INTRODUCE SOMETHING EXTRA It may sometimes be necessary to introduce something new, an auxiliary aid, to help make the connection between the given and the unknown. For instance, in a problem where a diagram is useful the auxiliary aid could be a new line drawn in a diagram. In a more algebraic problem it could be a new unknown that is related to the original unknown.

TAKE CASES We may sometimes have to split a problem into several cases and give a different argument for each of the cases. For instance, we often have to use this strategy in dealing with absolute value.

WORK BACKWARD Sometimes it is useful to imagine that your problem is solved and work backward, step by step, until you arrive at the given data. Then you may be able to reverse your steps and thereby construct a solution to the original problem. This procedure is commonly used in solving equations. For instance, in solving the equation $3x - 5 = 7$, we suppose that x is a number that satisfies $3x - 5 = 7$ and work backward. We add 5 to each side of the equation and then divide each side by 3 to get $x = 4$. Since each of these steps can be reversed, we have solved the problem.

ESTABLISH SUBGOALS In a complex problem it is often useful to set subgoals (in which the desired situation is only partially fulfilled). If we can first reach these subgoals, then we may be able to build on them to reach our final goal.

INDIRECT REASONING Sometimes it is appropriate to attack a problem indirectly. In using proof by contradiction to prove that P implies Q we assume that P is true and Q is false and try to see why this cannot happen. Somehow we have to use this information and arrive at a contradiction to what we absolutely know is true.

MATHEMATICAL INDUCTION In proving statements that involve a positive integer n, it is frequently helpful to use the Principle of Mathematical Induction.

STEP 3.
CARRY OUT THE PLAN

In Step 2 a plan was devised. In carrying out that plan we have to check each stage of the plan and write the details that prove that each stage is correct.

STEP 4.
LOOK BACK

Having completed our solution, it is wise to look back over it, partly to see if we have made errors in the solution and partly to see if we can think of an easier way to solve the problem. Another reason for looking back is that it will familiarize us with the method of solution and this may be useful for solving a future problem. Descartes said, "Every problem that I solved became a rule which served afterwards to solve other problems."

MULTIVARIABLE CALCULUS
THIRD EDITION

MULTIVARIABLE CALCULUS

THIRD EDITION

JAMES STEWART

McMaster University

BROOKS/COLE PUBLISHING COMPANY

I(T)P   An International Thomson Publishing Company

Pacific Grove ■ Albany ■ Bonn ■ Boston ■ Cincinnati ■ Detroit ■ London ■ Madrid ■ Melbourne
Mexico City ■ New York ■ Paris ■ San Francisco ■ Singapore ■ Tokyo ■ Toronto ■ Washington

Sponsoring Editor: *Jeremy Hayhurst*
Editorial Associate: *Elizabeth Rammel*
Marketing Team: *Patrick Farrant, Margaret Parks*
Production Services Manager: *Joan Marsh*
Production Coordination: *Kathi Townes, TECHarts*
Manuscript Editor: *Kathi Townes*
Interior Design: *Kathi Townes*

Cover Design: *Katherine Minerva, Vernon T. Boes*
Cover Sculpture: *Christian Haase*
Cover Photo: *Ed Young*
Interior Illustration: *TECHarts*
Typesetting: *Sandy Senter/Beacon Graphics*
Cover Printing: *Phoenix Color Corporation*
Printing and Binding: *Quebecor Printing/Hawkins*

For more information, contact:

BROOKS/COLE PUBLISHING COMPANY
511 Forest Lodge Road
Pacific Grove, CA 93950
USA

International Thomson Publishing Europe
Berkshire House 168-173
High Holborn
London WC1V 7AA
England

Thomas Nelson Australia
102 Dodds Street
South Melbourne, 3205
Victoria, Australia

Nelson Canada
1120 Birchmount Road
Scarborough, Ontario
Canada M1K 5G4

International Thomson Editores
Campos Eliseos 385, Piso 7
Col. Polanco
11560 México D. F. México

International Thomson Publishing GmbH
Königswinterer Strasse 418
53227 Bonn
Germany

International Thomson Publishing Asia
221 Henderson Road
#05–10 Henderson Building
Singapore 0315

International Thomson Publishing Japan
Hirakawacho Kyowa Building, 3F
2-2-1 Hirakawacho
Chiyoda-ku, Tokyo 102
Japan

Printed in the United States of America.

10 9 8 7 6 5 4 3

LIBRARY OF CONGRESS CATALOGING-IN-PUBLICATION DATA

Stewart, James
    Multivariable calculus / James Stewart.—3rd ed.
        p.   cm.
    Includes index.
    ISBN 0-534-25213-3
    1. Calculus.   I. Title
QA303.S88254 1994   94-41899
515–dc20   CIP

PREFACE

This book is a reprinting of the multivariable portion of my text *Calculus, Third Edition,* published by Brooks/Cole in 1995. The chapters reproduced here cover: infinite sequences and series; three-dimensional analytic geometry and vectors; partial derivatives; multiple integrals; vector calculus; and differential equations. These chapters are a direct continuation of my *Single Variable Calculus, Third Edition,* also published in 1995, which contains Chapters 1–10 of *Calculus.* Particularly important results from single variable calculus are summarized for at-a-glance review in the section titled *Key Definitions, Properties, and Theorems from Single Variable Calculus,* following the table of contents.

The last several years have seen much discussion about change in the calculus curriculum and in methods of teaching the subject. I have followed these discussions with great interest and have conducted experiments in my own calculus classes and listened to suggestions from colleagues and reviewers. What follows is a summary of how I have responded to these influences in preparing the third edition. You will see that the *spirit* of reform pervades the book, but within the context of a traditional curriculum.

TECHNOLOGY

For the past five years I have experimented with calculus laboratories for my own students, first with graphing software for computers, then with graphing calculators, and finally with computer algebra systems. Those of us who have watched our students use these machines know how enlivening such experiences can be. We have seen from the expressions on their faces how these devices can engage our students' attention and make them active learners.

Despite my enthusiasm for technology, I think there are potential dangers for misusing it. When I first started using technology, I tended to use it too much, but then I started to see where it is appropriate and where it is not. Many topics in calculus can be explained with chalk and blackboard (and reinforced with pencil and paper exercises) more simply, more quickly, and more clearly than with technology. Other topics cry out for the use of machines. What is important is the *appropriate* use of technology, which can be characterized as involving the *interaction* between technology and calculus. In short, technology is not a panacea, but, when used appropriately, it can be a powerful stimulus to learning.

This textbook can be used either with or without technology and I use three special symbols to indicate clearly when a particular type of machine is required. The symbol ⊞ means that an ordinary scientific calculator is needed for the calculations in an exercise. The icon ⟋⊏ indicates an example or exercise that requires the use of either a graphing calculator or a computer with graphing software. The symbol CAS is reserved for problems in which the full resources of a computer algebra system (like Derive, Maple, or Mathematica) are required. In all cases we assume that the student knows how to use the machine—we rarely give explicit commands.

Some of the exercises designated by ▲ or ⌨ are, in effect, calculus laboratories and require considerable time for their completion. Instructors should therefore consult the solutions manual to determine the complexity of a problem before assigning it. Some of those problems explore the shape of a family of surfaces depending on one or more parameters (see page 770). Other such projects involve technology in very different ways. See, for instance, pages 608 (logistic sequences), 740, 820, and 896.

VISUALIZATION

One of the themes of the calculus reform movement is the Rule of Three: Topics should be presented numerically, graphically, and symbolically, wherever possible. I believe that, even in its first and second editions, my calculus text has had a stronger focus on numerical and graphical points of view than other traditional books. In the third edition I have taken this principle farther. See page 611 for an example of how the Rule of Three comes into play. You will also see that I have included more work with tabular functions and more numerical estimates of sums of series.

I have added many examples and exercises that promote visual thinking. See pages 689, 731, 785, 896, and 915 for examples of exercises that test students' visual understanding.

In addition, I have added hundreds of new computer-generated figures to illustrate existing examples. These are not just pretty pictures—they constantly remind students of the geometric meaning behind the result of a calculation. (See, for instance, pages 804, 824, 840, 904, and 912.)

INCREASED EMPHASIS ON PROBLEM SOLVING

My educational philosophy was strongly influenced by attending the lectures of George Polya and Gabor Szego when I was a student at Stanford University. Both Polya and Szego consistently introduced a topic by relating it to something concrete or familiar. Wherever practical, I have introduced topics with an intuitive geometrical or physical description and attempted to tie mathematical concepts to the students' experience.

I found Polya's lectures on problem solving very inspirational and his books *How To Solve It, Mathematical Discovery,* and *Mathematics and Plausible Reasoning* have become the core text material for a mathematical problem-solving course that I instituted and teach at McMaster University. I have adapted these problem-solving strategies to the study of calculus both explicitly, by outlining strategies, and implicitly, by illustration and example.

Students usually have difficulties in situations that involve no single well-defined procedure for obtaining the answer. I think nobody has improved very much on Polya's four-stage problem-solving strategy and, accordingly, I have included in this edition a version of Polya's strategy on the front endpaper. In addition, I have retained from prior editions the separate special sections devoted to problem solving: 10.7 (Strategy for Testing Series) and 15.4 (Strategy for Solving First-Order Differential Equations).

In the second edition I included what I call *Problems Plus* after even-numbered chapters. These are problems that go beyond the usual exercises in one way or another and require a higher level of problem-solving ability. The very fact that they do not occur in the context of any particular chapter makes them a little more challenging. For instance, a problem that occurs after Chapter 14 need not have anything to do with Chapter 14. I particularly value problems in which a student has to combine methods from two or three different chapters. I have added a number of good new problems, including some with a geometric flavor (see Problem 32 after Chapter 10, Problems 3 and 15 after Chapter 12, and Problems 12 and 14 after Chapter 14). I have been testing these Problems Plus on my own students by putting them on assignments, tests, and exams. Because of their challenging nature I grade these problems in a different way. Here I reward a student significantly for ideas toward a solution and for recognizing which problem-solving principles are relevant. My aim is to teach my students to be unafraid to tackle a problem the likes of which they have never seen before.

The *Applications Plus* sections, which occur after odd-numbered chapters, are a counterpart to the Problems Plus. (Again the idea is often to combine ideas and techniques from different parts of the book.) See Problem 7 on page 891 for a good new applied problem. The Applications Plus, like the Problems Plus, are extended problems that would make good projects. There are many new applied problems in the ordinary sections of the book as well. (See, for instance, Exercises 29–33 on page 674.)

OTHER CHANGES

■ I have added historical and biographical margin notes, some of them fairly extensive, in order to enliven the course and to show students that mathematics was created by living, breathing human beings.

■ Chapter 10 contains more changes than any other chapter. I have added material on numerical estimates of sums of series based on which test was used to prove convergence: the Integral Test (page 621), the Comparison Test (page 627), or the Ratio Test (page 640). The last half of the chapter, on power series, has been completely reorganized and rewritten: Taylor's Formula occurs earlier, error estimates now include those from graphing devices, and applications of Taylor polynomials to physics are emphasized.

■ Euler's method for the numerical solution of differential equations has been added to Section 15.1.

■ About 20% of the exercises are new. In most cases, a relatively standard exercise has been replaced by one that uses technology or stimulates visual thinking without technology. Some of the new exercises encourage the development of communication skills by explicitly requesting descriptions, conjectures, and explanations. Many of these exercises are suitable as extended projects.

ACKNOWLEDGMENTS

The preparation of this and previous editions has involved much time spent reading the reasoned (but sometimes contradictory) advice from a large number of astute reviewers. I greatly appreciate the time they spent to understand my motivation for the approach taken. I have learned something from each of them.

FIRST EDITION REVIEWERS

John Alberghini,
 Manchester Community College
Daniel Anderson, *University of Iowa*
David Berman, *University of New Orleans*
Richard Biggs,
 University of Western Ontario
Stephen Brown
David Buchthal, *University of Akron*
James Choike, *Oklahoma State University*
Carl Cowen, *Purdue University*
Daniel Cyphert, *Armstrong State College*
Robert Dahlin
Daniel DiMaria, *Suffolk Community College*
Daniel Drucker, *Wayne State University*
Dennis Dunninger, *Michigan State University*
Bruce Edwards, *University of Florida*
Garret Etgen, *University of Houston*
Frederick Gass, *Miami University of Ohio*

Bruce Gilligan, *University of Regina*
Stuart Goldenberg,
 California Polytechnic State University
Michael Gregory, *University of North Dakota*
Charles Groetsch, *University of Cincinnati*
D. W. Hall, *Michigan State University*
Allen Hesse, *Rochester Community College*
Matt Kaufman
David Leeming, *University of Victoria*
Mark Pinsky, *Northwestern University*
Lothar Redlin,
 The Pennsylvania State University
Eric Schreiner, *Western Michigan University*
Wayne Skrapek, *University of Saskatchewan*
William Smith, *University of North Carolina*
Richard St. Andre,
 Central Michigan University
Steven Willard, *University of Alberta*

SECOND EDITION REVIEWERS

Michael Albert,
Carnegie-Mellon University
Jorge Cassio,
Miami-Dade Community College
Jack Ceder,
University of California, Santa Barbara
Seymour Ditor,
University of Western Ontario
Kenn Dunn, *Dalhousie University*
John Ellison, *Grove City College*
William Francis,
Michigan Technological University
Gerald Goff, *Oklahoma State University*
Stuart Goldenberg,
California Polytechnic State University
Richard Grassl, *University of New Mexico*
Melvin Hausner,
New York University/Courant Institute

Clement Jeske,
University of Wisconsin, Platteville
Jerry Johnson, *Oklahoma State University*
Virgil Kowalik, *Texas A & I University*
Sam Lesseig,
Northeast Missouri State University
Phil Locke, *University of Maine*
Phil McCartney,
Northern Kentucky University
Mary Martin, *Colgate University*
Igor Malyshev, *San José State University*
Richard Nowakowski, *Dalhousie University*
Vincent Panico, *University of the Pacific*
Tom Rishel, *Cornell University*
David Ryeburn, *Simon Fraser University*
Ricardo Salinas, *San Antonio College*
Stan Ver Nooy, *University of Oregon*
Jack Weiner, *University of Guelph*

THIRD EDITION REVIEWERS

B. D. Aggarwala, *University of Calgary*
Donna J. Bailey,
Northeast Missouri State University
Wayne Barber,
Chemeketa Community College
Neil Berger,
University of Illinois, Chicago
Robert Blumenthal,
Oglethorpe University
Barbara Bohannon, *Hofstra University*
Stephen W. Brady, *Wichita State University*
Jack Ceder,
University of California, Santa Barbara
Kenn Dunn, *Dalhousie University*
David Ellis,
San Francisco State University
Theodore Faticoni, *Fordham University*
Patrick Gallagher,
Columbia University–New York
Paul Garrett,
University of Minnesota–Minneapolis
Salim M. Haïdar,
Grand Valley State University
Melvin Hausner,
New York University/Courant Institute

Curtis Herink, *Mercer University*
John H. Jenkins,
Embry-Riddle Aeronautical University,
Prescott Campus
Matthias Kawski, *Arizona State University*
Kevin Kreider, *University of Akron*
Larry Mansfield, *Queens College*
Nathaniel F. G. Martin,
University of Virginia
Tom Metzger, *University of Pittsburgh*
Wayne N. Palmer, *Utica College*
Tom Rishel, *Cornell University*
Richard Rockwell, *Pacific Union College*
Robert Schmidt,
South Dakota State University
Mihr J. Shah,
Kent State University–Trumbull
Theodore Shifrin, *University of Georgia*
M. B. Tavakoli, *Chaffey College*
Andrei Verona,
California State University–Los Angeles
Theodore W. Wilcox,
Rochester Institute of Technology
Mary Wright,
Southern Illinois University–Carbondale

I also thank the authors of the supplements to this text: Richard St. Andre, *Central Michigan University,* for the **Study Guide;** Daniel Anderson, *University of Iowa,* Daniel Drucker, *Wayne State University,* and Barbara Frank, *St. Andrews Presbyterian College,* for volumes I and II of the **Student Solutions Manual;** and *Laurel Technical Services* and Edward Spitznagel and Joan Thomas of *Engineering Press, Inc.,* for their contributions to volumes I and II of the **Test Items.**

In addition, I would like to thank Ed Barbeau, Dan Drucker, Garret Etgen, Chris Fisher, E. L. Koh, Ron Lancaster, Lee Minor, and Saleem Watson for the use of problems that they devised; and McGill University students Andy Bulman-Fleming and Alex Taler for checking the accuracy of the manuscript and solving all of the exercises.

Finally, I thank Kathi Townes and the staff of TECHarts for their production coordination and interior illustration, Christian Haase for the cover sculpture, Ed Young for the cover photograph, and the following Brooks/Cole staff: Joan Marsh, production services manager; Katherine Minerva and Vernon T. Boes, cover designers; Patrick Farrant and Margaret Parks, marketing team; Elizabeth Rammel and Audra Silverie, supplements coordinators. I have been extremely fortunate to have worked with some of the best mathematics editors in the business over the past 15 years: Ron Munro, Harry Campbell, Craig Barth, Jeremy Hayhurst, and Gary W. Ostedt. Special thanks go to all of them.

JAMES STEWART

CONTENTS

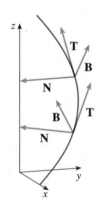

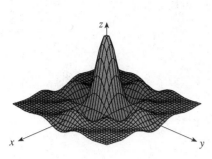

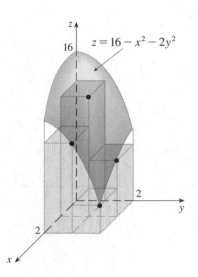

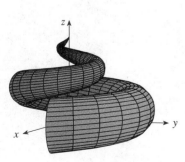

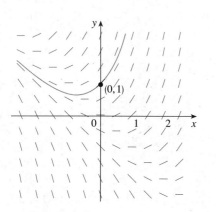

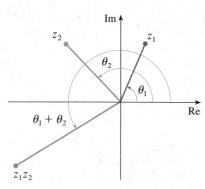

KEY DEFINITIONS, PROPERTIES, AND THEOREMS
from *Single Variable Calculus*

Blue numbers refer to *Single Variable Calculus, Third Edition: Early Transcendentals*

SECTION 1

A **function** f is a rule that assigns to each element x in a set A exactly one element, called $f(x)$, in a set B.

SECTION 1.2

(1) DEFINITION We write

$$\lim_{x \to a} f(x) = L$$

and say   "the limit of $f(x)$, as x approaches a, equals L"

if we can make the values of $f(x)$ arbitrarily close to L (as close to L as we like) by taking x to be sufficiently close to a but not equal to a.

SECTION 1.3

LIMIT LAWS   Suppose that c is a constant and the limits

$$\lim_{x \to a} f(x) \qquad \text{and} \qquad \lim_{x \to a} g(x)$$

exist. Then

1. $\lim_{x \to a} [f(x) + g(x)] = \lim_{x \to a} f(x) + \lim_{x \to a} g(x)$

2. $\lim_{x \to a} [f(x) - g(x)] = \lim_{x \to a} f(x) - \lim_{x \to a} g(x)$

3. $\lim_{x \to a} [cf(x)] = c \lim_{x \to a} f(x)$

4. $\lim_{x \to a} [f(x)g(x)] = \lim_{x \to a} f(x) \cdot \lim_{x \to a} g(x)$

5. $\lim_{x \to a} \dfrac{f(x)}{g(x)} = \dfrac{\lim_{x \to a} f(x)}{\lim_{x \to a} g(x)} \quad$ if $\lim_{x \to a} g(x) \neq 0$

FURTHER PROPERTIES OF LIMITS

6. $\lim\limits_{x \to a} [f(x)]^n = \left[\lim\limits_{x \to a} f(x)\right]^n$ where n is a positive integer

7. $\lim\limits_{x \to a} c = c$ 　　　　　　　**8.** $\lim\limits_{x \to a} x = a$

9. $\lim\limits_{x \to a} x^n = a^n$ where n is a positive integer

10. $\lim\limits_{x \to a} \sqrt[n]{x} = \sqrt[n]{a}$ where n is a positive integer

(If n is even, we assume that $a > 0$.)

11. $\lim\limits_{x \to a} \sqrt[n]{f(x)} = \sqrt[n]{\lim\limits_{x \to a} f(x)}$ where n is a positive integer

[If n is even, we assume that $\lim\limits_{x \to a} f(x) > 0$.]

---

(3) THE SQUEEZE THEOREM   If $f(x) \le g(x) \le h(x)$ for all x in an open interval that contains a (except possibly at a) and

$$\lim_{x \to a} f(x) = \lim_{x \to a} h(x) = L$$

then
$$\lim_{x \to a} g(x) = L$$

SECTION 1.4

(2) DEFINITION   Let f be a function defined on some open interval that contains the number a, except possibly at a itself. Then we say that the **limit of $f(x)$ as x approaches a is L**, and we write

$$\lim_{x \to a} f(x) = L$$

if for every number $\varepsilon > 0$ there is a corresponding number $\delta > 0$ such that

$$|f(x) - L| < \varepsilon \qquad \text{whenever} \qquad 0 < |x - a| < \delta$$

SECTION 1.5

(1) DEFINITION   A function f is **continuous at a number a** if

$$\lim_{x \to a} f(x) = f(a)$$

---

(4) THEOREM   If f and g are continuous at a and c is a constant, then the following functions are also continuous at a:

1. $f + g$　　　　**2.** $f - g$　　　　**3.** cf　　　　**4.** fg　　　　**5.** $\dfrac{f}{g}$ if $g(a) \ne 0$

---

(8) THEOREM   If g is continuous at a and f is continuous at $g(a)$, then $(f \circ g)(x) = f(g(x))$ is continuous at a.

SECTION 2.1

> **(2) DEFINITION**   The **derivative of a function f at a number a**, denoted by $f'(a)$, is
>
> $$f'(a) = \lim_{h \to 0} \frac{f(a + h) - f(a)}{h}$$
>
> if this limit exists.

(3) $$f'(a) = \lim_{x \to a} \frac{f(x) - f(a)}{x - a}$$

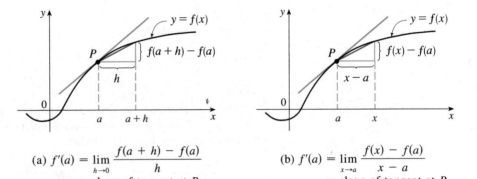

Geometric interpretation of the derivative

(a) $f'(a) = \lim_{h \to 0} \dfrac{f(a + h) - f(a)}{h}$ = slope of tangent at P

(b) $f'(a) = \lim_{x \to a} \dfrac{f(x) - f(a)}{x - a}$ = slope of tangent at P

> **(8) THEOREM**   If f is differentiable at a, then f is continuous at a.

SECTION 2.2

> **TABLE OF DIFFERENTIATION FORMULAS**
>
> $(cf)' = cf'$ $(f + g)' = f' + g'$ $(f - g)' = f' - g'$
>
> $(fg)' = f'g + fg'$ $\left(\dfrac{f}{g}\right)' = \dfrac{f'g - fg'}{g^2}$ $\dfrac{d}{dx} c = 0$
>
> $\dfrac{d}{dx}(x^n) = nx^{n-1}$

SECTION 2.3

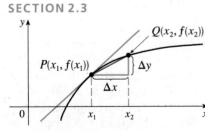

m_{PQ} = average rate of change
$m = f'(x_1)$ = instantaneous rate of change

The difference quotient $\dfrac{\Delta y}{\Delta x} = \dfrac{f(x_2) - f(x_1)}{x_2 - x_1}$

is the **average rate of change of y with respect to x** over the interval $[x_1, x_2]$ and can be interpreted as the slope of the secant line PQ in the figure. Its limit as $\Delta x \to 0$ is the derivative $f'(x_1)$, which can therefore be interpreted as the instantaneous rate of change of y with respect to x or the slope of the tangent line at $P(x_1, f(x_1))$. Using Leibniz notation, we write the process in the form

$$\frac{dy}{dx} = \lim_{\Delta x \to 0} \frac{\Delta y}{\Delta x}$$

SECTION 2.4

(4) THEOREM $$\lim_{\theta \to 0} \frac{\sin \theta}{\theta} = 1$$

TABLE OF DERIVATIVES OF TRIGONOMETRIC FUNCTIONS

$$\frac{d}{dx}(\sin x) = \cos x \qquad\qquad \frac{d}{dx}(\csc x) = -\csc x \cot x$$

$$\frac{d}{dx}(\cos x) = -\sin x \qquad\qquad \frac{d}{dx}(\sec x) = \sec x \tan x$$

$$\frac{d}{dx}(\tan x) = \sec^2 x \qquad\qquad \frac{d}{dx}(\cot x) = -\csc^2 x$$

SECTION 2.5

THE CHAIN RULE If the derivatives $g'(x)$ and $f'(g(x))$ both exist, and $F = f \circ g$ is the composite function defined by $F(x) = f(g(x))$, then $F'(x)$ exists and is given by the product

$$F'(x) = f'(g(x))g'(x)$$

In Leibniz notation, if $y = f(u)$ and $u = g(x)$ are both differentiable functions, then

$$\frac{dy}{dx} = \frac{dy}{du}\frac{du}{dx}$$

SECTION 2.10

(1) NEWTON'S METHOD

$$x_{n+1} = x_n - \frac{f(x_n)}{f'(x_n)}$$

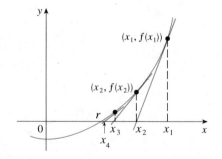

SECTION 3.1
SECTION 4.1

(1) DEFINITION A function f has an **absolute maximum** at c if $f(c) \geqslant f(x)$ for all x in D, where D is the domain of f. The number $f(c)$ is called the **maximum value** of f on D. Similarly, f has an **absolute minimum** at c if $f(c) \leqslant f(x)$ for all x in D and the number $f(c)$ is called the **minimum value** of f on D. The maximum and minimum values of f are called the **extreme values** of f.

(2) DEFINITION   A function f has a **local maximum** (or **relative maximum**) at c if there is an open interval I containing c such that $f(c) \geq f(x)$ for all x in I. Similarly, f has a **local minimum** at c if there is an open interval I containing c such that $f(c) \leq f(x)$ for all x in I.

(3) THE EXTREME VALUE THEOREM   If f is continuous on a closed interval $[a, b]$, then f attains an absolute maximum value $f(c)$ and an absolute minimum value $f(d)$ at some numbers c and d in $[a, b]$.

(4) FERMAT'S THEOREM   If f has a local extremum (that is, maximum or minimum) at c, and if $f'(c)$ exists, then $f'(c) = 0$.

(6) DEFINITION   A **critical number** of a function f is a number c in the domain of f such that either $f'(c) = 0$ or $f'(c)$ does not exist.

(8)   To find the *absolute* maximum and minimum values of a continuous function f on a closed interval $[a, b]$:

1. Find the values of f at the critical numbers of f in (a, b).

2. Find the values of $f(a)$ and $f(b)$.

3. The largest of the values from steps 1 and 2 is the absolute maximum value; the smallest of these values is the absolute minimum value.

SECTION 3.2
SECTION 4.2

ROLLE'S THEOREM   Let f be a function that satisfies the following three hypotheses:

1. f is continuous on the closed interval $[a, b]$.

2. f is differentiable on the open interval (a, b).

3. $f(a) = f(b)$

Then there is a number c in (a, b) such that $f'(c) = 0$.

THE MEAN VALUE THEOREM   Let f be a function that satisfies the following hypotheses:

1. f is continuous on the closed interval $[a, b]$.

2. f is differentiable on the open interval (a, b).

Then there is a number c in (a, b) such that

(1)
$$f'(c) = \frac{f(b) - f(a)}{b - a}$$

or, equivalently,

(2)
$$f(b) - f(a) = f'(c)(b - a)$$

SECTION 3.3
SECTION 4.3

(1) DEFINITION A function f is called **increasing** on an interval I if

$$f(x_1) < f(x_2) \qquad \text{whenever } x_1 < x_2 \text{ in } I$$

It is called **decreasing** on I if

$$f(x_1) > f(x_2) \qquad \text{whenever } x_1 < x_2 \text{ in } I$$

A function that is increasing or decreasing on I is called **monotonic** on I.

TEST FOR MONOTONIC FUNCTIONS Suppose f is continuous on $[a, b]$ and differentiable on (a, b).

(a) If $f'(x) > 0$ for all x in (a, b), then f is increasing on $[a, b]$.

(b) If $f'(x) < 0$ for all x in (a, b), then f is decreasing on $[a, b]$.

THE FIRST DERIVATIVE TEST Suppose that c is a critical number of a continuous function f.

(a) If f' changes from positive to negative at c, then f has a local maximum at c.

(b) If f' changes from negative to positive at c, then f has a local minimum at c.

(c) If f' does not change sign at c (that is, f' is positive on both sides of c or negative on both sides), then f has no local extremum at c.

SECTION 3.4
SECTION 4.4

(1) DEFINITION If the graph of f lies above all of its tangents on an interval I, then it is called **concave upward** on I. If the graph of f lies below all of its tangents on I, it is called **concave downward** on I.

THE TEST FOR CONCAVITY Suppose f is twice differentiable on an interval I.

(a) If $f''(x) > 0$ for all x in I, then the graph of f is concave upward on I.

(b) If $f''(x) < 0$ for all x in I, then the graph of f is concave downward on I.

THE SECOND DERIVATIVE TEST Suppose f'' is continuous on an open interval that contains c.

(a) If $f'(c) = 0$ and $f''(c) > 0$, then f has a local minimum at c.

(b) If $f'(c) = 0$ and $f''(c) < 0$, then f has a local maximum at c.

SECTION 3.5
SECTION 1.6

(1) DEFINITION Let f be a function defined on some interval (a, ∞). Then

$$\lim_{x \to \infty} f(x) = L$$

means that the values of $f(x)$ can be made arbitrarily close to L by taking x sufficiently large.

(5) DEFINITION Let f be a function defined on some interval (a, ∞). Then

$$\lim_{x \to \infty} f(x) = L$$

means that for every $\varepsilon > 0$ there is a corresponding number N such that

$$| f(x) - L | < \varepsilon \qquad \text{whenever} \qquad x > N$$

(7) DEFINITION Let f be a function defined on some interval (a, ∞). Then

$$\lim_{x \to \infty} f(x) = \infty$$

means that for every positive number M there is a corresponding number $N > 0$ such that

$$f(x) > M \qquad \text{whenever} \qquad x > N$$

SECTION 4.1
SECTION 5.1

(2) THEOREM If c is any constant (that is, it does not depend on i), then

(a) $\displaystyle\sum_{i=m}^{n} ca_i = c \sum_{i=m}^{n} a_i$

(b) $\displaystyle\sum_{i=m}^{n} (a_i + b_i) = \sum_{i=m}^{n} a_i + \sum_{i=m}^{n} b_i$

(c) $\displaystyle\sum_{i=m}^{n} (a_i - b_i) = \sum_{i=m}^{n} a_i - \sum_{i=m}^{n} b_i$

SECTION 4.3
SECTION 5.3

(2) DEFINITION OF A DEFINITE INTEGRAL If f is a function defined on a closed interval $[a, b]$, let P be a partition of $[a, b]$ with partition points $x_0, x_1, \ldots, x_n$, where

$$a = x_0 < x_1 < x_2 < \cdots < x_n = b$$

Choose points x_i^* in $[x_{i-1}, x_i]$ and let $\Delta x_i = x_i - x_{i-1}$ and $\|P\| = \max\{\Delta x_i\}$. Then the **definite integral of f from a to b** is

$$\int_a^b f(x)\, dx = \lim_{\|P\| \to 0} \sum_{i=1}^{n} f(x_i^*)\, \Delta x_i$$

if this limit exists. If the limit does exist, then f is called **integrable** on the interval $[a, b]$.

(5) THEOREM If f is integrable on $[a, b]$, then

$$\int_a^b f(x)\, dx = \lim_{n \to \infty} \frac{b - a}{n} \sum_{i=1}^{n} f\!\left(a + i\frac{b - a}{n} \right)$$

PROPERTIES OF THE INTEGRAL Suppose that all of the following integrals exist. Then

1. $\displaystyle\int_a^b c\,dx = c(b-a),$ where c is any constant

2. $\displaystyle\int_a^b [f(x) + g(x)]\,dx = \int_a^b f(x)\,dx + \int_a^b g(x)\,dx$

3. $\displaystyle\int_a^b cf(x)\,dx = c\int_a^b f(x)\,dx,$ where c is any constant

4. $\displaystyle\int_a^b [f(x) - g(x)]\,dx = \int_a^b f(x)\,dx - \int_a^b g(x)\,dx$

5. $\displaystyle\int_a^b f(x)\,dx = \int_a^c f(x)\,dx + \int_c^b f(x)\,dx$

ORDER PROPERTIES OF THE INTEGRAL Suppose the following integrals exist and $a \leq b$.

6. If $f(x) \geq 0$ for $a \leq x \leq b$, then $\int_a^b f(x)\,dx \geq 0$.

7. If $f(x) \geq g(x)$ for $a \leq x \leq b$, then $\int_a^b f(x)\,dx \geq \int_a^b g(x)\,dx$.

8. If $m \leq f(x) \leq M$ for $a \leq x \leq b$, then

$$m(b-a) \leq \int_a^b f(x)\,dx \leq M(b-a)$$

9. $\left|\int_a^b f(x)\,dx\right| \leq \int_a^b |f(x)|\,dx$

SECTION 4.4
SECTION 5.4

THE FUNDAMENTAL THEOREM OF CALCULUS Suppose f is continuous on $[a, b]$.

1. If $g(x) = \int_a^x f(t)\,dt$, then $g'(x) = f(x)$.

2. $\int_a^b f(x)\,dx = F(b) - F(a)$, where F is any antiderivative of f, that is, $F' = f$.

SECTION 4.5
SECTION 5.5

(5) THE SUBSTITUTION RULE FOR DEFINITE INTEGRALS If g' is continuous on $[a, b]$ and f is continuous on the range of g, then

$$\int_a^b f(g(x))g'(x)\,dx = \int_{g(a)}^{g(b)} f(u)\,du$$

---

(6) INTEGRALS OF SYMMETRIC FUNCTIONS Suppose f is continuous on $[-a, a]$.

(a) If f is even $[f(-x) = f(x)]$, then $\int_{-a}^a f(x)\,dx = 2\int_0^a f(x)\,dx$.

(b) If f is odd $[f(-x) = -f(x)]$, then $\int_{-a}^a f(x)\,dx = 0$.

SECTION 5.1
SECTION 6.1

(2)   The area of the region bounded by the curves $y = f(x)$, $y = g(x)$, and the lines $x = a$ and $x = b$, where f and g are continuous and $f(x) \geqslant g(x)$ for all x in $[a, b]$, is

$$A = \int_a^b [f(x) - g(x)]\,dx$$

SECTION 5.2
SECTION 6.2

(1) **DEFINITION OF VOLUME**   Let S be a solid that lies between the planes P_a and P_b. If the cross-sectional area of S in the plane P_x is $A(x)$, where A is an integrable function, then the **volume** of S is

$$V = \lim_{\|P\| \to 0} \sum_{i=1}^n A(x_i^*)\,\Delta x_i = \int_a^b A(x)\,dx$$

SECTION 5.5
SECTION 6.5

MEAN VALUE THEOREM FOR INTEGRALS   If f is continuous on $[a, b]$, then there exists a number c in $[a, b]$ such that

$$\int_a^b f(x)\,dx = f(c)\,(b - a)$$

SECTION 6.2
SECTION 3.1

(3)   If $a > 1$, then $\lim\limits_{x \to \infty} a^x = \infty$ and $\lim\limits_{x \to -\infty} a^x = 0$

If $0 < a < 1$, then $\lim\limits_{x \to \infty} a^x = 0$ and $\lim\limits_{x \to -\infty} a^x = \infty$

(7)

$$\frac{d}{dx}\,e^x = e^x$$

(9) **PROPERTIES OF THE EXPONENTIAL FUNCTION**

$$\lim_{x \to -\infty} e^x = 0 \qquad \lim_{x \to \infty} e^x = \infty$$

$$\lim_{x \to \infty} \ln x = \infty \qquad \lim_{x \to 0^+} \ln x = -\infty$$

SECTION 6.3
SECTION 3.3

(5)

$$\ln x = y \iff e^y = x$$

(6)

$$\ln(e^x) = x \qquad x \in \mathbb{R}$$

$$e^{\ln x} = x \qquad x > 0$$

SECTION 6.4
SECTION 3.4

(1)

$$\frac{d}{dx}(\ln x) = \frac{1}{x}$$

(7)

$$\frac{d}{dx}a^x = a^x \ln a$$

(8)

$$\lim_{x \to 0}(1 + x)^{1/x} = e$$

(9)

$$e = \lim_{n \to \infty}\left(1 + \frac{1}{n}\right)^n$$

SECTION 6.5
SECTION 3.5

(2) **THEOREM**   The only solutions of the differential equation $dy/dt = ky$ are the exponential functions

$$y(t) = y(0)e^{kt}$$

SECTION 6.6
SECTION 3.6

(11) **TABLE OF DERIVATIVES OF INVERSE TRIGONOMETRIC FUNCTIONS**

$$\frac{d}{dx}(\sin^{-1}x) = \frac{1}{\sqrt{1 - x^2}} \qquad \frac{d}{dx}(\csc^{-1}x) = -\frac{1}{x\sqrt{x^2 - 1}}$$

$$\frac{d}{dx}(\cos^{-1}x) = -\frac{1}{\sqrt{1 - x^2}} \qquad \frac{d}{dx}(\sec^{-1}x) = \frac{1}{x\sqrt{x^2 - 1}}$$

$$\frac{d}{dx}(\tan^{-1}x) = \frac{1}{1 + x^2} \qquad \frac{d}{dx}(\cot^{-1}x) = -\frac{1}{1 + x^2}$$

SECTION 6.7
SECTION 3.7

DEFINITION OF THE HYPERBOLIC FUNCTIONS

$$\sinh x = \frac{e^x - e^{-x}}{2} \qquad \cosh x = \frac{e^x + e^{-x}}{2} \qquad \tanh x = \frac{\sinh x}{\cosh x}$$

$$\operatorname{csch} x = \frac{1}{\sinh x} \qquad \operatorname{sech} x = \frac{1}{\cosh x} \qquad \coth x = \frac{\cosh x}{\sinh x}$$

(1) **TABLE OF DERIVATIVES OF HYPERBOLIC FUNCTIONS**

$$\frac{d}{dx}\sinh x = \cosh x \qquad \frac{d}{dx}\cosh x = \sinh x \qquad \frac{d}{dx}\tanh x = \operatorname{sech}^2 x$$

$$\frac{d}{dx}\operatorname{csch} x = -\operatorname{csch} x \coth x \qquad \frac{d}{dx}\operatorname{sech} x = -\operatorname{sech} x \tanh x \qquad \frac{d}{dx}\coth x = -\operatorname{csch}^2 x$$

SECTION 6.8
SECTION 3.8

(3) L'HOSPITAL'S RULE Suppose f and g are differentiable and $g'(x) \neq 0$ on an open interval I that contains a (except possibly at a). Suppose that

$$\lim_{x \to a} f(x) = 0 \qquad \text{and} \qquad \lim_{x \to a} g(x) = 0$$

or that

$$\lim_{x \to a} f(x) = \pm\infty \qquad \text{and} \qquad \lim_{x \to a} g(x) = \pm\infty$$

(In other words, we have an indeterminate form of type $\frac{0}{0}$ or ∞/∞.) Then

$$\lim_{x \to a} \frac{f(x)}{g(x)} = \lim_{x \to a} \frac{f'(x)}{g'(x)}$$

if the limit on the right side exists (or is ∞ or $-\infty$).

SECTION 7.9

(1) DEFINITION OF AN IMPROPER INTEGRAL OF TYPE 1

(a) If $\int_a^t f(x)\,dx$ exists for every number $t \geqslant a$, then

$$\int_a^\infty f(x)\,dx = \lim_{t \to \infty} \int_a^t f(x)\,dx$$

provided this limit exists (as a finite number).

(b) If $\int_t^b f(x)\,dx$ exists for every number $t \leqslant b$, then

$$\int_{-\infty}^b f(x)\,dx = \lim_{t \to -\infty} \int_t^b f(x)\,dx$$

provided this limit exists (as a finite number).

The improper integrals in (a) and (b) are called **convergent** if the limit exists and **divergent** if the limit does not exist.

(c) If both $\int_a^\infty f(x)\,dx$ and $\int_{-\infty}^a f(x)\,dx$ are convergent, then we define

$$\int_{-\infty}^\infty f(x)\,dx = \int_{-\infty}^a f(x)\,dx + \int_a^\infty f(x)\,dx$$

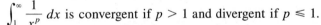

(2) $\qquad\qquad \int_1^\infty \dfrac{1}{x^p}\,dx$ is convergent if $p > 1$ and divergent if $p \leqslant 1$.

SECTION 8.2

(2) THE ARC LENGTH FORMULA If f' is continuous on $[a, b]$, then the length of the curve $y = f(x)$, $a \leqslant x \leqslant b$, is

$$L = \int_a^b \sqrt{1 + [f'(x)]^2}\,dx$$

SECTION 8.3

The **surface area** of the surface obtained by rotating the curve $y = f(x)$, $a \leqslant x \leqslant b$, about the x-axis is

(4) $$S = \int_a^b 2\pi f(x) \sqrt{1 + [f'(x)]^2}\,dx$$

SECTION 8.4

The centroid of the region $\mathcal{R}$ that lies between the two curves $y = f(x)$ and $y = g(x)$, $f(x) \geq g(x)$, is $(\bar{x}, \bar{y})$, where

(11)
$$\bar{x} = \frac{1}{A} \int_a^b x[\,f(x) - g(x)\,]\,dx$$

$$\bar{y} = \frac{1}{A} \int_a^b \tfrac{1}{2}\{[\,f(x)\,]^2 - [\,g(x)\,]^2\}\,dx$$

SECTION 9.3

(4) THEOREM   If a curve C is described by the parametric equations $x = f(t)$, $y = g(t)$, $\alpha \leq t \leq \beta$, where f' and g' are continuous on $[\alpha, \beta]$ and C is traversed exactly once as t increases from α to β, then the length of C is

$$L = \int_\alpha^\beta \sqrt{\left(\frac{dx}{dt}\right)^2 + \left(\frac{dy}{dt}\right)^2}\,dt$$

SECTION 9.4

(1)
$$x = r\cos\theta \qquad y = r\sin\theta$$

(2)
$$r^2 = x^2 + y^2 \qquad \tan\theta = \frac{y}{x}$$

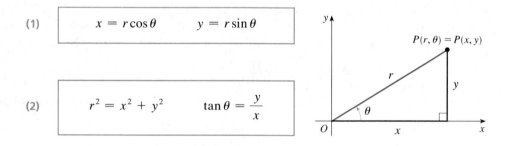

SECTION 9.5

The area of the region bounded by the polar curve $r = f(\theta) \geq 0$ and the rays $\theta = a$ and $\theta = b$ is

(3)
$$A = \int_a^b \tfrac{1}{2}[\,f(\theta)\,]^2\,d\theta$$

Formula 3 is often written as

(4)
$$A = \int_a^b \tfrac{1}{2}r^2\,d\theta$$

SECTION 9.6

(1)   The equation of a parabola with focus $(0, p)$ and directrix $y = -p$ is

$$x^2 = 4py$$

If we write $a = 1/(4p)$, then the standard equation of a parabola (1) becomes $y = ax^2$. It opens upward if $p > 0$ and downward if $p < 0$ [see parts (a) and (b) of the following figure]. The graph is symmetric with respect to the y-axis because (1) is unchanged when x is replaced by $-x$.

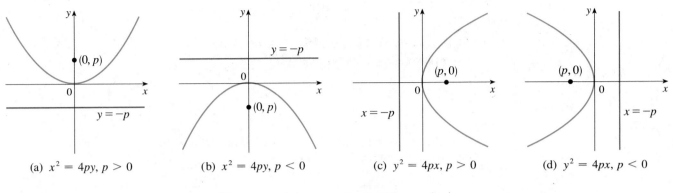

(a) $x^2 = 4py,\ p > 0$ (b) $x^2 = 4py,\ p < 0$ (c) $y^2 = 4px,\ p > 0$ (d) $y^2 = 4px,\ p < 0$

If we interchange x and y in (1), we obtain

(2)
$$y^2 = 4px$$

which is the equation of a parabola with focus $(p, 0)$ and directrix $x = -p$ [see parts (c) and (d) of the preceding figure].

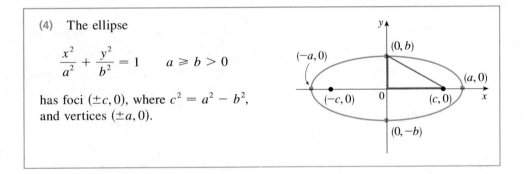

(4)   The ellipse

$$\frac{x^2}{a^2} + \frac{y^2}{b^2} = 1 \qquad a \geq b > 0$$

has foci $(\pm c, 0)$, where $c^2 = a^2 - b^2$, and vertices $(\pm a, 0)$.

If the foci of an ellipse are located on the y-axis at $(0, \pm c)$, then we can find its equation by interchanging x and y in (4).

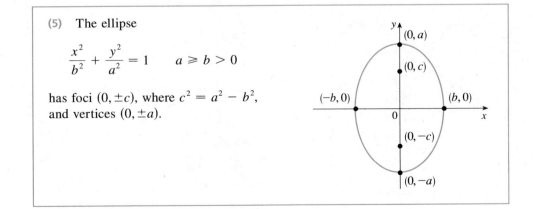

(5)   The ellipse

$$\frac{x^2}{b^2} + \frac{y^2}{a^2} = 1 \qquad a \geq b > 0$$

has foci $(0, \pm c)$, where $c^2 = a^2 - b^2$, and vertices $(0, \pm a)$.

(7)   The hyperbola

$$\frac{x^2}{a^2} - \frac{y^2}{b^2} = 1$$

has foci $(\pm c, 0)$, where $c^2 = a^2 + b^2$, vertices $(\pm a, 0)$, and asymptotes $y = \pm(b/a)x$.

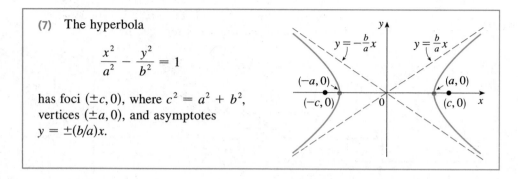

If the foci of a hyperbola are on the y-axis, then by reversing the roles of x and y we obtain the following information.

(8)   The hyperbola

$$\frac{y^2}{a^2} - \frac{x^2}{b^2} = 1$$

has foci $(0, \pm c)$, where $c^2 = a^2 + b^2$, vertices $(0, \pm a)$, and asymptotes $y = \pm(a/b)x$.

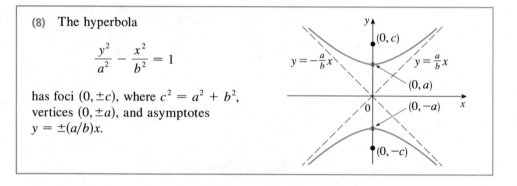

SECTION 9.7

(6) THEOREM   A polar equation of the form

$$r = \frac{ed}{1 \pm e \cos\theta} \qquad \text{or} \qquad r = \frac{ed}{1 \pm e \sin\theta}$$

represents a conic section with eccentricity e. The conic is an ellipse if $e < 1$, a parabola if $e = 1$, or a hyperbola if $e > 1$.

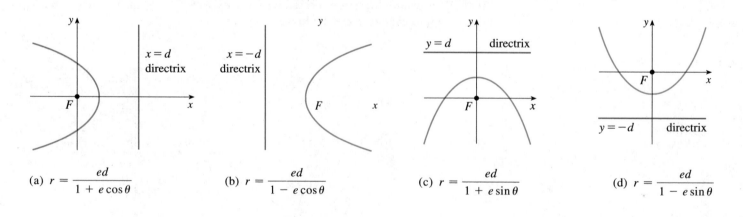

(a) $r = \dfrac{ed}{1 + e\cos\theta}$ (b) $r = \dfrac{ed}{1 - e\cos\theta}$ (c) $r = \dfrac{ed}{1 + e\sin\theta}$ (d) $r = \dfrac{ed}{1 - e\sin\theta}$

MULTIVARIABLE CALCULUS
THIRD EDITION

10

INFINITE SEQUENCES AND SERIES

■ Our minds are finite, and yet even in those circumstances of finitude, we are surrounded by possibilities that are infinite, and the purpose of human life is to grasp as much as we can out of that infinitude.

ALFRED NORTH WHITEHEAD

Infinite sequences and series were briefly introduced in the Preview of Calculus (Section 5 in Review and Preview) in connection with Zeno's paradoxes and the decimal representation of numbers. Their importance in calculus stems from Newton's idea of representing functions as sums of infinite series. For instance, in finding areas he often integrated a function by first expressing it as a series and then integrating each term of the series. We pursue this idea in Section 10.10 in order to integrate such functions as e^{-x^2}. (Recall that we have previously been unable to do this.) Many of the functions that arise in mathematical physics and chemistry, such as Bessel functions, are defined as sums of series, so it is important to be familiar with the basic concepts of convergence of infinite sequences and series.

10.1 SEQUENCES

A **sequence** can be thought of as a list of numbers written in a definite order:

$$a_1, a_2, a_3, a_4, \ldots, a_n, \ldots$$

The number a_1 is called the *first term,* a_2 is the *second term,* and in general a_n is the *nth term.* We will deal exclusively with infinite sequences and so each term a_n will have a successor a_{n+1}.

Notice that for every positive integer n there is a corresponding number a_n and so a sequence can be defined as a function whose domain is the set of positive integers. But we usually write a_n instead of the function notation $f(n)$ for the value of the function at the number n.

NOTATION: The sequence $\{a_1, a_2, a_3, \ldots\}$ is also denoted by

$$\{a_n\} \qquad \text{or} \qquad \{a_n\}_{n=1}^{\infty}$$

EXAMPLE 1   Some sequences can be defined by giving a formula for the nth term. In the following examples we give three descriptions of the sequence: one by using the preceding notation, another by using the defining formula, and a third by writing out the terms of the sequence. Notice that n doesn't have to start at 1.

(a) $\left\{\dfrac{n}{n+1}\right\}_{n=1}^{\infty}$ $a_n = \dfrac{n}{n+1}$ $\left\{\dfrac{1}{2}, \dfrac{2}{3}, \dfrac{3}{4}, \dfrac{4}{5}, \ldots, \dfrac{n}{n+1}, \ldots\right\}$

(b) $\left\{\dfrac{(-1)^n(n+1)}{3^n}\right\}$ $a_n = \dfrac{(-1)^n(n+1)}{3^n}$ $\left\{-\dfrac{2}{3}, \dfrac{3}{9}, -\dfrac{4}{27}, \dfrac{5}{81}, \ldots, \dfrac{(-1)^n(n+1)}{3^n}, \ldots\right\}$

(c) $\{\sqrt{n-3}\,\}_{n=3}^{\infty}$ $a_n = \sqrt{n-3}, \; n \geq 3$ $\{0, 1, \sqrt{2}, \sqrt{3}, \ldots, \sqrt{n-3}, \ldots\}$

(d) $\left\{\cos\dfrac{n\pi}{6}\right\}_{n=0}^{\infty}$ $a_n = \cos\dfrac{n\pi}{6}, \; n \geq 0$ $\left\{1, \dfrac{\sqrt{3}}{2}, \dfrac{1}{2}, 0, \ldots, \cos\dfrac{n\pi}{6}, \ldots\right\}$ ∎

EXAMPLE 2   Here are some sequences that do not have a simple defining equation.
(a) The sequence $\{p_n\}$, where p_n is the population of the world as of January 1 in the year n.
(b) If we let a_n be the digit in the nth decimal place of the number e, then $\{a_n\}$ is a well-defined sequence whose first few terms are

$$\{7, 1, 8, 2, 8, 1, 8, 2, 8, 4, 5, \ldots\}$$

(c) The **Fibonacci sequence** $\{f_n\}$ is defined recursively by the conditions

$$f_1 = 1 \qquad f_2 = 1 \qquad f_n = f_{n-1} + f_{n-2} \qquad n \geq 3$$

Each term is the sum of the two preceding terms. The first few terms are

$$\{1, 1, 2, 3, 5, 8, 13, 21, \ldots\}$$

This sequence arose when the 13th-century Italian mathematician known as Fibonacci solved a problem concerning the breeding of rabbits (see Exercise 63).   ∎

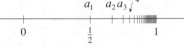

FIGURE 1

A sequence such as the one in Example 1(a), $a_n = n/(n+1)$, can be pictured either by plotting its terms on a number line as in Figure 1 or by plotting its graph as in Figure 2. Note that, since a sequence is a function whose domain is the set of positive integers, its graph consists of isolated points with coordinates

$$(1, a_1) \qquad (2, a_2) \qquad (3, a_3) \qquad \ldots \qquad (n, a_n) \qquad \ldots$$

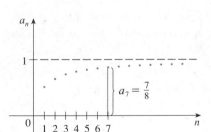

FIGURE 2

From Figure 1 or 2 it appears that the terms of the sequence $a_n = n/(n+1)$ are approaching 1 as n becomes large. In fact, the difference

$$1 - \frac{n}{n+1} = \frac{1}{n+1}$$

can be made as small as we like by taking n sufficiently large. We indicate this by writing

$$\lim_{n\to\infty} \frac{n}{n+1} = 1$$

In general, the notation

$$\lim_{n\to\infty} a_n = L$$

means that the terms of the sequence $\{a_n\}$ can be made arbitrarily close to L by taking n sufficiently large. Notice that the following precise definition of the limit of a sequence is very similar to the definition of a limit of a function at infinity given in Section 3.5.

---

(1) DEFINITION   A sequence $\{a_n\}$ has the **limit** L and we write

$$\lim_{n \to \infty} a_n = L \qquad \text{or} \qquad a_n \to L \text{ as } n \to \infty$$

if for every $\varepsilon > 0$ there is a corresponding integer N such that

$$|a_n - L| < \varepsilon \qquad \text{whenever} \qquad n > N$$

If $\lim_{n \to \infty} a_n$ exists, we say the sequence **converges** (or is **convergent**). Otherwise, we say the sequence **diverges** (or is **divergent**).

Definition 1 is illustrated by Figure 3, in which the terms $a_1, a_2, a_3, \ldots$ are plotted on a number line. No matter how small an interval $(L - \varepsilon, L + \varepsilon)$ is chosen, there exists an N such that all terms of the sequence from a_{N+1} onward must lie in that interval.

FIGURE 3
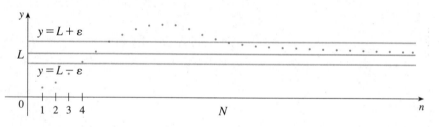

Another illustration of Definition 1 is given in Figure 4. The points on the graph of $\{a_n\}$ must lie between the horizontal lines $y = L + \varepsilon$ and $y = L - \varepsilon$ if $n > N$. This picture must be valid no matter how small ε is chosen, but usually a smaller ε requires a larger N.

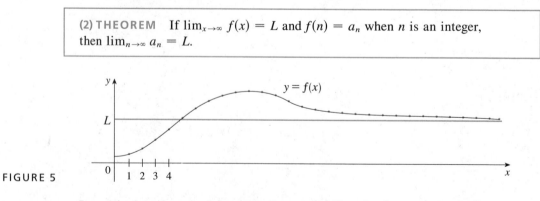

FIGURE 4

Comparison of Definition 1 and Definition 3.5.5 shows that the only difference between $\lim_{n \to \infty} a_n = L$ and $\lim_{x \to \infty} f(x) = L$ is that n is required to be an integer. Thus we have the following theorem, which is illustrated by Figure 5.

---

(2) THEOREM   If $\lim_{x \to \infty} f(x) = L$ and $f(n) = a_n$ when n is an integer, then $\lim_{n \to \infty} a_n = L$.

FIGURE 5

In particular, since we know that $\lim_{x \to \infty} (1/x^r) = 0$ when $r > 0$ (Theorem 3.5.4), we have

(3)
$$\lim_{n \to \infty} \frac{1}{n^r} = 0 \qquad \text{if } r > 0$$

The analogue of Definition 3.5.7 is the following:

---

(4) DEFINITION $\lim_{n \to \infty} a_n = \infty$ means that for every positive number M there is an integer N such that

$$a_n > M \quad \text{whenever} \quad n > N$$

If $\lim_{n \to \infty} a_n = \infty$, then the sequence $\{a_n\}$ is divergent but in a special way. We say that $\{a_n\}$ diverges to ∞.

The Limit Laws given in Section 1.3 also hold for the limits of sequences and their proofs are similar.

Limit Laws for Sequences

If $\{a_n\}$ and $\{b_n\}$ are convergent sequences and c is a constant, then

$$\lim_{n \to \infty} (a_n + b_n) = \lim_{n \to \infty} a_n + \lim_{n \to \infty} b_n$$

$$\lim_{n \to \infty} (a_n - b_n) = \lim_{n \to \infty} a_n - \lim_{n \to \infty} b_n$$

$$\lim_{n \to \infty} c a_n = c \lim_{n \to \infty} a_n$$

$$\lim_{n \to \infty} (a_n b_n) = \lim_{n \to \infty} a_n \cdot \lim_{n \to \infty} b_n$$

$$\lim_{n \to \infty} \frac{a_n}{b_n} = \frac{\lim_{n \to \infty} a_n}{\lim_{n \to \infty} b_n} \quad \text{if } \lim_{n \to \infty} b_n \neq 0$$

$$\lim_{n \to \infty} c = c$$

The Squeeze Theorem can also be adapted for sequences as follows.

Squeeze Theorem for Sequences

If $a_n \leq b_n \leq c_n$ for $n \geq n_0$ and $\lim_{n \to \infty} a_n = \lim_{n \to \infty} c_n = L$, then $\lim_{n \to \infty} b_n = L$.

Another useful fact about limits of sequences is given by the following theorem, whose proof is left as Exercise 67.

---

(5) THEOREM If $\lim_{n \to \infty} |a_n| = 0$, then $\lim_{n \to \infty} a_n = 0$.

EXAMPLE 3   Find $\lim\limits_{n\to\infty} \dfrac{n}{n+1}$.

SOLUTION   The method is similar to the one we used in Section 3.5: Divide numerator and denominator by the highest power of n and then use the Limit Laws.

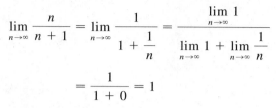

$$\lim_{n\to\infty} \frac{n}{n+1} = \lim_{n\to\infty} \frac{1}{1+\dfrac{1}{n}} = \frac{\lim\limits_{n\to\infty} 1}{\lim\limits_{n\to\infty} 1 + \lim\limits_{n\to\infty} \dfrac{1}{n}}$$

$$= \frac{1}{1+0} = 1$$

This shows that the guess we made earlier from Figures 1 and 2 was correct.

Here we used Equation 3 with $r = 1$.   ∎

EXAMPLE 4   Calculate $\lim\limits_{n\to\infty} \dfrac{\ln n}{n}$.

SOLUTION   Notice that both numerator and denominator approach infinity as $n\to\infty$. We cannot apply l'Hospital's Rule directly because it applies not to sequences but to functions of a real variable. However, we can apply l'Hospital's Rule to the related function $f(x) = (\ln x)/x$ and obtain

$$\lim_{x\to\infty} \frac{\ln x}{x} = \lim_{x\to\infty} \frac{1/x}{1} = 0$$

Therefore, by Theorem 2 we have

$$\lim_{n\to\infty} \frac{\ln n}{n} = 0$$

∎

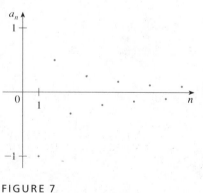

FIGURE 6

The graph of the sequence in Example 6 is shown in Figure 7 and supports the answer.

EXAMPLE 5   Determine whether the sequence $a_n = (-1)^n$ is convergent or divergent.

SOLUTION   If we write out the terms of the sequence, we obtain

$$\{-1, 1, -1, 1, -1, 1, -1, \dots\}$$

The graph of this sequence is shown in Figure 6. Since the terms oscillate between 1 and -1 infinitely often, a_n does not approach any number. Thus $\lim_{n\to\infty}(-1)^n$ does not exist; that is, the sequence $\{(-1)^n\}$ is divergent.   ∎

EXAMPLE 6   Evaluate $\lim\limits_{n\to\infty} \dfrac{(-1)^n}{n}$ if it exists.

SOLUTION

$$\lim_{n\to\infty} \left| \frac{(-1)^n}{n} \right| = \lim_{n\to\infty} \frac{1}{n} = 0$$

Therefore, by Theorem 5,

$$\lim_{n\to\infty} \frac{(-1)^n}{n} = 0$$

∎

FIGURE 7

EXAMPLE 7   Discuss the convergence of the sequence $a_n = n!/n^n$, where $n! = 1 \cdot 2 \cdot 3 \cdots \cdot n$.

CREATING GRAPHS OF SEQUENCES

Some computer algebra systems have special commands that enable us to create sequences and graph them directly. With most graphing calculators, however, sequences can be graphed by using parametric equations. For instance, the sequence in Example 7 can be graphed by entering the parametric equations

$$x = t \qquad y = t!/t^t$$

and graphing in dot mode starting with $t = 1$, setting the t-step equal to 1. The result is shown in Figure 8.

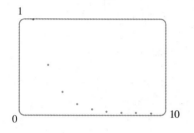

FIGURE 8

SOLUTION   Both numerator and denominator approach infinity as $n \to \infty$ but here we have no corresponding function for use with l'Hospital's Rule ($x!$ is not defined when x is not an integer). Let us write out a few terms to get a feeling for what happens to a_n as n gets large:

$$a_1 = 1 \qquad a_2 = \frac{1 \cdot 2}{2 \cdot 2} \qquad a_3 = \frac{1 \cdot 2 \cdot 3}{3 \cdot 3 \cdot 3}$$

(6)
$$a_n = \frac{1 \cdot 2 \cdot 3 \cdot \cdots \cdot n}{n \cdot n \cdot n \cdot \cdots \cdot n}$$

It appears from these expressions and the graph in Figure 8 that the terms are decreasing and perhaps approach 0. To confirm this, observe from Equation 6 that

$$a_n = \frac{1}{n}\left(\frac{2 \cdot 3 \cdot \cdots \cdot n}{n \cdot n \cdot \cdots \cdot n}\right)$$

so
$$0 < a_n \le \frac{1}{n}$$

We know that $1/n \to 0$ as $n \to \infty$. Therefore, $a_n \to 0$ as $n \to \infty$ by the Squeeze Theorem. ∎

EXAMPLE 8   For what values of r is the sequence $\{r^n\}$ convergent?

SOLUTION   We know from Section 6.2 that $\lim_{x \to \infty} a^x = \infty$ for $a > 1$ and $\lim_{x \to \infty} a^x = 0$ for $0 < a < 1$. Therefore, putting $a = r$ and using Theorem 2, we have

$$\lim_{n \to \infty} r^n = \begin{cases} \infty & \text{if } r > 1 \\ 0 & \text{if } 0 < r < 1 \end{cases}$$

It is obvious that

$$\lim_{n \to \infty} 1^n = 1 \qquad \text{and} \qquad \lim_{n \to \infty} 0^n = 0$$

If $-1 < r < 0$, then $0 < |r| < 1$, so

$$\lim_{n \to \infty} |r^n| = \lim_{n \to \infty} |r|^n = 0$$

and therefore $\lim_{n \to \infty} r^n = 0$ by Theorem 5. If $r \le -1$, then $\{r^n\}$ diverges as in Example 5. Figure 9 shows the graphs for various values of r. (The case $r = -1$ is shown in Figure 6.) ∎

FIGURE 9
The sequence $a_n = r^n$

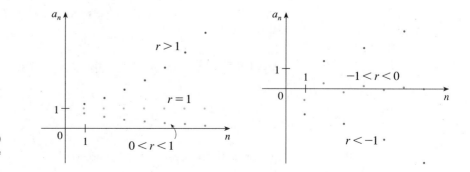

The results of Example 8 are summarized for future use as follows:

---

(7)   The sequence $\{r^n\}$ is convergent if $-1 < r \leq 1$ and divergent for all other values of r.

$$\lim_{n \to \infty} r^n = \begin{cases} 0 & \text{if } -1 < r < 1 \\ 1 & \text{if } r = 1 \end{cases}$$

---

(8) DEFINITION   A sequence $\{a_n\}$ is called **increasing** if $a_n \leq a_{n+1}$ for all $n \geq 1$, that is, $a_1 \leq a_2 \leq a_3 \leq \cdots$. It is called **decreasing** if $a_n \geq a_{n+1}$ for all $n \geq 1$. It is called **monotonic** if it is either increasing or decreasing.

EXAMPLE 9   The sequence $\left\{\dfrac{3}{n+5}\right\}$ is decreasing because

$$\frac{3}{n+5} > \frac{3}{n+6}$$

for all $n \geq 1$. (The right side is smaller because it has a larger denominator.)   ∎

EXAMPLE 10   Show that the sequence $a_n = \dfrac{n}{n^2 + 1}$ is decreasing.

SOLUTION 1   We must show that $a_{n+1} \leq a_n$, that is,

$$\frac{n+1}{(n+1)^2 + 1} \leq \frac{n}{n^2 + 1}$$

This inequality is equivalent to the one we get by cross-multiplication:

$$\frac{n+1}{(n+1)^2 + 1} \leq \frac{n}{n^2 + 1} \quad \Longleftrightarrow \quad (n+1)(n^2+1) \leq n[(n+1)^2 + 1]$$

$$\Longleftrightarrow \quad n^3 + n^2 + n + 1 \leq n^3 + 2n^2 + 2n$$

$$\Longleftrightarrow \quad 1 \leq n^2 + n$$

It is obvious that $n^2 + n \geq 1$ is true for $n \geq 1$. Therefore, $a_{n+1} \leq a_n$ and so $\{a_n\}$ is decreasing.

SOLUTION 2   Consider the function $f(x) = \dfrac{x}{x^2 + 1}$:

$$f'(x) = \frac{x^2 + 1 - 2x^2}{(x^2 + 1)^2} = \frac{1 - x^2}{(x^2 + 1)^2} < 0 \qquad \text{whenever } x^2 > 1$$

Thus f is decreasing on $[1, \infty)$ and so $f(n) > f(n+1)$. Therefore $\{a_n\}$ is decreasing.   ∎

> **(9) DEFINITION**   A sequence $\{a_n\}$ is **bounded above** if there is a number M such that
>
> $$a_n \leq M \qquad \text{for all } n \geq 1$$
>
> It is **bounded below** if there is a number m such that
>
> $$m \leq a_n \qquad \text{for all } n \geq 1$$
>
> If it is bounded above and below, then $\{a_n\}$ is a **bounded sequence.**

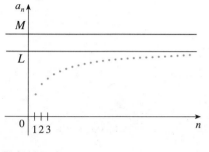

FIGURE 10

For instance, the sequence $a_n = n$ is bounded below ($a_n > 0$) but not above. The sequence $a_n = n/(n + 1)$ is bounded because $0 < a_n < 1$ for all n.

We know that not every bounded sequence is convergent [$a_n = (-1)^n$ satisfies $-1 \leq a_n \leq 1$ but is divergent from Example 5] and not every monotonic sequence is convergent ($a_n = n \to \infty$). But if a sequence is both bounded *and* monotonic, then it must be convergent. This fact is proved as Theorem 10, but intuitively you can understand why it is true by looking at Figure 10. If $\{a_n\}$ is increasing and $a_n \leq M$ for all n, then the terms are forced to crowd together and approach some number L.

The proof of Theorem 10 is based on the **Completeness Axiom** for the set $\mathbb{R}$ of real numbers, which says that if S is a nonempty set of real numbers that has an upper bound M ($x \leq M$ for all x in S), then S has a **least upper bound** b. (This means that b is an upper bound for S, but if M is any other upper bound, then $b \leq M$.) The Completeness Axiom is an expression of the fact that there is no gap or hole in the real number line.

> **(10) THEOREM**   Every bounded, monotonic sequence is convergent.

PROOF   Suppose $\{a_n\}$ is an increasing sequence. Since $\{a_n\}$ is bounded, the set $S = \{a_n \mid n \geq 1\}$ has an upper bound. By the Completeness Axiom it has a least upper bound L. Given $\varepsilon > 0$, $L - \varepsilon$ is *not* an upper bound for S (since L is the *least* upper bound). Therefore

$$a_N > L - \varepsilon \qquad \text{for some integer } N$$

But the sequence is increasing so $a_n \geq a_N$ for every $n > N$. Thus if $n > N$ we have

$$a_n > L - \varepsilon$$

so

$$0 \leq L - a_n < \varepsilon$$

since $a_n \leq L$. Thus

$$|L - a_n| < \varepsilon \qquad \text{whenever} \qquad n > N$$

so $\lim_{n \to \infty} a_n = L$.

A similar proof (using the greatest lower bound) works if $\{a_n\}$ is decreasing. $\square$

The proof of Theorem 10 shows that a sequence that is increasing and bounded above is convergent. (Likewise, a decreasing sequence that is bounded below is convergent.) This fact is used many times in dealing with infinite series.

EXAMPLE 11   Investigate the sequence $\{a_n\}$ defined by the recurrence relation

$$a_1 = 2 \qquad a_{n+1} = \tfrac{1}{2}(a_n + 6) \qquad \text{for } n = 1, 2, 3, \dots$$

SOLUTION   We begin by computing the first few terms:

$$a_1 = 2 \qquad a_2 = \tfrac{1}{2}(2 + 6) = 4 \qquad a_3 = \tfrac{1}{2}(4 + 6) = 5 \qquad a_4 = \tfrac{1}{2}(5 + 6) = 5.5$$

$$a_5 = 5.75 \qquad a_6 = 5.875 \qquad a_7 = 5.9375 \qquad a_8 = 5.96875$$

These initial terms suggest that the sequence is increasing and the terms are approaching 6. To confirm that the sequence is increasing we use mathematical induction to show that $a_{n+1} \geq a_n$ for all $n \geq 1$. This is true for $n = 1$ because $a_2 = 4 > a_1$. If we assume that it is true for $n = k$, then we have

$$a_{k+1} \geq a_k$$

so

$$a_{k+1} + 6 \geq a_k + 6$$

and

$$\tfrac{1}{2}(a_{k+1} + 6) \geq \tfrac{1}{2}(a_k + 6)$$

Thus

$$a_{k+2} \geq a_{k+1}$$

> Mathematical induction is often used in dealing with recursive sequences. See Appendix E for a discussion of the Principle of Mathematical Induction.

We have deduced that $a_{n+1} \geq a_n$ is true for $n = k + 1$. Therefore, the inequality is true for all n by induction.

Next we verify that $\{a_n\}$ is bounded by showing that $a_n < 6$ for all n. (Since the sequence is increasing, we already know that it has a lower bound: $a_n \geq a_1 = 2$ for all n.) We know that $a_1 < 6$, so the assertion is true for $n = 1$. Suppose it is true for $n = k$. Then

$$a_k < 6$$

so

$$a_k + 6 < 12$$

$$\tfrac{1}{2}(a_k + 6) < \tfrac{1}{2}(12) = 6$$

Thus

$$a_{k+1} < 6$$

This shows, by mathematical induction, that $a_n < 6$ for all n.

Since the sequence $\{a_n\}$ is increasing and bounded, Theorem 10 guarantees that it has a limit. The theorem doesn't tell us what the value of the limit is. But now that we know $L = \lim_{n \to \infty} a_n$ exists, we can use the recurrence relation to write

$$\lim_{n \to \infty} a_{n+1} = \lim_{n \to \infty} \tfrac{1}{2}(a_n + 6) = \tfrac{1}{2}\left(\lim_{n \to \infty} a_n + 6\right) = \tfrac{1}{2}(L + 6)$$

> A proof of this fact is requested in Exercise 50.

Since $a_n \to L$, it follows that $a_{n+1} \to L$, too (as $n \to \infty$, $n + 1 \to \infty$, too). So we have

$$L = \tfrac{1}{2}(L + 6)$$

Solving this equation for L, we get $L = 6$, as predicted.   ■

EXERCISES 10.1

1–6 ■ List the first five terms of the sequence.

1. $a_n = \dfrac{n}{2n + 1}$
  2. $a_n = \left(-\dfrac{2}{3}\right)^n$
  3. $a_n = \dfrac{1 \cdot 3 \cdot 5 \cdot \cdots \cdot (2n - 1)}{n!}$
  4. $\left\{\dfrac{(-7)^{n+1}}{n!}\right\}$

5. $\left\{ \sin \dfrac{n\pi}{2} \right\}$

6. $a_1 = 1,\ a_{n+1} = \dfrac{1}{1 + a_n}$

7–12 ■ Find a formula for the general term a_n of the sequence, assuming that the pattern of the first few terms continues.

7. $\left\{ \frac{1}{2}, \frac{1}{4}, \frac{1}{8}, \frac{1}{16}, \ldots \right\}$

8. $\left\{ \frac{1}{2}, \frac{1}{4}, \frac{1}{6}, \frac{1}{8}, \ldots \right\}$

9. $\{1, 4, 7, 10, \ldots\}$

10. $\left\{ \frac{3}{16}, \frac{4}{25}, \frac{5}{36}, \frac{6}{49}, \ldots \right\}$

11. $\left\{ \frac{3}{2}, -\frac{9}{4}, \frac{27}{8}, -\frac{81}{16}, \ldots \right\}$

12. $\{0, 2, 0, 2, 0, 2, \ldots\}$

13–40 ■ Determine whether the sequence converges or diverges. If it converges, find the limit.

13. $a_n = \dfrac{1}{4n^2}$

14. $a_n = 4\sqrt{n}$

15. $a_n = \dfrac{n^2 - 1}{n^2 + 1}$

16. $a_n = \dfrac{4n - 3}{3n + 4}$

17. $a_n = \dfrac{n^2}{n + 1}$

18. $a_n = \dfrac{\sqrt[3]{n} + \sqrt[4]{n}}{\sqrt{n} + \sqrt[5]{n}}$

19. $a_n = (-1)^n \dfrac{n^2}{1 + n^3}$

20. $a_n = \dfrac{1}{5^n}$

21. $a_n = \cos(n\pi/2)$

22. $a_n = \sin(n\pi/2)$

23. $\left\{ \dfrac{\pi^n}{3^n} \right\}$

24. $\{\arctan 2n\}$

25. $\left\{ \dfrac{3 + (-1)^n}{n^2} \right\}$

26. $\left\{ \dfrac{n!}{(n + 2)!} \right\}$

27. $\left\{ \dfrac{\ln(n^2)}{n} \right\}$

28. $\{(-1)^n \sin(1/n)\}$

29. $\{\sqrt{n + 2} - \sqrt{n}\}$

30. $\left\{ \dfrac{\ln(2 + e^n)}{3n} \right\}$

31. $a_n = n2^{-n}$

32. $a_n = \ln(n + 1) - \ln n$

33. $a_n = n^{-1/n}$

34. $a_n = (1 + 3n)^{1/n}$

35. $a_n = \dfrac{\cos^2 n}{2^n}$

36. $a_n = \dfrac{n \cos n}{n^2 + 1}$

37. $a_n = \dfrac{1}{n^2} + \dfrac{2}{n^2} + \cdots + \dfrac{n}{n^2}$

38. $a_n = (\sqrt{n + 1} - \sqrt{n})\sqrt{n + \frac{1}{2}}$

39. $a_n = \dfrac{n!}{2^n}$

40. $a_n = \dfrac{(-3)^n}{n!}$

41–48 ■ Use a graph of the sequence to decide whether the sequence is convergent or divergent. If the sequence is convergent, guess the value of the limit from the graph and then prove your guess. (See the margin note on page 603 for advice on graphing sequences.)

41. $a_n = (-1)^n \dfrac{n + 1}{n}$

42. $a_n = 2 + (-2/\pi)^n$

43. $\left\{ \arctan\left(\dfrac{2n}{2n + 1} \right) \right\}$

44. $\left\{ \dfrac{\sin n}{\sqrt{n}} \right\}$

45. $a_n = \dfrac{n^3}{n!}$

46. $a_n = \sqrt[n]{3^n + 5^n}$

47. $a_n = \dfrac{1 \cdot 3 \cdot 5 \cdot \cdots \cdot (2n - 1)}{(2n)^n}$

48. $a_n = \dfrac{1 \cdot 3 \cdot 5 \cdot \cdots \cdot (2n - 1)}{n!}$

49. For what values of r is the sequence $\{nr^n\}$ convergent?

50. (a) If $\{a_n\}$ is convergent, show that

$$\lim_{n \to \infty} a_{n+1} = \lim_{n \to \infty} a_n$$

(b) A sequence $\{a_n\}$ is defined by $a_1 = 1$ and $a_{n+1} = 1/(1 + a_n)$ for $n \geq 1$. Assuming that $\{a_n\}$ is convergent, find its limit.

51–58 ■ Determine whether the given sequence is increasing, decreasing, or not monotonic.

51. $a_n = \dfrac{1}{3n + 5}$

52. $a_n = \dfrac{1}{5^n}$

53. $a_n = \dfrac{n - 2}{n + 2}$

54. $a_n = \dfrac{3n + 4}{2n + 5}$

55. $a_n = \cos(n\pi/2)$

56. $a_n = 3 + (-1)^n/n$

57. $a_n = \dfrac{n}{n^2 + n - 1}$

58. $a_n = \dfrac{\sqrt{n + 1}}{5n + 3}$

59. Find the limit of the sequence

$$\left\{ \sqrt{2},\ \sqrt{2\sqrt{2}},\ \sqrt{2\sqrt{2\sqrt{2}}},\ \ldots \right\}$$

60. A sequence $\{a_n\}$ is given by $a_1 = \sqrt{2}$, $a_{n+1} = \sqrt{2 + a_n}$.
(a) By induction, or otherwise, show that $\{a_n\}$ is increasing and bounded above by 3. Apply Theorem 10 to show that $\lim_{n \to \infty} a_n$ exists.
(b) Find $\lim_{n \to \infty} a_n$.

61. Show that the sequence defined by $a_1 = 1$, $a_{n+1} = 3 - 1/a_n$ is increasing and $a_n < 3$ for all n. Deduce that $\{a_n\}$ is convergent and find its limit.

62. Show that the sequence defined by $a_1 = 2$, $a_{n+1} = 1/(3 - a_n)$ satisfies $0 < a_n \leq 2$ and is decreasing. Deduce that the sequence is convergent and find its limit.

63. (a) Fibonacci posed the following problem: Suppose that rabbits live forever and that every month each pair produces a new pair which becomes productive at age 2 months. If we start with one newborn pair, how many pairs of rabbits will we have in the nth month? Show that the answer is f_n, where $\{f_n\}$ is the Fibonacci sequence defined in Example 2(c).
(b) Let $a_n = f_{n+1}/f_n$ and show that $a_{n-1} = 1 + 1/a_{n-2}$. Assuming that $\{a_n\}$ is convergent, find its limit.

64. (a) Let $a_1 = a$, $a_2 = f(a)$, $a_3 = f(a_2) = f(f(a))$, ...,
$a_{n+1} = f(a_n)$, where f is a continuous function. If
$\lim_{n \to \infty} a_n = L$, show that $f(L) = L$.

(b) Illustrate part (a) by taking $f(x) = \cos x$, $a = 1$, and
estimating the value of L to five decimal places.

65. (a) Use a graph to guess the value of the limit

$$\lim_{n \to \infty} \frac{n^5}{n!}$$

(b) Use a graph of the sequence in part (a) to find the
smallest values of N that correspond to $\varepsilon = 0.1$ and
$\varepsilon = 0.001$ in Definition 1.

66. Use Definition 1 directly to prove that $\lim_{n \to \infty} r^n = 0$
when $|r| < 1$.

67. Prove Theorem 5.
[*Hint:* Use either Definition 1 or the Squeeze Theorem.]

68. Let $a_n = \left(1 + \dfrac{1}{n}\right)^n$.

(a) Show that if $0 \le a < b$, then

$$\frac{b^{n+1} - a^{n+1}}{b - a} < (n + 1)b^n$$

(b) Deduce that $b^n[(n + 1)a - nb] < a^{n+1}$.

(c) Use $a = 1 + 1/(n + 1)$ and $b = 1 + 1/n$ in part (b) to
show that $\{a_n\}$ is increasing.

(d) Use $a = 1$ and $b = 1 + 1/(2n)$ in part (b) to show that
$a_{2n} < 4$.

(e) Use parts (c) and (d) to show that $a_n < 4$ for all n.

(f) Use Theorem 10 to show that $\lim_{n \to \infty} (1 + 1/n)^n$ exists.
(The limit is e. See Equation 6.4.9.)

69. Let a and b be positive numbers with $a > b$. Let a_1 be their
arithmetic mean and b_1 their geometric mean:

$$a_1 = \frac{a + b}{2} \qquad b_1 = \sqrt{ab}$$

Repeat this process so that, in general,

$$a_{n+1} = \frac{a_n + b_n}{2} \qquad b_{n+1} = \sqrt{a_n b_n}$$

(a) Use mathematical induction to show that

$$a_n > a_{n+1} > b_{n+1} > b_n$$

(b) Deduce that both $\{a_n\}$ and $\{b_n\}$ are convergent.

(c) Show that $\lim_{n \to \infty} a_n = \lim_{n \to \infty} b_n$.
Gauss called the common value of these limits the
arithmetic-geometric mean of the numbers a and b.

70. (a) Show that if $\lim_{n \to \infty} a_{2n} = L$ and $\lim_{n \to \infty} a_{2n+1} = L$,
then $\{a_n\}$ is convergent and $\lim_{n \to \infty} a_n = L$.

(b) If $a_1 = 1$ and

$$a_{n+1} = 1 + \frac{1}{1 + a_n}$$

find the first eight terms of the sequence $\{a_n\}$. Then use
part (a) to show that $\lim_{n \to \infty} a_n = \sqrt{2}$. This gives the
continued fraction expansion

$$\sqrt{2} = 1 + \cfrac{1}{2 + \cfrac{1}{2 + \cdots}}$$

71. (a) Show that

$$2 \cos \theta - 1 = \frac{1 + 2 \cos 2\theta}{1 + 2 \cos \theta}$$

(b) Let $a_n = 2 \cos(\theta/2^n) - 1$ and $b_n = a_1 a_2 \cdots a_n$. Find a
formula for b_n that does not involve a product of n
terms and deduce that

$$\lim_{n \to \infty} b_n = \tfrac{1}{3}(1 + 2 \cos \theta)$$

72. A sequence that arises in ecology as a model for
population growth is defined by the **logistic difference
equation**

$$p_{n+1} = k p_n(1 - p_n)$$

where p_n measures the size of the population of the nth
generation of a single species. To keep the numbers
manageable, p_n is a fraction of the maximal size of the
population, so $0 \le p_n \le 1$. (Notice that the form of this
equation is similar to the logistic differential equation in
Section 8.1.) An ecologist is interested in predicting the
size of the population as time goes on and asks the
questions: Will it stabilize at a limiting value? Will it
change in a cyclical fashion? Or will it exhibit random
behavior?

Write a program to compute the first n terms of this
sequence starting with an initial population p_0, where
$0 < p_0 < 1$. Use this program to do the following.

(a) Calculate 20 or 30 terms of the sequence for $p_0 = \tfrac{1}{2}$ and
for two values of k such that $1 < k < 3$. Graph the
sequences. Do they appear to converge? Repeat for a
different value of p_0 between 0 and 1. Does the limit
depend on the choice of p_0? Does it depend on the
choice of k?

(b) Calculate terms of the sequence for a value of k
between 3 and 3.4 and plot them. What do you notice
about the behavior of the terms?

(c) Experiment with values of k between 3.4 and 3.5.
What happens to the terms?

(d) For values of k between 3.6 and 4, compute and plot at
least 100 terms and comment on the behavior of the
sequence. What happens if you change p_0 by 0.001?
This type of behavior is called *chaotic* and is exhibited
by insect populations under certain conditions.

10.2 SERIES

If we try to add the terms of an infinite sequence $\{a_n\}_{n=1}^{\infty}$ we get an expression of the form

$$(1) \qquad a_1 + a_2 + a_3 + \cdots + a_n + \cdots$$

which is called an **infinite series** (or just a **series**) and is denoted, for short, by the symbol

$$\sum_{n=1}^{\infty} a_n \qquad \text{or} \qquad \sum a_n$$

But does it make sense to talk about the sum of infinitely many terms?

It would be impossible to find a finite sum for the series

$$1 + 2 + 3 + 4 + 5 + \cdots + n + \cdots$$

because if we start adding the terms we get the cumulative sums 1, 3, 6, 10, 15, 21, ... and, after the nth term, $n(n + 1)/2$, which becomes very large as n increases.

However, if we start to add the terms of the series

$$\frac{1}{2} + \frac{1}{4} + \frac{1}{8} + \frac{1}{16} + \frac{1}{32} + \frac{1}{64} + \cdots + \frac{1}{2^n} + \cdots$$

we get $\frac{1}{2}, \frac{3}{4}, \frac{7}{8}, \frac{15}{16}, \frac{31}{32}, \frac{63}{64}, \ldots, 1 - 1/2^n, \ldots$. The table in the margin shows that as we add more and more terms, these partial sums become closer and closer to 1. (See also Figure 11 in Section 5 of Review and Preview.) In fact, by adding sufficiently many terms of the series we can make the partial sums as close as we like to 1. So it seems reasonable to say that the sum of this infinite series is 1 and to write

$$\sum_{n=1}^{\infty} \frac{1}{2^n} = \frac{1}{2} + \frac{1}{4} + \frac{1}{8} + \frac{1}{16} + \cdots + \frac{1}{2^n} + \cdots = 1$$

n	Sum of first n terms
1	0.50000000
2	0.75000000
3	0.87500000
4	0.93750000
5	0.96875000
6	0.98437500
7	0.99218750
10	0.99902344
15	0.99996948
20	0.99999905
25	0.99999997

We use a similar idea to determine whether or not a general series (1) has a sum. We consider the **partial sums**

$$s_1 = a_1$$

$$s_2 = a_1 + a_2$$

$$s_3 = a_1 + a_2 + a_3$$

$$s_4 = a_1 + a_2 + a_3 + a_4$$

and, in general,

$$s_n = a_1 + a_2 + a_3 + \cdots + a_n = \sum_{i=1}^{n} a_i$$

These partial sums form a new sequence $\{s_n\}$, which may or may not have a limit. If $\lim_{n \to \infty} s_n = s$ exists (as a finite number), then, as in the preceding example, we call it the sum of the infinite series $\sum a_n$.

(2) DEFINITION   Given a series $\sum_{n=1}^{\infty} a_n = a_1 + a_2 + a_3 + \cdots$, let s_n denote its nth partial sum:

$$s_n = \sum_{i=1}^{n} a_i = a_1 + a_2 + \cdots + a_n$$

If the sequence $\{s_n\}$ is convergent and $\lim_{n \to \infty} s_n = s$ exists as a real number, then the series $\sum a_n$ is called **convergent** and we write

$$a_1 + a_2 + \cdots + a_n + \cdots = s \qquad \text{or} \qquad \sum_{n=1}^{\infty} a_n = s$$

The number s is called the **sum** of the series. Otherwise, the series is called **divergent.**

Thus when we write $\sum_{n=1}^{\infty} a_n = s$ we mean that by adding sufficiently many terms of the series we can get as close as we like to the number s. Notice that

$$\sum_{n=1}^{\infty} a_n = \lim_{n \to \infty} \sum_{i=1}^{n} a_i$$

EXAMPLE 1   An important example of an infinite series is the **geometric series**

$$a + ar + ar^2 + ar^3 + \cdots + ar^{n-1} + \cdots = \sum_{n=1}^{\infty} ar^{n-1} \qquad a \neq 0$$

Each term is obtained from the preceding one by multiplying it by the common ratio r. (We have already considered the special case where $a = \frac{1}{2}$ and $r = \frac{1}{2}$.)

If $r = 1$, then $s_n = a + a + \cdots + a = na \to \pm\infty$. Since $\lim_{n \to \infty} s_n$ does not exist, the geometric series diverges in this case.

If $r \neq 1$, we have

$$s_n = a + ar + ar^2 + \cdots + ar^{n-1}$$

and
$$rs_n = \qquad ar + ar^2 + \cdots + ar^{n-1} + ar^n$$

Subtracting these equations, we get

$$s_n - rs_n = a - ar^n$$

(3)
$$s_n = \frac{a(1 - r^n)}{1 - r}$$

If $-1 < r < 1$, we know from (10.1.7) that $r^n \to 0$ as $n \to \infty$, so

$$\lim_{n \to \infty} s_n = \lim_{n \to \infty} \frac{a(1 - r^n)}{1 - r} = \frac{a}{1 - r} - \frac{a}{1 - r} \lim_{n \to \infty} r^n = \frac{a}{1 - r}$$

Thus when $|r| < 1$ the geometric series is convergent and its sum is $a/(1 - r)$.

If $r \leq -1$ or $r > 1$, the sequence $\{r^n\}$ is divergent by (10.1.7) and so, by Equation 3, $\lim_{n \to \infty} s_n$ does not exist. Therefore, the geometric series diverges in those cases.   ∎

We summarize the results of Example 1 as follows:

---

(4) The geometric series

$$\sum_{n=1}^{\infty} ar^{n-1} = a + ar + ar^2 + \cdots$$

is convergent if $|r| < 1$ and its sum is

$$\sum_{n=1}^{\infty} ar^{n-1} = \frac{a}{1-r} \qquad |r| < 1$$

If $|r| \geq 1$, the geometric series is divergent.

EXAMPLE 2 Find the sum of the geometric series

$$5 - \tfrac{10}{3} + \tfrac{20}{9} - \tfrac{40}{27} + \cdots$$

SOLUTION The first term is $a = 5$ and the common ratio is $r = -\tfrac{2}{3}$. Since $|r| = \tfrac{2}{3} < 1$, the series is convergent by (4) and its sum is

$$5 - \frac{10}{3} + \frac{20}{9} - \frac{40}{27} + \cdots = \frac{5}{1 - \left(-\frac{2}{3}\right)} = \frac{5}{\frac{5}{3}} = 3 \qquad \blacksquare$$

What do we really mean when we say that the sum of the series in Example 2 is 3? Of course, we can't literally add an infinite number of terms one by one. But, according to Definition 2, the total sum is the *limit* of the sequence of partial sums. So by taking the sum of sufficiently many terms we can get as close as we like to the number 3. The table shows the first 10 partial sums s_n and the graph in Figure 1 shows how they approach 3.

n	s_n
1	5.000000
2	1.666667
3	3.888889
4	2.407407
5	3.395062
6	2.736626
7	3.175583
8	2.882945
9	3.078037
10	2.947975

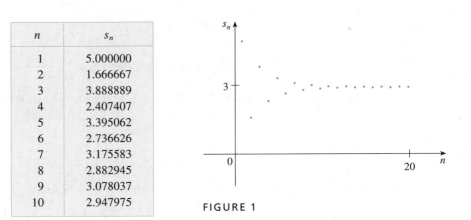

FIGURE 1

EXAMPLE 3 Is the series $\displaystyle\sum_{n=1}^{\infty} 2^{2n}3^{1-n}$ convergent or divergent?

SOLUTION

$$\sum_{n=1}^{\infty} 2^{2n}3^{1-n} = \sum_{n=1}^{\infty} \frac{4^n}{3^{n-1}} = \sum_{n=1}^{\infty} 4\left(\tfrac{4}{3}\right)^{n-1}$$

We recognize this series as a geometric series with $a = 4$ and $r = \tfrac{4}{3}$. Since $r > 1$, the series diverges by (4). $\qquad \blacksquare$

EXAMPLE 4   Write the number $2.3\overline{17} = 2.3171717\ldots$ as a ratio of integers.

SOLUTION
$$2.3171717\ldots = 2.3 + \frac{17}{10^3} + \frac{17}{10^5} + \frac{17}{10^7} + \cdots$$

After the first term we have a geometric series with $a = 17/10^3$ and $r = 1/10^2$. Therefore

$$2.3\overline{17} = 2.3 + \frac{\dfrac{17}{10^3}}{1 - \dfrac{1}{10^2}} = 2.3 + \frac{\dfrac{17}{1000}}{\dfrac{99}{100}}$$

$$= \frac{23}{10} + \frac{17}{990} = \frac{1147}{495}$$

∎

EXAMPLE 5   Find the sum of the series $\displaystyle\sum_{n=0}^{\infty} x^n$, where $|x| < 1$.

SOLUTION   Notice that this series starts with $n = 0$ and so the first term is $x^0 = 1$. Thus

$$\sum_{n=0}^{\infty} x^n = 1 + x + x^2 + x^3 + x^4 + \cdots$$

This is a geometric series with $a = 1$ and $r = x$. Since $|r| = |x| < 1$, it converges and (4) gives

(5)
$$\sum_{n=0}^{\infty} x^n = \frac{1}{1 - x}$$

∎

EXAMPLE 6   Show that the series $\displaystyle\sum_{n=1}^{\infty} \frac{1}{n(n + 1)}$ is convergent and find its sum.

SOLUTION   This is not a geometric series, so we go back to the definition of a convergent series and compute the partial sums

$$s_n = \sum_{i=1}^{n} \frac{1}{i(i + 1)} = \frac{1}{1 \cdot 2} + \frac{1}{2 \cdot 3} + \frac{1}{3 \cdot 4} + \cdots + \frac{1}{n(n + 1)}$$

We can simplify this expression if we use the partial fraction decomposition

$$\frac{1}{i(i + 1)} = \frac{1}{i} - \frac{1}{i + 1}$$

(see Section 7.4). Thus we have

$$s_n = \sum_{i=1}^{n} \frac{1}{i(i + 1)} = \sum_{i=1}^{n} \left(\frac{1}{i} - \frac{1}{i + 1} \right)$$

$$= \left(1 - \frac{1}{2} \right) + \left(\frac{1}{2} - \frac{1}{3} \right) + \left(\frac{1}{3} - \frac{1}{4} \right) + \cdots + \left(\frac{1}{n} - \frac{1}{n + 1} \right)$$

$$= 1 - \frac{1}{n + 1}$$

Notice that the terms cancel in pairs. This is an example of a **telescoping sum**: Because of all the cancellations, the sum collapses (like an old-fashioned collapsing telescope) into just two terms.

Figure 2 illustrates Example 6 by showing the graphs of the sequence of terms $a_n = 1/[n(n + 1)]$ and the sequence $\{s_n\}$ of partial sums. Notice that $a_n \to 0$ and $s_n \to 1$. See Exercises 56 and 57 for two geometric interpretations of Example 6.

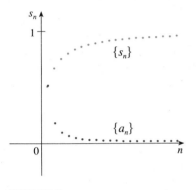

FIGURE 2

and so

$$\lim_{n \to \infty} s_n = \lim_{n \to \infty} \left(1 - \frac{1}{n + 1}\right) = 1 - 0 = 1$$

Therefore, the given series is convergent and

$$\sum_{n=1}^{\infty} \frac{1}{n(n + 1)} = 1$$

EXAMPLE 7 Show that the **harmonic series**

$$\sum_{n=1}^{\infty} \frac{1}{n} = 1 + \frac{1}{2} + \frac{1}{3} + \frac{1}{4} + \cdots$$

is divergent.

SOLUTION

$$s_1 = 1$$

$$s_2 = 1 + \tfrac{1}{2}$$

$$s_4 = 1 + \tfrac{1}{2} + \left(\tfrac{1}{3} + \tfrac{1}{4}\right) > 1 + \tfrac{1}{2} + \left(\tfrac{1}{4} + \tfrac{1}{4}\right) = 1 + \tfrac{2}{2}$$

$$s_8 = 1 + \tfrac{1}{2} + \left(\tfrac{1}{3} + \tfrac{1}{4}\right) + \left(\tfrac{1}{5} + \tfrac{1}{6} + \tfrac{1}{7} + \tfrac{1}{8}\right)$$

$$> 1 + \tfrac{1}{2} + \left(\tfrac{1}{4} + \tfrac{1}{4}\right) + \left(\tfrac{1}{8} + \tfrac{1}{8} + \tfrac{1}{8} + \tfrac{1}{8}\right)$$

$$= 1 + \tfrac{1}{2} + \tfrac{1}{2} + \tfrac{1}{2} = 1 + \tfrac{3}{2}$$

$$s_{16} = 1 + \tfrac{1}{2} + \left(\tfrac{1}{3} + \tfrac{1}{4}\right) + \left(\tfrac{1}{5} + \cdots + \tfrac{1}{8}\right) + \left(\tfrac{1}{9} + \cdots + \tfrac{1}{16}\right)$$

$$> 1 + \tfrac{1}{2} + \left(\tfrac{1}{4} + \tfrac{1}{4}\right) + \left(\tfrac{1}{8} + \cdots + \tfrac{1}{8}\right) + \left(\tfrac{1}{16} + \cdots + \tfrac{1}{16}\right)$$

$$= 1 + \tfrac{1}{2} + \tfrac{1}{2} + \tfrac{1}{2} + \tfrac{1}{2} = 1 + \tfrac{4}{2}$$

Similarly, $s_{32} > 1 + \tfrac{5}{2}$, $s_{64} > 1 + \tfrac{6}{2}$, and in general

$$s_{2^n} > 1 + \frac{n}{2}$$

The method used in Example 7 for showing that the harmonic series diverges is due to the French scholar Nicole Oresme (1323–1382).

This shows that $s_{2^n} \to \infty$ as $n \to \infty$ and so $\{s_n\}$ is divergent. Therefore the harmonic series diverges.

---

(6) THEOREM If the series $\displaystyle\sum_{n=1}^{\infty} a_n$ is convergent, then $\displaystyle\lim_{n \to \infty} a_n = 0$.

PROOF Let $s_n = a_1 + a_2 + \cdots + a_n$. Then $a_n = s_n - s_{n-1}$. Since $\Sigma\, a_n$ is convergent, the sequence $\{s_n\}$ is convergent. Let $\lim_{n \to \infty} s_n = s$. Since $n - 1 \to \infty$ as $n \to \infty$, we also have $\lim_{n \to \infty} s_{n-1} = s$. Therefore

$$\lim_{n \to \infty} a_n = \lim_{n \to \infty} (s_n - s_{n-1}) = \lim_{n \to \infty} s_n - \lim_{n \to \infty} s_{n-1}$$

$$= s - s = 0$$

NOTE 1: With any *series* $\Sigma\, a_n$ we associate two *sequences*: the sequence $\{s_n\}$ of its partial sums and the sequence $\{a_n\}$ of its terms. If $\Sigma\, a_n$ is convergent, then the limit of the sequence $\{s_n\}$ is s and, as Theorem 6 asserts, the limit of the sequence $\{a_n\}$ is 0.

⊘   **NOTE 2:**   The converse of Theorem 6 is not true in general. If $\lim_{n \to \infty} a_n = 0$, we cannot conclude that $\Sigma\ a_n$ is convergent. Observe that for the harmonic series $\Sigma\ 1/n$ we have $a_n = 1/n \to 0$ as $n \to \infty$, but we showed in Example 7 that $\Sigma\ 1/n$ is divergent.

---

(7) THE TEST FOR DIVERGENCE If $\lim\limits_{n \to \infty} a_n$ does not exist or if $\lim\limits_{n \to \infty} a_n \neq 0$, then the series $\sum\limits_{n=1}^{\infty} a_n$ is divergent.

PROOF   This follows immediately from Theorem 6.   □

EXAMPLE 8   Show that the series $\sum\limits_{n=1}^{\infty} \dfrac{n^2}{5n^2 + 4}$ diverges.

SOLUTION

$$\lim_{n \to \infty} a_n = \lim_{n \to \infty} \frac{n^2}{5n^2 + 4} = \lim_{n \to \infty} \frac{1}{5 + 4/n^2} = \frac{1}{5} \neq 0$$

So the series diverges by the Test for Divergence.   ■

NOTE 3:   If we find that $\lim_{n \to \infty} a_n \neq 0$, we know that $\Sigma\ a_n$ is divergent. If we find that $\lim_{n \to \infty} a_n = 0$, we know *nothing* about the convergence or divergence of $\Sigma\ a_n$. Remember the warning in Note 2: If $\lim_{n \to \infty} a_n = 0$, the series $\Sigma\ a_n$ might converge or it might diverge.

---

(8) THEOREM   If $\Sigma\ a_n$ and $\Sigma\ b_n$ are convergent series, then so are the series $\Sigma\ ca_n$ (where c is a constant), $\Sigma\ (a_n + b_n)$, and $\Sigma\ (a_n - b_n)$, and

(i) $\sum\limits_{n=1}^{\infty} ca_n = c \sum\limits_{n=1}^{\infty} a_n$ (ii) $\sum\limits_{n=1}^{\infty} (a_n + b_n) = \sum\limits_{n=1}^{\infty} a_n + \sum\limits_{n=1}^{\infty} b_n$

(iii) $\sum\limits_{n=1}^{\infty} (a_n - b_n) = \sum\limits_{n=1}^{\infty} a_n - \sum\limits_{n=1}^{\infty} b_n$

PROOF
(i) This proof is left as Exercise 61.
(ii) Let

$$s_n = \sum_{i=1}^{n} a_i \qquad s = \sum_{n=1}^{\infty} a_n \qquad t_n = \sum_{i=1}^{n} b_i \qquad t = \sum_{n=1}^{\infty} b_n$$

The nth partial sum for the series $\Sigma\ (a_n + b_n)$ is

$$u_n = \sum_{i=1}^{n} (a_i + b_i)$$

and, using Theorem 4.1.2, we have

$$\lim_{n \to \infty} u_n = \lim_{n \to \infty} \sum_{i=1}^{n} (a_i + b_i) = \lim_{n \to \infty} \left(\sum_{i=1}^{n} a_i + \sum_{i=1}^{n} b_i \right)$$

$$= \lim_{n \to \infty} \sum_{i=1}^{n} a_i + \lim_{n \to \infty} \sum_{i=1}^{n} b_i$$

$$= \lim_{n \to \infty} s_n + \lim_{n \to \infty} t_n = s + t$$

Therefore, $\Sigma (a_n + b_n)$ is convergent and its sum is

$$\sum_{n=1}^{\infty} (a_n + b_n) = s + t = \sum_{n=1}^{\infty} a_n + \sum_{n=1}^{\infty} b_n$$

(iii) This equation is proved like part (ii) or can be deduced from parts (i) and (ii).

$\square$

EXAMPLE 9 Find the sum of the series $\displaystyle\sum_{n=1}^{\infty} \left(\frac{3}{n(n + 1)} + \frac{1}{2^n} \right)$.

SOLUTION   The series $\Sigma \, 1/2^n$ is a geometric series with $a = \frac{1}{2}$ and $r = \frac{1}{2}$, so

$$\sum_{n=1}^{\infty} \frac{1}{2^n} = \frac{\frac{1}{2}}{1 - \frac{1}{2}} = 1$$

In Example 6 we found that

$$\sum_{n=1}^{\infty} \frac{1}{n(n + 1)} = 1$$

So, by Theorem 8, the given series is convergent and

$$\sum_{n=1}^{\infty} \left(\frac{3}{n(n + 1)} + \frac{1}{2^n} \right) = 3 \sum_{n=1}^{\infty} \frac{1}{n(n + 1)} + \sum_{n=1}^{\infty} \frac{1}{2^n}$$

$$= 3 \cdot 1 + 1 = 4 \qquad \blacksquare$$

NOTE 4:   A finite number of terms cannot affect the convergence of a series. For instance, suppose that we were able to show that the series

$$\sum_{n=4}^{\infty} \frac{n}{n^3 + 1}$$

is convergent. Since

$$\sum_{n=1}^{\infty} \frac{n}{n^3 + 1} = \frac{1}{2} + \frac{2}{9} + \frac{3}{28} + \sum_{n=4}^{\infty} \frac{n}{n^3 + 1}$$

it follows that the entire series $\sum_{n=1}^{\infty} n/(n^3 + 1)$ is convergent. Similarly, if it is known that the series $\sum_{n=N+1}^{\infty} a_n$ converges, then the full series

$$\sum_{n=1}^{\infty} a_n = \sum_{n=1}^{N} a_n + \sum_{n=N+1}^{\infty} a_n$$

is also convergent.

EXERCISES 10.2

1–6 ■ Find at least 10 partial sums of the series. Graph both the sequence of terms and the sequence of partial sums on the same screen. Does it appear that the series is convergent or divergent? If it is convergent, find the sum. If it is divergent, explain why.

1. $\displaystyle\sum_{n=1}^{\infty} \frac{10}{3^n}$

2. $\displaystyle\sum_{n=1}^{\infty} \sin n$

3. $\displaystyle\sum_{n=1}^{\infty} \frac{n}{n+1}$

4. $\displaystyle\sum_{n=4}^{\infty} \frac{3}{n(n-1)}$

5. $\displaystyle\sum_{n=1}^{\infty} \left(\frac{1}{n^{1.5}} - \frac{1}{(n+1)^{1.5}}\right)$

6. $\displaystyle\sum_{n=1}^{\infty} \left(-\frac{2}{7}\right)^{n-1}$

7–36 ■ Determine whether the series is convergent or divergent. If it is convergent, find its sum.

7. $4 + \frac{8}{5} + \frac{16}{25} + \frac{32}{125} + \cdots$

8. $1 - \frac{1}{2} + \frac{1}{4} - \frac{1}{8} + \cdots$

9. $\frac{2}{3} - \frac{2}{9} + \frac{2}{27} - \frac{2}{81} + \cdots$

10. $-\frac{81}{100} + \frac{9}{10} - 1 + \frac{10}{9} - \cdots$

11. $\displaystyle\sum_{n=1}^{\infty} 2\left(\frac{3}{4}\right)^{n-1}$

12. $\displaystyle\sum_{n=1}^{\infty} \left(-\frac{3}{\pi}\right)^{n-1}$

13. $\displaystyle\sum_{n=1}^{\infty} 5\left(\frac{e}{3}\right)^{n}$

14. $\displaystyle\sum_{n=1}^{\infty} \frac{1}{e^{2n}}$

15. $\displaystyle\sum_{n=0}^{\infty} \frac{5^n}{8^n}$

16. $\displaystyle\sum_{n=0}^{\infty} \frac{4^{n+1}}{5^n}$

17. $\displaystyle\sum_{n=1}^{\infty} 3^{-n}8^{n+1}$

18. $\displaystyle\sum_{n=1}^{\infty} (-1)^{n-1}\frac{3^{2n}}{2^{3n+1}}$

19. $\displaystyle\sum_{n=1}^{\infty} \frac{1}{2n}$

20. $\displaystyle\sum_{n=1}^{\infty} \frac{n^2}{3(n+1)(n+2)}$

21. $\displaystyle\sum_{n=1}^{\infty} \frac{1}{(3n-2)(3n+1)}$

22. $\displaystyle\sum_{n=1}^{\infty} \left(\frac{1}{2^{n-1}} + \frac{2}{3^{n-1}}\right)$

23. $\displaystyle\sum_{n=1}^{\infty} [2(0.1)^n + (0.2)^n]$

24. $\displaystyle\sum_{n=1}^{\infty} \left(\frac{1}{n} + 2^n\right)$

25. $\displaystyle\sum_{n=1}^{\infty} \frac{n}{\sqrt{1+n^2}}$

26. $\displaystyle\sum_{n=1}^{\infty} \frac{1}{4n^2-1}$

27. $\displaystyle\sum_{n=1}^{\infty} \frac{1}{n(n+2)}$

28. $\displaystyle\sum_{n=1}^{\infty} \ln\left(\frac{n}{2n+5}\right)$

29. $\displaystyle\sum_{n=1}^{\infty} \frac{3^n + 2^n}{6^n}$

30. $\displaystyle\sum_{n=1}^{\infty} \frac{2n+1}{n^2(n+1)^2}$

31. $\displaystyle\sum_{n=1}^{\infty} \left[\sin\left(\frac{1}{n}\right) - \sin\left(\frac{1}{n+1}\right)\right]$

32. $\displaystyle\sum_{n=1}^{\infty} \frac{1}{5+2^{-n}}$

33. $\displaystyle\sum_{n=1}^{\infty} \arctan n$

34. $\displaystyle\sum_{n=1}^{\infty} \frac{1}{n(n+1)(n+2)}$

35. $\displaystyle\sum_{n=1}^{\infty} \ln \frac{n}{n+1}$

36. $\displaystyle\sum_{n=2}^{\infty} \ln \frac{n^2-1}{n^2}$

37–42 ■ Express the number as a ratio of integers.

37. $0.\overline{5} = 0.5555\ldots$

38. $0.\overline{15} = 0.15151515\ldots$

39. $0.\overline{307} = 0.307307307307\ldots$

40. $1.\overline{123}$

41. $0.12\overline{3456}$

42. $4.1\overline{570}$

43–48 ■ Find the values of x for which the series converges. Find the sum of the series for those values of x.

43. $\displaystyle\sum_{n=0}^{\infty} (x-3)^n$

44. $\displaystyle\sum_{n=0}^{\infty} 3^n x^n$

45. $\displaystyle\sum_{n=2}^{\infty} \frac{x^n}{5^n}$

46. $\displaystyle\sum_{n=0}^{\infty} \frac{1}{x^n}$

47. $\displaystyle\sum_{n=0}^{\infty} 2^n \sin^n x$

48. $\displaystyle\sum_{n=0}^{\infty} \tan^n x$

CAS **49–50** ■ Use the partial fraction command on your CAS to find a convenient expression for the partial sum, and then use this expression to find the sum of the series. Check your answer by using the CAS to sum the series directly.

49. $\displaystyle\sum_{n=1}^{\infty} \frac{1}{(4n+1)(4n-3)}$

50. $\displaystyle\sum_{n=1}^{\infty} \frac{n^2+3n+1}{(n^2+n)^2}$

51. If the nth partial sum of a series $\sum_{n=1}^{\infty} a_n$ is

$$s_n = \frac{n-1}{n+1}$$

find a_n and $\sum_{n=1}^{\infty} a_n$.

52. If the nth partial sum of a series $\sum_{n=1}^{\infty} a_n$ is

$$s_n = 3 - n2^{-n}$$

find a_n and $\sum_{n=1}^{\infty} a_n$.

53. When money is spent on goods and services, those that receive the money also spend some of it. The people receiving some of the twice-spent money will spend some of that, and so on. Economists call this chain reaction the *multiplier effect*. In a hypothetical isolated community, the local government begins the process by spending D dollars. Suppose that each recipient of spent money spends $100c\%$ and saves $100s\%$ of the money that he or she receives. The values c and s are called the *marginal propensity to consume* and the *marginal propensity to save* and, of course, $c + s = 1$.
(a) Let S_n be the total spending that has been generated after n transactions. Find an equation for S_n.
(b) Show that $\lim_{n\to\infty} S_n = kD$, where $k = 1/s$. The number k is called the *multiplier*. What is the multiplier if the marginal propensity to consume is 80%?

Note: The federal government uses this principle to justify deficit spending. Banks use this principle to justify lending out a large percentage of the money that they receive in deposits.

54. A certain ball has the property that each time it falls from a height h onto a hard, level surface, it rebounds to a height rh, where $0 < r < 1$. Suppose that the ball is dropped from an initial height of H meters.
(a) Assuming that the ball continues to bounce indefinitely, find the total distance that it travels.
(b) Calculate the total time that the ball travels.
(c) Suppose that each time the ball strikes the surface with velocity v it rebounds with velocity kv, where $0 < k < 1$. How long will it take for the ball to come to rest?

55. What is the value of c if $\sum_{n=2}^{\infty} (1 + c)^{-n} = 2$?

56. Graph the curves $y = x^n$, $0 \le x \le 1$, for $n = 0, 1, 2, 3, 4, \ldots$ on a common screen. By finding the areas between successive curves, give a geometric demonstration of the fact, shown in Example 6, that
$$\sum_{n=1}^{\infty} \frac{1}{n(n + 1)} = 1$$

57. The figure shows two circles C and D of radius 1 that touch at P. T is a common tangent line; C_1 is the circle that touches C, D, and T; C_2 is the circle that touches C, D, and C_1; C_3 is the circle that touches C, D, and C_2. This procedure can be continued indefinitely and produces an infinite sequence of circles $\{C_n\}$. Find an expression for the diameter of C_n and thus provide another geometric demonstration of Example 6.

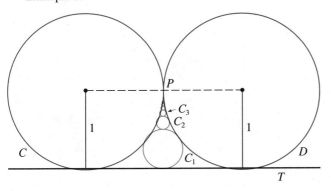

58. A right triangle ABC is given with $\angle A = \theta$ and $|AC| = b$. CD is drawn perpendicular to AB, DE is drawn perpen-

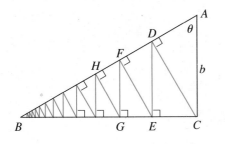

dicular to BC, $EF \perp AB$, and this process is continued indefinitely as in the figure. Find the total length of all the perpendiculars
$$|CD| + |DE| + |EF| + |FG| + \cdots$$
in terms of b and θ.

59. What is wrong with the following calculation?
$$\begin{aligned}
0 &= 0 + 0 + 0 + \cdots \\
&= (1 - 1) + (1 - 1) + (1 - 1) + \cdots \\
&= 1 - 1 + 1 - 1 + 1 - 1 + \cdots \\
&= 1 + (-1 + 1) + (-1 + 1) + (-1 + 1) + \cdots \\
&= 1 + 0 + 0 + 0 + \cdots = 1
\end{aligned}$$

(Guido Ubaldus thought that this proved the existence of God because "something has been created out of nothing.")

60. Suppose that $\sum_{n=1}^{\infty} a_n$ ($a_n \ne 0$) is known to be a convergent series. Prove that $\sum_{n=1}^{\infty} 1/a_n$ is a divergent series.

61. Prove part (i) of Theorem 8.

62. If $\Sigma\, a_n$ is divergent and $c \ne 0$, show that $\Sigma\, ca_n$ is divergent.

63. If $\Sigma\, a_n$ is convergent and $\Sigma\, b_n$ is divergent, show that the series $\Sigma\, (a_n + b_n)$ is divergent. [*Hint:* Argue by contradiction.]

64. If $\Sigma\, a_n$ and $\Sigma\, b_n$ are both divergent, is $\Sigma\, (a_n + b_n)$ necessarily divergent?

65. Suppose that a series $\Sigma\, a_n$ has positive terms and its partial sums s_n satisfy the inequality $s_n \le 1000$ for all n. Explain why $\Sigma\, a_n$ must be convergent.

66. The Fibonacci sequence was defined in Section 10.1 by the equations
$$f_1 = 1, \quad f_2 = 1, \quad f_n = f_{n-1} + f_{n-2} \qquad n \ge 3$$
Show that each of the following statements is true.
(a) $\dfrac{1}{f_{n-1} f_{n+1}} = \dfrac{1}{f_{n-1} f_n} - \dfrac{1}{f_n f_{n+1}}$
(b) $\displaystyle\sum_{n=2}^{\infty} \frac{1}{f_{n-1} f_{n+1}} = 1$
(c) $\displaystyle\sum_{n=2}^{\infty} \frac{f_n}{f_{n-1} f_{n+1}} = 2$

67. The **Cantor set,** named after the German mathematician Georg Cantor (1845–1918), is constructed as follows. We start with the closed interval $[0, 1]$ and remove the open interval $\left(\frac{1}{3}, \frac{2}{3}\right)$. That leaves the two intervals $\left[0, \frac{1}{3}\right]$ and $\left[\frac{2}{3}, 1\right]$ and we remove the open middle third of each. Four intervals remain and again we remove the open middle third of each of them. We continue this procedure indefinitely, at each step removing the open middle third of every interval that remains from the preceding step. The

Cantor set consists of the numbers that remain in $[0, 1]$ after all those intervals have been deleted.

(a) Show that the total length of all the intervals that are removed is 1. Despite that, the Cantor set contains infinitely many numbers. Give examples of some numbers in the Cantor set.

(b) The **Sierpinski carpet** is a two-dimensional analogue of the Cantor set. It is constructed by removing the center one-ninth of a square of side 1, then removing the centers of the eight smaller remaining squares, and so on. (The figure shows the first three steps of the construction.) Show that the sum of the areas of the removed squares is 1. This implies that the Sierpinski carpet has area 0.

68. (a) A sequence $\{a_n\}$ is defined recursively by the equation $a_n = \frac{1}{2}(a_{n-1} + a_{n-2})$ for $n \geq 3$, where a_1 and a_2 can be any real numbers. Experiment with various values of a_1 and a_2 and use your calculator to guess the limit of the sequence.

(b) Find $\lim_{n \to \infty} a_n$ in terms of a_1 and a_2 by expressing $a_{n+1} - a_n$ in terms of $a_2 - a_1$ and summing a series.

69. Consider the series

$$\sum_{n=1}^{\infty} \frac{n}{(n+1)!}$$

(a) Find the partial sums s_1, s_2, s_3, and s_4. Do you recognize the denominators? Use the pattern to guess a formula for s_n.

(b) Use mathematical induction to prove your guess.

(c) Show that the given infinite series is convergent and find its sum.

70. In the figure there are infinitely many circles approaching the vertices of an equilateral triangle, each circle touching other circles and sides of the triangle. If the triangle has sides of length 1, find the total area occupied by the circles.

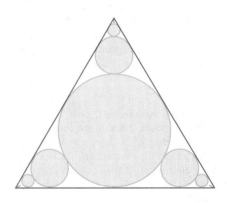

10.3   THE INTEGRAL TEST

In general it is difficult to find the exact sum of a series. We were able to accomplish this for geometric series and the series $\sum 1/[n(n+1)]$ because in each of those cases we could find a simple formula for the nth partial sum s_n. But usually it is not easy to compute $\lim_{n \to \infty} s_n$. Therefore, in the next few sections we develop several tests that enable us to determine whether a series is convergent or divergent without explicitly finding its sum. (In some cases, however, our methods will enable us to find good estimates of the sum.) Our first test involves improper integrals.

THE INTEGRAL TEST   Suppose f is a continuous, positive, decreasing function on $[1, \infty)$ and let $a_n = f(n)$. Then the series $\sum_{n=1}^{\infty} a_n$ is convergent if and only if the improper integral $\int_1^{\infty} f(x)\, dx$ is convergent. In other words:

(a) If $\int_1^{\infty} f(x)\, dx$ is convergent, then $\displaystyle\sum_{n=1}^{\infty} a_n$ is convergent.

(b) If $\int_1^{\infty} f(x)\, dx$ is divergent, then $\displaystyle\sum_{n=1}^{\infty} a_n$ is divergent.

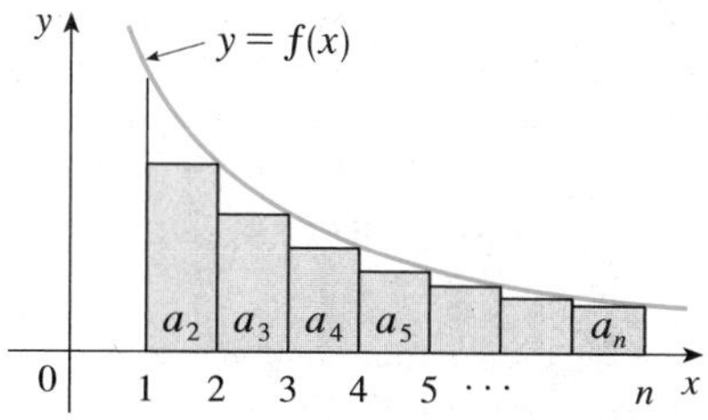

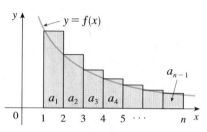

FIGURE 1

FIGURE 2

PROOF   The basic idea behind the Integral Test can be seen by looking at Figures 1 and 2. The area of the first shaded rectangle in Figure 1 is the value of f at the right endpoint of $[1, 2]$, that is, $f(2) = a_2$. So, comparing the areas of the shaded rectangles with the area under $y = f(x)$ from 1 to n, we see that

(1)
$$a_2 + a_3 + \cdots + a_n \le \int_1^n f(x)\, dx$$

(Notice that this inequality depends on the fact that f is decreasing.) Likewise, Figure 2 shows that

(2)
$$\int_1^n f(x)\, dx \le a_1 + a_2 + \cdots + a_{n-1}$$

(a) If $\int_1^\infty f(x)\, dx$ is convergent, then (1) gives

$$\sum_{i=2}^n a_i \le \int_1^n f(x)\, dx \le \int_1^\infty f(x)\, dx$$

since $f(x) \ge 0$. Therefore

$$s_n = a_1 + \sum_{i=2}^n a_i \le a_1 + \int_1^\infty f(x)\, dx = M, \text{ say}$$

Since $s_n \le M$ for all n, the sequence $\{s_n\}$ is bounded above. Also

$$s_{n+1} = s_n + a_{n+1} \ge s_n$$

since $a_{n+1} = f(n + 1) \ge 0$. Thus $\{s_n\}$ is an increasing bounded sequence and so it is convergent by Theorem 10.1.10. This means that $\sum a_n$ is convergent.

(b) If $\int_1^\infty f(x)\, dx$ is divergent, then $\int_1^n f(x)\, dx \to \infty$ as $n \to \infty$ because $f(x) \ge 0$. But (2) gives

$$\int_1^n f(x)\, dx \le \sum_{i=1}^{n-1} a_i = s_{n-1}$$

and so $s_{n-1} \to \infty$. This implies that $s_n \to \infty$ and so $\sum a_n$ diverges.   □

NOTE: When we use the Integral Test it is not necessary to start the series or the integral at $n = 1$. For instance, in testing the series

$$\sum_{n=4}^\infty \frac{1}{(n - 3)^2} \qquad \text{we use} \qquad \int_4^\infty \frac{1}{(x - 3)^2}\, dx$$

Also, it is not necessary that f be always decreasing. What is important is that f be *ultimately* decreasing, that is, decreasing for x larger than some number N. Then $\sum_{n=N}^\infty a_n$ is convergent, so $\sum_{n=1}^\infty a_n$ is convergent by Note 4 of Section 10.2.

EXAMPLE 1   Test the series $\displaystyle\sum_{n=1}^\infty \frac{1}{n^2 + 1}$ for convergence or divergence.

SOLUTION   The function $f(x) = 1/(x^2 + 1)$ is continuous, positive, and decreasing on $[1, \infty)$ so we use the Integral Test:

$$\int_1^\infty \frac{1}{x^2 + 1}\, dx = \lim_{t \to \infty} \int_1^t \frac{1}{x^2 + 1}\, dx = \lim_{t \to \infty} \tan^{-1}x \Big]_1^t$$

$$= \lim_{t \to \infty} \left(\tan^{-1}t - \frac{\pi}{4} \right) = \frac{\pi}{2} - \frac{\pi}{4} = \frac{\pi}{4}$$

Thus $\int_1^\infty 1/(x^2 + 1)\,dx$ is a convergent integral and so, by the Integral Test, the series $\Sigma\, 1/(n^2 + 1)$ is convergent.   ∎

EXAMPLE 2   For what values of p is the series $\displaystyle\sum_{n=1}^{\infty} \frac{1}{n^p}$ convergent?

SOLUTION   If $p < 0$, then $\lim_{n\to\infty} (1/n^p) = \infty$. If $p = 0$, then $\lim_{n\to\infty} (1/n^p) = 1$. In either case $\lim_{n\to\infty} (1/n^p) \neq 0$, so the given series diverges by the Test for Divergence (10.2.7).

If $p > 0$, then the function $f(x) = 1/x^p$ is clearly continuous, positive, and decreasing on $[1, \infty)$. We found in Chapter 7 [see (7.9.2)] that

$$\int_1^\infty \frac{1}{x^p}\,dx \text{ converges if } p > 1 \text{ and diverges if } p \leq 1$$

It follows from the Integral Test that the series $\Sigma\, 1/n^p$ converges if $p > 1$ and diverges if $0 < p \leq 1$. (For $p = 1$, this series is the harmonic series discussed in Example 7 in Section 10.2.)   ∎

The series in Example 2 is called the **p-series.** It is important in the rest of this chapter, so we summarize the results of Example 2 for future reference as follows:

(3)   The p-series $\displaystyle\sum_{n=1}^{\infty} \frac{1}{n^p}$ is convergent if $p > 1$ and divergent if $p \leq 1$.

EXAMPLE 3   The series $\Sigma\, 1/n^2$ is convergent because it is a p-series with $p = 2 > 1$. The exact sum of this series was found by the Swiss mathematician Leonhard Euler (1707–1783) to be

$$\sum_{n=1}^{\infty} \frac{1}{n^2} = \frac{\pi^2}{6}$$

but the proof of this fact is beyond the scope of this book.   ∎

NOTE: We should *not* infer from the Integral Test that the sum of the series is equal to the value of the integral. In fact, we know from Example 3 and from Section 7.9 that

$$\sum_{n=1}^{\infty} \frac{1}{n^2} = \frac{\pi^2}{6} \qquad \text{whereas} \qquad \int_1^\infty \frac{1}{x^2}\,dx = 1$$

Therefore, in general,

$$\sum_{n=1}^{\infty} a_n \neq \int_1^\infty f(x)\,dx$$

EXAMPLE 4   The series $\Sigma\, 1/\sqrt{n}$ is divergent because it can be rewritten as $\Sigma\, 1/n^{1/2}$, which is a p-series with $p = \frac{1}{2} < 1$.   ∎

EXAMPLE 5   Determine whether the series $\sum_{n=1}^{\infty} \dfrac{\ln n}{n}$ converges or diverges.

SOLUTION   The function $f(x) = (\ln x)/x$ is positive and continuous for $x > 1$ because the logarithm function is continuous. But it is not obvious whether or not f is decreasing, so we compute its derivative:

$$f'(x) = \frac{(1/x)x - \ln x}{x^2} = \frac{1 - \ln x}{x^2}$$

Thus $f'(x) < 0$ when $\ln x > 1$, that is, $x > e$. It follows that f is decreasing when $x > e$ and so we can apply the Integral Test:

$$\int_1^{\infty} \frac{\ln x}{x} \, dx = \lim_{t \to \infty} \int_1^t \frac{\ln x}{x} \, dx = \lim_{t \to \infty} \frac{(\ln x)^2}{2} \bigg]_1^t$$

$$= \lim_{t \to \infty} \frac{(\ln t)^2}{2} = \infty$$

Since this improper integral is divergent, the series $\sum (\ln n)/n$ is also divergent by the Integral Test. ∎

ESTIMATING THE SUM OF A SERIES

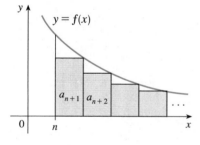

Suppose we have been able to use the Integral Test to show that a series $\sum a_n$ is convergent and we now want to find an approximation to the sum s of the series. Of course, any partial sum s_n is an approximation to s because $\lim_{n \to \infty} s_n = s$. But how good is such an approximation? To find out, we need to estimate the size of the **remainder**

$$R_n = s - s_n = a_{n+1} + a_{n+2} + a_{n+3} + \cdots$$

The remainder R_n is the error made when s_n, the sum of the first n terms, is used as an approximation to the total sum.

We use the same notation and ideas as in the Integral Test. Comparing the areas of the rectangles with the area under $y = f(x)$ for $x > n$ in Figure 3, we see that

$$R_n = a_{n+1} + a_{n+2} + \cdots \leqslant \int_n^{\infty} f(x) \, dx$$

FIGURE 3

Similarly, we see from Figure 4 that

$$R_n = a_{n+1} + a_{n+2} + \cdots \geqslant \int_{n+1}^{\infty} f(x) \, dx$$

So we have proved the following error estimate.

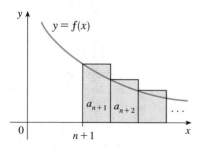

FIGURE 4

> **(4) REMAINDER ESTIMATE FOR THE INTEGRAL TEST**   If $\sum a_n$ converges by the Integral Test and $R_n = s - s_n$, then
>
> $$\int_{n+1}^{\infty} f(x) \, dx \leqslant R_n \leqslant \int_n^{\infty} f(x) \, dx$$

EXAMPLE 6

(a) Approximate the sum of the series $\Sigma\, 1/n^3$ by using the sum of the first 10 terms. Estimate the error involved in this approximation.

(b) How many terms are required to ensure that the sum is accurate to within 0.0005?

SOLUTION   In both parts (a) and (b) we need to know $\int_n^\infty f(x)\,dx$. With $f(x) = 1/x^3$, we have

$$\int_n^\infty \frac{1}{x^3}\,dx = \lim_{t\to\infty}\left[-\frac{1}{2x^2}\right]_n^t = \lim_{t\to\infty}\left(-\frac{1}{2t^2} + \frac{1}{2n^2}\right) = \frac{1}{2n^2}$$

(a)

$$\sum_{n=1}^{\infty} \frac{1}{n^3} \approx s_{10} = \frac{1}{1^3} + \frac{1}{2^3} + \frac{1}{3^3} + \cdots + \frac{1}{10^3} \approx 1.1975$$

According to the remainder estimate in (4), we have

$$R_{10} \leqslant \int_{10}^\infty \frac{1}{x^3}\,dx = \frac{1}{2(10)^2} = \frac{1}{200}$$

So the size of the error is at most 0.005.

(b) Accuracy to within 0.0005 means that we have to find a value of n such that $R_n \leqslant 0.0005$. Since

$$R_n \leqslant \int_n^\infty \frac{1}{x^3}\,dx = \frac{1}{2n^2}$$

we want

$$\frac{1}{2n^2} < 0.0005$$

Solving this inequality, we get

$$n^2 > \frac{1}{0.001} = 1000 \qquad \text{or} \qquad n > \sqrt{1000} \approx 31.6$$

We need 32 terms to ensure accuracy to within 0.0005. ∎

If we add s_n to each side of the inequalities in (4), we get

(5)

$$\boxed{\; s_n + \int_{n+1}^\infty f(x)\,dx \leqslant s \leqslant s_n + \int_n^\infty f(x)\,dx \;}$$

because $s_n + R_n = s$. The inequalities in (5) give a lower bound and an upper bound for s. They provide a more accurate approximation to the sum of the series than the partial sum s_n does.

EXAMPLE 7   Use (5) with $n = 10$ to estimate the sum of the series $\displaystyle\sum_{n=1}^{\infty} \frac{1}{n^3}$.

SOLUTION   The inequalities in (5) become

$$s_{10} + \int_{11}^\infty \frac{1}{x^3}\,dx \leqslant s \leqslant s_{10} + \int_{10}^\infty \frac{1}{x^3}\,dx$$

From Example 6 we know that

$$\int_n^\infty \frac{1}{x^3}\, dx = \frac{1}{2n^2}$$

so

$$s_{10} + \frac{1}{2(11)^2} \leqslant s \leqslant s_{10} + \frac{1}{2(10)^2}$$

Using $s_{10} \approx 1.197532$, we get

$$1.201664 \leqslant s \leqslant 1.202532$$

If we approximate s by the midpoint of this interval, then the error is at most half the length of the interval. So

$$\sum_{n=1}^\infty \frac{1}{n^3} \approx 1.2021 \qquad \text{with error} < 0.0005$$

■

If we compare Example 7 with Example 6, we see that the improved estimate in (5) can be much better than the estimate $s \approx s_n$. To make the error smaller than 0.0005 we had to use 32 terms in Example 6 but only 10 terms in Example 7.

EXERCISES 10.3

1–18 ■ Test the series for convergence or divergence.

1. $\displaystyle\sum_{n=1}^\infty \frac{2}{\sqrt[3]{n}}$

2. $\displaystyle\sum_{n=1}^\infty \left(\frac{2}{n\sqrt{n}} + \frac{3}{n^3} \right)$

3. $\displaystyle\sum_{n=5}^\infty \frac{1}{n^{1.0001}}$

4. $\displaystyle\sum_{n=1}^\infty n^{-0.99}$

5. $\displaystyle\sum_{n=5}^\infty \frac{1}{(n-4)^2}$

6. $\displaystyle\sum_{n=1}^\infty \frac{1}{2n+3}$

7. $\displaystyle\sum_{n=1}^\infty \frac{1}{\sqrt{n}+1}$

8. $\displaystyle\sum_{n=2}^\infty \frac{1}{n^2-1}$

9. $\displaystyle\sum_{n=1}^\infty n e^{-n^2}$

10. $\displaystyle\sum_{n=1}^\infty \frac{n}{2^n}$

11. $\displaystyle\sum_{n=1}^\infty \frac{n}{n^2+1}$

12. $\displaystyle\sum_{n=2}^\infty \frac{1}{2n^2-n-1}$

13. $\displaystyle\sum_{n=2}^\infty \frac{1}{n\ln n}$

14. $\displaystyle\sum_{n=1}^\infty \frac{1}{4n^2+1}$

15. $\displaystyle\sum_{n=1}^\infty \frac{\arctan n}{1+n^2}$

16. $\displaystyle\sum_{n=1}^\infty \frac{\ln n}{n^2}$

17. $\displaystyle\sum_{n=1}^\infty \frac{1}{n^2+2n+2}$

18. $\displaystyle\sum_{n=3}^\infty \frac{1}{n\ln n \ln(\ln n)}$

19–22 ■ Find the values of p for which the series is convergent.

19. $\displaystyle\sum_{n=2}^\infty \frac{1}{n(\ln n)^p}$

20. $\displaystyle\sum_{n=3}^\infty \frac{1}{n\ln n[\ln(\ln n)]^p}$

21. $\displaystyle\sum_{n=1}^\infty n(1+n^2)^p$

22. $\displaystyle\sum_{n=1}^\infty \frac{\ln n}{n^p}$

23. The Riemann zeta-function ζ is defined by

$$\zeta(x) = \sum_{n=1}^\infty \frac{1}{n^x}$$

and is used in number theory to study the distribution of prime numbers. What is the domain of ζ?

24. (a) Find the partial sum s_{10} of the series $\sum_{n=1}^\infty 1/n^4$. Estimate the error in using s_{10} as an approximation to the sum of the series.
 (b) Use (5) with $n = 10$ to give an improved estimate of the sum.
 (c) Find a value of n so that s_n is within 0.00001 of the sum.

25. (a) Use the sum of the first 10 terms to estimate the sum of the series $\sum_{n=1}^\infty 1/n^2$. How good is this estimate?
 (b) Improve this estimate using (5) with $n = 10$.
 (c) Find a value of n that will ensure that the error in the approximation $s \approx s_n$ is less than 0.001.

26. Find the sum of the series $\sum_{n=1}^\infty 1/n^5$ correct to three decimal places.

27. Estimate $\sum_{n=1}^\infty n^{-3/2}$ to within 0.01.

28. How many terms of the series $\sum_{n=2}^\infty 1/[n(\ln n)^2]$ would you need to add to find its sum to within 0.01?

29. (a) Use (1) to show that if s_n is the nth partial sum of the harmonic series, then

$$s_n \leq 1 + \ln n$$

(b) The harmonic series diverges but very slowly. Use part (a) to show that the sum of the first million terms is less than 15 and the sum of the first billion terms is less than 22.

CAS **30.** (a) Show that the series $\sum_{n=1}^{\infty} (\ln n)^2/n^2$ is convergent.
(b) Find an upper bound for the error in the approximation $s \approx s_n$.
(c) What is the smallest value of n such that this upper bound is less than 0.05?
(d) Find s_n for this value of n.

31. Find all positive values of b for which the series $\sum_{n=1}^{\infty} b^{\ln n}$ converges.

32. Use the following steps to show that the sequence

$$t_n = 1 + \frac{1}{2} + \frac{1}{3} + \cdots + \frac{1}{n} - \ln n$$

has a limit. (The value of the limit is denoted by γ and is called Euler's constant.)
(a) Draw a picture like Figure 2 with $f(x) = 1/x$ and interpret t_n as an area [or use (2)] to show that $t_n > 0$ for all n.
(b) Interpret

$$t_n - t_{n+1} = [\ln(n+1) - \ln n] - \frac{1}{n+1}$$

as a difference of areas to show that $t_n - t_{n+1} > 0$. Therefore $\{t_n\}$ is a decreasing sequence.
(c) Use Theorem 10.1.10 to show that $\{t_n\}$ is convergent.

10.4   THE COMPARISON TESTS

In the comparison tests the idea is to compare a given series with a series that is known to be convergent or divergent.

> **THE COMPARISON TEST**   Suppose that $\Sigma \, a_n$ and $\Sigma \, b_n$ are series with positive terms.
> (a) If $\Sigma \, b_n$ is convergent and $a_n \leq b_n$ for all n, then $\Sigma \, a_n$ is also convergent.
> (b) If $\Sigma \, b_n$ is divergent and $a_n \geq b_n$ for all n, then $\Sigma \, a_n$ is also divergent.

It is important to keep in mind the distinction between a sequence and a series. A sequence is a list of numbers, whereas a series is a sum. With every series $\Sigma \, a_n$ there are associated two sequences: the sequence $\{a_n\}$ of terms and the sequence $\{s_n\}$ of partial sums.

PROOF
(a) Let

$$s_n = \sum_{i=1}^{n} a_i \qquad t_n = \sum_{i=1}^{n} b_i \qquad t = \sum_{n=1}^{\infty} b_n$$

Since both series have positive terms, the sequences $\{s_n\}$ and $\{t_n\}$ are increasing $(s_{n+1} = s_n + a_{n+1} \geq s_n)$. Also $t_n \to t$, so $t_n \leq t$ for all n. Since $a_i \leq b_i$, we have $s_n \leq t_n$. Thus $s_n \leq t$ for all n. This means that $\{s_n\}$ is increasing and bounded above and therefore converges by Theorem 10.1.10. Thus $\Sigma \, a_n$ converges.

(b) If $\Sigma \, b_n$ is divergent, then $t_n \to \infty$ (since $\{t_n\}$ is increasing). But $a_i \geq b_i$ so $s_n \geq t_n$. Thus $s_n \to \infty$. Therefore, $\Sigma \, a_n$ diverges. $\square$

Standard series for use with the Comparison Test

In using the Comparison Test we must, of course, have some known series $\Sigma \, b_n$ for the purpose of comparison. Most of the time we use either a p-series [$\Sigma \, 1/n^p$ converges if $p > 1$ and diverges if $p \leq 1$; see (10.3.3)] or a geometric series [$\Sigma \, ar^{n-1}$ converges if $|r| < 1$ and diverges if $|r| \geq 1$; see (10.2.4)].

EXAMPLE 1   Determine whether the series $\displaystyle\sum_{n=1}^{\infty} \frac{5}{2n^2 + 4n + 3}$ converges or diverges.

SOLUTION   For large n the dominant term in the denominator is $2n^2$ so we compare the given series with the series $\Sigma\, 5/(2n^2)$. Observe that

$$\frac{5}{2n^2 + 4n + 3} < \frac{5}{2n^2}$$

because the left side has a bigger denominator. (In the notation of the Comparison Test, a_n is the left side and b_n is the right side.) We know that

$$\sum_{n=1}^{\infty} \frac{5}{2n^2} = \frac{5}{2} \sum_{n=1}^{\infty} \frac{1}{n^2}$$

is convergent (p-series with $p = 2 > 1$). Therefore

$$\sum_{n=1}^{\infty} \frac{5}{2n^2 + 4n + 3}$$

is convergent by part (a) of the Comparison Test.   ∎

Although the condition $a_n \le b_n$ or $a_n \ge b_n$ in the Comparison Test is given for all n, we need verify only that it holds for $n \ge N$, where N is some fixed integer, because the convergence of a series is not affected by a finite number of terms. This is illustrated in the next example.

EXAMPLE 2   Test the series $\displaystyle\sum_{n=1}^{\infty} \frac{\ln n}{n}$ for convergence or divergence.

SOLUTION   This series was tested (using the Integral Test) in Example 5 in Section 10.3, but it is also possible to test it by comparing it with the harmonic series. Observe that $\ln n > 1$ for $n \ge 3$ and so

$$\frac{\ln n}{n} > \frac{1}{n} \qquad n \ge 3$$

We know that $\Sigma\, 1/n$ is divergent (p-series with $p = 1$). Thus the given series is divergent by the Comparison Test.   ∎

EXAMPLE 3   Test the series $\displaystyle\sum_{n=1}^{\infty} \frac{1}{2^n + 1}$ for convergence or divergence.

SOLUTION   Notice that

$$\frac{1}{2^n + 1} < \frac{1}{2^n} = \left(\tfrac{1}{2}\right)^n \qquad n \ge 1$$

The series $\Sigma\, \left(\tfrac{1}{2}\right)^n$ is convergent (geometric series with $r = \tfrac{1}{2}$) and so the given series converges by the Comparison Test.   ∎

NOTE: The terms of the series being tested must be smaller than those of a convergent series or larger than those of a divergent series. If the terms are larger than the terms of a convergent series or smaller than those of a divergent series, then the Comparison Test does not apply. For instance, suppose that in Example 3 we had been given the similar series

$$\sum_{n=1}^{\infty} \frac{1}{2^n - 1}$$

The inequality

$$\frac{1}{2^n - 1} > \frac{1}{2^n}$$

is useless as far as the Comparison Test is concerned because $\Sigma\, b_n = \Sigma \left(\frac{1}{2}\right)^n$ is convergent and $a_n > b_n$. Nonetheless we have the feeling that $\Sigma\, 1/(2^n - 1)$ ought to be convergent since it is very similar to the convergent geometric series $\Sigma \left(\frac{1}{2}\right)^n$. In such cases the following test can be used.

THE LIMIT COMPARISON TEST Suppose that $\Sigma\, a_n$ and $\Sigma\, b_n$ are series with positive terms.

(a) If $\lim\limits_{n \to \infty} \dfrac{a_n}{b_n} = c > 0$, then either both series converge or both diverge.

(b) If $\lim\limits_{n \to \infty} \dfrac{a_n}{b_n} = 0$ and $\Sigma\, b_n$ converges, then $\Sigma\, a_n$ also converges.

(c) If $\lim\limits_{n \to \infty} \dfrac{a_n}{b_n} = \infty$ and $\Sigma\, b_n$ diverges, then $\Sigma\, a_n$ also diverges.

PROOF To prove part (a) we take $\varepsilon = c/2$ in Definition 10.1.1 and see that, since $\lim_{n \to \infty} (a_n/b_n) = c$, there is an integer N such that

$$\left| \frac{a_n}{b_n} - c \right| < \frac{c}{2} \qquad \text{when } n > N$$

Thus

$$\frac{c}{2} < \frac{a_n}{b_n} < \frac{3c}{2} \qquad \text{when } n > N$$

(1)

$$\left(\frac{c}{2}\right) b_n < a_n < \left(\frac{3c}{2}\right) b_n \qquad \text{when } n > N$$

If $\Sigma\, b_n$ converges, so does $\Sigma\, (3c/2)b_n$. The right half of (1) then shows that $\Sigma_N^{\infty}\, a_n$ converges by the Comparison Test. It follows that $\Sigma_1^{\infty}\, a_n$ converges. If $\Sigma\, b_n$ diverges, so does $\Sigma\, (c/2)b_n$, and the left half of (1) together with part (b) of the Comparison Test shows that $\Sigma\, a_n$ diverges.

The proofs of parts (b) and (c) are similar to that of part (a) and are left as Exercises 40 and 41.   □

EXAMPLE 4 Test the series $\displaystyle\sum_{n=1}^{\infty} \frac{1}{2^n - 1}$ for convergence or divergence.

SOLUTION We use the Limit Comparison Test with

$$a_n = \frac{1}{2^n - 1} \qquad\qquad b_n = \frac{1}{2^n}$$

$$\lim_{n \to \infty} \frac{a_n}{b_n} = \lim_{n \to \infty} \frac{2^n}{2^n - 1} = \lim_{n \to \infty} \frac{1}{1 - 1/2^n} = 1$$

Since this limit exists and $\Sigma\, 1/2^n$ is a convergent geometric series, the given series converges by the Limit Comparison Test.   ■

EXAMPLE 5 Solve Example 2 using the Limit Comparison Test.

SOLUTION Taking $a_n = (\ln n)/n$ and $b_n = 1/n$, we have

$$\lim_{n \to \infty} \frac{a_n}{b_n} = \lim_{n \to \infty} \frac{\dfrac{\ln n}{n}}{\dfrac{1}{n}} = \lim_{n \to \infty} \ln n = \infty$$

We know that the harmonic series $\Sigma \, 1/n$ is divergent so, by part (c) of the Limit Comparison Test, $\Sigma \, (\ln n)/n$ is also divergent. ∎

EXAMPLE 6 Determine whether the series $\displaystyle\sum_{n=1}^{\infty} \frac{2n^2 + 3n}{\sqrt{5 + n^7}}$ converges or diverges.

SOLUTION The dominant part of the numerator is $2n^2$ and the dominant part of the denominator is $\sqrt{n^7} = n^{7/2}$. This suggests taking

$$a_n = \frac{2n^2 + 3n}{\sqrt{5 + n^7}} \qquad b_n = \frac{2n^2}{n^{7/2}} = \frac{2}{n^{3/2}}$$

$$\lim_{n \to \infty} \frac{a_n}{b_n} = \lim_{n \to \infty} \frac{2n^2 + 3n}{\sqrt{5 + n^7}} \cdot \frac{n^{3/2}}{2} = \lim_{n \to \infty} \frac{2n^{7/2} + 3n^{5/2}}{2\sqrt{5 + n^7}}$$

$$= \lim_{n \to \infty} \frac{2 + \dfrac{3}{n}}{2\sqrt{\dfrac{5}{n^7} + 1}} = \frac{2 + 0}{2\sqrt{0 + 1}} = 1$$

Since $\Sigma \, b_n = 2 \, \Sigma \, 1/n^{3/2}$ is convergent (p-series with $p = \frac{3}{2} > 1$), the given series converges by the Limit Comparison Test. ∎

Notice that in testing many series we find a suitable comparison series $\Sigma \, b_n$ by keeping only the highest powers in the numerator and denominator.

ESTIMATING SUMS

If we have used the Comparison Test to show that a series $\Sigma \, a_n$ converges by comparison with a series $\Sigma \, b_n$, then we may be able to estimate the sum $\Sigma \, a_n$ by comparing remainders. As in Section 10.3, we consider the remainder

$$R_n = s - s_n = a_{n+1} + a_{n+2} + \cdots$$

For the comparison series $\Sigma \, b_n$ we consider the corresponding remainder

$$T_n = t - t_n = b_{n+1} + b_{n+2} + \cdots$$

Since $a_n \leq b_n$ for all n, we have $R_n \leq T_n$. If $\Sigma \, b_n$ is a p-series, we can estimate its remainder T_n as in Section 10.3. If $\Sigma \, b_n$ is a geometric series, then T_n is the sum of a geometric series and we can sum it exactly (see Exercises 35 and 36). In either case we know that R_n is smaller than T_n.

EXAMPLE 7 Use the sum of the first 100 terms to approximate the sum of the series $\Sigma \, 1/(n^3 + 1)$. Estimate the error involved in this approximation.

SOLUTION Since

$$\frac{1}{n^3 + 1} < \frac{1}{n^3}$$

the given series is convergent by the Comparison Test. The remainder T_n for the comparison series $\Sigma\ 1/n^3$ was estimated in Example 6 in Section 10.3 using the Remainder Estimate for the Integral Test. There we found that

$$T_n \leq \int_n^\infty \frac{1}{x^3}\ dx = \frac{1}{2n^2}$$

Therefore, the remainder R_n for the given series satisfies

$$R_n \leq T_n \leq \frac{1}{2n^2}$$

With $n = 100$ we have

$$R_{100} \leq \frac{1}{2(100)^2} = 0.00005$$

Using a programmable calculator or a computer, we find that

$$\sum_{n=1}^\infty \frac{1}{n^3 + 1} \approx \sum_{n=1}^{100} \frac{1}{n^3 + 1} \approx 0.6864538$$

with error less than 0.00005.

EXERCISES 10.4

1–32 ■ Determine whether the series converges or diverges.

1. $\displaystyle\sum_{n=1}^\infty \frac{1}{n^3 + n^2}$

2. $\displaystyle\sum_{n=1}^\infty \frac{3}{4^n + 5}$

3. $\displaystyle\sum_{n=1}^\infty \frac{3}{n2^n}$

4. $\displaystyle\sum_{n=2}^\infty \frac{1}{\sqrt{n} - 1}$

5. $\displaystyle\sum_{n=0}^\infty \frac{1 + 5^n}{4^n}$

6. $\displaystyle\sum_{n=1}^\infty \frac{\sin^2 n}{n\sqrt{n}}$

7. $\displaystyle\sum_{n=1}^\infty \frac{3}{n(n + 3)}$

8. $\displaystyle\sum_{n=1}^\infty \frac{1}{\sqrt{n(n + 1)(n + 2)}}$

9. $\displaystyle\sum_{n=2}^\infty \frac{\sqrt{n}}{n - 1}$

10. $\displaystyle\sum_{n=1}^\infty \frac{1}{\sqrt[3]{n(n + 1)(n + 2)}}$

11. $\displaystyle\sum_{n=1}^\infty \frac{n - 1}{n^3 + 1}$

12. $\displaystyle\sum_{n=1}^\infty \frac{n}{(n + 1)2^n}$

13. $\displaystyle\sum_{n=1}^\infty \frac{3 + \cos n}{3^n}$

14. $\displaystyle\sum_{n=1}^\infty \frac{5n}{2n^2 - 5}$

15. $\displaystyle\sum_{n=1}^\infty \frac{n}{\sqrt{n^5 + 4}}$

16. $\displaystyle\sum_{n=1}^\infty \frac{\arctan n}{n^4}$

17. $\displaystyle\sum_{n=1}^\infty \frac{2^n}{1 + 3^n}$

18. $\displaystyle\sum_{n=1}^\infty \frac{1 + 2^n}{1 + 3^n}$

19. $\displaystyle\sum_{n=1}^\infty \frac{1}{1 + \sqrt{n}}$

20. $\displaystyle\sum_{n=3}^\infty \frac{1}{n^2 - 4}$

21. $\displaystyle\sum_{n=1}^\infty \frac{n^2 + 1}{n^4 + 1}$

22. $\displaystyle\sum_{n=1}^\infty \frac{3n^3 - 2n^2}{n^4 + n^2 + 1}$

23. $\displaystyle\sum_{n=1}^\infty \frac{n^2 - n + 2}{\sqrt[4]{n^{10} + n^5 + 3}}$

24. $\displaystyle\sum_{n=1}^\infty \frac{n^2 - 3n}{\sqrt[3]{n^{10} - 4n^2}}$

25. $\displaystyle\sum_{n=1}^\infty \frac{n + 1}{n2^n}$

26. $\displaystyle\sum_{n=1}^\infty \frac{2n^2 + 7n}{3^n(n^2 + 5n - 1)}$

27. $\displaystyle\sum_{n=1}^\infty \frac{\ln n}{n^3}$

28. $\displaystyle\sum_{n=2}^\infty \frac{1}{\ln n}$

29. $\displaystyle\sum_{n=1}^\infty \frac{1}{n!}$

30. $\displaystyle\sum_{n=1}^\infty \frac{n!}{n^n}$

31. $\displaystyle\sum_{n=1}^\infty \sin\left(\frac{1}{n}\right)$

32. $\displaystyle\sum_{n=1}^\infty \frac{1}{n^{1+1/n}}$

33–36 ■ Use the sum of the first 10 terms to approximate the sum of the series. Estimate the error.

33. $\displaystyle\sum_{n=1}^\infty \frac{1}{n^4 + n^2}$

34. $\displaystyle\sum_{n=1}^\infty \frac{1 + \cos n}{n^5}$

35. $\displaystyle\sum_{n=1}^\infty \frac{1}{1 + 2^n}$

36. $\displaystyle\sum_{n=1}^\infty \frac{n}{(n + 1)3^n}$

37. The meaning of the decimal representation of a number $0.d_1 d_2 d_3 \ldots$ (where the digit d_i is one of the numbers 0, 1, 2, $\ldots$, 9) is that

$$0.d_1 d_2 d_3 d_4 \ldots = \frac{d_1}{10} + \frac{d_2}{10^2} + \frac{d_3}{10^3} + \frac{d_4}{10^4} + \cdots$$

Show that this series always converges.

38. For what values of p does the series $\sum_{n=2}^{\infty} 1/(n^p \ln n)$ converge?

39. Prove that if $a_n \geqslant 0$ and $\Sigma\, a_n$ converges, then $\Sigma\, a_n^2$ also converges.

40. Prove part (b) of the Limit Comparison Test.

41. Prove part (c) of the Limit Comparison Test.

42. Give an example of a pair of series $\Sigma\, a_n$ and $\Sigma\, b_n$ with positive terms where $\lim_{n \to \infty} (a_n/b_n) = 0$ and $\Sigma\, b_n$ diverges, but $\Sigma\, a_n$ converges. [Compare with part (b) of the Limit Comparison Test.]

43. Show that if $a_n > 0$ and $\lim_{n \to \infty} n a_n \neq 0$, then $\Sigma\, a_n$ is divergent.

44. Show that if $a_n > 0$ and $\Sigma\, a_n$ is convergent, then $\Sigma \ln(1 + a_n)$ is convergent.

45. If $\Sigma\, a_n$ is a convergent series with positive terms, is it true that $\Sigma \sin(a_n)$ is also convergent?

46. If $\Sigma\, a_n$ and $\Sigma\, b_n$ are both convergent series with positive terms, is it true that $\Sigma\, a_n b_n$ is also convergent?

10.5 ALTERNATING SERIES

An **alternating series** is a series whose terms are alternately positive and negative. Here are two examples:

$$1 - \frac{1}{2} + \frac{1}{3} - \frac{1}{4} + \frac{1}{5} - \frac{1}{6} + \cdots = \sum_{n=1}^{\infty} \frac{(-1)^{n-1}}{n}$$

$$-\frac{1}{2} + \frac{2}{3} - \frac{3}{4} + \frac{4}{5} - \frac{5}{6} + \frac{6}{7} - \cdots = \sum_{n=1}^{\infty} (-1)^n \frac{n}{n+1}$$

We see from these examples that the nth term of an alternating series is of the form

$$a_n = (-1)^{n-1} b_n \qquad \text{or} \qquad a_n = (-1)^n b_n$$

where b_n is a positive number. (In fact, $b_n = |a_n|$.)

The following test says that if the terms of an alternating series decrease to 0 in absolute value, then the series converges.

THE ALTERNATING SERIES TEST If the alternating series

$$\sum_{n=1}^{\infty} (-1)^{n-1} b_n = b_1 - b_2 + b_3 - b_4 + b_5 - b_6 + \cdots \qquad b_n > 0$$

satisfies

$$\text{(a)} \quad b_{n+1} \leqslant b_n \qquad \text{for all } n$$

$$\text{(b)} \quad \lim_{n \to \infty} b_n = 0$$

then the series is convergent.

Before giving the proof let us look at Figure 1, which gives a picture of the idea behind the proof. We first plot $s_1 = b_1$ on a number line. To find s_2 we subtract b_2, so s_2 is to the left of s_1. Then to find s_3 we add b_3, so s_3 is to the right of s_2. But, since $b_3 < b_2$, s_3 is to the left of s_1. Continuing in this manner, we see that the partial sums oscillate back and forth. Since $b_n \to 0$, the successive steps are becoming smaller and smaller. The even partial sums $s_2, s_4, s_6, \ldots$ are increasing and the odd partial sums $s_1, s_3, s_5, \ldots$ are decreasing. Thus it seems plausible that both are converging to some number s. Therefore, in the following proof we consider the even and odd partial sums separately.

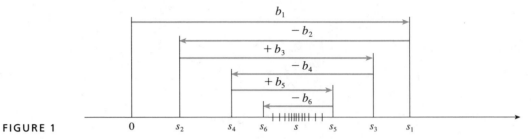

FIGURE 1

PROOF OF THE ALTERNATING SERIES TEST We first consider the even partial sums:

$$s_2 = b_1 - b_2 \geq 0 \qquad \text{since } b_2 \leq b_1$$

$$s_4 = s_2 + (b_3 - b_4) \geq s_2 \qquad \text{since } b_4 \leq b_3$$

In general

$$s_{2n} = s_{2n-2} + (b_{2n-1} - b_{2n}) \geq s_{2n-2} \qquad \text{since } b_{2n} \leq b_{2n-1}$$

Thus
$$0 \leq s_2 \leq s_4 \leq s_6 \leq \cdots \leq s_{2n} \leq \cdots$$

But we can also write

$$s_{2n} = b_1 - (b_2 - b_3) - (b_4 - b_5) - \cdots - (b_{2n-2} - b_{2n-1}) - b_{2n}$$

Every term in brackets is positive, so $s_{2n} \leq b_1$ for all n. Therefore, the sequence $\{s_{2n}\}$ of even partial sums is increasing and bounded above. It is therefore convergent by Theorem 10.1.10. Let us call its limit s, that is,

$$\lim_{n \to \infty} s_{2n} = s$$

Now we compute the limit of the odd partial sums:

$$\lim_{n \to \infty} s_{2n+1} = \lim_{n \to \infty} (s_{2n} + b_{2n+1})$$

$$= \lim_{n \to \infty} s_{2n} + \lim_{n \to \infty} b_{2n+1}$$

$$= s + 0 \qquad \text{[by condition (b)]}$$

$$= s$$

Since both the even and odd partial sums converge to s, we have $\lim_{n \to \infty} s_n = s$ (see Exercise 70 in Section 10.1) and so the series is convergent.   □

Figure 2 illustrates Example 1 by showing the graphs of the terms $a_n = (-1)^{n-1}/n$ and the partial sums s_n. Notice how the values of s_n zigzag across the limiting value, which appears to be about 0.7. In fact, the exact sum of the series is $\ln 2 \approx 0.693$ (see Exercise 35).

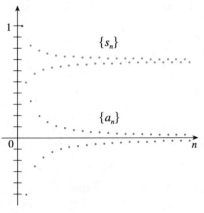

FIGURE 2

EXAMPLE 1   The alternating harmonic series

$$1 - \frac{1}{2} + \frac{1}{3} - \frac{1}{4} + \cdots = \sum_{n=1}^{\infty} \frac{(-1)^{n-1}}{n}$$

satisfies

(a) $b_{n+1} < b_n$ because $\dfrac{1}{n+1} < \dfrac{1}{n}$

(b) $\displaystyle\lim_{n\to\infty} b_n = \lim_{n\to\infty} \frac{1}{n} = 0$

so the series is convergent by the Alternating Series Test. ∎

EXAMPLE 2   The series $\displaystyle\sum_{n=1}^{\infty} \frac{(-1)^n 3n}{4n-1}$ is alternating but

$$\lim_{n\to\infty} b_n = \lim_{n\to\infty} \frac{3n}{4n-1} = \lim_{n\to\infty} \frac{3}{4 - \dfrac{1}{n}} = \frac{3}{4}$$

so condition (b) is not satisfied. Instead, we look at the limit of the nth term of the series:

$$\lim_{n\to\infty} a_n = \lim_{n\to\infty} \frac{(-1)^n 3n}{4n-1}$$

This limit does not exist, so the series diverges by the Test for Divergence. ∎

EXAMPLE 3   Test the series $\displaystyle\sum_{n=1}^{\infty} (-1)^{n+1} \frac{n^2}{n^3+1}$ for convergence or divergence.

SOLUTION   The given series is alternating so we try to verify conditions (a) and (b) of the Alternating Series Test.

Unlike the situation in Example 1, it is not obvious that the sequence given by $b_n = n^2/(n^3+1)$ is decreasing. However, if we consider the related function $f(x) = x^2/(x^3+1)$, we find that

$$f'(x) = \frac{x(2-x^3)}{(x^3+1)^2}$$

Since we are considering only positive x, we see that $f'(x) < 0$ if $2 - x^3 < 0$, that is, $x > \sqrt[3]{2}$. Thus f is decreasing on the interval $[\sqrt[3]{2}, \infty)$. This means that $f(n+1) < f(n)$ and therefore $b_{n+1} < b_n$ when $n \geq 2$. (The inequality $b_2 < b_1$ can be verified directly but all that really matters is that the sequence $\{b_n\}$ is eventually decreasing.)

Condition (b) is readily verified:

$$\lim_{n\to\infty} b_n = \lim_{n\to\infty} \frac{n^2}{n^3+1} = \lim_{n\to\infty} \frac{\dfrac{1}{n}}{1 + \dfrac{1}{n^3}} = 0$$

Thus the given series is convergent by the Alternating Series Test. ∎

NOTE: Instead of verifying condition (a) of the Alternating Series Test by computing a derivative as in Example 3, it is possible to verify that $b_{n+1} < b_n$ directly by using the technique of Solution 1 of Example 10 in Section 10.1.

ESTIMATING SUMS

A partial sum s_n of any convergent series can be used as an approximation to the total sum s but this is not of much use unless we can estimate the accuracy of the approximation. The error involved in using $s \approx s_n$ is the remainder $R_n = s - s_n$. The next theorem says that for series that satisfy the conditions of the Alternating Series Test, the size of the error is smaller than b_{n+1}, which is the magnitude of the first neglected term.

---

(1) ALTERNATING SERIES ESTIMATION THEOREM   If $s = \Sigma\, (-1)^{n-1}b_n$ is the sum of an alternating series that satisfies

$$\text{(a)}\ \ 0 \leqslant b_{n+1} \leqslant b_n \qquad \text{and} \qquad \text{(b)}\ \lim_{n\to\infty} b_n = 0$$

then

$$|R_n| = |s - s_n| \leqslant b_{n+1}$$

PROOF   The idea is similar to the one for the proof of the Alternating Series Test. (Indeed the result of Theorem 1 can be seen geometrically by looking at Figure 1.) We have

$$s - s_n = (-1)^n b_{n+1} + (-1)^{n+1}b_{n+2} + (-1)^{n+2}b_{n+3} + \cdots$$
$$= (-1)^n[b_{n+1} - b_{n+2} + b_{n+3} - \cdots]$$

and so

$$|s - s_n| = (b_{n+1} - b_{n+2}) + (b_{n+3} - b_{n+4}) + \cdots$$
$$= b_{n+1} - (b_{n+2} - b_{n+3}) - (b_{n+4} - b_{n+5}) - \cdots$$

Every term in brackets is positive, so $|s - s_n| \leqslant b_{n+1}$. $\square$

EXAMPLE 4   Find the sum of the series $\displaystyle\sum_{n=0}^{\infty} \frac{(-1)^n}{n!}$ correct to three decimal places. (By definition, $0! = 1$.)

SOLUTION   We first observe that the series is convergent by the Alternating Series Test because

$$\text{(a)}\ \ \frac{1}{(n + 1)!} = \frac{1}{n!(n + 1)} < \frac{1}{n!}$$

$$\text{(b)}\ \ 0 < \frac{1}{n!} < \frac{1}{n} \to 0 \quad \text{so}\quad \frac{1}{n!} \to 0 \text{ as } n \to \infty$$

To get a feel for how many terms we need to use in our approximation, let us write out the first few terms of the series:

$$s = \frac{1}{0!} - \frac{1}{1!} + \frac{1}{2!} - \frac{1}{3!} + \frac{1}{4!} - \frac{1}{5!} + \frac{1}{6!} - \frac{1}{7!} + \cdots$$
$$= 1 - 1 + \tfrac{1}{2} - \tfrac{1}{6} + \tfrac{1}{24} - \tfrac{1}{120} + \tfrac{1}{720} - \tfrac{1}{5040} + \cdots$$

Notice that

$$b_7 = \tfrac{1}{5040} < \tfrac{1}{5000} = 0.0002$$

and $\qquad s_6 = 1 - 1 + \frac{1}{2} - \frac{1}{6} + \frac{1}{24} - \frac{1}{120} + \frac{1}{720} \approx 0.368056$

By the Alternating Series Estimation Theorem we know that

$$|s - s_6| \leq b_7 < 0.0002$$

This error of less than 0.0002 does not affect the third decimal place, so we have

$$s \approx 0.368$$

correct to three decimal places.

In Section 10.10 we will prove that $e^x = \sum_{n=0}^{\infty} x^n/n!$ for all x, so what we have obtained in this example is actually an approximation to the number e^{-1}.   ■

⊘ **NOTE:** The rule that the error (in using s_n to approximate s) is smaller than the first neglected term is, in general, valid only for alternating series that satisfy the conditions of the Alternating Series Estimation Theorem. The rule does not apply to other types of series. (See the example in Appendix G.)

EXERCISES 10.5

1–20 ■ Test the series for convergence or divergence.

1. $\frac{3}{5} - \frac{3}{6} + \frac{3}{7} - \frac{3}{8} + \frac{3}{9} - \cdots$

2. $-5 - \frac{5}{2} + \frac{5}{5} - \frac{5}{8} + \frac{5}{11} - \frac{5}{14} + \cdots$

3. $-\frac{1}{2} + \frac{2}{3} - \frac{3}{4} + \frac{4}{5} - \frac{5}{6} + \frac{6}{7} - \cdots$

4. $\dfrac{1}{\ln 2} - \dfrac{1}{\ln 3} + \dfrac{1}{\ln 4} - \dfrac{1}{\ln 5} + \dfrac{1}{\ln 6} - \cdots$

5. $\displaystyle\sum_{n=1}^{\infty} \frac{(-1)^{n-1}}{n^2}$

6. $\displaystyle\sum_{n=1}^{\infty} \frac{(-1)^n}{\sqrt{n+3}}$

7. $\displaystyle\sum_{n=1}^{\infty} (-1)^{n+1} \frac{n}{5n+1}$

8. $\displaystyle\sum_{n=2}^{\infty} \frac{(-1)^{n-1}}{n \ln n}$

9. $\displaystyle\sum_{n=1}^{\infty} (-1)^n \frac{n}{n^2+1}$

10. $\displaystyle\sum_{n=1}^{\infty} (-1)^n \frac{n^2}{n^2+1}$

11. $\displaystyle\sum_{n=1}^{\infty} (-1)^{n-1} \frac{\sqrt{n}}{n+4}$

12. $\displaystyle\sum_{n=1}^{\infty} (-1)^{n+1} \frac{n}{2^n}$

13. $\displaystyle\sum_{n=2}^{\infty} (-1)^n \frac{n}{\ln n}$

14. $\displaystyle\sum_{n=1}^{\infty} (-1)^{n-1} \frac{\ln n}{n}$

15. $\displaystyle\sum_{n=1}^{\infty} \frac{\cos n\pi}{n^{3/4}}$

16. $\displaystyle\sum_{n=1}^{\infty} \frac{\sin(n\pi/2)}{n!}$

17. $\displaystyle\sum_{n=1}^{\infty} (-1)^n \sin\left(\frac{\pi}{n}\right)$

18. $\displaystyle\sum_{n=1}^{\infty} (-1)^n \cos\left(\frac{\pi}{n}\right)$

19. $\displaystyle\sum_{n=1}^{\infty} (-1)^n \frac{n^n}{n!}$

20. $\displaystyle\sum_{n=2}^{\infty} \frac{(-1)^{n-1}}{\sqrt[3]{\ln n}}$

21. Show that the series $\sum (-1)^{n-1} b_n$, where $b_n = 1/n$ if n is odd and $b_n = 1/n^2$ if n is even, is divergent. Why does the Alternating Series Test not apply?

22–24 ■ For what values of p is each series convergent?

22. $\displaystyle\sum_{n=1}^{\infty} \frac{(-1)^{n-1}}{n^p}$

23. $\displaystyle\sum_{n=1}^{\infty} \frac{(-1)^n}{n+p}$

24. $\displaystyle\sum_{n=1}^{\infty} (-1)^{n-1} \frac{(\ln n)^p}{n}$

25–32 ■ Approximate the sum of the series to the indicated accuracy.

25. $\displaystyle\sum_{n=1}^{\infty} \frac{(-1)^{n-1}}{n^2}$ (error < 0.01)

26. $\displaystyle\sum_{n=1}^{\infty} \frac{(-1)^{n+1}}{n^4}$ (error < 0.001)

27. $\displaystyle\sum_{n=0}^{\infty} \frac{(-2)^n}{n!}$ (error < 0.01)

28. $\displaystyle\sum_{n=0}^{\infty} \frac{(-1)^n n}{4^n}$ (error < 0.002)

29. $\displaystyle\sum_{n=1}^{\infty} \frac{(-1)^{n-1}}{(2n-1)!}$ (four decimal places)

30. $\displaystyle\sum_{n=0}^{\infty} \frac{(-1)^n}{(2n)!}$ (four decimal places)

31. $\displaystyle\sum_{n=0}^{\infty} \frac{(-1)^n}{2^n n!}$ (four decimal places)

32. $\displaystyle\sum_{n=1}^{\infty} \frac{(-1)^{n-1}}{n^6}$ (five decimal places)

33. Is the 50th partial sum s_{50} of the alternating series $\sum_{n=1}^{\infty} (-1)^{n-1}/n$ an overestimate or an underestimate of the total sum? Explain.

34. Calculate the first 10 partial sums of the series

$$\sum_{n=1}^{\infty} \frac{(-1)^{n-1}}{n^3}$$

and graph both the sequence of terms and the sequence of partial sums on the same screen. Estimate the error in using the 10th partial sum to approximate the total sum.

35. Use the following steps to show that

$$\sum_{n=1}^{\infty} \frac{(-1)^{n-1}}{n} = \ln 2$$

Let h_n and s_n be the partial sums of the harmonic and alternating harmonic series.
(a) Show that $s_{2n} = h_{2n} - h_n$.
(b) From Exercise 32 in Section 10.3 we have

$$h_n - \ln n \to \gamma \quad \text{as } n \to \infty$$

and therefore

$$h_{2n} - \ln(2n) \to \gamma \quad \text{as } n \to \infty$$

Use these facts together with part (a) to show that $s_{2n} \to \ln 2$ as $n \to \infty$.

10.6 ABSOLUTE CONVERGENCE AND THE RATIO AND ROOT TESTS

Given any series $\Sigma\, a_n$, we can consider the corresponding series

$$\sum_{n=1}^{\infty} |a_n| = |a_1| + |a_2| + |a_3| + \cdots$$

whose terms are the absolute values of the terms of the original series.

> **(1) DEFINITION**   A series $\Sigma\, a_n$ is called **absolutely convergent** if the series of absolute values $\Sigma\, |a_n|$ is convergent.

Notice that if $\Sigma\, a_n$ is a series with positive terms, then $|a_n| = a_n$ and so absolute convergence is the same as convergence.

EXAMPLE 1   The series

$$\sum_{n=1}^{\infty} \frac{(-1)^{n-1}}{n^2} = 1 - \frac{1}{2^2} + \frac{1}{3^2} - \frac{1}{4^2} + \cdots$$

is absolutely convergent because

$$\sum_{n=1}^{\infty} \left| \frac{(-1)^{n-1}}{n^2} \right| = \sum_{n=1}^{\infty} \frac{1}{n^2} = 1 + \frac{1}{2^2} + \frac{1}{3^2} + \frac{1}{4^2} + \cdots$$

is a convergent p-series ($p = 2$).   ∎

EXAMPLE 2   We know that the alternating harmonic series

$$\sum_{n=1}^{\infty} \frac{(-1)^{n-1}}{n} = 1 - \frac{1}{2} + \frac{1}{3} - \frac{1}{4} + \cdots$$

is convergent (see Example 1 in Section 10.5), but it is not absolutely convergent because the corresponding series of absolute values is

$$\sum_{n=1}^{\infty} \left| \frac{(-1)^{n-1}}{n} \right| = \sum_{n=1}^{\infty} \frac{1}{n} = 1 + \frac{1}{2} + \frac{1}{3} + \frac{1}{4} + \cdots$$

which is the harmonic series (p-series with $p = 1$) and is therefore divergent.   ∎

(2) **DEFINITION**   A series $\Sigma\, a_n$ is called **conditionally convergent** if it is convergent but not absolutely convergent.

Example 2 shows that the alternating harmonic series is conditionally convergent. Thus it is possible for a series to be convergent but not absolutely convergent. However, the next theorem shows that absolute convergence implies convergence.

(3) **THEOREM**   If a series $\Sigma\, a_n$ is absolutely convergent, then it is convergent.

PROOF   Observe that the inequality

$$-|a_n| \leqslant a_n \leqslant |a_n|$$

is true because a_n is either $-|a_n|$ or $|a_n|$. If we now add $|a_n|$ to each side of this inequality, we get

$$0 \leqslant a_n + |a_n| \leqslant 2|a_n|$$

Let $b_n = a_n + |a_n|$. Then $0 \leqslant b_n \leqslant 2|a_n|$. If $\Sigma\, a_n$ is absolutely convergent, then $\Sigma\, |a_n|$ is convergent, so $\Sigma\, 2|a_n|$ is convergent by part (a) of Theorem 10.2.8. Therefore, $\Sigma\, b_n$ is convergent by the Comparison Test. Since $a_n = b_n - |a_n|$,

$$\Sigma\, a_n = \Sigma\, b_n - \Sigma\, |a_n|$$

is convergent by part (c) of Theorem 10.2.8.   □

EXAMPLE 3   Determine whether the series

$$\sum_{n=1}^{\infty} \frac{\cos n}{n^2} = \frac{\cos 1}{1^2} + \frac{\cos 2}{2^2} + \frac{\cos 3}{3^2} + \cdots$$

is convergent or divergent.

SOLUTION   This series has both positive and negative terms, but it is not alternating. (The first term is positive, the next three are negative, and the following three are positive. The signs change irregularly.) We can apply the Comparison Test to the series of absolute values

$$\sum_{n=1}^{\infty} \left| \frac{\cos n}{n^2} \right| = \sum_{n=1}^{\infty} \frac{|\cos n|}{n^2}$$

Since $|\cos n| \leqslant 1$ for all n, we have

$$\frac{|\cos n|}{n^2} \leqslant \frac{1}{n^2}$$

We know that $\Sigma\, 1/n^2$ is convergent (p-series with $p = 2$) and therefore $\Sigma\, |\cos n|/n^2$ is convergent by the Comparison Test. Thus the given series $\Sigma\, (\cos n)/n^2$ is absolutely convergent and therefore convergent by Theorem 3.   ∎

Figure 1 shows the graphs of the terms a_n and partial sums s_n of the series in Example 3. Notice that the series is not alternating but has positive and negative terms.

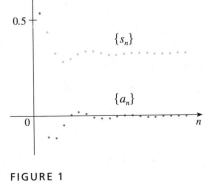

FIGURE 1

The following test is very useful in determining whether a given series is absolutely convergent.

THE RATIO TEST

(a) If $\lim\limits_{n \to \infty} \left| \dfrac{a_{n+1}}{a_n} \right| = L < 1$, then the series $\sum\limits_{n=1}^{\infty} a_n$ is absolutely convergent (and therefore convergent).

(b) If $\lim\limits_{n \to \infty} \left| \dfrac{a_{n+1}}{a_n} \right| = L > 1$ or $\lim\limits_{n \to \infty} \left| \dfrac{a_{n+1}}{a_n} \right| = \infty$, then the series $\sum\limits_{n=1}^{\infty} a_n$ is divergent.

PROOF

(a) The idea is to compare the given series with a convergent geometric series. Since $L < 1$, we can choose a number r such that $L < r < 1$. Since

$$\lim_{n \to \infty} \left| \frac{a_{n+1}}{a_n} \right| = L \qquad \text{and} \qquad L < r$$

the ratio $|a_{n+1}/a_n|$ will eventually be less than r; that is, there exists an integer N such that

$$\left| \frac{a_{n+1}}{a_n} \right| < r \qquad \text{whenever} \qquad n \geq N$$

or, equivalently,

(4) $$|a_{n+1}| < |a_n| r \qquad \text{whenever} \qquad n \geq N$$

Putting n successively equal to $N, N + 1, N + 2, \ldots$ in (4), we obtain

$$|a_{N+1}| < |a_N| r$$

$$|a_{N+2}| < |a_{N+1}| r < |a_N| r^2$$

$$|a_{N+3}| < |a_{N+2}| r < |a_N| r^3$$

and, in general,

(5) $$|a_{N+k}| < |a_N| r^k \qquad \text{for all } k \geq 1$$

Now the series

$$\sum_{k=1}^{\infty} |a_N| r^k = |a_N| r + |a_N| r^2 + |a_N| r^3 + \cdots$$

is convergent because it is a geometric series with $0 < r < 1$. So the inequality (5), together with the Comparison Test, shows that the series

$$\sum_{n=N+1}^{\infty} |a_n| = \sum_{k=1}^{\infty} |a_{N+k}| = |a_{N+1}| + |a_{N+2}| + |a_{N+3}| + \cdots$$

is also convergent. It follows that the series $\sum_{n=1}^{\infty} |a_n|$ is convergent. (Recall that a finite number of terms cannot affect convergence.) Therefore, $\Sigma\, a_n$ is absolutely convergent.

(b) If $|a_{n+1}/a_n| \to L > 1$ or $|a_{n+1}/a_n| \to \infty$, then the ratio $|a_{n+1}/a_n|$ will eventually be greater than 1; that is, there exists an integer N such that

$$\left| \frac{a_{n+1}}{a_n} \right| > 1 \qquad \text{whenever} \qquad n \geq N$$

This means that $|a_{n+1}| > |a_n|$ whenever $n \geq N$ and so

$$\lim_{n \to \infty} a_n \neq 0$$

Therefore, $\Sigma\, a_n$ diverges by the Test for Divergence.   □

NOTE: If $\lim_{n \to \infty} |a_{n+1}/a_n| = 1$, the Ratio Test gives no information. For instance, for the convergent series $\Sigma\, 1/n^2$ we have

$$\left| \frac{a_{n+1}}{a_n} \right| = \frac{\dfrac{1}{(n+1)^2}}{\dfrac{1}{n^2}} = \frac{n^2}{(n+1)^2} = \frac{1}{\left(1 + \dfrac{1}{n}\right)^2} \to 1 \qquad \text{as } n \to \infty$$

whereas for the divergent series $\Sigma\, 1/n$ we have

$$\left| \frac{a_{n+1}}{a_n} \right| = \frac{\dfrac{1}{n+1}}{\dfrac{1}{n}} = \frac{n}{n+1} = \frac{1}{1 + \dfrac{1}{n}} \to 1 \qquad \text{as } n \to \infty$$

Therefore, if $\lim_{n \to \infty} |a_{n+1}/a_n| = 1$, the series $\Sigma\, a_n$ might converge or it might diverge. In this case the Ratio Test fails and we must use some other test.

EXAMPLE 4   Test the series $\displaystyle\sum_{n=1}^{\infty} (-1)^n \frac{n^3}{3^n}$ for absolute convergence.

SOLUTION   We use the Ratio Test with $a_n = (-1)^n n^3/3^n$:

$$\left| \frac{a_{n+1}}{a_n} \right| = \left| \frac{\dfrac{(-1)^{n+1}(n+1)^3}{3^{n+1}}}{\dfrac{(-1)^n n^3}{3^n}} \right| = \frac{(n+1)^3}{3^{n+1}} \cdot \frac{3^n}{n^3}$$

$$= \frac{1}{3}\left(\frac{n+1}{n}\right)^3 = \frac{1}{3}\left(1 + \frac{1}{n}\right)^3 \to \frac{1}{3} < 1$$

ESTIMATING SUMS

In the last three sections we used various methods for estimating the sum of a series—the method depended on which test was used to prove convergence. What about series for which the Ratio Test works? There are two possibilities: If the series happens to be an alternating series, as in Example 4, then it is best to use the methods of Section 10.5. If the terms are all positive, then use the special methods explained in Exercise 38.

Thus, by the Ratio Test, the given series is absolutely convergent and therefore convergent.   ■

EXAMPLE 5   Test the convergence of the series $\sum_{n=1}^{\infty} \dfrac{n^n}{n!}$.

SOLUTION   Since the terms $a_n = n^n/n!$ are positive, we do not need the absolute value signs.

$$\frac{a_{n+1}}{a_n} = \frac{(n + 1)^{n+1}}{(n + 1)!} \cdot \frac{n!}{n^n} = \frac{(n + 1)(n + 1)^n}{(n + 1)n!} \cdot \frac{n!}{n^n}$$

$$= \left(\frac{n + 1}{n}\right)^n = \left(1 + \frac{1}{n}\right)^n \to e \qquad \text{as } n \to \infty$$

(see Equation 6.4.9). Since $e > 1$, the given series is divergent by the Ratio Test.   ∎

NOTE: Although the Ratio Test works in Example 5, an easier method is to use the Test for Divergence. Since

$$a_n = \frac{n^n}{n!} = \frac{n \cdot n \cdot n \cdots \cdot n}{1 \cdot 2 \cdot 3 \cdots \cdot n} \geq n$$

it follows that a_n does not approach 0 as $n \to \infty$. Therefore, the given series is divergent by the Test for Divergence.

The following test is convenient to apply when nth powers occur. Its proof is similar to the proof of the Ratio Test and is left as Exercise 42.

THE ROOT TEST

(a) If $\lim\limits_{n \to \infty} \sqrt[n]{|a_n|} = L < 1$, then the series $\sum_{n=1}^{\infty} a_n$ is absolutely convergent (and therefore convergent).

(b) If $\lim\limits_{n \to \infty} \sqrt[n]{|a_n|} = L > 1$ or $\lim\limits_{n \to \infty} \sqrt[n]{|a_n|} = \infty$, then the series $\sum_{n=1}^{\infty} a_n$ is divergent.

If $\lim_{n \to \infty} \sqrt[n]{|a_n|} = 1$, then the Root Test gives no information. The series $\sum a_n$ could converge or diverge. (If $L = 1$ in the Ratio Test, do not try the Root Test because L will again be 1.)

EXAMPLE 6   Test the convergence of the series $\sum_{n=1}^{\infty} \left(\dfrac{2n + 3}{3n + 2}\right)^n$.

SOLUTION $$a_n = \left(\frac{2n + 3}{3n + 2}\right)^n$$

$$\sqrt[n]{|a_n|} = \frac{2n + 3}{3n + 2} = \frac{2 + \dfrac{3}{n}}{3 + \dfrac{2}{n}} \to \frac{2}{3} < 1$$

Thus the given series converges by the Root Test.   ∎

REARRANGEMENTS

The question of whether a given convergent series is absolutely convergent or conditionally convergent has a bearing on the question of whether infinite sums behave like finite sums.

If we rearrange the order of the terms in a finite sum, then of course the value of the sum remains unchanged. But this is not always the case for an infinite series. By a **rearrangement** of an infinite series $\sum a_n$ we mean a series obtained by simply changing the order of the terms. For instance, a rearrangement of $\sum a_n$ could start as follows:

$$a_1 + a_2 + a_5 + a_3 + a_4 + a_{15} + a_6 + a_7 + a_{20} + \cdots$$

It turns out that **if $\sum a_n$ is an absolutely convergent series with sum s, then any rearrangement of $\sum a_n$ has the same sum s.** However, any conditionally convergent series can be rearranged to give a different sum. To illustrate this fact let us consider the alternating harmonic series

(6) $$1 - \tfrac{1}{2} + \tfrac{1}{3} - \tfrac{1}{4} + \tfrac{1}{5} - \tfrac{1}{6} + \tfrac{1}{7} - \tfrac{1}{8} + \cdots = \ln 2$$

(See Exercise 35 in Section 10.5.) If we multiply this series by $\tfrac{1}{2}$, we get

$$\tfrac{1}{2} - \tfrac{1}{4} + \tfrac{1}{6} - \tfrac{1}{8} + \cdots = \tfrac{1}{2} \ln 2$$

Inserting zeros between the terms of this series, we have

Adding these zeros does not affect the sum of the series; each term in the sequence of partial sums is repeated, but the limit is the same.

(7) $$0 + \tfrac{1}{2} + 0 - \tfrac{1}{4} + 0 + \tfrac{1}{6} + 0 - \tfrac{1}{8} + \cdots = \tfrac{1}{2} \ln 2$$

Now we add the series in Equations 6 and 7 using Theorem 10.2.8:

(8) $$1 + \tfrac{1}{3} - \tfrac{1}{2} + \tfrac{1}{5} + \tfrac{1}{7} - \tfrac{1}{4} + \cdots = \tfrac{3}{2} \ln 2$$

Notice that the series in (8) contains the same terms as in (6), but rearranged so that one negative term occurs after each pair of positive terms. The sums of these series, however, are different. In fact, Riemann proved that **if $\sum a_n$ is a conditionally convergent series and r is any real number whatsoever, then there is a rearrangement of $\sum a_n$ that has a sum equal to r.** A proof of this fact is outlined in Exercise 44.

EXERCISES 10.6

1–32 ■ Determine whether the series is absolutely convergent, conditionally convergent, or divergent.

1. $\displaystyle\sum_{n=1}^{\infty} \frac{(-1)^{n-1}}{n\sqrt{n}}$

2. $\displaystyle\sum_{n=1}^{\infty} \frac{(-1)^n}{\sqrt{n}}$

3. $\displaystyle\sum_{n=1}^{\infty} \frac{(-3)^n}{n^3}$

4. $\displaystyle\sum_{n=0}^{\infty} \frac{(-3)^n}{n!}$

5. $\displaystyle\sum_{n=1}^{\infty} \frac{(-1)^{n+1}}{2n+1}$

6. $\displaystyle\sum_{n=1}^{\infty} \frac{(-1)^{n-1}}{n^2+1}$

7. $\displaystyle\sum_{n=1}^{\infty} \frac{(-1)^{n-1}}{(2n-1)!}$

8. $\displaystyle\sum_{n=1}^{\infty} e^{-n} n!$

9. $\displaystyle\sum_{n=1}^{\infty} (-1)^n \frac{n}{n^2+4}$

10. $\displaystyle\sum_{n=1}^{\infty} (-1)^{n-1} \frac{\sqrt{n}}{n+1}$

11. $\displaystyle\sum_{n=1}^{\infty} (-1)^n \frac{2n}{3n-4}$

12. $\displaystyle\sum_{n=1}^{\infty} (-1)^n \frac{2^n}{n^2+1}$

13. $\displaystyle\sum_{n=1}^{\infty} \frac{\sin 2n}{n^2}$

14. $\displaystyle\sum_{n=1}^{\infty} \frac{(-1)^n \arctan n}{n^3}$

15. $\displaystyle\sum_{n=1}^{\infty} \frac{(-2)^n}{n3^{n+1}}$

16. $\displaystyle\sum_{n=1}^{\infty} \frac{(-1)^{n+1} 5^{n-1}}{(n+1)^2 4^{n+2}}$

17. $\displaystyle\sum_{n=1}^{\infty} \frac{(n+1)5^n}{n3^{2n}}$

18. $\displaystyle\sum_{n=1}^{\infty} \frac{\cos(n\pi/6)}{n\sqrt{n}}$

19. $\displaystyle\sum_{n=1}^{\infty} \frac{n!}{(-10)^n}$

20. $\displaystyle\sum_{n=1}^{\infty} \frac{n!}{n^n}$

21. $\displaystyle\sum_{n=1}^{\infty} \frac{\cos(n\pi/3)}{n!}$

22. $\displaystyle\sum_{n=2}^{\infty} \frac{(-1)^n}{(\ln n)^n}$

23. $\displaystyle\sum_{n=1}^{\infty} \frac{(-n)^n}{5^{2n+3}}$

24. $\displaystyle\sum_{n=2}^{\infty} \frac{(-1)^n}{n \ln n}$

25. $\displaystyle\sum_{n=1}^{\infty} \left(\frac{1-3n}{3+4n}\right)^n$

26. $\displaystyle\sum_{n=1}^{\infty} \frac{(-2)^n n^2}{(n+2)!}$

27. $1 - \dfrac{2!}{1 \cdot 3} + \dfrac{3!}{1 \cdot 3 \cdot 5} - \dfrac{4!}{1 \cdot 3 \cdot 5 \cdot 7} + \cdots$

$\qquad + \dfrac{(-1)^{n-1}n!}{1 \cdot 3 \cdot 5 \cdot \cdots \cdot (2n-1)} + \cdots$

28. $\dfrac{1}{3} + \dfrac{1 \cdot 4}{3 \cdot 5} + \dfrac{1 \cdot 4 \cdot 7}{3 \cdot 5 \cdot 7} + \dfrac{1 \cdot 4 \cdot 7 \cdot 10}{3 \cdot 5 \cdot 7 \cdot 9} + \cdots$

$\qquad + \dfrac{1 \cdot 4 \cdot 7 \cdot \cdots \cdot (3n-2)}{3 \cdot 5 \cdot 7 \cdot \cdots \cdot (2n+1)} + \cdots$

29. $\displaystyle\sum_{n=1}^{\infty} \dfrac{2 \cdot 4 \cdot 6 \cdot \cdots \cdot (2n)}{n!}$

30. $\displaystyle\sum_{n=1}^{\infty} (-1)^n \dfrac{2^n n!}{5 \cdot 8 \cdot 11 \cdot \cdots \cdot (3n+2)}$

31. $\displaystyle\sum_{n=1}^{\infty} \dfrac{(n+2)!}{n!\,10^n}$

32. $\displaystyle\sum_{n=1}^{\infty} \dfrac{(-1)^n}{(\arctan n)^n}$

33. The terms of a series are defined recursively by the equations

$$a_1 = 2, \qquad a_{n+1} = \dfrac{5n+1}{4n+3} a_n$$

Determine whether $\Sigma\, a_n$ converges or diverges.

34. A series $\Sigma\, a_n$ is defined by the equations

$$a_1 = 1, \qquad a_{n+1} = \dfrac{2 + \cos n}{\sqrt{n}} a_n$$

Determine whether $\Sigma\, a_n$ converges or diverges.

35. For which of the following series is the Ratio Test inconclusive (that is, it fails to give a definite answer)?

(a) $\displaystyle\sum_{n=1}^{\infty} \dfrac{1}{n^3}$ $\qquad$ (b) $\displaystyle\sum_{n=1}^{\infty} \dfrac{n}{2^n}$

(c) $\displaystyle\sum_{n=1}^{\infty} \dfrac{(-3)^{n-1}}{\sqrt{n}}$ $\qquad$ (d) $\displaystyle\sum_{n=1}^{\infty} \dfrac{\sqrt{n}}{1 + n^2}$

36. For which positive integers k is the series

$$\sum_{n=1}^{\infty} \dfrac{(n!)^2}{(kn)!}$$

convergent?

37. (a) Show that $\sum_{n=0}^{\infty} x^n/n!$ converges for all x.
(b) Deduce that $\lim_{n \to \infty} x^n/n! = 0$ for all x.

38. Let $\Sigma\, a_n$ be a series with positive terms and let $r_n = a_{n+1}/a_n$. Suppose that $\lim_{n \to \infty} r_n = L < 1$, so $\Sigma\, a_n$ converges by the Ratio Test. As usual, we let R_n be the remainder after n terms, that is,

$$R_n = a_{n+1} + a_{n+2} + a_{n+3} + \cdots$$

(a) If $\{r_n\}$ is a decreasing sequence and $r_{n+1} < 1$, show, by summing a geometric series, that

$$R_n \leqslant \dfrac{a_{n+1}}{1 - r_{n+1}}$$

(b) If $\{r_n\}$ is an increasing sequence, show that

$$R_n \leqslant \dfrac{a_{n+1}}{1 - L}$$

39. (a) Find the partial sum s_5 of the series

$$\sum_{n=1}^{\infty} \dfrac{1}{n2^n}$$

Use Exercise 38 to estimate the error in using s_5 as an approximation to the sum of the series.
(b) Find a value of n so that s_n is within 0.00005 of the sum. Use this value of n to approximate the sum of the series.

40. Use the sum of the first 10 terms to approximate the sum of the series

$$\sum_{n=1}^{\infty} \dfrac{n}{2^n}$$

Use Exercise 38 to estimate the error.

41. Prove that if $\Sigma\, a_n$ is absolutely convergent, then

$$\left| \sum_{n=1}^{\infty} a_n \right| \leqslant \sum_{n=1}^{\infty} |a_n|$$

42. Prove the Root Test. [*Hint for part (a):* Take any number r such that $L < r < 1$ and use the fact that there is an integer N such that $\sqrt[n]{|a_n|} < r$ whenever $n \geqslant N$.]

43. Given any series $\Sigma\, a_n$ we define a series $\Sigma\, a_n^+$ whose terms are all the positive terms of $\Sigma\, a_n$ and a series $\Sigma\, a_n^-$ whose terms are all the negative terms of $\Sigma\, a_n$. To be specific, we let

$$a_n^+ = \dfrac{a_n + |a_n|}{2} \qquad a_n^- = \dfrac{a_n - |a_n|}{2}$$

Notice that if $a_n > 0$, then $a_n^+ = a_n$ and $a_n^- = 0$, whereas if $a_n < 0$, then $a_n^- = a_n$ and $a_n^+ = 0$.
(a) If $\Sigma\, a_n$ is absolutely convergent, show that both of the series $\Sigma\, a_n^+$ and $\Sigma\, a_n^-$ are convergent.
(b) If $\Sigma\, a_n$ is conditionally convergent, show that both of the series $\Sigma\, a_n^+$ and $\Sigma\, a_n^-$ are divergent.

44. Prove that if $\Sigma\, a_n$ is a conditionally convergent series and r is any real number, then there is a rearrangement of $\Sigma\, a_n$ whose sum is r. [*Hints:* Use the notation of Exercise 43. Take just enough positive terms a_n^+ so that their sum is greater than r. Then add just enough negative terms a_n^- so that the cumulative sum is less than r. Continue in this manner and use Theorem 10.2.6.]

10.7 | STRATEGY FOR TESTING SERIES

We now have several ways of testing a series for convergence or divergence; the problem is to decide which test to use on which series. In this respect testing series is similar to integrating functions. Again there are no hard and fast rules about which test to apply to a given series, but you may find the following advice of some use.

It is not wise to apply a list of the tests in a specific order until one finally works. That would be a waste of time and effort. Instead, as with integration, the main strategy is to classify the series according to its *form*.

1. If the series is of the form $\Sigma\, 1/n^p$, it is a *p*-series, which we know to be convergent if $p > 1$ and divergent if $p \leqslant 1$.

2. If the series has the form $\Sigma\, ar^{n-1}$ or $\Sigma\, ar^n$, it is a geometric series, which converges if $|r| < 1$ and diverges if $|r| \geqslant 1$. Some preliminary algebraic manipulation may be required to bring the series into this form.

3. If the series has a form that is similar to a *p*-series or a geometric series, then one of the comparison tests should be considered. In particular, if a_n is a rational function or algebraic function of n (involving roots of polynomials), then the series should be compared with a *p*-series. Notice that most of the series in Exercises 10.4 have this form. (The value of p should be chosen as in Section 10.4 by keeping only the highest powers of n in the numerator and denominator.) The comparison tests apply only to series with positive terms, but if $\Sigma\, a_n$ has some negative terms, then we can apply the Comparison Test to $\Sigma\, |a_n|$ and test for absolute convergence.

4. If you can see at a glance that $\lim_{n \to \infty} a_n \neq 0$, then the Test for Divergence should be used.

5. If the series is of the form $\Sigma\, (-1)^{n-1}b_n$ or $\Sigma\, (-1)^n b_n$, then the Alternating Series Test is an obvious possibility.

6. Series that involve factorials or other products (including a constant raised to the nth power) are often conveniently tested using the Ratio Test. Bear in mind that $|a_{n+1}/a_n| \to 1$ as $n \to \infty$ for all *p*-series and therefore all rational or algebraic functions of n. Thus the Ratio Test should not be used for such series.

7. If a_n is of the form $(b_n)^n$, then the Root Test may be useful.

8. If $a_n = f(n)$, where $\int_1^\infty f(x)\, dx$ is easily evaluated, then the Integral Test is effective (assuming the hypotheses of this test are satisfied).

In the following examples we do not work out all the details but simply indicate which tests should be used.

EXAMPLE 1 $\displaystyle\sum_{n=1}^{\infty} \frac{n-1}{2n+1}$

Since $a_n \to \frac{1}{2} \neq 0$ as $n \to \infty$, we should use the Test for Divergence.

EXAMPLE 2 $\displaystyle\sum_{n=1}^{\infty} \frac{\sqrt{n^3+1}}{3n^3+4n^2+2}$

Since a_n is an algebraic function of n, we compare the given series with a *p*-series.

The comparison series is $\Sigma \, b_n$, where

$$b_n = \frac{\sqrt{n^3}}{3n^3} = \frac{n^{3/2}}{3n^3} = \frac{1}{3n^{3/2}}$$

$\blacksquare$

EXAMPLE 3 $\displaystyle\sum_{n=1}^{\infty} ne^{-n^2}$

Since the integral $\int_1^{\infty} xe^{-x^2}\, dx$ is easily evaluated, we use the Integral Test. The Ratio Test also works.

$\blacksquare$

EXAMPLE 4 $\displaystyle\sum_{n=1}^{\infty} (-1)^n \frac{n^3}{n^4 + 1}$

Since the series is alternating, we use the Alternating Series Test.

$\blacksquare$

EXAMPLE 5 $\displaystyle\sum_{n=1}^{\infty} \frac{2^n}{n!}$

Since the series involves $n!$, we use the Ratio Test.

$\blacksquare$

EXAMPLE 6 $\displaystyle\sum_{n=1}^{\infty} \frac{1}{2 + 3^n}$

Since the series is closely related to the geometric series $\Sigma \, 1/3^n$, we use the Comparison Test.

$\blacksquare$

EXERCISES 10.7

1–40 ■ Test the series for convergence or divergence.

1. $\displaystyle\sum_{n=1}^{\infty} \frac{\sqrt{n}}{n^2 + 1}$

2. $\displaystyle\sum_{n=1}^{\infty} \cos n$

3. $\displaystyle\sum_{n=1}^{\infty} \frac{4^n}{3^{2n-1}}$

4. $\displaystyle\sum_{i=1}^{\infty} \frac{i^4}{4^i}$

5. $\displaystyle\sum_{n=2}^{\infty} \frac{(-1)^n}{(\ln n)^2}$

6. $\displaystyle\sum_{n=1}^{\infty} n^2 e^{-n^3}$

7. $\displaystyle\sum_{k=1}^{\infty} k^{-1.7}$

8. $\displaystyle\sum_{n=0}^{\infty} \frac{10^n}{n!}$

9. $\displaystyle\sum_{n=1}^{\infty} \frac{n}{e^n}$

10. $\displaystyle\sum_{m=1}^{\infty} \frac{2m}{8m - 5}$

11. $\displaystyle\sum_{n=2}^{\infty} \frac{n^3 + 1}{n^4 - 1}$

12. $\displaystyle\sum_{n=1}^{\infty} \left(\frac{n^2 + 1}{2n^2 + 1}\right)^n$

13. $\displaystyle\sum_{n=2}^{\infty} \frac{2}{n(\ln n)^3}$

14. $\displaystyle\sum_{n=1}^{\infty} \frac{\sqrt{n}}{e^{\sqrt{n}}}$

15. $\displaystyle\sum_{n=1}^{\infty} \frac{3^n n^2}{n!}$

16. $\displaystyle\sum_{n=1}^{\infty} \frac{3}{4n - 5}$

17. $\displaystyle\sum_{n=1}^{\infty} \frac{3^n}{5^n + n}$

18. $\displaystyle\sum_{k=1}^{\infty} \frac{k + 5}{5^k}$

19. $\displaystyle\sum_{n=0}^{\infty} \frac{n!}{2 \cdot 5 \cdot 8 \cdot \cdots \cdot (3n + 2)}$

20. $\displaystyle\sum_{n=1}^{\infty} \frac{(-1)^n n}{(n + 1)(n + 2)}$

21. $\displaystyle\sum_{i=1}^{\infty} \frac{1}{\sqrt{i(i + 1)}}$

22. $\displaystyle\sum_{n=1}^{\infty} \frac{n^2}{\sqrt{n^5 + n^2 + 2}}$

23. $\displaystyle\sum_{n=1}^{\infty} (-1)^n 2^{1/n}$

24. $\displaystyle\sum_{n=1}^{\infty} \frac{\cos(n/2)}{n^2 + 4n}$

25. $\displaystyle\sum_{n=1}^{\infty} (-1)^n \frac{\ln n}{\sqrt{n}}$

26. $\displaystyle\sum_{n=1}^{\infty} \frac{\tan(1/n)}{n}$

27. $\displaystyle\sum_{n=0}^{\infty} (-\pi)^n$

28. $\displaystyle\sum_{n=1}^{\infty} \frac{\sqrt[3]{n} + 1}{n(\sqrt{n} + 1)}$

29. $\displaystyle\sum_{n=1}^{\infty} \frac{(-2)^{2n}}{n^n}$

30. $\displaystyle\sum_{n=1}^{\infty} \frac{2^{3n-1}}{n^2 + 1}$

31. $\displaystyle\sum_{k=1}^{\infty} \frac{k \ln k}{(k + 1)^3}$

32. $\displaystyle\sum_{n=1}^{\infty} \frac{e^{1/n}}{n^2}$

33. $\displaystyle\sum_{n=1}^{\infty} \frac{2^n}{(2n + 1)!}$

34. $\displaystyle\sum_{j=1}^{\infty} (-1)^j \frac{\sqrt{j}}{j + 5}$

35. $\displaystyle\sum_{n=1}^{\infty} \frac{\tan^{-1} n}{n \sqrt{n}}$

36. $\displaystyle\sum_{n=1}^{\infty} \frac{(2n)^n}{n^{2n}}$

37. $\displaystyle\sum_{n=1}^{\infty} \left(\frac{n}{n + 1}\right)^{n^2}$

38. $\displaystyle\sum_{n=2}^{\infty} \frac{1}{(\ln n)^{\ln n}}$

39. $\displaystyle\sum_{n=1}^{\infty} (\sqrt[n]{2} - 1)^n$

40. $\displaystyle\sum_{n=1}^{\infty} (\sqrt[n]{2} - 1)$

10.8 POWER SERIES

A **power series** is a series of the form

$$(1) \qquad \sum_{n=0}^{\infty} c_n x^n = c_0 + c_1 x + c_2 x^2 + c_3 x^3 + \cdots$$

where x is a variable and the c_n's are constants called the **coefficients** of the series. For each fixed x, the series (1) is a series of constants that we can test for convergence or divergence. A power series may converge for some values of x and diverge for other values of x. The sum of the series is a function

$$f(x) = c_0 + c_1 x + c_2 x^2 + \cdots + c_n x^n + \cdots$$

whose domain is the set of all x for which the series converges. Notice that f resembles a polynomial. The only difference is that f has infinitely many terms.

For instance, if we take $c_n = 1$ for all n, the power series becomes the geometric series

$$\sum_{n=0}^{\infty} x^n = 1 + x + x^2 + \cdots + x^n + \cdots = \frac{1}{1 - x}$$

which converges when $-1 < x < 1$ and diverges when $|x| \geq 1$ (see Equation 10.2.5).

More generally, a series of the form

$$(2) \qquad \sum_{n=0}^{\infty} c_n (x - a)^n = c_0 + c_1 (x - a) + c_2 (x - a)^2 + \cdots$$

is called a **power series in $(x - a)$** or a **power series centered at a** or a **power series about a.** Notice that in writing out the term corresponding to $n = 0$ in Equations 1 and 2 we have adopted the convention that $(x - a)^0 = 1$ even when $x = a$. Notice also that when $x = a$ all of the terms are 0 for $n \geq 1$ and so the power series (2) always converges when $x = a$.

EXAMPLE 1 For what values of x is the series $\displaystyle\sum_{n=0}^{\infty} n! x^n$ convergent?

SOLUTION   We use the Ratio Test. If we let a_n, as usual, denote the nth term of the series, then $a_n = n! x^n$. If $x \neq 0$, we have

$$\lim_{n \to \infty} \left| \frac{a_{n+1}}{a_n} \right| = \lim_{n \to \infty} \left| \frac{(n + 1)! x^{n+1}}{n! x^n} \right|$$
$$= \lim_{n \to \infty} (n + 1)|x| = \infty$$

By the Ratio Test, the series diverges when $x \neq 0$. Thus the given series converges only when $x = 0$. ∎

EXAMPLE 2 For what values of x does the series $\displaystyle\sum_{n=1}^{\infty} \frac{(x - 3)^n}{n}$ converge?

SOLUTION   Let $a_n = (x - 3)^n / n$. Then

$$\left| \frac{a_{n+1}}{a_n} \right| = \left| \frac{(x - 3)^{n+1}}{n + 1} \cdot \frac{n}{(x - 3)^n} \right|$$
$$= \frac{1}{1 + \dfrac{1}{n}} |x - 3| \to |x - 3| \qquad \text{as } n \to \infty$$

By the Ratio Test, the given series is absolutely convergent, and therefore convergent, when $|x - 3| < 1$ and divergent when $|x - 3| > 1$. Now

$$|x - 3| < 1 \iff -1 < x - 3 < 1 \iff 2 < x < 4$$

so the series converges when $2 < x < 4$ and diverges when $x < 2$ or $x > 4$.

The Ratio Test gives no information when $|x - 3| = 1$ so we must consider $x = 2$ and $x = 4$ separately. If we put $x = 4$ in the series, it becomes $\Sigma \ 1/n$, the harmonic series, which is divergent. If $x = 2$, the series is $\Sigma \ (-1)^n/n$, which converges by the Alternating Series Test. Thus the given power series converges for $2 \leq x < 4$. ∎

We will see that the main use of a power series is that it provides a way to represent some of the most important functions that arise in mathematics, physics, and chemistry. In particular, the sum of the power series in the next example is called a **Bessel function,** after the German astronomer Friedrich Bessel (1784–1846), and the function given in Exercise 31 is another example of a Bessel function. In fact, Bessel functions first arose in solving Kepler's equation that describes planetary motion. Since that time, these functions have been applied in many different physical situations—the temperature distribution in a circular plate is one example.

EXAMPLE 3   Find the domain of the Bessel function of order 0 defined by

$$J_0(x) = \sum_{n=0}^{\infty} \frac{(-1)^n x^{2n}}{2^{2n}(n!)^2}$$

SOLUTION   Let $a_n = (-1)^n x^{2n}/[2^{2n}(n!)^2]$. Then

$$\left| \frac{a_{n+1}}{a_n} \right| = \left| \frac{(-1)^{n+1} x^{2(n+1)}}{2^{2(n+1)}[(n + 1)!]^2} \cdot \frac{2^{2n}(n!)^2}{(-1)^n x^{2n}} \right|$$

$$= \frac{x^2}{4(n + 1)^2} \to 0 < 1 \qquad \text{for all } x$$

Thus, by the Ratio Test, the given series converges for all values of x. In other words, the domain of the Bessel function J_0 is $(-\infty, \infty) = \mathbb{R}$. ∎

Recall that the sum of a series is equal to the limit of the sequence of partial sums. So when we define the Bessel function in Example 3 as the sum of a series we mean that, for every real number x,

$$J_0(x) = \lim_{n \to \infty} s_n(x) \qquad \text{where} \qquad s_{2n}(x) = \sum_{i=0}^{n} \frac{(-1)^i x^{2i}}{2^{2i}(i!)^2}$$

The first few partial sums are

$$s_0(x) = 1 \qquad s_2(x) = 1 - \frac{x^2}{4} \qquad s_4(x) = 1 - \frac{x^2}{4} + \frac{x^4}{64}$$

$$s_6(x) = 1 - \frac{x^2}{4} + \frac{x^4}{64} - \frac{x^6}{2304} \qquad s_8(x) = 1 - \frac{x^2}{4} + \frac{x^4}{64} - \frac{x^6}{2304} + \frac{x^8}{147456}$$

FIGURE 1
Partial sums of the Bessel function J_0

Figure 1 shows the graph of these partial sums, which are polynomials. They are all

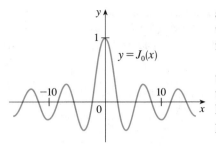

y = J₀(x) indicated on graph

FIGURE 2

approximations to the function J_0, but notice that the approximations become better when more terms are included. Figure 2 shows a more complete graph of the Bessel function.

For the power series that we have looked at so far, the set of values of x for which the series is convergent has always turned out to be an interval [a finite interval for the geometric series and the series in Example 2, the infinite interval $(-\infty, \infty)$ in Example 3, and a collapsed interval $[0, 0] = \{0\}$ in Example 1]. The following theorem, proved in Appendix F, says that this is true in general.

---

(3) THEOREM For a given power series $\sum_{n=0}^{\infty} c_n(x - a)^n$ there are only three possibilities:

(i) The series converges only when $x = a$.

(ii) The series converges for all x.

(iii) There is a positive number R such that the series converges if $|x - a| < R$ and diverges if $|x - a| > R$.

The number R in case (iii) is called the **radius of convergence** of the power series. By convention, the radius of convergence is $R = 0$ in case (i) and $R = \infty$ in case (ii). The **interval of convergence** of a power series is the interval that consists of all values of x for which the series converges. In case (i) the interval consists of just a single point a. In case (ii) the interval is $(-\infty, \infty)$. In case (iii) note that the inequality $|x - a| < R$ can be rewritten as $a - R < x < a + R$. When x is an *endpoint* of the interval, that is, $x = a \pm R$, anything can happen—the series might converge at one or both endpoints or it might diverge at both endpoints. Thus in case (iii) there are four possibilities for the interval of convergence:

$$(a - R, a + R) \qquad (a - R, a + R] \qquad [a - R, a + R) \qquad [a - R, a + R]$$

The situation is illustrated in Figure 3.

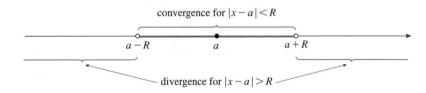

FIGURE 3

We summarize here the radius and interval of convergence for each of the examples already considered in this section.

	Series	Radius of convergence	Interval of convergence
Geometric series	$\sum_{n=0}^{\infty} x^n$	$R = 1$	$(-1, 1)$
Example 1	$\sum_{n=0}^{\infty} n! x^n$	$R = 0$	$\{0\}$
Example 2	$\sum_{n=1}^{\infty} \dfrac{(x - 3)^n}{n}$	$R = 1$	$[2, 4)$
Example 3	$\sum_{n=0}^{\infty} \dfrac{(-1)^n x^{2n}}{2^{2n}(n!)^2}$	$R = \infty$	$(-\infty, \infty)$

In general, the Ratio Test (or sometimes the Root Test) should be used to determine the radius of convergence R. The Ratio and Root Tests always fail when x is an endpoint of the interval of convergence, so the endpoints should be checked using some other test.

EXAMPLE 4   Find the radius of convergence and interval of convergence of the series

$$\sum_{n=0}^{\infty} \frac{(-3)^n x^n}{\sqrt{n+1}}$$

SOLUTION   Let $a_n = (-3)^n x^n / \sqrt{n+1}$. Then

$$\left| \frac{a_{n+1}}{a_n} \right| = \left| \frac{(-3)^{n+1} x^{n+1}}{\sqrt{n+2}} \cdot \frac{\sqrt{n+1}}{(-3)^n x^n} \right|$$

$$= 3 \sqrt{\frac{1+(1/n)}{1+(2/n)}} \, |x| \rightarrow 3|x| \qquad \text{as } n \rightarrow \infty$$

By the Ratio Test, the given series converges if $3|x| < 1$ and diverges if $3|x| > 1$. Thus it converges if $|x| < \frac{1}{3}$ and diverges if $|x| > \frac{1}{3}$. This means that the radius of convergence is $R = \frac{1}{3}$.

We know the series converges in the interval $\left(-\frac{1}{3}, \frac{1}{3}\right)$ but we must now test for convergence at the endpoints of this interval. If $x = -\frac{1}{3}$, the series becomes

$$\sum_{n=0}^{\infty} \frac{(-3)^n \left(-\frac{1}{3}\right)^n}{\sqrt{n+1}} = \sum_{n=0}^{\infty} \frac{1}{\sqrt{n+1}} = \frac{1}{\sqrt{1}} + \frac{1}{\sqrt{2}} + \frac{1}{\sqrt{3}} + \frac{1}{\sqrt{4}} + \cdots$$

which diverges. (Use the Integral Test or simply observe that it is a p-series with $p = \frac{1}{2} < 1$.) If $x = \frac{1}{3}$, the series is

$$\sum_{n=0}^{\infty} \frac{(-3)^n \left(\frac{1}{3}\right)^n}{\sqrt{n+1}} = \sum_{n=0}^{\infty} \frac{(-1)^n}{\sqrt{n+1}}$$

which converges by the Alternating Series Test. Therefore, the given power series converges when $-\frac{1}{3} < x \leq \frac{1}{3}$, so the interval of convergence is $\left(-\frac{1}{3}, \frac{1}{3}\right]$.   ∎

EXAMPLE 5   Find the radius of convergence and interval of convergence of the series

$$\sum_{n=0}^{\infty} \frac{n(x+2)^n}{3^{n+1}}$$

SOLUTION   If $a_n = n(x+2)^n / 3^{n+1}$, then

$$\left| \frac{a_{n+1}}{a_n} \right| = \left| \frac{(n+1)(x+2)^{n+1}}{3^{n+2}} \cdot \frac{3^{n+1}}{n(x+2)^n} \right|$$

$$= \left(1 + \frac{1}{n}\right) \frac{|x+2|}{3} \rightarrow \frac{|x+2|}{3} \qquad \text{as } n \rightarrow \infty$$

Using the Ratio Test, we see that the series converges if $|x+2|/3 < 1$ and it diverges if $|x+2|/3 > 1$. So it converges if $|x+2| < 3$ and diverges if $|x+2| > 3$. Thus the radius of convergence is $R = 3$.

The inequality $|x + 2| < 3$ can be written as $-5 < x < 1$, so we test the series at the endpoints -5 and 1. When $x = -5$, the series is

$$\sum_{n=0}^{\infty} \frac{n(-3)^n}{3^{n+1}} = \tfrac{1}{3} \sum_{n=0}^{\infty} (-1)^n n$$

which diverges by the Test for Divergence $[(-1)^n n$ does not converge to $0]$. When $x = 1$, the series is

$$\sum_{n=0}^{\infty} \frac{n(3)^n}{3^{n+1}} = \tfrac{1}{3} \sum_{n=0}^{\infty} n$$

which also diverges by the Test for Divergence. Thus the series converges only when $-5 < x < 1$, so the interval of convergence is $(-5, 1)$. ∎

EXERCISES 10.8

1. If $\sum_{n=0}^{\infty} c_n 4^n$ is convergent, does it follow that the following series are convergent?

(a) $\sum_{n=0}^{\infty} c_n(-2)^n$ (b) $\sum_{n=0}^{\infty} c_n(-4)^n$

2. Suppose that $\sum_{n=0}^{\infty} c_n x^n$ converges when $x = -4$ and diverges when $x = 6$. What can be said about the convergence or divergence of the following series?

(a) $\sum_{n=0}^{\infty} c_n$ (b) $\sum_{n=0}^{\infty} c_n 8^n$

(c) $\sum_{n=0}^{\infty} c_n(-3)^n$ (d) $\sum_{n=0}^{\infty} (-1)^n c_n 9^n$

3–28 ■ Find the radius of convergence and interval of convergence of the series.

3. $\sum_{n=0}^{\infty} \frac{x^n}{n+2}$ **4.** $\sum_{n=1}^{\infty} \frac{(-1)^n x^n}{\sqrt[3]{n}}$

5. $\sum_{n=0}^{\infty} nx^n$ **6.** $\sum_{n=1}^{\infty} \frac{x^n}{n^2}$

7. $\sum_{n=0}^{\infty} \frac{x^n}{n!}$ **8.** $\sum_{n=1}^{\infty} n^n x^n$

9. $\sum_{n=1}^{\infty} \frac{(-1)^n x^n}{n 2^n}$ **10.** $\sum_{n=1}^{\infty} n 5^n x^n$

11. $\sum_{n=0}^{\infty} \frac{3^n x^n}{(n+1)^2}$ **12.** $\sum_{n=0}^{\infty} \frac{n^2 x^n}{10^n}$

13. $\sum_{n=2}^{\infty} \frac{x^n}{\ln n}$ **14.** $\sum_{n=0}^{\infty} \sqrt{n}(3x + 2)^n$

15. $\sum_{n=0}^{\infty} \frac{n}{4^n}(2x - 1)^n$ **16.** $\sum_{n=1}^{\infty} \frac{(-1)^n x^{2n-1}}{(2n - 1)!}$

17. $\sum_{n=1}^{\infty} (-1)^n \frac{(x - 1)^n}{\sqrt{n}}$ **18.** $\sum_{n=1}^{\infty} \frac{(x - 4)^n}{n 5^n}$

19. $\sum_{n=1}^{\infty} \frac{(x - 2)^n}{n^n}$ **20.** $\sum_{n=0}^{\infty} \frac{(-3)^n(x - 1)^n}{\sqrt{n + 1}}$

21. $\sum_{n=0}^{\infty} \frac{2^n(x - 3)^n}{n + 3}$ **22.** $\sum_{n=1}^{\infty} \frac{(x + 1)^n}{n(n + 1)}$

23. $\sum_{n=1}^{\infty} (n/2)^n(x + 6)^n$ **24.** $\sum_{n=1}^{\infty} \frac{nx^n}{1 \cdot 3 \cdot 5 \cdot \cdots \cdot (2n - 1)}$

25. $\sum_{n=1}^{\infty} \frac{(2x - 1)^n}{n^3}$ **26.** $\sum_{n=2}^{\infty} (-1)^n \frac{(2x + 3)^n}{n \ln n}$

27. $\sum_{n=2}^{\infty} \frac{x^n}{(\ln n)^n}$ **28.** $\sum_{n=1}^{\infty} \frac{2 \cdot 4 \cdot 6 \cdot \cdots \cdot (2n)}{1 \cdot 3 \cdot 5 \cdot \cdots \cdot (2n - 1)} x^n$

29. If k is a positive integer, find the radius of convergence of the series

$$\sum_{n=0}^{\infty} \frac{(n!)^k}{(kn)!} x^n$$

30. Graph the first several partial sums $s_n(x)$ of the series $\sum_{n=0}^{\infty} x^n$, together with the sum function $f(x) = 1/(1 - x)$, on a common screen. On what interval do these partial sums appear to be converging to $f(x)$?

31. The function J_1 defined by

$$J_1(x) = \sum_{n=0}^{\infty} \frac{(-1)^n x^{2n+1}}{n!(n + 1)! 2^{2n+1}}$$

is called the *Bessel function of order 1*.
(a) Find its domain.
(b) Graph the first several partial sums on a common screen.
(c) If your CAS has built-in Bessel functions, graph J_1 on the same screen as the partial sums in part (b) and observe how the partial sums approximate J_1.

32. The function A defined by

$$A(x) = 1 + \frac{x^3}{2 \cdot 3} + \frac{x^6}{2 \cdot 3 \cdot 5 \cdot 6} + \frac{x^9}{2 \cdot 3 \cdot 5 \cdot 6 \cdot 8 \cdot 9} + \cdots$$

is called the *Airy function* after the English mathematician and astronomer Sir George Airy (1801–1892).

(a) Find the domain of the Airy function.

(b) Graph the first several partial sums $s_n(x)$ on a common screen.

(c) If your CAS has built-in Airy functions, graph A on the same screen as the partial sums in part (b) and observe how the partial sums approximate A.

33. A function f is defined by

$$f(x) = 1 + 2x + x^2 + 2x^3 + x^4 + \cdots$$

that is, its coefficients are $c_{2n} = 1$ and $c_{2n+1} = 2$ for all $n \geq 0$. Find the interval of convergence of the series and find an explicit formula for $f(x)$.

34. If $f(x) = \sum_{n=0}^{\infty} c_n x^n$, where $c_{n+4} = c_n$ for all $n \geq 0$, find the interval of convergence of the series and a formula for $f(x)$.

35. Show that if $\lim_{n \to \infty} \sqrt[n]{|c_n|} = c$, then the radius of convergence of the power series $\sum c_n x^n$ is $R = 1/c$.

36. Suppose that the radius of convergence of the power series $\sum c_n x^n$ is R. What is the radius of convergence of the power series $\sum c_n x^{2n}$?

37. Suppose the series $\sum c_n x^n$ has radius of convergence 2 and the series $\sum d_n x^n$ has radius of convergence 3. What can you say about the radius of convergence of the series $\sum (c_n + d_n)x^n$?

10.9   REPRESENTATION OF FUNCTIONS AS POWER SERIES

In this section we learn how to represent a function as a sum of a power series by manipulating geometric series or by differentiating or integrating such a series. You might wonder why we would ever want to express a known function as a sum of infinitely many terms. We will see later that this strategy is useful for integrating functions that are otherwise intractable, for solving differential equations, and for approximating functions by polynomials. (Scientists do this to simplify the expressions they deal with; computer scientists do this to represent functions on calculators and computers.)

We start with an equation that we have seen before:

A geometric illustration of Equation 1 is shown in Figure 1. Because the sum of a series is the limit of the sequence of partial sums, we have

$$\frac{1}{1-x} = \lim_{n \to \infty} s_n(x)$$

where

$$s_n(x) = 1 + x + x^2 + \cdots + x^n$$

is the nth partial sum. Notice that as n increases, $s_n(x)$ becomes a better approximation to $f(x)$ for $-1 < x < 1$.

(1)
$$\frac{1}{1-x} = 1 + x + x^2 + x^3 + \cdots = \sum_{n=0}^{\infty} x^n \qquad |x| < 1$$

We first encountered this equation in Example 5 in Section 10.2, where we obtained it by observing that it is a geometric series with $a = 1$ and $r = x$. But here our point of view is different. We now regard Equation 1 as expressing the function $f(x) = 1/(1-x)$ as a sum of a power series.

EXAMPLE 1   Express $1/(1 + x^2)$ as the sum of a power series and find the interval of convergence.

SOLUTION   Replacing x by $-x^2$ in Equation 1, we have

$$\frac{1}{1+x^2} = \frac{1}{1-(-x^2)} = \sum_{n=0}^{\infty} (-x^2)^n$$

$$= \sum_{n=0}^{\infty} (-1)^n x^{2n} = 1 - x^2 + x^4 - x^6 + x^8 - \cdots$$

Because this is a geometric series it converges when $|-x^2| < 1$, that is, $x^2 < 1$, or $|x| < 1$. Therefore, the interval of convergence is $(-1, 1)$.   ■

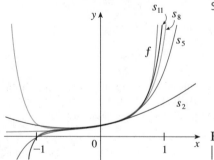

FIGURE 1

$$f(x) = \frac{1}{1-x} \text{ and some partial sums}$$

EXAMPLE 2   Find a power series representation for $1/(x + 2)$.

SOLUTION   In order to put this function in the form of the left side of Equation 1 we first factor a 2 from the denominator:

$$\frac{1}{2 + x} = \frac{1}{2\left(1 + \frac{x}{2}\right)} = \frac{1}{2\left[1 - \left(-\frac{x}{2}\right)\right]}$$

$$= \frac{1}{2} \sum_{n=0}^{\infty} \left(-\frac{x}{2}\right)^n = \sum_{n=0}^{\infty} \frac{(-1)^n}{2^{n+1}} x^n$$

This series converges when $\left|-x/2\right| < 1$, that is, $\left|x\right| < 2$. So the interval of convergence is $(-2, 2)$. ∎

EXAMPLE 3 Find a power series representation of $x^3/(x + 2)$.

SOLUTION Since this function is just x^3 times the function in Example 2, all we have to do is to multiply that series by x^3:

$$\frac{x^3}{x + 2} = x^3 \sum_{n=0}^{\infty} \frac{(-1)^n}{2^{n+1}} x^n = \sum_{n=0}^{\infty} \frac{(-1)^n}{2^{n+1}} x^{n+3}$$

$$= \tfrac{1}{2} x^3 - \tfrac{1}{4} x^4 + \tfrac{1}{8} x^5 - \tfrac{1}{16} x^6 + \cdots$$

Another way of writing this series is as follows:

$$\frac{x^3}{x + 2} = \sum_{n=3}^{\infty} \frac{(-1)^{n-1}}{2^{n-2}} x^n$$

As in Example 2, the interval of convergence is $(-2, 2)$. ∎

DIFFERENTIATION AND INTEGRATION OF POWER SERIES

The sum of a power series is a function $f(x) = \sum_{n=0}^{\infty} c_n(x - a)^n$ whose domain is the interval of convergence of the series. We would like to be able to differentiate and integrate such functions, and the following theorem says that we can do so by differentiating or integrating each individual term in the series, just as we would for a polynomial. This is called **term-by-term differentiation and integration.** The proof is lengthy and is therefore omitted.

---

(2) THEOREM If the power series $\sum c_n(x - a)^n$ has radius of convergence $R > 0$, then the function f defined by

$$f(x) = c_0 + c_1(x - a) + c_2(x - a)^2 + \cdots = \sum_{n=0}^{\infty} c_n(x - a)^n$$

is differentiable (and therefore continuous) on the interval $(a - R, a + R)$ and

(a) $f'(x) = c_1 + 2c_2(x - a) + 3c_3(x - a)^2 + \cdots = \sum_{n=1}^{\infty} n c_n(x - a)^{n-1}$

(b) $\displaystyle\int f(x)\, dx = C + c_0(x - a) + c_1 \frac{(x - a)^2}{2} + c_2 \frac{(x - a)^3}{3} + \cdots$

$$= C + \sum_{n=0}^{\infty} c_n \frac{(x - a)^{n+1}}{n + 1}$$

The radii of convergence of the power series in Equations (a) and (b) are both R.

NOTE 1:   Equations (a) and (b) can be rewritten in the form

(c) $\dfrac{d}{dx}\left[\displaystyle\sum_{n=0}^{\infty} c_n(x-a)^n\right] = \displaystyle\sum_{n=0}^{\infty} \dfrac{d}{dx}[c_n(x-a)^n]$

(d) $\displaystyle\int\left[\displaystyle\sum_{n=0}^{\infty} c_n(x-a)^n\right]dx = \displaystyle\sum_{n=0}^{\infty} \int c_n(x-a)^n\,dx$

We know that, for finite sums, the derivative of a sum is the sum of the derivatives and the integral of a sum is the sum of the integrals. Equations (c) and (d) assert that the same is true for infinite sums provided we are dealing with *power series*. (For other types of series of functions the situation is not as simple; see Exercise 34.)

NOTE 2:   Although Theorem 2 says that the radius of convergence remains the same when a power series is differentiated or integrated, this does not mean that the *interval* of convergence remains the same. It may happen that the original series converges at an endpoint, whereas the differentiated series diverges there. (See Exercise 35.)

NOTE 3:   The idea of differentiating a power series term by term is the basis for a powerful method for solving differential equations. We will discuss this method in Chapter 15.

EXAMPLE 4   In Example 3 in Section 10.8 we saw that the Bessel function

$$J_0(x) = \sum_{n=0}^{\infty} \frac{(-1)^n x^{2n}}{2^{2n}(n!)^2}$$

is defined for all x. Thus, by Theorem 2, J_0 is differentiable for all x and its derivative is found by term-by-term differentiation as follows:

$$J_0'(x) = \sum_{n=0}^{\infty} \frac{d}{dx}\frac{(-1)^n x^{2n}}{2^{2n}(n!)^2} = \sum_{n=1}^{\infty} \frac{(-1)^n 2nx^{2n-1}}{2^{2n}(n!)^2}$$ ∎

EXAMPLE 5   Express $1/(1-x)^2$ as a power series by differentiating Equation 1. What is the radius of convergence?

SOLUTION   Differentiating both sides of the equation

$$\frac{1}{1-x} = 1 + x + x^2 + x^3 + \cdots = \sum_{n=0}^{\infty} x^n$$

we get $$\frac{1}{(1-x)^2} = 1 + 2x + 3x^2 + \cdots = \sum_{n=1}^{\infty} nx^{n-1}$$

An equivalent answer is

$$\frac{1}{(1-x)^2} = \sum_{n=0}^{\infty} (n+1)x^n$$

According to Theorem 2, the radius of convergence of the differentiated series is the same as the radius of convergence of the original series, namely, $R = 1$.   ∎

EXAMPLE 6   Find a power series representation for $\ln(1-x)$ and its radius of convergence.

SOLUTION We notice that, except for a factor of -1, the derivative of this function is $1/(1-x)$. So we integrate both sides of Equation 1:

$$-\ln(1-x) = \int \frac{1}{1-x}\,dx = C + x + \frac{x^2}{2} + \frac{x^3}{3} + \cdots$$

$$= C + \sum_{n=0}^{\infty} \frac{x^{n+1}}{n+1} = C + \sum_{n=1}^{\infty} \frac{x^n}{n} \qquad |x| < 1$$

To determine the value of C we put $x = 0$ in this equation and obtain $-\ln(1-0) = C$. Thus $C = 0$ and

$$\ln(1-x) = -x - \frac{x^2}{2} - \frac{x^3}{3} - \cdots = -\sum_{n=1}^{\infty} \frac{x^n}{n} \qquad |x| < 1$$

The radius of convergence is the same as for the original series: $R = 1$.   ∎

Notice what happens if we put $x = \frac{1}{2}$ in the result of Example 6. Since $\ln \frac{1}{2} = -\ln 2$, we see that

$$\ln 2 = \frac{1}{2} + \frac{1}{8} + \frac{1}{24} + \frac{1}{64} + \cdots = \sum_{n=1}^{\infty} \frac{1}{n2^n}$$

EXAMPLE 7 Find a power series representation for $f(x) = \tan^{-1}x$.

The power series for $\tan^{-1}x$ obtained in Example 7 is called *Gregory's series* after the Scottish mathematician James Gregory (1638–1675), who had anticipated some of Newton's discoveries. We have shown that Gregory's series is valid when $-1 < x < 1$, but it turns out (although it is not easy to prove) that it is also valid when $x = \pm 1$. Notice that when $x = 1$ the series becomes

$$\frac{\pi}{4} = 1 - \frac{1}{3} + \frac{1}{5} - \frac{1}{7} + \cdots$$

This beautiful result is known as the Leibniz formula for π.

SOLUTION We observe that $f'(x) = 1/(1+x^2)$ and find the required series by integrating the power series for $1/(1+x^2)$ found in Example 1.

$$\tan^{-1}x = \int \frac{1}{1+x^2}\,dx = \int (1 - x^2 + x^4 - x^6 + \cdots)\,dx$$

$$= C + x - \frac{x^3}{3} + \frac{x^5}{5} - \frac{x^7}{7} + \cdots$$

To find C we put $x = 0$ and obtain $C = \tan^{-1}0 = 0$. Therefore

$$\tan^{-1}x = x - \frac{x^3}{3} + \frac{x^5}{5} - \frac{x^7}{7} + \cdots = \sum_{n=0}^{\infty} (-1)^n \frac{x^{2n+1}}{2n+1}$$

Since the radius of convergence of the series for $1/(1+x^2)$ is 1, the radius of convergence of this series for $\tan^{-1}x$ is also 1.   ∎

EXAMPLE 8
(a) Evaluate $\int [1/(1+x^7)]\,dx$ as a power series.
(b) Use part (a) to approximate $\int_0^{0.5} [1/(1+x^7)]\,dx$ correct to within 10^{-7}.

SOLUTION
(a) The first step is to express the integrand, $1/(1+x^7)$, as the sum of a power series. As in Example 1, we start with Equation 1 and replace x by $-x^7$:

$$\frac{1}{1+x^7} = \frac{1}{1-(-x^7)} = \sum_{n=0}^{\infty} (-x^7)^n = \sum_{n=0}^{\infty} (-1)^n x^{7n} = 1 - x^7 + x^{14} - \cdots$$

This example demonstrates one way in which power series representations are useful. Integrating $1/(1 + x^7)$ by hand is incredibly difficult. Different computer algebra systems return different forms of the answer, but they are all extremely complicated. (If you have a CAS, try it yourself.) The infinite series answer that we obtain in Example 8(a) is actually much easier to deal with than the finite answer provided by a CAS.

Now we integrate term by term:

$$\int \frac{1}{1 + x^7}\, dx = \int \sum_{n=0}^{\infty} (-1)^n x^{7n}\, dx = C + \sum_{n=0}^{\infty} (-1)^n \frac{x^{7n+1}}{7n + 1}$$

$$= C + x - \frac{x^8}{8} + \frac{x^{15}}{15} - \frac{x^{22}}{22} + \cdots$$

This series converges for $|-x^7| < 1$, that is, for $|x| < 1$.

(b) Using the antiderivative given by the power series in part (a) with $C = 0$, we have

$$\int_0^{0.5} \frac{1}{1 + x^7}\, dx = \left[x - \frac{x^8}{8} + \frac{x^{15}}{15} - \frac{x^{22}}{22} + \cdots \right]_0^{1/2}$$

$$= \frac{1}{2} - \frac{1}{8 \cdot 2^8} + \frac{1}{15 \cdot 2^{15}} - \frac{1}{22 \cdot 2^{22}} + \cdots + \frac{(-1)^n}{(7n + 1)2^{7n+1}} + \cdots$$

This infinite series is the exact value of the definite integral, but since it is an alternating series, we can approximate the sum using Theorem 10.5.1. If we stop after the term with $n = 3$, the error is smaller than

$$\frac{1}{29 \cdot 2^{29}} \approx 6.4 \times 10^{-11}$$

and we have

$$\int_0^{0.5} \frac{1}{1 + x^7}\, dx \approx \frac{1}{2} - \frac{1}{8 \cdot 2^8} + \frac{1}{15 \cdot 2^{15}} - \frac{1}{22 \cdot 2^{22}} \approx 0.49951374 \quad \blacksquare$$

EXERCISES 10.9

1–8 ■ Find a power series representation for the function and determine the interval of convergence.

1. $f(x) = \dfrac{1}{1 + x}$

2. $f(x) = \dfrac{x}{1 - x}$

3. $f(x) = \dfrac{1}{1 + 4x^2}$

4. $f(x) = \dfrac{1}{x^4 + 16}$

5. $f(x) = \dfrac{1}{4 + x^2}$

6. $f(x) = \dfrac{1 + x^2}{1 - x^2}$

7. $f(x) = \dfrac{x}{x - 3}$

8. $f(x) = \dfrac{2}{3x + 4}$

9–10 ■ Express the function as the sum of a power series by first using partial fractions. Find the interval of convergence.

9. $f(x) = \dfrac{3x - 2}{2x^2 - 3x + 1}$

10. $f(x) = \dfrac{x}{x^2 - 3x + 2}$

11–18 ■ Find a power series representation for the function and determine the radius of convergence.

11. $f(x) = \dfrac{1}{(1 + x)^2}$

12. $f(x) = \ln(1 + x)$

13. $f(x) = \dfrac{1}{(1 + x)^3}$

14. $f(x) = x \ln(1 + x)$

15. $f(x) = \ln(5 - x)$

16. $f(x) = \tan^{-1}(2x)$

17. $f(x) = \ln\left(\dfrac{1 + x}{1 - x}\right)$

18. $f(x) = \dfrac{x^2}{(1 - 2x)^2}$

19–20 ■ Find a power series representation for f and graph f and several partial sums $s_n(x)$ on the same screen. What happens as n increases?

19. $f(x) = \ln(3 + x)$

20. $f(x) = \dfrac{1}{x^2 + 25}$

21–24 ■ Evaluate the indefinite integral as a power series.

21. $\displaystyle\int \frac{1}{1 + x^4}\, dx$

22. $\displaystyle\int \frac{x}{1 + x^5}\, dx$

23. $\displaystyle\int \frac{\arctan x}{x}\, dx$

24. $\displaystyle\int \tan^{-1}(x^2)\, dx$

25–28 ▪ Use a power series to approximate the definite integral to six decimal places.

25. $\displaystyle\int_0^{0.2} \frac{1}{1+x^4}\,dx$

26. $\displaystyle\int_0^{1/2} \tan^{-1}(x^2)\,dx$

27. $\displaystyle\int_0^{1/3} x^2\tan^{-1}(x^4)\,dx$

28. $\displaystyle\int_0^{0.5} \frac{dx}{1+x^6}$

29. Use the result of Example 6 to compute $\ln 1.1$ correct to five decimal places.

30. Show that the function
$$f(x) = \sum_{n=0}^{\infty} \frac{(-1)^n x^{2n}}{(2n)!}$$
is a solution of the differential equation $f''(x) + f(x) = 0$.

31. (a) Show that J_0 (the Bessel function of order 0 given in Example 4) satisfies the differential equation
$$x^2 J_0''(x) + x J_0'(x) + x^2 J_0(x) = 0$$
 (b) Evaluate $\int_0^1 J_0(x)\,dx$ correct to three decimal places.

32. The Bessel function of order 1 is defined by
$$J_1(x) = \sum_{n=0}^{\infty} \frac{(-1)^n x^{2n+1}}{n!(n+1)!2^{2n+1}}$$
 (a) Show that J_1 satisfies the differential equation
$$x^2 J_1''(x) + x J_1'(x) + (x^2 - 1) J_1(x) = 0$$

 (b) Show that $J_0'(x) = -J_1(x)$.

33. (a) Show that the function $\displaystyle f(x) = \sum_{n=0}^{\infty} \frac{x^n}{n!}$ is a solution of the differential equation $f'(x) = f(x)$.
 (b) Show that $f(x) = e^x$.

34. Let $f_n(x) = (\sin nx)/n^2$. Show that the series $\sum f_n(x)$ converges for all values of x but the series of derivatives $\sum f_n'(x)$ diverges when $x = 2n\pi$, n an integer. For what values of x does the series $\sum f_n''(x)$ converge?

35. Let
$$f(x) = \sum_{n=1}^{\infty} \frac{x^n}{n^2}$$
Find the intervals of convergence for f, f', and f''.

36. (a) Starting with the geometric series $\sum_{n=0}^{\infty} x^n$, find the sum of the series
$$\sum_{n=1}^{\infty} nx^{n-1} \qquad |x| < 1$$
 (b) Find the sums of the following series.
  (i) $\displaystyle \sum_{n=1}^{\infty} nx^n, \quad |x| < 1$ (ii) $\displaystyle \sum_{n=1}^{\infty} \frac{n}{2^n}$
 (c) Find the sums of the following series.
  (i) $\displaystyle \sum_{n=2}^{\infty} n(n-1)x^n, \quad |x| < 1$ (ii) $\displaystyle \sum_{n=2}^{\infty} \frac{n^2-n}{2^n}$
  (iii) $\displaystyle \sum_{n=1}^{\infty} \frac{n^2}{2^n}$

10.10 TAYLOR AND MACLAURIN SERIES

In the preceding section we were able to find power series representations for a certain restricted class of functions. Here we investigate more general problems: Which functions have power series representations? How can we find such representations?

We start by supposing that f is any function that can be represented by a power series

(1) $\quad f(x) = c_0 + c_1(x-a) + c_2(x-a)^2 + c_3(x-a)^3 + c_4(x-a)^4 + \cdots$
$$|x-a| < R$$

and let us try to determine what the coefficients c_n must be in terms of f. To begin, notice that if we put $x = a$ in Equation 1, then all terms after the first one are 0 and we get

$$f(a) = c_0$$

If we apply Theorem 10.9.2 to Equation 1, we obtain

(2) $\quad f'(x) = c_1 + 2c_2(x-a) + 3c_3(x-a)^2 + 4c_4(x-a)^3 + \cdots \qquad |x-a| < R$

and substitution of $x = a$ in Equation 2 gives

$$f'(a) = c_1$$

Now we apply Theorem 10.9.2 a second time, this time to Equation 2, and obtain

(3) $f''(x) = 2c_2 + 2 \cdot 3c_3(x - a) + 3 \cdot 4c_4(x - a)^2 + \cdots$ $|x - a| < R$

Again we put $x = a$ in Equation 3. The result is

$$f''(a) = 2c_2$$

Let us apply the procedure one more time. Differentiation of the series in Equation 3 gives

(4) $f'''(x) = 2 \cdot 3c_3 + 2 \cdot 3 \cdot 4c_4(x - a) + 3 \cdot 4 \cdot 5c_5(x - a)^2 + \cdots$ $|x - a| < R$

and substitution of $x = a$ in Equation 4 gives

$$f'''(a) = 2 \cdot 3c_3 = 3!c_3$$

By now you can see the pattern. If we continue to differentiate and substitute $x = a$, we obtain

$$f^{(n)}(a) = 2 \cdot 3 \cdot 4 \cdot \cdots \cdot nc_n = n!c_n$$

Solving this equation for the nth coefficient c_n, we get

$$c_n = \frac{f^{(n)}(a)}{n!}$$

This formula remains valid even for $n = 0$ if we adopt the conventions that $0! = 1$ and $f^{(0)} = f$. Thus we have proved the following theorem:

---

(5) THEOREM   If f has a power series representation (expansion) at a, that is, if

$$f(x) = \sum_{n=0}^{\infty} c_n(x - a)^n \qquad |x - a| < R$$

then its coefficients are given by the formula

$$c_n = \frac{f^{(n)}(a)}{n!}$$

Substituting this formula for c_n back into the series, we see that *if f has a power series expansion at a, then it must be of the following form:*

---

(6) $f(x) = \displaystyle\sum_{n=0}^{\infty} \frac{f^{(n)}(a)}{n!} (x - a)^n$

$\qquad = f(a) + \dfrac{f'(a)}{1!}(x - a) + \dfrac{f''(a)}{2!}(x - a)^2 + \dfrac{f'''(a)}{3!}(x - a)^3 + \cdots$

The series in Equation 6 is called the **Taylor series of the function f at a** (or **about a** or **centered at a**). For the special case $a = 0$ the Taylor series becomes

The Taylor series is named after the English mathematician Brook Taylor (1685–1731) and the Maclaurin series is named in honor of the Scottish mathematician Colin Maclaurin (1698–1746) despite the fact that the Maclaurin series is really just a special case of the Taylor series. But the idea of representing particular functions as sums of power series goes back to Sir Isaac Newton, and the general Taylor series was known to the Scottish mathematician James Gregory in 1668 and to the Swiss mathematician John Bernoulli in the 1690s. Taylor was apparently unaware of the work of Gregory and Bernoulli when he published his discoveries on series in 1715 in his book *Methodus incrementorum directa et inversa*. Maclaurin series are named after Colin Maclaurin because he popularized them in his calculus textbook *Treatise of Fluxions* published in 1742.

$$(7) \qquad f(x) = \sum_{n=0}^{\infty} \frac{f^{(n)}(0)}{n!} x^n = f(0) + \frac{f'(0)}{1!} x + \frac{f''(0)}{2!} x^2 + \cdots$$

This case arises frequently enough that it is given the special name **Maclaurin series.**

NOTE: We have shown that *if* f can be represented as a power series about a (such functions are called **analytic at a**), then f is equal to the sum of its Taylor series. Theorem 10.9.2 shows that analytic functions are infinitely differentiable at a; that is, they have derivatives of all orders at a. However, not all infinitely differentiable functions are analytic. Exercise 56 gives an example of an infinitely differentiable function that is not analytic at 0. This function is therefore not equal to the sum of its Taylor series.

EXAMPLE 1   Find the Maclaurin series of the function $f(x) = e^x$ and its radius of convergence.

SOLUTION   If $f(x) = e^x$, then $f^{(n)}(x) = e^x$, so $f^{(n)}(0) = e^0 = 1$ for all n. Therefore, the Taylor series for f at 0 (that is, the Maclaurin series) is

$$\sum_{n=0}^{\infty} \frac{f^{(n)}(0)}{n!} x^n = \sum_{n=0}^{\infty} \frac{x^n}{n!} = 1 + \frac{x}{1!} + \frac{x^2}{2!} + \frac{x^3}{3!} + \cdots$$

To find the radius of convergence we let $a_n = x^n/n!$. Then

$$\left| \frac{a_{n+1}}{a_n} \right| = \left| \frac{x^{n+1}}{(n+1)!} \cdot \frac{n!}{x^n} \right| = \frac{|x|}{n+1} \to 0 < 1$$

so, by the Ratio Test, the series converges for all x and the radius of convergence is $R = \infty$.   ∎

The conclusion we can draw from Theorem 5 and Example 1 is that *if* e^x has a power series expansion at 0, then

$$e^x = \sum_{n=0}^{\infty} \frac{x^n}{n!}$$

So how can we determine whether e^x *does* have a power series representation?

Let us investigate the more general question: Under what circumstances is a function $f(x)$ equal to the sum of its Taylor series? In other words, if f has derivatives of all orders, when is it true that

$$f(x) = \sum_{n=0}^{\infty} \frac{f^{(n)}(a)}{n!} (x-a)^n$$

As with any convergent series, this means that $f(x)$ is the limit of the sequence of partial sums. In the case of the Taylor series the partial sums are

$$T_n(x) = \sum_{i=0}^{n} \frac{f^{(i)}(a)}{i!} (x-a)^i$$

$$= f(a) + \frac{f'(a)}{1!} (x-a) + \frac{f''(a)}{2!} (x-a)^2 + \cdots + \frac{f^{(n)}(a)}{n!} (x-a)^n$$

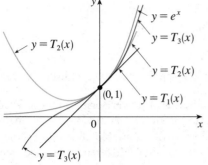

FIGURE 1

As n increases, $T_n(x)$ appears to approach e^x in Figure 1. This suggests that e^x is equal to the sum of its Taylor series.

Notice that T_n is a polynomial of degree n called the **nth-degree Taylor polynomial of f at a.** For instance, for the exponential function $f(x) = e^x$, the result of Example 1 shows that its first three Taylor polynomials at 0 (or Maclaurin polynomials) are

$$T_1(x) = 1 + x \qquad T_2(x) = 1 + x + \frac{x^2}{2!} \qquad T_3(x) = 1 + x + \frac{x^2}{2!} + \frac{x^3}{3!}$$

The graphs of the exponential function and these three Taylor polynomials are drawn in Figure 1.

In general, $f(x)$ is the sum of its Taylor series if

$$f(x) = \lim_{n \to \infty} T_n(x)$$

If we let $R_n(x)$ be the remainder of the series, then

$$R_n(x) = f(x) - T_n(x) \qquad \text{and} \qquad f(x) = T_n(x) + R_n(x)$$

If we can somehow show that $\lim_{n \to \infty} R_n(x) = 0$, then it follows that

$$\lim_{n \to \infty} T_n(x) = \lim_{n \to \infty} [f(x) - R_n(x)] = f(x) - \lim_{n \to \infty} R_n(x) = f(x)$$

We have therefore proved the following theorem.

---

(8) THEOREM   If $f(x) = T_n(x) + R_n(x)$, where T_n is the nth-degree Taylor polynomial of f at a and

$$\lim_{n \to \infty} R_n(x) = 0$$

for $|x - a| < R$, then f is equal to the sum of its Taylor series on the interval $|x - a| < R$; that is, f is analytic at a.

In trying to show that $\lim_{n \to \infty} R_n(x) = 0$ for a specific function f, we usually use the expression in the next theorem.

---

(9) TAYLOR'S FORMULA   If f has $n + 1$ derivatives in an interval I that contains the number a, then for x in I there is a number z strictly between x and a such that the remainder term in the Taylor series can be expressed as

$$R_n(x) = \frac{f^{(n+1)}(z)}{(n + 1)!} (x - a)^{n+1}$$

NOTE 1:   For the special case $n = 0$, if we put $x = b$ and $z = c$ in Taylor's Formula, we get $f(b) = f(a) + f'(c)(b - a)$, which is the Mean Value Theorem. In fact, Theorem 9 can be proved by a method similar to the proof of the Mean Value Theorem. The proof is given at the end of this section.

NOTE 2:   Notice that the remainder term

(10) $$R_n(x) = \frac{f^{(n+1)}(z)}{(n + 1)!} (x - a)^{n+1}$$

is very similar to the terms in the Taylor series except that $f^{(n+1)}$ is evaluated at z instead of at a. All we can say about the number z is that it lies somewhere between x and a. The expression for $R_n(x)$ in Equation 10 is known as **Lagrange's form of the remainder term.**

NOTE 3:   In Section 10.12 we will explore the use of Taylor's Formula in approximating functions. Our immediate use of it is in conjunction with Theorem 8.

In applying Theorems 8 and 9 it is often helpful to make use of the following fact:

(11)
$$\lim_{n \to \infty} \frac{x^n}{n!} = 0 \qquad \text{for every real number } x$$

This is true because we know from Example 1 that the series $\Sigma \, x^n/n!$ converges for all x and so its nth term approaches 0.

EXAMPLE 2   Prove that e^x is equal to the sum of its Taylor series.

SOLUTION   If $f(x) = e^x$, then $f^{(n+1)}(x) = e^x$, so the remainder term in Taylor's Formula is

$$R_n(x) = \frac{e^z}{(n + 1)!} x^{n+1}$$

where z lies between 0 and x. (Note, however, that z depends on n.) If $x > 0$, then $0 < z < x$, so $e^z < e^x$. Therefore

$$0 < R_n(x) = \frac{e^z}{(n + 1)!} x^{n+1} < e^x \frac{x^{n+1}}{(n + 1)!} \to 0$$

by Equation 11, so $R_n(x) \to 0$ as $n \to \infty$ by the Squeeze Theorem. If $x < 0$, then $x < z < 0$, so $e^z < e^0 = 1$ and

$$\left| R_n(x) \right| < \frac{\left| x \right|^{n+1}}{(n + 1)!} \to 0$$

Again $R_n(x) \to 0$. Thus, by Theorem 8, e^x is equal to the sum of its Taylor series, that is,

(12)
$$e^x = \sum_{n=0}^{\infty} \frac{x^n}{n!} \qquad \text{for all } x$$

In particular if we put $x = 1$ in Equation 12, we obtain the following expression for the number e as a sum of an infinite series:

(13)
$$e = \sum_{n=0}^{\infty} \frac{1}{n!} = 1 + \frac{1}{1!} + \frac{1}{2!} + \frac{1}{3!} + \cdots$$

EXAMPLE 3   Find the Taylor series for $f(x) = e^x$ at $a = 2$.

SOLUTION   We have $f^{(n)}(2) = e^2$ and so, putting $a = 2$ in the definition of a Taylor series (6), we get

$$\sum_{n=0}^{\infty} \frac{f^{(n)}(2)}{n!}(x - 2)^n = \sum_{n=0}^{\infty} \frac{e^2}{n!}(x - 2)^n$$

Again it can be verified, as in Example 1, that the radius of convergence is $R = \infty$. As in Example 2 we can verify that $\lim_{n \to \infty} R_n(x) = 0$, so

(14) $$e^x = \sum_{n=0}^{\infty} \frac{e^2}{n!}(x - 2)^n \qquad \text{for all } x$$

We have two power series expansions for e^x, the Maclaurin series in Equation 12 and the Taylor series in Equation 14. The first is better if we are interested in values of x near 0 and the second is better if x is near 2.

EXAMPLE 4   Find the Maclaurin series for $\sin x$ and prove that it represents $\sin x$ for all x.

SOLUTION   We arrange our computation in two columns as follows:

$$f(x) = \sin x \qquad f(0) = 0$$
$$f'(x) = \cos x \qquad f'(0) = 1$$
$$f''(x) = -\sin x \qquad f''(0) = 0$$
$$f'''(x) = -\cos x \qquad f'''(0) = -1$$
$$f^{(4)}(x) = \sin x \qquad f^{(4)}(0) = 0$$

Since the derivatives repeat in a cycle of four, we can write the Maclaurin series as follows:

$$f(0) + \frac{f'(0)}{1!}x + \frac{f''(0)}{2!}x^2 + \frac{f'''(0)}{3!}x^3 + \cdots$$
$$= x - \frac{x^3}{3!} + \frac{x^5}{5!} - \frac{x^7}{7!} + \cdots = \sum_{n=0}^{\infty} (-1)^n \frac{x^{2n+1}}{(2n + 1)!}$$

Using the remainder term (10) with $a = 0$, we have

$$R_n(x) = \frac{f^{(n+1)}(z)}{(n + 1)!} x^{n+1}$$

where $f(x) = \sin x$ and z lies between 0 and x. But $f^{(n+1)}(z)$ is $\pm \sin z$ or $\pm \cos z$. In any case, $|f^{(n+1)}(z)| \leq 1$ and so

(15) $$|R_n(x)| = \frac{|f^{(n+1)}(z)|}{(n + 1)!} |x^{n+1}| \leq \frac{|x|^{n+1}}{(n + 1)!}$$

By Equation 11 the right side of this inequality approaches 0 as $n \to \infty$, so $|R_n(x)| \to 0$ by the Squeeze Theorem. It follows that $R_n(x) \to 0$ as $n \to \infty$, so $\sin x$ is equal to the sum of its Maclaurin series by Theorem 8.

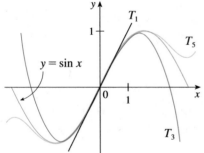

FIGURE 2

Figure 2 shows the graph of $\sin x$ together with its Taylor (or Maclaurin) polynomials

$$T_1(x) = x$$
$$T_3(x) = x - \frac{x^3}{3!}$$
$$T_5(x) = x - \frac{x^3}{3!} + \frac{x^5}{5!}$$

Notice that, as n increases, $T_n(x)$ becomes a better approximation to $\sin x$.

We state the result of Example 4 for future reference:

(16)

$$
\sin x = x - \frac{x^3}{3!} + \frac{x^5}{5!} - \frac{x^7}{7!} + \cdots
$$

$$
= \sum_{n=0}^{\infty} (-1)^n \frac{x^{2n+1}}{(2n+1)!} \qquad \text{for all } x
$$

EXAMPLE 5   Find the Maclaurin series for $\cos x$.

SOLUTION   We could proceed directly as in Example 4 but it is easier to use Theorem 10.9.2 to differentiate the Maclaurin series for $\sin x$ given by Equation 16:

$$
\cos x = \frac{d}{dx} (\sin x) = \frac{d}{dx} \left(x - \frac{x^3}{3!} + \frac{x^5}{5!} - \frac{x^7}{7!} + \cdots \right)
$$

$$
= 1 - \frac{3x^2}{3!} + \frac{5x^4}{5!} - \frac{7x^6}{7!} + \cdots = 1 - \frac{x^2}{2!} + \frac{x^4}{4!} - \frac{x^6}{6!} + \cdots
$$

Since the Maclaurin series for $\sin x$ converges for all x, Theorem 10.9.2 tells us that the differentiated series for $\cos x$ also converges for all x. Thus

(17)

$$
\cos x = 1 - \frac{x^2}{2!} + \frac{x^4}{4!} - \frac{x^6}{6!} + \cdots
$$

$$
= \sum_{n=0}^{\infty} (-1)^n \frac{x^{2n}}{(2n)!} \qquad \text{for all } x
$$

EXAMPLE 6   Find the Maclaurin series for the function $f(x) = x \cos x$.

SOLUTION   Instead of computing derivatives and substituting in Equation 7, it is easier to multiply the series for $\cos x$ (Equation 17) by x:

$$
x \cos x = x \sum_{n=0}^{\infty} (-1)^n \frac{x^{2n}}{(2n)!} = \sum_{n=0}^{\infty} (-1)^n \frac{x^{2n+1}}{(2n)!}
$$

EXAMPLE 7   Represent $f(x) = \sin x$ as the sum of its Taylor series centered at $\pi/3$.

SOLUTION   Arranging our work in columns, we have

$$
f(x) = \sin x \qquad\qquad f\left(\frac{\pi}{3}\right) = \frac{\sqrt{3}}{2}
$$

$$
f'(x) = \cos x \qquad\qquad f'\left(\frac{\pi}{3}\right) = \frac{1}{2}
$$

$$
f''(x) = -\sin x \qquad\qquad f''\left(\frac{\pi}{3}\right) = -\frac{\sqrt{3}}{2}
$$

$$
f'''(x) = -\cos x \qquad\qquad f'''\left(\frac{\pi}{3}\right) = -\frac{1}{2}
$$

We have obtained two different series representations for $\sin x$, the Maclaurin series in Example 4 and the Taylor series in Example 7. It is best to use the Maclaurin series for values of x near 0 and the Taylor series for x near $\pi/3$. Notice that the third Taylor polynomial T_3 in Figure 3 is a good approximation to $\sin x$ near $\pi/3$ but not as good near 0. Compare it with the third Maclaurin polynomial T_3 in Figure 2, where the opposite is true.

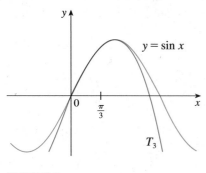

FIGURE 3

and this pattern repeats indefinitely. Therefore, the Taylor series at $\pi/3$ is

$$f\left(\frac{\pi}{3}\right) + \frac{f'\left(\frac{\pi}{3}\right)}{1!}\left(x - \frac{\pi}{3}\right) + \frac{f''\left(\frac{\pi}{3}\right)}{2!}\left(x - \frac{\pi}{3}\right)^2 + \frac{f'''\left(\frac{\pi}{3}\right)}{3!}\left(x - \frac{\pi}{3}\right)^3 + \cdots$$

$$= \frac{\sqrt{3}}{2} + \frac{1}{2 \cdot 1!}\left(x - \frac{\pi}{3}\right) - \frac{\sqrt{3}}{2 \cdot 2!}\left(x - \frac{\pi}{3}\right)^2 - \frac{1}{2 \cdot 3!}\left(x - \frac{\pi}{3}\right)^3 + \cdots$$

The proof that this series represents $\sin x$ for all x is very similar to that in Example 4. [Just replace x by $x - \pi/3$ in (15).] We can write the series in sigma notation if we separate the terms that contain $\sqrt{3}$:

$$\sin x = \sum_{n=0}^{\infty} \frac{(-1)^n \sqrt{3}}{2(2n)!}\left(x - \frac{\pi}{3}\right)^{2n} + \sum_{n=0}^{\infty} \frac{(-1)^n}{2(2n+1)!}\left(x - \frac{\pi}{3}\right)^{2n+1} \quad\blacksquare$$

The power series that we obtained by indirect methods in Examples 5 and 6 and in Section 10.9 are indeed the Taylor or Maclaurin series of the given functions because Theorem 5 asserts that, no matter how a power series representation $f(x) = \Sigma\, c_n(x - a)^n$ is obtained, it is always true that $c_n = f^{(n)}(a)/n!$. In other words, the coefficients are uniquely determined.

We collect in the following table, for future reference, some important Maclaurin series that we have derived in this section and the preceding one.

Important Maclaurin series and their intervals of convergence

$$\frac{1}{1-x} = \sum_{n=0}^{\infty} x^n = 1 + x + x^2 + x^3 + \cdots \qquad (-1, 1)$$

$$e^x = \sum_{n=0}^{\infty} \frac{x^n}{n!} = 1 + \frac{x}{1!} + \frac{x^2}{2!} + \frac{x^3}{3!} + \cdots \qquad (-\infty, \infty)$$

$$\sin x = \sum_{n=0}^{\infty} (-1)^n \frac{x^{2n+1}}{(2n+1)!} = x - \frac{x^3}{3!} + \frac{x^5}{5!} - \frac{x^7}{7!} + \cdots \qquad (-\infty, \infty)$$

$$\cos x = \sum_{n=0}^{\infty} (-1)^n \frac{x^{2n}}{(2n)!} = 1 - \frac{x^2}{2!} + \frac{x^4}{4!} - \frac{x^6}{6!} + \cdots \qquad (-\infty, \infty)$$

$$\tan^{-1}x = \sum_{n=0}^{\infty} (-1)^n \frac{x^{2n+1}}{2n+1} = x - \frac{x^3}{3} + \frac{x^5}{5} - \frac{x^7}{7} + \cdots \qquad [-1, 1]$$

One reason that Taylor series are important is that they enable us to integrate functions that we could not previously handle. In fact, in the introduction to this chapter we mentioned that Newton often integrated functions by first expressing them as power series and then integrating the series term by term. The function $f(x) = e^{-x^2}$ cannot be integrated in the usual way because its antiderivative is not an elementary function (see Section 7.6). In the following example we use Newton's idea to integrate this function.

EXAMPLE 8
(a) Evaluate $\int e^{-x^2}\, dx$ as an infinite series.
(b) Evaluate $\int_0^1 e^{-x^2}\, dx$ correct to within an error of 0.001.

SOLUTION

(a) First we find the Maclaurin series for $f(x) = e^{-x^2}$. Although it is possible to use the direct method, let us find it simply by replacing x with $-x^2$ in the series for e^x in the table of Maclaurin series. Thus

$$e^{-x^2} = \sum_{n=0}^{\infty} \frac{(-x^2)^n}{n!} = \sum_{n=0}^{\infty} (-1)^n \frac{x^{2n}}{n!} = 1 - \frac{x^2}{1!} + \frac{x^4}{2!} - \frac{x^6}{3!} + \cdots$$

Now we integrate term by term:

$$\int e^{-x^2} dx = \int \left(1 - \frac{x^2}{1!} + \frac{x^4}{2!} - \frac{x^6}{3!} + \cdots + (-1)^n \frac{x^{2n}}{n!} + \cdots \right) dx$$

$$= C + x - \frac{x^3}{3 \cdot 1!} + \frac{x^5}{5 \cdot 2!} - \frac{x^7}{7 \cdot 3!} + \cdots + (-1)^n \frac{x^{2n+1}}{(2n+1)n!} + \cdots$$

This series converges for all x because the original series for e^{-x^2} converges for all x.

(b) The Fundamental Theorem of Calculus gives

$$\int_0^1 e^{-x^2} dx = \left[x - \frac{x^3}{3 \cdot 1!} + \frac{x^5}{5 \cdot 2!} - \frac{x^7}{7 \cdot 3!} + \frac{x^9}{9 \cdot 4!} - \cdots \right]_0^1$$

$$= 1 - \frac{1}{3} + \frac{1}{10} - \frac{1}{42} + \frac{1}{216} - \cdots$$

$$\approx 0.7475$$

Since this series is alternating, Theorem 10.5.1 shows that the error involved in this approximation is less than

$$\frac{1}{11 \cdot 5!} = \frac{1}{1320} < 0.001$$

MULTIPLICATION AND DIVISION OF POWER SERIES

If power series are added or subtracted, they behave like polynomials (Theorem 10.2.8 shows this). In fact, as the following example illustrates, they can also be multiplied and divided like polynomials. We find only the first few terms because the calculations for the later terms become tedious and the initial terms are the most important ones.

EXAMPLE 9 Find the first three nonzero terms in the Maclaurin series for
(a) $e^x \sin x$ and (b) $\tan x$.

SOLUTION

(a) Using the Maclaurin series for e^x and $\sin x$ in the table, we have

$$e^x \sin x = \left(1 + \frac{x}{1!} + \frac{x^2}{2!} + \frac{x^3}{3!} + \cdots \right) \left(x - \frac{x^3}{3!} + \cdots \right)$$

We multiply these expressions, collecting like terms just as for polynomials:

$$1 + x + \tfrac{1}{2}x^2 + \tfrac{1}{6}x^3 + \cdots$$
$$x \qquad\qquad - \tfrac{1}{6}x^3 + \cdots$$
$$\overline{\qquad\qquad\qquad\qquad\qquad\qquad\qquad\qquad}$$
$$x + x^2 + \tfrac{1}{2}x^3 + \tfrac{1}{6}x^4 + \cdots$$
$$- \tfrac{1}{6}x^3 - \tfrac{1}{6}x^4 - \cdots$$
$$\overline{\qquad\qquad\qquad\qquad\qquad\qquad\qquad\qquad}$$
$$x + x^2 + \tfrac{1}{3}x^3 + \cdots$$

Thus
$$e^x \sin x = x + x^2 + \tfrac{1}{3}x^3 + \cdots$$

(b) Using the Maclaurin series in the table, we have

$$\tan x = \frac{\sin x}{\cos x} = \frac{x - \dfrac{x^3}{3!} + \dfrac{x^5}{5!} - \cdots}{1 - \dfrac{x^2}{2!} + \dfrac{x^4}{4!} - \cdots}$$

We use a procedure like long division:

$$
\begin{array}{r}
x + \tfrac{1}{3}x^3 + \tfrac{2}{15}x^5 + \cdots \\
1 - \tfrac{1}{2}x^2 + \tfrac{1}{24}x^4 - \cdots \overline{)\, x - \tfrac{1}{6}x^3 + \tfrac{1}{120}x^5 - \cdots} \\
x - \tfrac{1}{2}x^3 + \tfrac{1}{24}x^5 - \cdots \\
\hline
\tfrac{1}{3}x^3 - \tfrac{1}{30}x^5 + \cdots \\
\tfrac{1}{3}x^3 - \tfrac{1}{6}x^5 + \cdots \\
\hline
\tfrac{2}{15}x^5 + \cdots
\end{array}
$$

Thus
$$\tan x = x + \tfrac{1}{3}x^3 + \tfrac{2}{15}x^5 + \cdots \qquad\blacksquare$$

Although we have not attempted to justify the formal manipulations used in Example 9, they are legitimate. There is a theorem which states that if $f(x) = \Sigma\, c_n x^n$ and $g(x) = \Sigma\, b_n x^n$ both converge for $|x| < R$ and the series are multiplied as if they were polynomials, then the resulting series also converges for $|x| < R$ and represents $f(x)g(x)$. For division we require $b_0 \neq 0$; the resulting series converges for sufficiently small $|x|$.

PROOF OF TAYLOR'S FORMULA

We conclude this section by giving the promised proof of Theorem 9.

Let $R_n(x) = f(x) - T_n(x)$, where T_n is the nth-degree Taylor polynomial of f at a. The idea for the proof is the same as that for the Mean Value Theorem: We apply Rolle's Theorem to a specially constructed function. We think of x as a constant, $x \neq a$, and we define a function g on I by

$$g(t) = f(x) - f(t) - f'(t)(x - t) - \frac{f''(t)}{2!}(x - t)^2 - \cdots$$

$$- \frac{f^{(n)}(t)}{n!}(x - t)^n - R_n(x)\frac{(x - t)^{n+1}}{(x - a)^{n+1}}$$

Then

$$g(x) = f(x) - f(x) - 0 - \cdots - 0 = 0$$

$$g(a) = f(x) - [T_n(x) + R_n(x)] = f(x) - f(x) = 0$$

Thus, by Rolle's Theorem (applied to g on the interval from a to x), there is a number z between x and a such that $g'(z) = 0$. If we differentiate the expression for g, then most terms cancel. We leave it to you to verify that the expression for $g'(t)$ simplifies to

$$g'(t) = -\frac{f^{(n+1)}(t)}{n!}(x - t)^n + (n + 1)R_n(x)\frac{(x - t)^n}{(x - a)^{n+1}}$$

Thus we have

$$g'(z) = -\frac{f^{(n+1)}(z)}{n!}(x - z)^n + (n + 1)R_n(x)\frac{(x - z)^n}{(x - a)^{n+1}} = 0$$

and so

$$R_n(x) = \frac{f^{(n+1)}(z)}{(n + 1)!}(x - a)^{n+1}$$

$\square$

EXERCISES 10.10

1–6 ■ Find the Maclaurin series for $f(x)$ using the definition of a Maclaurin series. [Assume that f has a power series expansion. Do not show that $R_n(x) \to 0$.] Also find the associated radius of convergence.

1. $f(x) = \cos x$

2. $f(x) = \sin 2x$

3. $f(x) = \dfrac{1}{(1 + x)^2}$

4. $f(x) = \dfrac{x}{1 - x}$

5. $f(x) = \sinh x$

6. $f(x) = \cosh x$

7–12 ■ Find the Taylor series for $f(x)$ at the given value of a. [Assume that f has a power series expansion. Do not show that $R_n(x) \to 0$.]

7. $f(x) = \sin x, \quad a = \pi/4$

8. $f(x) = \cos x, \quad a = -\pi/4$

9. $f(x) = 1/x, \quad a = 1$

10. $f(x) = \sqrt{x}, \quad a = 4$

11. $f(x) = e^x, \quad a = 3$

12. $f(x) = \ln x, \quad a = 2$

13. Prove that the series obtained in Exercise 1 represents $\cos x$ for all x.

14. Prove that the series obtained in Exercise 7 represents $\sin x$ for all x.

15. Prove that the series obtained in Exercise 5 represents $\sinh x$ for all x.

16. Prove that the series obtained in Exercise 6 represents $\cosh x$ for all x.

17–26 ■ Use a Maclaurin series derived in this section to obtain the Maclaurin series for the given function.

17. $f(x) = e^{3x}$

18. $f(x) = \sin 2x$

19. $f(x) = x^2 \cos x$

20. $f(x) = \cos(x^3)$

21. $f(x) = x \sin(x/2)$

22. $f(x) = xe^{-x}$

23. $f(x) = \sin^2 x$ [*Hint*: Use $\sin^2 x = \frac{1}{2}(1 - \cos 2x)$.]

24. $f(x) = \cos^2 x$

25. $f(x) = \begin{cases} \dfrac{\sin x}{x} & \text{if } x \neq 0 \\ 1 & \text{if } x = 0 \end{cases}$

26. $f(x) = \begin{cases} \dfrac{1 - \cos x}{x^2} & \text{if } x \neq 0 \\ \frac{1}{2} & \text{if } x = 0 \end{cases}$

27–30 ■ Find the Maclaurin series of f (by any method) and its radius of convergence. Graph f and its first few Taylor polynomials on the same screen.

27. $f(x) = \sqrt{1 + x}$

28. $f(x) = 1/\sqrt{1 + 2x}$

29. $f(x) = (1 + x)^{-3}$

30. $f(x) = 2^x$

31. Find the Maclaurin series for $\ln(1 + x)$ and use it to calculate $\ln 1.1$ correct to five decimal places.

32. Use the Maclaurin series for $\sin x$ to compute $\sin 3°$ correct to five decimal places.

33–36 ■ Evaluate the indefinite integral as an infinite series.

33. $\displaystyle\int \sin(x^2)\, dx$

34. $\displaystyle\int \frac{\sin x}{x}\, dx$

35. $\displaystyle\int \sqrt{x^3 + 1}\, dx$

36. $\displaystyle\int e^{x^3}\, dx$

37–40 ■ Use series to approximate the definite integral to within the indicated accuracy.

37. $\displaystyle\int_0^1 \sin(x^2)\, dx$ (three decimal places)

38. $\displaystyle\int_0^{0.5} \cos(x^2)\, dx$ (three decimal places)

39. $\displaystyle\int_0^{0.1} \frac{dx}{\sqrt{1 + x^3}}$ (error $< 10^{-8}$)

40. $\displaystyle\int_0^{0.5} x^2 e^{-x^2}\, dx$ (error < 0.001)

41–44 ■ Use multiplication or division of power series to find the first three nonzero terms in the Maclaurin series for each function.

41. $y = e^{-x^2}\cos x$

42. $y = \sec x$

43. $y = \dfrac{\ln(1 - x)}{e^x}$

44. $y = e^x \ln(1 - x)$

45–50 ■ Find the sum of the series.

45. $\displaystyle\sum_{n=0}^{\infty} (-1)^n \frac{x^{4n}}{n!}$

46. $\displaystyle\sum_{n=0}^{\infty} \frac{(-1)^n \pi^{2n}}{6^{2n}(2n)!}$

47. $\sum_{n=0}^{\infty} \dfrac{(-1)^n \pi^{2n+1}}{4^{2n+1}(2n+1)!}$

48. $\sum_{n=2}^{\infty} \dfrac{x^{3n+1}}{n!}$

49. $\sum_{n=0}^{\infty} \dfrac{x^{n+1}}{(n+1)!}$

50. $\sum_{n=0}^{\infty} \dfrac{x^n}{2^n(n+1)!}$

51. Show that $e^x > 1 + x$ for all $x > 0$.

52. Show that $\cosh x \geqslant 1 + \frac{1}{2}x^2$ for all x.

53. The limit

$$\lim_{x \to 0} \frac{\sin x - x + \frac{1}{6}x^3}{x^5}$$

could be evaluated using l'Hospital's Rule. Instead, use a series to evaluate it.

54. Use series to evaluate the limit

$$\lim_{x \to 0} \frac{1 - \cos x}{1 + x - e^x}$$

55. If f has derivatives of all orders on an interval $I = (a - R, a + R)$ and these derivatives have a common bound M [$|f^{(n)}(x)| \leqslant M$ for all x in I and all $n = 1, 2, 3, \ldots$], prove that f is analytic at a.

56. (a) Show that the function defined in Exercise 101 in Section 6.8 is not equal to its Maclaurin series.

(b) Graph the function in part (a) and comment on its behavior near the origin.

10.11 THE BINOMIAL SERIES

You may be acquainted with the Binomial Theorem, which states that if a and b are any real numbers and k is a positive integer, then

$$(a + b)^k = a^k + ka^{k-1}b + \frac{k(k-1)}{2!}a^{k-2}b^2 + \frac{k(k-1)(k-2)}{3!}a^{k-3}b^3$$

$$+ \cdots + \frac{k(k-1)(k-2)\cdots(k-n+1)}{n!}a^{k-n}b^n$$

$$+ \cdots + kab^{k-1} + b^k$$

The traditional notation for the binomial coefficients is

$$\binom{k}{0} = 1 \qquad \binom{k}{n} = \frac{k(k-1)(k-2)\cdots(k-n+1)}{n!} \qquad n = 1, 2, \ldots, k$$

which enables us to write the Binomial Theorem in the abbreviated form

$$(a + b)^k = \sum_{n=0}^{k} \binom{k}{n} a^{k-n}b^n$$

In particular, if we put $a = 1$ and $b = x$, we get

(1)
$$(1 + x)^k = \sum_{n=0}^{k} \binom{k}{n} x^n$$

One of Newton's accomplishments was to extend the Binomial Theorem (Equation 1) to the case in which k is no longer a positive integer. In this case the expression for $(1 + x)^k$ is no longer a finite sum; it becomes an infinite series. To find this series we compute the Maclaurin series of $(1 + x)^k$ in the usual way:

$$f(x) = (1 + x)^k \qquad\qquad f(0) = 1$$

$$f'(x) = k(1 + x)^{k-1} \qquad\qquad f'(0) = k$$

$$f''(x) = k(k - 1)(1 + x)^{k-2} \qquad\qquad f''(0) = k(k - 1)$$

$$f'''(x) = k(k - 1)(k - 2)(1 + x)^{k-3} \qquad\qquad f'''(0) = k(k - 1)(k - 2)$$

$$\vdots \qquad\qquad\qquad\qquad\qquad\qquad \vdots$$

$$f^{(n)}(x) = k(k - 1)\cdots(k - n + 1)(1 + x)^{k-n} \qquad f^{(n)}(0) = k(k - 1)\cdots(k - n + 1)$$

Therefore, the Maclaurin series of $f(x) = (1 + x)^k$ is

$$\sum_{n=0}^{\infty} \frac{f^{(n)}(0)}{n!}x^n = \sum_{n=0}^{\infty} \frac{k(k - 1)\cdots(k - n + 1)}{n!}x^n$$

This series is called the **binomial series.** If its nth term is a_n, then

$$\left| \frac{a_{n+1}}{a_n} \right| = \left| \frac{k(k - 1)\cdots(k - n + 1)(k - n)x^{n+1}}{(n + 1)!} \cdot \frac{n!}{k(k - 1)\cdots(k - n + 1)x^n} \right|$$

$$= \frac{|k - n|}{n + 1}|x| = \frac{\left| 1 - \dfrac{k}{n} \right|}{1 + \dfrac{1}{n}}|x| \to |x| \qquad \text{as } n \to \infty$$

Thus, by the Ratio Test, the binomial series converges if $|x| < 1$ and diverges if $|x| > 1$.

The following theorem states that $(1 + x)^k$ is equal to the sum of its Maclaurin series. It is possible to prove this by showing that the remainder term $R_n(x)$ approaches 0, but that turns out to be quite difficult. The proof outlined in Exercise 21 is much easier.

---

(2) THE BINOMIAL SERIES If k is any real number and $|x| < 1$, then

$$(1 + x)^k = 1 + kx + \frac{k(k - 1)}{2!}x^2 + \frac{k(k - 1)(k - 2)}{3!}x^3 + \cdots$$

$$= \sum_{n=0}^{\infty} \binom{k}{n} x^n$$

where $\displaystyle \binom{k}{n} = \frac{k(k - 1)\cdots(k - n + 1)}{n!} \quad (n \geqslant 1)$ and $\displaystyle \binom{k}{0} = 1$

Although the binomial series always converges when $|x| < 1$, the question of whether or not it converges at the endpoints, ± 1, depends on the value of k. It turns out that the series converges at 1 if $-1 < k \leqslant 0$ and at both endpoints if $k \geqslant 0$. Notice that if k is a positive integer and $n > k$, then the expression for $\binom{k}{n}$ contains a factor $(k - k)$, so $\binom{k}{n} = 0$ for $n > k$. This means that the series terminates and reduces to the ordinary Binomial Theorem (Equation 1) when k is a positive integer.

Although, as we have seen, the binomial series is just a special case of the Maclaurin series, it occurs frequently and so it is worth remembering.

EXAMPLE 1   Expand $\dfrac{1}{(1+x)^2}$ as a power series.

SOLUTION   We use the binomial series with $k = -2$. The binomial coefficient is

$$\binom{-2}{n} = \frac{(-2)(-3)(-4)\cdots(-2-n+1)}{n!}$$

$$= \frac{(-1)^n 2 \cdot 3 \cdot 4 \cdot \cdots \cdot n(n+1)}{n!} = (-1)^n(n+1)$$

and so, when $|x| < 1$,

$$\frac{1}{(1+x)^2} = (1+x)^{-2} = \sum_{n=0}^{\infty} \binom{-2}{n} x^n$$

$$= \sum_{n=0}^{\infty} (-1)^n(n+1)x^n$$

■

EXAMPLE 2   Find the Maclaurin series for the function $f(x) = \dfrac{1}{\sqrt{4-x}}$ and its radius of convergence.

SOLUTION   As given, $f(x)$ is not quite of the form $(1+x)^k$ so we rewrite it as follows:

$$\frac{1}{\sqrt{4-x}} = \frac{1}{\sqrt{4\left(1-\dfrac{x}{4}\right)}} = \frac{1}{2\sqrt{1-\dfrac{x}{4}}} = \frac{1}{2}\left(1-\frac{x}{4}\right)^{-1/2}$$

Using the binomial series with $k = -\frac{1}{2}$ and with x replaced by $-x/4$, we have

$$\frac{1}{\sqrt{4-x}} = \frac{1}{2}\left(1-\frac{x}{4}\right)^{-1/2} = \frac{1}{2}\sum_{n=0}^{\infty}\binom{-\frac{1}{2}}{n}\left(-\frac{x}{4}\right)^n$$

$$= \frac{1}{2}\left[1 + \left(-\frac{1}{2}\right)\left(-\frac{x}{4}\right) + \frac{\left(-\frac{1}{2}\right)\left(-\frac{3}{2}\right)}{2!}\left(-\frac{x}{4}\right)^2 + \frac{\left(-\frac{1}{2}\right)\left(-\frac{3}{2}\right)\left(-\frac{5}{2}\right)}{3!}\left(-\frac{x}{4}\right)^3\right.$$

$$\left. + \cdots + \frac{\left(-\frac{1}{2}\right)\left(-\frac{3}{2}\right)\left(-\frac{5}{2}\right)\cdots\left(-\frac{1}{2}-n+1\right)}{n!}\left(-\frac{x}{4}\right)^n + \cdots\right]$$

$$= \frac{1}{2}\left[1 + \frac{1}{8}x + \frac{1\cdot 3}{2!\,8^2}x^2 + \frac{1\cdot 3\cdot 5}{3!\,8^3}x^3\right.$$

$$\left. + \cdots + \frac{1\cdot 3\cdot 5\cdot\cdots\cdot(2n-1)}{n!\,8^n}x^n + \cdots\right]$$

We know from (2) that this series converges when $|-x/4| < 1$, that is, $|x| < 4$, so the radius of convergence is $R = 4$.

■

A binomial series is a special case of a Taylor series. Figure 1 shows the graphs of the first three Taylor polynomials computed from the answer to Example 2.

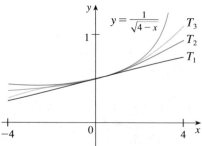

FIGURE 1

EXERCISES 10.11

1–10 ■ Use the binomial series to expand the given function as a power series. State the radius of convergence.

1. $\sqrt{1 + x}$

2. $\dfrac{1}{(1 + x)^3}$

3. $\dfrac{1}{(1 + 2x)^4}$

4. $\sqrt[3]{1 + x^2}$

5. $\dfrac{x}{\sqrt{1 - x}}$

6. $\dfrac{1}{\sqrt{2 + x}}$

7. $\sqrt[4]{1 - x^4}$

8. $\dfrac{x^2}{\sqrt{1 - x^3}}$

9. $\left(\dfrac{x}{1 - x}\right)^5$

10. $\sqrt[5]{x - 1}$

11–12 ■ Use the binomial series to expand the given function as a Maclaurin series and to find the first three Taylor polynomials T_1, T_2, and T_3. Graph the function and these Taylor polynomials in the interval of convergence.

11. $\dfrac{1}{\sqrt[3]{8 + x}}$

12. $(4 + x)^{3/2}$

13. (a) Use the binomial series to expand $1/\sqrt{1 - x^2}$.
(b) Use part (a) to find the Maclaurin series for $\sin^{-1}x$.

14. (a) Use the binomial series to expand $1/\sqrt{1 + x^2}$.
(b) Use part (a) to find the Maclaurin series for $\sinh^{-1}x$.

15. (a) Expand $1/\sqrt{1 + x}$ as a power series.
(b) Use part (a) to estimate $1/\sqrt{1.1}$ correct to three decimal places.

16. (a) Expand $\sqrt[3]{8 + x}$ as a power series.
(b) Use part (a) to estimate $\sqrt[3]{8.2}$ correct to four decimal places.

17. (a) Expand $f(x) = x/(1 - x)^2$ as a power series.
(b) Use part (a) to find the sum of the series

$$\sum_{n=1}^{\infty} \frac{n}{2^n}$$

18. (a) Expand $f(x) = (x + x^2)/(1 - x)^3$ as a power series.
(b) Use part (a) to find the sum of the series

$$\sum_{n=1}^{\infty} \frac{n^2}{2^n}$$

19. (a) Use the binomial series to find the Maclaurin series of $f(x) = \sqrt{1 + x^2}$.
(b) Use part (a) to evaluate $f^{(10)}(0)$.

20. (a) Use the binomial series to find the Maclaurin series of $f(x) = 1/\sqrt{1 + x^3}$.
(b) Use part (a) to evaluate $f^{(9)}(0)$.

21. Use the following steps to prove (2).
(a) Let $g(x) = \sum_{n=0}^{\infty} \binom{k}{n}x^n$. Differentiate this series to show that

$$g'(x) = \frac{kg(x)}{1 + x} \qquad -1 < x < 1$$

(b) Let $h(x) = (1 + x)^{-k}g(x)$ and show that $h'(x) = 0$.
(c) Deduce that $g(x) = (1 + x)^k$.

22. The period of a pendulum with length L that makes a maximum angle θ_0 with the vertical is

$$T = 4\sqrt{\frac{L}{g}} \int_0^{\pi/2} \frac{dx}{\sqrt{1 - k^2 \sin^2 x}}$$

where $k = \sin(\tfrac{1}{2}\theta_0)$ and g is the acceleration due to gravity. (In Exercise 35 in Section 7.8 we approximated this integral using Simpson's Rule.)
(a) Expand the integrand as a binomial series and use the result of Exercise 40 in Section 7.1 to show that

$$T = 2\pi\sqrt{\frac{L}{g}}\left[1 + \frac{1^2}{2^2}k^2 + \frac{1^2 3^2}{2^2 4^2}k^4 + \frac{1^2 3^2 5^2}{2^2 4^2 6^2}k^6 + \cdots\right]$$

If θ_0 is not too large, the approximation $T \approx 2\pi\sqrt{L/g}$, obtained by using only the first term in the series, is often used. A better approximation is obtained by using two terms:

$$T \approx 2\pi\sqrt{\frac{L}{g}}\left(1 + \tfrac{1}{4}k^2\right)$$

(b) Notice that all the terms in the series after the first one have coefficients that are at most $\tfrac{1}{4}$. Use this fact to compare this series with a geometric series and show that

$$2\pi\sqrt{\frac{L}{g}}\left(1 + \tfrac{1}{4}k^2\right) \leqslant T \leqslant 2\pi\sqrt{\frac{L}{g}\,\frac{4 - 3k^2}{4 - 4k^2}}$$

(c) Use the inequalities in part (b) to estimate the period of a pendulum with $L = 1$ meter and $\theta_0 = 10°$. How does it compare with the estimate $T \approx 2\pi\sqrt{L/g}$? What if $\theta_0 = 42°$?

10.12   APPLICATIONS OF TAYLOR POLYNOMIALS

Suppose that $f(x)$ is equal to the sum of its Taylor series at a:

$$f(x) = \sum_{n=0}^{\infty} \frac{f^{(n)}(a)}{n!}(x - a)^n$$

In Section 10.10 we introduced the notation $T_n(x)$ for the nth partial sum of this series and called it the nth-degree Taylor polynomial of f at a. Thus

$$T_n(x) = \sum_{i=0}^{n} \frac{f^{(i)}(a)}{i!}(x - a)^i$$

$$= f(a) + \frac{f'(a)}{1!}(x - a) + \frac{f''(a)}{2!}(x - a)^2 + \cdots + \frac{f^{(n)}(a)}{n!}(x - a)^n$$

Since f is the sum of its Taylor series, we know that $T_n(x) \to f(x)$ as $n \to \infty$ and so T_n can be used as an approximation to f: $f(x) \approx T_n(x)$. It is useful to be able to approximate a function by a polynomial because polynomials are the simplest of functions. In this section we explore the use of such approximations by physical scientists and computer scientists.

Notice that the first-degree Taylor polynomial

$$T_1(x) = f(a) + f'(a)(x - a)$$

is the same as the linear approximation (or tangent line approximation) and the second-degree Taylor polynomial

$$T_2(x) = f(a) + \frac{f'(a)}{1!}(x - a) + \frac{f''(a)}{2!}(x - a)^2$$

is the same as the quadratic approximation to f discussed in Section 2.9. Recall that the quadratic approximation was constructed so that it and its first two derivatives have the same values at a that f, f', and f'' have. In general, it can be shown that the derivatives of T_n at a agree with those of f up to and including derivatives of order n (see Exercise 34).

To illustrate these ideas let's take another look at the graphs of $y = e^x$ and its first few Taylor polynomials, as shown in Figure 1. The graph of T_1 is the tangent line to $y = e^x$ at $(0, 1)$, which is the best linear approximation to e^x near $(0, 1)$. The graph of T_2 is the parabola $y = 1 + x + x^2/2$ and the graph of T_3 is the cubic curve $y = 1 + x + x^2/2 + x^3/6$, which is a closer fit to the exponential curve $y = e^x$ than T_2. The next Taylor polynomial T_4 would be an even better approximation, and so on.

When using a Taylor polynomial T_n to approximate a function f, we have to ask the question: How good an approximation is it? or How large should we take n to be in order to achieve a desired accuracy? To answer these questions we need to look at the magnitude of the remainder:

$$|R_n(x)| = |f(x) - T_n(x)|$$

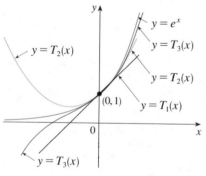

FIGURE 1

There are three possible methods for estimating the size of the error:

1. If a graphing device is available, we can use it to graph $|R_n(x)|$ and thereby estimate the error.

2. If the series happens to be an alternating series, we can use Theorem 10.5.1 (the Alternating Series Estimation Theorem).

3. In all cases we can use Taylor's Formula (10.10.9), which says that

$$R_n(x) = \frac{f^{(n+1)}(z)}{(n+1)!}(x-a)^{n+1}$$

where z is a number that lies between x and a.

EXAMPLE 1
(a) Approximate the function $f(x) = \sqrt[3]{x}$ by a Taylor polynomial of degree 2 at $a = 8$.
(b) How accurate is this approximation when $7 \le x \le 9$?

SOLUTION

(a)
$$f(x) = \sqrt[3]{x} = x^{1/3} \qquad f(8) = 2$$

$$f'(x) = \tfrac{1}{3}x^{-2/3} \qquad f'(8) = \tfrac{1}{12}$$

$$f''(x) = -\tfrac{2}{9}x^{-5/3} \qquad f''(8) = -\tfrac{1}{144}$$

$$f'''(x) = \tfrac{10}{27}x^{-8/3}$$

Thus the second-degree Taylor polynomial is

$$T_2(x) = f(8) + \frac{f'(8)}{1!}(x-8) + \frac{f''(8)}{2!}(x-8)^2$$

$$= 2 + \tfrac{1}{12}(x-8) - \tfrac{1}{288}(x-8)^2$$

The desired approximation is

$$\sqrt[3]{x} \approx T_2(x) = 2 + \tfrac{1}{12}(x-8) - \tfrac{1}{288}(x-8)^2$$

(b) The Taylor series is not alternating when $x < 8$, so we can't use the Alternating Series Estimation Theorem in this example. But using Taylor's Formula we can write

$$R_2(x) = \frac{f'''(z)}{3!}(x-8)^3 = \tfrac{10}{27}z^{-8/3}\frac{(x-8)^3}{3!} = \frac{5(x-8)^3}{81z^{8/3}}$$

where z lies between 8 and x. In order to estimate the error we note that if $7 \le x \le 9$, then $-1 \le x - 8 \le 1$, so $|x-8| \le 1$ and therefore $|x-8|^3 \le 1$. Also, since $z > 7$, we have

$$z^{8/3} > 7^{8/3} > 179$$

and so
$$|R_2(x)| = \frac{5|x-8|^3}{81z^{8/3}} < \frac{5 \cdot 1}{81 \cdot 179} < 0.0004$$

Thus if $7 \le x \le 9$, the approximation in part (a) is accurate to within 0.0004. ∎

2.5

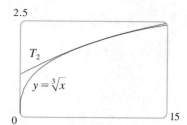

FIGURE 2

0.0003

FIGURE 3

Let's use a graphing device to check the calculation in Example 1. Figure 2 shows that the graphs of $y = \sqrt[3]{x}$ and $y = T_2(x)$ are very close to each other when x is near 8. Figure 3 shows the graph of $|R_2(x)|$ computed from the expression

$$|R_2(x)| = |\sqrt[3]{x} - T_2(x)|$$

We see from the graph that

$$|R_2(x)| < 0.0003$$

when $7 \leqslant x \leqslant 9$. Thus the error estimate from graphical methods is slightly better than the error estimate from Taylor's Formula in this case.

EXAMPLE 2
(a) What is the maximum error possible in using the approximation

$$\sin x \approx x - \frac{x^3}{3!} + \frac{x^5}{5!}$$

when $-0.3 \leqslant x \leqslant 0.3$? Use this approximation to find $\sin 12°$ correct to six decimal places.
(b) For what values of x is this approximation accurate to within 0.00005?

SOLUTION
(a) Notice that the Maclaurin series

$$\sin x = x - \frac{x^3}{3!} + \frac{x^5}{5!} - \frac{x^7}{7!} + \cdots$$

is alternating for all nonzero values of x, so we can use the Alternating Series Estimation Theorem (10.5.1). The error in approximating $\sin x$ by the first three terms of its Maclaurin series is at most

$$\left| \frac{x^7}{7!} \right| = \frac{|x|^7}{5040}$$

If $-0.3 \leqslant x \leqslant 0.3$, then $|x| \leqslant 0.3$, so the error is smaller than

$$\frac{(0.3)^7}{5040} \approx 4.3 \times 10^{-8}$$

To find $\sin 12°$ we first convert to radian measure.

$$\sin 12° = \sin\left(\frac{12\pi}{180}\right) = \sin\left(\frac{\pi}{15}\right)$$

$$\approx \frac{\pi}{15} - \left(\frac{\pi}{15}\right)^3 \frac{1}{3!} + \left(\frac{\pi}{15}\right)^5 \frac{1}{5!}$$

$$\approx 0.20791169$$

Thus, correct to six decimal places, $\sin 12° \approx 0.207912$.
(b) The error will be smaller than 0.00005 if

$$\frac{|x|^7}{5040} < 0.00005$$

Solving this inequality for x, we get

$$|x|^7 < 0.252 \qquad \text{or} \qquad |x| < (0.252)^{1/7} \approx 0.821$$

So the given approximation is accurate to within 0.00005 when $|x| < 0.82$.   ∎

What if we had used Taylor's Formula to solve Example 2? The remainder term is

$$R_6(x) = \frac{f^{(7)}(z)}{7!}x^7 = -\cos z \frac{x^7}{7!}$$

But $|-\cos z| \leqslant 1$, so $|R_6(x)| \leqslant |x|^7/7!$ and we get the same estimates as with the Alternating Series Estimation Theorem.

What about graphical methods? Figure 4 shows the graph of

$$|R_6(x)| = \left| \sin x - \left(x - \tfrac{1}{6}x^3 + \tfrac{1}{120}x^5\right)\right|$$

and we see from it that $|R_6(x)| < 4.3 \times 10^{-8}$ when $|x| \leqslant 0.3$. This is the same estimate that we obtained in Example 2. For part (b) we want $|R_6(x)| < 0.00005$, so we graph both $y = |R_6(x)|$ and $y = 0.00005$ in Figure 5. By placing the cursor on the right intersection point we find that the inequality is satisfied when $|x| < 0.82$. Again this is the same estimate as we obtained in the solution to Example 2.

If we had been asked to approximate $\sin 72°$ instead of $\sin 12°$ in Example 2, it would have been wise to use the Taylor polynomials at $a = \pi/3$ (instead of $a = 0$) because they are better approximations to $\sin x$ for values of x close to $\pi/3$. Notice that $72°$ is close to $60°$ (or $\pi/3$ radians) and the derivatives of $\sin x$ are easy to compute at $\pi/3$.

Figure 6 shows the graphs of the Taylor polynomial approximations

$$T_1(x) = x \qquad\qquad T_3(x) = x - \frac{x^3}{3!}$$

$$T_5(x) = x - \frac{x^3}{3!} + \frac{x^5}{5!} \qquad T_7(x) = x - \frac{x^3}{3!} + \frac{x^5}{5!} - \frac{x^7}{7!}$$

to the sine curve. You can see that as n increases, $T_n(x)$ is a good approximation to $\sin x$ on a larger and larger interval.

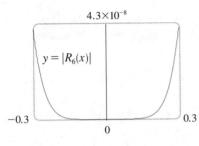

4.3×10^{-8}

$y = |R_6(x)|$

-0.3 0 0.3

FIGURE 4

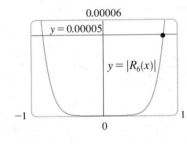

0.00006

$y = 0.00005$

$y = |R_6(x)|$

-1 0 1

FIGURE 5

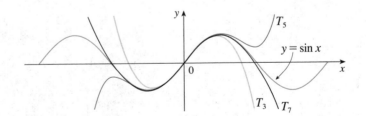

FIGURE 6

One use of the type of calculation done in Examples 1 and 2 occurs in calculators and computers. For instance, when you press the sin or e^x key on your calculator, or when a computer programmer uses a subroutine for a trigonometric or exponential or Bessel function, in many machines a polynomial approximation is calculated. The polynomial is often a Taylor polynomial that has been modified so that the error is spread more evenly throughout an interval.

Taylor polynomials are also used frequently in physics. In order to gain insight into an equation, a physicist often simplifies a function by considering only the first two or three terms in its Taylor series. In other words, the physicist uses a Taylor polynomial as an approximation to the function. When this happens, Taylor's Formula can be used to gauge the accuracy of the approximation. The following example shows one way in which this idea is used in special relativity. In Exercises 27–31 various other applications are explored.

EXAMPLE 3   In Einstein's theory of special relativity the mass of an object moving with velocity v is

$$m = \frac{m_0}{\sqrt{1 - v^2/c^2}}$$

where m_0 is the mass of the object when at rest and c is the speed of light. The kinetic energy of the object is the difference between its total energy and its energy at rest:

$$K = mc^2 - m_0 c^2$$

(a) Show that when v is very small compared with c, this expression for K agrees with classical Newtonian physics: $K = \frac{1}{2} m_0 v^2$.

(b) Use Taylor's Formula to estimate the difference in these expressions for K when $|v| \le 100$ m/s.

SOLUTION

(a) Using the expressions given for K and m, we get

$$K = mc^2 - m_0 c^2 = \frac{m_0 c^2}{\sqrt{1 - v^2/c^2}} - m_0 c^2$$

$$= m_0 c^2 \left[\left(1 - \frac{v^2}{c^2} \right)^{-1/2} - 1 \right]$$

With $x = -v^2/c^2$ the Maclaurin series for $(1 + x)^{-1/2}$ is most easily computed as a binomial series with $k = -\frac{1}{2}$. (Notice that $|x| < 1$ because $v < c$.) Therefore, we have

$$(1 + x)^{-1/2} = 1 - \frac{1}{2}x + \frac{\left(-\frac{1}{2}\right)\left(-\frac{3}{2}\right)}{2!}x^2 + \frac{\left(-\frac{1}{2}\right)\left(-\frac{3}{2}\right)\left(-\frac{5}{2}\right)}{3!}x^3 + \cdots$$

$$= 1 - \frac{1}{2}x + \frac{3}{8}x^2 - \frac{5}{16}x^3 + \cdots$$

and

$$K = m_0 c^2 \left[\left(1 + \frac{1}{2}\frac{v^2}{c^2} + \frac{3}{8}\frac{v^4}{c^4} + \frac{5}{16}\frac{v^6}{c^6} + \cdots \right) - 1 \right]$$

$$= m_0 c^2 \left(\frac{1}{2}\frac{v^2}{c^2} + \frac{3}{8}\frac{v^4}{c^4} + \frac{5}{16}\frac{v^6}{c^6} + \cdots \right)$$

If v is much smaller than c, then all terms after the first are very small when compared with the first term. If we omit them, we get

$$K \approx m_0 c^2 \left(\frac{1}{2}\frac{v^2}{c^2} \right) = \frac{1}{2} m_0 v^2$$

(b) By Taylor's Formula we can write the remainder term as

$$R_1(x) = \frac{f''(z)}{2!} x^2$$

where $f(x) = m_0 c^2 [(1 + x)^{-1/2} - 1]$ and $x = -v^2/c^2$. Since $f''(x) = \frac{3}{4} m_0 c^2 (1 + x)^{-5/2}$, we get

$$R_1(x) = \frac{3m_0 c^2}{8(1 + z)^{5/2}} \cdot \frac{v^4}{c^4}$$

where z lies between 0 and $-v^2/c^2$. We have $c = 3 \times 10^8$ m/s and $|v| \leq 100$ m/s, so

$$R_1(x) \leq \frac{\frac{3}{8} m_0 (9 \times 10^{16})(100/c)^4}{(1 - 100^2/c^2)^{5/2}} < (4.17 \times 10^{-10}) m_0$$

Thus when $|v| \leq 100$ m/s, the magnitude of the error in using the Newtonian expression for kinetic energy is at most $(4.2 \times 10^{-10}) m_0$. ∎

EXERCISES 10.12

1–8 ■ Find the Taylor polynomial $T_n(x)$ for the function f at the number a.
If you have a graphing device, graph f and T_n on the same screen.

1. $f(x) = \sin x$, $a = \pi/6$, $n = 3$

2. $f(x) = \cos x$, $a = 2\pi/3$, $n = 4$

3. $f(x) = \tan x$, $a = 0$, $n = 4$

4. $f(x) = \tan x$, $a = \pi/4$, $n = 4$

5. $f(x) = e^x \sin x$, $a = 0$, $n = 3$

6. $f(x) = \sqrt{x}$, $a = 9$, $n = 3$

7. $f(x) = 1/\sqrt[3]{x}$, $a = 8$, $n = 3$

8. $f(x) = \sec x$, $a = \pi/3$, $n = 3$

9–10 ■ Find the Taylor polynomials $T_n(x)$ at a for the given function. Then plot the graphs of f and these approximating polynomials.

9. $f(x) = \cos x$, $a = 0$, $n = 1, 2, 3, 4$

10. $f(x) = 1/x$, $a = 1$, $n = 1, 2, 3$

CAS 11–12 ■ Use a computer algebra system to find the Taylor polynomials T_n at $a = 0$ for the given values of n. Then graph these polynomials and f on the same screen.

11. $f(x) = \sec x$, $n = 2, 4, 6, 8$

12. $f(x) = \tan x$, $n = 1, 3, 5, 7, 9$

13–22 ■
(a) Approximate f by a Taylor polynomial with degree n at the number a.
(b) Use Taylor's formula to estimate the accuracy of the approximation $f(x) \approx T_n(x)$ when x lies in the given interval.

(c) Check your result in part (b) by graphing $|R_n(x)|$.

13. $f(x) = \sqrt{1 + x}$, $a = 0$, $n = 1$, $0 \leq x \leq 0.1$

14. $f(x) = 1/x$, $a = 1$, $n = 3$, $0.8 \leq x \leq 1.2$

15. $f(x) = \sin x$, $a = \pi/4$, $n = 5$, $0 \leq x \leq \pi/2$

16. $f(x) = \cos x$, $a = \pi/3$, $n = 4$, $0 \leq x \leq 2\pi/3$

17. $f(x) = \tan x$, $a = 0$, $n = 3$, $0 \leq x \leq \pi/6$

18. $f(x) = \sqrt[3]{1 + x^2}$, $a = 0$, $n = 2$, $|x| \leq 0.5$

19. $f(x) = e^{x^2}$, $a = 0$, $n = 3$, $0 \leq x \leq 0.1$

20. $f(x) = \cosh x$, $a = 0$, $n = 5$, $|x| \leq 1$

21. $f(x) = x^{3/4}$, $a = 16$, $n = 3$, $15 \leq x \leq 17$

22. $f(x) = \ln x$, $a = 4$, $n = 3$, $3 \leq x \leq 5$

23. Use the information from Exercise 1 to estimate $\sin 35°$ correct to five decimal places.

24. Use the information from Exercise 16 to estimate $\cos 69°$ correct to five decimal places.

25. Use Taylor's Formula to determine the number of terms of the Maclaurin series for e^x that should be used to estimate $e^{0.1}$ to within 0.00001.

26. How many terms of the Maclaurin series for $\ln(1 + x)$ do you need to use to estimate $\ln 1.4$ to within 0.001?

27–28 ■ Use the Alternating Series Estimation Theorem or Taylor's Formula to estimate the range of values of x for which the given approximation is accurate to within the stated error.
If you have a graphing device, check your answer graphically.

27. $\sin x \approx x - \frac{x^3}{6}$, error < 0.01

28. $\cos x \approx 1 - \dfrac{x^2}{2} + \dfrac{x^4}{24}$,   error < 0.005

29. A car is moving with speed 20 m/s and acceleration 2 m/s^2 at a given instant. Using a second-degree Taylor polynomial, estimate how far the car moves in the next second. Would it be reasonable to use this polynomial to estimate the distance traveled during the next minute?

30. In the late 19th century, the Rayleigh-Jeans Law expressed the energy density of blackbody radiation of wavelength λ as

$$f(\lambda) = \frac{8\pi kT}{\lambda^4}$$

where λ is measured in meters, T is the temperature in kelvins, and k is Boltzmann's constant. (See Problem 8 on page 501 for background information on blackbody radiation.) The Rayleigh-Jeans Law agrees with experimental measurements for long wavelengths but disagrees drastically for short wavelengths. [The law predicts that $f(\lambda) \to \infty$ as $\lambda \to 0^+$ but experiments have shown that $f(\lambda) \to 0$.] This fact is known as the *ultraviolet catastrophe*.

In 1900 Max Planck found a better model (known now as Planck's Law) for blackbody radiation:

$$f(\lambda) = \frac{8\pi hc\lambda^{-5}}{e^{hc/(\lambda kT)} - 1}$$

where h is Planck's constant and c is the speed of light.
(a) Use a Taylor polynomial to show that, for large wavelengths, Planck's Law gives approximately the same values as the Rayleigh-Jeans Law.
(b) Graph f as given by both laws on the same screen and comment on the similarities and differences. Use $T = 5700$ K (the temperature of the sun), $h = 6.6262 \times 10^{-34}$ J-s, $c = 2.997925 \times 10^8$ m/s, and $k = 1.3807 \times 10^{-23}$ J/K. (You may want to change from meters to the more convenient unit of micrometers: 1 μm $= 10^{-6}$ m.)

31. An electric dipole consists of two electric charges of equal magnitude and opposite signs. If the charges are q and $-q$ and are located at a distance d from each other, then the electric field E at the point P in the figure is

$$E = \frac{q}{D^2} - \frac{q}{(D+d)^2}$$

By expanding this expression for E as a series in powers of d/D, show that E is approximately proportional to $1/D^3$ when P is far away from the dipole.

32. The resistivity ρ of a conducting wire is the reciprocal of the conductivity and is measured in units of ohm-meters

(Ω-m). The resistivity of a given metal depends on the temperature according to the equation

$$\rho(t) = \rho_{20}e^{\alpha(t-20)}$$

where t is the temperature in °C. There are tables that list the values of α (called the temperature coefficient) and ρ_{20} (the resistivity at 20 °C) for various metals. Except at very low temperatures, the resistivity varies almost linearly with temperature and so it is common to approximate the expression for $\rho(t)$ by its first- or second-degree Taylor polynomial at $t = 20$.
(a) Find expressions for these linear and quadratic approximations.
(b) For copper, the tables give $\alpha = 0.0039/°C$ and $\rho_{20} = 1.7 \times 10^{-8}$ Ω-m. Graph the resistivity of copper and the linear and quadratic approximations for $-250\,°C \le t \le 1000\,°C$.
(c) For what values of t does the linear approximation agree with the exponential expression to within one percent?

33. If a water wave with length L moves with velocity v across a body of water with depth d, as in the figure, then

$$v^2 = \frac{gL}{2\pi} \tanh \frac{2\pi d}{L}$$

(a) If the water is deep, show that $v \approx \sqrt{gL/(2\pi)}$.
(b) If the water is shallow, use the Maclaurin series for tanh to show that $v \approx \sqrt{gd}$. (Thus in shallow water the velocity of a wave tends to be independent of the length of the wave.)
(c) Use the Alternating Series Estimation Theorem to show that if $L > 10d$, then the estimate $v^2 \approx gd$ is accurate to within $0.014gL$.

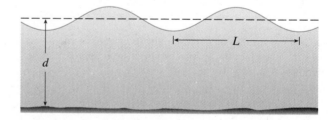

34. Show that T_n and f have the same derivatives at a up to order n.

35. In Section 2.10 we considered Newton's method for approximating a root r of the equation $f(x) = 0$, and from an initial approximation x_1 we obtained successive approximations $x_2, x_3, \ldots$, where

$$x_{n+1} = x_n - \frac{f(x_n)}{f'(x_n)}$$

Use Taylor's Formula with $n = 1$, $a = x_n$, and $x = r$ to show that if $f''(x)$ exists on an interval I containing r, x_n, and x_{n+1}, and $|f''(x)| \leq M$, $|f'(x)| \geq K$ for all $x \in I$, then

$$|x_{n+1} - r| \leq \frac{M}{2K}|x_n - r|^2$$

[This means that if x_n is accurate to d decimal places, then x_{n+1} is accurate to about $2d$ decimal places. More precisely, if the error at stage n is at most 10^{-m}, then the error at stage $n + 1$ is at most $(M/2K)10^{-2m}$.]

36. Use the following outline to prove that e is an irrational number.
   (a) If e were rational, then it would be of the form $e = p/q$, where p and q are positive integers and $q > 2$. Use Taylor's Formula to write

$$\frac{p}{q} = e = 1 + \frac{1}{1!} + \frac{1}{2!} + \cdots + \frac{1}{q!} + \frac{e^z}{(q + 1)!}$$

$$= s_q + \frac{e^z}{(q + 1)!}$$

   where $0 < z < 1$.
   (b) Show that $q!(e - s_q)$ is an integer.
   (c) Show that $0 < q!(e - s_q) < 1$.
   (d) Use parts (b) and (c) to deduce that e is irrational.

---

10 REVIEW

KEY TOPICS ■ Define, state, or discuss the following.

1. Sequence
2. Limit of a sequence
3. Convergent sequence; divergent sequence
4. Increasing, decreasing, and monotonic sequences
5. Bounded sequence
6. Completeness Axiom
7. Convergence of bounded, monotonic sequences
8. Series
9. Partial sums
10. Convergent series; divergent series
11. Sum of a series
12. Geometric series
13. Harmonic series
14. Test for Divergence
15. Integral Test
16. Convergence of a p-series
17. Remainder Estimate for the Integral Test
18. Comparison Test
19. Limit Comparison Test

20. Alternating Series Test
21. Alternating Series Estimation Theorem
22. Absolute convergence
23. Conditional convergence
24. Relation between convergence and absolute convergence
25. Ratio Test
26. Root Test
27. Power series
28. Radius of convergence
29. Interval of convergence
30. The power series for $1/(1 - x)$
31. Differentiation and integration of power series
32. Taylor series
33. Maclaurin series
34. Taylor polynomial
35. Taylor's Formula
36. Maclaurin series for e^x, $\sin x$, $\cos x$
37. Binomial series

EXERCISES

1–18 ■ Determine whether the statement is true or false.

1. If $\lim_{n \to \infty} a_n = 0$, then $\Sigma\, a_n$ is convergent.

2. If $\Sigma\, c_n 6^n$ is convergent, then $\Sigma\, c_n(-2)^n$ is convergent.

3. If $\Sigma\, c_n 6^n$ is convergent, then $\Sigma\, c_n(-6)^n$ is convergent.

4. If $\Sigma\, c_n x^n$ diverges when $x = 6$, then it diverges when $x = 10$.

5. The Ratio Test can be used to determine whether $\Sigma\, 1/n^3$ converges.

6. The Ratio Test can be used to determine whether $\Sigma\, 1/n!$ converges.

7. If $0 \le a_n \le b_n$ and $\Sigma\, b_n$ diverges, then $\Sigma\, a_n$ diverges.

8. $\displaystyle\sum_{n=0}^{\infty} \frac{(-1)^n}{n!} = \frac{1}{e}$

9. $1^x + 2^x + 3^x + \cdots$ is a power series.

10. If f has infinitely many derivatives on $(-\infty, \infty)$, then
$$f(x) = \sum_{n=0}^{\infty} \frac{f^{(n)}(0)}{n!} x^n \text{ for all } x.$$

11. If $-1 < \alpha < 1$, then $\lim_{n\to\infty} \alpha^n = 0$.

12. If $\Sigma\, a_n$ is divergent, then $\Sigma\, |a_n|$ is divergent.

13. If $f(x) = 2x - x^2 + \frac{1}{3}x^3 - \cdots$ converges for all x, then $f'''(0) = 2$.

14. If $\{a_n\}$ and $\{b_n\}$ are divergent, then $\{a_n + b_n\}$ is divergent.

15. If $\{a_n\}$ and $\{b_n\}$ are divergent, then $\{a_n b_n\}$ is divergent.

16. If $\{a_n\}$ is decreasing and $a_n > 0$ for all n, then $\{a_n\}$ is convergent.

17. If $a_n > 0$ and $\Sigma\, a_n$ converges, then $\Sigma\, (-1)^n a_n$ converges.

18. If $a_n > 0$ and $\lim_{n\to\infty} (a_{n+1}/a_n) < 1$, then $\lim_{n\to\infty} a_n = 0$.

19–26 ■ Determine whether the sequence is convergent or divergent. If it is convergent, find its limit.

19. $a_n = \dfrac{n}{2n + 5}$

20. $a_n = 5 - (0.9)^n$

21. $a_n = 2n + 5$

22. $a_n = n/\ln n$

23. $a_n = \sin n$

24. $a_n = (\sin n)/n$

25. $\{(1 + 3/n)^{4n}\}$

26. $\{(-10)^n/n!\}$

27. A sequence is defined recursively by the equations $a_1 = 1$, $a_{n+1} = \frac{1}{3}(a_n + 4)$. Show that $\{a_n\}$ is increasing and $a_n < 2$ for all n. Deduce that $\{a_n\}$ is convergent and find its limit.

28. Show that $\lim_{n\to\infty} n^4 e^{-n} = 0$ and use a graph to find the smallest value of N that corresponds to $\varepsilon = 0.1$ in the definition of a limit.

29–40 ■ Determine whether the series is convergent or divergent.

29. $\displaystyle\sum_{n=1}^{\infty} \frac{n^2}{n^3 + 1}$

30. $\displaystyle\sum_{n=1}^{\infty} \frac{n + n^2}{n + n^4}$

31. $\displaystyle\sum_{n=1}^{\infty} \frac{(-1)^n}{\sqrt[4]{n}}$

32. $\displaystyle\sum_{n=1}^{\infty} \frac{n^2}{3^n}$

33. $\displaystyle\sum_{n=1}^{\infty} \left(\frac{n}{3n + 1}\right)^n$

34. $\displaystyle\sum_{n=1}^{\infty} \sqrt{\frac{n - 1}{n}}$

35. $\displaystyle\sum_{n=1}^{\infty} \frac{\sin n}{1 + n^2}$

36. $\displaystyle\sum_{n=2}^{\infty} \frac{1}{n(\ln n)^2}$

37. $\displaystyle\sum_{n=1}^{\infty} \frac{1 \cdot 3 \cdot 5 \cdot \cdots \cdot (2n - 1)}{5^n n!}$

38. $\displaystyle\sum_{n=1}^{\infty} (-1)^{n+1} \frac{\ln n}{\sqrt{n}}$

39. $\displaystyle\sum_{n=1}^{\infty} \frac{4^n}{n3^n}$

40. $\displaystyle\sum_{n=1}^{\infty} \frac{\sqrt{n + 1} - \sqrt{n - 1}}{n}$

41–44 ■ Determine whether the series is conditionally convergent, absolutely convergent, or divergent.

41. $\displaystyle\sum_{n=1}^{\infty} (-1)^{n-1} n^{-1/3}$

42. $\displaystyle\sum_{n=1}^{\infty} (-1)^{n-1} n^{-3}$

43. $\displaystyle\sum_{n=1}^{\infty} \frac{(-1)^n (n + 1)3^n}{2^{2n+1}}$

44. $\displaystyle\sum_{n=1}^{\infty} \frac{(-1)^n \sqrt{n}}{\ln n}$

45–48 ■ Find the sum of the series.

45. $\displaystyle\sum_{n=1}^{\infty} \frac{2^{2n+1}}{5^n}$

46. $\displaystyle\sum_{n=1}^{\infty} \frac{1}{n(n + 3)}$

47. $\displaystyle\sum_{n=1}^{\infty} [\tan^{-1}(n + 1) - \tan^{-1}n]$

48. $\displaystyle\sum_{n=0}^{\infty} \frac{(-1)^n x^n}{2^{2n} n!}$

49. Express the repeating decimal $1.2345345345\ldots$ as a fraction.

50. For what values of x does the series $\sum_{n=1}^{\infty} (\ln x)^n$ converge?

51. Find the sum of the series $\displaystyle\sum_{n=1}^{\infty} \frac{(-1)^{n+1}}{n^5}$ correct to four decimal places.

52. (a) Find the partial sum s_5 of the series $\sum_{n=1}^{\infty} 1/n^6$ and estimate the error in using it as an approximation to the sum of the series.
(b) Find the sum of this series correct to five decimal places.

53. Use the sum of the first eight terms to approximate the sum of the series $\sum_{n=1}^{\infty} (2 + 5^n)^{-1}$. Estimate the error involved in this approximation.

54. (a) Show that the series $\displaystyle\sum_{n=1}^{\infty} \frac{n^n}{(2n)!}$ is convergent.
(b) Deduce that $\displaystyle\lim_{n\to\infty} \frac{n^n}{(2n)!} = 0$.

55. Prove that if the series $\sum_{n=1}^{\infty} a_n$ is absolutely convergent, then the series
$$\sum_{n=1}^{\infty} \left(\frac{n + 1}{n}\right) a_n$$
is also absolutely convergent.

56–59 ■ Find the radius of convergence and interval of convergence of the series.

56. $\displaystyle\sum_{n=0}^{\infty} \frac{(-3)^n x^{2n}}{n + 1}$

57. $\displaystyle\sum_{n=1}^{\infty} \frac{x^n}{3^n n^3}$

58. $\displaystyle\sum_{n=1}^{\infty} \frac{(x+1)^n}{n^n}$

59. $\displaystyle\sum_{n=0}^{\infty} \frac{2^n(x-3)^n}{\sqrt{n+3}}$

60. Find the radius of convergence of the series

$$\sum_{n=1}^{\infty} \frac{(2n)!}{(n!)^2} x^n$$

61. Find the Taylor series of $f(x) = \sin x$ at $a = \pi/6$.

62. Find the Taylor series of $f(x) = \cos x$ at $a = \pi/3$.

63–70 ■ Find the Maclaurin series for f and its radius of convergence. You may use either the direct method (definition of a Maclaurin series) or known series such as geometric series, binomial series, or the Maclaurin series for e^x and $\sin x$.

63. $f(x) = \dfrac{x^2}{1+x}$

64. $f(x) = \sqrt{1-x^2}$

65. $f(x) = \ln(1-x)$

66. $f(x) = xe^{2x}$

67. $f(x) = \sin(x^4)$

68. $f(x) = 10^x$

69. $f(x) = 1/\sqrt[4]{16-x}$

70. $f(x) = (1-3x)^{-5}$

71. Evaluate $\displaystyle\int \frac{e^x}{x}\, dx$ as an infinite series.

72. Use series to approximate $\int_0^1 \sqrt{1+x^4}\, dx$ correct to two decimal places.

73–74 ■
(a) Approximate f by a Taylor polynomial with degree n at the number a.
(b) Graph f and T_n on a common screen.
(c) Use Taylor's Formula to estimate the accuracy of the approximation $f(x) \approx T_n(x)$ when x lies in the given interval.
(d) Check your result in part (c) by graphing $|R_n(x)|$.

73. $f(x) = \sqrt{x}$, $\quad a = 1$, $\quad n = 3$, $\quad 0.9 \le x \le 1.1$

74. $f(x) = \sec x$, $\quad a = 0$, $\quad n = 2$, $\quad 0 \le x \le \pi/6$

75. Use series to evaluate $\lim_{x\to\infty} x^2(1 - e^{-1/x^2})$.

76. The force due to gravity on an object with mass m at a height h above the surface of the earth is

$$F = \frac{mgR^2}{(R+h)^2}$$

where R is the radius of the earth and g is the acceleration due to gravity.
(a) Express F as a series in powers of h/R.
(b) Observe that if we approximate F by the first term in the series, we get the expression $F \approx mg$ that is usually used when h is much smaller than R. Use the Alternating Series Estimation Theorem to estimate the range of values of h for which the approximation $F \approx mg$ is accurate to within 1%. (Use $R = 6400$ km.)

77. Suppose that $f(x) = \sum_{n=0}^{\infty} c_n x^n$ for all x.
(a) If f is an odd function, show that

$$c_0 = c_2 = c_4 = \cdots = 0$$

(b) If f is an even function, show that

$$c_1 = c_3 = c_5 = \cdots = 0$$

78. If $f(x) = e^{x^2}$, show that $f^{(2n)}(0) = \dfrac{(2n)!}{n!}$.

79. If $f(x) = \sum_{m=0}^{\infty} c_m x^m$ has positive radius of convergence and $e^{f(x)} = \sum_{n=0}^{\infty} d_n x^n$, show that

$$nd_n = \sum_{i=1}^{n} i c_i d_{n-i} \qquad n \ge 1$$

PROBLEMS PLUS

Cover up the solution to the example
and try it yourself first.

EXAMPLE Find the sum of the series $\displaystyle\sum_{n=0}^{\infty} \frac{(x + 2)^n}{(n + 3)!}$.

SOLUTION The problem-solving principle that is relevant here is: *Try to recognize something familiar.* Does the given series look anything like a series that we already know? Well, it does have some ingredients in common with the Maclaurin series for the exponential function:

$$e^x = \sum_{n=0}^{\infty} \frac{x^n}{n!} = 1 + x + \frac{x^2}{2!} + \frac{x^3}{3!} + \cdots$$

We can make this series look more like our given series by replacing x by $x + 2$:

$$e^{x+2} = \sum_{n=0}^{\infty} \frac{(x + 2)^n}{n!} = 1 + (x + 2) + \frac{(x + 2)^2}{2!} + \frac{(x + 2)^3}{3!} + \cdots$$

But here the exponent in the numerator matches the number in the denominator whose factorial is taken. To make that happen in the given series, let's multiply and divide by $(x + 2)^3$:

$$\sum_{n=0}^{\infty} \frac{(x + 2)^n}{(n + 3)!} = \frac{1}{(x + 2)^3} \sum_{n=0}^{\infty} \frac{(x + 2)^{n+3}}{(n + 3)!}$$

$$= (x + 2)^{-3}\left[\frac{(x + 2)^3}{3!} + \frac{(x + 2)^4}{4!} + \cdots \right]$$

We see that the series between brackets is just the series for e^{x+2} with the first three terms missing. So

$$\sum_{n=0}^{\infty} \frac{(x + 2)^n}{(n + 3)!} = (x + 2)^{-3}\left[e^{x+2} - 1 - (x + 2) - \frac{(x + 2)^2}{2!} \right] \qquad\blacksquare$$

PROBLEMS

1. If $f(x) = \sin(x^3)$, find $f^{(15)}(0)$.

2. A function f is defined by

$$f(x) = \lim_{n \to \infty} \frac{x^{2n} - 1}{x^{2n} + 1}$$

Where is f continuous?

3. (a) Show that $\tan \frac{1}{2}x = \cot \frac{1}{2}x - 2 \cot x$.
   (b) Find the sum of the series

$$\sum_{n=1}^{\infty} \frac{1}{2^n} \tan \frac{x}{2^n}$$

4. A curve is defined by the parametric equations

$$x = \int_1^t \frac{\cos u}{u}\, du \qquad y = \int_1^t \frac{\sin u}{u}\, du$$

Find the length of the arc of the curve from the origin to the nearest point where there is a vertical tangent line.

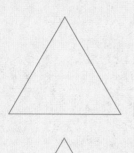

1

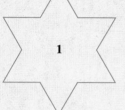

2

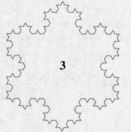

3

FIGURE FOR PROBLEM 5

5. To construct the **snowflake curve**, start with an equilateral triangle with sides of length 1. Step 1 in the construction is to divide each side into three equal parts, construct an equilateral triangle on the middle part, and then delete the middle part (see the figure). Step 2 is to repeat step 1 for each side of the resulting polygon. This process is repeated at each succeeding step. The snowflake curve is the curve that results from repeating this process indefinitely.

(a) Let s_n, l_n, and p_n represent the number of sides, the length of a side, and the total length of the nth approximating curve (the curve obtained after step n of the construction), respectively. Find formulas for s_n, l_n, and p_n.

(b) Show that $p_n \to \infty$ as $n \to \infty$.

(c) Sum an infinite series to find the area enclosed by the snowflake curve. Parts (b) and (c) show that the snowflake curve is infinitely long but encloses only a finite area.

6. (a) Find the highest and lowest points on the curve $x^4 + y^4 = x^2 + y^2$.

(b) Sketch the curve. (Notice that it is symmetric with respect to both axes and both of the lines $y = \pm x$, so it suffices to consider $y \geq x \geq 0$ initially.)

(c) Find the area enclosed by the curve. (You may want to use polar coordinates and use a CAS for this part.)

7. Find the area of the region $S = \{(x, y) \mid x \geq 0, y \leq 1, x^2 + y^2 \leq 4y\}$.

8. (a) Show that, for $n = 1, 2, 3, \ldots$,

$$\sin \theta = 2^n \sin \frac{\theta}{2^n} \cos \frac{\theta}{2} \cos \frac{\theta}{4} \cos \frac{\theta}{8} \cdots \cos \frac{\theta}{2^n}$$

(b) Deduce that

$$\frac{\sin \theta}{\theta} = \cos \frac{\theta}{2} \cos \frac{\theta}{4} \cos \frac{\theta}{8} \cdots$$

The meaning of this infinite product is that we take the product of the first n factors and then we take the limit of these partial products as $n \to \infty$.

(c) Show that

$$\frac{2}{\pi} = \frac{\sqrt{2}}{2} \frac{\sqrt{2 + \sqrt{2}}}{2} \frac{\sqrt{2 + \sqrt{2 + \sqrt{2}}}}{2} \cdots$$

This infinite product is due to the French mathematician François Viète (1540–1603). Notice that it expresses π in terms of just the number 2 and repeated square roots.

9. If $a_1 = \cos \theta$, $-\pi/2 \leq \theta \leq \pi/2$, $b_1 = 1$, and

$$a_{n+1} = \tfrac{1}{2}(a_n + b_n) \qquad b_{n+1} = \sqrt{b_n a_{n+1}}$$

show that

$$\lim_{n \to \infty} a_n = \lim_{n \to \infty} b_n = \frac{\sin \theta}{\theta}$$

10. If the curve $y = e^{-x/10} \sin x$, $x \geq 0$, is rotated about the x-axis, the resulting solid looks like an infinite decreasing string of beads. (If you have a graphing device, take a look at the graph of the curve.)

(a) Find the exact volume of the nth bead. (Use either a computer algebra system or a table of integrals.)

(b) Find the total volume of the beads.

11. Find the interval of convergence of $\sum_{n=1}^{\infty} n^3 x^n$ and find its sum.

12. Let $\{P_n\}$ be a sequence of points determined as in the figure. Thus $|AP_1| = 1$, $|P_n P_{n+1}| = 2^{n-1}$, and angle $AP_n P_{n+1}$ is a right angle. Find $\lim_{n \to \infty} \angle P_n A P_{n+1}$.

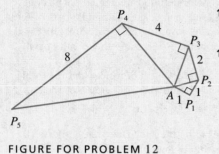

FIGURE FOR PROBLEM 12

13. (a) Show that for $xy \neq -1$,

$$\arctan x - \arctan y = \arctan \frac{x - y}{1 + xy}$$

if the left side lies between $-\pi/2$ and $\pi/2$.

(b) Show that

$$\arctan \tfrac{120}{119} - \arctan \tfrac{1}{239} = \frac{\pi}{4}$$

(c) Deduce the following formula of John Machin (1680–1751):

$$4 \arctan \tfrac{1}{5} - \arctan \tfrac{1}{239} = \frac{\pi}{4}$$

(d) Use the Maclaurin series for arctan to show that

$$0.197395560 < \arctan \tfrac{1}{5} < 0.197395562$$

(e) Show that

$$0.004184075 < \arctan \tfrac{1}{239} < 0.004184077$$

(f) Deduce that, correct to seven decimal places,

$$\pi \approx 3.1415927$$

Machin used this method in 1706 to find π correct to 100 decimal places. In this century, with the aid of computers, the value of π has been computed to increasingly greater accuracy. In 1989 Gregory and David Chudnovsky of Columbia University used supercomputers to find the value of π correct to more than a billion decimal places!

14. (a) Prove a formula similar to the one in Problem 13(a) but involving arccot instead of arctan.

(b) Find the sum of the series

$$\sum_{n=0}^{\infty} \operatorname{arccot}(n^2 + n + 1)$$

15. Let

$$u = 1 + \frac{x^3}{3!} + \frac{x^6}{6!} + \frac{x^9}{9!} + \cdots$$

$$v = x + \frac{x^4}{4!} + \frac{x^7}{7!} + \frac{x^{10}}{10!} + \cdots$$

$$w = \frac{x^2}{2!} + \frac{x^5}{5!} + \frac{x^8}{8!} + \cdots$$

Show that $u^3 + v^3 + w^3 - 3uvw = 1$.

16. A curve called the **folium of Descartes** is defined by the parametric equations

$$x = \frac{3t}{1 + t^3} \qquad y = \frac{3t^2}{1 + t^3}$$

(a) Show that if (a, b) lies on the curve, then so does (b, a); that is, the curve is symmetric with respect to the line $y = x$. Where does the curve intersect this line?
(b) Find the points on the curve where the tangent lines are horizontal or vertical.
(c) Show that the line $y = -x - 1$ is a slant asymptote.
(d) Sketch the curve.
(e) Show that a Cartesian equation of this curve is $x^3 + y^3 = 3xy$.
(f) Show that the polar equation can be written in the form

$$r = \frac{3 \sec \theta \tan \theta}{1 + \tan^3 \theta}$$

(g) Find the area enclosed by the loop of this curve.
[CAS] (h) Show that the area of the loop is the same as the area that lies between the asymptote and the infinite branches of the curve.

17. If $0 < a \le b \le c$, show that $\lim_{n \to \infty} (a^n + b^n + c^n)^{1/n} = c$.

18. Four bugs are placed at the four corners of a square with side length a. The bugs crawl counterclockwise at the same speed and each bug crawls directly toward the next bug at all times. They approach the center of the square along spiral paths.
(a) Find the polar equation of a bug's path assuming the pole is at the center of the square. (Use the fact that the line joining one bug to the next is tangent to the bug's path.)
(b) Find the distance traveled by a bug by the time it meets the other bugs at the center.

19. If the value of x^x at $x = 0$ is taken to be 1, show that

$$\int_0^1 x^x \, dx = \sum_{n=1}^{\infty} \frac{(-1)^n}{n^n}$$

20. Find the sum of the series

$$1 + \frac{1}{2} + \frac{1}{3} + \frac{1}{4} + \frac{1}{6} + \frac{1}{8} + \frac{1}{9} + \frac{1}{12} + \cdots$$

where the terms are the reciprocals of the positive integers whose only prime factors are 2s and 3s.

21. Consider the series whose terms are the reciprocals of the positive integers that can be written in base 10 notation without using the digit 0. Show that this series is convergent and the sum is less than 90.

22. If $p > 1$, evaluate the expression

$$\frac{1 + \dfrac{1}{2^p} + \dfrac{1}{3^p} + \dfrac{1}{4^p} + \cdots}{1 - \dfrac{1}{2^p} + \dfrac{1}{3^p} - \dfrac{1}{4^p} + \cdots}$$

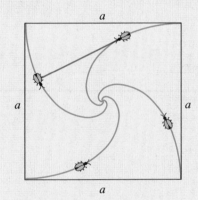

a

a a

a

FIGURE FOR PROBLEM 18

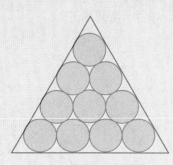

FIGURE FOR PROBLEM 23

23. Suppose that circles of equal diameter are packed tightly in n rows inside an equilateral triangle. (The figure illustrates the case $n = 4$.) If A is the area of the triangle and A_n is the total area occupied by the n rows of circles, show that

$$\lim_{n \to \infty} \frac{A_n}{A} = \frac{\pi}{2\sqrt{3}}$$

24. If $a_0 + a_1 + a_2 + \cdots + a_k = 0$, show that

$$\lim_{n \to \infty} \left(a_0 \sqrt{n} + a_1 \sqrt{n+1} + a_2 \sqrt{n+2} + \cdots + a_k \sqrt{n+k} \right) = 0$$

If you don't see how to prove this, try the problem-solving strategy of analogy (see Section 4 of Review and Preview). Try the special cases $k = 1$ and $k = 2$ first. If you can see how to prove the assertion for these cases, then you will probably see how to prove it in general.

25. Let $f(x) = x \sin(\pi/x)$, $-1 \le x \le 1$, $x \ne 0$, and $f(0) = 0$.
   (a) Show that f is continuous on $(-1, 1)$.
   (b) Sketch the graph of f, showing the local maxima and minima.
   (c) Use your graph from part (b) to show that the length of the graph from $x = 1/n$ to $x = 1/(n-1)$ is greater than $2/(2n-1)$. Deduce that the graph of f has infinite length. (This shows that it is possible for a continuous function on a finite interval to have a graph that is infinitely long.)

26. A sequence $\{a_n\}$ is defined recursively by the equations

$$a_0 = a_1 = 1 \qquad n(n-1)a_n = (n-1)(n-2)a_{n-1} - (n-3)a_{n-2}$$

Find the sum of the series $\sum_{n=0}^{\infty} a_n$.

27. Show that

$$1 + \frac{1}{2} - \frac{2}{3} + \frac{1}{4} + \frac{1}{5} - \frac{2}{6} + \frac{1}{7} + \frac{1}{8} - \frac{2}{9} + \cdots = \ln 3$$

[*Hint:* See Exercise 10.5.35.]

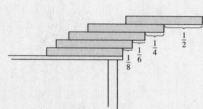

FIGURE FOR PROBLEM 28

28. Suppose you have a large supply of books, all the same size, and you stack them at the edge of a table, with each book extending farther beyond the edge of the table than the one beneath it. Show that it is possible to do this so that the top book extends entirely beyond the table. In fact, show that the top book can extend any distance at all beyond the edge of the table if the stack is high enough. Use the following method of stacking: The top book extends half its length beyond the second book. The second book extends a quarter of its length beyond the third. The third extends one-sixth of its length beyond the fourth, and so on. (Try it yourself with a deck of cards.) Consider centers of mass.

29. Evaluate

$$\int_0^1 \binom{-x-1}{100} \left(\frac{1}{x+1} + \frac{1}{x+2} + \frac{1}{x+3} + \cdots + \frac{1}{x+100} \right) dx$$

where the first factor in the integrand denotes a binomial coefficient.

30. For which numbers c is it true that $\cosh x \le e^{cx^2}$ for all x?

31. Find the sum of the double series

$$\sum_{n=1}^{\infty} \sum_{m=1}^{\infty} \frac{1}{m^2 n + mn^2 + 2mn}$$

[*Hint:* Use partial fractions.]

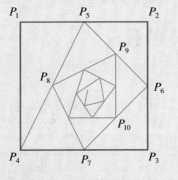

P_1 P_5 P_2

P_9

P_8

P_6

P_{10}

P_4 P_7 P_3

FIGURE FOR PROBLEM 32

32. Starting with the vertices $P_1(0,1)$, $P_2(1,1)$, $P_3(1,0)$, $P_4(0,0)$ of a square, we construct further points as shown in the figure: P_5 is the midpoint of $P_1 P_2$, P_6 is the midpoint of $P_2 P_3$, P_7 is the midpoint of $P_3 P_4$, and so on. The polygonal spiral path $P_1 P_2 P_3 P_4 P_5 P_6 P_7 \ldots$ approaches a point P inside the square.
    (a) If the coordinates of P_n are (x_n, y_n), show that

$$\tfrac{1}{2} x_n + x_{n+1} + x_{n+2} + x_{n+3} = 2$$

and find a similar equation for the y-coordinates.
    (b) Find the coordinates of P.

33. Let $a_1 = \tfrac{3}{2}$ and $a_{n+1} = \dfrac{3a_n^2 + 4a_n - 3}{4a_n^2}$ for $n = 1, 2, 3, \ldots$.
    (a) Show that $a_n > 1$ and $a_{n+1} < a_n$ for all n.
    (b) Find $\lim_{n \to \infty} a_n$.
    (c) Find $\lim_{n \to \infty} (a_1 a_2 a_3 \cdots a_n)$.

11

THREE-DIMENSIONAL ANALYTIC GEOMETRY AND VECTORS

I now demonstrate the frame of the System of the World.
SIR ISAAC NEWTON

The analytic geometry of three-dimensional space is important not only in its own right, but also because it will be needed in the next chapter to study the calculus of functions of several variables. We use vectors in our treatment of three-dimensional analytic geometry because vectors give particularly simple descriptions of lines, planes, and curves in space. We also see that vector-valued functions can be used to describe the motion of objects through space. In particular, we use them to derive Kepler's laws of planetary motion.

11.1 THREE-DIMENSIONAL COORDINATE SYSTEMS

To locate a point in a plane, two numbers are necessary. We know that any point in the plane can be represented as an ordered pair (a, b) of real numbers, where a is the x-coordinate and b is the y-coordinate. For this reason, a plane is called two-dimensional. To locate a point in space, three numbers are required. We represent any point in space by an ordered triple (a, b, c) of real numbers.

In order to represent points in space, we first choose a fixed point O (the origin) and three directed lines through O that are perpendicular to each other, called the **coordinate axes** and labeled the x-axis, y-axis, and z-axis. Usually we think of the x- and y-axes as being horizontal and the z-axis as being vertical, and we draw the orientation of the axes as in Figure 1.

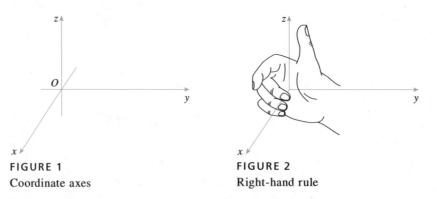

FIGURE 1
Coordinate axes

FIGURE 2
Right-hand rule

In looking at Figure 2 you can think of the y- and z-axes as lying in the plane of the paper and the x-axis as coming out of the paper toward you. The direction of the z-axis is determined by the **right-hand rule:** If you curl the fingers of your right hand

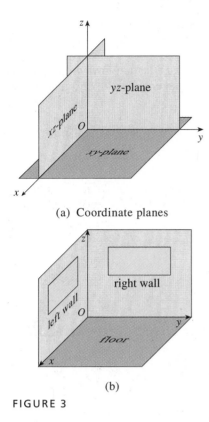

(a) Coordinate planes

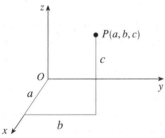

(b)

FIGURE 3

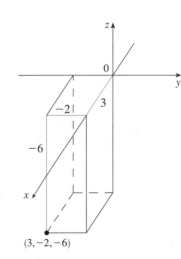

FIGURE 4

around the z-axis in the direction of a 90° counterclockwise rotation from the positive x-axis to the positive y-axis, then your thumb points in the positive direction of the z-axis.

The three coordinate axes determine the three **coordinate planes** illustrated in Figure 3(a). The xy-plane is the plane that contains the x- and y-axes; the yz-plane contains the y- and z-axes; the xz-plane contains the x- and z-axes. These three coordinate planes divide space into eight parts, called **octants.** The **first octant,** in the foreground, is determined by the positive axes.

Because many people have some difficulty visualizing diagrams of three-dimensional figures, you may find it helpful to do the following [see Figure 3(b)]. Look at any bottom corner of a room and call the corner the origin. The wall on your left is in the xz-plane, the wall on your right is in the yz-plane, and the floor is in the xy-plane. The x-axis runs along the intersection of the floor and the left wall. The y-axis runs along the intersection of the floor and the right wall. The z-axis runs up from the floor toward the ceiling along the intersection of the two walls. You are situated in the first octant, and you can now imagine seven other rooms situated in the other seven octants (three on the same floor and four on the floor below), all connected by the common corner point O.

Now if P is any point in space, let a be the (directed) distance from the yz-plane to P, let b be the distance from the xz-plane to P, and let c be the distance from the xy-plane to P. We represent the point P by the ordered triple (a, b, c) of real numbers and we call a, b, and c the **coordinates** of P; a is the x-coordinate, b is the y-coordinate, and c is the z-coordinate. Thus, to locate the point (a, b, c) we can start at the origin O and move a units along the x-axis, then b units parallel to the y-axis, and then c units parallel to the z-axis as in Figure 4.

The point $P(a, b, c)$ determines a rectangular box as in Figure 5. If we drop a perpendicular from P to the xy-plane, we get a point Q with coordinates $(a, b, 0)$ called the **projection** of P on the xy-plane. Similarly, $R(0, b, c)$ and $S(a, 0, c)$ are the projections of P on the yz-plane and xz-plane, respectively.

As numerical illustrations, the points $(-4, 3, -5)$ and $(3, -2, -6)$ are plotted in Figure 6.

The Cartesian product $\mathbb{R} \times \mathbb{R} \times \mathbb{R} = \{(x, y, z) \mid x, y, z \in \mathbb{R}\}$ is the set of all ordered triples of real numbers and is denoted by $\mathbb{R}^3$. We have given a one-to-one correspondence between points P in space and ordered triples (a, b, c) in $\mathbb{R}^3$. It is called a

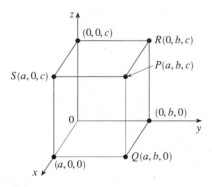

FIGURE 5

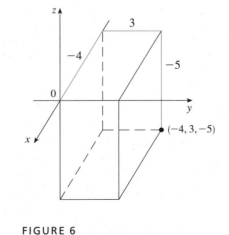

FIGURE 6

three-dimensional rectangular coordinate system. Notice that, in terms of coordinates, the first octant can be described as the set of points whose coordinates are all positive.

In two-dimensional analytic geometry, the graph of an equation involving x and y is a curve in $\mathbb{R}^2$. In three-dimensional analytic geometry, an equation in x, y, and z represents a **surface** in $\mathbb{R}^3$.

EXAMPLE 1   What surfaces in $\mathbb{R}^3$ are represented by the following equations?
(a) $z = 3$ (b) $y = 5$

SOLUTION

(a) The equation $z = 3$ represents the set $\{(x, y, z) \mid z = 3\}$, which is the set of all points in $\mathbb{R}^3$ whose z-coordinate is 3. This is the horizontal plane that is parallel to the xy-plane and three units above it as in Figure 7(a).

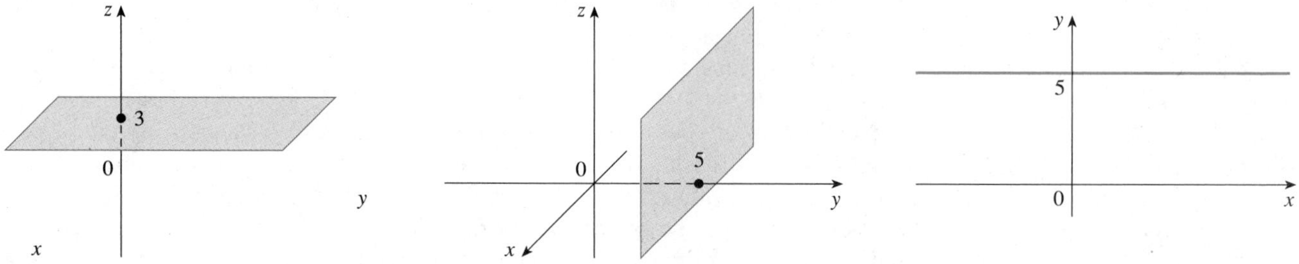

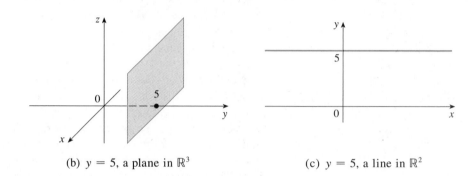

(a) $z = 3$, a plane in $\mathbb{R}^3$ (b) $y = 5$, a plane in $\mathbb{R}^3$ (c) $y = 5$, a line in $\mathbb{R}^2$

FIGURE 7

(b) The equation $y = 5$ represents the set of all points in $\mathbb{R}^3$ whose y-coordinate is 5. This is the vertical plane that is parallel to the xz-plane and five units to the right of it as in Figure 7(b).   ∎

NOTE: When an equation is given, it must be understood from the context whether it represents a curve in $\mathbb{R}^2$ or a surface in $\mathbb{R}^3$. In Example 1, $y = 5$ represents a plane in $\mathbb{R}^3$, but of course $y = 5$ can also represent a line in $\mathbb{R}^2$ if we are dealing with two-dimensional analytic geometry. See Figure 7(b) and (c).

In general, if k is a constant, then $x = k$ represents a plane parallel to the yz-plane, $y = k$ is a plane parallel to the xz-plane, and $z = k$ is a plane parallel to the xy-plane. In Figure 5, the faces of the rectangular box are formed by the three coordinate planes $x = 0$ (the yz-plane), $y = 0$ (the xz-plane), and $z = 0$ (the xy-plane), and the planes $x = a$, $y = b$, and $z = c$.

EXAMPLE 2   Describe and sketch the surface in $\mathbb{R}^3$ represented by the equation $y = x$.

SOLUTION   The equation represents the set of all points in $\mathbb{R}^3$ whose x- and y-coordinates are equal, that is, $\{(x, x, z) \mid x \in \mathbb{R}, z \in \mathbb{R}\}$. This is a vertical plane that intersects the xy-plane in the line $y = x$, $z = 0$. The portion of this plane that lies in the first octant is sketched in Figure 8.   ∎

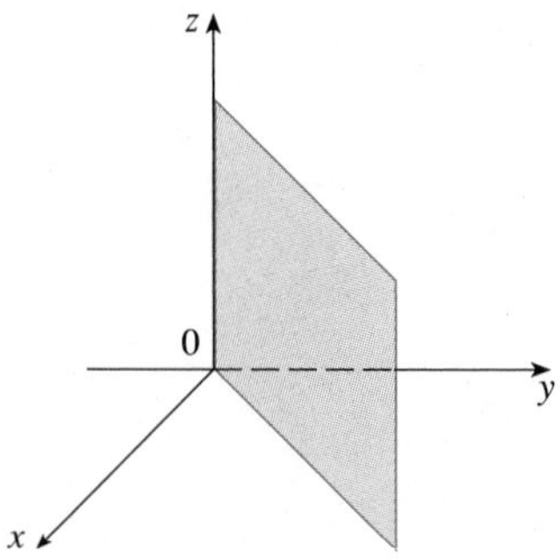

FIGURE 8
The plane $y = x$

The familiar formula for the distance between two points in a plane is easily extended to the following three-dimensional formula.

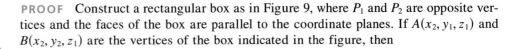

DISTANCE FORMULA IN THREE DIMENSIONS The distance $|P_1P_2|$ between the points $P_1(x_1, y_1, z_1)$ and $P_2(x_2, y_2, z_2)$ is

$$|P_1P_2| = \sqrt{(x_2 - x_1)^2 + (y_2 - y_1)^2 + (z_2 - z_1)^2}$$

FIGURE 9

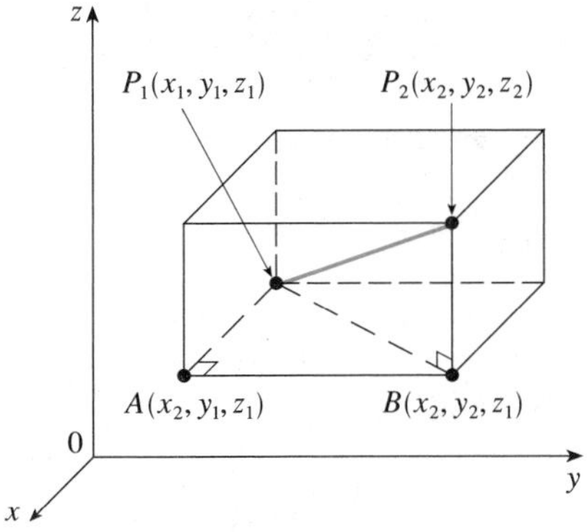

PROOF Construct a rectangular box as in Figure 9, where P_1 and P_2 are opposite vertices and the faces of the box are parallel to the coordinate planes. If $A(x_2, y_1, z_1)$ and $B(x_2, y_2, z_1)$ are the vertices of the box indicated in the figure, then

$$|P_1A| = |x_2 - x_1| \qquad |AB| = |y_2 - y_1| \qquad |BP_2| = |z_2 - z_1|$$

Because triangles P_1BP_2 and P_1AB are both right-angled, two applications of the Pythagorean Theorem give

$$|P_1P_2|^2 = |P_1B|^2 + |BP_2|^2 \qquad \text{and} \qquad |P_1B|^2 = |P_1A|^2 + |AB|^2$$

Combining these equations, we get

$$\begin{aligned} |P_1P_2|^2 &= |P_1A|^2 + |AB|^2 + |BP_2|^2 \\ &= |x_2 - x_1|^2 + |y_2 - y_1|^2 + |z_2 - z_1|^2 \\ &= (x_2 - x_1)^2 + (y_2 - y_1)^2 + (z_2 - z_1)^2 \end{aligned}$$

Therefore $$|P_1P_2| = \sqrt{(x_2 - x_1)^2 + (y_2 - y_1)^2 + (z_2 - z_1)^2}$$ □

EXAMPLE 3 The distance from the point $P(2, -1, 7)$ to the point $Q(1, -3, 5)$ is

$$\begin{aligned} |PQ| &= \sqrt{(1 - 2)^2 + (-3 + 1)^2 + (5 - 7)^2} \\ &= \sqrt{1 + 4 + 4} = 3 \end{aligned}$$ ■

EXAMPLE 4 Find an equation of a sphere with radius r and center $C(h, k, l)$.

FIGURE 10

SOLUTION By definition, a sphere is the set of all points $P(x, y, z)$ whose distance from C is r (see Figure 10). Thus P is on the sphere if and only if $|PC| = r$. Squaring both sides, we have $|PC|^2 = r^2$ or

$$(x - h)^2 + (y - k)^2 + (z - l)^2 = r^2$$ ■

This result is worth remembering:

EQUATION OF A SPHERE An equation of a sphere with center $C(h, k, l)$ and radius r is

$$(x - h)^2 + (y - k)^2 + (z - l)^2 = r^2$$

In particular, if the center is the origin O, then an equation of the sphere is

$$x^2 + y^2 + z^2 = r^2$$

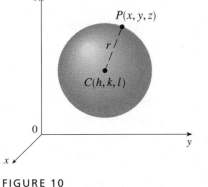

EXAMPLE 5 Show that $x^2 + y^2 + z^2 + 4x - 6y + 2z + 6 = 0$ is the equation of a sphere, and find its center and radius.

SOLUTION We can rewrite the given equation in the form of an equation of a sphere if we complete squares:

$$(x^2 + 4x + 4) + (y^2 - 6y + 9) + (z^2 + 2z + 1) = -6 + 4 + 9 + 1$$

$$(x + 2)^2 + (y - 3)^2 + (z + 1)^2 = 8$$

Comparing this equation with the standard form, we see that it is the equation of a sphere with center $(-2, 3, -1)$ and radius $\sqrt{8} = 2\sqrt{2}$. ∎

EXERCISES 11.1

1–4 ■ Draw a rectangular box that has P and Q as opposite vertices and has its faces parallel to the coordinate planes. Then find (a) the coordinates of the other six vertices of the box and (b) the length of the diagonal of the box.

1. $P(0, 0, 0)$, $Q(2, 3, 5)$ **2.** $P(0, 0, 0)$, $Q(-4, -1, 2)$

3. $P(1, 1, 2)$, $Q(3, 4, 5)$ **4.** $P(4, 3, 0)$, $Q(1, 6, -4)$

5–8 ■ Find the lengths of the sides of the triangle ABC and determine whether the triangle is isosceles, a right triangle, both, or neither.

5. $A(2, 1, 0)$, $B(3, 3, 4)$, $C(5, 4, 3)$

6. $A(5, 5, 1)$, $B(3, 3, 2)$, $C(1, 4, 4)$

7. $A(-2, 6, 1)$, $B(5, 4, -3)$, $C(2, -6, 4)$

8. $A(3, -4, 1)$, $B(5, -3, 0)$, $C(6, -7, 4)$

9–10 ■ Determine whether the given points are collinear.

9. $P(1, 2, 3)$, $Q(0, 3, 7)$, $R(3, 5, 11)$

10. $K(0, 3, -4)$, $L(1, 2, -2)$, $M(3, 0, 1)$

11–14 ■ Find the equation of the sphere with center C and radius r.

11. $C(0, 1, -1)$, $r = 4$ **12.** $C(-1, 2, 4)$, $r = \frac{1}{2}$

13. $C(-6, -1, 2)$, $r = 2\sqrt{3}$ **14.** $C(1, 2, -3)$, $r = 7$

15–20 ■ Show that the given equation represents a sphere, and find its center and radius.

15. $x^2 + y^2 + z^2 + 2x + 8y - 4z = 28$

16. $x^2 + y^2 + z^2 = 6x + 4y + 10z$

17. $x^2 + y^2 + z^2 + x - 2y + 6z - 2 = 0$

18. $2x^2 + 2y^2 + 2z^2 + 4y - 2z = 1$

19. $x^2 + y^2 + z^2 = x$

20. $x^2 + y^2 + z^2 + ax + by + cz + d = 0$, where $a^2 + b^2 + c^2 > 4d$

21. Prove that the midpoint of the line segment from $P_1(x_1, y_1, z_1)$ to $P_2(x_2, y_2, z_2)$ is

$$\left(\frac{x_1 + x_2}{2}, \frac{y_1 + y_2}{2}, \frac{z_1 + z_2}{2} \right)$$

22. Find an equation of a sphere if one of its diameters has endpoints $(2, 1, 4)$ and $(4, 3, 10)$.

23. Find the lengths of the medians of the triangle with vertices $A(1, 2, 3)$, $B(-2, 0, 5)$, and $C(4, 1, 5)$.

24. Find an equation of the sphere that has center $(1, 2, 3)$ and passes through the point $(-1, 1, -2)$.

25. Find equations of the spheres with center $(2, -3, 6)$ that touch (a) the xy-plane, (b) the yz-plane, (c) the xz-plane.

26. Consider the points P such that the distance from P to $A(-1, 5, 3)$ is twice the distance from P to $B(6, 2, -2)$. Show that the set of all such points is a sphere, and find its center and radius.

27. Find an equation of the set of all points equidistant from the points $A(-1, 5, 3)$ and $B(6, 2, -2)$. Describe the set.

28–43 ■ Describe in words the region of $\mathbb{R}^3$ represented by the equation or inequality.

28. $x = 0$ **29.** $x = 9$

30. $z = -8$ **31.** $y > 2$

32. $z \leq 0$ **33.** $z = x$

34. $y = z$ **35.** $x^2 + y^2 = 1$

36. $y^2 + z^2 \leq 4$ **37.** $x^2 + y^2 + z^2 > 1$

38. $1 \leq x^2 + y^2 + z^2 \leq 25$ **39.** $x^2 + y^2 + z^2 - 2z < 3$

40. $xy = 0$ **41.** $xy = 1$

42. $xyz = 0$ **43.** $|z| \leq 2$

44–47 ■ Write inequalities to describe the given region.

44. The solid rectangular box in the first octant bounded by the planes $x = 1$, $y = 2$, and $z = 3$.

45. The half-space consisting of all points to the left of the *xz*-plane

46. The solid upper hemisphere of the sphere of radius 2 centered at the origin

47. The region consisting of all points between (but not on) the spheres of radius *r* and *R* centered at the origin, where $r < R$

48. Use a computer with three-dimensional graphing software to graph the sphere with center the origin and radius 1. (You may need to graph the upper and lower hemispheres separately.)

49. The figure shows a line L_1 in space and a second line L_2, which is the projection of L_1 on the *xy*-plane. (In other words, the points on L_2 are directly beneath, or above, the points on L_1.)
(a) Find the coordinates of the point *P*.
(b) Locate on the diagram the points *A*, *B*, and *C*, where the line L_1 intersects the *xy*-plane, the *yz*-plane, and the *xz*-plane, respectively.

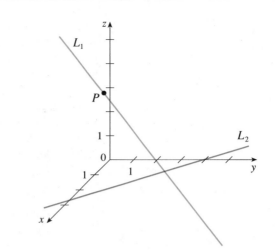

50. Find the volume of the intersection of the spheres $x^2 + y^2 + z^2 + 4x - 2y + 4z + 5 = 0$ and $x^2 + y^2 + z^2 = 4$.

11.2 VECTORS

The term **vector** is used by scientists to indicate a quantity (such as velocity or force) that has both magnitude and direction. A vector is often represented by an arrow or a directed line segment. The length of the arrow represents the magnitude of the vector and the arrow points in the direction of the vector. For instance, Figure 1(a) shows a particle moving along a path in the plane and its velocity vector **v** at a specific location of the particle. Here the length of the arrow represents the speed of the particle and it points in the direction that the particle is moving.

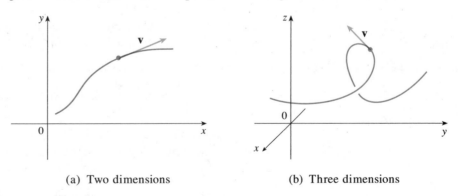

(a) Two dimensions   (b) Three dimensions

FIGURE 1 The velocity vector of a particle

Figure 1(b) shows the path of a particle moving in space. Here the velocity vector **v** is a three-dimensional vector. (This application of vectors will be studied in detail in Section 11.9.)

Notice that all of the arrows in Figure 2 are equivalent in the sense that they have the same length and point in the same direction even though they are in different locations. All of the directed line segments have the property that the terminal point is

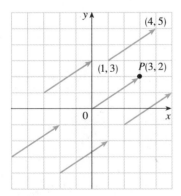

FIGURE 2
Representations of the vector
$\mathbf{v} = \langle 3, 2 \rangle$

reached from the initial point by a displacement of three units to the right and two upward. We regard each of the directed line segments as equivalent representations of a single entity called a **vector.** In other words, we can regard a vector **v** as a set of equivalent directed line segments. These line segments are characterized by the numbers 3 and 2, and we symbolize this situation by writing **v** = $\langle 3, 2 \rangle$. Thus a two-dimensional vector can be thought of as an ordered pair of real numbers. We use the notation $\langle a, b \rangle$ for the ordered pair that refers to a vector so as not to confuse it with the ordered pair (a, b) that refers to a point in the plane. A vector can be indicated by printing a letter in boldface (**v**) or by putting an arrow above the letter ($\vec{v}$).

> **(1) DEFINITION**   A **two-dimensional vector** is an ordered pair $\mathbf{a} = \langle a_1, a_2 \rangle$ of real numbers. A **three-dimensional vector** is an ordered triple $\mathbf{a} = \langle a_1, a_2, a_3 \rangle$ of real numbers. The numbers a_1, a_2, and a_3 are called the **components** of **a**.

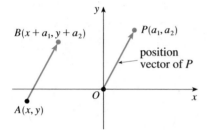

(a) Representations of $\mathbf{a} = \langle a_1, a_2 \rangle$

A **representation** of the vector $\mathbf{a} = \langle a_1, a_2 \rangle$ is a directed line segment $\overrightarrow{AB}$ from any point $A(x, y)$ to the point $B(x + a_1, y + a_2)$. A particular representation of **a** is the directed line segment $\overrightarrow{OP}$ from the origin to the point $P(a_1, a_2)$, and $\langle a_1, a_2 \rangle$ is called the **position vector** of the point $P(a_1, a_2)$. Likewise, in three dimensions, the vector $\mathbf{a} = \langle a_1, a_2, a_3 \rangle$ is the position vector of the point $P(a_1, a_2, a_3)$ (see Figure 3).

Observe that if $\mathbf{a} = \langle a_1, a_2, a_3 \rangle$ is a vector that has the representation $\overrightarrow{AB}$, where the initial point is $A(x_1, y_1, z_1)$ and the terminal point is $B(x_2, y_2, z_2)$, then we must have $x_1 + a_1 = x_2, y_1 + a_2 = y_2$, and $z_1 + a_3 = z_2$ and so $a_1 = x_2 - x_1, a_2 = y_2 - y_1$, and $a_3 = z_2 - z_1$. Thus we have the following:

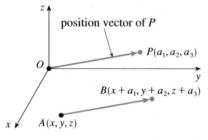

(b) Representations of $\mathbf{a} = \langle a_1, a_2, a_3 \rangle$

FIGURE 3

> **(2)**   Given the points $A(x_1, y_1, z_1)$ and $B(x_2, y_2, z_2)$, the vector **a** with representation $\overrightarrow{AB}$ is
>
> $$\mathbf{a} = \langle x_2 - x_1, y_2 - y_1, z_2 - z_1 \rangle$$

EXAMPLE 1   Find the vector represented by the directed line segment with initial point $A(2, -3, 4)$ and terminal point $B(-2, 1, 1)$.

SOLUTION   By (2), the vector corresponding to $\overrightarrow{AB}$ is

$$\mathbf{a} = \langle -2 - 2, 1 - (-3), 1 - 4 \rangle = \langle -4, 4, -3 \rangle \qquad \blacksquare$$

The **magnitude** or **length** of the vector **v** is the length of any of its representations and is denoted by the symbol $|\mathbf{v}|$ or $\|\mathbf{v}\|$. By using the distance formula to compute the length of a segment OP, we obtain the following:

> **(3)**   The length of the two-dimensional vector $\mathbf{a} = \langle a_1, a_2 \rangle$ is
>
> $$|\mathbf{a}| = \sqrt{a_1^2 + a_2^2}$$
>
> The length of the three-dimensional vector $\mathbf{a} = \langle a_1, a_2, a_3 \rangle$ is
>
> $$|\mathbf{a}| = \sqrt{a_1^2 + a_2^2 + a_3^2}$$

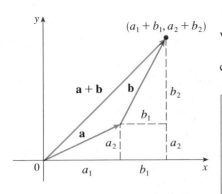

FIGURE 4
Triangle Law

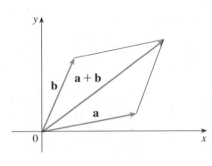

FIGURE 5
Parallelogram Law

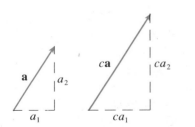

FIGURE 6

The only vector with length 0 is the **zero vector 0** $= \langle 0, 0 \rangle$ (or $\mathbf{0} = \langle 0, 0, 0 \rangle$). This vector is also the only vector with no specific direction.

According to the following definition, we add vectors by adding the corresponding components of two vectors.

---

(4) VECTOR ADDITION   If $\mathbf{a} = \langle a_1, a_2 \rangle$ and $\mathbf{b} = \langle b_1, b_2 \rangle$, then the vector $\mathbf{a} + \mathbf{b}$ is defined by

$$\mathbf{a} + \mathbf{b} = \langle a_1 + b_1, a_2 + b_2 \rangle$$

Similarly, for three-dimensional vectors,

$$\langle a_1, a_2, a_3 \rangle + \langle b_1, b_2, b_3 \rangle = \langle a_1 + b_1, a_2 + b_2, a_3 + b_3 \rangle$$

Definition 4 is illustrated geometrically in Figure 4 for the two-dimensional case. You can see why the definition of vector addition is sometimes called the **Triangle Law.** Alternatively, another interpretation of vector addition is shown in Figure 5 and is called the **Parallelogram Law.**

It is possible to multiply a vector by a real number c. (In this context we call the real number c a **scalar** to distinguish it from a vector.) For instance, we want $2\mathbf{a}$ to be the same vector as $\mathbf{a} + \mathbf{a}$, so

$$2\langle a_1, a_2 \rangle = \langle a_1, a_2 \rangle + \langle a_1, a_2 \rangle = \langle 2a_1, 2a_2 \rangle$$

In general, we multiply a vector by a scalar by multiplying each component by that scalar.

---

(5) MULTIPLICATION OF A VECTOR BY A SCALAR   If c is a scalar and $\mathbf{a} = \langle a_1, a_2 \rangle$, then the vector $c\mathbf{a}$ is defined by

$$c\mathbf{a} = \langle ca_1, ca_2 \rangle$$

Similarly, for three-dimensional vectors,

$$c\langle a_1, a_2, a_3 \rangle = \langle ca_1, ca_2, ca_3 \rangle$$

Definition 5 is illustrated by Figure 6.

Let us see how the scalar multiple $c\mathbf{a}$ compares with the original vector $\mathbf{a}$. If $\mathbf{a} = \langle a_1, a_2 \rangle$, then

$$|c\mathbf{a}| = \sqrt{(ca_1)^2 + (ca_2)^2} = \sqrt{c^2(a_1^2 + a_2^2)}$$
$$= \sqrt{c^2}\sqrt{a_1^2 + a_2^2} = |c||\mathbf{a}|$$

so the length of $c\mathbf{a}$ is $|c|$ times the length of $\mathbf{a}$.

If $a_1 \neq 0$, we can talk about the slope of $\mathbf{a}$ as being a_2/a_1. But, if $c \neq 0$, then the slope of $c\mathbf{a}$ is $ca_2/ca_1 = a_2/a_1$, the same as the slope of $\mathbf{a}$. If $c > 0$, then a_1 and ca_1 have the same sign. Also, a_2 and ca_2 have the same sign. This means that $\mathbf{a}$ and $c\mathbf{a}$ have the same direction. On the other hand, if $c < 0$, then a_1 and ca_1 have opposite signs, as do a_2 and ca_2, so $\mathbf{a}$ and $c\mathbf{a}$ have opposite directions. In particular, the vector $-\mathbf{a} = (-1)\mathbf{a}$

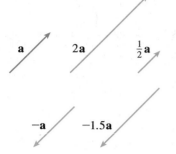

a   **2a** $\frac{1}{2}\mathbf{a}$

−a   **−1.5a**

FIGURE 7
Scalar multiples of **a**

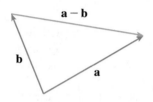

a − b

b

a

FIGURE 8

has the same length as **a** but points in the opposite direction. Illustrations of representations are shown in Figure 7.

Although we have been considering two-dimensional vectors, it is also true for three-dimensional vectors that $c\mathbf{a}$ is a vector that is $|c|$ times as long as **a** and has the same direction as **a** if $c > 0$ and the opposite direction if $c < 0$. Two vectors **a** and **b** are called **parallel** if $\mathbf{b} = c\mathbf{a}$ for some scalar c.

By the **difference a − b** of two vectors, we mean

$$\mathbf{a} - \mathbf{b} = \mathbf{a} + (-\mathbf{b})$$

so if $\mathbf{a} = \langle a_1, a_2 \rangle$ and $\mathbf{b} = \langle b_1, b_2 \rangle$, then

$$\mathbf{a} - \mathbf{b} = \langle a_1 - b_1, a_2 - b_2 \rangle$$

Since $(\mathbf{a} - \mathbf{b}) + \mathbf{b} = \mathbf{a}$, the vector $\mathbf{a} - \mathbf{b}$, when added to **b**, gives **a**. This is illustrated in Figure 8 by means of the Triangle Law.

EXAMPLE 2   If $\mathbf{a} = \langle 4, 0, 3 \rangle$ and $\mathbf{b} = \langle -2, 1, 5 \rangle$, find $|\mathbf{a}|$ and the vectors $\mathbf{a} + \mathbf{b}$, $\mathbf{a} - \mathbf{b}$, $3\mathbf{b}$, and $2\mathbf{a} + 5\mathbf{b}$.

SOLUTION $$|\mathbf{a}| = \sqrt{4^2 + 0^2 + 3^2} = \sqrt{25} = 5$$

$$\mathbf{a} + \mathbf{b} = \langle 4, 0, 3 \rangle + \langle -2, 1, 5 \rangle$$
$$= \langle 4 - 2, 0 + 1, 3 + 5 \rangle = \langle 2, 1, 8 \rangle$$

$$\mathbf{a} - \mathbf{b} = \langle 4, 0, 3 \rangle - \langle -2, 1, 5 \rangle$$
$$= \langle 4 - (-2), 0 - 1, 3 - 5 \rangle = \langle 6, -1, -2 \rangle$$

$$3\mathbf{b} = 3\langle -2, 1, 5 \rangle = \langle 3(-2), 3(1), 3(5) \rangle = \langle -6, 3, 15 \rangle$$

$$2\mathbf{a} + 5\mathbf{b} = 2\langle 4, 0, 3 \rangle + 5\langle -2, 1, 5 \rangle$$
$$= \langle 8, 0, 6 \rangle + \langle -10, 5, 25 \rangle = \langle -2, 5, 31 \rangle$$ ∎

We denote by V_2 the set of all two-dimensional vectors and by V_3 the set of all three-dimensional vectors. More generally, we will later need to consider the set V_n of all n-dimensional vectors. An n-dimensional vector is an ordered n-tuple:

$$\mathbf{a} = \langle a_1, a_2, \ldots, a_n \rangle$$

where $a_1, a_2, \ldots, a_n$ are real numbers that are called the components of **a**. Addition and scalar multiplication are defined in terms of components just as for the cases $n = 2$ and $n = 3$.

---

(6) PROPERTIES OF VECTORS   If **a**, **b**, and **c** are vectors in V_n and c and d are scalars, then

1. $\mathbf{a} + \mathbf{b} = \mathbf{b} + \mathbf{a}$ 2. $\mathbf{a} + (\mathbf{b} + \mathbf{c}) = (\mathbf{a} + \mathbf{b}) + \mathbf{c}$
3. $\mathbf{a} + \mathbf{0} = \mathbf{a}$ 4. $\mathbf{a} + (-\mathbf{a}) = \mathbf{0}$
5. $c(\mathbf{a} + \mathbf{b}) = c\mathbf{a} + c\mathbf{b}$ 6. $(c + d)\mathbf{a} = c\mathbf{a} + d\mathbf{a}$
7. $(cd)\mathbf{a} = c(d\mathbf{a})$ 8. $1\mathbf{a} = \mathbf{a}$

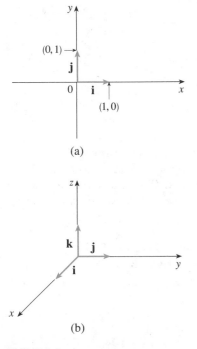

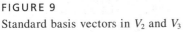

FIGURE 9
Standard basis vectors in V_2 and V_3

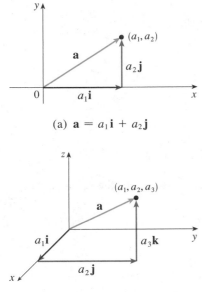

(a) $\mathbf{a} = a_1\mathbf{i} + a_2\mathbf{j}$

(b) $\mathbf{a} = a_1\mathbf{i} + a_2\mathbf{j} + a_3\mathbf{k}$

FIGURE 10

The eight properties of vectors in Theorem 6 can be readily verified using Definitions 4 and 5. For instance, here is the verification of Property 1 for the case $n = 2$:

$$\mathbf{a} + \mathbf{b} = \langle a_1, a_2 \rangle + \langle b_1, b_2 \rangle = \langle a_1 + b_1, a_2 + b_2 \rangle$$
$$= \langle b_1 + a_1, b_2 + a_2 \rangle = \langle b_1, b_2 \rangle + \langle a_1, a_2 \rangle$$
$$= \mathbf{b} + \mathbf{a}$$

The remaining proofs are left as exercises.

Three vectors in V_3 play a special role. Let

$$\mathbf{i} = \langle 1, 0, 0 \rangle \qquad \mathbf{j} = \langle 0, 1, 0 \rangle \qquad \mathbf{k} = \langle 0, 0, 1 \rangle$$

Then $\mathbf{i}$, $\mathbf{j}$, and $\mathbf{k}$ are vectors that have length 1 and point in the directions of the positive x-, y-, and z-axes. Similarly, in two dimensions we define $\mathbf{i} = \langle 1, 0 \rangle$ and $\mathbf{j} = \langle 0, 1 \rangle$ (see Figure 9).

If $\mathbf{a} = \langle a_1, a_2, a_3 \rangle$, then we can write

$$\mathbf{a} = \langle a_1, a_2, a_3 \rangle = \langle a_1, 0, 0 \rangle + \langle 0, a_2, 0 \rangle + \langle 0, 0, a_3 \rangle$$
$$= a_1\langle 1, 0, 0 \rangle + a_2\langle 0, 1, 0 \rangle + a_3\langle 0, 0, 1 \rangle$$

(7)
$$\mathbf{a} = a_1\mathbf{i} + a_2\mathbf{j} + a_3\mathbf{k}$$

Thus any vector in V_3 can be expressed in terms of the **standard basis vectors i, j**, and **k**. For instance,

$$\langle 1, -2, 6 \rangle = \mathbf{i} - 2\mathbf{j} + 6\mathbf{k}$$

Similarly, in two dimensions, we can write

(8)
$$\mathbf{a} = \langle a_1, a_2 \rangle = a_1\mathbf{i} + a_2\mathbf{j}$$

See Figure 10 for the geometric interpretation of Equations 8 and 7 and compare with Figure 9.

EXAMPLE 3 If $\mathbf{a} = \mathbf{i} + 2\mathbf{j} - 3\mathbf{k}$ and $\mathbf{b} = 4\mathbf{i} + 7\mathbf{k}$, express the vector $2\mathbf{a} + 3\mathbf{b}$ in terms of $\mathbf{i}$, $\mathbf{j}$, and $\mathbf{k}$.

SOLUTION   Using Properties 1, 2, 5, 6, and 7 of Theorem 6, we have

$$2\mathbf{a} + 3\mathbf{b} = 2(\mathbf{i} + 2\mathbf{j} - 3\mathbf{k}) + 3(4\mathbf{i} + 7\mathbf{k})$$
$$= 2\mathbf{i} + 4\mathbf{j} - 6\mathbf{k} + 12\mathbf{i} + 21\mathbf{k} = 14\mathbf{i} + 4\mathbf{j} + 15\mathbf{k} \qquad \blacksquare$$

A **unit vector** is a vector whose length is 1. For instance, $\mathbf{i}$, $\mathbf{j}$, and $\mathbf{k}$ are all unit vectors. In general, if $\mathbf{a} \neq \mathbf{0}$, then the unit vector that has the same direction as $\mathbf{a}$ is

(9)
$$\mathbf{u} = \frac{1}{|\mathbf{a}|}\,\mathbf{a} = \frac{\mathbf{a}}{|\mathbf{a}|}$$

In order to verify this, we let $c = 1/|\mathbf{a}|$. Then $\mathbf{u} = c\mathbf{a}$ and c is a positive scalar, so $\mathbf{u}$ has the same direction as $\mathbf{a}$. Also

$$|\mathbf{u}| = |c\mathbf{a}| = |c||\mathbf{a}| = \frac{1}{|\mathbf{a}|}|\mathbf{a}| = 1$$

EXAMPLE 4   Find the unit vector in the direction of the vector $2\,\mathbf{i} - \mathbf{j} - 2\,\mathbf{k}$.

SOLUTION   The given vector has length

$$|2\,\mathbf{i} - \mathbf{j} - 2\,\mathbf{k}| = \sqrt{2^2 + (-1)^2 + (-2)^2} = \sqrt{9} = 3$$

so, by Equation 9, the unit vector with the same direction is

$$\tfrac{1}{3}(2\,\mathbf{i} - \mathbf{j} - 2\,\mathbf{k}) = \tfrac{2}{3}\mathbf{i} - \tfrac{1}{3}\mathbf{j} - \tfrac{2}{3}\mathbf{k}$$ ■

We conclude this section by considering one of the many applications of vectors in physics and engineering. A force is represented by a vector because it has both a magnitude (measured in pounds or newtons) and a direction. If several forces are acting on an object, the **resultant force** experienced by the object is the vector sum of these forces.

EXAMPLE 5   A 100-lb weight hangs from two wires as shown in Figure 11. Find the tensions (forces) $\mathbf{T}_1$ and $\mathbf{T}_2$ in both wires and their magnitudes.

SOLUTION   We first express $\mathbf{T}_1$ and $\mathbf{T}_2$ in terms of their horizontal and vertical components. From Figure 12 we see that

(10) $$\mathbf{T}_1 = -|\mathbf{T}_1|\cos 50°\,\mathbf{i} + |\mathbf{T}_1|\sin 50°\,\mathbf{j}$$

(11) $$\mathbf{T}_2 = |\mathbf{T}_2|\cos 32°\,\mathbf{i} + |\mathbf{T}_2|\sin 32°\,\mathbf{j}$$

The resultant $\mathbf{T}_1 + \mathbf{T}_2$ of the tensions counterbalances the weight $\mathbf{w}$ and so we must have

$$\mathbf{T}_1 + \mathbf{T}_2 = -\mathbf{w} = 100\,\mathbf{j}$$

Thus

$$(-|\mathbf{T}_1|\cos 50° + |\mathbf{T}_2|\cos 32°)\,\mathbf{i} + (|\mathbf{T}_1|\sin 50° + |\mathbf{T}_2|\sin 32°)\,\mathbf{j} = 100\,\mathbf{j}$$

Equating components, we get

$$-|\mathbf{T}_1|\cos 50° + |\mathbf{T}_2|\cos 32° = 0$$

$$|\mathbf{T}_1|\sin 50° + |\mathbf{T}_2|\sin 32° = 100$$

Solving the first of these equations for $|\mathbf{T}_2|$ and substituting into the second, we get

$$|\mathbf{T}_1|\sin 50° + \frac{|\mathbf{T}_1|\cos 50°}{\cos 32°}\sin 32° = 100$$

So the magnitudes of the tensions are

$$|\mathbf{T}_1| = \frac{100}{\sin 50° + \tan 32°\cos 50°} \approx 85.64 \text{ lb}$$

and

$$|\mathbf{T}_2| = \frac{|\mathbf{T}_1|\cos 50°}{\cos 32°} \approx 64.91 \text{ lb}$$

Substituting these values in (10) and (11), we obtain the tension vectors

$$\mathbf{T}_1 \approx -55.05\,\mathbf{i} + 65.60\,\mathbf{j} \qquad \mathbf{T}_2 \approx 55.05\,\mathbf{i} + 34.40\,\mathbf{j}$$ ■

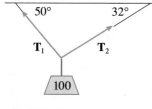

FIGURE 11

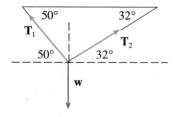

FIGURE 12

EXERCISES 11.2

1–6 ■ Find a vector **a** with representation given by the directed line segment $\overrightarrow{AB}$. Draw $\overrightarrow{AB}$ and the equivalent representation starting at the origin.

1. $A(1, 3)$, $B(4, 4)$

2. $A(-3, 4)$, $B(-1, 0)$

3. $A(3, -1)$, $B(3, -3)$

4. $A(4, -1)$, $B(1, 2)$

5. $A(0, 3, 1)$, $B(2, 3, -1)$

6. $A(1, -2, 0)$, $B(1, -2, 3)$

7–10 ■ Find the sum of the given vectors and illustrate geometrically.

7. $\langle 2, 3 \rangle$, $\langle 3, -4 \rangle$

8. $\langle -1, 2 \rangle$, $\langle 5, 3 \rangle$

9. $\langle 1, 0, 1 \rangle$, $\langle 0, 0, 1 \rangle$

10. $\langle 0, 3, 2 \rangle$, $\langle 1, 0, -3 \rangle$

11–18 ■ Find $|\mathbf{a}|$, $\mathbf{a} + \mathbf{b}$, $\mathbf{a} - \mathbf{b}$, $2\mathbf{a}$, and $3\mathbf{a} + 4\mathbf{b}$.

11. $\mathbf{a} = \langle 5, -12 \rangle$, $\mathbf{b} = \langle -2, 8 \rangle$

12. $\mathbf{a} = \langle -1, 2 \rangle$, $\mathbf{b} = \langle 4, 3 \rangle$

13. $\mathbf{a} = \langle 2, -3, 6 \rangle$, $\mathbf{b} = \langle 1, 1, 4 \rangle$

14. $\mathbf{a} = \langle 3, 2, -1 \rangle$, $\mathbf{b} = \langle 0, 6, 7 \rangle$

15. $\mathbf{a} = \mathbf{i} - \mathbf{j}$, $\mathbf{b} = \mathbf{i} + \mathbf{j}$

16. $\mathbf{a} = 2\mathbf{i} + 3\mathbf{j}$, $\mathbf{b} = 3\mathbf{i} - 2\mathbf{j}$

17. $\mathbf{a} = \mathbf{i} + \mathbf{j} + \mathbf{k}$, $\mathbf{b} = 2\mathbf{i} - \mathbf{j} + 3\mathbf{k}$

18. $\mathbf{a} = 6\mathbf{i} + \mathbf{k}$, $\mathbf{b} = \mathbf{i} - 2\mathbf{j} + 7\mathbf{k}$

19–24 ■ Find a unit vector that has the same direction as the given vector.

19. $\langle 1, 2 \rangle$

20. $\langle 3, -5 \rangle$

21. $\langle -2, 4, 3 \rangle$

22. $\langle 1, -4, 8 \rangle$

23. $\mathbf{i} + \mathbf{j}$

24. $2\mathbf{i} - 4\mathbf{j} + 7\mathbf{k}$

25–26 ■ Express **i** and **j** in terms of **a** and **b**.

25. $\mathbf{a} = 2\mathbf{i} + 3\mathbf{j}$, $\mathbf{b} = \mathbf{i} - \mathbf{j}$

26. $\mathbf{a} = \mathbf{i} - 2\mathbf{j}$, $\mathbf{b} = 3\mathbf{i} + \mathbf{j}$

27. If A, B, and C are the vertices of a triangle, find $\overrightarrow{AB} + \overrightarrow{BC} + \overrightarrow{CA}$.

28. Let C be the point on the line segment AB that is twice as far from B as it is from A. If $\mathbf{a} = \overrightarrow{OA}$, $\mathbf{b} = \overrightarrow{OB}$, and $\mathbf{c} = \overrightarrow{OC}$, show that $\mathbf{c} = \frac{2}{3}\mathbf{a} + \frac{1}{3}\mathbf{b}$.

29. (a) Draw the vectors $\mathbf{a} = \langle 3, 2 \rangle$, $\mathbf{b} = \langle 2, -1 \rangle$, and $\mathbf{c} = \langle 7, 1 \rangle$.
(b) Show, by means of a sketch, that there are scalars s and t such that $\mathbf{c} = s\mathbf{a} + t\mathbf{b}$.
(c) Use the sketch to estimate the values of s and t.
(d) Find the exact values of s and t.

30. Suppose that **a** and **b** are nonzero vectors that are not parallel and **c** is any vector in the plane determined by **a** and **b**. Give a geometric argument to show that **c** can be

written as $\mathbf{c} = s\mathbf{a} + t\mathbf{b}$ for suitable scalars s and t. Then give an argument using components.

31. Two forces $\mathbf{F}_1$ and $\mathbf{F}_2$ with magnitudes 10 lb and 12 lb act on an object at a point P as shown in the figure. Find the resultant force $\mathbf{F}$ acting at P as well as its magnitude and its direction. (Indicate the direction by finding the angle θ shown in the figure.)

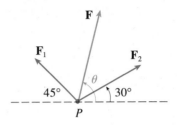

32. Velocities have both direction and magnitude and thus are vectors. The magnitude of a velocity vector is called *speed*. Suppose that a wind is blowing from the direction N45°W at a speed of 50 km/h. (This means that the direction from which the wind blows is 45° west of the northerly direction.) A pilot is steering a plane in the direction N60°E at an airspeed (speed in still air) of 250 km/h. The *true course*, or *track*, of the plane is the direction of the resultant of the velocity vectors of the plane and the wind. The *ground speed* of the plane is the magnitude of the resultant. Find the true course and the ground speed of the plane.

33. A woman walks due west on the deck of a ship at 3 mi/h. The ship is moving north at a speed of 22 mi/h. Find the speed and direction of the woman relative to the surface of the water.

34. Ropes 3 m and 5 m in length are fastened to a holiday decoration that is suspended over a town square. The decoration has a mass of 5 kg. The ropes, fastened at different heights, make angles of 52° and 40° with the horizontal. Find the tension in each wire and the magnitude of each tension.

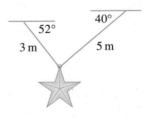

35. If $\mathbf{r} = \langle x, y, z \rangle$ and $\mathbf{r}_0 = \langle x_0, y_0, z_0 \rangle$, describe the set of all points (x, y, z) such that $|\mathbf{r} - \mathbf{r}_0| = 1$.

36. If $\mathbf{r} = \langle x, y \rangle$, $\mathbf{r}_1 = \langle x_1, y_1 \rangle$, and $\mathbf{r}_2 = \langle x_2, y_2 \rangle$, describe the set of all points (x, y) such that $|\mathbf{r} - \mathbf{r}_1| + |\mathbf{r} - \mathbf{r}_2| = k$, where $k > |\mathbf{r}_1 - \mathbf{r}_2|$.

37. Prove Property 2 of Theorem 6 for the case $n = 2$.

38. Prove Property 5 of Theorem 6 for the case $n = 3$.

39. Prove Property 6 of Theorem 6 for the case $n = 3$.

40. Use vectors to prove that the line joining the midpoints of two sides of a triangle is parallel to the third side and half its length.

41. A quadrilateral has one pair of opposite sides parallel and of equal length. Use vectors to prove that the other pair of opposite sides is parallel and of equal length.

42. Suppose the three coordinate planes are all mirrored and a light ray given by the vector $\mathbf{a} = \langle a_1, a_2, a_3 \rangle$ first strikes the xz-plane, as shown in the figure. Use the fact that the angle of incidence equals the angle of reflection to show that the direction of the reflected ray is given by $\mathbf{b} = \langle a_1, -a_2, a_3 \rangle$. Deduce that, after being reflected by all

three mutually perpendicular mirrors, the resulting ray is parallel to the initial ray. (American space scientists used this principle, together with laser beams and an array of corner mirrors on the moon, to calculate very precisely the distance from the earth to the moon.)

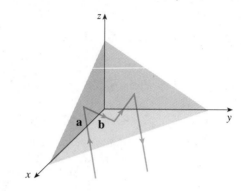

11.3 THE DOT PRODUCT

So far we have added two vectors and multiplied a vector by a scalar. The question arises: Is it possible to multiply two vectors so that their product is a useful quantity? One such product is the dot product, whose definition follows. Another is the cross product, which is discussed in the next section.

> **(1) DEFINITION**   If $\mathbf{a} = \langle a_1, a_2, a_3 \rangle$ and $\mathbf{b} = \langle b_1, b_2, b_3 \rangle$, then the **dot product** of $\mathbf{a}$ and $\mathbf{b}$ is the number $\mathbf{a} \cdot \mathbf{b}$ given by
>
> $$\mathbf{a} \cdot \mathbf{b} = a_1 b_1 + a_2 b_2 + a_3 b_3$$

Thus, to find the dot product of $\mathbf{a}$ and $\mathbf{b}$ we multiply corresponding components and add. The result is not a vector. It is a real number, that is, a scalar. For this reason, the dot product is sometimes called the **scalar product** (or **inner product**). Although Definition 1 is given for three-dimensional vectors, the dot product of two-dimensional vectors is defined in a similar fashion:

$$\langle a_1, a_2 \rangle \cdot \langle b_1, b_2 \rangle = a_1 b_1 + a_2 b_2$$

EXAMPLE 1

$$\langle 2, 4 \rangle \cdot \langle 3, -1 \rangle = 2(3) + 4(-1) = 2$$

$$\langle -1, 7, 4 \rangle \cdot \langle 6, 2, -\tfrac{1}{2} \rangle = (-1)(6) + 7(2) + 4(-\tfrac{1}{2}) = 6$$

$$(\mathbf{i} + 2\mathbf{j} - 3\mathbf{k}) \cdot (2\mathbf{j} - \mathbf{k}) = 1(0) + 2(2) + (-3)(-1) = 7 \qquad \blacksquare$$

The dot product obeys many of the laws that hold for ordinary products of real numbers. These are stated in the following theorem.

> **(2) PROPERTIES OF THE DOT PRODUCT** If $\mathbf{a}$, $\mathbf{b}$, and $\mathbf{c}$ are vectors in V_3 and c is a scalar, then
>
> 1. $\mathbf{a} \cdot \mathbf{a} = |\mathbf{a}|^2$
> 2. $\mathbf{a} \cdot \mathbf{b} = \mathbf{b} \cdot \mathbf{a}$
> 3. $\mathbf{a} \cdot (\mathbf{b} + \mathbf{c}) = \mathbf{a} \cdot \mathbf{b} + \mathbf{a} \cdot \mathbf{c}$
> 4. $(c\mathbf{a}) \cdot \mathbf{b} = c(\mathbf{a} \cdot \mathbf{b}) = \mathbf{a} \cdot (c\mathbf{b})$
> 5. $\mathbf{0} \cdot \mathbf{a} = 0$

These properties are easily proved using Definition 1. For instance, here are the proofs of Properties 1 and 3:

1. $\mathbf{a} \cdot \mathbf{a} = a_1^2 + a_2^2 + a_3^2 = |\mathbf{a}|^2$
3. $\mathbf{a} \cdot (\mathbf{b} + \mathbf{c}) = \langle a_1, a_2, a_3 \rangle \cdot \langle b_1 + c_1, b_2 + c_2, b_3 + c_3 \rangle$
$= a_1(b_1 + c_1) + a_2(b_2 + c_2) + a_3(b_3 + c_3)$
$= a_1 b_1 + a_1 c_1 + a_2 b_2 + a_2 c_2 + a_3 b_3 + a_3 c_3$
$= (a_1 b_1 + a_2 b_2 + a_3 b_3) + (a_1 c_1 + a_2 c_2 + a_3 c_3)$
$= \mathbf{a} \cdot \mathbf{b} + \mathbf{a} \cdot \mathbf{c}$

The proofs of the remaining properties are left as exercises.

The dot product $\mathbf{a} \cdot \mathbf{b}$ can be given a geometric interpretation in terms of the **angle θ between $\mathbf{a}$ and $\mathbf{b}$,** which is defined to be the angle between the representations of $\mathbf{a}$ and $\mathbf{b}$ that start at the origin, where $0 \le \theta \le \pi$. In other words, θ is the angle between the line segments $\overrightarrow{OA}$ and $\overrightarrow{OB}$ in Figure 1. Note that if $\mathbf{a}$ and $\mathbf{b}$ are parallel vectors, then $\theta = 0$ or π.

The formula in the following theorem is used by physicists as the *definition* of the dot product.

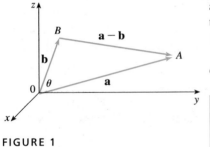

FIGURE 1

> **(3) THEOREM** If θ is the angle between the vectors $\mathbf{a}$ and $\mathbf{b}$, then
>
> $$\mathbf{a} \cdot \mathbf{b} = |\mathbf{a}||\mathbf{b}| \cos\theta$$

PROOF If we apply the Law of Cosines to triangle OAB in Figure 1, we get

(4) $$|AB|^2 = |OA|^2 + |OB|^2 - 2|OA||OB| \cos\theta$$

(Observe that the Law of Cosines still applies in the limiting cases when $\theta = 0$ or π, or $\mathbf{a} = \mathbf{0}$ or $\mathbf{b} = \mathbf{0}$.) But $|OA| = |\mathbf{a}|$, $|OB| = |\mathbf{b}|$, and $|AB| = |\mathbf{a} - \mathbf{b}|$, so Equation 4 becomes

(5) $$|\mathbf{a} - \mathbf{b}|^2 = |\mathbf{a}|^2 + |\mathbf{b}|^2 - 2|\mathbf{a}||\mathbf{b}| \cos\theta$$

Using Properties 1, 2, and 3 of the dot product, we can rewrite the left side of this equation as follows:

$$|\mathbf{a} - \mathbf{b}|^2 = (\mathbf{a} - \mathbf{b}) \cdot (\mathbf{a} - \mathbf{b})$$
$$= \mathbf{a} \cdot \mathbf{a} - \mathbf{a} \cdot \mathbf{b} - \mathbf{b} \cdot \mathbf{a} + \mathbf{b} \cdot \mathbf{b}$$
$$= |\mathbf{a}|^2 - 2\mathbf{a} \cdot \mathbf{b} + |\mathbf{b}|^2$$

Therefore, Equation 5 gives

$$|\mathbf{a}|^2 - 2\mathbf{a} \cdot \mathbf{b} + |\mathbf{b}|^2 = |\mathbf{a}|^2 + |\mathbf{b}|^2 - 2|\mathbf{a}||\mathbf{b}| \cos\theta$$

Thus $$-2\mathbf{a} \cdot \mathbf{b} = -2|\mathbf{a}||\mathbf{b}| \cos\theta$$

or $$\mathbf{a} \cdot \mathbf{b} = |\mathbf{a}||\mathbf{b}| \cos\theta$$

(6) COROLLARY   If θ is the angle between the nonzero vectors **a** and **b**, then

$$\cos\theta = \frac{\mathbf{a}\cdot\mathbf{b}}{|\mathbf{a}||\mathbf{b}|}$$

EXAMPLE 2   Find the angle between the vectors $\mathbf{a} = \langle 2, 2, -1 \rangle$ and $\mathbf{b} = \langle 5, -3, 2 \rangle$.
SOLUTION   Since

$$|\mathbf{a}| = \sqrt{2^2 + 2^2 + (-1)^2} = 3 \quad \text{and} \quad |\mathbf{b}| = \sqrt{5^2 + (-3)^2 + 2^2} = \sqrt{38}$$

and since $\mathbf{a}\cdot\mathbf{b} = 2(5) + 2(-3) + (-1)(2) = 2$

we have, from Corollary 6,

$$\cos\theta = \frac{\mathbf{a}\cdot\mathbf{b}}{|\mathbf{a}||\mathbf{b}|} = \frac{2}{3\sqrt{38}}$$

So the angle between **a** and **b** is

$$\theta = \cos^{-1}\left(\frac{2}{3\sqrt{38}}\right) \approx 1.46 \quad \text{(or } 84°\text{)}$$

Two nonzero vectors **a** and **b** are called **perpendicular** or **orthogonal** if the angle between them is $\theta = \pi/2$. Then Theorem 3 gives

$$\mathbf{a}\cdot\mathbf{b} = |\mathbf{a}||\mathbf{b}|\cos(\pi/2) = 0$$

and conversely if $\mathbf{a}\cdot\mathbf{b} = 0$, then $\cos\theta = 0$, so $\theta = \pi/2$. The zero vector **0** is considered to be perpendicular to all vectors. Therefore

(7)   **a** and **b** are orthogonal if and only if $\mathbf{a}\cdot\mathbf{b} = 0$.

EXAMPLE 3   Show that $2\mathbf{i} + 2\mathbf{j} - \mathbf{k}$ is perpendicular to $5\mathbf{i} - 4\mathbf{j} + 2\mathbf{k}$.
SOLUTION   Since

$$(2\mathbf{i} + 2\mathbf{j} - \mathbf{k})\cdot(5\mathbf{i} - 4\mathbf{j} + 2\mathbf{k}) = 2(5) + 2(-4) + (-1)(2) = 0$$

these vectors are perpendicular by (7).

**DIRECTION ANGLES
AND DIRECTION COSINES**

The **direction angles** of a nonzero vector **a** are the angles α, β, and γ in the interval $[0, \pi]$ that **a** makes with the positive x-, y-, and z-axes (see Figure 2).
The cosines of these direction angles, $\cos\alpha$, $\cos\beta$, and $\cos\gamma$, are called the **direction cosines** of the vector **a**. Using Corollary 6 with **b** replaced by **i**, we obtain

(8)
$$\cos\alpha = \frac{\mathbf{a}\cdot\mathbf{i}}{|\mathbf{a}||\mathbf{i}|} = \frac{a_1}{|\mathbf{a}|}$$

(This can also be seen directly from Figure 2.) Similarly, we also have

(9)
$$\cos\beta = \frac{a_2}{|\mathbf{a}|} \qquad \cos\gamma = \frac{a_3}{|\mathbf{a}|}$$

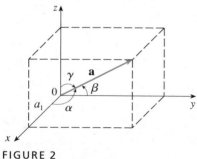

FIGURE 2

By squaring the expressions in Equations 8 and 9 and adding, we see that

(10)
$$\cos^2\alpha + \cos^2\beta + \cos^2\gamma = 1$$

We can also use Equations 8 and 9 to write

$$\mathbf{a} = \langle a_1, a_2, a_3 \rangle = \langle |\mathbf{a}|\cos\alpha, |\mathbf{a}|\cos\beta, |\mathbf{a}|\cos\gamma \rangle$$
$$= |\mathbf{a}|\langle \cos\alpha, \cos\beta, \cos\gamma \rangle$$

Therefore

(11)
$$\frac{1}{|\mathbf{a}|}\mathbf{a} = \langle \cos\alpha, \cos\beta, \cos\gamma \rangle$$

which says that the direction cosines of **a** are the components of the unit vector in the direction of **a**.

EXAMPLE 4 Find the direction angles of the vector $\mathbf{a} = \langle 1, 2, 3 \rangle$.

SOLUTION Since $|\mathbf{a}| = \sqrt{1^2 + 2^2 + 3^2} = \sqrt{14}$, Equations 8 and 9 give

$$\cos\alpha = \frac{1}{\sqrt{14}} \qquad \cos\beta = \frac{2}{\sqrt{14}} \qquad \cos\gamma = \frac{3}{\sqrt{14}}$$

and so

$$\alpha = \cos^{-1}\left(\frac{1}{\sqrt{14}}\right) \approx 74° \qquad \beta = \cos^{-1}\left(\frac{2}{\sqrt{14}}\right) \approx 58° \qquad \gamma = \cos^{-1}\left(\frac{3}{\sqrt{14}}\right) \approx 37°$$

PROJECTIONS

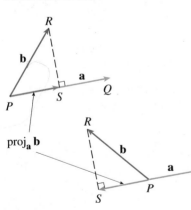

FIGURE 3
Vector projections

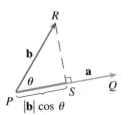

FIGURE 4
Scalar projection

Figure 3 shows representations $\overrightarrow{PQ}$ and $\overrightarrow{PR}$ of two vectors **a** and **b** with the same initial point P. If S is the foot of the perpendicular from R to the line containing $\overrightarrow{PQ}$, then the vector with representation $\overrightarrow{PS}$ is called the **vector projection** of **b** onto **a** and is denoted by $\text{proj}_{\mathbf{a}}\,\mathbf{b}$. The **scalar projection** of **b** onto **a** (also called the **component of b along a**) is defined to be the number $|\mathbf{b}|\cos\theta$, where θ is the angle between **a** and **b** (see Figure 4). This is denoted by $\text{comp}_{\mathbf{a}}\,\mathbf{b}$. Observe that it is negative if $\pi/2 < \theta \le \pi$. The equation

$$\mathbf{a} \cdot \mathbf{b} = |\mathbf{a}||\mathbf{b}|\cos\theta = |\mathbf{a}|(|\mathbf{b}|\cos\theta)$$

shows that the dot product of **a** and **b** can be interpreted as the length of **a** times the scalar projection of **b** onto **a**. Since

$$|\mathbf{b}|\cos\theta = \frac{\mathbf{a} \cdot \mathbf{b}}{|\mathbf{a}|} = \frac{\mathbf{a}}{|\mathbf{a}|} \cdot \mathbf{b}$$

the component of **b** along **a** can be computed by taking the dot product of **b** with the unit vector in the direction of **a**. To summarize:

Scalar projection of **b** onto **a**: $\text{comp}_{\mathbf{a}}\,\mathbf{b} = \dfrac{\mathbf{a} \cdot \mathbf{b}}{|\mathbf{a}|}$

Vector projection of **b** onto **a**: $\text{proj}_{\mathbf{a}}\,\mathbf{b} = \left(\dfrac{\mathbf{a} \cdot \mathbf{b}}{|\mathbf{a}|}\right)\dfrac{\mathbf{a}}{|\mathbf{a}|} = \dfrac{\mathbf{a} \cdot \mathbf{b}}{|\mathbf{a}|^2}\mathbf{a}$

EXAMPLE 5   Find the scalar projection and vector projection of $\mathbf{b} = \langle 1, 1, 2 \rangle$ onto $\mathbf{a} = \langle -2, 3, 1 \rangle$.

SOLUTION   Since $|\mathbf{a}| = \sqrt{(-2)^2 + 3^2 + 1^2} = \sqrt{14}$, the scalar projection of $\mathbf{b}$ onto $\mathbf{a}$ is

$$\text{comp}_{\mathbf{a}}\,\mathbf{b} = \frac{\mathbf{a} \cdot \mathbf{b}}{|\mathbf{a}|} = \frac{(-2)(1) + 3(1) + 1(2)}{\sqrt{14}} = \frac{3}{\sqrt{14}}$$

The vector projection is this scalar projection times the unit vector in the direction of $\mathbf{a}$:

$$\text{proj}_{\mathbf{a}}\,\mathbf{b} = \frac{3}{\sqrt{14}}\frac{\mathbf{a}}{|\mathbf{a}|} = \frac{3}{14}\mathbf{a} = \left\langle -\frac{3}{7}, \frac{9}{14}, \frac{3}{14} \right\rangle$$ ∎

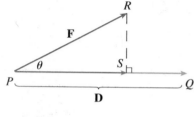

FIGURE 5

One use of projections occurs in physics in calculating work. In Section 5.4 we defined the work done by a constant force F in moving an object through a distance d as $W = Fd$, but this applies only when the force is directed along the line of motion of the object. Suppose, however, that the constant force is a vector $\mathbf{F} = \overrightarrow{PR}$ pointing in some other direction as in Figure 5. If the force moves the object from P to Q, then the **displacement vector** is $\mathbf{D} = \overrightarrow{PQ}$. The work done by this force is defined to be the product of the component of the force along $\mathbf{D}$ and the distance moved:

$$W = (|\mathbf{F}|\cos\theta)|\mathbf{D}|$$

But then, from Theorem 3, we have

(12)
$$W = |\mathbf{F}||\mathbf{D}|\cos\theta = \mathbf{F} \cdot \mathbf{D}$$

Thus the work done by a constant force $\mathbf{F}$ is the dot product $\mathbf{F} \cdot \mathbf{D}$, where $\mathbf{D}$ is the displacement vector.

EXAMPLE 6   A force is given by a vector $\mathbf{F} = 3\mathbf{i} + 4\mathbf{j} + 5\mathbf{k}$ and moves a particle from the point $P(2, 1, 0)$ to the point $Q(4, 6, 2)$. Find the work done.

SOLUTION   The displacement vector is $\mathbf{D} = \overrightarrow{PQ} = \langle 2, 5, 2 \rangle$, so by Equation 12, the work done is

$$W = \mathbf{F} \cdot \mathbf{D} = \langle 3, 4, 5 \rangle \cdot \langle 2, 5, 2 \rangle$$
$$= 6 + 20 + 10 = 36$$

If the unit of length is meters and the magnitude of the force is measured in newtons, then the work done is 36 joules. ∎

EXERCISES 11.3

1–8 ■ Find $\mathbf{a} \cdot \mathbf{b}$.

1. $\mathbf{a} = \langle 2, 5 \rangle$, $\mathbf{b} = \langle -3, 1 \rangle$

2. $\mathbf{a} = \langle -2, -8 \rangle$, $\mathbf{b} = \langle 6, -4 \rangle$

3. $\mathbf{a} = \langle 4, 7, -1 \rangle$, $\mathbf{b} = \langle -2, 1, 4 \rangle$

4. $\mathbf{a} = \langle -1, -2, -3 \rangle$, $\mathbf{b} = \langle 2, 8, -6 \rangle$

5. $\mathbf{a} = 2\mathbf{i} + 3\mathbf{j} - 4\mathbf{k}$, $\mathbf{b} = \mathbf{i} - 3\mathbf{j} + \mathbf{k}$

6. $\mathbf{a} = \mathbf{i} - \mathbf{k}$, $\mathbf{b} = \mathbf{i} + 2\mathbf{j}$

7. $|\mathbf{a}| = 2$, $|\mathbf{b}| = 3$,   the angle between $\mathbf{a}$ and $\mathbf{b}$ is $\pi/3$

8. $|\mathbf{a}| = 6$, $|\mathbf{b}| = \frac{1}{3}$,   the angle between $\mathbf{a}$ and $\mathbf{b}$ is $\pi/4$

9. If $\mathbf{a} = \langle a_1, a_2, a_3 \rangle$, show that $\mathbf{a} \cdot \mathbf{i} = a_1$, $\mathbf{a} \cdot \mathbf{j} = a_2$, and $\mathbf{a} \cdot \mathbf{k} = a_3$.

10. (a) Show that $\mathbf{i} \cdot \mathbf{j} = \mathbf{j} \cdot \mathbf{k} = \mathbf{k} \cdot \mathbf{i} = 0$.
(b) Show that $\mathbf{i} \cdot \mathbf{i} = \mathbf{j} \cdot \mathbf{j} = \mathbf{k} \cdot \mathbf{k} = 1$.

11–16 ■ Find the angle between the vectors. (First find an exact expression and then approximate to the nearest degree.)

11. $\mathbf{a} = \langle 1, 2, 2 \rangle$, $\mathbf{b} = \langle 3, 4, 0 \rangle$

12. $\mathbf{a} = \langle 6, 0, 2 \rangle$, $\mathbf{b} = \langle 5, 3, -2 \rangle$

13. $\mathbf{a} = \langle 1, 2 \rangle$, $\mathbf{b} = \langle 12, -5 \rangle$

14. $\mathbf{a} = \langle 3, 1 \rangle$, $\mathbf{b} = \langle 2, 4 \rangle$

15. $\mathbf{a} = 6\mathbf{i} - 2\mathbf{j} - 3\mathbf{k}$, $\mathbf{b} = \mathbf{i} + \mathbf{j} + \mathbf{k}$

16. $\mathbf{a} = \mathbf{i} + \mathbf{j} + 2\mathbf{k}$, $\mathbf{b} = 2\mathbf{j} - 3\mathbf{k}$

17–18 ■ Find, correct to the nearest degree, the three angles of the triangle with the given vertices.

17. $A(1, 2, 3)$, $B(6, 1, 5)$, $C(-1, -2, 0)$

18. $P(0, -1, 6)$, $Q(2, 1, -3)$, $R(5, 4, 2)$

19–24 ■ Determine whether the given vectors are orthogonal, parallel, or neither.

19. $\mathbf{a} = \langle 2, -4 \rangle$, $\mathbf{b} = \langle -1, 2 \rangle$

20. $\mathbf{a} = \langle 2, -4 \rangle$, $\mathbf{b} = \langle 4, 2 \rangle$

21. $\mathbf{a} = \langle 2, 8, -3 \rangle$, $\mathbf{b} = \langle -1, 2, 5 \rangle$

22. $\mathbf{a} = \langle -1, 5, 2 \rangle$, $\mathbf{b} = \langle 4, 2, -3 \rangle$

23. $\mathbf{a} = 3\mathbf{i} + \mathbf{j} - \mathbf{k}$, $\mathbf{b} = \mathbf{i} - \mathbf{j} + 2\mathbf{k}$

24. $\mathbf{a} = 2\mathbf{i} + 6\mathbf{j} - 4\mathbf{k}$, $\mathbf{b} = -3\mathbf{i} - 9\mathbf{j} + 6\mathbf{k}$

25–28 ■ Find the values of x such that the given vectors are orthogonal.

25. $x\mathbf{i} - 2\mathbf{j}$, $x\mathbf{i} + 8\mathbf{j}$

26. $x\mathbf{i} + 2x\mathbf{j}$, $x\mathbf{i} - 2\mathbf{j}$

27. $\langle x, 1, 2 \rangle$, $\langle 3, 4, x \rangle$

28. $\langle x, x, -1 \rangle$, $\langle 1, x, 6 \rangle$

29. Find a unit vector that is orthogonal to both $\mathbf{i} + \mathbf{j}$ and $\mathbf{i} + \mathbf{k}$.

30. For what values of c is the angle between the vectors $\langle 1, 2, 1 \rangle$ and $\langle 1, 0, c \rangle$ equal to $60°$?

31–35 ■ Find the direction cosines and direction angles of the vector. (Give the direction angles correct to the nearest degree.)

31. $\langle 1, 2, 2 \rangle$

32. $\langle -4, -1, 2 \rangle$

33. $-8\mathbf{i} + 3\mathbf{j} + 2\mathbf{k}$

34. $3\mathbf{i} + 5\mathbf{j} - 4\mathbf{k}$

35. $\langle 2, 1.2, 0.8 \rangle$

36. If a vector has direction angles $\alpha = \pi/4$ and $\beta = \pi/3$, find the third direction angle γ.

37–42 ■ Find the scalar and vector projections of $\mathbf{b}$ onto $\mathbf{a}$.

37. $\mathbf{a} = \langle 2, 3 \rangle$, $\mathbf{b} = \langle 4, 1 \rangle$

38. $\mathbf{a} = \langle 3, -1 \rangle$, $\mathbf{b} = \langle 2, 3 \rangle$

39. $\mathbf{a} = \langle 4, 2, 0 \rangle$, $\mathbf{b} = \langle 1, 1, 1 \rangle$

40. $\mathbf{a} = \langle -1, -2, 2 \rangle$, $\mathbf{b} = \langle 3, 3, 4 \rangle$

41. $\mathbf{a} = \mathbf{i} + \mathbf{k}$, $\mathbf{b} = \mathbf{i} - \mathbf{j}$

42. $\mathbf{a} = 2\mathbf{i} - 3\mathbf{j} + \mathbf{k}$, $\mathbf{b} = \mathbf{i} + 6\mathbf{j} - 2\mathbf{k}$

43. Show that the vector

$$\text{orth}_a \mathbf{b} = \mathbf{b} - \text{proj}_a \mathbf{b}$$

is orthogonal to $\mathbf{a}$. (It is called an **orthogonal projection** of $\mathbf{b}$.)

44. For the vectors in Exercise 38, find $\text{orth}_a \mathbf{b}$ and illustrate by drawing the vectors $\mathbf{a}$, $\mathbf{b}$, $\text{proj}_a \mathbf{b}$, and $\text{orth}_a \mathbf{b}$.

45. If $\mathbf{a} = \langle 3, 0, -1 \rangle$, find a vector $\mathbf{b}$ such that $\text{comp}_a \mathbf{b} = 2$.

46. Suppose that $\mathbf{a}$ and $\mathbf{b}$ are nonzero vectors.
(a) Under what circumstances is $\text{comp}_a \mathbf{b} = \text{comp}_b \mathbf{a}$?
(b) Under what circumstances is $\text{proj}_a \mathbf{b} = \text{proj}_b \mathbf{a}$?

47. A constant force with vector representation $\mathbf{F} = 10\mathbf{i} + 18\mathbf{j} - 6\mathbf{k}$ moves an object along a straight line from the point $(2, 3, 0)$ to the point $(4, 9, 15)$. Find the work done if the distance is measured in meters and the magnitude of the force is measured in newtons.

48. Find the work done by a force of 20 lb acting in the direction N50°W in moving an object 4 ft due west.

49. A woman exerts a horizontal force of 25 lb on a crate as she pushes it up a ramp that is 10 ft long and inclined at an angle of 20° above the horizontal. Find the work done on the box.

50. A wagon is pulled a distance of 100 m along a horizontal path by a constant force of 50 N. The handle of the wagon is at an angle of 30° above the horizontal. How much work is done?

51. Which of the following expressions have no meaning?
(a) $(\mathbf{a} \cdot \mathbf{b}) \cdot \mathbf{c}$
(b) $(\mathbf{a} \cdot \mathbf{b})\mathbf{c}$
(c) $|\mathbf{a}|(\mathbf{b} \cdot \mathbf{c})$
(d) $\mathbf{a} \cdot (\mathbf{b} + \mathbf{c})$
(e) $\mathbf{a} \cdot \mathbf{b} + \mathbf{c}$
(f) $|\mathbf{a}| \cdot (\mathbf{b} + \mathbf{c})$

52. Suppose that all sides of a quadrilateral are equal in length and opposite sides are parallel. Use vector methods to show that the diagonals are perpendicular.

53. Use a scalar projection to show that the distance from a point $P_1(x_1, y_1)$ to the line with equation $ax + by + c = 0$ is

$$\frac{|ax_1 + by_1 + c|}{\sqrt{a^2 + b^2}}$$

Use this formula to find the distance from the point $(-2, 3)$ to the line $3x - 4y + 5 = 0$.

54. If $\mathbf{r} = \langle x, y, z \rangle$, $\mathbf{a} = \langle a_1, a_2, a_3 \rangle$, and $\mathbf{b} = \langle b_1, b_2, b_3 \rangle$, show that the vector equation $(\mathbf{r} - \mathbf{a}) \cdot (\mathbf{r} - \mathbf{b}) = 0$ represents a sphere, and find its center and radius.

55. Find the angle between a diagonal of a cube and one of its edges.

56. Find the angle between a diagonal of a cube and a diagonal of one of its faces.

57. A molecule of methane, CH_4, is structured with the four hydrogen atoms at the vertices of a regular tetrahedron and the carbon atom at the centroid. The *bond angle* is the angle formed by the H—C—H combination; it is the angle between the lines that join the carbon atom to two of the hydrogen atoms. Show that the bond angle is about $109.5°$. [*Hint:* Take the vertices of the tetrahedron to be the points $(1, 0, 0)$, $(0, 1, 0)$, $(0, 0, 1)$, and $(1, 1, 1)$ as shown in the figure. Then the centroid is $\left(\frac{1}{2}, \frac{1}{2}, \frac{1}{2}\right)$.]

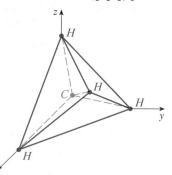

58. If $\mathbf{c} = |\mathbf{a}|\mathbf{b} + |\mathbf{b}|\mathbf{a}$, where $\mathbf{a}$, $\mathbf{b}$, and $\mathbf{c}$ are all nonzero vectors, show that $\mathbf{c}$ bisects the angle between $\mathbf{a}$ and $\mathbf{b}$.

59. Prove Properties 2 and 5 of the dot product (Theorem 2).

60. Prove Property 4 of the dot product (Theorem 2).

61. Use Theorem 3 to prove the Cauchy-Schwarz Inequality:

$$|\mathbf{a} \cdot \mathbf{b}| \le |\mathbf{a}||\mathbf{b}|$$

62. The Triangle Inequality for vectors is

$$|\mathbf{a} + \mathbf{b}| \le |\mathbf{a}| + |\mathbf{b}|$$

(a) Give a geometric interpretation of the Triangle Inequality.
(b) Use the Cauchy-Schwarz Inequality from Exercise 61 to prove the Triangle Inequality. [*Hint:* Use the fact that $|\mathbf{a} + \mathbf{b}|^2 = (\mathbf{a} + \mathbf{b}) \cdot (\mathbf{a} + \mathbf{b})$ and use Property 3 of the dot product.]

63. The Parallelogram Law states that

$$|\mathbf{a} + \mathbf{b}|^2 + |\mathbf{a} - \mathbf{b}|^2 = 2|\mathbf{a}|^2 + 2|\mathbf{b}|^2$$

(a) Give a geometric interpretation of the Parallelogram Law.
(b) Prove the Parallelogram Law. (See the hint in Exercise 62.)

11.4   THE CROSS PRODUCT

The **cross product** $\mathbf{a} \times \mathbf{b}$ of two vectors $\mathbf{a}$ and $\mathbf{b}$, unlike the dot product, is a vector. For this reason it is also called the **vector product**. Note that $\mathbf{a} \times \mathbf{b}$ is defined only when $\mathbf{a}$ and $\mathbf{b}$ are *three-dimensional* vectors.

> **(1) DEFINITION**   If $\mathbf{a} = \langle a_1, a_2, a_3 \rangle$ and $\mathbf{b} = \langle b_1, b_2, b_3 \rangle$, then the **cross product** of $\mathbf{a}$ and $\mathbf{b}$ is the vector
>
> $$\mathbf{a} \times \mathbf{b} = \langle a_2 b_3 - a_3 b_2, a_3 b_1 - a_1 b_3, a_1 b_2 - a_2 b_1 \rangle$$

This may seem like a strange way of defining a product. The reason for the particular form of Definition 1 is that the cross product defined in this way has many useful properties, as we will soon see. In particular, we will show that the vector $\mathbf{a} \times \mathbf{b}$ is perpendicular to both $\mathbf{a}$ and $\mathbf{b}$.

In order to make Definition 1 easier to remember, we use the notation of determinants. A **determinant of order 2** is defined by

$$\begin{vmatrix} a & b \\ c & d \end{vmatrix} = ad - bc$$

For example,

$$\begin{vmatrix} 2 & 1 \\ -6 & 4 \end{vmatrix} = 2(4) - 1(-6) = 14$$

A **determinant of order 3** can be defined in terms of second-order determinants as follows:

(2)
$$\begin{vmatrix} a_1 & a_2 & a_3 \\ b_1 & b_2 & b_3 \\ c_1 & c_2 & c_3 \end{vmatrix} = a_1 \begin{vmatrix} b_2 & b_3 \\ c_2 & c_3 \end{vmatrix} - a_2 \begin{vmatrix} b_1 & b_3 \\ c_1 & c_3 \end{vmatrix} + a_3 \begin{vmatrix} b_1 & b_2 \\ c_1 & c_2 \end{vmatrix}$$

Observe that each term on the right side of Equation 2 involves a number a_i in the first row of the determinant, and a_i is multiplied by the second-order determinant obtained from the left side by deleting the row and column in which a_i appears. Notice also the minus sign in the second term. For example,

$$\begin{vmatrix} 1 & 2 & -1 \\ 3 & 0 & 1 \\ -5 & 4 & 2 \end{vmatrix} = 1 \begin{vmatrix} 0 & 1 \\ 4 & 2 \end{vmatrix} - 2 \begin{vmatrix} 3 & 1 \\ -5 & 2 \end{vmatrix} + (-1) \begin{vmatrix} 3 & 0 \\ -5 & 4 \end{vmatrix}$$

$$= 1(0 - 4) - 2(6 + 5) + (-1)(12 - 0) = -38$$

If we now rewrite Definition 1 using second-order determinants and the standard basis vectors $\mathbf{i}$, $\mathbf{j}$, and $\mathbf{k}$, we see that the cross product of $\mathbf{a} = a_1\mathbf{i} + a_2\mathbf{j} + a_3\mathbf{k}$ and $\mathbf{b} = b_1\mathbf{i} + b_2\mathbf{j} + b_3\mathbf{k}$ is

(3)
$$\mathbf{a} \times \mathbf{b} = \begin{vmatrix} a_2 & a_3 \\ b_2 & b_3 \end{vmatrix} \mathbf{i} - \begin{vmatrix} a_1 & a_3 \\ b_1 & b_3 \end{vmatrix} \mathbf{j} + \begin{vmatrix} a_1 & a_2 \\ b_1 & b_2 \end{vmatrix} \mathbf{k}$$

In view of the similarity between Equations 2 and 3, we often write

(4)
$$\mathbf{a} \times \mathbf{b} = \begin{vmatrix} \mathbf{i} & \mathbf{j} & \mathbf{k} \\ a_1 & a_2 & a_3 \\ b_1 & b_2 & b_3 \end{vmatrix}$$

Although the first row of the symbolic determinant in Equation 4 consists of vectors, if we expand it as if it were an ordinary determinant using the rule in Equation 2, we obtain Equation 3. The symbolic formula in Equation 4 is probably the easiest way of remembering and computing cross products.

EXAMPLE 1   If $\mathbf{a} = \langle 1, 3, 4 \rangle$ and $\mathbf{b} = \langle 2, 7, -5 \rangle$, then

$$\mathbf{a} \times \mathbf{b} = \begin{vmatrix} \mathbf{i} & \mathbf{j} & \mathbf{k} \\ 1 & 3 & 4 \\ 2 & 7 & -5 \end{vmatrix}$$

$$= \begin{vmatrix} 3 & 4 \\ 7 & -5 \end{vmatrix} \mathbf{i} - \begin{vmatrix} 1 & 4 \\ 2 & -5 \end{vmatrix} \mathbf{j} + \begin{vmatrix} 1 & 3 \\ 2 & 7 \end{vmatrix} \mathbf{k}$$

$$= (-15 - 28)\mathbf{i} - (-5 - 8)\mathbf{j} + (7 - 6)\mathbf{k} = -43\mathbf{i} + 13\mathbf{j} + \mathbf{k} \qquad \blacksquare$$

EXAMPLE 2   Show that $\mathbf{a} \times \mathbf{a} = \mathbf{0}$ for any vector $\mathbf{a}$ in V_3.

SOLUTION   If $\mathbf{a} = \langle a_1, a_2, a_3 \rangle$, then

$$\mathbf{a} \times \mathbf{a} = \begin{vmatrix} \mathbf{i} & \mathbf{j} & \mathbf{k} \\ a_1 & a_2 & a_3 \\ a_1 & a_2 & a_3 \end{vmatrix}$$

$$= (a_2 a_3 - a_3 a_2)\mathbf{i} - (a_1 a_3 - a_3 a_1)\mathbf{j} + (a_1 a_2 - a_2 a_1)\mathbf{k}$$

$$= 0\mathbf{i} - 0\mathbf{j} + 0\mathbf{k} = \mathbf{0} \qquad \blacksquare$$

One of the most important properties of the cross product is given by the following theorem.

---

(5) THEOREM   The vector $\mathbf{a} \times \mathbf{b}$ is orthogonal to both $\mathbf{a}$ and $\mathbf{b}$.

PROOF   In order to show that $\mathbf{a} \times \mathbf{b}$ is orthogonal to $\mathbf{a}$, we compute their dot product as follows:

$$(\mathbf{a} \times \mathbf{b}) \cdot \mathbf{a} = \begin{vmatrix} a_2 & a_3 \\ b_2 & b_3 \end{vmatrix} a_1 - \begin{vmatrix} a_1 & a_3 \\ b_1 & b_3 \end{vmatrix} a_2 + \begin{vmatrix} a_1 & a_2 \\ b_1 & b_2 \end{vmatrix} a_3$$

$$= a_1(a_2 b_3 - a_3 b_2) - a_2(a_1 b_3 - a_3 b_1) + a_3(a_1 b_2 - a_2 b_1)$$

$$= a_1 a_2 b_3 - a_1 b_2 a_3 - a_1 a_2 b_3 + b_1 a_2 a_3 + a_1 b_2 a_3 - b_1 a_2 a_3$$

$$= 0$$

A similar computation shows that $(\mathbf{a} \times \mathbf{b}) \cdot \mathbf{b} = 0$. Therefore, $\mathbf{a} \times \mathbf{b}$ is orthogonal to both $\mathbf{a}$ and $\mathbf{b}$.   □

If $\mathbf{a}$ and $\mathbf{b}$ are represented by directed line segments with the same initial point (as in Figure 1), then Theorem 5 says that the cross product $\mathbf{a} \times \mathbf{b}$ points in a direction perpendicular to the plane through $\mathbf{a}$ and $\mathbf{b}$. It turns out that the direction of $\mathbf{a} \times \mathbf{b}$ is given by the *right-hand rule:* If the fingers of your right hand curl in the direction of a rotation (through an angle less than 180°) from $\mathbf{a}$ to $\mathbf{b}$, then your thumb points in the direction of $\mathbf{a} \times \mathbf{b}$.

Now that we know the direction of the vector $\mathbf{a} \times \mathbf{b}$, the remaining thing we need to complete its geometric description is its length $|\mathbf{a} \times \mathbf{b}|$. This is given by the following theorem.

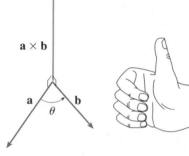

FIGURE 1

---

(6) THEOREM   If θ is the angle between $\mathbf{a}$ and $\mathbf{b}$ (so $0 \leq \theta \leq \pi$), then

$$|\mathbf{a} \times \mathbf{b}| = |\mathbf{a}||\mathbf{b}|\sin\theta$$

PROOF   From the definitions of the cross product and length of a vector, we have

$$|\mathbf{a} \times \mathbf{b}|^2 = (a_2 b_3 - a_3 b_2)^2 + (a_3 b_1 - a_1 b_3)^2 + (a_1 b_2 - a_2 b_1)^2$$

$$= a_2^2 b_3^2 - 2a_2 a_3 b_2 b_3 + a_3^2 b_2^2 + a_3^2 b_1^2 - 2a_1 a_3 b_1 b_3 + a_1^2 b_3^2$$

$$+ a_1^2 b_2^2 - 2a_1 a_2 b_1 b_2 + a_2^2 b_1^2$$

$$= (a_1^2 + a_2^2 + a_3^2)(b_1^2 + b_2^2 + b_3^2) - (a_1 b_1 + a_2 b_2 + a_3 b_3)^2$$

$$= |\mathbf{a}|^2 |\mathbf{b}|^2 - (\mathbf{a} \cdot \mathbf{b})^2$$

$$= |\mathbf{a}|^2 |\mathbf{b}|^2 - |\mathbf{a}|^2 |\mathbf{b}|^2 \cos^2\theta \quad \text{(by Theorem 11.3.3)}$$

$$= |\mathbf{a}|^2 |\mathbf{b}|^2 (1 - \cos^2\theta)$$

$$= |\mathbf{a}|^2 |\mathbf{b}|^2 \sin^2\theta$$

Taking square roots and observing that $\sqrt{\sin^2\theta} = \sin\theta$ because $\sin\theta \geq 0$ when $0 \leq \theta \leq \pi$, we have

$$|\mathbf{a} \times \mathbf{b}| = |\mathbf{a}||\mathbf{b}|\sin\theta$$ □

Geometric characterization of $\mathbf{a} \times \mathbf{b}$

Since a vector is completely determined by its magnitude and direction, we can now say that $\mathbf{a} \times \mathbf{b}$ is the vector that is perpendicular to both $\mathbf{a}$ and $\mathbf{b}$, whose orientation is determined by the right-hand rule, and whose length is $|\mathbf{a}||\mathbf{b}|\sin\theta$. In fact, that is exactly how physicists *define* $\mathbf{a} \times \mathbf{b}$.

---

(7) COROLLARY   Two nonzero vectors $\mathbf{a}$ and $\mathbf{b}$ are parallel if and only if $\mathbf{a} \times \mathbf{b} = \mathbf{0}$.

PROOF   Two nonzero vectors $\mathbf{a}$ and $\mathbf{b}$ are parallel if and only if $\theta = 0$ or π. In either case $\sin\theta = 0$. $\square$

The geometric interpretation of Theorem 6 can be seen by looking at Figure 2. If $\mathbf{a}$ and $\mathbf{b}$ are represented by directed line segments with the same initial point, then they determine a parallelogram with base $|\mathbf{a}|$, altitude $|\mathbf{b}|\sin\theta$, and area

$$A = |\mathbf{a}|(|\mathbf{b}|\sin\theta) = |\mathbf{a} \times \mathbf{b}|$$

Thus **the length of the cross product $\mathbf{a} \times \mathbf{b}$ is equal to the area of the parallelogram determined by $\mathbf{a}$ and $\mathbf{b}$.**

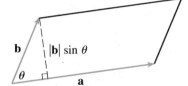

FIGURE 2

EXAMPLE 3   Find the area of the triangle with vertices $P(1, 4, 6)$, $Q(-2, 5, -1)$, and $R(1, -1, 1)$.

SOLUTION   By (11.2.2) the vectors that correspond to the directed line segments $\overrightarrow{PQ}$ and $\overrightarrow{PR}$ are $\mathbf{a} = \langle -3, 1, -7 \rangle$ and $\mathbf{b} = \langle 0, -5, -5 \rangle$. We compute the cross product of these vectors:

$$\mathbf{a} \times \mathbf{b} = \begin{vmatrix} \mathbf{i} & \mathbf{j} & \mathbf{k} \\ -3 & 1 & -7 \\ 0 & -5 & -5 \end{vmatrix}$$

$$= (-5 - 35)\mathbf{i} - (15 - 0)\mathbf{j} + (15 - 0)\mathbf{k} = -40\mathbf{i} - 15\mathbf{j} + 15\mathbf{k}$$

The area A of triangle PQR is half the area of the parallelogram with adjacent sides PQ and PR. Thus

$$A = \tfrac{1}{2}|\mathbf{a} \times \mathbf{b}| = \tfrac{1}{2}\sqrt{(-40)^2 + (-15)^2 + 15^2} = \frac{5\sqrt{82}}{2}$$ ∎

If we apply Theorems 5 and 6 to the standard basis vectors $\mathbf{i}$, $\mathbf{j}$, and $\mathbf{k}$ using $\theta = \pi/2$, we obtain

$$\mathbf{i} \times \mathbf{j} = \mathbf{k} \qquad \mathbf{j} \times \mathbf{k} = \mathbf{i} \qquad \mathbf{k} \times \mathbf{i} = \mathbf{j}$$

$$\mathbf{j} \times \mathbf{i} = -\mathbf{k} \qquad \mathbf{k} \times \mathbf{j} = -\mathbf{i} \qquad \mathbf{i} \times \mathbf{k} = -\mathbf{j}$$

Observe that

$$\mathbf{i} \times \mathbf{j} \neq \mathbf{j} \times \mathbf{i}$$

Thus the cross product is not commutative. Also

$$\mathbf{i} \times (\mathbf{i} \times \mathbf{j}) = \mathbf{i} \times \mathbf{k} = -\mathbf{j}$$

whereas

$$(\mathbf{i} \times \mathbf{i}) \times \mathbf{j} = \mathbf{0} \times \mathbf{j} = \mathbf{0}$$

So the associative law for multiplication does not usually hold; that is, in general,

$$(\mathbf{a} \times \mathbf{b}) \times \mathbf{c} \neq \mathbf{a} \times (\mathbf{b} \times \mathbf{c})$$

However, some of the usual laws of algebra do hold for cross products. The following theorem summarizes the properties of vector products.

---

(8) THEOREM   If $\mathbf{a}$, $\mathbf{b}$, and $\mathbf{c}$ are vectors and c is a scalar, then

1. $\mathbf{a} \times \mathbf{b} = -\mathbf{b} \times \mathbf{a}$
2. $(c\mathbf{a}) \times \mathbf{b} = c(\mathbf{a} \times \mathbf{b}) = \mathbf{a} \times (c\mathbf{b})$
3. $\mathbf{a} \times (\mathbf{b} + \mathbf{c}) = \mathbf{a} \times \mathbf{b} + \mathbf{a} \times \mathbf{c}$
4. $(\mathbf{a} + \mathbf{b}) \times \mathbf{c} = \mathbf{a} \times \mathbf{c} + \mathbf{b} \times \mathbf{c}$
5. $\mathbf{a} \cdot (\mathbf{b} \times \mathbf{c}) = (\mathbf{a} \times \mathbf{b}) \cdot \mathbf{c}$
6. $\mathbf{a} \times (\mathbf{b} \times \mathbf{c}) = (\mathbf{a} \cdot \mathbf{c})\mathbf{b} - (\mathbf{a} \cdot \mathbf{b})\mathbf{c}$

These properties can be proved by writing the vectors in terms of their components and using the definition of a cross product. We give the proof of Property 5 and leave the remaining proofs as exercises.

PROOF OF PROPERTY 5   If $\mathbf{a} = \langle a_1, a_2, a_3 \rangle$, $\mathbf{b} = \langle b_1, b_2, b_3 \rangle$, and $\mathbf{c} = \langle c_1, c_2, c_3 \rangle$, then

$$
\begin{aligned}
(9) \quad \mathbf{a} \cdot (\mathbf{b} \times \mathbf{c}) &= a_1(b_2c_3 - b_3c_2) + a_2(b_3c_1 - b_1c_3) + a_3(b_1c_2 - b_2c_1) \\
&= a_1b_2c_3 - a_1b_3c_2 + a_2b_3c_1 - a_2b_1c_3 + a_3b_1c_2 - a_3b_2c_1 \\
&= (a_2b_3 - a_3b_2)c_1 + (a_3b_1 - a_1b_3)c_2 + (a_1b_2 - a_2b_1)c_3 \\
&= (\mathbf{a} \times \mathbf{b}) \cdot \mathbf{c} \qquad \qquad \square
\end{aligned}
$$

The product $\mathbf{a} \cdot (\mathbf{b} \times \mathbf{c})$ that occurs in Property 5 is called the **scalar triple product** of the vectors $\mathbf{a}$, $\mathbf{b}$, and $\mathbf{c}$. Notice from Equation 9 that we can write the scalar triple product as a determinant:

$$
(10) \qquad \mathbf{a} \cdot (\mathbf{b} \times \mathbf{c}) = \begin{vmatrix} a_1 & a_2 & a_3 \\ b_1 & b_2 & b_3 \\ c_1 & c_2 & c_3 \end{vmatrix}
$$

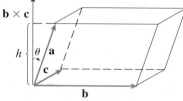

The geometric significance of the scalar triple product can be seen by considering the parallelepiped determined by the vectors $\mathbf{a}$, $\mathbf{b}$, and $\mathbf{c}$ (Figure 3). The area of the base parallelogram is $A = |\mathbf{b} \times \mathbf{c}|$. If θ is the angle between $\mathbf{a}$ and $\mathbf{b} \times \mathbf{c}$, then the height h of the parallelepiped is $h = |\mathbf{a}||\cos\theta|$. (We must use $|\cos\theta|$ instead of $\cos\theta$ in case $\theta > \pi/2$.) Thus the volume of the parallelepiped is

$$V = Ah = |\mathbf{b} \times \mathbf{c}||\mathbf{a}||\cos\theta| = |\mathbf{a} \cdot (\mathbf{b} \times \mathbf{c})|$$

FIGURE 3

Thus we have proved the following:

---

(11)   The volume of the parallelepiped determined by the vectors $\mathbf{a}$, $\mathbf{b}$, and $\mathbf{c}$ is the magnitude of their scalar triple product:

$$V = |\mathbf{a} \cdot (\mathbf{b} \times \mathbf{c})|$$

EXAMPLE 4 Use the scalar triple product to show that the vectors $\mathbf{a} = \langle 1, 4, -7 \rangle$, $\mathbf{b} = \langle 2, -1, 4 \rangle$, and $\mathbf{c} = \langle 0, -9, 18 \rangle$ are coplanar; that is, they lie in the same plane.

SOLUTION We use Equation 10 to compute their scalar triple product:

$$\mathbf{a} \cdot (\mathbf{b} \times \mathbf{c}) = \begin{vmatrix} 1 & 4 & -7 \\ 2 & -1 & 4 \\ 0 & -9 & 18 \end{vmatrix}$$

$$= 1 \begin{vmatrix} -1 & 4 \\ -9 & 18 \end{vmatrix} - 4 \begin{vmatrix} 2 & 4 \\ 0 & 18 \end{vmatrix} - 7 \begin{vmatrix} 2 & -1 \\ 0 & -9 \end{vmatrix}$$

$$= 1(18) - 4(36) - 7(-18) = 0$$

Therefore, by (11) the volume of the parallelepiped determined by $\mathbf{a}$, $\mathbf{b}$, and $\mathbf{c}$ is 0. This means that $\mathbf{a}$, $\mathbf{b}$, and $\mathbf{c}$ are coplanar. ∎

The idea of a cross product occurs often in physics. In particular, we consider a force $\mathbf{F}$ acting on a rigid body at a point given by a position vector $\mathbf{r}$ (see Figure 4). The **torque** $\boldsymbol{\tau}$ (relative to the origin) is defined to be the cross product of the position and force vectors

$$\boldsymbol{\tau} = \mathbf{r} \times \mathbf{F}$$

and measures the tendency of the body to rotate about the origin. The direction of the torque vector indicates the axis of rotation. According to Theorem 6, the magnitude of the torque vector is

$$|\boldsymbol{\tau}| = |\mathbf{r} \times \mathbf{F}| = |\mathbf{r}||\mathbf{F}| \sin \theta$$

where θ is the angle between the position and force vectors. Observe that the only component of $\mathbf{F}$ that can cause a rotation is the one perpendicular to $\mathbf{r}$, that is, $|\mathbf{F}| \sin \theta$. The magnitude of the torque is equal to the area of the parallelogram determined by $\mathbf{r}$ and $\mathbf{F}$.

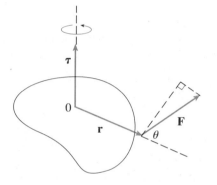

FIGURE 4

EXAMPLE 5 A bolt is tightened by applying a 40-N force to a 0.25-m wrench as shown in Figure 5. Find the magnitude of the torque about the center of the bolt.

SOLUTION The magnitude of the torque vector is

$$|\boldsymbol{\tau}| = |\mathbf{r} \times \mathbf{F}| = |\mathbf{r}||\mathbf{F}| \sin 75° = (0.25)(40) \sin 75°$$

$$= 10 \sin 75° \approx 9.66 \text{ N-m} = 9.66 \text{ J}$$

If the bolt is right-threaded, then the torque vector itself is

$$\boldsymbol{\tau} = |\boldsymbol{\tau}| \mathbf{n} \approx 9.66 \mathbf{n}$$

where $\mathbf{n}$ is a unit vector directed down into the page. ∎

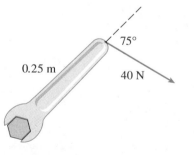

FIGURE 5

EXERCISES 11.4

1–7 ■ Find the cross product $\mathbf{a} \times \mathbf{b}$.

1. $\mathbf{a} = \langle 1, 0, 1 \rangle$, $\mathbf{b} = \langle 0, 1, 0 \rangle$

2. $\mathbf{a} = \langle 2, 4, 0 \rangle$, $\mathbf{b} = \langle -3, 1, 6 \rangle$

3. $\mathbf{a} = \langle -2, 3, 4 \rangle$, $\mathbf{b} = \langle 3, 0, 1 \rangle$

4. $\mathbf{a} = \langle 1, 2, -3 \rangle$, $\quad \mathbf{b} = \langle 5, -1, -2 \rangle$

5. $\mathbf{a} = \mathbf{i} + \mathbf{j} + \mathbf{k}$, $\quad \mathbf{b} = \mathbf{i} + \mathbf{j} - \mathbf{k}$

6. $\mathbf{a} = \mathbf{i} + 2\mathbf{j} - \mathbf{k}$, $\quad \mathbf{b} = 3\mathbf{i} - \mathbf{j} + 7\mathbf{k}$

7. $\mathbf{a} = 2\mathbf{i} - \mathbf{k}$, $\quad \mathbf{b} = \mathbf{i} + 2\mathbf{j}$

8. The figure shows a vector $\mathbf{a}$ in the xy-plane and a vector $\mathbf{b}$ in the direction of $\mathbf{k}$. Their lengths are $|\mathbf{a}| = 3$ and $|\mathbf{b}| = 2$.
(a) Find $|\mathbf{a} \times \mathbf{b}|$.
(b) Use the right-hand rule to decide whether the components of $\mathbf{a} \times \mathbf{b}$ are positive, negative, or 0.

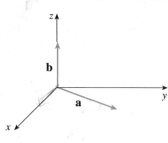

9. If $\mathbf{a} = \langle 0, 1, 2 \rangle$ and $\mathbf{b} = \langle 3, 1, 0 \rangle$, find $\mathbf{a} \times \mathbf{b}$ and $\mathbf{b} \times \mathbf{a}$.

10. If $\mathbf{a} = \langle -4, 0, 3 \rangle$, $\mathbf{b} = \langle 2, -1, 0 \rangle$, and $\mathbf{c} = \langle 0, 2, 5 \rangle$, show that $\mathbf{a} \times (\mathbf{b} \times \mathbf{c}) \neq (\mathbf{a} \times \mathbf{b}) \times \mathbf{c}$.

11. Find two unit vectors orthogonal to both $\langle 1, -1, 1 \rangle$ and $\langle 0, 4, 4 \rangle$.

12. Find two unit vectors orthogonal to both $\mathbf{i} + \mathbf{j}$ and $\mathbf{i} - \mathbf{j} + \mathbf{k}$.

13. Show that $\mathbf{0} \times \mathbf{a} = \mathbf{0} = \mathbf{a} \times \mathbf{0}$ for any vector $\mathbf{a}$ in V_3.

14. Show that $(\mathbf{a} \times \mathbf{b}) \cdot \mathbf{b} = 0$ for all vectors $\mathbf{a}$ and $\mathbf{b}$ in V_3.

15. Prove Property 1 of Theorem 8.

16. Prove Property 2 of Theorem 8.

17. Prove Property 3 of Theorem 8.

18. Prove Property 4 of Theorem 8.

19. Find the area of the parallelogram with vertices $A(0, 1)$, $B(3, 0)$, $C(5, -2)$, and $D(2, -1)$.

20. Find the area of the parallelogram with vertices $P(0, 0, 0)$, $Q(5, 0, 0)$, $R(2, 6, 6)$, and $S(7, 6, 6)$.

21–24 ■ (a) Find a vector orthogonal to the plane through the points P, Q, and R, and (b) find the area of triangle PQR.

21. $P(1, 0, 0)$, $\quad Q(0, 2, 0)$, $\quad R(0, 0, 3)$

22. $P(1, 0, -1)$, $\quad Q(2, 4, 5)$, $\quad R(3, 1, 7)$

23. $P(0, 0, 0)$, $\quad Q(1, -1, 1)$, $\quad R(4, 3, 7)$

24. $P(-4, -4, -4)$, $\quad Q(0, 5, -1)$, $\quad R(3, 1, 2)$

25–26 ■ Find the volume of the parallelepiped determined by the vectors $\mathbf{a}$, $\mathbf{b}$, and $\mathbf{c}$.

25. $\mathbf{a} = \langle 1, 0, 6 \rangle$, $\quad \mathbf{b} = \langle 2, 3, -8 \rangle$, $\quad \mathbf{c} = \langle 8, -5, 6 \rangle$

26. $\mathbf{a} = 2\mathbf{i} + 3\mathbf{j} - 2\mathbf{k}$, $\quad \mathbf{b} = \mathbf{i} - \mathbf{j}$, $\quad \mathbf{c} = 2\mathbf{i} + 3\mathbf{k}$

27–28 ■ Find the volume of the parallelepiped with adjacent edges PQ, PR, and PS.

27. $P(1, 1, 1)$, $\quad Q(2, 0, 3)$, $\quad R(4, 1, 7)$, $\quad S(3, -1, -2)$

28. $P(0, 1, 2)$, $\quad Q(2, 4, 5)$, $\quad R(-1, 0, 1)$, $\quad S(6, -1, 4)$

29. Use the scalar triple product to verify that the vectors $\mathbf{a} = 2\mathbf{i} + 3\mathbf{j} + \mathbf{k}$, $\mathbf{b} = \mathbf{i} - \mathbf{j}$, and $\mathbf{c} = 7\mathbf{i} + 3\mathbf{j} + 2\mathbf{k}$ are coplanar.

30. Use the scalar triple product to verify that the points $P(1, 0, 1)$, $Q(2, 4, 6)$, $R(3, -1, 2)$, and $S(6, 2, 8)$ are coplanar.

31. A bicycle pedal is pushed by a foot with a 60-N force as shown. The shaft of the pedal is 18 cm long. Find the magnitude of the torque about P.

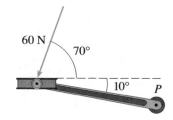

32. Find the magnitude of the torque about P if a 36-lb force is applied as shown.

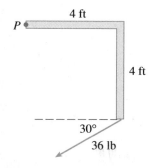

33. (a) Let P be a point not on the line L that passes through the points Q and R. Show that the distance d from the point P to the line L is

$$d = \frac{|\mathbf{a} \times \mathbf{b}|}{|\mathbf{a}|}$$

where $\mathbf{a} = \overrightarrow{QR}$ and $\mathbf{b} = \overrightarrow{QP}$.
(b) Use the formula in part (a) to find the distance from the point $P(1, 1, 1)$ to the line through $Q(0, 6, 8)$ and $R(-1, 4, 7)$.

34. (a) Let P be a point not on the plane that passes through the points Q, R, and S. Show that the distance d from P to the plane is

$$d = \frac{|\mathbf{a} \cdot (\mathbf{b} \times \mathbf{c})|}{|\mathbf{a} \times \mathbf{b}|}$$

where $\mathbf{a} = \overrightarrow{QR}$, $\mathbf{b} = \overrightarrow{QS}$, and $\mathbf{c} = \overrightarrow{QP}$.

(b) Use the formula in part (a) to find the distance from the point $P(2, 1, 4)$ to the plane through the points $Q(1, 0, 0)$, $R(0, 2, 0)$, and $S(0, 0, 3)$.

35. Prove that $(\mathbf{a} - \mathbf{b}) \times (\mathbf{a} + \mathbf{b}) = 2(\mathbf{a} \times \mathbf{b})$.

36. The product $\mathbf{a} \times (\mathbf{b} \times \mathbf{c})$ is called the **vector triple product** of $\mathbf{a}$, $\mathbf{b}$, and $\mathbf{c}$. Prove the following formula for the vector triple product:

$$\mathbf{a} \times (\mathbf{b} \times \mathbf{c}) = (\mathbf{a} \cdot \mathbf{c})\mathbf{b} - (\mathbf{a} \cdot \mathbf{b})\mathbf{c}$$

37. Prove that

$$\mathbf{a} \times (\mathbf{b} \times \mathbf{c}) + \mathbf{b} \times (\mathbf{c} \times \mathbf{a}) + \mathbf{c} \times (\mathbf{a} \times \mathbf{b}) = \mathbf{0}$$

38. Prove that

$$(\mathbf{a} \times \mathbf{b}) \cdot (\mathbf{c} \times \mathbf{d}) = \begin{vmatrix} \mathbf{a} \cdot \mathbf{c} & \mathbf{b} \cdot \mathbf{c} \\ \mathbf{a} \cdot \mathbf{d} & \mathbf{b} \cdot \mathbf{d} \end{vmatrix}$$

39. Suppose that $\mathbf{a} \neq \mathbf{0}$.
(a) If $\mathbf{a} \cdot \mathbf{b} = \mathbf{a} \cdot \mathbf{c}$, does it follow that $\mathbf{b} = \mathbf{c}$?
(b) If $\mathbf{a} \times \mathbf{b} = \mathbf{a} \times \mathbf{c}$, does it follow that $\mathbf{b} = \mathbf{c}$?
(c) If $\mathbf{a} \cdot \mathbf{b} = \mathbf{a} \cdot \mathbf{c}$ and $\mathbf{a} \times \mathbf{b} = \mathbf{a} \times \mathbf{c}$, does it follow that $\mathbf{b} = \mathbf{c}$?

40. If $\mathbf{v}_1$, $\mathbf{v}_2$, and $\mathbf{v}_3$ are noncoplanar vectors, let

$$\mathbf{k}_1 = \frac{\mathbf{v}_2 \times \mathbf{v}_3}{\mathbf{v}_1 \cdot (\mathbf{v}_2 \times \mathbf{v}_3)} \qquad \mathbf{k}_2 = \frac{\mathbf{v}_3 \times \mathbf{v}_1}{\mathbf{v}_1 \cdot (\mathbf{v}_2 \times \mathbf{v}_3)}$$

$$\mathbf{k}_3 = \frac{\mathbf{v}_1 \times \mathbf{v}_2}{\mathbf{v}_1 \cdot (\mathbf{v}_2 \times \mathbf{v}_3)}$$

(These vectors occur in the study of crystallography. Vectors of the form $n_1\mathbf{v}_1 + n_2\mathbf{v}_2 + n_3\mathbf{v}_3$, where each n_i is an integer, form a *lattice* for a crystal. Vectors written similarly in terms of $\mathbf{k}_1$, $\mathbf{k}_2$, and $\mathbf{k}_3$ form the *reciprocal lattice*.)

(a) Show that $\mathbf{k}_i$ is perpendicular to $\mathbf{v}_j$ if $i \neq j$.
(b) Show that $\mathbf{k}_i \cdot \mathbf{v}_i = 1$ for $i = 1, 2, 3$.
(c) Show that $\mathbf{k}_1 \cdot (\mathbf{k}_2 \times \mathbf{k}_3) = \dfrac{1}{\mathbf{v}_1 \cdot (\mathbf{v}_2 \times \mathbf{v}_3)}$.

11.5 EQUATIONS OF LINES AND PLANES

A line in the xy-plane is determined when a point on the line and the direction of the line (its slope or angle of inclination) are given. The equation of the line can then be written using the point-slope form.

Likewise, a line L in three-dimensional space is determined when we know a point $P_0(x_0, y_0, z_0)$ on L and the direction of L. In three dimensions the direction of a line is conveniently described by a vector, so we let $\mathbf{v}$ be a vector parallel to L. Let $P(x, y, z)$ be an arbitrary point on L and let $\mathbf{r}_0$ and $\mathbf{r}$ be the position vectors of P_0 and P (that is, they have representations $\overrightarrow{OP_0}$ and $\overrightarrow{OP}$). If $\mathbf{a}$ is the vector with representation $\overrightarrow{P_0P}$, as in Figure 1, then the Triangle Law for vector addition gives $\mathbf{r} = \mathbf{r}_0 + \mathbf{a}$. But, since $\mathbf{a}$ and $\mathbf{v}$ are parallel vectors, there is a scalar t such that $\mathbf{a} = t\mathbf{v}$. Thus

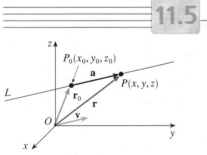

FIGURE 1

$$(1) \qquad \boxed{\mathbf{r} = \mathbf{r}_0 + t\mathbf{v}}$$

which is a **vector equation** of L. Each value of the **parameter** t gives the position vector $\mathbf{r}$ of a point on L. As Figure 2 indicates, positive values of t correspond to points on L that lie on one side of P_0, whereas negative values of t correspond to points that lie on the other side of P_0.

If the vector $\mathbf{v}$ that gives the direction of L is written in component form as $\mathbf{v} = \langle a, b, c \rangle$, then we have $t\mathbf{v} = \langle ta, tb, tc \rangle$. We can also write $\mathbf{r} = \langle x, y, z \rangle$ and $\mathbf{r}_0 = \langle x_0, y_0, z_0 \rangle$, so the vector equation (1) becomes

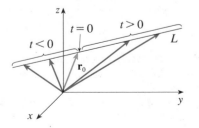

FIGURE 2

$$\langle x, y, z \rangle = \langle x_0 + ta, \ y_0 + tb, \ z_0 + tc \rangle$$

Two vectors are equal if and only if corresponding components are equal. Therefore, we have the three scalar equations:

(2)

$$x = x_0 + at \qquad y = y_0 + bt \qquad z = z_0 + ct$$

where $t \in \mathbb{R}$. These equations are called **parametric equations** of the line L through the point $P_0(x_0, y_0, z_0)$ and parallel to the vector $\mathbf{v} = \langle a, b, c \rangle$. Each value of the parameter t gives a point (x, y, z) on L.

EXAMPLE 1

(a) Find a vector equation and parametric equations for the line that passes through the point $(5, 1, 3)$ and is parallel to the vector $\mathbf{i} + 4\mathbf{j} - 2\mathbf{k}$.
(b) Find two other points on the line.

Figure 3 shows the line L in Example 1 and its relation to the given point and to the vector that gives its direction.

SOLUTION
(a) Here $\mathbf{r}_0 = \langle 5, 1, 3 \rangle = 5\mathbf{i} + \mathbf{j} + 3\mathbf{k}$ and $\mathbf{v} = \mathbf{i} + 4\mathbf{j} - 2\mathbf{k}$, so the vector equation (1) becomes

$$\mathbf{r} = (5\mathbf{i} + \mathbf{j} + 3\mathbf{k}) + t(\mathbf{i} + 4\mathbf{j} - 2\mathbf{k})$$

or

$$\mathbf{r} = (5 + t)\mathbf{i} + (1 + 4t)\mathbf{j} + (3 - 2t)\mathbf{k}$$

Parametric equations are

$$x = 5 + t \qquad y = 1 + 4t \qquad z = 3 - 2t$$

(b) Choosing the parameter value $t = 1$ gives $x = 6$, $y = 5$, and $z = 1$, so $(6, 5, 1)$ is a point on the line. Similarly, $t = -1$ gives the point $(4, -3, 5)$. ∎

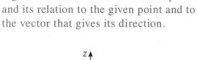

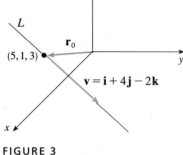

FIGURE 3

The vector equation and parametric equations of a line are not unique. If we change the point or the parameter or choose a different parallel vector, then the equations change. For instance, if, instead of $(5, 1, 3)$, we choose the point $(6, 5, 1)$ in Example 1, then the parametric equations of the line become

$$x = 6 + t \qquad y = 5 + 4t \qquad z = 1 - 2t$$

Or, if we stay with the point $(5, 1, 3)$ but choose the parallel vector $2\mathbf{i} + 8\mathbf{j} - 4\mathbf{k}$, we arrive at the equations

$$x = 5 + 2t \qquad y = 1 + 8t \qquad z = 3 - 4t$$

In general, if a vector $\mathbf{v} = \langle a, b, c \rangle$ is used to describe the direction of a line L, then the numbers a, b, and c are called **direction numbers** of L. Since any vector parallel to $\mathbf{v}$ could also be used, we see that any three numbers proportional to a, b, and c could also be used as a set of direction numbers for L.

Another way of describing a line L is to eliminate the parameter t from Equations 2. If none of a, b, or c is 0, we can solve each of these equations for t, equate the results, and obtain

(3)

$$\frac{x - x_0}{a} = \frac{y - y_0}{b} = \frac{z - z_0}{c}$$

These equations are called **symmetric equations** of L. Notice that the numbers a, b, and c that appear in the denominators of Equations 3 are direction numbers of L, that is, components of a vector parallel to L. If one of a, b, or c is 0, we can still eliminate t. For instance, if $a = 0$, we could write the equations of L as

$$x = x_0 \qquad \frac{y - y_0}{b} = \frac{z - z_0}{c}$$

This means that L lies in the vertical plane $x = x_0$.

EXAMPLE 2

(a) Find parametric equations and symmetric equations of the line that passes through the points $A(2, 4, -3)$ and $B(3, -1, 1)$.
(b) At what point does this line intersect the xy-plane?

SOLUTION
(a) We are not explicitly given a vector parallel to the line, but observe that the vector $\mathbf{v}$ with representation $\overrightarrow{AB}$ is parallel to the line and

$$\mathbf{v} = \langle 3 - 2, -1 - 4, 1 - (-3) \rangle = \langle 1, -5, 4 \rangle$$

Thus direction numbers are $a = 1$, $b = -5$, and $c = 4$. Taking the point $(2, 4, -3)$ as P_0, we see that parametric equations (2) are

$$x = 2 + t \qquad y = 4 - 5t \qquad z = -3 + 4t$$

Figure 4 shows the line L in Example 2 and the point P where it intersects the xy-plane.

and symmetric equations (3) are

$$\frac{x - 2}{1} = \frac{y - 4}{-5} = \frac{z + 3}{4}$$

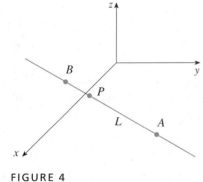

FIGURE 4

(b) The line intersects the xy-plane when $z = 0$, so we put $z = 0$ in the symmetric equations and obtain

$$\frac{x - 2}{1} = \frac{y - 4}{-5} = \frac{3}{4}$$

This gives $x = \frac{11}{4}$ and $y = \frac{1}{4}$, so the line intersects the xy-plane at the point $\left(\frac{11}{4}, \frac{1}{4}, 0\right)$. ∎

In general, the procedure of Example 2 shows that direction numbers of the line L through the points $P_0(x_0, y_0, z_0)$ and $P_1(x_1, y_1, z_1)$ are $x_1 - x_0$, $y_1 - y_0$, and $z_1 - z_0$ and so symmetric equations of L are

$$\frac{x - x_0}{x_1 - x_0} = \frac{y - y_0}{y_1 - y_0} = \frac{z - z_0}{z_1 - z_0}$$

The lines L_1 and L_2 in Example 3, shown in Figure 5, are skew lines.

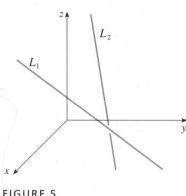

FIGURE 5

EXAMPLE 3   Show that the lines L_1 and L_2 with parametric equations

$$x = 1 + t \qquad y = -2 + 3t \qquad z = 4 - t$$

$$x = 2s \qquad y = 3 + s \qquad z = -3 + 4s$$

are **skew lines;** that is, they do not intersect and are not parallel (and therefore do not lie in the same plane).

SOLUTION   The lines are not parallel because the corresponding vectors $\langle 1, 3, -1 \rangle$ and $\langle 2, 1, 4 \rangle$ are not parallel. (Their components are not proportional.) If L_1 and L_2 had a point of intersection, there would be values of t and s such that

$$1 + \ t = 2s$$

$$-2 + 3t = 3 + \ s$$

$$4 - \ t = -3 + 4s$$

But if we solve the first two equations, we get $t = \frac{11}{5}$ and $s = \frac{8}{5}$, and these values do not satisfy the third equation. Therefore, there are no values of t and s that satisfy the three equations. Thus L_1 and L_2 do not intersect. Hence L_1 and L_2 are skew lines.   ∎

PLANES

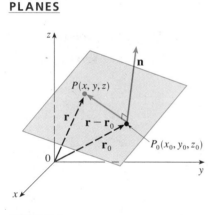

FIGURE 6

A plane in space is determined by a point $P_0(x_0, y_0, z_0)$ in the plane and a vector $\mathbf{n}$ that is orthogonal to the plane. This orthogonal vector $\mathbf{n}$ is called a **normal vector.** Let $P(x, y, z)$ be an arbitrary point in the plane and let $\mathbf{r}_0$ and $\mathbf{r}$ be the position vectors of P_0 and P. Then the vector $\mathbf{r} - \mathbf{r}_0$ is represented by $\overrightarrow{P_0 P}$ (see Figure 6). The normal vector $\mathbf{n}$ is orthogonal to every vector in the given plane. In particular, $\mathbf{n}$ is orthogonal to $\mathbf{r} - \mathbf{r}_0$ and so we have

(4)
$$\mathbf{n} \cdot (\mathbf{r} - \mathbf{r}_0) = 0$$

which can be rewritten as

(5)
$$\mathbf{n} \cdot \mathbf{r} = \mathbf{n} \cdot \mathbf{r}_0$$

Either Equation 4 or Equation 5 is called a **vector equation of the plane.**

To obtain a scalar equation for the plane, we write $\mathbf{n} = \langle a, b, c \rangle$, $\mathbf{r} = \langle x, y, z \rangle$, and $\mathbf{r}_0 = \langle x_0, y_0, z_0 \rangle$. Then the vector equation (4) becomes

$$\langle a, b, c \rangle \cdot \langle x - x_0, y - y_0, z - z_0 \rangle = 0$$

or

(6)
$$a(x - x_0) + b(y - y_0) + c(z - z_0) = 0$$

Equation 6 is the **scalar equation of the plane through** $P_0(x_0, y_0, z_0)$ **with normal vector** $\mathbf{n} = \langle a, b, c \rangle$.

EXAMPLE 4   Find an equation of the plane through the point $(2, 4, -1)$ with normal vector $\mathbf{n} = \langle 2, 3, 4 \rangle$. Find the intercepts and sketch the plane.

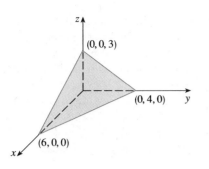

FIGURE 7

SOLUTION   Putting $a = 2$, $b = 3$, $c = 4$, $x_0 = 2$, $y_0 = 4$, and $z_0 = -1$ in Equation 6, we see that an equation of the plane is

$$2(x - 2) + 3(y - 4) + 4(z + 1) = 0$$

or

$$2x + 3y + 4z = 12$$

To find the x-intercept we set $y = z = 0$ in this equation and obtain $x = 6$. Similarly, the y-intercept is 4 and the z-intercept is 3. This enables us to sketch the portion of the plane that lies in the first octant (see Figure 7).   ∎

By collecting terms in Equation 6 as we did in Example 4, we can rewrite the equation of a plane as

(7)
$$ax + by + cz = d$$

where $d = ax_0 + by_0 + cz_0$. Equation 7 is called a **linear equation** in x, y, and z. Conversely, it can be shown that if a, b, and c are not all 0, then the linear equation (7) represents a plane with normal vector $\langle a, b, c \rangle$. (See Exercise 71.)

Figure 8 shows the portion of the plane in Example 5 that is enclosed by triangle PQR.

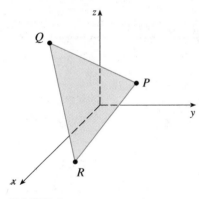

FIGURE 8

EXAMPLE 5   Find an equation of the plane that passes through the points $P(1, 3, 2)$, $Q(3, -1, 6)$, and $R(5, 2, 0)$.

SOLUTION   The vectors $\mathbf{a}$ and $\mathbf{b}$ corresponding to $\overrightarrow{PQ}$ and $\overrightarrow{PR}$ are

$$\mathbf{a} = \langle 2, -4, 4 \rangle \qquad \mathbf{b} = \langle 4, -1, -2 \rangle$$

Since both $\mathbf{a}$ and $\mathbf{b}$ lie in the plane, their cross product $\mathbf{a} \times \mathbf{b}$ is orthogonal to the plane and can be taken as the normal vector. Thus

$$\mathbf{n} = \mathbf{a} \times \mathbf{b} = \begin{vmatrix} \mathbf{i} & \mathbf{j} & \mathbf{k} \\ 2 & -4 & 4 \\ 4 & -1 & -2 \end{vmatrix} = 12\mathbf{i} + 20\mathbf{j} + 14\mathbf{k}$$

and an equation of the plane is

$$12(x - 1) + 20(y - 3) + 14(z - 2) = 0$$

or

$$6x + 10y + 7z = 50$$ ∎

EXAMPLE 6   Find the point at which the line with parametric equations $x = 2 + 3t$, $y = -4t$, $z = 5 + t$ intersects the plane $4x + 5y - 2z = 18$.

SOLUTION   We substitute the expressions for x, y, and z from the parametric equations into the equation of the plane:

$$4(2 + 3t) + 5(-4t) - 2(5 + t) = 18$$

This simplifies to $-10t = 20$, so $t = -2$. Therefore, the point of intersection occurs when the parameter value is $t = -2$. Then $x = 2 + 3(-2) = -4$, $y = -4(-2) = 8$, $z = 5 - 2 = 3$, and so the point of intersection is $(-4, 8, 3)$.   ∎

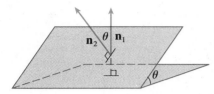

FIGURE 9

Figure 10 shows the planes in Example 7 and their line of intersection L.

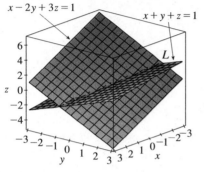

FIGURE 10

Another way to find the line of intersection is to solve the equations of the planes for two of the variables in terms of the third, which can be taken as the parameter.

Two planes are **parallel** if their normal vectors are parallel. For instance, the planes $x + 2y - 3z = 4$ and $2x + 4y - 6z = 3$ are parallel because their normal vectors are $\mathbf{n}_1 = \langle 1, 2, -3 \rangle$ and $\mathbf{n}_2 = \langle 2, 4, -6 \rangle$ and $\mathbf{n}_2 = 2\mathbf{n}_1$. If two planes are not parallel, then they intersect in a straight line and the angle between the two planes is defined as the acute angle between their normal vectors (see Figure 9).

EXAMPLE 7
(a) Find the angle between the planes $x + y + z = 1$ and $x - 2y + 3z = 1$.
(b) Find symmetric equations for the line of intersection L of these two planes.

SOLUTION
(a) The normal vectors of these planes are

$$\mathbf{n}_1 = \langle 1, 1, 1 \rangle \qquad \mathbf{n}_2 = \langle 1, -2, 3 \rangle$$

and so if θ is the angle between the planes, Corollary 11.3.6 gives

$$\cos\theta = \frac{\mathbf{n}_1 \cdot \mathbf{n}_2}{|\mathbf{n}_1||\mathbf{n}_2|} = \frac{1(1) + 1(-2) + 1(3)}{\sqrt{1+1+1}\sqrt{1+4+9}} = \frac{2}{\sqrt{42}}$$

$$\theta = \cos^{-1}\left(\frac{2}{\sqrt{42}}\right) \approx 72°$$

(b) We first need to find a point on L. For instance, we can find the point where the line intersects the xy-plane by setting $z = 0$ in the equations of both planes. This gives the equations $x + y = 1$ and $x - 2y = 1$, whose solution is $x = 1$, $y = 0$. So the point $(1, 0, 0)$ lies on L.

Now we observe that, since L lies in both planes, it is perpendicular to both of the normal vectors. Thus a vector $\mathbf{v}$ parallel to L is given by the cross product

$$\mathbf{v} = \mathbf{n}_1 \times \mathbf{n}_2 = \begin{vmatrix} \mathbf{i} & \mathbf{j} & \mathbf{k} \\ 1 & 1 & 1 \\ 1 & -2 & 3 \end{vmatrix} = 5\mathbf{i} - 2\mathbf{j} - 3\mathbf{k}$$

and so the symmetric equations of L can be written as

$$\frac{x-1}{5} = \frac{y}{-2} = \frac{z}{-3}$$

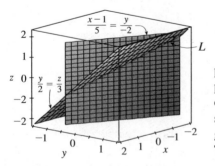

FIGURE 11

Figure 11 shows how the line L in Example 7 can also be regarded as the line of intersection of planes derived from its symmetric equations.

NOTE: Since a linear equation in x, y, and z represents a plane and two nonparallel planes intersect in a line, it follows that two linear equations can represent a line. The points (x, y, z) that satisfy both $a_1x + b_1y + c_1z = d_1$ and $a_2x + b_2y + c_2z = d_2$ lie on both of these planes and so the pair of linear equations represents the line of intersection of the planes (if they are not parallel). For instance, in Example 7 the line L was given as the line of intersection of the planes $x + y + z = 1$ and $x - 2y + 3z = 1$. The symmetric equations that we found for L could be written as

$$\frac{x-1}{5} = \frac{y}{-2} \quad \text{and} \quad \frac{y}{-2} = \frac{z}{-3}$$

which is again a pair of linear equations. They exhibit L as the line of intersection of the planes $(x - 1)/5 = y/(-2)$ and $y/(-2) = z/(-3)$. (See Figure 11.)

In general, when we write the equations of a line in the symmetric form

$$\frac{x - x_0}{a} = \frac{y - y_0}{b} = \frac{z - z_0}{c}$$

we can regard the line as the line of intersection of the two planes

$$\frac{x - x_0}{a} = \frac{y - y_0}{b} \qquad \text{and} \qquad \frac{y - y_0}{b} = \frac{z - z_0}{c}$$

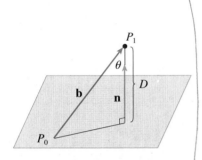

FIGURE 12

EXAMPLE 8 Find a formula for the distance D from a point $P_1(x_1, y_1, z_1)$ to the plane $ax + by + cz + d = 0$.

SOLUTION Let $P_0(x_0, y_0, z_0)$ be any point in the given plane and let $\mathbf{b}$ be the vector corresponding to $\overrightarrow{P_0 P_1}$. Then

$$\mathbf{b} = \langle x_1 - x_0, y_1 - y_0, z_1 - z_0 \rangle$$

From Figure 12 you can see that the distance D from P_1 to the plane is equal to the absolute value of the scalar projection of $\mathbf{b}$ onto the normal vector $\mathbf{n} = \langle a, b, c \rangle$. (See Section 11.3.) Thus

$$\begin{aligned} D = \text{comp}_{\mathbf{n}}\, \mathbf{b} &= \frac{|\mathbf{n} \cdot \mathbf{b}|}{|\mathbf{n}|} \\ &= \frac{|a(x_1 - x_0) + b(y_1 - y_0) + c(z_1 - z_0)|}{\sqrt{a^2 + b^2 + c^2}} \\ &= \frac{|(ax_1 + by_1 + cz_1) - (ax_0 + by_0 + cz_0)|}{\sqrt{a^2 + b^2 + c^2}} \end{aligned}$$

Since P_0 lies in the plane, its coordinates satisfy the equation of the plane and so we have $ax_0 + by_0 + cz_0 + d = 0$. Thus the formula for D can be written as

(8)
$$D = \frac{|ax_1 + by_1 + cz_1 + d|}{\sqrt{a^2 + b^2 + c^2}}$$

∎

EXAMPLE 9 Find the distance between the parallel planes $10x + 2y - 2z = 5$ and $5x + y - z = 1$.

SOLUTION First we note that the planes are parallel because their normal vectors $\langle 10, 2, -2 \rangle$ and $\langle 5, 1, -1 \rangle$ are parallel. To find the distance D between the planes, we choose any point on one plane and calculate its distance to the other plane. In particular, if we put $y = z = 0$ in the equation of the first plane, we get $10x = 5$ and so $\left(\tfrac{1}{2}, 0, 0\right)$ is a point in this plane. By Formula 8, the distance between $\left(\tfrac{1}{2}, 0, 0\right)$ and the plane $5x + y - z - 1 = 0$ is

$$D = \frac{\left|5\left(\tfrac{1}{2}\right) + 1(0) - 1(0) - 1\right|}{\sqrt{5^2 + 1^2 + (-1)^2}} = \frac{\tfrac{3}{2}}{3\sqrt{3}} = \frac{\sqrt{3}}{6}$$

So the distance between the planes is $\sqrt{3}/6$.

∎

EXAMPLE 10   In Example 3 we showed that the lines

$$L_1: \quad x = 1 + t \qquad y = -2 + 3t \qquad z = 4 - t$$

$$L_2: \quad x = 2s \qquad y = 3 + s \qquad z = -3 + 4s$$

are skew. Find the distance between them.

SOLUTION   Since the two lines L_1 and L_2 are skew, they can be viewed as lying on two parallel planes P_1 and P_2. The distance between L_1 and L_2 is the same as the distance between P_1 and P_2, which can be computed as in Example 9. The common normal vector to both planes must be orthogonal to both $\mathbf{v}_1 = \langle 1, 3, -1 \rangle$ (the direction of L_1) and $\mathbf{v}_2 = \langle 0, 1, 4 \rangle$ (the direction of L_2). So a normal vector is

$$\mathbf{n} = \mathbf{v}_1 \times \mathbf{v}_2 = \begin{vmatrix} \mathbf{i} & \mathbf{j} & \mathbf{k} \\ 1 & 3 & -1 \\ 0 & 1 & 4 \end{vmatrix} = 13\mathbf{i} - 4\mathbf{j} + \mathbf{k}$$

If we put $s = 0$ in the equations of L_2, we get the point $(0, 3, -3)$ on L_2 and so an equation for P_2 is

$$13(x - 0) - 4(y - 3) + 1(z + 3) = 0 \qquad \text{or} \qquad 13x - 4y + z + 15 = 0$$

If we now set $t = 0$ in the equations for L_1, we get the point $(1, -2, 4)$ on P_1. So the distance between L_1 and L_2 is the same as the distance from $(1, -2, 4)$ to $13x - 4y + z + 15 = 0$. By Formula 8, this distance is

$$D = \frac{|13(1) - 4(-2) + 1(4) + 15|}{\sqrt{13^2 + (-4)^2 + 1^2}} = \frac{40}{\sqrt{186}} \approx 2.9 \qquad \blacksquare$$

EXERCISES 11.5

1–4 ■ Find the vector equation and parametric equations for the line passing through the given point and parallel to the vector **a**.

1. $(3, -1, 8)$, $\mathbf{a} = \langle 2, 3, 5 \rangle$

2. $(-2, 4, 5)$, $\mathbf{a} = \langle 3, -1, 6 \rangle$

3. $(0, 1, 2)$, $\mathbf{a} = 6\mathbf{i} + 3\mathbf{j} + 2\mathbf{k}$

4. $(1, -1, -2)$, $\mathbf{a} = 2\mathbf{i} - 7\mathbf{k}$

5–10 ■ Find parametric equations and symmetric equations for the line passing through the given points.

5. $(2, 1, 8)$, $(6, 0, 3)$ **6.** $(-1, 0, 5)$, $(4, -3, 3)$

7. $(3, 1, -1)$, $(3, 2, -6)$ **8.** $(3, 1, \frac{1}{2})$, $(-1, 4, 1)$

9. $(-\frac{1}{3}, 1, 1)$, $(0, 5, -8)$ **10.** $(2, -7, 5)$, $(-4, 2, 5)$

11. Show that the line through the points $(2, -1, -5)$ and $(8, 8, 7)$ is parallel to the line through the points $(4, 2, -6)$ and $(8, 8, 2)$.

12. Show that the line through the points $(0, 1, 1)$ and $(1, -1, 6)$ is perpendicular to the line through the points $(-4, 2, 1)$ and $(-1, 6, 2)$.

13. (a) Find symmetric equations for the line that passes through the point $(0, 2, -1)$ and is parallel to the line with parametric equations $x = 1 + 2t$, $y = 3t$, and $z = 5 - 7t$.

(b) Find the points in which the required line in part (a) intersects the coordinate planes.

14. (a) Find parametric equations for the line through $(5, 1, 0)$ that is perpendicular to the plane $2x - y + z = 1$.

(b) In what points does this line intersect the coordinate planes?

15–18 ■ Determine whether the lines L_1 and L_2 are parallel, skew, or intersecting. If they intersect, find the point of intersection.

15. L_1: $\dfrac{x - 4}{2} = \dfrac{y + 5}{4} = \dfrac{z - 1}{-3}$, L_2: $\dfrac{x - 2}{1} = \dfrac{y + 1}{3} = \dfrac{z}{2}$

16. L_1: $\dfrac{x - 1}{2} = \dfrac{y}{1} = \dfrac{z - 1}{4}$, L_2: $\dfrac{x}{1} = \dfrac{y + 2}{2} = \dfrac{z + 2}{3}$

17. L_1: $x = -6t$, $y = 1 + 9t$, $z = -3t$
L_2: $x = 1 + 2s$, $y = 4 - 3s$, $z = s$

18. L_1: $x = 1 + t$, $y = 2 - t$, $z = 3t$
L_2: $x = 2 - s$, $y = 1 + 2s$, $z = 4 + s$

19–22 ■ Find an equation of the plane through the given point and with the specified normal vector.

19. $(1, 4, 5)$, $\mathbf{n} = \langle 7, 1, 4 \rangle$

20. $(-5, 1, 2)$, $\mathbf{n} = \langle 3, -5, 2 \rangle$

21. $(1, 2, 3)$, $\mathbf{n} = 15\mathbf{i} + 9\mathbf{j} - 12\mathbf{k}$

22. $(-1, -6, -4)$, $\mathbf{n} = -5\mathbf{i} + 2\mathbf{j} - 2\mathbf{k}$

23–26 ■ Find an equation of the plane passing through the given point and parallel to the indicated plane.

23. $(6, 5, -2)$, $x + y - z + 1 = 0$

24. $(3, 0, 8)$, $2x + 5y + 8z = 17$

25. $(-1, 3, -8)$, $3x - 4y - 6z = 9$

26. $(2, -4, 5)$, $z = 2x + 3y$

27–30 ■ Find an equation of the plane passing through the three given points.

27. $(0, 0, 0)$, $(1, 1, 1)$, $(1, 2, 3)$

28. $(-1, 1, -1)$, $(1, -1, 2)$, $(4, 0, 3)$

29. $(1, 0, -3)$, $(0, -2, -4)$, $(4, 1, 6)$

30. $(2, 1, -3)$, $(5, -1, 4)$, $(2, -2, 4)$

31–34 ■ Find an equation of the plane that passes through the given point and contains the indicated line.

31. $(1, 6, -4)$; $x = 1 + 2t$, $y = 2 - 3t$, $z = 3 - t$

32. $(-1, -3, 2)$; $x = -1 - 2t$, $y = 4t$, $z = 2 + t$

33. $(0, 1, 2)$; $x = y = z$

34. $(-1, 0, 1)$; $x = 5t$, $y = 1 + t$, $z = -t$

35–38 ■ Find the point at which the given line intersects the specified plane.

35. $x = 1 + t$, $y = 2t$, $z = 3t$; $x + y + z = 1$

36. $x = 5$, $y = 4 - t$, $z = 2t$; $2x - y + z = 5$

37. $x = 1 + 2t$, $y = -1$, $z = t$; $2x + y - z + 5 = 0$

38. $x = 1 - t$, $y = t$, $z = 1 + t$; $z = 1 - 2x + y$

39. Find direction numbers for the line of intersection of the planes $x + y + z = 1$ and $x + z = 0$.

40. Find the cosine of the angle between the planes $x + y + z = 0$ and $x + 2y + 3z = 1$.

41–46 ■ Determine whether the planes are parallel, perpendicular, or neither. If neither, find the angle between them.

41. $x + z = 1$, $y + z = 1$

42. $-8x - 6y + 2z = 1$, $z = 4x + 3y$

43. $x + 4y - 3z = 1$, $-3x + 6y + 7z = 0$

44. $2x + 2y - z = 4$, $6x - 3y + 2z = 5$

45. $2x + 4y - 2z = 1$, $-3x - 6y + 3z = 10$

46. $2x - 5y + z = 3$, $4x + 2y + 2z = 1$

47–48 ■ (a) Find symmetric equations for the line of intersection of the planes and (b) find the angle between the planes.

47. $x + y - z = 2$, $3x - 4y + 5z = 6$

48. $x - 2y + z = 1$, $2x + y + z = 1$

49–50 ■ Find parametric equations for the line of intersection of the planes.

49. $z = x + y$, $2x - 5y - z = 1$

50. $2x + 5z + 3 = 0$, $x - 3y + z + 2 = 0$

51. Find an equation for the plane consisting of all points that are equidistant from the points $(1, 1, 0)$ and $(0, 1, 1)$.

52. Find an equation for the plane consisting of all points that are equidistant from the points $(-4, 2, 1)$ and $(2, -4, 3)$.

53. Find an equation of the plane that passes through the line of intersection of the planes $x + y - z = 2$ and $2x - y + 3z = 1$ and passes through the point $(-1, 2, 1)$.

54. Find an equation of the plane that passes through the line of intersection of the planes $x - z = 1$ and $y + 2z = 3$ and is perpendicular to the plane $x + y - 2z = 1$.

55. Find an equation of the plane with x-intercept a, y-intercept b, and z-intercept c.

56. (a) Find the point at which the lines
$\mathbf{r} = \langle 1, 1, 0 \rangle + t\langle 1, -1, 2 \rangle$ and $\mathbf{r} = \langle 2, 0, 2 \rangle + s\langle -1, 1, 0 \rangle$
intersect.
(b) Find an equation of the plane that contains these lines.

57. Find parametric equations for the line through the point $(0, 1, 2)$ that is parallel to the plane $x + y + z = 2$ and perpendicular to the line $x = 1 + t$, $y = 1 - t$, $z = 2t$.

58. Find parametric equations for the line through the point $(0, 1, 2)$ that is perpendicular to the line $x = 1 + t$, $y = 1 - t$, $z = 2t$ and intersects this line.

59. Which of the following four planes are parallel? Are any of them identical?

P_1: $4x - 2y + 6z = 3$ P_2: $4x - 2y - 2z = 6$

P_3: $-6x + 3y - 9z = 5$ P_4: $z = 2x - y - 3$

60. Which of the following four lines are parallel? Are any of them identical?

L_1: $x = 1 + t$, $y = t$, $z = 2 - 5t$

L_2: $x + 1 = y - 2 = 1 - z$

L_3: $x = 1 + t$, $y = 4 + t$, $z = 1 - t$

L_4: $\mathbf{r} = \langle 2, 1, -3 \rangle + t\langle 2, 2, -10 \rangle$

61–62 ■ Use the formula in Exercise 33 in Section 11.4 to find the distance from the point to the given line.

61. $(1, 2, 3)$; $x = 2 + t, y = 2 - 3t, z = 5t$

62. $(1, 0, -1)$; $x = 5 - t, y = 3t, z = 1 + 2t$

63–64 ■ Find the distance from the point to the given plane.

63. $(2, 8, 5)$, $x - 2y - 2z = 1$

64. $(3, -2, 7)$, $4x - 6y + z = 5$

65–66 ■ Find the distance between the given parallel planes.

65. $z = x + 2y + 1$, $3x + 6y - 3z = 4$

66. $3x + 6y - 9z = 4$, $x + 2y - 3z = 1$

67. Show that the distance between the parallel planes $ax + by + cz = d_1$ and $ax + by + cz = d_2$ is

$$D = \frac{|d_1 - d_2|}{\sqrt{a^2 + b^2 + c^2}}$$

68. Find equations of the planes that are parallel to the plane $x + 2y - 2z = 1$ and two units away from it.

69. Show that the lines with symmetric equations $x = y = z$ and $x + 1 = y/2 = z/3$ are skew lines, and find the distance between these lines.

70. Find the distance between the skew lines with parametric equations $x = 1 + t, y = 1 + 6t, z = 2t$ and $x = 1 + 2s, y = 5 + 15s, z = -2 + 6s$.

71. If a, b, and c are not all 0, show that the equation $ax + by + cz = d$ represents a plane and $\langle a, b, c \rangle$ is a normal vector to the plane.   *Hint:* Suppose $a \neq 0$ and rewrite the equation in the form

$$a\left(x - \frac{d}{a}\right) + b(y - 0) + c(z - 0) = 0$$

72. Give a geometric description of each family of planes.
(a) $x + y + z = c$
(b) $x + y + cz = 1$
(c) $y \cos \theta + z \sin \theta = 1$

11.6   QUADRIC SURFACES

A **quadric surface** is the graph of a second-degree equation in three variables x, y, and z. The most general such equation is

$$Ax^2 + By^2 + Cz^2 + Dxy + Eyz + Fxz + Gx + Hy + Iz + J = 0$$

where $A, B, C, \ldots, J$ are constants, but by translation and rotation it can be brought into one of the two standard forms

$$Ax^2 + By^2 + Cz^2 + J = 0 \qquad \text{or} \qquad Ax^2 + By^2 + Iz = 0$$

Quadric surfaces are the analogues in three dimensions of the conic sections in the plane. (See Section 9.6 for a review of conic sections.)

In order to sketch the graph of a quadric surface (or any surface), it is useful to determine the curves of intersection of the surface with planes parallel to the coordinate planes. These curves are called **traces** (or cross-sections) of the surface.

ELLIPSOIDS   The quadric surface with equation

(1)
$$\frac{x^2}{a^2} + \frac{y^2}{b^2} + \frac{z^2}{c^2} = 1$$

is called an **ellipsoid** because its traces are ellipses. For instance, the horizontal plane $z = k$ (where $-c < k < c$) intersects the surface in the ellipse

$$\frac{x^2}{a^2} + \frac{y^2}{b^2} = 1 - \frac{k^2}{c^2} \qquad z = k$$

and, in particular, the trace in the xy-plane is just the ellipse $x^2/a^2 + y^2/b^2 = 1$, $z = 0$. Similarly, the traces in the other coordinate planes are the ellipses with equations

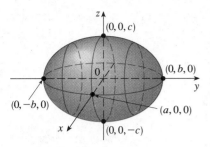

FIGURE 1

The ellipsoid $\dfrac{x^2}{a^2} + \dfrac{y^2}{b^2} + \dfrac{z^2}{c^2} = 1$

$y^2/b^2 + z^2/c^2 = 1$, $x = 0$, and $x^2/a^2 + z^2/c^2 = 1$, $y = 0$. Figure 1 shows how drawing some traces indicates the shape of the ellipsoid.

The six intercepts of the ellipsoid are $(\pm a, 0, 0)$, $(0, \pm b, 0)$, and $(0, 0, \pm c)$ and the ellipsoid lies in the box

$$|x| \leqslant a \qquad |y| \leqslant b \qquad |z| \leqslant c$$

Since the equation involves only even powers of x, y, and z, the ellipsoid is symmetric with respect to each coordinate plane.

If two of the three semiaxes a, b, and c are equal, then the ellipsoid is a surface of revolution. For instance, if $c = a$, then the ellipsoid could be obtained by revolving the ellipse $x^2/a^2 + y^2/b^2 = 1$, $z = 0$, around the y-axis. If $a = b = c$, the ellipsoid is a sphere.

HYPERBOLOIDS   The quadric surface

(2)
$$\frac{x^2}{a^2} + \frac{y^2}{b^2} - \frac{z^2}{c^2} = 1$$

is also symmetric with respect to the coordinate planes. The trace in any horizontal plane $z = k$ is the ellipse

$$\frac{x^2}{a^2} + \frac{y^2}{b^2} = 1 + \frac{k^2}{c^2} \qquad z = k$$

but the traces in the xz- and yz-planes are the hyperbolas

$$\frac{x^2}{a^2} - \frac{z^2}{c^2} = 1, \quad y = 0 \qquad \text{and} \qquad \frac{y^2}{b^2} - \frac{z^2}{c^2} = 1, \quad x = 0$$

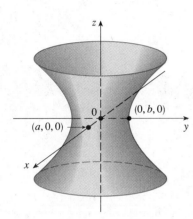

FIGURE 2

Hyperboloid of one sheet
$\dfrac{x^2}{a^2} + \dfrac{y^2}{b^2} - \dfrac{z^2}{c^2} = 1$

Furthermore, in contrast to the next example, this surface consists of just one piece, so it is called a **hyperboloid of one sheet** (see Figure 2). The z-axis is called the **axis** of this hyperboloid. (If the minus sign in Equation 2 occurs in front of the first or second term instead of the third term, then the axis is the x- or y-axis, respectively.)

If $a = b$ in Equation 2, the surface is a hyperboloid of revolution and is obtained by rotating a hyperbola about the z-axis.

Now consider the surface

(3)
$$-\frac{x^2}{a^2} - \frac{y^2}{b^2} + \frac{z^2}{c^2} = 1$$

Traces in the xz- and yz-planes are the hyperbolas

$$-\frac{x^2}{a^2} + \frac{z^2}{c^2} = 1, \quad y = 0 \qquad \text{and} \qquad -\frac{y^2}{b^2} + \frac{z^2}{c^2} = 1, \quad x = 0$$

If $|k| > c$, the horizontal plane $z = k$ intersects the surface in the ellipse

$$\frac{x^2}{a^2} + \frac{y^2}{b^2} = \frac{k^2}{c^2} - 1 \qquad z = k$$

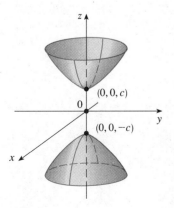

FIGURE 3

Hyperboloid of two sheets
$-\dfrac{x^2}{a^2} - \dfrac{y^2}{b^2} + \dfrac{z^2}{c^2} = 1$

whereas if $|k| < c$, the plane $z = k$ does not intersect the surface at all. Thus the surface consists of two parts, one above the plane $z = c$ and one below the plane $z = -c$, and is called a **hyperboloid of two sheets** whose axis is the z-axis. (See Figure 3.)

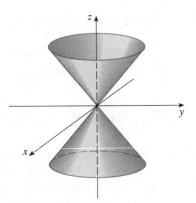

FIGURE 4

The cone $\dfrac{z^2}{c^2} = \dfrac{x^2}{a^2} + \dfrac{y^2}{b^2}$

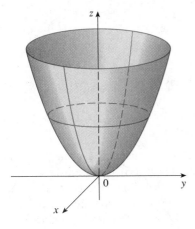
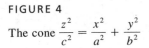
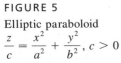

FIGURE 5

Elliptic paraboloid

$\dfrac{z}{c} = \dfrac{x^2}{a^2} + \dfrac{y^2}{b^2}$, $c > 0$

Notice, in comparing Equations 2 and 3, that the number of minus signs in the equation indicates the number of sheets of the hyperboloid.

CONES   If we replace the right side of Equation 2 or 3 by 0, we get the surface

(4)
$$\frac{z^2}{c^2} = \frac{x^2}{a^2} + \frac{y^2}{b^2}$$

which is a **cone.** This surface has the property that if P is any point on the cone, then the line OP lies entirely on the cone.

You can verify that traces in horizontal planes $z = k$ are ellipses and traces in vertical planes $x = k$ or $y = k$ are hyperbolas if $k \neq 0$ but are pairs of lines if $k = 0$.

The cone given by Equation 4 is asymptotic to both of the hyperboloids given by Equations 2 and 3. (Compare Figures 2, 3, and 4.)

PARABOLOIDS   The surface

(5)
$$\frac{z}{c} = \frac{x^2}{a^2} + \frac{y^2}{b^2}$$

is called an **elliptic paraboloid** because its traces in horizontal planes $z = k$ are ellipses, whereas its traces in vertical planes $x = k$ or $y = k$ are parabolas. For instance, its trace in the yz-plane is the parabola

$$z = \frac{c}{b^2} y^2 \qquad x = 0$$

The **axis** of the paraboloid given by Equation 5 is the z-axis and its **vertex** is the origin. The case where $c > 0$ is illustrated in Figure 5. If $a = b$, the surface is a **circular paraboloid,** also called a paraboloid of revolution.

The **hyperbolic paraboloid**

(6)
$$\frac{z}{c} = \frac{x^2}{a^2} - \frac{y^2}{b^2}$$

also has parabolas as its vertical traces, but it has hyperbolas as its horizontal traces. The case where $c < 0$ is illustrated in Figure 6. Notice that the shape of the surface near the origin resembles that of a saddle. This surface will be investigated further in Section 12.7 when we discuss saddle points.

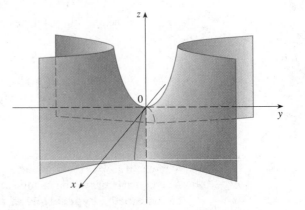

FIGURE 6

Hyperbolic paraboloid

$\dfrac{z}{c} = \dfrac{x^2}{a^2} - \dfrac{y^2}{b^2}$, $c < 0$

The idea of using traces to draw a surface is employed in three-dimensional graphing software for computers. Figure 7 shows two paraboloids drawn by a computer using vertical traces.

(a) The paraboloid $z = x^2 + y^2$

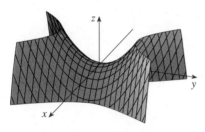

(b) The paraboloid $z = y^2 - x^2$

FIGURE 7

QUADRIC CYLINDERS   When one of the variables x, y, or z is missing from the equation of a surface, then the surface is a cylinder. For instance, the equation

(7)
$$\frac{x^2}{a^2} + \frac{y^2}{b^2} = 1$$

represents the **elliptic cylinder** $\{(x, y, z) \mid x^2/a^2 + y^2/b^2 = 1\}$. All horizontal traces are congruent ellipses and the generators of the cylinder are vertical lines. The **parabolic cylinder** $y = ax^2$ is illustrated in Figure 8.

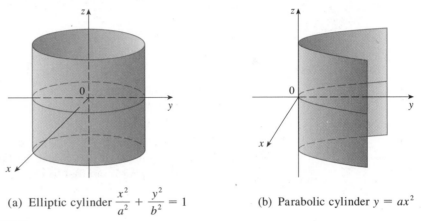

(a) Elliptic cylinder $\dfrac{x^2}{a^2} + \dfrac{y^2}{b^2} = 1$ (b) Parabolic cylinder $y = ax^2$

FIGURE 8

EXAMPLE 1   Identify and sketch the surface $4x^2 - y^2 + 2z^2 + 4 = 0$.

SOLUTION   Dividing by -4, we first put the equation in standard form:

$$-x^2 + \frac{y^2}{4} - \frac{z^2}{2} = 1$$

Comparing this equation with Equation 3, we see that it represents a hyperboloid of two sheets, the only difference being that in this case the axis of the hyperboloid is the y-axis. The traces in the xy- and yz-planes are the hyperbolas

$$-x^2 + \frac{y^2}{4} = 1, \quad z = 0 \quad \text{and} \quad \frac{y^2}{4} - \frac{z^2}{2} = 1, \quad x = 0$$

The surface has no trace in the xz-plane, but traces in the vertical planes $y = k$ for $|k| > 2$ are the ellipses

$$x^2 + \frac{z^2}{2} = \frac{k^2}{4} - 1 \qquad y = k$$

which can be written as

$$\frac{x^2}{\dfrac{k^2}{4} - 1} + \frac{z^2}{2\left(\dfrac{k^2}{4} - 1\right)} = 1 \qquad y = k$$

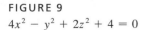

FIGURE 9
$4x^2 - y^2 + 2z^2 + 4 = 0$

These traces are used to make the sketch in Figure 9.   ∎

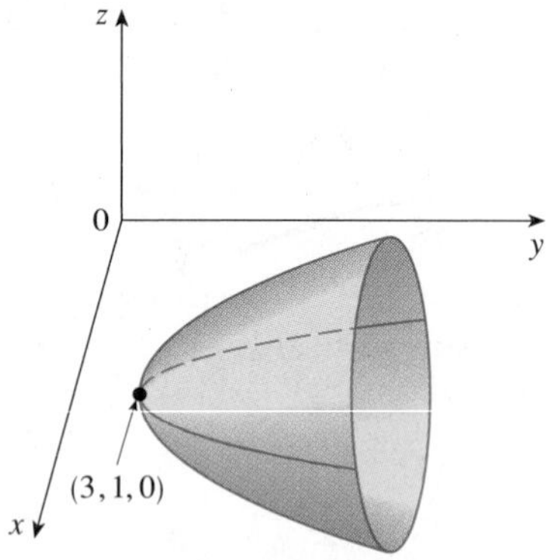

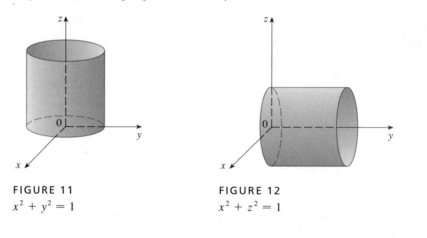

FIGURE 10
$x^2 + 2z^2 - 6x - y + 10 = 0$

EXAMPLE 2   Classify the quadric surface $x^2 + 2z^2 - 6x - y + 10 = 0$.

SOLUTION   By completing the square we rewrite the equation as

$$y - 1 = (x - 3)^2 + 2z^2$$

Comparing this equation with Equation 5, we see that it represents an elliptic paraboloid. Here, however, the axis of the paraboloid is parallel to the y-axis and its vertex is the point $(3, 1, 0)$. The traces in the plane $y = k$ ($k > 1$) are the ellipses $(x - 3)^2 + 2z^2 = k - 1$, $y = k$. The trace in the xy-plane is the parabola with equation $y = 1 + (x - 3)^2$, $z = 0$. The paraboloid is sketched in Figure 10.   ■

EXAMPLE 3   Identify and sketch the surfaces:
(a) $x^2 + y^2 = 1$ (b) $x^2 + z^2 = 1$

SOLUTION

(a) Since z is missing and the equations $x^2 + y^2 = 1$, $z = k$, represent a circle with radius 1 in the plane $z = k$, the surface $x^2 + y^2 = 1$ is a circular cylinder whose axis is the z-axis (see Figure 11).

(b) In this case y is missing and the surface is a circular cylinder whose axis is the y-axis (see Figure 12). It is obtained by taking the circle $x^2 + z^2 = 1$, $y = 0$, in the xz-plane and moving it parallel to the y-axis.

FIGURE 11
$x^2 + y^2 = 1$

FIGURE 12
$x^2 + z^2 = 1$
■

⊘   **NOTE:**   When you are dealing with surfaces, it is important to recognize that an equation like $x^2 + y^2 = 1$ represents a cylinder and not a circle. The trace of the cylinder $x^2 + y^2 = 1$ in the xy-plane is the circle with equations $x^2 + y^2 = 1$, $z = 0$.

EXERCISES 11.6

1–16 ■ Find the traces of the given surface in the planes $x = k$, $y = k$, $z = k$. Then identify the surface and sketch it.

1. $x^2 - y^2 + z^2 = 1$

2. $x = y^2 + z^2$

3. $4x^2 + 9y^2 + 36z^2 = 36$

4. $2x^2 + z^2 = 4$

5. $4z^2 - x^2 - y^2 = 1$

6. $z = x^2 - y^2$

7. $z = y^2$

8. $25y^2 + z^2 = 100 + 4x^2$

9. $y^2 = x^2 + z^2$

10. $9x^2 - y^2 - z^2 = 9$

11. $x^2 + 4z^2 - y = 0$

12. $x^2 - y^2 = 1$

13. $y^2 + 9z^2 = 9$

14. $x^2 + 4y^2 + z^2 = 4$

15. $y = z^2 - x^2$

16. $16x^2 = y^2 + 4z^2$

17–24 ■ Match the equation with its graph (labeled I–VIII). Give reasons for your choice.

17. $x^2 + 4y^2 + 9z^2 = 1$

18. $9x^2 + 4y^2 + z^2 = 1$

19. $x^2 - y^2 + z^2 = 1$

20. $-x^2 + y^2 - z^2 = 1$

21. $y = 2x^2 + z^2$

22. $y^2 = x^2 + 2z^2$

23. $x^2 + 2z^2 = 1$

24. $y = x^2 - z^2$

I

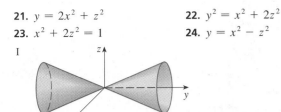

II

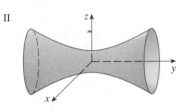

III

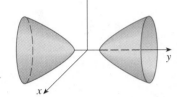

IV

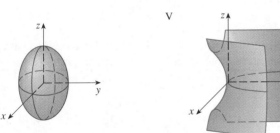

V

VI

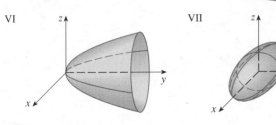

VII

VIII

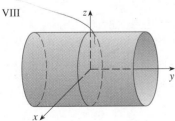

25–34 ■ Reduce the equation to one of the standard forms, classify the surface, and sketch it.

25. $z^2 = 3x^2 + 4y^2 - 12$

26. $4x^2 - 9y^2 + z^2 + 36 = 0$

27. $z = x^2 + y^2 + 1$

28. $x^2 + 4y^2 + z^2 - 2x = 0$

29. $x^2 + y^2 - 4z^2 + 4x - 6y - 8z = 13$

30. $4x = y^2 - 2z^2$

31. $x^2 + 4y^2 = 100$

32. $9x^2 + y^2 - z^2 - 2y + 2z = 0$

33. $x^2 - y^2 + 4y + z = 4$

34. $4x^2 - y^2 + z^2 + 8x + 8z + 24 = 0$

35–38 ■ Use a computer with three-dimensional graphing software to graph the surface.

35. $z = 3x^2 - 5y^2$

36. $8x^2 + 15y^2 + 5z^2 = 100$

37. $z^2 = x^2 + 4y^2$

38. $z = y^2 + xy$

39. Sketch the region bounded by the surfaces $z = \sqrt{x^2 + y^2}$ and $x^2 + y^2 = 1$ for $1 \leq z \leq 2$.

40. Sketch the region bounded by the paraboloids $z = x^2 + y^2$ and $z = 2 - x^2 - y^2$.

41. Find an equation for the surface obtained by rotating the parabola $y = x^2$ about the y-axis.

42. Find an equation for the surface obtained by rotating the line $x = 3y$ about the x-axis.

43. Find an equation for the surface consisting of all points that are equidistant from the point $(-1, 0, 0)$ and the plane $x = 1$. Identify the surface.

44. Find an equation for the surface consisting of all points P for which the distance from P to the x-axis is twice the distance from P to the yz-plane. Identify the surface.

45. Show that if the point (a, b, c) lies on the hyperbolic paraboloid $z = y^2 - x^2$, then the lines with parametric equations $x = a + t$, $y = b + t$, $z = c + 2(b - a)t$ and $x = a + t$, $y = b - t$, $z = c - 2(b + a)t$ both lie entirely on this paraboloid. (This shows that the hyperbolic paraboloid is what is called a **ruled surface;** that is, it can be generated by the motion of a straight line. In fact, this exercise shows that through each point on the hyperbolic paraboloid there are two generating lines. The only other quadric surfaces that are ruled surfaces are cylinders, cones, and hyperboloids of one sheet.)

46. Show that the curve of intersection of the surfaces $x^2 + 2y^2 - z^2 + 3x = 1$ and $2x^2 + 4y^2 - 2z^2 - 5y = 0$ lies in a plane.

47. Graph the surfaces $z = x^2 + y^2$ and $z = 1 - y^2$ on a common screen using the domain $|x| \leq 1.2$, $|y| \leq 1.2$ and observe the curve of intersection of these surfaces. Show that the projection of this curve onto the xy-plane is an ellipse.

48. Investigate the family of surfaces $z = x^2 + y^2 + cxy$. In particular, you should determine the transitional values of c for which the surface changes from one type of quadric surface to another.

11.7 VECTOR FUNCTIONS AND SPACE CURVES

The functions that we have used so far have been real-valued functions. We now study functions whose values are vectors because such functions are needed to describe curves in space and the motion of particles in space.

In general, a function is defined as a rule that assigns to each element in the domain an element in the range. A **vector-valued function**, or **vector function**, is simply a function whose domain is a set of real numbers and whose range is a set of vectors. We are most interested in vector functions $\mathbf{r}$ whose values are three-dimensional vectors. This means that for every number t in the domain of $\mathbf{r}$ there is a unique vector in V_3 denoted by $\mathbf{r}(t)$. If $f(t)$, $g(t)$, and $h(t)$ are the components of the vector $\mathbf{r}(t)$, then f, g, and h are real-valued functions called the **component functions** of $\mathbf{r}$ and we can write

$$\mathbf{r}(t) = \langle f(t), g(t), h(t) \rangle = f(t)\,\mathbf{i} + g(t)\,\mathbf{j} + h(t)\,\mathbf{k}$$

We use the letter t to denote the independent variable because it represents time in most applications of vector functions.

EXAMPLE 1 If

$$\mathbf{r}(t) = \langle t^3, \ln(3 - t), \sqrt{t}\, \rangle$$

then the component functions are

$$f(t) = t^3 \qquad g(t) = \ln(3 - t) \qquad h(t) = \sqrt{t}$$

By our usual convention, the domain of $\mathbf{r}$ consists of all values of t for which the expression for $\mathbf{r}(t)$ is defined. The expressions t^3, $\ln(3 - t)$, and $\sqrt{t}$ are all defined when $3 - t > 0$ and $t \geqslant 0$. Therefore, the domain of $\mathbf{r}$ is the interval $[0, 3)$. ∎

The **limit** of a vector function $\mathbf{r}$ is defined by taking the limits of its component functions as follows:

---

(1) If $\mathbf{r}(t) = \langle f(t), g(t), h(t) \rangle$, then

$$\lim_{t \to a} \mathbf{r}(t) = \left\langle \lim_{t \to a} f(t), \lim_{t \to a} g(t), \lim_{t \to a} h(t) \right\rangle$$

provided the limits of the component functions exist.

Equivalently, we could have used an ε, δ definition (see Exercise 64). Limits of vector functions obey the same rules as limits of real-valued functions. (See Exercise 63.)

EXAMPLE 2 Find $\displaystyle\lim_{t \to 0} \mathbf{r}(t)$ where $\mathbf{r}(t) = (1 + t^3)\,\mathbf{i} + te^{-t}\mathbf{j} + \dfrac{\sin t}{t}\,\mathbf{k}$.

SOLUTION

$$\lim_{t \to 0} \mathbf{r}(t) = \left[\lim_{t \to 0}(1 + t^3)\right]\mathbf{i} + \left[\lim_{t \to 0} te^{-t}\right]\mathbf{j} + \left[\lim_{t \to 0} \frac{\sin t}{t}\right]\mathbf{k}$$

$$= \mathbf{i} + \mathbf{k} \qquad \text{[by (2.4.4)]}$$ ∎

A vector function **r** is **continuous at *a*** if

$$\lim_{t \to a} \mathbf{r}(t) = \mathbf{r}(a)$$

In view of Definition 1, we see that **r** is continuous at *a* if and only if its component functions *f*, *g*, and *h* are continuous at *a*.

There is a close connection between continuous vector functions and space curves. Suppose that *f*, *g*, and *h* are continuous real-valued functions on an interval *I*. Then the set *C* of all points (x, y, z) in space, where

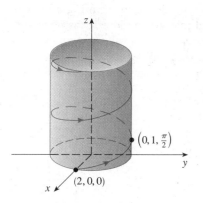

z

C

$P(f(t), g(t), h(t))$

0

y

$\mathbf{r}(t) = \langle f(t), g(t), h(t) \rangle$

x

FIGURE 1

(2) $$x = f(t) \qquad y = g(t) \qquad z = h(t)$$

and *t* varies throughout the interval *I*, is called a **space curve**. The equations in (2) are called **parametric equations of *C*** and *t* is called a **parameter**. We can think of *C* as being traced out by a moving particle whose position at time *t* is $(f(t), g(t), h(t))$. If we now consider the vector function $\mathbf{r}(t) = \langle f(t), g(t), h(t) \rangle$, then $\mathbf{r}(t)$ is the position vector of the point $P(f(t), g(t), h(t))$ on *C*. Thus any continuous vector function **r** defines a space curve *C* that is traced out by the tip of the moving vector $\mathbf{r}(t)$, as shown in Figure 1.

EXAMPLE 3   Describe the curve defined by the vector function

$$\mathbf{r}(t) = \langle 1 + t, 2 + 5t, -1 + 6t \rangle$$

SOLUTION   The corresponding parametric equations are

$$x = 1 + t \qquad y = 2 + 5t \qquad z = -1 + 6t$$

which we recognize from Equations 11.5.2 as parametric equations of a line passing through the point $(1, 2, -1)$ and parallel to the vector $\langle 1, 5, 6 \rangle$. Alternatively, we could observe that the function can be written as $\mathbf{r} = \mathbf{r}_0 + t\mathbf{v}$, where $\mathbf{r}_0 = \langle 1, 2, -1 \rangle$ and $\mathbf{v} = \langle 1, 5, 6 \rangle$, and this is the vector equation of a line as given by Equation 11.5.1.   ∎

EXAMPLE 4   Sketch the curve whose vector equation is

$$\mathbf{r}(t) = 2 \cos t \, \mathbf{i} + \sin t \, \mathbf{j} + t \, \mathbf{k}$$

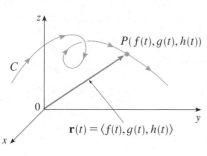

z

$(0, 1, \frac{\pi}{2})$

y

x $(2, 0, 0)$

FIGURE 2

SOLUTION   The parametric equations for this curve are

$$x = 2 \cos t \qquad y = \sin t \qquad z = t$$

Since $(x/2)^2 + y^2 = \cos^2 t + \sin^2 t = 1$, the curve must lie on the elliptical cylinder $x^2/4 + y^2 = 1$. Since $z = t$, the curve spirals upward around the cylinder as *t* increases. The curve, shown in Figure 2, is called a **helix**.   ∎

The corkscrew shape of the helix in Example 4 is familiar from its occurrence in coiled springs. It also occurs in the model of DNA (deoxyribonucleic acid, the genetic material of living cells). In 1953 James Watson and Francis Crick showed that the structure of the DNA molecule is that of two linked, parallel helices that are intertwined as in Figure 3.

Plane curves can also be represented in vector notation. For instance, the curve given by the parametric equations $x = \sin t$ and $y = \sin^2 t$ in Example 4 in Section 9.1

FIGURE 3

could also be described by the vector equation

$$\mathbf{r}(t) = \langle \sin t, \sin^2 t \rangle = \sin t\,\mathbf{i} + \sin^2 t\,\mathbf{j}$$

where $\mathbf{i} = \langle 1, 0 \rangle$ and $\mathbf{j} = \langle 0, 1 \rangle$.

USING COMPUTERS TO DRAW SPACE CURVES

Space curves are inherently more difficult to draw by hand than plane curves; for an accurate representation we need to use technology. Although most graphing calculators are presently unable to graph space curves, computer graphing programs with this capability are now widely available. For instance, Figure 4 shows a computer-generated graph of the curve with parametric equations

$$x = (4 + \sin 20t)\cos t \qquad y = (4 + \sin 20t)\sin t \qquad z = \cos 20t$$

It is called a **toroidal spiral** because it lies on a torus. Another interesting curve, the **trefoil knot,** with equations

$$x = (2 + \cos 1.5t)\cos t \qquad y = (2 + \cos 1.5t)\sin t \qquad z = \sin 1.5t$$

is graphed in Figure 5. It would not be easy to plot either of these curves by hand.

Even when a computer is used to draw a space curve, optical illusions make it difficult to get a good impression of what the curve really looks like. (This is especially true in Figure 5.) The next example shows how to cope with this problem.

EXAMPLE 5   Use a computer to sketch the curve with vector equation $\mathbf{r}(t) = \langle t, t^2, t^3 \rangle$. This curve is called a **twisted cubic.**

SOLUTION   We start by using the computer to plot the curve with parametric equations $x = t$, $y = t^2$, $z = t^3$ for $-2 \le t \le 2$. The result is shown in Figure 6(a), but it's hard to see the true nature of the curve from that graph alone. Most three-dimensional computer graphing programs allow the user to enclose a curve or surface in a box instead of displaying the coordinate axes. When we look at the same curve in a box in Figure 6(b), we have a much clearer picture of the curve. We

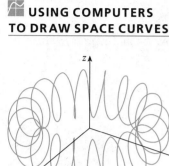

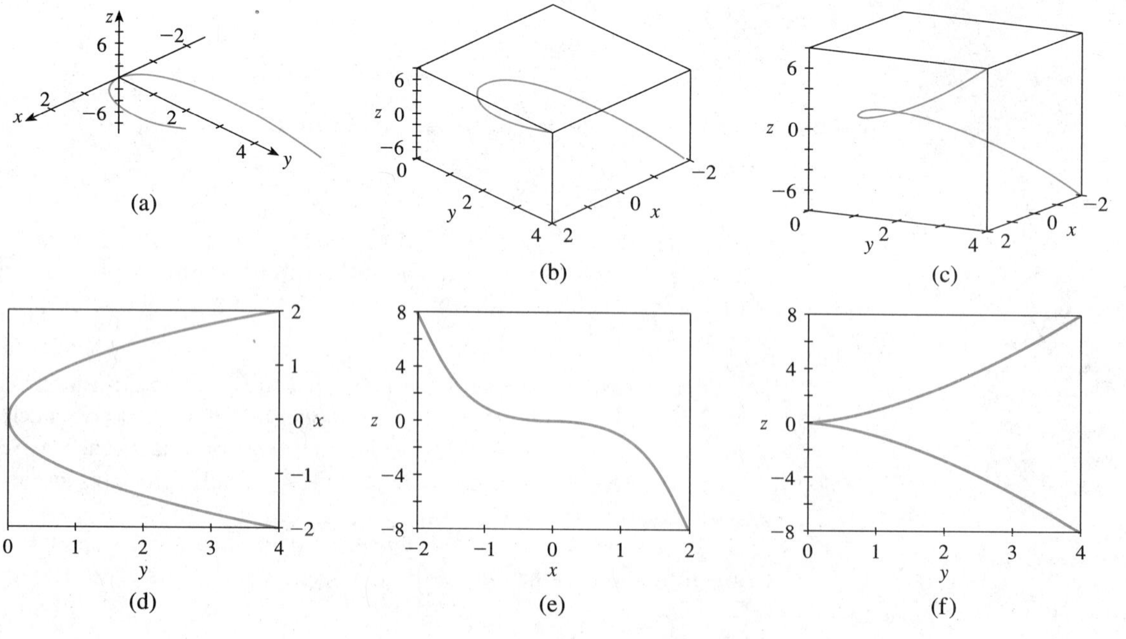

FIGURE 4
A toroidal spiral

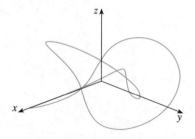

FIGURE 5
A trefoil knot

FIGURE 6
Views of the
twisted cubic

(a)

(b)

(c)

(d)

(e)

(f)

can see that it climbs from a lower corner of the box to the upper corner nearest us, and it twists as it climbs.

We get an even better idea of the curve when we view it from different vantage points. Part (c) shows the result of rotating the box to give another viewpoint. Parts (d), (e), and (f) show the views we get when we look directly at a face of the box. In particular, part (d) shows the view from directly above the box. It is the projection of the curve on the xy-plane, namely, the parabola $y = x^2$. Part (e) shows the projection on the xz-plane, the cubic curve $z = x^3$. It is now obvious why the given curve is called a twisted cubic. ∎

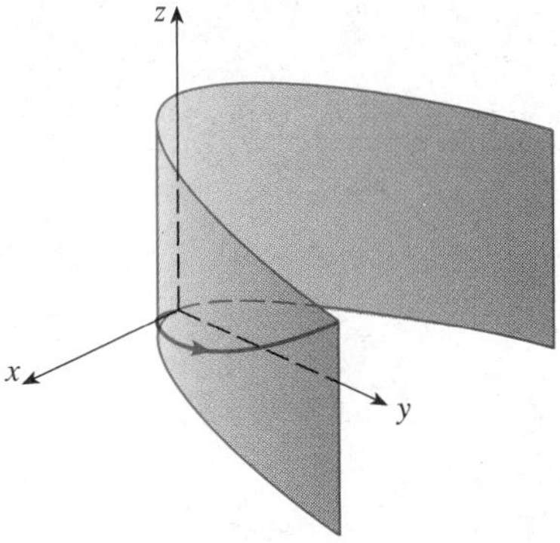

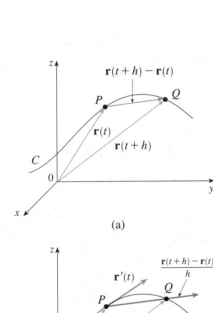

FIGURE 7

Another method of visualizing a space curve is to draw it on a surface. For instance, the twisted cubic in Example 5 lies on the parabolic cylinder $y = x^2$. (Eliminate the parameter from the first two parametric equations, $x = t$ and $y = t^2$.) Figure 7 shows both the cylinder and the twisted cubic, and we see that the curve moves upward from the origin along the surface of the cylinder.

A third method for visualizing the twisted cubic is to realize that it also lies on the cylinder $z = x^3$. So it can be viewed as the curve of intersection of the cylinders $y = x^2$ and $z = x^3$.

DERIVATIVES AND INTEGRALS

The **derivative r′** of a vector function **r** is defined just as for a real-valued function:

(3)
$$\frac{d\mathbf{r}}{dt} = \mathbf{r}'(t) = \lim_{h \to 0} \frac{\mathbf{r}(t + h) - \mathbf{r}(t)}{h}$$

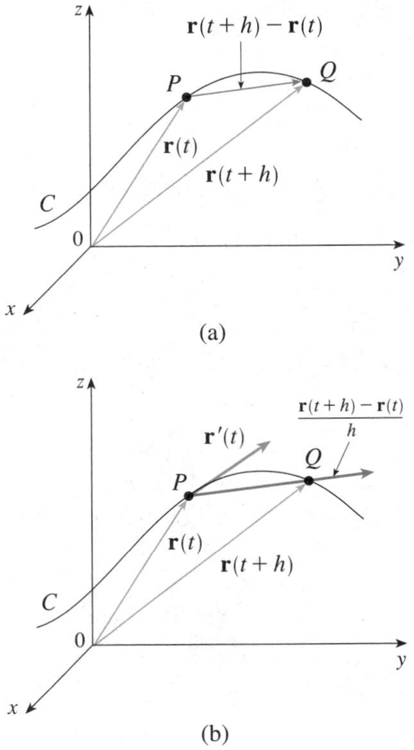

if this limit exists. The geometric significance of this definition is shown in Figure 8. If P and Q have position vectors $\mathbf{r}(t)$ and $\mathbf{r}(t + h)$, then $\overrightarrow{PQ}$ represents the vector $\mathbf{r}(t + h) - \mathbf{r}(t)$, which can therefore be regarded as a secant vector. If $h > 0$, the scalar multiple $(1/h)(\mathbf{r}(t + h) - \mathbf{r}(t))$ has the same direction as $\mathbf{r}(t + h) - \mathbf{r}(t)$. As $h \to 0$, it appears that this vector approaches a vector that lies on the tangent line. For this reason, the vector $\mathbf{r}'(t)$ is called the **tangent vector** to the curve defined by **r** at the point P, provided that $\mathbf{r}'(t)$ exists and $\mathbf{r}'(t) \neq \mathbf{0}$. The **tangent line** to C at P is defined to be the line through P parallel to the tangent vector $\mathbf{r}'(t)$. We will also have occasion to consider the **unit tangent vector,** which is

$$\mathbf{T}(t) = \frac{\mathbf{r}'(t)}{|\mathbf{r}'(t)|}$$

The following theorem gives us a convenient method for computing the derivative of a vector function **r**; just differentiate each component of **r**.

(a)

(b)

FIGURE 8

> **(4) THEOREM** If $\mathbf{r}(t) = \langle f(t), g(t), h(t) \rangle = f(t)\,\mathbf{i} + g(t)\,\mathbf{j} + h(t)\,\mathbf{k}$, where f, g, and h are differentiable functions, then
>
> $$\mathbf{r}'(t) = \langle f'(t), g'(t), h'(t) \rangle = f'(t)\,\mathbf{i} + g'(t)\,\mathbf{j} + h'(t)\,\mathbf{k}$$

PROOF

$$\mathbf{r}'(t) = \lim_{\Delta t \to 0} \frac{1}{\Delta t}[\mathbf{r}(t + \Delta t) - \mathbf{r}(t)]$$

$$= \lim_{\Delta t \to 0} \frac{1}{\Delta t}[\langle f(t + \Delta t), g(t + \Delta t), h(t + \Delta t)\rangle - \langle f(t), g(t), h(t)\rangle]$$

$$= \lim_{\Delta t \to 0} \left\langle \frac{f(t + \Delta t) - f(t)}{\Delta t}, \frac{g(t + \Delta t) - g(t)}{\Delta t}, \frac{h(t + \Delta t) - h(t)}{\Delta t} \right\rangle$$

$$= \left\langle \lim_{\Delta t \to 0} \frac{f(t + \Delta t) - f(t)}{\Delta t}, \lim_{\Delta t \to 0} \frac{g(t + \Delta t) - g(t)}{\Delta t}, \lim_{\Delta t \to 0} \frac{h(t + \Delta t) - h(t)}{\Delta t} \right\rangle$$

$$= \langle f'(t), g'(t), h'(t)\rangle \qquad \square$$

EXAMPLE 6
(a) Find the derivative of $\mathbf{r}(t) = (1 + t^3)\mathbf{i} + te^{-t}\mathbf{j} + \sin 2t\,\mathbf{k}$.
(b) Find the unit tangent vector at the point where $t = 0$.

SOLUTION
(a) Theorem 4 gives

$$\mathbf{r}'(t) = 3t^2\mathbf{i} + (1 - t)e^{-t}\mathbf{j} + 2\cos 2t\,\mathbf{k}$$

(b) Since $\mathbf{r}(0) = \mathbf{i}$ and $\mathbf{r}'(0) = \mathbf{j} + 2\,\mathbf{k}$, the unit tangent vector at the point $(1, 0, 0)$ is

$$\mathbf{T}(0) = \frac{\mathbf{r}'(0)}{|\mathbf{r}'(0)|} = \frac{\mathbf{j} + 2\,\mathbf{k}}{\sqrt{1 + 4}} = \frac{1}{\sqrt{5}}\mathbf{j} + \frac{2}{\sqrt{5}}\mathbf{k} \qquad \blacksquare$$

EXAMPLE 7   For the curve $\mathbf{r}(t) = \sqrt{t}\,\mathbf{i} + (2 - t)\mathbf{j}$, find $\mathbf{r}'(t)$ and sketch the position vector $\mathbf{r}(1)$ and the tangent vector $\mathbf{r}'(1)$.

SOLUTION   We have

$$\mathbf{r}'(t) = \frac{1}{2\sqrt{t}}\mathbf{i} - \mathbf{j} \qquad \text{and} \qquad \mathbf{r}'(1) = \frac{1}{2}\mathbf{i} - \mathbf{j}$$

The curve is a plane curve and elimination of the parameter from the equations $x = \sqrt{t}$, $y = 2 - t$ gives $y = 2 - x^2$, $x \geq 0$. In Figure 9 we draw the position vector $\mathbf{r}(1) = \mathbf{i} + \mathbf{j}$ starting at the origin and the tangent vector $\mathbf{r}'(1)$ starting at the corresponding point $(1, 1)$. $\qquad \blacksquare$

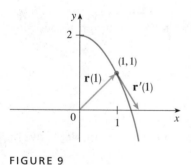

FIGURE 9

EXAMPLE 8   Find parametric equations for the tangent line to the helix with parametric equations

$$x = 2\cos t \qquad y = \sin t \qquad z = t$$

at the point $(0, 1, \pi/2)$.

The helix and the tangent line in Example 8 are shown in Figure 10.

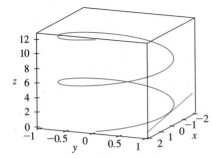

FIGURE 10

SOLUTION   The vector equation of the helix is $\mathbf{r}(t) = \langle 2\cos t, \sin t, t \rangle$, so

$$\mathbf{r}'(t) = \langle -2\sin t, \cos t, 1 \rangle$$

The parameter value corresponding to the point $(0, 1, \pi/2)$ is $t = \pi/2$, so the tangent vector there is $\mathbf{r}'(\pi/2) = \langle -2, 0, 1 \rangle$. The tangent line is the line through $(0, 1, \pi/2)$ parallel to the vector $\langle -2, 0, 1 \rangle$, so by Equations 11.5.2 its parametric equations are

$$x = -2t \qquad y = 1 \qquad z = \frac{\pi}{2} + t$$

∎

The next theorem shows that the differentiation formulas for real-valued functions have their counterparts for vector-valued functions.

---

(5) THEOREM   Suppose $\mathbf{u}$ and $\mathbf{v}$ are differentiable vector functions, c is a scalar, and f is a real-valued function. Then

1. $\dfrac{d}{dt}[\mathbf{u}(t) + \mathbf{v}(t)] = \mathbf{u}'(t) + \mathbf{v}'(t)$

2. $\dfrac{d}{dt}[c\mathbf{u}(t)] = c\mathbf{u}'(t)$

3. $\dfrac{d}{dt}[f(t)\mathbf{u}(t)] = f'(t)\mathbf{u}(t) + f(t)\mathbf{u}'(t)$

4. $\dfrac{d}{dt}[\mathbf{u}(t) \cdot \mathbf{v}(t)] = \mathbf{u}'(t) \cdot \mathbf{v}(t) + \mathbf{u}(t) \cdot \mathbf{v}'(t)$

5. $\dfrac{d}{dt}[\mathbf{u}(t) \times \mathbf{v}(t)] = \mathbf{u}'(t) \times \mathbf{v}(t) + \mathbf{u}(t) \times \mathbf{v}'(t)$

6. $\dfrac{d}{dt}[\mathbf{u}(f(t))] = f'(t)\mathbf{u}'(f(t))$ (Chain Rule)

This theorem can be proved either directly from Definition 3 or by using Theorem 4 and the corresponding differentiation formulas for real-valued functions. The proofs of Formulas 1, 2, 3, 5, and 6 are left as exercises.

PROOF OF FORMULA 4   Let

$$\mathbf{u}(t) = \langle f_1(t), f_2(t), f_3(t) \rangle \qquad \mathbf{v}(t) = \langle g_1(t), g_2(t), g_3(t) \rangle$$

Then $\mathbf{u}(t) \cdot \mathbf{v}(t) = f_1(t)g_1(t) + f_2(t)g_2(t) + f_3(t)g_3(t) = \displaystyle\sum_{i=1}^{3} f_i(t)g_i(t)$

so the ordinary Product Rule gives

$$\frac{d}{dt}[\mathbf{u}(t) \cdot \mathbf{v}(t)] = \frac{d}{dt}\sum_{i=1}^{3} f_i(t)g_i(t) = \sum_{i=1}^{3} \frac{d}{dt}[f_i(t)g_i(t)]$$

$$= \sum_{i=1}^{3} [f_i'(t)g_i(t) + f_i(t)g_i'(t)]$$

$$= \sum_{i=1}^{3} f_i'(t)g_i(t) + \sum_{i=1}^{3} f_i(t)g_i'(t)$$

$$= \mathbf{u}'(t) \cdot \mathbf{v}(t) + \mathbf{u}(t) \cdot \mathbf{v}'(t)$$

□

EXAMPLE 9   Show that if $|\mathbf{r}(t)| = c$ (a constant), then $\mathbf{r}'(t)$ is orthogonal to $\mathbf{r}(t)$ for all t.

SOLUTION   Since

$$\mathbf{r}(t) \cdot \mathbf{r}(t) = |\mathbf{r}(t)|^2 = c^2$$

and c^2 is a constant, Formula 4 of Theorem 5 gives

$$0 = \frac{d}{dt}[\mathbf{r}(t) \cdot \mathbf{r}(t)] = \mathbf{r}'(t) \cdot \mathbf{r}(t) + \mathbf{r}(t) \cdot \mathbf{r}'(t) = 2\mathbf{r}'(t) \cdot \mathbf{r}(t)$$

Thus $\mathbf{r}'(t) \cdot \mathbf{r}(t) = 0$, which says that $\mathbf{r}'(t)$ is orthogonal to $\mathbf{r}(t)$.

Geometrically, this result says that if a curve lies on a sphere, then the tangent vector $\mathbf{r}'(t)$ is always perpendicular to the position vector $\mathbf{r}(t)$. ∎

The **definite integral** of a continuous vector function $\mathbf{r}(t)$ can be defined in much the same way as for real-valued functions except that the integral is a vector. But then we can express the integral of $\mathbf{r}$ in terms of the integrals of its component functions f, g, and h as follows. (We use the notation of Chapter 4.)

$$\int_a^b \mathbf{r}(t)\, dt = \lim_{\|P\| \to 0} \sum_{i=1}^{n} \mathbf{r}(t_i^*)\, \Delta t_i$$

$$= \lim_{\|P\| \to 0} \left[\left(\sum_{i=1}^{n} f(t_i^*)\, \Delta t_i \right) \mathbf{i} + \left(\sum_{i=1}^{n} g(t_i^*)\, \Delta t_i \right) \mathbf{j} + \left(\sum_{i=1}^{n} h(t_i^*)\, \Delta t_i \right) \mathbf{k} \right]$$

and so

$$\int_a^b \mathbf{r}(t)\, dt = \left(\int_a^b f(t)\, dt \right) \mathbf{i} + \left(\int_a^b g(t)\, dt \right) \mathbf{j} + \left(\int_a^b h(t)\, dt \right) \mathbf{k}$$

This means that we can evaluate an integral of a vector function by integrating each component function.

The Fundamental Theorem of Calculus for continuous vector functions says that

$$\int_a^b \mathbf{r}(t)\, dt = \mathbf{R}(t) \Big]_a^b = \mathbf{R}(b) - \mathbf{R}(a)$$

where $\mathbf{R}$ is an antiderivative of $\mathbf{r}$, that is, $\mathbf{R}'(t) = \mathbf{r}(t)$. We use the notation $\int \mathbf{r}(t)\, dt$ for indefinite integrals (antiderivatives).

EXAMPLE 10   If $\mathbf{r}(t) = 2\cos t\, \mathbf{i} + \sin t\, \mathbf{j} + 2t\, \mathbf{k}$, then

$$\int \mathbf{r}(t)\, dt = \left(\int 2\cos t\, dt \right) \mathbf{i} + \left(\int \sin t\, dt \right) \mathbf{j} + \left(\int 2t\, dt \right) \mathbf{k}$$

$$= 2\sin t\, \mathbf{i} - \cos t\, \mathbf{j} + t^2 \mathbf{k} + \mathbf{C}$$

where $\mathbf{C}$ is a vector constant of integration, and

$$\int_0^{\pi/2} \mathbf{r}(t)\, dt = \left[2\sin t\, \mathbf{i} - \cos t\, \mathbf{j} + t^2 \mathbf{k} \right]_0^{\pi/2} = 2\mathbf{i} + \mathbf{j} + \frac{\pi^2}{4} \mathbf{k}$$

∎

EXERCISES 11.7

1–6 ■ Match the parametric equations with the graphs (labeled I–VI). Give reasons for your choices.

1. $x = \cos 4t, \quad y = t, \quad z = \sin 4t$

2. $x = t^2 - 2, \quad y = t^3, \quad z = t^4 + 1$

3. $x = t, \quad y = 1/(1 + t^2), \quad z = t^2$

4. $x = \sin 3t \cos t, \quad y = \sin 3t \sin t, \quad z = t$

5. $x = \cos t, \quad y = \sin t, \quad z = \sin 5t$

6. $x = \cos t, \quad y = \sin t, \quad z = \ln t$

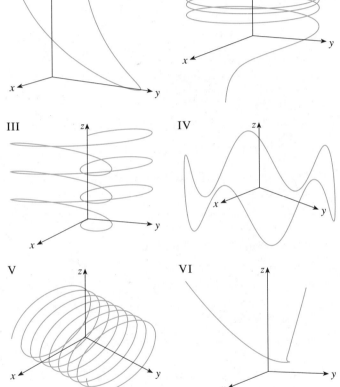

I

II

III

IV

V

VI

7–14 ■ Sketch the curve with the given vector equation. Indicate with an arrow the direction in which t increases.

7. $\mathbf{r}(t) = \langle t, -t, 2t \rangle$

8. $\mathbf{r}(t) = \langle t^2, t, 2 \rangle$

9. $\mathbf{r}(t) = \langle \sin t, 3, \cos t \rangle$

10. $\mathbf{r}(t) = \langle \sin t, t, \cos t \rangle$

11. $\mathbf{r}(t) = (t^4 + 1)\mathbf{i} + t\mathbf{k}$

12. $\mathbf{r}(t) = t\mathbf{i} + t\mathbf{j} + \cos t\mathbf{k}$

13. $\mathbf{r}(t) = t^2\mathbf{i} + t^4\mathbf{j} + t^6\mathbf{k}$

14. $\mathbf{r}(t) = \sin t\mathbf{i} + \sin t\mathbf{j} + \sqrt{2} \cos t\mathbf{k}$

15. Show that the curve with parametric equations $x = t \cos t$, $y = t \sin t$, $z = t$ lies on the cone $z^2 = x^2 + y^2$, and use this fact to help sketch the curve.

16. Show that the curve with parametric equations $x = \sin t$, $y = \cos t$, $z = \sin^2 t$ is the curve of intersection of the surfaces $z = x^2$ and $x^2 + y^2 = 1$. Use this fact to help sketch the curve.

17–20 ■ Use a computer to graph the curve with the given vector equation. Make sure you choose a parameter domain and viewpoints that reveal the true nature of the curve.

17. $\mathbf{r}(t) = \langle \sin t, \cos t, t^2 \rangle$

18. $\mathbf{r}(t) = \langle t^2, t^3 - t, t \rangle$

19. $\mathbf{r}(t) = \langle \sqrt{t}, t, t^2 - 2 \rangle$

20. $\mathbf{r}(t) = \langle \sin t, \sin 2t, \sin 3t \rangle$

21. Graph the curve with parametric equations $x = (1 + \cos 16t) \cos t$, $y = (1 + \cos 16t) \sin t$, $z = 1 + \cos 16t$. Explain the appearance of the graph by showing that it lies on a cone.

22. Graph the curve with parametric equations $x = \sqrt{1 - 0.25 \cos^2 10t} \cos t$, $y = \sqrt{1 - 0.25 \cos^2 10t} \sin t$, $z = 0.5 \cos 10t$. Explain the appearance of the graph by showing that it lies on a sphere.

23–26 ■ Find the limit.

23. $\lim\limits_{t \to 0} \langle t, \cos t, 2 \rangle$

24. $\lim\limits_{t \to 0} \left\langle \dfrac{1 - \cos t}{t}, t^3, e^{-1/t^2} \right\rangle$

25. $\lim\limits_{t \to 1} \left\langle \sqrt{t + 3}\,\mathbf{i} + \dfrac{t - 1}{t^2 - 1}\mathbf{j} + \dfrac{\tan t}{t}\mathbf{k} \right\rangle$

26. $\lim\limits_{t \to \infty} \left\langle e^{-t}\mathbf{i} + \dfrac{t - 1}{t + 1}\mathbf{j} + \tan^{-1}t\,\mathbf{k} \right\rangle$

27–34 ■ Find the domain and derivative of the vector function.

27. $\mathbf{r}(t) = \langle t, t^2, t^3 \rangle$

28. $\mathbf{r}(t) = \langle t^2 - 4, \sqrt{t - 4}, \sqrt{6 - t} \rangle$

29. $\mathbf{r}(t) = \mathbf{i} + \tan t\,\mathbf{j} + \sec t\,\mathbf{k}$

30. $\mathbf{r}(t) = te^{2t}\mathbf{i} + \dfrac{t - 1}{t + 1}\mathbf{j} + \tan^{-1}t\,\mathbf{k}$

31. $\mathbf{r}(t) = \ln(4 - t^2)\mathbf{i} + \sqrt{1 + t}\,\mathbf{j} - 4e^{3t}\mathbf{k}$

32. $\mathbf{r}(t) = e^{-t}\cos t\,\mathbf{i} + e^{-t}\sin t\,\mathbf{j} + \ln|t|\,\mathbf{k}$

33. $\mathbf{r}(t) = \mathbf{a} + t\mathbf{b} + t^2\mathbf{c}$

34. $\mathbf{r}(t) = t\mathbf{a} \times (\mathbf{b} + t\mathbf{c})$

35–40 ■
(a) Sketch the plane curve with the given vector equation.
(b) Find $\mathbf{r}'(t)$.
(c) Sketch the position vector $\mathbf{r}(t)$ and the tangent vector $\mathbf{r}'(t)$ for the given value of t.

35. $\mathbf{r}(t) = \langle \cos t, \sin t \rangle, \quad t = \pi/4$

36. $\mathbf{r}(t) = \langle t^3, t^2 \rangle, \quad t = 1$

37. $\mathbf{r}(t) = (1 + t)\mathbf{i} + t^2\mathbf{j}, \quad t = 1$

38. $\mathbf{r}(t) = 2\sin t\,\mathbf{i} + 3\cos t\,\mathbf{j}, \quad t = \pi/3$

39. $\mathbf{r}(t) = e^t\mathbf{i} + e^{-2t}\mathbf{j}, \quad t = 0$

40. $\mathbf{r}(t) = \sec t\,\mathbf{i} + \tan t\,\mathbf{j}, \quad t = \pi/4$

41–44 ■ Find the unit tangent vector $\mathbf{T}(t)$ at the point with the given value of the parameter t.

41. $\mathbf{r}(t) = \langle 2t, 3t^2, 4t^3 \rangle, \quad t = 1$

42. $\mathbf{r}(t) = \langle e^{2t}, e^{-2t}, te^{2t} \rangle, \quad t = 0$

43. $\mathbf{r}(t) = t\mathbf{i} + 2\sin t\,\mathbf{j} + 3\cos t\,\mathbf{k}, \quad t = \pi/6$

44. $\mathbf{r}(t) = e^{2t}\cos t\,\mathbf{i} + e^{2t}\sin t\,\mathbf{j} + e^{2t}\mathbf{k}, \quad t = \pi/2$

45–48 ■ Find parametric equations for the tangent line to the curve with the given parametric equations at the specified point.

45. $x = t, \ y = t^2, \ z = t^3; \quad (1, 1, 1)$

46. $x = 1 + 2t, \ y = 1 + t - t^2, \ z = 1 - t + t^2 - t^3; \quad (1, 1, 1)$

47. $x = t\cos 2\pi t, \ y = t\sin 2\pi t, \ z = 4t; \quad (0, \frac{1}{4}, 1)$

48. $x = \sin \pi t, \ y = \sqrt{t}, \ z = \cos \pi t; \quad (0, 1, -1)$

49–50 ■ Find parametric equations for the tangent line to the curve with the given parametric equations at the specified point. Illustrate by graphing both the curve and the tangent line on a common screen.

49. $x = t, \ y = \sqrt{2}\cos t, \ z = \sqrt{2}\sin t; \quad (\pi/4, 1, 1)$

50. $x = \cos t, \ y = 3e^{2t}, \ z = 3e^{-2t}; \quad (1, 3, 3)$

51. The curves $\mathbf{r}_1(t) = \langle t, t^2, t^3 \rangle$ and $\mathbf{r}_2(t) = \langle \sin t, \sin 2t, t \rangle$ intersect at the origin. Find their angle of intersection correct to the nearest degree.

52. At what point do the curves $\mathbf{r}_1(t) = \langle t, 1 - t, 3 + t^2 \rangle$ and $\mathbf{r}_2(s) = \langle 3 - s, s - 2, s^2 \rangle$ intersect? Find their angle of intersection correct to the nearest degree.

53. Show that the curve with parametric equations $x = t^2$, $y = 1 - 3t, \ z = 1 + t^3$ passes through the points $(1, 4, 0)$ and $(9, -8, 28)$ but not the point $(4, 7, -6)$.

54. (a) Find the point of intersection of the tangent lines to the curve $\mathbf{r}(t) = \langle \sin \pi t, 2\sin \pi t, \cos \pi t \rangle$ at the points where $t = 0$ and $t = 0.5$.
  (b) Illustrate by graphing the curve and both tangent lines.

55. Try to sketch by hand the curve of intersection of the circular cylinder $x^2 + y^2 = 4$ and the parabolic cylinder $z = x^2$. Then find parametric equations for this curve using the parameter t, where $x = 2\cos t$, and use these equations and a computer to graph the curve.

56. Try to sketch by hand the curve of intersection of the parabolic cylinder $y = x^2$ and the top half of the ellipsoid $x^2 + 4y^2 + 4z^2 = 16$. Then find parametric equations for

this curve using the parameter t, where $x = t$, and use these equations and a computer to graph the curve.

57–60 ■ Evaluate the integral.

57. $\int_0^1 (t\mathbf{i} + t^2\mathbf{j} + t^3\mathbf{k})\,dt$

58. $\int_1^2 [(1 + t^2)\mathbf{i} - 4t^4\mathbf{j} - (t^2 - 1)\mathbf{k}]\,dt$

59. $\int_0^{\pi/4} (\cos 2t\,\mathbf{i} + \sin 2t\,\mathbf{j} + t\sin t\,\mathbf{k})\,dt$

60. $\int_1^4 \left(\sqrt{t}\,\mathbf{i} + te^{-t}\mathbf{j} + \dfrac{1}{t^2}\,\mathbf{k} \right) dt$

61. Find $\mathbf{r}(t)$ if $\mathbf{r}'(t) = t^2\mathbf{i} + 4t^3\mathbf{j} - t^2\mathbf{k}$ and $\mathbf{r}(0) = \mathbf{j}$.

62. Find $\mathbf{r}(t)$ if $\mathbf{r}'(t) = \sin t\,\mathbf{i} - \cos t\,\mathbf{j} + 2t\,\mathbf{k}$ and $\mathbf{r}(0) = \mathbf{i} + \mathbf{j} + 2\mathbf{k}$.

63. Suppose $\mathbf{u}$ and $\mathbf{v}$ are vector functions that possess limits as $t \to a$ and let c be a constant. Prove the following properties of limits.
  (a) $\lim\limits_{t \to a} [\mathbf{u}(t) + \mathbf{v}(t)] = \lim\limits_{t \to a} \mathbf{u}(t) + \lim\limits_{t \to a} \mathbf{v}(t)$
  (b) $\lim\limits_{t \to a} c\mathbf{u}(t) = c\lim\limits_{t \to a} \mathbf{u}(t)$
  (c) $\lim\limits_{t \to a} [\mathbf{u}(t) \cdot \mathbf{v}(t)] = \lim\limits_{t \to a} \mathbf{u}(t) \cdot \lim\limits_{t \to a} \mathbf{v}(t)$
  (d) $\lim\limits_{t \to a} [\mathbf{u}(t) \times \mathbf{v}(t)] = \lim\limits_{t \to a} \mathbf{u}(t) \times \lim\limits_{t \to a} \mathbf{v}(t)$

64. Show that $\lim_{t \to a} \mathbf{r}(t) = \mathbf{b}$ if and only if for every $\varepsilon > 0$ there is a number $\delta > 0$ such that $|\mathbf{r}(t) - \mathbf{b}| < \varepsilon$ whenever $0 < |t - a| < \delta$.

65. Prove Formula 1 of Theorem 5.

66. Prove Formula 3 of Theorem 5.

67. Prove Formula 5 of Theorem 5.

68. Prove Formula 6 of Theorem 5.

69. If $\mathbf{u}(t) = \mathbf{i} - 2t^2\mathbf{j} + 3t^3\mathbf{k}$ and $\mathbf{v}(t) = t\mathbf{i} + \cos t\,\mathbf{j} + \sin t\,\mathbf{k}$, find $D_t[\mathbf{u}(t) \cdot \mathbf{v}(t)]$.

70. If $\mathbf{u}$ and $\mathbf{v}$ are the vector functions in Exercise 69, find $D_t[\mathbf{u}(t) \times \mathbf{v}(t)]$.

71. Show that if $\mathbf{r}$ is a vector function such that $\mathbf{r}'' = (\mathbf{r}')'$ exists, then

$$\frac{d}{dt} [\mathbf{r}(t) \times \mathbf{r}'(t)] = \mathbf{r}(t) \times \mathbf{r}''(t)$$

72. Find an expression for $\dfrac{d}{dt} [\mathbf{u}(t) \cdot (\mathbf{v}(t) \times \mathbf{w}(t))]$.

73. If $\mathbf{r}(t) \neq \mathbf{0}$, show that $\dfrac{d}{dt} |\mathbf{r}(t)| = \dfrac{1}{|\mathbf{r}(t)|} \mathbf{r}(t) \cdot \mathbf{r}'(t)$.

74. If a curve has the property that the position vector $\mathbf{r}(t)$ is always perpendicular to the tangent vector $\mathbf{r}'(t)$, show that the curve lies on a sphere with center the origin.

75. If $\mathbf{u}(t) = \mathbf{r}(t) \cdot [\mathbf{r}'(t) \times \mathbf{r}''(t)]$, show that

$$\mathbf{u}'(t) = \mathbf{r}(t) \cdot [\mathbf{r}'(t) \times \mathbf{r}'''(t)]$$

11.8 ARC LENGTH AND CURVATURE

Recall that we defined the length of a plane curve $x = f(t)$, $y = g(t)$, $a \leq t \leq b$, as the limit of lengths of inscribed polygons and, for the case where f' and g' are continuous, we arrived at the formula

(1)
$$L = \int_a^b \sqrt{[f'(t)]^2 + [g'(t)]^2} \, dt = \int_a^b \sqrt{\left(\frac{dx}{dt}\right)^2 + \left(\frac{dy}{dt}\right)^2} \, dt$$

in Theorem 9.3.4.

The length of a space curve is defined in exactly the same way (see Figure 1). Suppose that the curve has the vector equation $\mathbf{r}(t) = \langle f(t), g(t), h(t) \rangle$, $a \leq t \leq b$, or, equivalently, the parametric equations $x = f(t)$, $y = g(t)$, $z = h(t)$, where f', g', and h' are continuous. If the curve is traversed exactly once as t increases from a to b, then it can be shown that its length is

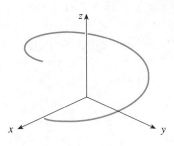

FIGURE 1

(2)
$$L = \int_a^b \sqrt{[f'(t)]^2 + [g'(t)]^2 + [h'(t)]^2} \, dt$$
$$= \int_a^b \sqrt{\left(\frac{dx}{dt}\right)^2 + \left(\frac{dy}{dt}\right)^2 + \left(\frac{dz}{dt}\right)^2} \, dt$$

Notice that both of the arc length formulas (1) and (2) can be put into the more compact form

(3)
$$L = \int_a^b |\mathbf{r}'(t)| \, dt$$

because, for plane curves $\mathbf{r}(t) = f(t)\,\mathbf{i} + g(t)\,\mathbf{j}$,

$$|\mathbf{r}'(t)| = |f'(t)\,\mathbf{i} + g'(t)\,\mathbf{j}| = \sqrt{[f'(t)]^2 + [g'(t)]^2}$$

whereas, for space curves $\mathbf{r}(t) = f(t)\,\mathbf{i} + g(t)\,\mathbf{j} + h(t)\,\mathbf{k}$,

$$|\mathbf{r}'(t)| = |f'(t)\,\mathbf{i} + g'(t)\,\mathbf{j} + h'(t)\,\mathbf{k}| = \sqrt{[f'(t)]^2 + [g'(t)]^2 + [h'(t)]^2}$$

Figure 2 shows the arc of the helix whose length is computed in Example 1.

EXAMPLE 1 Find the length of the arc of the circular helix with vector equation $\mathbf{r}(t) = \cos t\,\mathbf{i} + \sin t\,\mathbf{j} + t\,\mathbf{k}$ from the point $(1, 0, 0)$ to the point $(1, 0, 2\pi)$.

SOLUTION Since $\mathbf{r}'(t) = -\sin t\,\mathbf{i} + \cos t\,\mathbf{j} + \mathbf{k}$, we have

$$|\mathbf{r}'(t)| = \sqrt{(-\sin t)^2 + \cos^2 t + 1} = \sqrt{2}$$

The arc from $(1, 0, 0)$ to $(1, 0, 2\pi)$ is described by the parameter interval $0 \leq t \leq 2\pi$ and so, from Formula 3, we have

$$L = \int_0^{2\pi} |\mathbf{r}'(t)| \, dt = \int_0^{2\pi} \sqrt{2} \, dt = 2\sqrt{2}\,\pi$$

■

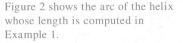

FIGURE 2

A curve given by a vector function $\mathbf{r}(t)$ on an interval I is called **smooth** if $\mathbf{r}'$ is continuous and $\mathbf{r}'(t) \neq \mathbf{0}$ (except possibly at any endpoints of I). The significance of

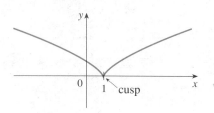

FIGURE 3
The curve $\mathbf{r}(t) = \langle 1 + t^3, t^2 \rangle$

the condition $\mathbf{r}'(t) \neq \mathbf{0}$ is illustrated by the graph of the semicubical parabola $\mathbf{r}(t) = \langle 1 + t^3, t^2 \rangle$ that is shown in Figure 3. Since $\mathbf{r}'(t) = \langle 3t^2, 2t \rangle$, we have $\mathbf{r}'(0) = \langle 0, 0 \rangle = \mathbf{0}$. The point that corresponds to $t = 0$ is $(1, 0)$ and here there is a sharp corner that is called a **cusp.** We can think of a smooth curve as a curve with no such cusp. A curve, such as the semicubical parabola, that is made up of a finite number of smooth pieces is called **piecewise smooth.** The arc length formula (3) holds for piecewise-smooth curves.

A single curve C can be represented by more than one vector function. For instance, the twisted cubic

$$(4) \qquad \mathbf{r}_1(t) = \langle t, t^2, t^3 \rangle \qquad 1 \leq t \leq 2$$

could also be represented by the function

$$(5) \qquad \mathbf{r}_2(u) = \langle e^u, e^{2u}, e^{3u} \rangle \qquad 0 \leq u \leq \ln 2$$

where the connection between the parameters t and u is given by $t = e^u$. We say that Equations 4 and 5 are **parametrizations** of the curve C. If we were to use Equation 3 to compute the length of C using Equations 4 and 5, we would get the same answer. In general, it can be shown (see Exercise 43) that when Equation 3 is used to compute the length of any piecewise-smooth curve, the arc length is independent of the parametrization that is used.

Now we suppose that C is a piecewise-smooth curve given by a vector function $\mathbf{r}(t) = f(t)\mathbf{i} + g(t)\mathbf{j} + h(t)\mathbf{k}$, $a \leq t \leq b$, and at least one of f, g, h is one-to-one on (a, b). As in Section 8.2, we define its **arc length function** s by

$$(6) \qquad s(t) = \int_a^t |\mathbf{r}'(u)| \, du = \int_a^t \sqrt{\left(\frac{dx}{du}\right)^2 + \left(\frac{dy}{du}\right)^2 + \left(\frac{dz}{du}\right)^2} \, du$$

Thus $s(t)$ is the length of the part of C between $\mathbf{r}(a)$ and $\mathbf{r}(t)$ (see Figure 4). If we differentiate both sides of Equation 6 using Part 1 of the Fundamental Theorem of Calculus, we obtain

$$(7) \qquad \frac{ds}{dt} = |\mathbf{r}'(t)|$$

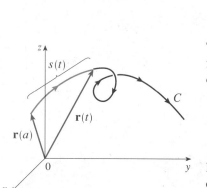

FIGURE 4

It is often useful to **parametrize a curve with respect to arc length** because arc length arises naturally from the shape of the curve and does not depend on a particular coordinate system. If a curve $\mathbf{r}(t)$ is already given in terms of a parameter t and $s(t)$ is the arc length function given by Equation 6, then we may be able to solve for t as a function of s: $t = t(s)$. Then the curve can be reparametrized in terms of s by substituting for t: $\mathbf{r} = \mathbf{r}(t(s))$.

EXAMPLE 2   Reparametrize the helix $\mathbf{r}(t) = \cos t\,\mathbf{i} + \sin t\,\mathbf{j} + t\,\mathbf{k}$ with respect to arc length measured from $(1, 0, 0)$ in the direction of increasing t.

SOLUTION   The initial point $(1, 0, 0)$ corresponds to the parameter value $t = 0$. From Example 1 we have

$$\frac{ds}{dt} = |\mathbf{r}'(t)| = \sqrt{2}$$

and so
$$s = s(t) = \int_0^t |\mathbf{r}'(u)| \, du = \int_0^t \sqrt{2} \, du = \sqrt{2}\, t$$

Therefore, $t = s/\sqrt{2}$ and the required reparametrization is obtained by substituting for t:

$$\mathbf{r}(t(s)) = \cos(s/\sqrt{2})\,\mathbf{i} + \sin(s/\sqrt{2})\,\mathbf{j} + (s/\sqrt{2})\,\mathbf{k}$$

CURVATURE

If C is a smooth curve defined by the vector function $\mathbf{r}$, then $\mathbf{r}'(t) \neq \mathbf{0}$. Recall that the unit tangent vector $\mathbf{T}(t)$ is given by

$$\mathbf{T}(t) = \frac{\mathbf{r}'(t)}{|\mathbf{r}'(t)|}$$

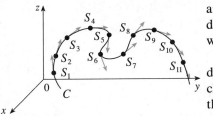

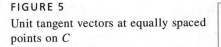

FIGURE 5
Unit tangent vectors at equally spaced points on C

and indicates the direction of the curve. From Figure 5 you can see that $\mathbf{T}(t)$ changes direction very slowly when C is fairly straight, but it changes direction more quickly when C bends or twists more sharply.

The curvature of C at a given point is a measure of how quickly the curve changes direction at that point. Specifically, we define it to be the magnitude of the rate of change of the unit tangent vector with respect to arc length. (We use arc length so that the curvature will be independent of the parametrization.)

> **(8) DEFINITION** The **curvature** of a curve is
> $$\kappa = \left| \frac{d\mathbf{T}}{ds} \right|$$
> where $\mathbf{T}$ is the unit tangent vector.

The curvature is easier to compute if it is expressed in terms of the parameter t instead of s, so we use the Chain Rule (Theorem 11.7.5, Formula 6) to write

$$\frac{d\mathbf{T}}{dt} = \frac{d\mathbf{T}}{ds}\frac{ds}{dt} \qquad \text{and} \qquad \kappa = \left| \frac{d\mathbf{T}}{ds} \right| = \left| \frac{d\mathbf{T}/dt}{ds/dt} \right|$$

But $ds/dt = |\mathbf{r}'(t)|$ from Equation 7, so

(9)
$$\kappa(t) = \frac{|\mathbf{T}'(t)|}{|\mathbf{r}'(t)|}$$

EXAMPLE 3   Show that the curvature of a circle of radius a is $1/a$.

SOLUTION   We can take the circle to have center the origin, and then a parametrization is

$$\mathbf{r}(t) = a\cos t\,\mathbf{i} + a\sin t\,\mathbf{j}$$

Therefore
$$\mathbf{r}'(t) = -a\sin t\,\mathbf{i} + a\cos t\,\mathbf{j} \qquad \text{and} \qquad |\mathbf{r}'(t)| = a$$

so
$$\mathbf{T}(t) = \frac{\mathbf{r}'(t)}{|\mathbf{r}'(t)|} = -\sin t\,\mathbf{i} + \cos t\,\mathbf{j}$$

and
$$\mathbf{T}'(t) = -\cos t\,\mathbf{i} - \sin t\,\mathbf{j}$$

This gives $|\mathbf{T}'(t)| = 1$, so using Equation 9, we have

$$\kappa(t) = \frac{|\mathbf{T}'(t)|}{|\mathbf{r}'(t)|} = \frac{1}{a}$$ ∎

The result of Example 3 shows that small circles have large curvature and large circles have small curvature, in accordance with our intuition. We can see directly from the definition of curvature that the curvature of a straight line is always 0 because the tangent vector is constant.

Although Formula 9 can be used in all cases to compute the curvature, the formula given by the following theorem is often more convenient to apply.

> **(10) THEOREM**   The curvature of the curve given by the vector function $\mathbf{r}$ is
>
> $$\kappa(t) = \frac{|\mathbf{r}'(t) \times \mathbf{r}''(t)|}{|\mathbf{r}'(t)|^3}$$

PROOF   Since $\mathbf{T} = \mathbf{r}'/|\mathbf{r}'|$ and $|\mathbf{r}'| = ds/dt$, we have

$$\mathbf{r}' = |\mathbf{r}'|\mathbf{T} = \frac{ds}{dt}\mathbf{T}$$

so the Product Rule (Theorem 11.7.5, Formula 3) gives

$$\mathbf{r}'' = \frac{d^2s}{dt^2}\mathbf{T} + \frac{ds}{dt}\mathbf{T}'$$

Using the fact that $\mathbf{T} \times \mathbf{T} = \mathbf{0}$ (Example 2 in Section 11.4), we have

$$\mathbf{r}' \times \mathbf{r}'' = \left(\frac{ds}{dt}\right)^2 (\mathbf{T} \times \mathbf{T}')$$

Now $|\mathbf{T}(t)| = 1$ for all t, so $\mathbf{T}$ and $\mathbf{T}'$ are orthogonal by Example 9 in Section 11.7. Therefore, by Theorem 11.4.6,

$$|\mathbf{r}' \times \mathbf{r}''| = \left(\frac{ds}{dt}\right)^2 |\mathbf{T} \times \mathbf{T}'| = \left(\frac{ds}{dt}\right)^2 |\mathbf{T}||\mathbf{T}'| = \left(\frac{ds}{dt}\right)^2 |\mathbf{T}'|$$

Thus $$|\mathbf{T}'| = \frac{|\mathbf{r}' \times \mathbf{r}''|}{(ds/dt)^2} = \frac{|\mathbf{r}' \times \mathbf{r}''|}{|\mathbf{r}'|^2}$$

and $$\kappa = \frac{|\mathbf{T}'|}{|\mathbf{r}'|} = \frac{|\mathbf{r}' \times \mathbf{r}''|}{|\mathbf{r}'|^3}$$ □

EXAMPLE 4   Find the curvature of the twisted cubic $\mathbf{r}(t) = \langle t, t^2, t^3 \rangle$ at a general point and at $(0,0,0)$.

SOLUTION   We first compute the required ingredients:

$$\mathbf{r}'(t) = \langle 1, 2t, 3t^2 \rangle \qquad \mathbf{r}''(t) = \langle 0, 2, 6t \rangle$$

$$|\mathbf{r}'(t)| = \sqrt{1 + 4t^2 + 9t^4}$$

$$\mathbf{r}'(t) \times \mathbf{r}''(t) = \begin{vmatrix} \mathbf{i} & \mathbf{j} & \mathbf{k} \\ 1 & 2t & 3t^2 \\ 0 & 2 & 6t \end{vmatrix} = 6t^2\mathbf{i} - 6t\mathbf{j} + 2\mathbf{k}$$

$$|\mathbf{r}'(t) \times \mathbf{r}''(t)| = \sqrt{36t^4 + 36t^2 + 4} = 2\sqrt{9t^4 + 9t^2 + 1}$$

Theorem 10 then gives

$$\kappa(t) = \frac{|\mathbf{r}'(t) \times \mathbf{r}''(t)|}{|\mathbf{r}'(t)|^3} = \frac{2\sqrt{1 + 9t^2 + 9t^4}}{(1 + 4t^2 + 9t^4)^{3/2}}$$

At the origin the curvature is $\kappa(0) = 2$.   ∎

For the special case of a plane curve with equation $y = f(x)$, we can choose x as the parameter and write $\mathbf{r}(x) = x\mathbf{i} + f(x)\mathbf{j}$. Then $\mathbf{r}'(x) = \mathbf{i} + f'(x)\mathbf{j}$, $\mathbf{r}''(x) = f''(x)\mathbf{j}$, and, since $\mathbf{i} \times \mathbf{j} = \mathbf{k}$ and $\mathbf{j} \times \mathbf{j} = \mathbf{0}$, we have $\mathbf{r}'(x) \times \mathbf{r}''(x) = f''(x)\mathbf{k}$. We also have $|\mathbf{r}'(x)| = \sqrt{1 + [f'(x)]^2}$, so, by Theorem 10,

(11)
$$\kappa(x) = \frac{|f''(x)|}{[1 + (f'(x))^2]^{3/2}}$$

EXAMPLE 5   Find the curvature of the parabola $y = x^2$ at the points $(0,0)$, $(1,1)$, and $(2,4)$.

SOLUTION   Since $y' = 2x$ and $y'' = 2$, Formula 11 gives

$$\kappa(x) = \frac{|y''|}{[1 + (y')^2]^{3/2}} = \frac{2}{(1 + 4x^2)^{3/2}}$$

The curvature at $(0,0)$ is $\kappa(0) = 2$. At $(1,1)$ it is $\kappa(1) = 2/5^{3/2} \approx 0.18$. At $(2,4)$ it is $\kappa(2) = 2/17^{3/2} \approx 0.03$. Observe that $\kappa(x) \to 0$ as $x \to \pm\infty$. This corresponds to the fact that the parabola appears to become flatter as $x \to \pm\infty$.   ∎

THE NORMAL AND BINORMAL VECTORS

At a given point on a smooth space curve $\mathbf{r}(t)$, there are many vectors that are orthogonal to the unit tangent vector $\mathbf{T}(t)$. We single out one by observing that, since $|\mathbf{T}(t)| = 1$ for all t, we have $\mathbf{T}(t) \cdot \mathbf{T}'(t) = 0$ by Example 9 in Section 11.7, so $\mathbf{T}'(t)$ is orthogonal to $\mathbf{T}(t)$. Note that $\mathbf{T}'(t)$ is itself not a unit vector. But if $\mathbf{r}'$ is also smooth, we can define the **principal unit normal vector** $\mathbf{N}(t)$ (or **unit normal** for short) as

$$\mathbf{N}(t) = \frac{\mathbf{T}'(t)}{|\mathbf{T}'(t)|}$$

The vector $\mathbf{B}(t) = \mathbf{T}(t) \times \mathbf{N}(t)$ is called the **binormal vector.** It is perpendicular to both $\mathbf{T}$ and $\mathbf{N}$ and is also a unit vector. (See Figure 6.)

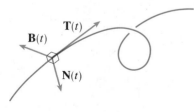

FIGURE 6

Figure 7 illustrates Example 6 by showing the vectors **T**, **N**, and **B** at two locations on the helix. In general, the vectors **T**, **N**, and **B**, starting at the various points on a curve, form a set of orthogonal vectors, called the **TNB** frame, that moves along the curve as t varies. This **TNB** frame plays an important role in the branch of mathematics known as differential geometry and in its applications to the motion of spacecraft.

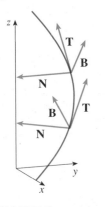

FIGURE 7

EXAMPLE 6   Find the unit normal and binormal vectors for the circular helix

$$\mathbf{r}(t) = \cos t\,\mathbf{i} + \sin t\,\mathbf{j} + t\,\mathbf{k}$$

SOLUTION $\mathbf{r}'(t) = -\sin t\,\mathbf{i} + \cos t\,\mathbf{j} + \mathbf{k}$ $|\mathbf{r}'(t)| = \sqrt{2}$

$$\mathbf{T}(t) = \frac{\mathbf{r}'(t)}{|\mathbf{r}'(t)|} = \frac{1}{\sqrt{2}}(-\sin t\,\mathbf{i} + \cos t\,\mathbf{j} + \mathbf{k})$$

$$\mathbf{T}'(t) = \frac{1}{\sqrt{2}}(-\cos t\,\mathbf{i} - \sin t\,\mathbf{j})\qquad |\mathbf{T}'(t)| = \frac{1}{\sqrt{2}}$$

$$\mathbf{N}(t) = \frac{\mathbf{T}'(t)}{|\mathbf{T}'(t)|} = -\cos t\,\mathbf{i} - \sin t\,\mathbf{j} = \langle -\cos t, -\sin t, 0\rangle$$

This shows that the normal vector at a point on the helix is horizontal and points toward the z-axis. The binormal vector is

$$\mathbf{B}(t) = \mathbf{T}(t) \times \mathbf{N}(t) = \frac{1}{\sqrt{2}}\begin{bmatrix} \mathbf{i} & \mathbf{j} & \mathbf{k} \\ -\sin t & \cos t & 1 \\ -\cos t & -\sin t & 0 \end{bmatrix} = \frac{1}{\sqrt{2}}\langle \sin t, -\cos t, 1\rangle \quad\blacksquare$$

The plane determined by the normal and binormal vectors **N** and **B** at a point P on a curve C is called the **normal plane** of C at P. It consists of all lines that are orthogonal to the tangent vector **T**. The plane determined by the vectors **T** and **N** is called the **osculating plane** of C at P. The name comes from the Latin *osculum*, meaning "kiss." It is the plane that comes closest to containing the part of the curve near P. (For a plane curve, the osculating plane is simply the plane that contains the curve.)

The circle that lies in the osculating plane of C at P, has the same tangent as C at P, lies on the concave side of C (toward which **N** points), and has radius $\rho = 1/\kappa$ (the reciprocal of the curvature) is called the **osculating circle** (or the **circle of curvature**) of C at P. It is the circle that best describes how C behaves near P; it shares the same tangent, normal, and curvature at P.

EXAMPLE 7   Find the equations of the normal plane and osculating plane of the helix in Example 6 at the point $P(0, 1, \pi/2)$.

SOLUTION   The normal plane at P has normal vector $\mathbf{r}'(\pi/2) = \langle -1, 0, 1\rangle$, so an equation is

Figure 8 shows the helix and the osculating plane in Example 7.

$$-1(x - 0) + 0(y - 1) + 1\left(z - \frac{\pi}{2}\right) = 0 \qquad\text{or}\qquad z = x + \frac{\pi}{2}$$

The osculating plane at P contains the vectors **T** and **N**, so its normal vector is $\mathbf{T} \times \mathbf{N} = \mathbf{B}$. From Example 6 we have

$$\mathbf{B}(t) = \frac{1}{\sqrt{2}}\langle \sin t, -\cos t, 1\rangle \qquad \mathbf{B}\!\left(\frac{\pi}{2}\right) = \left\langle \frac{1}{\sqrt{2}}, 0, \frac{1}{\sqrt{2}}\right\rangle$$

A simpler normal vector is $\langle 1, 0, 1\rangle$, so an equation of the osculating plane is

$$1(x - 0) + 0(y - 1) + 1\left(z - \frac{\pi}{2}\right) = 0 \qquad\text{or}\qquad z = -x + \frac{\pi}{2} \quad\blacksquare$$

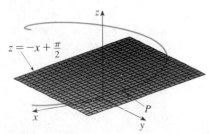

FIGURE 8

EXAMPLE 8 Find and graph the osculating circle of the parabola $y = x^2$ at the origin.

SOLUTION From Example 5 the curvature of the parabola at the origin is $\kappa(0) = 2$. So the radius of the osculating circle at the origin is $1/\kappa = \frac{1}{2}$ and its center is $(0, \frac{1}{2})$. Its equation is therefore

$$x^2 + \left(y - \tfrac{1}{2}\right)^2 = \tfrac{1}{4}$$

For the graph in Figure 9 we use parametric equations of this circle:

$$x = \tfrac{1}{2}\cos t \qquad y = \tfrac{1}{2} + \tfrac{1}{2}\sin t \qquad\blacksquare$$

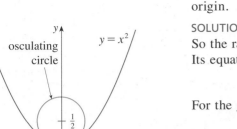

FIGURE 9

We summarize here the formulas for unit tangent, unit normal and binormal vectors, and curvature.

$$\mathbf{T}(t) = \frac{\mathbf{r}'(t)}{|\mathbf{r}'(t)|} \qquad \mathbf{N}(t) = \frac{\mathbf{T}'(t)}{|\mathbf{T}'(t)|} \qquad \mathbf{B}(t) = \mathbf{T}(t) \times \mathbf{N}(t)$$

$$\kappa = \left|\frac{d\mathbf{T}}{ds}\right| = \frac{|\mathbf{T}'(t)|}{|\mathbf{r}'(t)|} = \frac{|\mathbf{r}'(t) \times \mathbf{r}''(t)|}{|\mathbf{r}'(t)|^3}$$

EXERCISES 11.8

1–4 ■ Find the length of the given curve.

1. $\mathbf{r}(t) = \langle 2t, 3\sin t, 3\cos t\rangle, \quad a \leqslant t \leqslant b$

2. $\mathbf{r}(t) = \langle e^t, e^t\sin t, e^t\cos t\rangle, \quad 0 \leqslant t \leqslant 2\pi$

3. $\mathbf{r}(t) = 6t\,\mathbf{i} + 3\sqrt{2}\,t^2\mathbf{j} + 2t^3\mathbf{k}, \quad 0 \leqslant t \leqslant 1$

4. $\mathbf{r}(t) = t^2\mathbf{i} + 2t\,\mathbf{j} + \ln t\,\mathbf{k}, \quad 1 \leqslant t \leqslant e$

5. Use Simpson's Rule with $n = 10$ to estimate the length of the arc of the twisted cubic $x = t$, $y = t^2$, $z = t^3$ from the origin to the point $(2, 4, 8)$.

6. Use a computer to graph the curve with parametric equations $x = \cos t$, $y = \sin 3t$, $z = \sin t$. If your computer (or calculator) evaluates definite integrals, use it to find the total length of this curve correct to four decimal places. Otherwise, use Simpson's Rule.

7–10 ■ Reparametrize the curve with respect to arc length measured from the point where $t = 0$ in the direction of increasing t.

7. $\mathbf{r}(t) = e^t\sin t\,\mathbf{i} + e^t\cos t\,\mathbf{j}$

8. $\mathbf{r}(t) = (1 + 2t)\,\mathbf{i} + (3 + t)\,\mathbf{j} - 5t\,\mathbf{k}$

9. $\mathbf{r}(t) = 3\sin t\,\mathbf{i} + 4t\,\mathbf{j} + 3\cos t\,\mathbf{k}$

10. $\mathbf{r}(t) = \cos^3 t\,\mathbf{i} + \sin^3 t\,\mathbf{j} + \cos 2t\,\mathbf{k}, \quad 0 \leqslant t \leqslant \pi/2$

11–14 ■
(a) Find the unit tangent and unit normal vectors $\mathbf{T}(t)$ and $\mathbf{N}(t)$.
(b) Use Formula 9 to find the curvature.

11. $\mathbf{r}(t) = \langle \sin 4t, 3t, \cos 4t\rangle$

12. $\mathbf{r}(t) = \langle 6t, 3\sqrt{2}\,t^2, 2t^3\rangle$

13. $\mathbf{r}(t) = \langle\sqrt{2}\,\cos t, \sin t, \sin t\rangle$

14. $\mathbf{r}(t) = \langle t^2, 2t, \ln t\rangle$

15–19 ■ Use Theorem 10 to find the curvature.

15. $\mathbf{r}(t) = \mathbf{i} + t\,\mathbf{j} + t^2\mathbf{k}$

16. $\mathbf{r}(t) = (1 + t)\,\mathbf{i} + (1 - t)\,\mathbf{j} + 3t^2\mathbf{k}$

17. $\mathbf{r}(t) = 2t^3\mathbf{i} - 3t^2\mathbf{j} + 6t\,\mathbf{k}$

18. $\mathbf{r}(t) = (t^2 + 2)\,\mathbf{i} + (t^2 - 4t)\,\mathbf{j} + 2t\,\mathbf{k}$

19. $\mathbf{r}(t) = \sin t\,\mathbf{i} + \cos t\,\mathbf{j} + \sin t\,\mathbf{k}$

20. Graph the curve with parametric equations $x = t$, $y = 4t^{3/2}$, $z = -t^2$, and find the curvature at the point $(1, 4, -1)$.

21–24 ■ Use Formula 11 to find the curvature.

21. $y = x^3$

22. $y = \sqrt{x}$

23. $y = \sin x$

24. $y = \ln x$

25. At what point does the curve $y = e^x$ have maximum curvature?

26–27 ■ Use a graphing calculator or computer to graph both the curve and its curvature function $\kappa(x)$ on the same screen. Is the graph of κ what you would expect?

26. $y = xe^{-x}$

27. $y = x^4$

28. Use Theorem 10 to show that the curvature of a plane parametric curve $x = f(t)$, $y = g(t)$ is

$$\kappa = \frac{|\dot{x}\ddot{y} - \dot{y}\ddot{x}|}{[\dot{x}^2 + \dot{y}^2]^{3/2}}$$

where the dots indicate derivatives with respect to t.

29–30 ■ Use the formula in Exercise 28 to find the curvature.

29. $x = t^3$, $y = t^2$ **30.** $x = t\sin t$, $y = t\cos t$

31–32 ■ Find the vectors **T**, **N**, and **B** at the given point.

31. $\mathbf{r}(t) = \langle t^2, \frac{2}{3}t^3, t\rangle$, $(1, \frac{2}{3}, 1)$

32. $\mathbf{r}(t) = \langle e^t, e^t\sin t, e^t\cos t\rangle$, $(1, 0, 1)$

33–34 ■ Find equations of the normal plane and osculating plane of the curve at the given point.

33. $x = 2\sin 3t$, $y = t$, $z = 2\cos 3t$; $(0, \pi, -2)$

34. $x = t$, $y = t^2$, $z = t^3$; $(1, 1, 1)$

35. Find the equations of the osculating circles of the ellipse $9x^2 + 4y^2 = 36$ at the points $(2, 0)$ and $(0, 3)$. Use a graphing calculator or computer to graph the ellipse and both osculating circles on the same screen.

36. Find the equations of the osculating circles of the parabola $y = \frac{1}{2}x^2$ at the points $(0, 0)$ and $(1, \frac{1}{2})$. Graph both osculating circles and the parabola.

37. At what point on the curve $x = t^3$, $y = 3t$, $z = t^4$ is the normal plane parallel to the plane $6x + 6y - 8z = 1$?

38. Is there a point on the curve in Exercise 37 where the osculating plane is parallel to the plane $x + y + z = 1$? (*Note:* You will need a CAS for differentiating, for simplifying, and for computing a cross product.)

39. Show that the curvature κ is related to the tangent and normal vectors by the equation

$$\frac{d\mathbf{T}}{ds} = \kappa\mathbf{N}$$

40. Show that the curvature of a plane curve is $\kappa = |d\phi/ds|$, where ϕ is the angle between **T** and **i**; that is, ϕ is the angle of inclination of the tangent line. (This shows that the definition of curvature is consistent with the definition for plane curves given in Exercise 29 in Section 9.3.)

41. (a) Show that $d\mathbf{B}/ds$ is perpendicular to **B**.
    (b) Show that $d\mathbf{B}/ds$ is perpendicular to **T**.
    (c) Deduce from parts (a) and (b) that $d\mathbf{B}/ds = -\tau(s)\mathbf{N}$ for some number $\tau(s)$ called the **torsion** of the curve. (The torsion measures the degree of twisting of a curve.)

42. Show that for a plane curve the torsion is $\tau(s) = 0$.

43. Show that arc length is independent of parametrization. [*Hint:* Suppose C is given by $\mathbf{r}_1(t)$, $a \leq t \leq b$, and also by $\mathbf{r}_2(u)$, $\alpha \leq u \leq \beta$, where $t = g(u)$ and $g'(u) > 0$. If $L_1 = \int_a^b |\mathbf{r}_1'(t)|\,dt$ and $L_2 = \int_\alpha^\beta |\mathbf{r}_2'(u)|\,du$, show that $L_1 = L_2$.]

44. The following formulas, called the **Frenet-Serret formulas,** are of fundamental importance in differential geometry:
    1. $d\mathbf{T}/ds = \kappa\mathbf{N}$
    2. $d\mathbf{N}/ds = -\kappa\mathbf{T} + \tau\mathbf{B}$
    3. $d\mathbf{B}/ds = -\tau\mathbf{N}$
(Formula 1 comes from Exercise 39 and Formula 3 comes from Exercise 41.) Use the fact that $\mathbf{N} = \mathbf{B} \times \mathbf{T}$ to deduce Formula 2 from Formulas 1 and 3.

45. Use the Frenet-Serret formulas to prove each of the following. (Primes denote derivatives with respect to t. Start as in the proof of Theorem 10.)
    (a) $\mathbf{r}'' = s''\mathbf{T} + \kappa(s')^2\mathbf{N}$
    (b) $\mathbf{r}' \times \mathbf{r}'' = \kappa(s')^3\mathbf{B}$
    (c) $\mathbf{r}''' = [s''' - \kappa^2(s')^3]\mathbf{T} + [3\kappa s's'' + \kappa'(s')^2]\mathbf{N} + \kappa\tau(s')^3\mathbf{B}$
    (d) $\tau = \dfrac{(\mathbf{r}' \times \mathbf{r}'') \cdot \mathbf{r}'''}{|\mathbf{r}' \times \mathbf{r}''|^2}$

46–47 ■ Use the formula in Exercise 45(d) to find the torsion of the curve.

46. $\mathbf{r}(t) = \langle\cos t, \sin t, t\rangle$ **47.** $\mathbf{r}(t) = \langle t, \frac{1}{2}t^2, \frac{1}{3}t^3\rangle$

48. Find the curvature and torsion of the curve $x = \sinh t$, $y = \cosh t$, $z = t$ at the point $(0, 1, 0)$.

49. The DNA molecule has the shape of a double helix (see Figure 3 on page 725). The radius of each helix is about 10 angstroms (1 angstrom = 10^{-8} cm). Each helix rises about 34 angstroms during each complete turn and there are about 2.9×10^8 complete turns. Estimate the length of each helix.

50. Let us consider the problem of designing a railroad track to make a smooth transition between sections of straight track. Existing track along the negative x-axis is to be joined smoothly to a track along the line $y = 1$ for $x \geq 1$.
    (a) Find a polynomial $P = P(x)$ of degree 5 such that the function F defined by

$$F(x) = \begin{cases} 0 & \text{if } x \leq 0 \\ P(x) & \text{if } 0 < x < 1 \\ 1 & \text{if } x \geq 1 \end{cases}$$

is continuous and has continuous slope and continuous curvature.
    (b) Use a graphing calculator or computer to draw the graph of F.

11.9 MOTION IN SPACE: VELOCITY AND ACCELERATION

In this section we show how the ideas of tangent and normal vectors and curvature can be used in physics to study the motion of an object, including its velocity and acceleration, along a space curve. In particular, we follow in the footsteps of Newton by using these methods to derive Kepler's laws of planetary motion.

Suppose a particle moves through space so that its position vector at time t is $\mathbf{r}(t)$. Notice from Figure 1 that, for small values of h, the vector

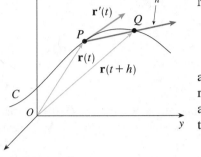

FIGURE 1

$$(1) \qquad \frac{\mathbf{r}(t + h) - \mathbf{r}(t)}{h}$$

approximates the direction of the particle moving along the curve $\mathbf{r}(t)$. Its magnitude measures the size of the displacement vector per unit time. The vector (1) gives the average velocity over a time interval of length h and its limit is the **velocity vector** $\mathbf{v}(t)$ at time t:

$$(2) \qquad \boxed{\mathbf{v}(t) = \lim_{h \to 0} \frac{\mathbf{r}(t + h) - \mathbf{r}(t)}{h} = \mathbf{r}'(t)}$$

Thus the velocity vector points in the direction of the unit tangent vector.

The **speed** of the particle at time t is the magnitude of the velocity vector, that is, $|\mathbf{v}(t)|$. This is appropriate because, from Equations 2 and 11.8.7, we have

$$|\mathbf{v}(t)| = |\mathbf{r}'(t)| = \frac{ds}{dt} = \text{rate of change of distance with respect to time}$$

As in the case of one-dimensional motion, the **acceleration** of the particle is defined as the derivative of the velocity:

$$\mathbf{a}(t) = \mathbf{v}'(t) = \mathbf{r}''(t)$$

EXAMPLE 1 The position vector of an object moving in a plane is given by $\mathbf{r}(t) = t^3\mathbf{i} + t^2\mathbf{j}$, $t \geq 0$. Find its velocity, speed, and acceleration when $t = 1$ and illustrate geometrically.

SOLUTION The velocity and acceleration at time t are

$$\mathbf{v}(t) = \mathbf{r}'(t) = 3t^2\mathbf{i} + 2t\mathbf{j}$$

$$\mathbf{a}(t) = \mathbf{r}''(t) = 6t\mathbf{i} + 2\mathbf{j}$$

and the speed is

$$|\mathbf{v}(t)| = \sqrt{(3t^2)^2 + (2t)^2} = \sqrt{9t^4 + 4t^2}$$

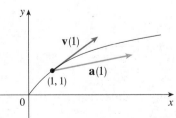
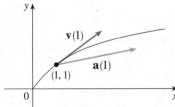

FIGURE 2

When $t = 1$, we have

$$\mathbf{v}(1) = 3\mathbf{i} + 2\mathbf{j} \qquad \mathbf{a}(1) = 6\mathbf{i} + 2\mathbf{j} \qquad |\mathbf{v}(1)| = \sqrt{13}$$

These velocity and acceleration vectors are shown in Figure 2. ∎

Figure 3 shows the path of the particle in Example 2 with the velocity and acceleration vectors when $t = 1$.

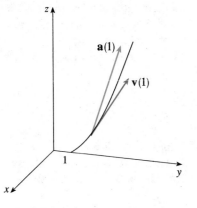

FIGURE 3

EXAMPLE 2   Find the velocity, acceleration, and speed of a particle with position vector $\mathbf{r}(t) = \langle t^2, e^t, te^t \rangle$.

SOLUTION

$$\mathbf{v}(t) = \mathbf{r}'(t) = \langle 2t, e^t, (1 + t)e^t \rangle$$

$$\mathbf{a}(t) = \mathbf{v}'(t) = \langle 2, e^t, (2 + t)e^t \rangle$$

$$|\mathbf{v}(t)| = \sqrt{4t^2 + e^{2t} + (1 + t)^2 e^{2t}}$$ ∎

The vector integrals that were introduced in Section 11.7 can be used to find position vectors when velocity or acceleration vectors are known, as in the following example.

EXAMPLE 3   A moving particle starts at an initial position $\mathbf{r}(0) = \langle 1, 0, 0 \rangle$ with initial velocity $\mathbf{v}(0) = \mathbf{i} - \mathbf{j} + \mathbf{k}$. Its acceleration is $\mathbf{a}(t) = 4t\,\mathbf{i} + 6t\,\mathbf{j} + \mathbf{k}$. Find its velocity and position at time t.

SOLUTION   Since $\mathbf{a}(t) = \mathbf{v}'(t)$, we have

$$\mathbf{v}(t) = \int \mathbf{a}(t)\,dt = \int (4t\,\mathbf{i} + 6t\,\mathbf{j} + \mathbf{k})\,dt$$

$$= 2t^2\,\mathbf{i} + 3t^2\,\mathbf{j} + t\,\mathbf{k} + \mathbf{C}$$

To determine the value of the constant vector $\mathbf{C}$, we use the fact that $\mathbf{v}(0) = \mathbf{i} - \mathbf{j} + \mathbf{k}$. The preceding equation gives $\mathbf{v}(0) = \mathbf{C}$, so $\mathbf{C} = \mathbf{i} - \mathbf{j} + \mathbf{k}$ and

$$\mathbf{v}(t) = 2t^2\,\mathbf{i} + 3t^2\,\mathbf{j} + t\,\mathbf{k} + \mathbf{i} - \mathbf{j} + \mathbf{k}$$

$$= (2t^2 + 1)\,\mathbf{i} + (3t^2 - 1)\,\mathbf{j} + (t + 1)\,\mathbf{k}$$

Since $\mathbf{v}(t) = \mathbf{r}'(t)$, we have

$$\mathbf{r}(t) = \int \mathbf{v}(t)\,dt$$

$$= \int [(2t^2 + 1)\,\mathbf{i} + (3t^2 - 1)\,\mathbf{j} + (t + 1)\,\mathbf{k}]\,dt$$

$$= \left(\frac{2t^3}{3} + t\right)\mathbf{i} + (t^3 - t)\,\mathbf{j} + \left(\frac{t^2}{2} + t\right)\mathbf{k} + \mathbf{D}$$

Putting $t = 0$, we find that $\mathbf{D} = \mathbf{r}(0) = \mathbf{i}$, so

$$\mathbf{r}(t) = \left(\frac{2t^3}{3} + t + 1\right)\mathbf{i} + (t^3 - t)\,\mathbf{j} + \left(\frac{t^2}{2} + t\right)\mathbf{k}$$ ∎

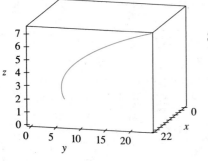

FIGURE 4

The expression for $\mathbf{r}(t)$ that we obtained in Example 3 was used to plot the path of the particle in Figure 4 for $0 \leq t \leq 3$.

In general, vector integrals allow us to recover velocity when acceleration is known and position when velocity is known:

$$\mathbf{v}(t) = \mathbf{v}(t_0) + \int_{t_0}^{t} \mathbf{a}(u)\,du \qquad \mathbf{r}(t) = \mathbf{r}(t_0) + \int_{t_0}^{t} \mathbf{v}(u)\,du$$

If the force that acts on a particle is known, then the acceleration can be found from **Newton's Second Law of Motion.** The vector version of this law states that if, at any time t, a force $\mathbf{F}(t)$ acts on an object of mass m producing an acceleration $\mathbf{a}(t)$, then

$$\mathbf{F}(t) = m\mathbf{a}(t)$$

EXAMPLE 4   An object with mass m that moves in an elliptical path with constant angular speed ω has position vector $\mathbf{r}(t) = a\cos\omega t\,\mathbf{i} + b\sin\omega t\,\mathbf{j}$. Find the force acting on the object and show that it is directed toward the origin.

SOLUTION $$\mathbf{v}(t) = \mathbf{r}'(t) = -a\omega\sin\omega t\,\mathbf{i} + b\omega\cos\omega t\,\mathbf{j}$$

$$\mathbf{a}(t) = \mathbf{v}'(t) = -a\omega^2\cos\omega t\,\mathbf{i} - b\omega^2\sin\omega t\,\mathbf{j}$$

Therefore Newton's Second Law gives the force as

$$\mathbf{F}(t) = m\mathbf{a}(t) = -m\omega^2(a\cos\omega t\,\mathbf{i} + b\sin\omega t\,\mathbf{j})$$

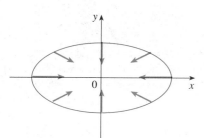

FIGURE 5

Notice that $\mathbf{F}(t) = -m\omega^2\mathbf{r}(t)$. This shows that the force acts in the direction opposite to the radius vector $\mathbf{r}(t)$ and therefore points toward the origin (see Figure 5). Such a force is called a *centripetal* (center-seeking) force. ∎

EXAMPLE 5   A projectile is fired with angle of elevation α and initial velocity $\mathbf{v}_0$ (see Figure 6). Assuming that air resistance is negligible and the only external force is due to gravity, find the position function $\mathbf{r}(t)$ of the projectile. What value of α maximizes the range (the horizontal distance traveled)?

SOLUTION   We set up the axes so that the projectile starts at the origin. Since the force due to gravity acts downward, we have

$$\mathbf{F} = m\mathbf{a} = -mg\,\mathbf{j}$$

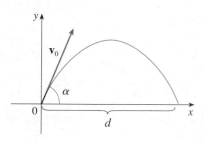

FIGURE 6

where $g = |\mathbf{a}| \approx 9.8$ m/s². Thus

$$\mathbf{a} = -g\,\mathbf{j}$$

Since $\mathbf{v}'(t) = \mathbf{a}$, we have

$$\mathbf{v}(t) = -gt\,\mathbf{j} + \mathbf{C}$$

where $\mathbf{C} = \mathbf{v}(0) = \mathbf{v}_0$. Therefore

$$\mathbf{r}'(t) = \mathbf{v}(t) = -gt\,\mathbf{j} + \mathbf{v}_0$$

Integrating again, we obtain

$$\mathbf{r}(t) = -\tfrac{1}{2}gt^2\,\mathbf{j} + t\mathbf{v}_0 + \mathbf{D}$$

But $\mathbf{D} = \mathbf{r}(0) = \mathbf{0}$, so the position vector of the projectile is given by

(3) $$\mathbf{r}(t) = -\tfrac{1}{2}gt^2\,\mathbf{j} + t\mathbf{v}_0$$

If we write $|\mathbf{v}_0| = v_0$ (the initial speed of the projectile), then

$$\mathbf{v}_0 = v_0\cos\alpha\,\mathbf{i} + v_0\sin\alpha\,\mathbf{j}$$

and Equation 3 becomes

$$\mathbf{r}(t) = (v_0\cos\alpha)t\,\mathbf{i} + \left[(v_0\sin\alpha)t - \tfrac{1}{2}gt^2\right]\mathbf{j}$$

The parametric equations of the trajectory are therefore

$$(4) \qquad x = (v_0 \cos \alpha)t \qquad y = (v_0 \sin \alpha)t - \tfrac{1}{2}gt^2$$

The horizontal distance d is the value of x when $y = 0$. Setting $y = 0$, we obtain $t = 0$ or $t = (2v_0 \sin \alpha)/g$. The latter value of t then gives

$$d = x = (v_0 \cos \alpha) \frac{2v_0 \sin \alpha}{g} = \frac{v_0^2 \sin 2\alpha}{g}$$

Clearly, d has its maximum value when $\sin 2\alpha = 1$, that is, $\alpha = \pi/4$. ■

TANGENTIAL AND NORMAL COMPONENTS OF ACCELERATION

When we study the motion of a particle, it is often useful to resolve the acceleration into two components, one in the direction of the tangent and the other in the direction of the normal. If we write $v = |\mathbf{v}|$ for the speed of the particle, then

$$\mathbf{T}(t) = \frac{\mathbf{r}'(t)}{|\mathbf{r}'(t)|} = \frac{\mathbf{v}(t)}{|\mathbf{v}(t)|} = \frac{\mathbf{v}}{v}$$

and so

$$\mathbf{v} = v\mathbf{T}$$

If we differentiate both sides of this equation with respect to t, we get

$$(5) \qquad \mathbf{a} = \mathbf{v}' = v'\mathbf{T} + v\mathbf{T}'$$

If we use the expression for the curvature given by Equation 11.8.9, then we have

$$(6) \qquad \kappa = \frac{|\mathbf{T}'|}{|\mathbf{r}'|} = \frac{|\mathbf{T}'|}{v} \qquad \text{so} \qquad |\mathbf{T}'| = \kappa v$$

The unit normal vector was defined in the preceding section as $\mathbf{N} = \mathbf{T}'/|\mathbf{T}'|$, so (6) gives

$$\mathbf{T}' = |\mathbf{T}'|\mathbf{N} = \kappa v \mathbf{N}$$

and Equation 5 becomes

$$(7) \qquad \boxed{\mathbf{a} = v'\mathbf{T} + \kappa v^2 \mathbf{N}}$$

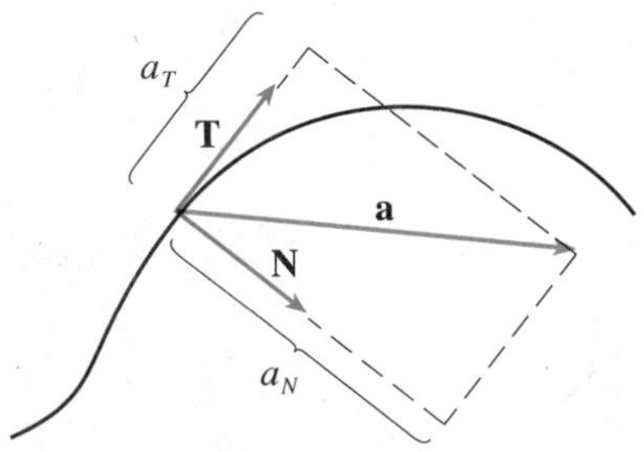

Writing a_T and a_N for the tangential and normal components of acceleration, we have

$$\mathbf{a} = a_T \mathbf{T} + a_N \mathbf{N}$$

where

$$(8) \qquad a_T = v' \qquad \text{and} \qquad a_N = \kappa v^2$$

FIGURE 7

This resolution is illustrated in Figure 7.

Although we have expressions for the tangential and normal components of acceleration in Equations 8, it is desirable to have expressions that depend only on $\mathbf{r}$, $\mathbf{r}'$, and $\mathbf{r}''$. To this end we take the dot product of $\mathbf{v} = v\mathbf{T}$ with $\mathbf{a}$ as given by Equation 7:

$$\mathbf{v} \cdot \mathbf{a} = v\mathbf{T} \cdot (v'\mathbf{T} + \kappa v^2 \mathbf{N})$$

$$= vv'\mathbf{T} \cdot \mathbf{T} + \kappa v^3 \mathbf{T} \cdot \mathbf{N}$$

$$= vv' \qquad\qquad \text{(since } \mathbf{T} \cdot \mathbf{T} = 1 \text{ and } \mathbf{T} \cdot \mathbf{N} = 0\text{)}$$

Therefore

(9)
$$a_T = v' = \frac{\mathbf{v} \cdot \mathbf{a}}{v} = \frac{\mathbf{r}'(t) \cdot \mathbf{r}''(t)}{|\mathbf{r}'(t)|}$$

Using the formula for curvature given by Theorem 11.8.10, we have

(10)
$$a_N = \kappa v^2 = \frac{|\mathbf{r}'(t) \times \mathbf{r}''(t)|}{|\mathbf{r}'(t)|^3} |\mathbf{r}'(t)|^2 = \frac{|\mathbf{r}'(t) \times \mathbf{r}''(t)|}{|\mathbf{r}'(t)|}$$

EXAMPLE 6   A particle moves with position function $\mathbf{r}(t) = \langle t^2, t^2, t^3 \rangle$. Find the tangential and normal components of acceleration.

SOLUTION

$$\mathbf{r}(t) = t^2 \mathbf{i} + t^2 \mathbf{j} + t^3 \mathbf{k}$$

$$\mathbf{r}'(t) = 2t\,\mathbf{i} + 2t\,\mathbf{j} + 3t^2\,\mathbf{k}$$

$$\mathbf{r}''(t) = 2\,\mathbf{i} + 2\,\mathbf{j} + 6t\,\mathbf{k}$$

$$|\mathbf{r}'(t)| = \sqrt{8t^2 + 9t^4}$$

Therefore, Equation 9 gives the tangential component as

$$a_T = \frac{\mathbf{r}'(t) \cdot \mathbf{r}''(t)}{|\mathbf{r}'(t)|} = \frac{8t + 18t^3}{\sqrt{8t^2 + 9t^4}}$$

Since
$$\mathbf{r}'(t) \times \mathbf{r}''(t) = \begin{vmatrix} \mathbf{i} & \mathbf{j} & \mathbf{k} \\ 2t & 2t & 3t^2 \\ 2 & 2 & 6t \end{vmatrix} = 6t^2 \mathbf{i} - 6t^2 \mathbf{j}$$

Equation 10 gives the normal component as

$$a_N = \frac{|\mathbf{r}'(t) \times \mathbf{r}''(t)|}{|\mathbf{r}'(t)|} = \frac{6\sqrt{2}\,t^2}{\sqrt{8t^2 + 9t^4}} \qquad \blacksquare$$

KEPLER'S LAWS OF PLANETARY MOTION

We now describe one of the great accomplishments of calculus by showing how the material of this chapter can be used to prove Kepler's laws of planetary motion. After 20 years of studying the astronomical observations of the Danish astronomer Tycho Brahe, the German mathematician and astronomer Johannes Kepler (1571–1630) formulated the following three laws.

KEPLER'S LAWS

1. A planet revolves around the sun in an elliptical orbit with the sun at one focus.
2. The line joining the sun to a planet sweeps out equal areas in equal times.
3. The square of the period of revolution of a planet is proportional to the cube of the length of the major axis of its orbit.

In his book *Principia Mathematica* of 1687, Sir Isaac Newton was able to show that these three laws are consequences of two of his own laws, the Second Law of Motion

and the Law of Universal Gravitation. In what follows we prove Kepler's First Law. The remaining laws are left as exercises (with hints).

Since the gravitational force of the sun on a planet is so much larger than the forces exerted by other celestial bodies, we can safely ignore all bodies in the universe except for the sun and one planet revolving about it. We use a coordinate system with the sun at the origin and we let $\mathbf{r} = \mathbf{r}(t)$ be the position vector of the planet. (Equally well, $\mathbf{r}$ could be the position vector of the moon or a satellite moving around the earth or a comet moving around a star.) The velocity vector is $\mathbf{v} = \mathbf{r}'$ and the acceleration vector is $\mathbf{a} = \mathbf{r}''$. We use the following laws of Newton:

$$\text{Second Law of Motion:} \quad \mathbf{F} = m\mathbf{a}$$

$$\text{Law of Gravitation:} \quad \mathbf{F} = -\frac{GMm}{r^3}\mathbf{r} = -\frac{GMm}{r^2}\mathbf{u}$$

where $\mathbf{F}$ is the gravitational force on the planet, m and M are the masses of the planet and the sun, G is the gravitational constant, $r = |\mathbf{r}|$, and $\mathbf{u} = (1/r)\mathbf{r}$ is the unit vector in the direction of $\mathbf{r}$.

We first show that the planet moves in one plane. By equating the expressions for $\mathbf{F}$ in Newton's two laws we find that

$$\mathbf{a} = -\frac{GM}{r^3}\mathbf{r}$$

and so $\mathbf{a}$ is parallel to $\mathbf{r}$. It follows that $\mathbf{r} \times \mathbf{a} = \mathbf{0}$. We use Formula 5 in Theorem 11.7.5 to write

$$\frac{d}{dt}(\mathbf{r} \times \mathbf{v}) = \mathbf{r}' \times \mathbf{v} + \mathbf{r} \times \mathbf{v}'$$

$$= \mathbf{v} \times \mathbf{v} + \mathbf{r} \times \mathbf{a} = \mathbf{0} + \mathbf{0} = \mathbf{0}$$

Therefore $\qquad\qquad\qquad \mathbf{r} \times \mathbf{v} = \mathbf{h}$

where $\mathbf{h}$ is a constant vector. (We may assume that $\mathbf{h} \neq \mathbf{0}$; that is, $\mathbf{r}$ and $\mathbf{v}$ are not parallel.) This means that the vector $\mathbf{r} = \mathbf{r}(t)$ is perpendicular to $\mathbf{h}$ for all values of t and so the planet always lies in the plane through the origin perpendicular to $\mathbf{h}$. Thus the orbit of the planet is a plane curve.

To prove Kepler's First Law we rewrite the vector $\mathbf{h}$ as follows:

$$\mathbf{h} = \mathbf{r} \times \mathbf{v} = \mathbf{r} \times \mathbf{r}' = r\mathbf{u} \times (r\mathbf{u})'$$

$$= r\mathbf{u} \times (r\mathbf{u}' + r'\mathbf{u}) = r^2(\mathbf{u} \times \mathbf{u}') + rr'(\mathbf{u} \times \mathbf{u})$$

$$= r^2(\mathbf{u} \times \mathbf{u}')$$

Then

$$\mathbf{a} \times \mathbf{h} = \frac{-GM}{r^2}\mathbf{u} \times (r^2\mathbf{u} \times \mathbf{u}') = -GM\mathbf{u} \times (\mathbf{u} \times \mathbf{u}')$$

$$= -GM[(\mathbf{u} \cdot \mathbf{u}')\mathbf{u} - (\mathbf{u} \cdot \mathbf{u})\mathbf{u}'] \quad \text{(by Theorem 11.4.8, Property 6)}$$

But $\mathbf{u} \cdot \mathbf{u} = |\mathbf{u}|^2 = 1$ and, since $|\mathbf{u}(t)| = 1$, it follows from Example 9 in Section 11.7

that $\mathbf{u} \cdot \mathbf{u}' = 0$. Therefore

$$\mathbf{a} \times \mathbf{h} = GM\,\mathbf{u}'$$

and so

$$(\mathbf{v} \times \mathbf{h})' = \mathbf{v}' \times \mathbf{h} = \mathbf{a} \times \mathbf{h} = GM\,\mathbf{u}'$$

Integrating both sides of this equation, we get

(11)
$$\mathbf{v} \times \mathbf{h} = GM\,\mathbf{u} + \mathbf{c}$$

where $\mathbf{c}$ is a constant vector.

At this point it is convenient to choose the coordinate axes so that the standard basis vector $\mathbf{k}$ points in the direction of the vector $\mathbf{h}$. Then the planet moves in the xy-plane. Since both $\mathbf{v} \times \mathbf{h}$ and $\mathbf{u}$ are perpendicular to $\mathbf{h}$, Equation 11 shows that $\mathbf{c}$ lies in the xy-plane. This means that we can choose the x- and y-axes so that the vector $\mathbf{i}$ lies in the direction of $\mathbf{c}$, as shown in Figure 8.

If θ is the angle between $\mathbf{c}$ and $\mathbf{r}$, then (r, θ) are polar coordinates of the planet. From Equation 11 we have

$$\mathbf{r} \cdot (\mathbf{v} \times \mathbf{h}) = \mathbf{r} \cdot (GM\,\mathbf{u} + \mathbf{c}) = GM\,\mathbf{r} \cdot \mathbf{u} + \mathbf{r} \cdot \mathbf{c}$$
$$= GMr\,\mathbf{u} \cdot \mathbf{u} + |\mathbf{r}||\mathbf{c}|\cos\theta = GMr + rc\cos\theta$$

where $c = |\mathbf{c}|$. Then

$$r = \frac{\mathbf{r} \cdot (\mathbf{v} \times \mathbf{h})}{GM + c\cos\theta} = \frac{1}{GM}\,\frac{\mathbf{r} \cdot (\mathbf{v} \times \mathbf{h})}{1 + e\cos\theta}$$

where $e = c/(GM)$. But

$$\mathbf{r} \cdot (\mathbf{v} \times \mathbf{h}) = (\mathbf{r} \times \mathbf{v}) \cdot \mathbf{h} = \mathbf{h} \cdot \mathbf{h} = |\mathbf{h}|^2 = h^2$$

where $h = |\mathbf{h}|$. So

$$r = \frac{h^2/(GM)}{1 + e\cos\theta} = \frac{eh^2/c}{1 + e\cos\theta}$$

Writing $d = h^2/c$, we obtain the equation

(12)
$$r = \frac{ed}{1 + e\cos\theta}$$

Comparing with Theorem 9.7.6, we see that Equation 12 is the polar equation of a conic section with focus at the origin and eccentricity e. We know that the orbit of a planet is a closed curve and so the conic must be an ellipse.

This completes the derivation of Kepler's First Law. We will guide you through the derivation of the second and third laws in Exercises 35 and 36. The proofs of these three laws show that the methods of this chapter provide a powerful tool for describing some of the laws of nature.

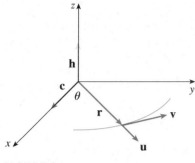

FIGURE 8

EXERCISES 11.9

1–6 ■ Find the velocity, acceleration, and speed of a particle with the given position function. Sketch the path of the particle and draw the velocity and acceleration vectors for the given value of t.

1. $\mathbf{r}(t) = \langle t^2 - 1, t \rangle, \quad t = 1$

2. $\mathbf{r}(t) = \langle \sqrt{t}, 1 - t \rangle, \quad t = 1$

3. $\mathbf{r}(t) = e^t\mathbf{i} + e^{-t}\mathbf{j}, \quad t = 0$

4. $\mathbf{r}(t) = \sin t\,\mathbf{i} + 2\cos t\,\mathbf{j}, \quad t = \pi/6$

5. $\mathbf{r}(t) = \sin t\,\mathbf{i} + t\,\mathbf{j} + \cos t\,\mathbf{k}, \quad t = 0$

6. $\mathbf{r}(t) = t\,\mathbf{i} + t^2\mathbf{j} + t^3\mathbf{k},\quad t = 1$

7–12 ■ Find the velocity, acceleration, and speed of a particle with the given position function.

7. $\mathbf{r}(t) = \langle t^3, t^2 + 1, t^3 - 1 \rangle$

8. $\mathbf{r}(t) = \langle \sqrt{t}, t, t\sqrt{t} \rangle$

9. $\mathbf{r}(t) = (1/t)\,\mathbf{i} + \mathbf{j} + t^2\mathbf{k}$

10. $\mathbf{r}(t) = e^t\mathbf{i} + 2t\,\mathbf{j} + e^{-t}\mathbf{k}$

11. $\mathbf{r}(t) = e^t(\cos t\,\mathbf{i} + \sin t\,\mathbf{j} + t\,\mathbf{k})$

12. $\mathbf{r}(t) = \cosh t\,\mathbf{i} + \sinh t\,\mathbf{j} + t\,\mathbf{k}$

13–14 ■ Find the velocity and position vectors of a particle that has the given acceleration and the given initial velocity and position.

13. $\mathbf{a}(t) = \mathbf{k},\quad \mathbf{v}(0) = \mathbf{i} - \mathbf{j},\quad \mathbf{r}(0) = \mathbf{0}$

14. $\mathbf{a}(t) = -10\,\mathbf{k},\quad \mathbf{v}(0) = \mathbf{i} + \mathbf{j} - \mathbf{k},\quad \mathbf{r}(0) = 2\,\mathbf{i} + 3\,\mathbf{j}$

15–16 ■
(a) Find the position vector of a particle that has the given acceleration and the given initial velocity and position.
(b) Use a computer to graph the path of the particle.

15. $\mathbf{a}(t) = \mathbf{i} + 2\,\mathbf{j} + 2t\,\mathbf{k},\quad \mathbf{v}(0) = \mathbf{0},\quad \mathbf{r}(0) = \mathbf{i} + \mathbf{k}$

16. $\mathbf{a}(t) = t\,\mathbf{i} + t^2\mathbf{j} + \cos 2t\,\mathbf{k},\quad \mathbf{v}(0) = \mathbf{i} + \mathbf{k},\quad \mathbf{r}(0) = \mathbf{j}$

17. The position function of a particle is given by $\mathbf{r}(t) = \langle t^2, 5t, t^2 - 16t \rangle$. When is the speed a minimum?

18. What force is required so that a particle of mass m has the position function $\mathbf{r}(t) = t^3\mathbf{i} + t^2\mathbf{j} + t^3\mathbf{k}$?

19. A force with magnitude 20 N acts directly upward from the xy-plane on an object with mass 4 kg. The object starts at the origin with initial velocity $\mathbf{v}(0) = \mathbf{i} - \mathbf{j}$. Find its position function and its speed at time t.

20. Show that if a particle moves with constant speed, then the velocity and acceleration vectors are orthogonal.

21. A projectile is fired with an initial speed of 500 m/s and angle of elevation 30°. Find (a) the range of the projectile, (b) the maximum height reached, and (c) the speed at impact.

22. Rework Exercise 21 if the projectile is fired from a position 200 m above the ground.

23. A ball is thrown at an angle of 45° to the ground. If the ball lands 90 m away, what was the initial speed of the ball?

24. A gun has muzzle speed 120 m/s. What angle of elevation should be used to hit an object 500 m away?

25. Show that the trajectory of a projectile is a parabola by eliminating the parameter from Equations 4.

26–31 ■ Find the tangential and normal components of the acceleration vector.

26. $\mathbf{r}(t) = (t^2 + 4)\,\mathbf{i} + (2t - 3)\,\mathbf{j}$

27. $\mathbf{r}(t) = (t - \sin t)\,\mathbf{i} + (1 - \cos t)\,\mathbf{j}$

28. $\mathbf{r}(t) = t\,\mathbf{i} + 4\sin t\,\mathbf{j} + 4\cos t\,\mathbf{k}$

29. $\mathbf{r}(t) = t^3\mathbf{i} + t^2\mathbf{j} + t\,\mathbf{k}$

30. $\mathbf{r}(t) = t\,\mathbf{i} + \cos^2 t\,\mathbf{j} + \sin^2 t\,\mathbf{k}$

31. $\mathbf{r}(t) = e^t\mathbf{i} + \sqrt{2}\,t\,\mathbf{j} + e^{-t}\mathbf{k}$

32. If a particle with mass m moves with position vector $\mathbf{r}(t)$, its **angular momentum** is defined as $\mathbf{L}(t) = m\mathbf{r}(t) \times \mathbf{v}(t)$ and its **torque** as $\boldsymbol{\tau}(t) = m\mathbf{r}(t) \times \mathbf{a}(t)$. Show that $\mathbf{L}'(t) = \boldsymbol{\tau}(t)$. Deduce that if $\boldsymbol{\tau}(t) = 0$ for all t, then $\mathbf{L}(t)$ is constant. (This is the law of conservation of angular momentum.)

33. The position function of a spaceship is

$$\mathbf{r}(t) = (3 + t)\,\mathbf{i} + (2 + \ln t)\,\mathbf{j} + \left(7 - \frac{4}{t^2 + 1}\right)\mathbf{k}$$

and the coordinates of a space station are $(6, 4, 9)$. The captain wants the spaceship to coast into the space station. When should the engines be turned off?

34. A rocket burning its onboard fuel while moving through space has velocity $\mathbf{v}(t)$ and mass $m(t)$ at time t. If the exhaust gases escape with velocity $\mathbf{v}_e$ relative to the rocket, it can be deduced from Newton's Second Law of Motion that

$$m\,\frac{d\mathbf{v}}{dt} = \frac{dm}{dt}\,\mathbf{v}_e$$

(a) Show that $\mathbf{v}(t) = \mathbf{v}(0) - \ln\dfrac{m(0)}{m(t)}\,\mathbf{v}_e$.

(b) For the rocket to accelerate in a straight line from rest to twice the speed of its own exhaust gases, what fraction of its initial mass would the rocket have to burn as fuel?

35. Use the following steps to prove Kepler's Second Law. The notation is the same as in the proof of the first law. In particular, use polar coordinates so that $\mathbf{r} = (r\cos\theta)\,\mathbf{i} + (r\sin\theta)\,\mathbf{j}$.

(a) Show that $\mathbf{h} = r^2\,\dfrac{d\theta}{dt}\,\mathbf{k}$.

(b) Deduce that $r^2\,\dfrac{d\theta}{dt} = h$.

(c) If $A = A(t)$ is the area swept out by the radius vector $\mathbf{r} = \mathbf{r}(t)$ in the time interval $[t_0, t]$, show that

$$\frac{dA}{dt} = \tfrac{1}{2}r^2\,\frac{d\theta}{dt}$$

(d) Deduce that

$$\frac{dA}{dt} = \tfrac{1}{2}h = \text{constant}$$

This says that the rate at which A is swept out is constant and proves Kepler's Second Law.

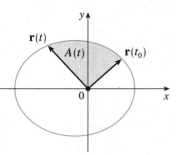

36. Let T be the period of a planet about the sun; that is, T is the time required for it to travel once around its elliptical orbit. Suppose that the lengths of the major and minor axes of the ellipse are $2a$ and $2b$.

(a) Use Exercise 35 part (d) to show that $T = 2\pi ab/h$.

(b) Show that $\dfrac{h^2}{GM} = ed = \dfrac{b^2}{a}$.

(c) Use parts (a) and (b) to show that $T^2 = \dfrac{4\pi^2}{GM} a^3$.

This proves Kepler's Third Law. [Notice that the proportionality constant $4\pi^2/(GM)$ is independent of the planet.]

37. The period of the earth's orbit is approximately 365.25 days. Use this fact and Exercise 36 to find the length of the major axis of the earth's orbit. You will need the mass of the sun, $M = 1.99 \times 10^{30}$ kg, and the gravitational constant, $G = 6.67 \times 10^{-11}$ N·m²/kg².

11.10 CYLINDRICAL AND SPHERICAL COORDINATES

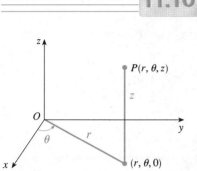

FIGURE 1

Recall that in plane geometry we introduced the polar coordinate system in order to give a more convenient description of certain curves and regions. In three dimensions there are two coordinate systems that are similar to polar coordinates and give convenient descriptions of some commonly occurring surfaces and solids. They will be especially useful in Chapter 13 when we compute volumes and triple integrals.

In the **cylindrical coordinate system,** a point P in three-dimensional space is represented by the ordered triple (r, θ, z), where r and θ are polar coordinates of the projection of P onto the xy-plane and z is the directed distance from the xy-plane to P (see Figure 1).

To convert from cylindrical to rectangular coordinates we use the equations

$$(1) \qquad x = r\cos\theta \qquad y = r\sin\theta \qquad z = z$$

whereas to convert from rectangular to cylindrical coordinates we use

$$(2) \qquad r^2 = x^2 + y^2 \qquad \tan\theta = \frac{y}{x} \qquad z = z$$

These equations follow from Equations 9.4.1 and 9.4.2.

EXAMPLE 1

(a) Plot the point with cylindrical coordinates $(2, 2\pi/3, 1)$ and find its rectangular coordinates.

(b) Find the cylindrical coordinates of the point with rectangular coordinates $(3, -3, -7)$.

SOLUTION

(a) The point with cylindrical coordinates $(2, 2\pi/3, 1)$ is plotted in Figure 2. From Equations 1, its rectangular coordinates are

$$x = 2\cos\frac{2\pi}{3} = 2\left(-\frac{1}{2}\right) = -1 \qquad y = 2\sin\frac{2\pi}{3} = 2\left(\frac{\sqrt{3}}{2}\right) = \sqrt{3} \qquad z = 1$$

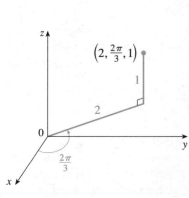

FIGURE 2

Thus the point is $\left(-1, \sqrt{3}, 1\right)$ in rectangular coordinates.

(b) From Equations 2 we have

$$r = \sqrt{3^2 + (-3)^2} = 3\sqrt{2}$$

$$\tan \theta = \frac{-3}{3} = -1 \qquad \text{so} \qquad \theta = \frac{7\pi}{4} + 2n\pi$$

$$z = -7$$

Therefore, one set of cylindrical coordinates is $(3\sqrt{2}, 7\pi/4, -7)$. Another is $(3\sqrt{2}, -\pi/4, -7)$. As with polar coordinates, there are infinitely many choices. ■

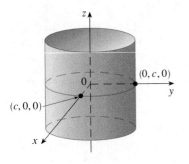

FIGURE 3
$r = c$, a cylinder

Cylindrical coordinates are useful in problems that involve symmetry about the z-axis. For instance, the axis of the circular cylinder with Cartesian equation $x^2 + y^2 = c^2$ is the z-axis. In cylindrical coordinates this cylinder has the very simple equation $r = c$ (see Figure 3). This is the reason for the name "cylindrical" coordinates.

EXAMPLE 2   Describe the surface whose equation in cylindrical coordinates is $z = r$.

SOLUTION   We first convert to an equation in rectangular coordinates. From the first equation in (2) we have

$$z^2 = r^2 = x^2 + y^2$$

We recognize the equation $z^2 = x^2 + y^2$ (by comparison with Equation 11.6.4) as being a circular cone whose axis is the z-axis (see Figure 4). ■

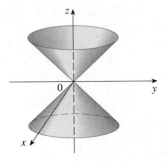

FIGURE 4
$z = r$, a cone

EXAMPLE 3   Find an equation in cylindrical coordinates for the ellipsoid $4x^2 + 4y^2 + z^2 = 1$.

SOLUTION   Since $r^2 = x^2 + y^2$ from Equations 2, we have

$$z^2 = 1 - 4(x^2 + y^2) = 1 - 4r^2$$

So the equation of the ellipsoid in cylindrical coordinates is $z^2 = 1 - 4r^2$. ■

The **spherical coordinates** (ρ, θ, ϕ) of a point P in space are shown in Figure 5, where $\rho = |OP|$ is the distance from the origin to P, θ is the same angle as in cylindrical coordinates, and ϕ is the angle between the positive z-axis and the line segment OP. Note that

$$\rho \geq 0 \qquad 0 \leq \phi \leq \pi$$

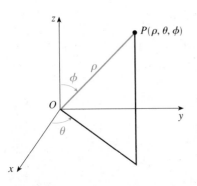

FIGURE 5

The spherical coordinate system is especially useful in problems where there is symmetry about the origin. For example, the sphere with center the origin and radius c has the simple equation $\rho = c$ (see Figure 6) and this is the reason for the name "spherical" coordinates. The graph of the equation $\theta = c$ is a vertical half-plane (see Figure 7), and the equation $\phi = c$ represents a half-cone with the z-axis as its axis (see Figure 8).

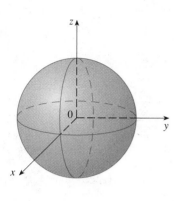

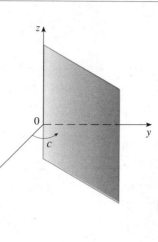

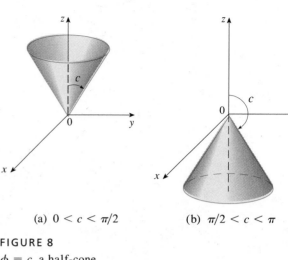

(a) $0 < c < \pi/2$ (b) $\pi/2 < c < \pi$

FIGURE 6
$\rho = c$, a sphere

FIGURE 7
$\theta = c$, a half-plane

FIGURE 8
$\phi = c$, a half-cone

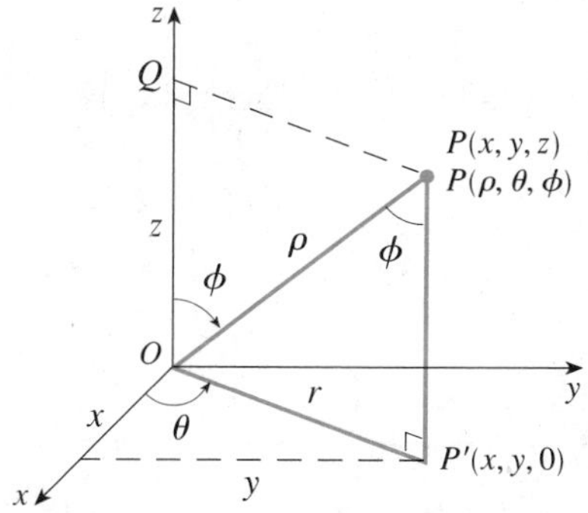

The relationship between rectangular and spherical coordinates can be seen from Figure 9. From triangles OPQ and OPP' we have

$$z = \rho \cos \phi \qquad r = \rho \sin \phi$$

But $x = r \cos \theta$ and $y = r \sin \theta$, so

(3) $$x = \rho \sin \phi \cos \theta \qquad y = \rho \sin \phi \sin \theta \qquad z = \rho \cos \phi$$

Also, the distance formula shows that

(4) $$\rho^2 = x^2 + y^2 + z^2$$

FIGURE 9

EXAMPLE 4   The point $(2, \pi/4, \pi/3)$ is given in spherical coordinates. Find its rectangular coordinates.

SOLUTION   From Equations 3 we have

$$x = \rho \sin \phi \cos \theta = 2 \sin \frac{\pi}{3} \cos \frac{\pi}{4} = 2\left(\frac{\sqrt{3}}{2}\right)\left(\frac{1}{\sqrt{2}}\right) = \sqrt{\frac{3}{2}}$$

$$y = \rho \sin \phi \sin \theta = 2 \sin \frac{\pi}{3} \sin \frac{\pi}{4} = 2\left(\frac{\sqrt{3}}{2}\right)\left(\frac{1}{\sqrt{2}}\right) = \sqrt{\frac{3}{2}}$$

$$z = \rho \cos \phi = 2 \cos \frac{\pi}{3} = 2(\tfrac{1}{2}) = 1$$

Thus the point $(2, \pi/4, \pi/3)$ is $(\sqrt{3/2}, \sqrt{3/2}, 1)$ in rectangular coordinates. ∎

EXAMPLE 5   The point $(0, 2\sqrt{3}, -2)$ is given in rectangular coordinates. Find its spherical coordinates.

SOLUTION   From Equation 4 we have

$$\rho = \sqrt{x^2 + y^2 + z^2} = \sqrt{0 + 12 + 4} = 4$$

and so Equations 3 give

$$\cos\phi = \frac{z}{\rho} = \frac{-2}{4} = -\frac{1}{2} \qquad \phi = \frac{2\pi}{3}$$

$$\cos\theta = \frac{x}{\rho\sin\phi} = 0 \qquad \theta = \frac{\pi}{2}$$

(Note that $\theta \neq 3\pi/2$ because $y = 2\sqrt{3} > 0$.) Therefore, the spherical coordinates of the given point are $(4, \pi/2, 2\pi/3)$. ■

EXAMPLE 6   Find an equation in spherical coordinates for the hyperboloid of two sheets with equation $x^2 - y^2 - z^2 = 1$.

SOLUTION   Substituting the expressions in Equations 3 into the given equation, we have

$$\rho^2\sin^2\phi\cos^2\theta - \rho^2\sin^2\phi\sin^2\theta - \rho^2\cos^2\phi = 1$$

$$\rho^2[\sin^2\phi\,(\cos^2\theta - \sin^2\theta) - \cos^2\phi] = 1$$

or $$\rho^2(\sin^2\phi\cos 2\theta - \cos^2\phi) = 1$$ ■

EXAMPLE 7   Find a rectangular equation for the surface whose spherical equation is $\rho = \sin\theta\sin\phi$.

SOLUTION   From Equations 4 and 3 we have

$$x^2 + y^2 + z^2 = \rho^2 = \rho\sin\theta\sin\phi = y$$

or $$x^2 + \left(y - \tfrac{1}{2}\right)^2 + z^2 = \tfrac{1}{4}$$

which is the equation of a sphere with center $\left(0, \tfrac{1}{2}, 0\right)$ and radius $\tfrac{1}{2}$. ■

EXAMPLE 8   Use a computer to draw a picture of the solid that remains when a hole of radius 3 is drilled through the center of a sphere of radius 4.

SOLUTION   To keep the equations simple, let's choose the coordinate system so that the center of the sphere is at the origin and the axis of the cylinder that forms the hole is the z-axis. We could use either cylindrical or spherical coordinates to describe the solid, but the description is somewhat simpler if we use cylindrical coordinates. Then the equation of the cylinder is $r = 3$ and the equation of the sphere is $x^2 + y^2 + z^2 = 16$ or $r^2 + z^2 = 16$. The points in the solid lie outside the cylinder and inside the sphere, so they satisfy the inequalities

$$3 \leq r \leq \sqrt{16 - z^2}$$

To ensure that the computer graphs only the appropriate parts of these surfaces, we find where they intersect by solving the equations $r = 3$ and $r = \sqrt{16 - z^2}$:

$$\sqrt{16 - z^2} = 3 \quad \Rightarrow \quad 16 - z^2 = 9 \quad \Rightarrow \quad z^2 = 7 \quad \Rightarrow \quad z = \pm\sqrt{7}$$

The solid lies between $z = -\sqrt{7}$ and $z = \sqrt{7}$, so we use the computer to graph the surfaces with the following equations and domains:

$$r = 3 \qquad\qquad 0 \leq \theta \leq 2\pi \qquad -\sqrt{7} \leq z \leq \sqrt{7}$$

$$r = \sqrt{16 - z^2} \qquad 0 \leq \theta \leq 2\pi \qquad -\sqrt{7} \leq z \leq \sqrt{7}$$

The resulting picture, shown in Figure 10, is exactly what we want. ■

Most three-dimensional graphing programs can graph surfaces whose equations are given in cylindrical or spherical coordinates. As Example 8 demonstrates, this is often the most convenient way of drawing a solid.

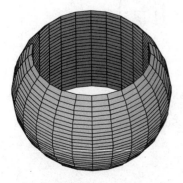

FIGURE 10

EXERCISES 11.10

1–6 ■ Change from cylindrical to rectangular coordinates.

1. $(3, \pi/2, 1)$ **2.** $(\sqrt{2}, \pi/4, \sqrt{2})$

3. $(2, 4\pi/3, 8)$ **4.** $(6, -\pi/6, 5)$

5. $(3, 0, -6)$ **6.** $(1, \pi, e)$

7–12 ■ Change from rectangular to cylindrical coordinates.

7. $(-1, 0, 0)$ **8.** $(1, 1, 1)$

9. $(\sqrt{3}, 1, 4)$ **10.** $(-\sqrt{2}, \sqrt{2}, 0)$

11. $(4, 4, 4)$ **12.** $(-1, \sqrt{3}, 2)$

13–18 ■ Change from spherical to rectangular coordinates.

13. $(1, 0, 0)$ **14.** $(3, 0, \pi)$

15. $(1, \pi/6, \pi/6)$ **16.** $(2, \pi/2, 3\pi/4)$

17. $(4, \pi/4, \pi/6)$ **18.** $(2, \pi/4, \pi/4)$

19–24 ■ Change from rectangular to spherical coordinates.

19. $(-3, 0, 0)$ **20.** $(1, 1, \sqrt{2})$

21. $(\sqrt{3}, 0, 1)$ **22.** $(-\sqrt{3}, -3, -2)$

23. $(1, -1, -\sqrt{2})$ **24.** $(\sqrt{3}, 1, 2\sqrt{3})$

25–28 ■ Change from cylindrical to spherical coordinates.

25. $(\sqrt{2}, \pi/4, 0)$ **26.** $(1, \pi/2, 1)$

27. $(4, \pi/3, 4)$ **28.** $(12, \pi, 5)$

29–32 ■ Change from spherical to cylindrical coordinates.

29. $(2, 0, 0)$ **30.** $(2\sqrt{2}, 3\pi/2, \pi/2)$

31. $(8, \pi/6, \pi/2)$ **32.** $(4, \pi/4, \pi/3)$

33–50 ■ Identify the surface whose equation is given.

33. $r = 3$ **34.** $\rho = 3$

35. $\phi = \pi/3$ **36.** $\theta = \pi/3$

37. $z = r^2$ **38.** $r = 4\sin\theta$

39. $\rho\cos\phi = 2$ **40.** $\rho\sin\phi = 2$

41. $\phi = 0$ **42.** $\phi = \pi/2$

43. $r = 2\cos\theta$ **44.** $\rho = 2\cos\phi$

45. $r^2 + z^2 = 25$ **46.** $r^2 - 2z^2 = 4$

47. $\rho^2(\sin^2\phi\cos^2\theta + \cos^2\phi) = 4$

48. $\rho^2(\sin^2\phi - 4\cos^2\phi) = 1$

49. $r^2 = r$ **50.** $\rho^2 - 6\rho + 8 = 0$

51–58 ■ Write the equation (a) in cylindrical coordinates and (b) in spherical coordinates.

51. $x^2 + y^2 + z^2 = 16$ **52.** $x^2 + y^2 - z^2 = 16$

53. $x + 2y + 3z = 6$ **54.** $x^2 + y^2 = 2z$

55. $x^2 - y^2 - 2z^2 = 4$ **56.** $y^2 + z^2 = 1$

57. $x^2 + y^2 = 2y$ **58.** $z = x^2 - y^2$

59–62 ■ Sketch the solid described by the given inequalities.

59. $r^2 \le z \le 2 - r^2$

60. $0 \le \theta \le \pi/2, \quad r \le z \le 2$

61. $-\pi/2 \le \theta \le \pi/2, \quad 0 \le \phi \le \pi/6, \quad 0 \le \rho \le \sec\phi$

62. $0 \le \phi \le \pi/3, \quad \rho \le 2$

63. A solid lies above the cone $z = \sqrt{x^2 + y^2}$ and below the sphere $x^2 + y^2 + z^2 = z$. Write a description of the solid in terms of inequalities involving spherical coordinates.

64. Use a computer to draw the solid enclosed by the paraboloids $z = x^2 + y^2$ and $z = 5 - x^2 - y^2$.

65. Use a computer to draw a silo consisting of a cylinder with radius 3 and height 10 surmounted by a hemisphere.

66. The latitude and longitude of a point P in the Northern Hemisphere are related to spherical coordinates ρ, θ, ϕ as follows. We take the origin to be the center of the earth and the positive z-axis to pass through the North Pole. The positive x-axis passes through the point where the prime meridian (the meridian through Greenwich, England) intersects the equator. Then the latitude of P is $\alpha = 90° - \phi°$ and the longitude is $\beta = 360° - \theta°$. Find the great-circle distance from Los Angeles (latitude 34.06° north, longitude 118.25° west) to Montreal (latitude 45.50° north, longitude 73.60° west). Take the radius of the earth to be 3960 mi. (A *great circle* is the circle of intersection of a sphere and a plane through the center of the sphere.)

═══════════════════
11 REVIEW
═══════════════════

KEY TOPICS ■ Define, state, or discuss the following.

1. $\mathbb{R}^3$

2. Distance formula in $\mathbb{R}^3$

3. Equation of a sphere

4. Vectors in V_2 and V_3

5. Position vector

6. Length of a vector

7. Addition of vectors

8. Multiplication of a vector by a scalar

9. Unit vector

10. Dot product and its properties

11. Angle between two vectors

12. Orthogonal vectors

13. Parallel vectors

14. Direction cosines

15. Vector and scalar projections

16. Work done by a constant force

17. Cross product and its properties

18. Scalar triple product

19. Vector, parametric, and symmetric equations of a line

20. Vector and scalar equations of a plane

21. Distance from a point to a plane

22. Standard equations of ellipsoids, hyperboloids, paraboloids, and cones

23. Vector functions and component functions

24. Limit of a vector function

25. Continuity of a vector function

26. Derivative of a vector function

27. Space curve

28. Tangent vector and tangent line

29. Differentiation formulas for vector functions

30. Integrals of vector functions

31. Length of a curve

32. Smooth curve

33. Arc length function

34. Curvature

35. Unit normal vector

36. Binormal vector

37. Normal plane

38. Osculating plane

39. Osculating circle

40. Velocity and acceleration along a curve

41. Tangential and normal components of acceleration

42. Kepler's Laws

43. Cylindrical coordinates

44. Spherical coordinates

EXERCISES

1–14 ■ Determine whether the statement is true or false.

1. For any vectors $\mathbf{u}$ and $\mathbf{v}$ in V_3, $\mathbf{u} \cdot \mathbf{v} = \mathbf{v} \cdot \mathbf{u}$.

2. For any vectors $\mathbf{u}$ and $\mathbf{v}$ in V_3, $\mathbf{u} \times \mathbf{v} = \mathbf{v} \times \mathbf{u}$.

3. For any vectors $\mathbf{u}$ and $\mathbf{v}$ in V_3, $|\mathbf{u} \times \mathbf{v}| = |\mathbf{v} \times \mathbf{u}|$.

4. For any vectors $\mathbf{u}$ and $\mathbf{v}$ in V_3 and any scalar k, $k(\mathbf{u} \cdot \mathbf{v}) = (k\mathbf{u}) \cdot \mathbf{v}$.

5. For any vectors $\mathbf{u}$ and $\mathbf{v}$ in V_3 and any scalar k, $k(\mathbf{u} \times \mathbf{v}) = (k\mathbf{u}) \times \mathbf{v}$.

6. For any vectors $\mathbf{u}$, $\mathbf{v}$, and $\mathbf{w}$ in V_3, $(\mathbf{u} + \mathbf{v}) \times \mathbf{w} = \mathbf{u} \times \mathbf{w} + \mathbf{v} \times \mathbf{w}$.

7. For any vectors $\mathbf{u}$, $\mathbf{v}$, and $\mathbf{w}$ in V_3, $\mathbf{u} \cdot (\mathbf{v} \times \mathbf{w}) = (\mathbf{u} \times \mathbf{v}) \cdot \mathbf{w}$.

8. For any vectors $\mathbf{u}$, $\mathbf{v}$, and $\mathbf{w}$ in V_3, $\mathbf{u} \times (\mathbf{v} \times \mathbf{w}) = (\mathbf{u} \times \mathbf{v}) \times \mathbf{w}$.

9. For any vectors $\mathbf{u}$ and $\mathbf{v}$ in V_3, $(\mathbf{u} \times \mathbf{v}) \cdot \mathbf{u} = 0$.

10. For any vectors $\mathbf{u}$ and $\mathbf{v}$ in V_3, $(\mathbf{u} + \mathbf{v}) \times \mathbf{v} = \mathbf{u} \times \mathbf{v}$.

11. The cross product of two unit vectors is a unit vector.

12. A linear equation $Ax + By + Cz + D = 0$ represents a line in space.

13. The set of points $\{(x, y, z) \mid x^2 + y^2 = 1\}$ is a circle.

14. The curve with vector equation $\mathbf{r}(t) = t^3\mathbf{i} + 2t^3\mathbf{j} + 3t^3\mathbf{k}$ is a line.

15. Find the lengths of the sides of the triangle with vertices $A(2, 6, -4)$, $B(-1, 2, 8)$, and $C(0, 1, 2)$.

16. Find an equation of the sphere with center $(1, -1, 2)$ and radius 3.

17. Find the center and radius of the sphere

$$x^2 + y^2 + z^2 + 4x + 6y - 10z + 2 = 0$$

18–30 ■ Calculate the given quantity if

$$\mathbf{a} = \mathbf{i} + \mathbf{j} - 2\mathbf{k} \qquad \mathbf{b} = 3\mathbf{i} - 2\mathbf{j} + \mathbf{k} \qquad \mathbf{c} = \mathbf{j} - 5\mathbf{k}$$

18. $2\mathbf{a} + 3\mathbf{b}$

19. $6\mathbf{a} - 5\mathbf{c}$

20. $|\mathbf{b}|$

21. $\mathbf{a} \cdot \mathbf{b}$

22. $\mathbf{a} \times \mathbf{b}$

23. $|\mathbf{b} \times \mathbf{c}|$

24. $\mathbf{a} \cdot (\mathbf{b} \times \mathbf{c})$

25. $\mathbf{c} \times \mathbf{c}$

26. $\mathbf{a} \times (\mathbf{b} \times \mathbf{c})$

27. The angle between $\mathbf{a}$ and $\mathbf{b}$ (correct to the nearest degree)

28. The direction cosines of $\mathbf{b}$

29. The scalar projection of $\mathbf{b}$ onto $\mathbf{a}$

30. The vector projection of $\mathbf{b}$ onto $\mathbf{a}$

31. Find the value of x such that the vectors $\langle 2, x, 4 \rangle$ and $\langle 2x, 3, -7 \rangle$ are orthogonal.

32. Find two unit vectors orthogonal to both $\langle 1, 0, 1 \rangle$ and $\langle 2, 3, 4 \rangle$.

33. Suppose that $\mathbf{u} \cdot (\mathbf{v} \times \mathbf{w}) = 2$. Find
(a) $(\mathbf{u} \times \mathbf{v}) \cdot \mathbf{w}$
(b) $\mathbf{u} \cdot (\mathbf{w} \times \mathbf{v})$
(c) $\mathbf{v} \cdot (\mathbf{u} \times \mathbf{w})$
(d) $(\mathbf{u} \times \mathbf{v}) \cdot \mathbf{v}$

34. Show that if $\mathbf{a}$, $\mathbf{b}$, and $\mathbf{c}$ are in V_3, then
$$(\mathbf{a} \times \mathbf{b}) \cdot [(\mathbf{b} \times \mathbf{c}) \times (\mathbf{c} \times \mathbf{a})] = [\mathbf{a} \cdot (\mathbf{b} \times \mathbf{c})]^2$$

35. Describe a method for determining whether three points P, Q, and R lie on the same line.

36. Describe a method for determining whether four points P, Q, R, and S lie in the same plane.

37. Find the acute angle between two diagonals of a cube.

38. Given the points $A(1, 0, 1)$, $B(2, 3, 0)$, $C(-1, 1, 4)$, and $D(0, 3, 2)$, find the volume of the parallelepiped with adjacent edges AB, AC, and AD.

39. (a) Find a vector perpendicular to the plane through the points $A(1, 0, 0)$, $B(2, 0, -1)$, and $C(1, 4, 3)$.
(b) Find the area of triangle ABC.

40. A constant force $\mathbf{F} = 3\mathbf{i} + 5\mathbf{j} + 10\mathbf{k}$ moves an object along the line segment from $(1, 0, 2)$ to $(5, 3, 8)$. Find the work done if the distance is measured in meters and the force in newtons.

41. A boat is pulled onto shore using two ropes, as shown in the diagram. If a force of 255 N is needed to pull the boat, find the magnitude of the force in each rope.

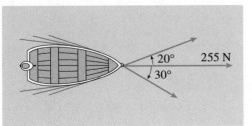

42. Find the magnitude of the torque about P if a 50-N force is applied as shown.

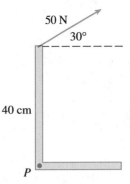

43–45 ■ Find parametric equations for the line that satisfies the given conditions.

43. Passing through $(1, 2, 4)$ and in the direction of $\mathbf{v} = 2\mathbf{i} - \mathbf{j} + 3\mathbf{k}$

44. Passing through $(-6, -1, 0)$ and $(2, -3, 5)$

45. Passing through $(1, 0, 1)$ and parallel to the line with parametric equations $x = 4t$, $y = 1 - 3t$, $z = 2 + 5t$

46–48 ■ Find an equation of the plane that satisfies the given conditions.

46. Passing through $(4, -1, -1)$ and with normal vector $\langle 2, 6, -3 \rangle$

47. Passing through $(-4, 1, 2)$ and parallel to the plane $x + 2y + 5z = 3$

48. Passing through $(-1, 2, 0)$, $(2, 0, 1)$, and $(-5, 3, 1)$

49. Find the point in which the line with parametric equations $x = 2 - t$, $y = 1 + 3t$, $z = 4t$ intersects the plane $2x - y + z = 2$.

50. Find the distance from the origin to the line $x = 1 + t$, $y = 2 - t$, $z = -1 + 2t$.

51. Determine whether the lines given by the symmetric equations
$$\frac{x - 1}{2} = \frac{y - 2}{3} = \frac{z - 3}{4}$$
and
$$\frac{x + 1}{6} = \frac{y - 3}{-1} = \frac{z + 5}{2}$$
are parallel, skew, or intersecting.

52. (a) Show that the planes $x + y - z = 1$ and $2x - 3y + 4z = 5$ are neither parallel nor perpendicular.
(b) Find, correct to the nearest degree, the angle between these planes.

53. Find the distance between the planes $3x + y - 4z = 2$ and $3x + y - 4z = 24$.

54–62 ■ Identify and sketch the graph of each surface.

54. $x = 6$

55. $y = z$

56. $6x + 4y + 3z = 12$

57. $y = x^2 + z^2$

58. $225x^2 + 9y^2 + 25z^2 = 225$

59. $x^2 = y^2 + 4z^2$

60. $y^2 + z^2 = 1 + x^2$

61. $-4x^2 + y^2 - 4z^2 = 4$

62. $y = z^2$

63. An ellipsoid is created by rotating the ellipse $4x^2 + y^2 = 16$ about the x-axis. Find an equation of the ellipsoid.

64. A surface consists of all points P such that the distance from P to the plane $y = 1$ is twice the distance from P to the point $(0, -1, 0)$. Find an equation for this surface and identify it.

65. (a) Sketch the curve with vector function
$$\mathbf{r}(t) = 2\mathbf{i} + \sin t\,\mathbf{j} + \cos t\,\mathbf{k}$$
(b) Find $\mathbf{r}'(t)$ and $\mathbf{r}''(t)$.

66. Find parametric equations for the tangent line to the curve $x = t^2$, $y = t^4$, $z = t^3$ at the point $(1, 1, 1)$. Graph the curve and the tangent line on a common screen.

67. If $\mathbf{r}(t) = (t + t^2)\mathbf{i} + (2 + t^3)\mathbf{j} + t^4\mathbf{k}$, evaluate $\int_0^1 \mathbf{r}(t)\,dt$.

68. Let C be the curve with equations $x = 2 - t^3$, $y = 2t - 1$, $z = \ln t$. Find (a) the point where C intersects the xz-plane, (b) parametric equations of the tangent line at $(1, 1, 0)$, and (c) an equation of the normal plane to C at $(1, 1, 0)$.

69. Use Simpson's Rule with $n = 4$ to estimate the length of the arc of the curve with equations $x = \sqrt{t}$, $y = 4/t$, $z = t^2 + 1$ from $(1, 4, 2)$ to $(2, 1, 17)$.

70. Find the length of the curve $\mathbf{r}(t) = \langle 2t^{3/2}, \cos 2t, \sin 2t \rangle$, $0 \le t \le 1$.

71. The helix $\mathbf{r}_1(t) = \cos t\,\mathbf{i} + \sin t\,\mathbf{j} + t\,\mathbf{k}$ intersects the curve $\mathbf{r}_2(t) = (1 + t)\mathbf{i} + t^2\mathbf{j} + t^3\mathbf{k}$ at the point $(1, 0, 0)$. Find the angle of intersection of these curves.

72. Reparametrize the curve $\mathbf{r}(t) = e^t\mathbf{i} + e^t\sin t\,\mathbf{j} + e^t\cos t\,\mathbf{k}$ with respect to arc length measured from the point $(1, 0, 1)$ in the direction of increasing t.

73. For the curve given by $\mathbf{r}(t) = \langle t^3/3, t^2/2, t \rangle$, find (a) the unit tangent vector, (b) the unit normal vector, and (c) the curvature.

74. Find the curvature of the ellipse $x = 3\cos t$, $y = 4\sin t$ at the points $(3, 0)$ and $(0, 4)$.

75. Find the curvature of the curve $y = x^4$ at the point $(1, 1)$.

76. Find an equation of the osculating circle of the curve $y = x^4 - x^2$ at the origin. Graph both the curve and its osculating circle.

77. Find an equation of the osculating plane of the curve $x = \sin 2t$, $y = t$, $z = \cos 2t$ at the point $(0, \pi, 1)$.

78. A particle moves with position function $\mathbf{r}(t) = 2\sqrt{2}\,t\,\mathbf{i} + e^{2t}\mathbf{j} + e^{-2t}\mathbf{k}$. Find the velocity, speed, and acceleration of the particle.

79. A particle starts at the origin with initial velocity $\mathbf{i} + 2\mathbf{j} + \mathbf{k}$. Its acceleration is $\mathbf{a}(t) = t\mathbf{i} + \mathbf{j} + t^2\mathbf{k}$. Find its position function.

80. Find the tangential and normal components of the acceleration vector of a particle with position function $\mathbf{r}(t) = t\mathbf{i} + 2t\mathbf{j} + t^2\mathbf{k}$.

81. The cylindrical coordinates of a point are $(2, \pi/6, 2)$. Find the rectangular and spherical coordinates of the point.

82. The rectangular coordinates of a point are $(2, 2, -1)$. Find the cylindrical and spherical coordinates of the point.

83. The spherical coordinates of a point are $(4, \pi/3, \pi/6)$. Find the rectangular and cylindrical coordinates of the point.

84–87 ■ Identify the surface whose equation is given.

84. $\phi = \pi/4$

85. $\theta = \pi/4$

86. $r = \cos\theta$

87. $\rho = 3\sec\phi$

88–90 ■ Write the given equation in cylindrical coordinates and in spherical coordinates.

88. $x^2 + y^2 = 4$

89. $x^2 + y^2 + z^2 = 4$

90. $x^2 + y^2 + z^2 = 2x$

91. The parabola $z = 4y^2$, $x = 0$ is rotated about the z-axis. Write an equation of the resulting surface in cylindrical coordinates.

92. Sketch the solid consisting of all points with spherical coordinates (ρ, θ, ϕ) such that $0 \le \theta \le \pi/2$, $0 \le \phi \le \pi/6$, and $0 \le \rho \le 2\cos\phi$.

93. A disk of radius 1 is rotating in the counterclockwise direction at a constant angular speed ω. A particle starts at the center of the disk and moves toward the edge along a fixed radius so that its position at time t, $t \ge 0$, is given by $\mathbf{r}(t) = t\mathbf{R}(t)$, where
$$\mathbf{R}(t) = \cos\omega t\,\mathbf{i} + \sin\omega t\,\mathbf{j}$$
(a) Show that the velocity $\mathbf{v}$ of the particle is
$$\mathbf{v} = \cos\omega t\,\mathbf{i} + \sin\omega t\,\mathbf{j} + t\mathbf{v}_d$$
where $\mathbf{v}_d = \mathbf{R}'(t)$ is the velocity of a point on the edge of the disk.
(b) Show that the acceleration $\mathbf{a}$ of the particle is
$$\mathbf{a} = 2\mathbf{v}_d + t\mathbf{a}_d$$
where $\mathbf{a}_d = \mathbf{R}''(t)$ is the acceleration of a point on the rim of the disk. The extra term $2\mathbf{v}_d$ is called the *Coriolis acceleration;* it is the result of the interaction of the rotation of the disk and the motion of the particle. One can obtain a physical demonstration of this acceleration by walking toward the edge of a moving merry-go-round.
(c) Determine the Coriolis acceleration of a particle that moves on a rotating disk according to
$$\mathbf{r}(t) = e^{-t}\cos\omega t\,\mathbf{i} + e^{-t}\sin\omega t\,\mathbf{j}$$

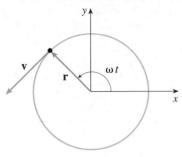

FIGURE FOR PROBLEM 1

1. A particle P moves with constant angular speed ω around a circle whose center is at the origin and whose radius is R. The particle is said to be in *uniform circular motion*. Assume that the motion is counterclockwise and that the particle is at the point $(R, 0)$ when $t = 0$. The position vector at time $t \geq 0$ is

$$\mathbf{r}(t) = R\cos\omega t\,\mathbf{i} + R\sin\omega t\,\mathbf{j}$$

(a) Find the velocity vector $\mathbf{v}$ and show that $\mathbf{v} \cdot \mathbf{r} = 0$. Conclude that $\mathbf{v}$ is tangent to the circle and points in the direction of the motion.

(b) Show that the speed $|\mathbf{v}|$ of the particle is the constant ωR. The *period T* of the particle is the time required for one complete revolution. Conclude that

$$T = \frac{2\pi R}{|\mathbf{v}|} = \frac{2\pi}{\omega}$$

(c) Find the acceleration vector $\mathbf{a}$. Show that it is proportional to $\mathbf{r}$ and that it points toward the origin. An acceleration with this property is called a *centripetal acceleration*. Show that the magnitude of the acceleration vector is $|\mathbf{a}| = R\omega^2$.

(d) Suppose that the particle has mass m. Show that the magnitude of the force $\mathbf{F}$ that is required to produce this motion, called a *centripetal force*, is

$$|\mathbf{F}| = \frac{m|\mathbf{v}|^2}{R}$$

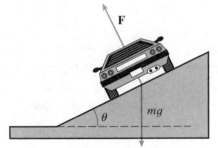

FIGURE FOR PROBLEM 2

2. A circular curve of radius R on a highway is banked at an angle θ so that a car can safely traverse the curve without skidding when there is no friction between the road and the tires. The loss of friction could occur, for example, if the road is covered with a film of water or ice. The rated speed v_R of the curve is the maximum speed that a car can attain without skidding. Suppose a car of mass m is traversing the curve at the rated speed v_R. Two forces are acting on the car: the vertical force, mg, due to the weight of the car, and a force $\mathbf{F}$ exerted by, and normal to, the road. (See the figure.)

   The vertical component of $\mathbf{F}$ balances the weight of the car, so that $|\mathbf{F}|\cos\theta = mg$. The horizontal component of $\mathbf{F}$ produces a centripetal force on the car so that, by Newton's Second Law and part (d) of Problem 1,

$$|\mathbf{F}|\sin\theta = \frac{mv_R^2}{R}$$

(a) Show that $v_R^2 = Rg\tan\theta$.

(b) Find the rated speed of a circular curve with radius 400 ft that is banked at an angle of $12°$.

(c) Suppose the design engineers want to keep the banking at $12°$, but wish to increase the rated speed by 50%. What should the radius of the curve be?

3. A projectile is fired from the origin with angle of elevation α and initial speed v_0. Assuming that air resistance is negligible and that the only force acting on the projectile is gravity, g, we showed in Example 5 in Section 11.9 that the position vector of the projectile is

$$\mathbf{r}(t) = (v_0\cos\alpha)t\,\mathbf{i} + \left[(v_0\sin\alpha)t - \tfrac{1}{2}gt^2\right]\mathbf{j}$$

We also showed that the maximum horizontal distance of the projectile is achieved when $\alpha = 45°$ and in this case the range is $R = v_0^2/g$.

(a) At what angle should the projectile be fired to achieve maximum height and what is the maximum height?

(b) Fix the initial speed v_0 and consider the parabola $x^2 + 2Ry - R^2 = 0$, whose graph is shown in the figure. Show that the projectile can hit any target inside or on the boundary of the region bounded by the parabola and the x-axis, and that it cannot hit any target outside this region.

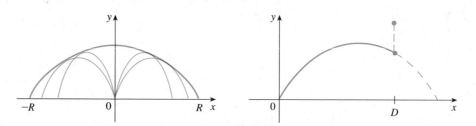

FIGURE FOR PROBLEM 3

(c) Suppose that the gun is elevated to an angle of inclination α in order to aim at a target that is suspended at a height h directly over a point D units downrange. The target is released at the instant the gun is fired. Show that the projectile always hits the target, regardless of the value v_0, provided the projectile does not hit the ground "before" D.

FIGURE FOR PROBLEM 4

4. A projectile is fired from the origin down an inclined plane that makes an angle θ with the horizontal. The angle of elevation of the gun and the initial velocity of the projectile are α and v_0, respectively.
   (a) Find the position vector of the projectile and the parametric equations of the path of the projectile as functions of the time t. (Ignore air resistance.)
   (b) Determine the angle of elevation α that will maximize the downhill range.
   (c) Suppose the projectile is fired up an inclined plane whose angle of inclination is θ. At what angle should the projectile be fired to maximize the (uphill) range?

5. A projectile of mass m is fired from the origin at an angle of elevation α. In addition to gravity, assume that air resistance provides a force that is proportional to the velocity and that opposes the motion. Then, by Newton's Second Law, the total force acting on the projectile satisfies the equation

(1) $$m\frac{d^2\mathbf{R}}{dt^2} = -mg\,\mathbf{j} - k\frac{d\mathbf{R}}{dt}$$

where $\mathbf{R}$ is the position vector and $k > 0$ is the constant of proportionality.
   (a) Show that Equation 1 can be integrated to obtain the equation

$$\frac{d\mathbf{R}}{dt} + \frac{k}{m}\mathbf{R} = \mathbf{v}_0 - gt\,\mathbf{j}$$

where $\mathbf{v}_0 = \mathbf{v}(0) = \dfrac{d\mathbf{R}}{dt}(0)$.
   (b) Multiply both sides of the equation in part (a) by $e^{(k/m)t}$ and show that the left-hand side of the resulting equation is the derivative of the product $e^{(k/m)t}\mathbf{R}(t)$. Then integrate to find an expression for the position vector $\mathbf{R}(t)$.

6. Based on clinical evidence, it can be assumed that the rate of decrease of the concentration C of a drug in the bloodstream is proportional to the concentration itself; that is, $dC/dt = -kC$, where k is a positive constant. Thus the concentration at time t is given by $C(t) = C_0e^{-kt}$, where C_0 is the initial dosage. Now suppose that the drug is to be administered in equal dosages C_0 at equally spaced time intervals of length T.

Assume that when the drug is administered it is diffused immediately throughout the bloodstream; that is, assume that there is an instantaneous rise in the concentration at each administration of the drug.

(a) Let C_{n-1} be the concentration of the drug immediately after the nth injection and let R_n be the concentration immediately before the $(n + 1)$th injection. Thus C_{n-1} and R_n are the concentrations at the beginning and end, respectively, of the nth interval. Find expressions for C_{n-1} and R_n.

(b) Calculate $\lim_{n\to\infty} C_{n-1}$ and $\lim_{n\to\infty} R_n$.

(c) Let $C = C(t)$ denote the concentration of the drug in the bloodstream at any time t. Sketch the graph of C.

(d) Suppose that the drug is ineffective below the concentration level C_L and is harmful above the concentration level C_H. Find a dosage C_0 and a time interval T so that, eventually

$$C_L \le C(t) \le C_H$$

7. In designing *transfer curves* to connect sections of straight railroad tracks, it is important to realize that the acceleration of the train should be continuous so that the reactive force exerted by the train on the track is also continuous. Because of the formulas for the components of acceleration in Section 11.9, this will be the case if the curvature varies continuously.

(a) A logical candidate for a transfer curve to join existing tracks given by $y = 1$ for $x \le 0$ and $y = \sqrt{2} - x$ for $x \ge 1/\sqrt{2}$ might be the function $f(x) = \sqrt{1 - x^2}$, $0 < x < 1/\sqrt{2}$, whose graph is the arc of the circle shown in the figure. It looks reasonable at first glance. Show that the function

$$F(x) = \begin{cases} 1 & \text{if } x \le 0 \\ \sqrt{1 - x^2} & \text{if } 0 < x < 1/\sqrt{2} \\ \sqrt{2} - x & \text{if } x \ge 1/\sqrt{2} \end{cases}$$

is continuous and has continuous slope, but does not have continuous curvature. Therefore f is not an appropriate transfer curve.

(b) Find a fifth-degree polynomial to serve as a transfer curve between the following straight line segments: $y = 0$ for $x \le 0$ and $y = x$ for $x \ge 1$. Could this be done with a fourth-degree polynomial? Use a graphing calculator or computer to sketch the graph of the "connected" function and check to see that it looks like the one in the figure.

8. A ball rolls off a table with a speed of 2 ft/s. The table is 3.5 ft high.

(a) Determine the point at which the ball hits the floor and find its speed at the instant of impact.

(b) Find the angle θ between the path of the ball and the vertical line drawn through the point of impact. (See the figure.)

(c) Suppose the ball rebounds from the floor at the same angle with which it hits the floor, but loses 20% of its speed due to energy absorbed by the ball on impact. Where does the ball strike the floor on the second bounce?

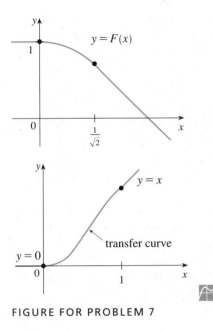

FIGURE FOR PROBLEM 7

3.5 ft

FIGURE FOR PROBLEM 8

12

PARTIAL DERIVATIVES

■ Strange as it may seem, the power of mathematics rests on its evasion of all unnecessary thought and on its wonderful savings of mental operations.

ERNST MACH

So far we have dealt with the calculus of functions of a single variable. But, in the real world, physical quantities usually depend on two or more variables, so in this chapter we turn our attention to functions of several variables and extend the basic ideas of differential calculus to such functions.

12.1 FUNCTIONS OF SEVERAL VARIABLES

The temperature T at a point on the surface of the earth at any given time depends on the longitude x and latitude y of the point. We can think of T as being a function of the two variables x and y, or as a function of the pair (x, y). We indicate this functional dependence by writing $T = f(x, y)$.

The volume V of a circular cylinder depends on its radius r and height h. In fact, we know from Chapter 5 that $V = \pi r^2 h$. We say that V is a function of r and h, and we write $V(r, h) = \pi r^2 h$.

---

(1) DEFINITION Let $D \subset \mathbb{R}^2$. A **function f of two variables** is a rule that assigns to each ordered pair (x, y) in D a unique real number denoted by $f(x, y)$. The set D is the **domain** of f and its **range** is the set of values that f takes on, that is, $\{ f(x, y) \,|\, (x, y) \in D \}$.

We often write $z = f(x, y)$ to make explicit the value taken on by f at the general point (x, y). The variables x and y are **independent variables** and z is the **dependent variable.** [Compare this with the notation $y = f(x)$ for functions of a single variable.]

The situation described in Definition 1 is indicated by the notation $f: D \to \mathbb{R}$. A function of two variables is just a special case of the general idea of a function $f: X \to Y$ (see Section 1 of Review and Preview) where $X = D \subset \mathbb{R}^2$ and $Y \subset \mathbb{R}$. One way of visualizing such a function is by means of an arrow diagram (see Figure 1), where the domain D is represented as a subset of the xy-plane.

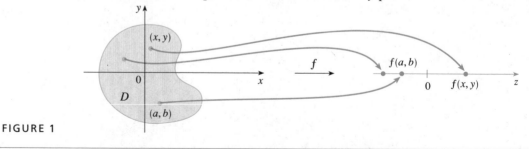

FIGURE 1

If a function f is given by a formula and no domain is specified, then the domain of f is understood to be the set of all pairs (x, y) for which the given expression is a well-defined real number.

EXAMPLE 1   Find the domain of each of the following functions and evaluate $f(3, 2)$.

(a) $f(x, y) = \dfrac{\sqrt{x + y + 1}}{x - 1}$ (b) $f(x, y) = x \ln(y^2 - x)$

SOLUTION

(a) $f(3, 2) = \dfrac{\sqrt{3 + 2 + 1}}{3 - 1} = \dfrac{\sqrt{6}}{2}$

The expression for f makes sense if the denominator is not 0 and the quantity under the square root sign is nonnegative. So the domain of f is

$$D = \{(x, y) \mid x + y + 1 \geq 0, \ x \neq 1\}$$

The inequality $x + y + 1 \geq 0$, or $y \geq -x - 1$, describes the points that lie on or above the line $y = -x - 1$, while $x \neq 1$ means that the points on the line $x = 1$ must be excluded from the domain. (See Figure 2.)

(b) $f(3, 2) = 3 \ln(2^2 - 3) = 3 \ln 1 = 0$

Since $\ln(y^2 - x)$ is defined only when $y^2 - x > 0$, that is, $x < y^2$, the domain of f is $D = \{(x, y) \mid x < y^2\}$. This is the set of points to the left of the parabola $x = y^2$. (See Figure 3.)   ■

> **(2) DEFINITION**   If f is a function of two variables with domain D, the **graph** of f is the set
>
> $$S = \{(x, y, z) \in \mathbb{R}^3 \mid z = f(x, y), (x, y) \in D\}$$

Just as the graph of a function f of one variable is a curve C with equation $y = f(x)$, so the graph of a function f of two variables is a surface S with equation $z = f(x, y)$. We can visualize the graph S of f as lying directly above or below its domain D in the xy-plane (see Figure 4).

EXAMPLE 2   Sketch the graph of the function $f(x, y) = 6 - 3x - 2y$.

SOLUTION   The graph of f has the equation $z = 6 - 3x - 2y$, or $3x + 2y + z = 6$, which is a plane. The portion of this graph that lies in the first octant is sketched in Figure 5.

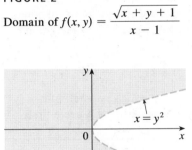

$x + y + 1 = 0$

$x = 1$

FIGURE 2

Domain of $f(x, y) = \dfrac{\sqrt{x + y + 1}}{x - 1}$

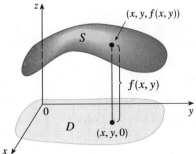

$x = y^2$

FIGURE 3

Domain of $f(x, y) = x \ln(y^2 - x)$

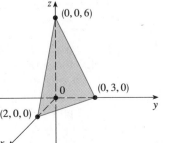

$(x, y, f(x, y))$

S

$f(x, y)$

D $(x, y, 0)$

FIGURE 4

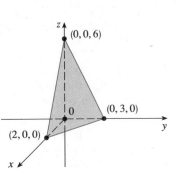

$(0, 0, 6)$

$(0, 3, 0)$

$(2, 0, 0)$

FIGURE 5   ■

EXAMPLE 3   Find the domain and range and sketch the graph of

$$g(x, y) = \sqrt{9 - x^2 - y^2}$$

SOLUTION   The domain of g is

$$D = \{(x, y) \mid 9 - x^2 - y^2 \geqslant 0\} = \{(x, y) \mid x^2 + y^2 \leqslant 9\}$$

which is the disk with center $(0, 0)$ and radius 3. The range of g is

$$\{z \mid z = \sqrt{9 - x^2 - y^2}, \ (x, y) \in D\}$$

Since z is a positive square root, $z \geqslant 0$. Also

$$9 - x^2 - y^2 \leqslant 9 \quad \Rightarrow \quad \sqrt{9 - x^2 - y^2} \leqslant 3$$

So the range is

$$\{z \mid 0 \leqslant z \leqslant 3\} = [0, 3]$$

The graph has equation $z = \sqrt{9 - x^2 - y^2}$. We square both sides of this equation to obtain $z^2 = 9 - x^2 - y^2$, or $x^2 + y^2 + z^2 = 9$, which we recognize as an equation of the sphere with center the origin and radius 3. But, since $z \geqslant 0$, the graph of g is just the top half of this sphere (see Figure 6).   ■

EXAMPLE 4   Sketch the graph of the function $h(x, y) = 4x^2 + y^2$.

SOLUTION   The graph of h has the equation $z = 4x^2 + y^2$, which is an elliptic paraboloid (see Equation 11.6.5). Traces in the horizontal planes $z = k$ are ellipses if $k > 0$. Vertical traces are parabolas. (See Figure 7.)   ■

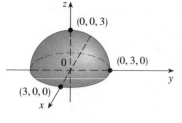

FIGURE 6
Graph of $g(x, y) = \sqrt{9 - x^2 - y^2}$

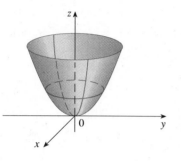

FIGURE 7
Graph of $h(x, y) = 4x^2 + y^2$

Computer programs are now readily available for graphing functions of two variables. In most such programs, traces in the vertical planes $x = k$ and $y = k$ are drawn for equally spaced values of k and parts of the graph are eliminated using hidden line removal. Figure 8 shows computer-generated graphs of several functions. Notice that we get an especially good picture of a function when rotation is used to give views from different vantage points.

LEVEL CURVES

So far we have two methods for visualizing functions: arrow diagrams and graphs. A third method, borrowed from mapmakers, is a contour map on which points of constant elevation are joined to form *contour curves,* or *level curves.*

> **(3) DEFINITION**   The **level curves** of a function f of two variables are the curves with equations $f(x, y) = k$, where k is a constant (in the range of f).

A level curve $f(x, y) = k$ is the locus of all points at which f takes on a given value k. In other words, it shows where the graph of f has height k.

You can see from Figure 9 the relation between level curves and horizontal traces. The level curves $f(x, y) = k$ are just the traces of the graph of f in the horizontal plane $z = k$ projected down to the xy-plane. So if you draw the level curves of a function and visualize them being lifted up to the surface at the indicated height, then you can mentally piece together a picture of the graph. The surface is steep where the level curves are close together. It is somewhat flatter where they are farther apart.

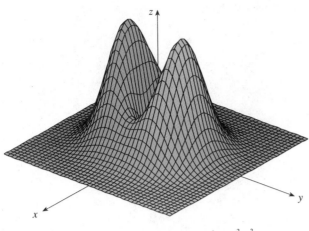

(a) $f(x, y) = (x^2 + 3y^2)e^{-x^2-y^2}$

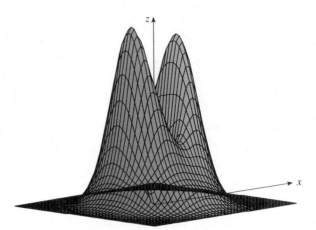

(b) $f(x, y) = (x^2 + 3y^2)e^{-x^2-y^2}$

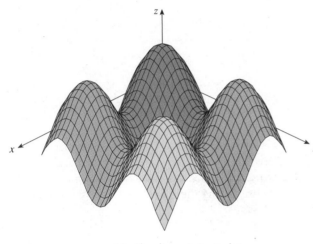

(c) $f(x, y) = \sin x + \sin y$

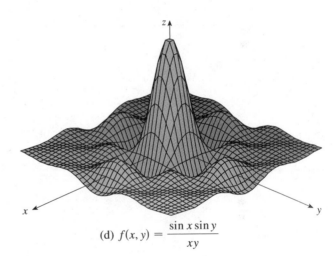

(d) $f(x, y) = \dfrac{\sin x \sin y}{xy}$

FIGURE 8

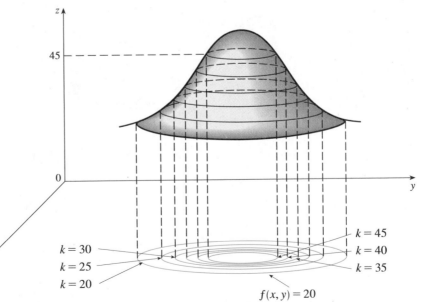

FIGURE 9

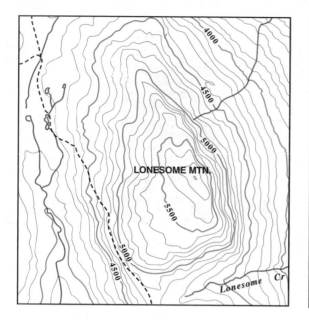

FIGURE 10

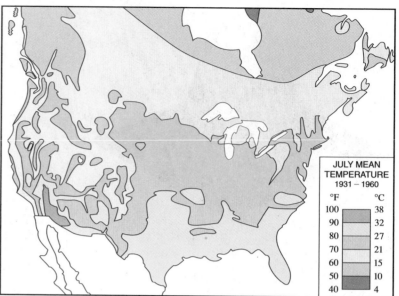

FIGURE 11

One common example of level curves occurs in topographic maps of mountainous regions, such as the map in Figure 10. The level curves are curves of constant elevation above sea level. If you walk along one of these contour lines you neither ascend nor descend. Another common example is the temperature function introduced in the opening paragraph of this section. Here the level curves are called **isothermals** and join locations with the same temperature. Figure 11 shows a weather map of North America indicating the average temperature during the month of July. The isothermals are the curves that separate the colored bands.

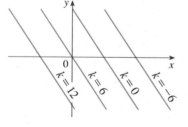

FIGURE 12

Contour map of
$f(x, y) = 6 - 3x - 2y$

EXAMPLE 5   Sketch the level curves of the function $f(x, y) = 6 - 3x - 2y$ for the values $k = -6, 0, 6, 12$.

SOLUTION   The level curves are

$$6 - 3x - 2y = k \qquad \text{or} \qquad 3x + 2y + (k - 6) = 0$$

This is a family of lines with slope $-\frac{3}{2}$. The four particular level curves with $k = -6, 0, 6,$ and 12 are $3x + 2y - 12 = 0$, $3x + 2y - 6 = 0$, $3x + 2y = 0$, and $3x + 2y + 6 = 0$. They are sketched in Figure 12.   ∎

EXAMPLE 6   Sketch the level curves of the function

$$g(x, y) = \sqrt{9 - x^2 - y^2} \qquad \text{for} \qquad k = 0, 1, 2, 3$$

SOLUTION   The level curves are

$$\sqrt{9 - x^2 - y^2} = k \qquad \text{or} \qquad x^2 + y^2 = 9 - k^2$$

This is a family of concentric circles with center $(0, 0)$ and radius $\sqrt{9 - k^2}$. The cases $k = 0, 1, 2, 3$ are shown in Figure 13. Try to visualize these level curves lifted up to form a surface and compare with the graph of g in Figure 6.

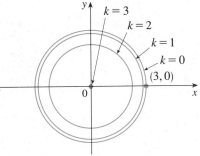

FIGURE 13

Contour map of

$g(x, y) = \sqrt{9 - x^2 - y^2}$

EXAMPLE 7   Sketch some level curves of the function $h(x, y) = 4x^2 + y^2$.

SOLUTION   The level curves are

$$4x^2 + y^2 = k \qquad \text{or} \qquad \frac{x^2}{k/4} + \frac{y^2}{k} = 1$$

which, for $k > 0$, describes a family of ellipses with semiaxes $\sqrt{k}/2$ and $\sqrt{k}$. Compare the sketches of these level curves in Figure 14 with the graph of h in Figure 7.

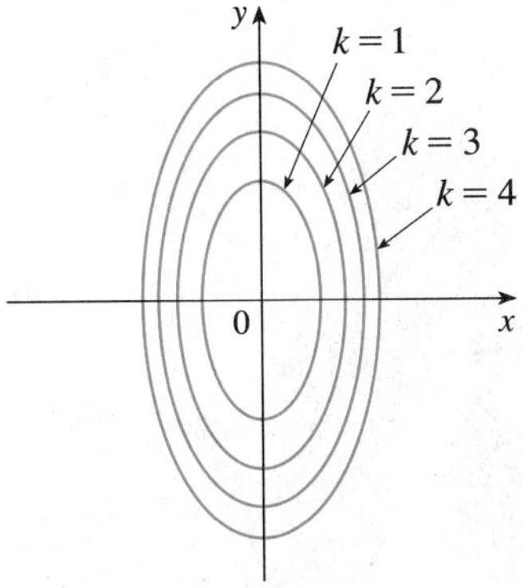

FIGURE 14

Contour map of $h(x, y) = 4x^2 + y^2$

EXAMPLE 8   Sketch the level curves of $f(x, y) = x + y^2$ for $k = -1, 0, 1, 2, 3$.

SOLUTION   The level curves are $x + y^2 = k$. These form a family of parabolas with x-intercept k [see Figure 15(a)]. We see from Figure 15(b) how the graph $z = x + y^2$ is put together from the level curves.

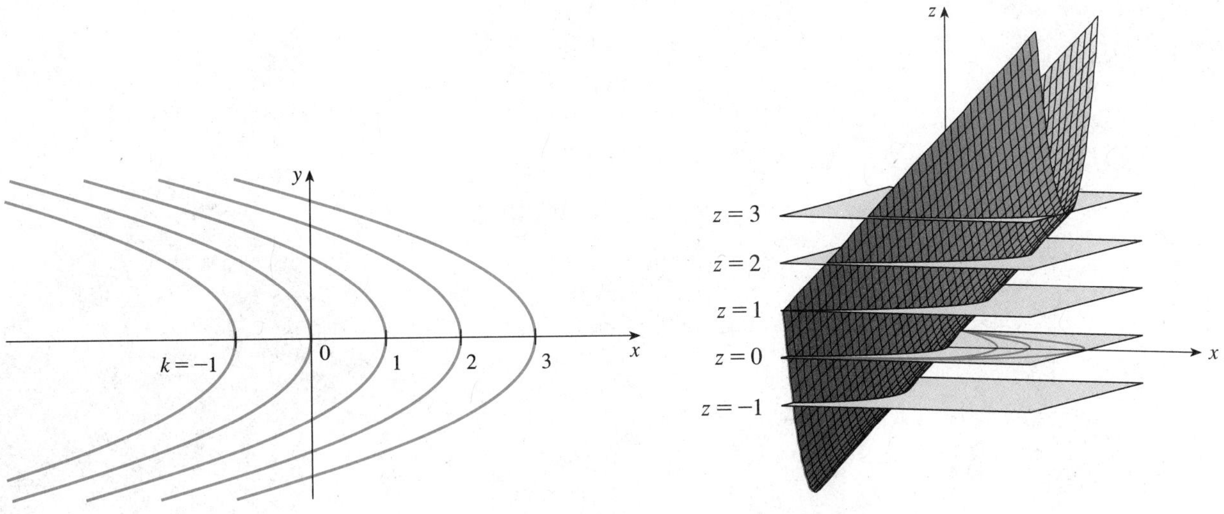

FIGURE 15   (a) Contour map of $f(x, y) = x + y^2$ (b) Graph of $f(x, y) = x + y^2$

Figure 16 on page 766 shows some computer-generated level curves together with the corresponding computer-generated graphs.

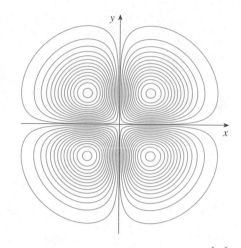

(a) Level curves of $f(x, y) = -xye^{-x^2-y^2}$

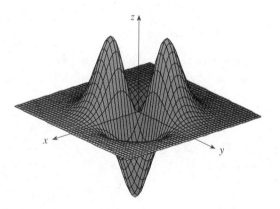
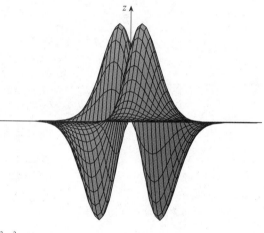

(b) Two views of $f(x, y) = -xye^{-x^2-y^2}$

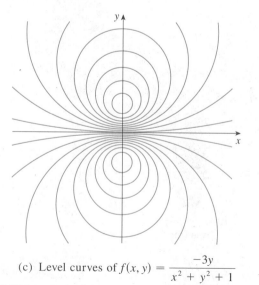

(c) Level curves of $f(x, y) = \dfrac{-3y}{x^2 + y^2 + 1}$

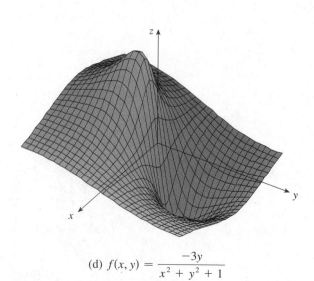

(d) $f(x, y) = \dfrac{-3y}{x^2 + y^2 + 1}$

FIGURE 16

FUNCTIONS OF THREE OR MORE VARIABLES

A **function of three variables**, f, is a rule that assigns to each ordered triple (x, y, z) in a domain $D \subset \mathbb{R}^3$ a unique real number denoted by $f(x, y, z)$. For instance, in the illustration at the beginning of this section the temperature T depends on the time t as well as on the longitude x and latitude y, so we could write $T = f(x, y, t)$.

It is possible to visualize a function f of three variables by an arrow diagram but not by its graph, since that would lie in a four-dimensional space. However we do gain some insight into f by examining its **level surfaces,** which are the surfaces with equations $f(x, y, z) = k$, where k is a constant. If the point (x, y, z) moves along a level surface, the value of $f(x, y, z)$ remains fixed.

EXAMPLE 9   Find the domain of f if $f(x, y, z) = \ln(z - y) + xy \sin z$.

SOLUTION   The expression for $f(x, y, z)$ is defined as long as $z - y > 0$, so the domain of f is

$$D = \{(x, y, z) \in \mathbb{R}^3 \mid z > y\}$$

This is a **half-space** consisting of all points that lie above the plane $z = y$. ∎

EXAMPLE 10   Find the level surfaces of the function

$$f(x, y, z) = x^2 + y^2 + z^2$$

SOLUTION   The level surfaces are $x^2 + y^2 + z^2 = k$, where $k \geqslant 0$. These form a family of concentric spheres with radius $\sqrt{k}$ (see Figure 17). Thus as (x, y, z) varies over any sphere with center O, the value of $f(x, y, z)$ remains fixed. ∎

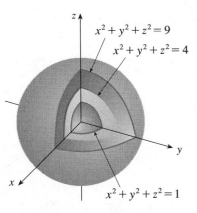

$x^2 + y^2 + z^2 = 9$
$x^2 + y^2 + z^2 = 4$
$x^2 + y^2 + z^2 = 1$

FIGURE 17

Functions of any number of variables can also be considered. A **function of n variables** is a rule that assigns a number $z = f(x_1, x_2, \ldots, x_n)$ to an n-tuple $(x_1, x_2, \ldots, x_n)$ of real numbers. For example, if a company uses n different ingredients in making a food product, c_i is the cost per unit of the ith ingredient, and x_i units of the ith ingredient are used, then the total cost C of the ingredients is a function of the n variables x_1, $x_2, \ldots, x_n$:

(4) $$C = f(x_1, x_2, \ldots, x_n) = c_1 x_1 + c_2 x_2 + \cdots + c_n x_n$$

The notation $$f : D \subset \mathbb{R}^n \to \mathbb{R}$$

is used to signify that f is a real-valued function whose domain D is a subset of $\mathbb{R}^n$. Sometimes we will use vector notation to write functions more compactly: If $\mathbf{x} = \langle x_1, x_2, \ldots, x_n \rangle$, we often write $f(\mathbf{x})$ in place of $f(x_1, x_2, \ldots, x_n)$. With this notation we can rewrite the function defined in Equation 4 as

$$f(\mathbf{x}) = \mathbf{c} \cdot \mathbf{x}$$

where $\mathbf{c} = \langle c_1, c_2, \ldots, c_n \rangle$.

In view of the one-to-one correspondence between points $(x_1, x_2, \ldots, x_n)$ in $\mathbb{R}^n$ and their position vectors $\mathbf{x} = \langle x_1, x_2, \ldots, x_n \rangle$ in V_n, we have three ways of looking at a function $f : \mathbb{R}^n \to \mathbb{R}$:

1. As a function of n real variables $x_1, x_2, \ldots, x_n$
2. As a function of a single point variable $(x_1, x_2, \ldots, x_n)$
3. As a function of a single vector variable $\mathbf{x} = \langle x_1, x_2, \ldots, x_n \rangle$

We will see that all three points of view are useful.

EXERCISES 12.1

1. If $f(x, y) = x^2 - y^2 + 4xy - 7x + 10$, find
  (a) $f(2, 1)$ (b) $f(-3, 5)$
  (c) $f(x + h, y)$ (d) $f(x, y + k)$
  (e) $f(x, x)$

2. If $g(x, y) = \ln(xy + y - 1)$, find
  (a) $g(1, 1)$ (b) $g(e, 1)$
  (c) $g(x, 1)$ (d) $g(x + h, y)$
  (e) $g(x, y + k)$

3. If $F(x, y) = 3xy/(x^2 + 2y^2)$, find
  (a) $F(1, 1)$ (b) $F(-1, 2)$
  (c) $F(t, 1)$ (d) $F(-1, y)$
  (e) $F(x, x^2)$

4. If $G(x, y, z) = x \sin y \cos z$, find
  (a) $G(2, \pi/6, \pi/3)$ (b) $G(4, \pi/4, 0)$
  (c) $G(t, t, t)$ (d) $G(u, v, 0)$
  (e) $G(x, x + y, x)$

5–14 ■ Find the domain and range of the function.

5. $f(x, y) = x + 2y - 5$ **6.** $f(x, y) = \sqrt{x - y}$

7. $f(x, y) = 2/(x + y)$ **8.** $f(x, y) = \tan^{-1}(y/x)$

9. $f(x, y) = e^{x^2 - y}$

10. $f(x, y) = \sqrt{36 - 9x^2 - 4y^2}$

11. $f(x, y, z) = x^2 \ln(x - y + z)$

12. $f(x, y, z) = x/(yz)$

13. $f(x, y, z) = x \sin(y + z)$

14. $f(x, y, z) = 1/\sqrt{x^2 + y^2 + z^2 - 1}$

15–30 ■ Find and sketch the domain of the function.

15. $f(x, y) = \sqrt[4]{y - 2x}$ **16.** $f(x, y) = \sqrt{x} + \sqrt{y}$

17. $f(x, y) = \dfrac{\sqrt{9 - x^2 - y^2}}{x + 2y}$ **18.** $f(x, y) = \dfrac{x^2 + y^2}{x^2 - y^2}$

19. $f(x, y) = xy\sqrt{x^2 + y}$ **20.** $f(x, y) = \tan(x - y)$

21. $f(x, y) = \ln(xy - 1)$ **22.** $f(x, y) = \ln(x^2 - y^2)$

23. $f(x, y) = x^2 \sec y$

24. $f(x, y) = \sqrt{x^2 + y^2 - 1} + \ln(4 - x^2 - y^2)$

25. $f(x, y) = \sin^{-1}(x + y)$

26. $f(x, y) = \sqrt{4 - 2x^2 - y^2}$

27. $f(x, y) = \ln x + \ln \sin y$

28. $f(x, y) = \sqrt{y - x} \ln(y + x)$

29. $f(x, y, z) = \sqrt{1 - x^2 - y^2 - z^2}$

30. $f(x, y, z) = \ln(16 - 4x^2 - 4y^2 - z^2)$

31–42 ■ Sketch the graph of the function.

31. $f(x, y) = 3$ **32.** $f(x, y) = x$

33. $f(x, y) = 1 - x - y$ **34.** $f(x, y) = y^2$

35. $f(x, y) = x^2 + 9y^2$ **36.** $f(x, y) = 3 - x^2 - y^2$

37. $f(x, y) = \sqrt{x^2 + y^2}$

38. $f(x, y) = \sqrt{16 - x^2 - 16y^2}$

39. $f(x, y) = y^2 - x^2$

40. $f(x, y) = \sin y$

41. $f(x, y) = 1 - x^2$

42. $f(x, y) = x^2 + y^2 - 4x - 2y + 5$

43–52 ■ Draw a contour map of the function showing several level curves.

43. $f(x, y) = xy$ **44.** $f(x, y) = x^2 - y^2$

45. $f(x, y) = x^2 + 9y^2$ **46.** $f(x, y) = e^{xy}$

47. $f(x, y) = \dfrac{x}{y}$ **48.** $f(x, y) = \dfrac{x + y}{x - y}$

49. $f(x, y) = \sqrt{x + y}$ **50.** $f(x, y) = y - \cos x$

51. $f(x, y) = x - y^2$ **52.** $f(x, y) = e^{1/(x^2 + y^2)}$

53–56 ■ Describe the level surfaces of the function.

53. $f(x, y, z) = x + 3y + 5z$

54. $f(x, y, z) = x^2 + 3y^2 + 5z^2$

55. $f(x, y, z) = x^2 - y^2 + z^2$

56. $f(x, y, z) = x^2 - y^2$

57. A thin metal plate, located in the xy-plane, has temperature $T(x, y)$ at the point (x, y). The level curves of T are called *isothermals* because at all points on an isothermal the temperature is the same. Sketch some isothermals if the temperature function is given by
$$T(x, y) = 100/(1 + x^2 + 2y^2)$$

58. If $V(x, y)$ is the electric potential at a point (x, y) in the xy-plane, then the level curves of V are called *equipotential curves* because at all points on such a curve the electric potential is the same. Sketch some equipotential curves if $V(x, y) = c/\sqrt{r^2 - x^2 - y^2}$, where c is a positive constant.

59–64 ■ Match the function (a) with its graph (labeled A–F) and (b) with its contour map (labeled I–VI). Give reasons for your choices.

59. $z = \sin\sqrt{x^2 + y^2}$ **60.** $z = x^2y^2e^{-x^2 - y^2}$

61. $z = \dfrac{1}{x^2 + 4y^2}$ **62.** $z = x^3 - 3xy^2$

63. $z = \sin x \sin y$ **64.** $z = \sin^2 x + \frac{1}{4}y^2$

A

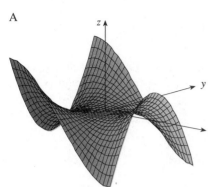

B

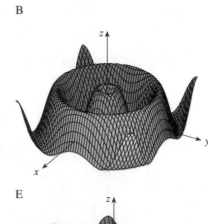

C

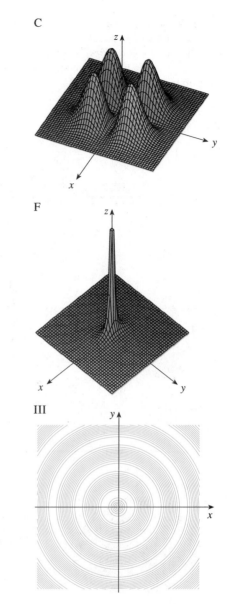

D

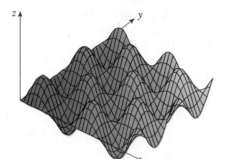

E

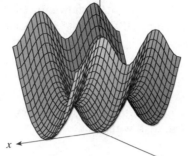

F

I

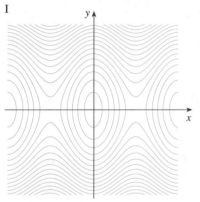

II

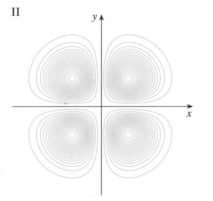

III

IV

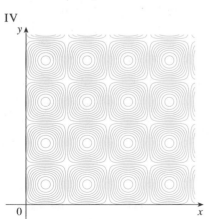

V

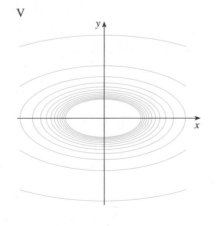

VI

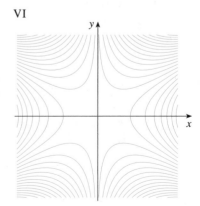

65–68 ■ Use a computer to graph the function using various domains and viewpoints. Get a printout of one that, in your opinion, gives a good view. If your software also produces level curves, then plot some contour lines of the same function and compare with the graph.

65. $f(x, y) = x^3 + y^3$

66. $f(x, y) = \sin(ye^{-x})$

67. $f(x, y) = xy^2 - x^3$ (monkey saddle)

68. $f(x, y) = xy^3 - yx^3$ (dog saddle)

69–70 ■ Use a computer to investigate the family of functions. How does the shape of the graph depend on the numbers a and b?

69. $f(x, y) = e^{ax^2 + by^2}$ **70.** $f(x, y) = (ax^2 + by^2)e^{-x^2 - y^2}$

12.2 LIMITS AND CONTINUITY

Consider the function $f(x, y) = \sqrt{9 - x^2 - y^2}$ whose domain is the closed disk $D = \{(x, y) \mid x^2 + y^2 \leq 9\}$ shown in Figure 1 and whose graph is the hemisphere shown in Figure 2. If the point (x, y) is close to the origin, then x and y are both close to 0, and so $f(x, y)$ is close to 3. In fact, if (x, y) lies in a small open disk $x^2 + y^2 < \delta^2$, then

$$f(x, y) = \sqrt{9 - (x^2 + y^2)} > \sqrt{9 - \delta^2}$$

Thus we can make the values of $f(x, y)$ as close to 3 as we like by taking (x, y) in a small enough disk with center $(0, 0)$. We describe this situation by using the notation

$$\lim_{(x, y) \to (0, 0)} \sqrt{9 - x^2 - y^2} = 3$$

In general, the notation

$$\lim_{(x, y) \to (a, b)} f(x, y) = L$$

means that the values of $f(x, y)$ can be made as close as we wish to the number L by taking the point (x, y) close enough to the point (a, b). A more precise definition follows.

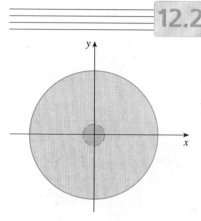

FIGURE 1
Domain of f

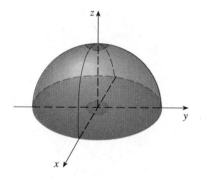

FIGURE 2
Graph of f

> **(1) DEFINITION**   Let f be a function of two variables defined on a disk with center (a, b), except possibly at (a, b). Then we say that the **limit of $f(x, y)$ as (x, y) approaches (a, b)** is L and we write
>
> $$\lim_{(x, y) \to (a, b)} f(x, y) = L$$
>
> if for every number $\varepsilon > 0$ there is a corresponding number $\delta > 0$ such that
>
> $$|f(x, y) - L| < \varepsilon \quad \text{whenever} \quad 0 < \sqrt{(x - a)^2 + (y - b)^2} < \delta$$

Other notations for the limit in Definition 1 are

$$\lim_{\substack{x \to a \\ y \to b}} f(x, y) = L \qquad \text{and} \qquad f(x, y) \to L \text{ as } (x, y) \to (a, b)$$

Since $|f(x, y) - L|$ is the distance between the numbers $f(x, y)$ and L, and $\sqrt{(x - a)^2 + (y - b)^2}$ is the distance between the point (x, y) and the point (a, b),

Definition 1 says that the distance between $f(x, y)$ and L can be made arbitrarily small by making the distance from (x, y) to (a, b) sufficiently small (but not 0). Figure 3 illustrates Definition 1 by means of an arrow diagram. If any small interval $(L - \varepsilon, L + \varepsilon)$ is given around L, then we can find a disk D_δ with center (a, b) and radius $\delta > 0$ such that f maps all the points in D_δ [except possibly (a, b)] into the interval $(L - \varepsilon, L + \varepsilon)$.

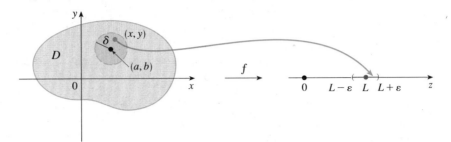

FIGURE 3

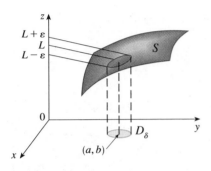

FIGURE 4

Another illustration of Definition 1 is given in Figure 4 where the surface S is the graph of f. If $\varepsilon > 0$ is given, we can find $\delta > 0$ such that if (x, y) is restricted to lie in the disk D_δ and $(x, y) \neq (a, b)$, then the corresponding part of S lies between the horizontal planes $z = L - \varepsilon$ and $z = L + \varepsilon$.

For functions of a single variable, when we let x approach a, there are only two possible directions of approach, from the left or from the right. We recall from Chapter 1 that if $\lim_{x \to a^-} f(x) \neq \lim_{x \to a^+} f(x)$, then $\lim_{x \to a} f(x)$ does not exist.

For functions of two variables the situation is not as simple because we can let (x, y) approach (a, b) from an infinite number of directions in any manner whatsoever (see Figure 5).

Definition 1 refers only to the *distance* between (x, y) and (a, b). It does not refer to the direction of approach. Therefore, if the limit exists, then $f(x, y)$ must approach the same limit no matter how (x, y) approaches (a, b). Thus, if we can find two different paths of approach along which $f(x, y)$ has different limits, then it follows that $\lim_{(x, y) \to (a, b)} f(x, y)$ does not exist.

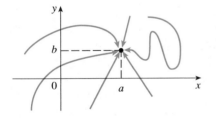

FIGURE 5

> If $f(x, y) \to L_1$ as $(x, y) \to (a, b)$ along a path C_1 and $f(x, y) \to L_2$ as $(x, y) \to (a, b)$ along a path C_2, where $L_1 \neq L_2$, then $\lim_{(x, y) \to (a, b)} f(x, y)$ does not exist.

EXAMPLE 1 Find $\displaystyle\lim_{(x, y) \to (0, 0)} \frac{x^2 - y^2}{x^2 + y^2}$ if it exists.

SOLUTION Let $f(x, y) = (x^2 - y^2)/(x^2 + y^2)$. First let us approach $(0, 0)$ along the x-axis. Then $y = 0$ gives $f(x, 0) = x^2/x^2 = 1$ for all $x \neq 0$, so

$$f(x, y) \to 1 \quad \text{as} \quad (x, y) \to (0, 0) \text{ along the } x\text{-axis}$$

We now approach along the y-axis by putting $x = 0$. Then $f(0, y) = -y^2/y^2 = -1$ for all $y \neq 0$, so

$$f(x, y) \to -1 \quad \text{as} \quad (x, y) \to (0, 0) \text{ along the } y\text{-axis}$$

(See Figure 6.) Since f has two different limits along two different lines, the given limit does not exist. ∎

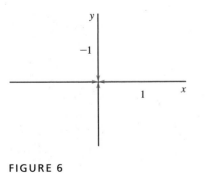

FIGURE 6

EXAMPLE 2   If $f(x, y) = xy/(x^2 + y^2)$, does $\lim\limits_{(x, y) \to (0,0)} f(x, y)$ exist?

SOLUTION   If $y = 0$, then $f(x, 0) = 0/x^2 = 0$. Therefore

$$f(x, y) \to 0 \qquad \text{as} \qquad (x, y) \to (0, 0) \text{ along the } x\text{-axis}$$

If $x = 0$, then $f(0, y) = 0/y^2 = 0$, so

$$f(x, y) \to 0 \qquad \text{as} \qquad (x, y) \to (0, 0) \text{ along the } y\text{-axis}$$

Although we have obtained identical limits along the axes, that does not show that the given limit is 0. Let us now approach $(0, 0)$ along another line, say $y = x$. For all $x \neq 0$,

$$f(x, x) = \frac{x^2}{x^2 + x^2} = \frac{1}{2}$$

Therefore $f(x, y) \to \frac{1}{2}$ as $(x, y) \to (0, 0)$ along $y = x$

(See Figure 7.) Since we obtained different limits along different paths, the given limit does not exist. ∎

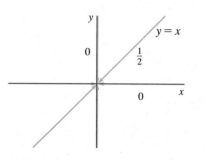

FIGURE 7

Figure 8 sheds some light on Example 2. The ridge that occurs above the line $y = x$ corresponds to the fact that $f(x, y) = \frac{1}{2}$ for all points (x, y) on that line except the origin.

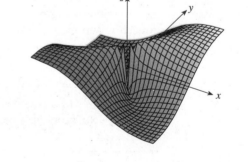

FIGURE 8

$$f(x, y) = \frac{xy}{x^2 + y^2}$$

EXAMPLE 3   If $f(x, y) = \dfrac{xy^2}{x^2 + y^4}$, does $\lim\limits_{(x, y) \to (0,0)} f(x, y)$ exist?

SOLUTION   With the solution of Example 2 in mind, let us try to save time by letting $(x, y) \to (0, 0)$ along any line through the origin. Then $y = mx$, where m is the slope, and if $m \neq 0$,

$$f(x, y) = f(x, mx) = \frac{x(mx)^2}{x^2 + (mx)^4} = \frac{m^2x^3}{x^2 + m^4x^4} = \frac{m^2x}{1 + m^4x^2}$$

So $f(x, y) \to 0$ as $(x, y) \to (0, 0)$ along $y = mx$

Thus f has the same limiting value along every line through the origin. But that does not show that the given limit is 0, for if we now let $(x, y) \to (0, 0)$ along the parabola

Figure 9 shows the graph of the function in Example 3. Notice the ridge above the parabola $x = y^2$.

FIGURE 9

$x = y^2$, we have

$$f(x, y) = f(y^2, y) = \frac{y^2 \cdot y^2}{(y^2)^2 + y^4} = \frac{y^4}{2y^4} = \frac{1}{2}$$

so $\qquad f(x, y) \to \frac{1}{2} \qquad$ as $\qquad (x, y) \to (0, 0)$ along $x = y^2$

Since different paths lead to different limiting values, the given limit does not exist. ∎

EXAMPLE 4 Find $\displaystyle\lim_{(x, y) \to (0,0)} \frac{3x^2 y}{x^2 + y^2}$ if it exists.

SOLUTION As in Example 3, one can show that the limit along any line through the origin is 0. This does not prove that the given limit is 0, but the limits along the parabolas $y = x^2$ and $x = y^2$ also turn out to be 0, so we begin to suspect that the limit does exist.

Let $\varepsilon > 0$. We want to find $\delta > 0$ such that

$$\left| \frac{3x^2 y}{x^2 + y^2} - 0 \right| < \varepsilon \qquad \text{whenever} \qquad 0 < \sqrt{x^2 + y^2} < \delta$$

that is, $\qquad \dfrac{3x^2 |y|}{x^2 + y^2} < \varepsilon \qquad$ whenever $\qquad 0 < \sqrt{x^2 + y^2} < \delta$

But $x^2 \leqslant x^2 + y^2$ since $y^2 \geqslant 0$, so

(2) $$\frac{3x^2 |y|}{x^2 + y^2} \leqslant 3|y| = 3\sqrt{y^2} \leqslant 3\sqrt{x^2 + y^2}$$

Thus if we choose $\delta = \varepsilon/3$ and let $0 < \sqrt{x^2 + y^2} < \delta$, then

$$\left| \frac{3x^2 y}{x^2 + y^2} - 0 \right| \leqslant 3\sqrt{x^2 + y^2} \leqslant 3\delta = 3\left(\frac{\varepsilon}{3} \right) = \varepsilon$$

Hence, by Definition 1,

$$\lim_{(x, y) \to (0,0)} \frac{3x^2 y}{x^2 + y^2} = 0$$

∎

Another way to do Example 4 is to use the Squeeze Theorem instead of Definition 1. From (4) it follows that

$$\lim_{(x, y) \to (0,0)} 3|y| = 0$$

and so the first inequality in (2) shows that the given limit is 0.

Just as for functions of one variable, the calculation of limits can be greatly simplified by the use of properties of limits and by the use of continuity.

The properties of limits listed in Section 1.3 can be extended to functions of two variables. The limit of a sum is the sum of the limits, the limit of a product is the product of the limits, and so on.

Recall that evaluating limits of *continuous* functions of a single variable is easy. It can be accomplished by direct substitution because the defining property of a continuous function is $\lim_{x \to a} f(x) = f(a)$. Continuous functions of two variables are also defined by the direct substitution property.

> **(3) DEFINITION** Let f be a function of two variables defined on a disk with center (a, b). Then f is called **continuous at** (a, b) if
>
> $$\lim_{(x, y) \to (a, b)} f(x, y) = f(a, b)$$

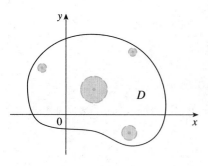

FIGURE 10
Interior points of D

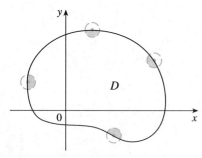

FIGURE 11
Boundary points of D

If the domain of f is a set $D \subset \mathbb{R}^2$, then Definition 3 defines the continuity of f at an **interior point** (a, b) of D, that is, a point that is contained in a disk $D_\delta \subset D$ (see Figure 10). But D may also contain a **boundary point,** that is, a point (a, b) such that every disk with center (a, b) contains points in D and also points not in D (see Figure 11).

If (a, b) is a boundary point of D, then Definition 1 is modified so that the last line reads

$$|f(x, y) - L| < \varepsilon \quad \text{whenever} \quad (x, y) \in D \quad \text{and} \quad 0 < \sqrt{(x - a)^2 + (y - b)^2} < \delta$$

With this convention, Definition 3 also applies when f is defined at a boundary point (a, b) of D.

Finally, we say f is **continuous on** D if f is continuous at every point (a, b) in D.

The intuitive meaning of continuity is that if the point (x, y) changes by a small amount, then the value of $f(x, y)$ changes by a small amount. This means that a surface that is the graph of a continuous function has no hole or break.

Using the properties of limits, you can see that sums, differences, products, and quotients of continuous functions are continuous on their domains. Let us use this fact to give examples of continuous functions.

A **polynomial function of two variables** (or polynomial, for short) is a sum of terms of the form $cx^m y^n$, where c is a constant and m and n are nonnegative integers. A **rational function** is a ratio of polynomials. For instance,

$$f(x, y) = x^4 + 5x^3 y^2 + 6xy^4 - 7y + 6$$

is a polynomial, whereas

$$g(x, y) = \frac{2xy + 1}{x^2 + y^2}$$

is a rational function.

From Definition 1 it can be shown (see Exercise 51) that

$$(4) \qquad \lim_{(x, y) \to (a, b)} x = a \qquad \lim_{(x, y) \to (a, b)} y = b \qquad \lim_{(x, y) \to (a, b)} c = c$$

These limits show that the functions $f(x, y) = x$, $g(x, y) = y$, and $h(x, y) = c$ are continuous. Since any polynomial can be built up out of the simple functions f, g, and h by multiplication and addition, it follows that *all polynomials are continuous on $\mathbb{R}^2$.* Likewise, any rational function is continuous on its domain since it is a quotient of continuous functions.

EXAMPLE 5   Evaluate $\lim\limits_{(x, y) \to (1, 2)} (x^2 y^3 - x^3 y^2 + 3x + 2y)$.

SOLUTION   Since $f(x, y) = x^2 y^3 - x^3 y^2 + 3x + 2y$ is a polynomial, it is continuous everywhere, so the limit can be found by direct substitution:

$$\lim_{(x, y) \to (1, 2)} (x^2 y^3 - x^3 y^2 + 3x + 2y) = 1^2 \cdot 2^3 - 1^3 \cdot 2^2 + 3 \cdot 1 + 2 \cdot 2 = 11 \qquad \blacksquare$$

EXAMPLE 6   Where is the function $f(x, y) = \dfrac{x^2 - y^2}{x^2 + y^2}$ continuous?

SOLUTION   The function f is discontinuous at $(0, 0)$ because it is not defined there. Since f is a rational function, it is continuous on its domain, which is the set $D = \{(x, y) \mid (x, y) \neq (0, 0)\}$.   ■

EXAMPLE 7   Let

$$g(x, y) = \begin{cases} \dfrac{x^2 - y^2}{x^2 + y^2} & \text{if } (x, y) \neq (0, 0) \\ 0 & \text{if } (x, y) = (0, 0) \end{cases}$$

Here g is defined at $(0, 0)$ but g is still discontinuous at 0 because $\lim_{(x, y) \to (0, 0)} g(x, y)$ does not exist (see Example 1).   ■

Figure 12 shows the graph of the continuous function in Example 8.

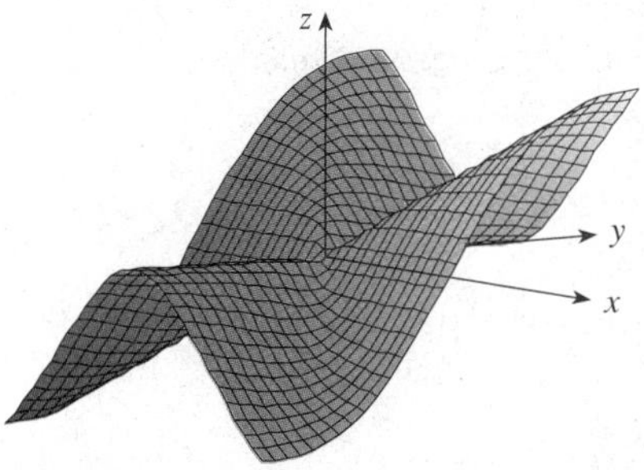

EXAMPLE 8   Let

$$f(x, y) = \begin{cases} \dfrac{3x^2 y}{x^2 + y^2} & \text{if } (x, y) \neq (0, 0) \\ 0 & \text{if } (x, y) = (0, 0) \end{cases}$$

We know f is continuous for $(x, y) \neq (0, 0)$ since it is equal to a rational function there. Also, from Example 4, we have

$$\lim_{(x, y) \to (0, 0)} f(x, y) = \lim_{(x, y) \to (0, 0)} \frac{3x^2 y}{x^2 + y^2} = 0 = f(0, 0)$$

Therefore, f is continuous at $(0, 0)$, and so it is continuous on $\mathbb{R}^2$.   ■

FIGURE 12

EXAMPLE 9   Let

$$g(x, y) = \begin{cases} \dfrac{3x^2 y}{x^2 + y^2} & \text{if } (x, y) \neq (0, 0) \\ 17 & \text{if } (x, y) = (0, 0) \end{cases}$$

Again from Example 4, we have

$$\lim_{(x, y) \to (0, 0)} g(x, y) = \lim_{(x, y) \to (0, 0)} \frac{3x^2 y}{x^2 + y^2} = 0 \neq 17 = g(0, 0)$$

and this shows that g is discontinuous at $(0, 0)$. However, g is continuous on the set $S = \{(x, y) \mid (x, y) \neq (0, 0)\}$ since it is equal to a rational function on S.   ■

Composition is another way of combining two continuous functions to get a third. The proof of the following theorem is similar to that of Theorem 1.5.8.

(5) THEOREM   If f is continuous at (a, b) and g is a function of a single variable that is continuous at $f(a, b)$, then the composite function $h = g \circ f$ defined by $h(x, y) = g(f(x, y))$ is continuous at (a, b).

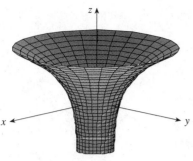

FIGURE 13

EXAMPLE 10   On what set is the function $h(x, y) = \ln(x^2 + y^2 - 1)$ continuous?

SOLUTION   Let $f(x, y) = x^2 + y^2 - 1$ and $g(t) = \ln t$. Then

$$g(f(x, y)) = \ln(x^2 + y^2 - 1) = h(x, y)$$

so $h = g \circ f$. Now f is continuous everywhere since it is a polynomial, and g is continuous on its domain $\{t \mid t > 0\}$. Thus, by Theorem 5, h is continuous on its domain

$$D = \{(x, y) \mid x^2 + y^2 - 1 > 0\} = \{(x, y) \mid x^2 + y^2 > 1\}$$

which consists of all points outside the circle $x^2 + y^2 = 1$. A computer-drawn graph of h is shown in Figure 13. ∎

Everything that we have done in this section can be extended to functions of three or more variables. The distance between two points (x, y, z) and (a, b, c) in $\mathbb{R}^3$ is given by $\sqrt{(x - a)^2 + (y - b)^2 + (z - c)^2}$, so the definitions of limit and continuity of a function of three variables are as follows.

---

(6) DEFINITION   Let f be a function of three variables defined inside a sphere with center (a, b, c), except possibly at (a, b, c).

(a)
$$\lim_{(x, y, z) \to (a, b, c)} f(x, y, z) = L$$

means that for every number $\varepsilon > 0$ there is a corresponding number $\delta > 0$ such that

$$|f(x, y, z) - L| < \varepsilon \quad \text{whenever} \quad 0 < \sqrt{(x - a)^2 + (y - b)^2 + (z - c)^2} < \delta$$

(b) f is **continuous** at (a, b, c) if

$$\lim_{(x, y, z) \to (a, b, c)} f(x, y, z) = f(a, b, c)$$

If we use the vector notation introduced at the end of Section 12.1, then we can write the definitions of a limit for functions of two or three variables in a single compact form as follows.

---

(7)   If $f : D \subset \mathbb{R}^n \to \mathbb{R}$, then $\lim_{\mathbf{x} \to \mathbf{a}} f(\mathbf{x}) = L$ means that for every number $\varepsilon > 0$ there is a corresponding number $\delta > 0$ such that

$$|f(\mathbf{x}) - L| < \varepsilon \quad \text{whenever} \quad 0 < |\mathbf{x} - \mathbf{a}| < \delta$$

Notice that if $n = 1$, then $\mathbf{x} = x$ and $\mathbf{a} = a$, and (7) is just the definition of a limit for functions of a single variable. For the case $n = 2$, we have $\mathbf{x} = \langle x, y \rangle$, $\mathbf{a} = \langle a, b \rangle$, and $|\mathbf{x} - \mathbf{a}| = \sqrt{(x - a)^2 + (y - b)^2}$, so (7) becomes Definition 1. If $n = 3$, then $\mathbf{x} = \langle x, y, z \rangle$, $\mathbf{a} = \langle a, b, c \rangle$, and (7) becomes part (a) of Definition 6. In each case the definition of continuity can be written as

$$\lim_{\mathbf{x} \to \mathbf{a}} f(\mathbf{x}) = f(\mathbf{a})$$

EXERCISES 12.2

1–26 ■ Find the limit, if it exists, or show that the limit does not exist.

1. $\displaystyle\lim_{(x,y)\to(2,3)} (x^2y^2 - 2xy^5 + 3y)$

2. $\displaystyle\lim_{(x,y)\to(-3,4)} (x^3 + 3x^2y^2 - 5y^3 + 1)$

3. $\displaystyle\lim_{(x,y)\to(0,0)} \frac{x^2y^3 + x^3y^2 - 5}{2 - xy}$

4. $\displaystyle\lim_{(x,y)\to(-2,1)} \frac{x^2 + xy + y^2}{x^2 - y^2}$

5. $\displaystyle\lim_{(x,y)\to(\pi,\pi)} x\sin\!\left(\frac{x+y}{4}\right)$

6. $\displaystyle\lim_{(x,y)\to(1,4)} e^{\sqrt{x+2y}}$

7. $\displaystyle\lim_{(x,y)\to(0,0)} \frac{x-y}{x^2+y^2}$

8. $\displaystyle\lim_{(x,y)\to(0,0)} \frac{x^2}{x^2+y^2}$

9. $\displaystyle\lim_{(x,y)\to(0,0)} \frac{8x^2y^2}{x^4+y^4}$

10. $\displaystyle\lim_{(x,y)\to(0,0)} \frac{x^3 + xy^2}{x^2+y^2}$

11. $\displaystyle\lim_{(x,y)\to(0,0)} \frac{2xy}{x^2+2y^2}$

12. $\displaystyle\lim_{(x,y)\to(0,0)} \frac{(x+y)^2}{x^2+y^2}$

13. $\displaystyle\lim_{(x,y)\to(0,0)} \frac{xy}{\sqrt{x^2+y^2}}$

14. $\displaystyle\lim_{(x,y)\to(0,0)} \frac{xy+1}{x^2+y^2+1}$

15. $\displaystyle\lim_{(x,y)\to(0,0)} \frac{2x^2y}{x^4+y^2}$

16. $\displaystyle\lim_{(x,y)\to(0,0)} \frac{x^3y^2}{x^2+y^2}$

17. $\displaystyle\lim_{(x,y)\to(0,0)} \frac{x^2+y^2}{\sqrt{x^2+y^2+1}-1}$

18. $\displaystyle\lim_{(x,y)\to(0,0)} \frac{\sqrt{x^2y^2+1}-1}{x^2+y^2}$

19. $\displaystyle\lim_{(x,y)\to(0,1)} \frac{xy-x}{x^2+y^2-2y+1}$

20. $\displaystyle\lim_{(x,y)\to(1,-1)} \frac{x^2+y^2-2x-2y}{x^2+y^2-2x+2y+2}$

21. $\displaystyle\lim_{(x,y,z)\to(1,2,3)} \frac{xz^2-y^2z}{xyz-1}$

22. $\displaystyle\lim_{(x,y,z)\to(2,3,0)} [xe^z + \ln(2x-y)]$

23. $\displaystyle\lim_{(x,y,z)\to(0,0,0)} \frac{x^2-y^2-z^2}{x^2+y^2+z^2}$

24. $\displaystyle\lim_{(x,y,z)\to(0,0,0)} \frac{xy+yz+zx}{x^2+y^2+z^2}$

25. $\displaystyle\lim_{(x,y,z)\to(0,0,0)} \frac{xy+yz^2+xz^2}{x^2+y^2+z^4}$

26. $\displaystyle\lim_{(x,y,z)\to(0,0,0)} \frac{x^2y^2z^2}{x^2+y^2+z^2}$

27–28 ■ Use a computer graph of the function to explain why the limit does not exist.

27. $\displaystyle\lim_{(x,y)\to(0,0)} \frac{2x^2+3xy+4y^2}{3x^2+5y^2}$

28. $\displaystyle\lim_{(x,y)\to(0,0)} \frac{xy^3}{x^2+y^6}$

29–32 ■ Find $h(x,y) = g(f(x,y))$ and the set on which h is continuous.

29. $g(t) = e^{-t}\cos t, \quad f(x,y) = x^4 + x^2y^2 + y^4$

30. $g(t) = \dfrac{\sqrt{t}-1}{\sqrt{t}+1}, \quad f(x,y) = x^2 - y$

31. $g(t) = t^2 + \sqrt{t}, \quad f(x,y) = 2x + 3y - 6$

32. $g(z) = \sin z, \quad f(x,y) = y\ln x$

33–46 ■ Determine the largest set on which the function is continuous.

33. $F(x,y) = \dfrac{x^2+y^2+1}{x^2+y^2-1}$

34. $F(x,y) = \dfrac{x^6 + x^3y^3 + y^6}{x^3 + y^3}$

35. $F(x,y) = \tan(x^4 - y^4)$

36. $F(x,y) = \ln(2x + 3y)$

37. $G(x,y) = e^{xy}\sin(x+y)$

38. $G(x,y) = \sin^{-1}(x^2 + y^2)$

39. $G(x,y) = \sqrt{x+y} - \sqrt{x-y}$

40. $G(x,y) = 2^{x\tan y}$

41. $f(x,y,z) = x\ln(yz)$

42. $f(x,y,z) = x + y\sqrt{x+z}$

43. $f(x,y) = \begin{cases} \dfrac{2x^2-y^2}{2x^2+y^2} & \text{if } (x,y)\neq(0,0) \\ 0 & \text{if } (x,y)=(0,0) \end{cases}$

44. $f(x,y) = \begin{cases} \dfrac{x^2y^3}{2x^2+y^2} & \text{if } (x,y)\neq(0,0) \\ 0 & \text{if } (x,y)=(0,0) \end{cases}$

45. $f(x,y) = \begin{cases} \dfrac{x^2y^3}{2x^2+y^2} & \text{if } (x,y)\neq(0,0) \\ 1 & \text{if } (x,y)=(0,0) \end{cases}$

46. $f(x,y) = \begin{cases} \dfrac{xy}{x^2+xy+y^2} & \text{if } (x,y)\neq(0,0) \\ 0 & \text{if } (x,y)=(0,0) \end{cases}$

47–48 ■ Use polar coordinates to find the limit. [If (r,θ) are polar coordinates of the point (x,y) with $r \geqslant 0$, note that $r \to 0^+$ as $(x,y) \to (0,0)$.]

47. $\displaystyle\lim_{(x,y)\to(0,0)} \frac{x^3+y^3}{x^2+y^2}$

48. $\displaystyle\lim_{(x,y)\to(0,0)} (x^2+y^2)\ln(x^2+y^2)$

49. Use spherical coordinates to find

$$\lim_{(x,y,z)\to(0,0,0)} \frac{xyz}{x^2+y^2+z^2}$$

50. Use a computer to graph the function

$$f(x, y) = \frac{\sin(x^2 + y^2)}{x^2 + y^2}$$

and use polar coordinates to find the limit as $(x, y) \to (0, 0)$.

51. Prove, using Definition 1, that
(a) $\displaystyle\lim_{(x, y) \to (a, b)} x = a$
(b) $\displaystyle\lim_{(x, y) \to (a, b)} y = b$
(c) $\displaystyle\lim_{(x, y) \to (a, b)} c = c$

52. For what values of the number r is the function

$$f(x, y, z) = \begin{cases} \dfrac{(x + y + z)^r}{x^2 + y^2 + z^2} & \text{if } (x, y, z) \neq (0, 0, 0) \\ 0 & \text{if } (x, y, z) = (0, 0, 0) \end{cases}$$

continuous on $\mathbb{R}^3$?

53. Show that the function $f : \mathbb{R}^n \to \mathbb{R}$ given by $f(\mathbf{x}) = |\mathbf{x}|$ is continuous on $\mathbb{R}^n$.
[*Hint*: Consider $|\mathbf{x} - \mathbf{a}|^2 = (\mathbf{x} - \mathbf{a}) \cdot (\mathbf{x} - \mathbf{a})$.]

54. If $\mathbf{c} \in V_n$, show that the function $f : \mathbb{R}^n \to \mathbb{R}$ given by $f(\mathbf{x}) = \mathbf{c} \cdot \mathbf{x}$ is continuous on $\mathbb{R}^n$.

12.3   PARTIAL DERIVATIVES

If f is a function of two variables x and y, suppose we let only x vary while keeping y fixed, say $y = b$, where b is a constant. Then we are really considering a function of a single variable x, namely, $g(x) = f(x, b)$. If g has a derivative at a, then we call it the **partial derivative of f with respect to x at (a, b)** and denote it by $f_x(a, b)$. Thus

(1) $$f_x(a, b) = g'(a) \qquad \text{where} \qquad g(x) = f(x, b)$$

By Definition 2.1.2 we know that

$$g'(a) = \lim_{h \to 0} \frac{g(a + h) - g(a)}{h}$$

and so Equation 1 becomes

(2) $$f_x(a, b) = \lim_{h \to 0} \frac{f(a + h, b) - f(a, b)}{h}$$

Similarly, the **partial derivative of f with respect to y at (a, b)**, denoted by $f_y(a, b)$, is obtained by keeping x fixed ($x = a$) and finding the ordinary derivative at b of the function $G(y) = f(a, y)$:

(3) $$f_y(a, b) = \lim_{h \to 0} \frac{f(a, b + h) - f(a, b)}{h}$$

If we now let the point (a, b) vary, f_x and f_y become functions of two variables.

(4)   If f is a function of two variables, its **partial derivatives** are the functions f_x and f_y defined by

$$f_x(x, y) = \lim_{h \to 0} \frac{f(x + h, y) - f(x, y)}{h}$$

$$f_y(x, y) = \lim_{h \to 0} \frac{f(x, y + h) - f(x, y)}{h}$$

There are many alternative notations for partial derivatives. For instance, instead of f_x we can write f_1 or $D_1 f$ (to indicate differentiation with respect to the *first* variable) or $\partial f / \partial x$. But here $\partial f / \partial x$ cannot be interpreted as a ratio of differentials.

NOTATIONS FOR PARTIAL DERIVATIVES If $z = f(x, y)$, we write

$$f_x(x, y) = f_x = \frac{\partial f}{\partial x} = \frac{\partial}{\partial x} f(x, y) = \frac{\partial z}{\partial x} = f_1 = D_1 f = D_x f$$

$$f_y(x, y) = f_y = \frac{\partial f}{\partial y} = \frac{\partial}{\partial y} f(x, y) = \frac{\partial z}{\partial y} = f_2 = D_2 f = D_y f$$

To compute partial derivatives, all we have to do is remember from Equation 1 that the partial derivative with respect to x is just the *ordinary* derivative of the function g of a single variable that we get by keeping y fixed. Thus we have the following rule:

RULE FOR FINDING PARTIAL DERIVATIVES OF $z = f(x, y)$

1. To find f_x, regard y as a constant and differentiate $f(x, y)$ with respect to x.
2. To find f_y, regard x as a constant and differentiate $f(x, y)$ with respect to y.

EXAMPLE 1 If $f(x, y) = x^3 + x^2 y^3 - 2y^2$, find $f_x(2, 1)$ and $f_y(2, 1)$.

SOLUTION Holding y constant and differentiating with respect to x, we get

$$f_x(x, y) = 3x^2 + 2xy^3$$

and so

$$f_x(2, 1) = 3 \cdot 2^2 + 2 \cdot 2 \cdot 1^3 = 16$$

Holding x constant and differentiating with respect to y, we get

$$f_y(x, y) = 3x^2 y^2 - 4y$$

$$f_y(2, 1) = 3 \cdot 2^2 \cdot 1^2 - 4 \cdot 1 = 8 \qquad \blacksquare$$

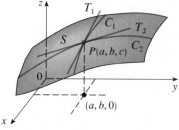

FIGURE 1

To give a geometric interpretation of partial derivatives, we recall that the equation $z = f(x, y)$ represents a surface S (the graph of f). If $f(a, b) = c$, then the point $P(a, b, c)$ lies on S. The vertical plane $y = b$ intersects S in a curve C_1. (In other words, C_1 is the trace of S in the plane $y = b$.) Likewise, the vertical plane $x = a$ intersects S in a curve C_2. Both of the curves C_1 and C_2 pass through the point P (see Figure 1).

Notice that the curve C_1 is the graph of the function $g(x) = f(x, b)$, so the slope of its tangent T_1 at P is $g'(a) = f_x(a, b)$. The curve C_2 is the graph of the function $G(y) = f(a, y)$, so the slope of its tangent T_2 at P is $G'(b) = f_y(a, b)$.

Thus the partial derivatives $f_x(a, b)$ and $f_y(a, b)$ can be interpreted geometrically as the slopes of the tangent lines at $P(a, b, c)$ to the traces C_1 and C_2 of S in the planes $y = b$ and $x = a$.

Partial derivatives can also be interpreted as rates of change. If $z = f(x, y)$, then $\partial z / \partial x$ represents the rate of change of z with respect to x when y is fixed. Similarly, $\partial z / \partial y$ represents the rate of change of z with respect to y when x is fixed. For instance, referring to the temperature function T at the beginning of Section 12.1, we see that $\partial T / \partial x$ is the rate at which the temperature changes in the east-west direction and $\partial T / \partial y$ is the rate at which it changes in the north-south direction.

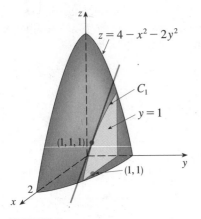

FIGURE 2

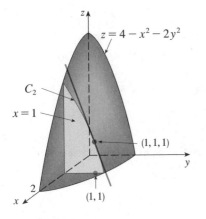

FIGURE 3

Some computer algebra systems can plot surfaces defined by implicit equations in three variables. Figure 4 shows such a plot of the surface defined by the equation in Example 4.

FIGURE 4

EXAMPLE 2 If $f(x, y) = 4 - x^2 - 2y^2$, find $f_x(1, 1)$ and $f_y(1, 1)$ and interpret these numbers as slopes.

SOLUTION   We have

$$f_x(x, y) = -2x \qquad f_y(x, y) = -4y$$

$$f_x(1, 1) = -2 \qquad f_y(1, 1) = -4$$

The graph of f is the paraboloid $z = 4 - x^2 - 2y^2$ and the vertical plane $y = 1$ intersects it in the parabola $z = 2 - x^2$, $y = 1$. (As in the preceding discussion, we label it C_1 in Figure 2.) The slope of the tangent line to this parabola at the point $(1, 1, 1)$ is $f_x(1, 1) = -2$. Similarly, the curve C_2 in which the plane $x = 1$ intersects the paraboloid is the parabola $z = 3 - 2y^2$, $x = 1$ and the slope of the tangent line at $(1, 1, 1)$ is $f_y(1, 1) = -4$ (see Figure 3). ∎

EXAMPLE 3 If $f(x, y) = \sin\left(\dfrac{x}{1 + y}\right)$, calculate $\dfrac{\partial f}{\partial x}$ and $\dfrac{\partial f}{\partial y}$.

SOLUTION   Using the Chain Rule for functions of one variable, we have

$$\frac{\partial f}{\partial x} = \cos\left(\frac{x}{1 + y}\right) \cdot \frac{\partial}{\partial x}\left(\frac{x}{1 + y}\right) = \cos\left(\frac{x}{1 + y}\right) \cdot \frac{1}{1 + y}$$

$$\frac{\partial f}{\partial y} = \cos\left(\frac{x}{1 + y}\right) \cdot \frac{\partial}{\partial y}\left(\frac{x}{1 + y}\right) = -\cos\left(\frac{x}{1 + y}\right) \cdot \frac{x}{(1 + y)^2}$$ ∎

EXAMPLE 4 Find $\partial z/\partial x$ and $\partial z/\partial y$ if z is defined implicitly as a function of x and y by the equation

$$x^3 + y^3 + z^3 + 6xyz = 1$$

SOLUTION   To find $\partial z/\partial x$, we differentiate implicitly with respect to x, being careful to treat y as a constant:

$$3x^2 + 3z^2 \frac{\partial z}{\partial x} + 6yz + 6xy \frac{\partial z}{\partial x} = 0$$

Solving this equation for $\partial z/\partial x$, we obtain

$$\frac{\partial z}{\partial x} = -\frac{x^2 + 2yz}{z^2 + 2xy}$$

Similarly, implicit differentiation with respect to y gives

$$\frac{\partial z}{\partial y} = -\frac{y^2 + 2xz}{z^2 + 2xy}$$ ∎

FUNCTIONS OF MORE THAN TWO VARIABLES

Partial derivatives can also be defined for functions of three or more variables. For example, if f is a function of three variables x, y, and z, then its partial derivative with respect to x is defined as

$$f_x(x, y, z) = \lim_{h \to 0} \frac{f(x + h, y, z) - f(x, y, z)}{h}$$

and it is found by regarding y and z as constants and differentiating $f(x, y, z)$ with respect to x.

In general, if u is a function of n variables, $u = f(x_1, x_2, \ldots, x_n)$, its partial derivative with respect to the ith variable x_i is

$$\frac{\partial u}{\partial x_i} = \lim_{h \to 0} \frac{f(x_1, \ldots, x_{i-1}, x_i + h, x_{i+1}, \ldots, x_n) - f(x_1, \ldots, x_i, \ldots, x_n)}{h}$$

and we also write

$$\frac{\partial u}{\partial x_i} = \frac{\partial f}{\partial x_i} = f_{x_i} = f_i = D_i f$$

EXAMPLE 5 Find f_x, f_y, and f_z if $f(x, y, z) = e^{xy} \ln z$.

SOLUTION Holding y and z constant and differentiating with respect to x, we have

$$f_x = y e^{xy} \ln z$$

Similarly, $\qquad\qquad f_y = x e^{xy} \ln z \qquad \text{and} \qquad f_z = \frac{e^{xy}}{z}$ ∎

HIGHER DERIVATIVES

If f is a function of two variables, then its partial derivatives f_x and f_y are also functions of two variables, so we can consider their partial derivatives $(f_x)_x$, $(f_x)_y$, $(f_y)_x$, and $(f_y)_y$, which are called the **second partial derivatives** of f. If $z = f(x, y)$, we use the following notation:

$$(f_x)_x = f_{xx} = f_{11} = \frac{\partial}{\partial x}\left(\frac{\partial f}{\partial x}\right) = \frac{\partial^2 f}{\partial x^2} = \frac{\partial^2 z}{\partial x^2}$$

$$(f_x)_y = f_{xy} = f_{12} = \frac{\partial}{\partial y}\left(\frac{\partial f}{\partial x}\right) = \frac{\partial^2 f}{\partial y\, \partial x} = \frac{\partial^2 z}{\partial y\, \partial x}$$

$$(f_y)_x = f_{yx} = f_{21} = \frac{\partial}{\partial x}\left(\frac{\partial f}{\partial y}\right) = \frac{\partial^2 f}{\partial x\, \partial y} = \frac{\partial^2 z}{\partial x\, \partial y}$$

$$(f_y)_y = f_{yy} = f_{22} = \frac{\partial}{\partial y}\left(\frac{\partial f}{\partial y}\right) = \frac{\partial^2 f}{\partial y^2} = \frac{\partial^2 z}{\partial y^2}$$

Thus the notation f_{xy} (or $\partial^2 f / \partial y\, \partial x$) means that we first differentiate with respect to x and then with respect to y, whereas in computing f_{yx} the order is reversed.

EXAMPLE 6   Find the second partial derivatives of

$$f(x, y) = x^3 + x^2y^3 - 2y^2$$

SOLUTION   In Example 1 we found that

$$f_x(x, y) = 3x^2 + 2xy^3 \qquad f_y(x, y) = 3x^2y^2 - 4y$$

Figure 5 shows the graph of the function f in Example 6 and the graphs of its first- and second-order partial derivatives for $-2 \le x \le 2$, $-2 \le y \le 2$. Notice that these graphs are consistent with our interpretations of f_x and f_y as slopes of tangent lines to traces of the graph of f. For instance, the graph of f decreases if we start at $(0, -2)$ and move in the positive x-direction. This is reflected in the negative values of f_x. You should compare the graphs of f_{yx} and f_{yy} with the graph of f_y to see the relationships.

Therefore

$$f_{xx} = \frac{\partial}{\partial x}(3x^2 + 2xy^3) = 6x + 2y^3$$

$$f_{xy} = \frac{\partial}{\partial y}(3x^2 + 2xy^3) = 6xy^2$$

$$f_{yx} = \frac{\partial}{\partial x}(3x^2y^2 - 4y) = 6xy^2$$

$$f_{yy} = \frac{\partial}{\partial y}(3x^2y^2 - 4y) = 6x^2y - 4$$ ∎

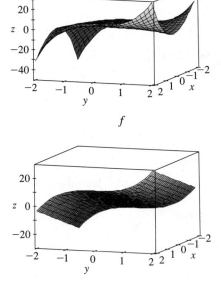

f

f_x

f_y

f_{xx}

$f_{xy} = f_{yx}$

f_{yy}

FIGURE 5

Notice that $f_{xy} = f_{yx}$ in Example 6. This is not just a coincidence. It turns out that the mixed partial derivatives f_{xy} and f_{yx} are equal for most functions that one meets in practice. The following theorem, discovered by the French mathematician Alexis Clairaut (1713–1765), gives conditions under which we can assert that $f_{xy} = f_{yx}$. The proof is given in Appendix F.

---

(5) CLAIRAUT'S THEOREM   Suppose f is defined on a disk D that contains the point (a, b). If the functions f_{xy} and f_{yx} are both continuous on D, then

$$f_{xy}(a, b) = f_{yx}(a, b)$$

Alexis Clairaut was a child prodigy in mathematics, having read l'Hospital's textbook on calculus when he was ten and presented a paper on geometry to the French Academy of Sciences when he was 13. At the age of 18, Clairaut published *Recherches sur les courbes à double courbure,* which was the first systematic treatise on three-dimensional analytic geometry and included the calculus of space curves.

Partial derivatives of order 3 or higher can also be defined. For instance,

$$f_{xyy} = (f_{xy})_y = \frac{\partial}{\partial y}\left(\frac{\partial^2 z}{\partial y\,\partial x}\right) = \frac{\partial^3 z}{\partial y^2\,\partial x}$$

and using Clairaut's Theorem it can be shown that $f_{xyy} = f_{yxy} = f_{yyx}$ if these functions are continuous.

EXAMPLE 7 Calculate f_{xxyz} if $f(x, y, z) = \sin(3x + yz)$.

SOLUTION

$$f_x = 3\cos(3x + yz)$$

$$f_{xx} = -9\sin(3x + yz)$$

$$f_{xxy} = -9z\cos(3x + yz)$$

$$f_{xxyz} = -9\cos(3x + yz) + 9yz\sin(3x + yz) \qquad \blacksquare$$

PARTIAL DIFFERENTIAL EQUATIONS

Partial derivatives occur in *partial differential equations* that express certain physical laws. For instance, the partial differential equation

$$\frac{\partial^2 u}{\partial x^2} + \frac{\partial^2 u}{\partial y^2} = 0$$

is called **Laplace's equation** after Pierre Laplace (1749–1827). Solutions of this equation are called **harmonic functions** and play a role in problems of heat conduction, fluid flow, and electric potential.

EXAMPLE 8 Show that the function $u(x, y) = e^x \sin y$ is a solution of Laplace's equation.

SOLUTION

$$u_x = e^x \sin y \qquad\qquad u_y = e^x \cos y$$

$$u_{xx} = e^x \sin y \qquad\qquad u_{yy} = -e^x \sin y$$

$$u_{xx} + u_{yy} = e^x \sin y - e^x \sin y = 0$$

Therefore, u satisfies Laplace's equation. $\qquad \blacksquare$

The **wave equation**

$$\frac{\partial^2 u}{\partial t^2} = a^2\,\frac{\partial^2 u}{\partial x^2}$$

describes the motion of a waveform, which could be an ocean wave, a sound wave, a light wave, or a wave traveling along a vibrating string. For instance, if $u(x, t)$ represents the displacement of a vibrating violin string at time t and at a distance x from one end of the string (as in Figure 6), then $u(x, t)$ satisfies the wave equation. Here the constant a depends on the density of the string and on the tension in the string.

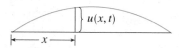

FIGURE 6

EXAMPLE 9   Verify that the function $u(x, t) = \sin(x - at)$ satisfies the wave equation.

SOLUTION

$$u_x = \cos(x - at) \qquad\qquad u_{xx} = -\sin(x - at)$$

$$u_t = -a\cos(x - at) \qquad u_{tt} = -a^2\sin(x - at) = a^2 u_{xx}$$

So u satisfies the wave equation.   ∎

EXERCISES 12.3

1. If $f(x, y) = 16 - 4x^2 - y^2$, find $f_x(1, 2)$ and $f_y(1, 2)$ and interpret these numbers as slopes. Illustrate with sketches.

2. If $f(x, y) = \sqrt{4 - x^2 - 4y^2}$, find $f_x(1, 0)$ and $f_y(1, 0)$ and interpret these numbers as slopes. Illustrate with sketches.

3–16 ■ Find the indicated partial derivatives.

3. $f(x, y) = x^3 y^5$; $f_x(3, -1)$

4. $f(x, y) = \sqrt{2x + 3y}$; $f_y(2, 4)$

5. $f(x, y) = xe^{-y} + 3y$; $\dfrac{\partial f}{\partial y}(1, 0)$

6. $f(x, y) = \sin(y - x)$; $\dfrac{\partial f}{\partial x}(3, 3)$

7. $z = \dfrac{x^3 + y^3}{x^2 + y^2}$; $\dfrac{\partial z}{\partial x}, \dfrac{\partial z}{\partial y}$

8. $z = x\sqrt{y} - \dfrac{y}{\sqrt{x}}$; $\dfrac{\partial z}{\partial x}, \dfrac{\partial z}{\partial y}$

9. $xy + yz = xz$; $\dfrac{\partial z}{\partial x}, \dfrac{\partial z}{\partial y}$

10. $xyz = \cos(x + y + z)$; $\dfrac{\partial z}{\partial x}, \dfrac{\partial z}{\partial y}$

11. $x^2 + y^2 - z^2 = 2x(y + z)$; $\dfrac{\partial z}{\partial x}, \dfrac{\partial z}{\partial y}$

12. $xy^2 z^3 + x^3 y^2 z = x + y + z$; $\dfrac{\partial z}{\partial x}, \dfrac{\partial z}{\partial y}$

13. $f(x, y, z) = xyz$; $f_y(0, 1, 2)$

14. $f(x, y, z) = \sqrt{x^2 + y^2 + z^2}$; $f_z(0, 3, 4)$

15. $u = xy + yz + zx$; u_x, u_y, u_z

16. $u = x^2 y^3 t^4$; u_x, u_y, u_t

17–44 ■ Find the first partial derivatives of the function.

17. $f(x, y) = x^3 y^5 - 2x^2 y + x$

18. $f(x, y) = x^2 y^2(x^4 + y^4)$

19. $f(x, y) = x^4 + x^2 y^2 + y^4$

20. $f(x, y) = \ln(x^2 + y^2)$

21. $f(x, y) = \dfrac{x - y}{x + y}$

22. $f(x, y) = x^y$

23. $f(x, y) = e^x \tan(x - y)$

24. $f(s, t) = s/\sqrt{s^2 + t^2}$

25. $f(u, v) = \tan^{-1}(u/v)$

26. $f(x, t) = e^{\sin(t/x)}$

27. $g(x, y) = y\tan(x^2 y^3)$

28. $g(x, y) = \ln(x + \ln y)$

29. $z = \ln(x + \sqrt{x^2 + y^2})$

30. $z = x^{xy}$

31. $f(x, y) = \displaystyle\int_x^y e^{t^2}\, dt$

32. $f(x, y) = \displaystyle\int_y^x \dfrac{e^t}{t}\, dt$

33. $f(x, y, z) = x^2 yz^3 + xy - z$

34. $f(x, y, z) = x\sqrt{yz}$

35. $f(x, y, z) = x^{yz}$

36. $f(x, y, z) = xe^y + ye^z + ze^x$

37. $u = z\sin\dfrac{y}{x + z}$

38. $u = x^{y/z}$

39. $u = xy^2 z^3 \ln(x + 2y + 3z)$

40. $u = x^{y^z}$

41. $f(x, y, z, t) = \dfrac{x - y}{z - t}$

42. $f(x, y, z, t) = xy^2 z^3 t^4$

43. $u = \sqrt{x_1^2 + x_2^2 + \cdots + x_n^2}$

44. $u = \sin(x_1 + 2x_2 + \cdots + nx_n)$

45–46 ■ Use the definition of partial derivatives as limits (4) to find $f_x(x, y)$ and $f_y(x, y)$.

45. $f(x, y) = x^2 - xy + 2y^2$

46. $f(x, y) = \sqrt{3x - y}$

47–48 ■ Find f_x and f_y and graph f, f_x, and f_y with domains and viewpoints that enable you to see the relationships between them.

47. $f(x, y) = x^2 + y^2 + x^2 y$

48. $f(x, y) = xe^{-x^2 - y^2}$

49. The following surfaces, labeled a, b, and c, are graphs of a function f and its partial derivatives f_x and f_y. Identify each surface and give reasons for your choices.

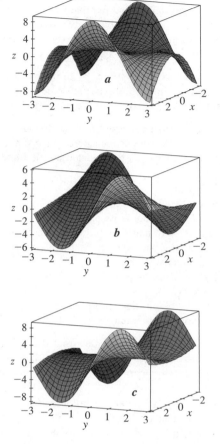

50. A contour map is given for a function f. Use it to estimate $f_x(2, 1)$ and $f_y(2, 1)$.

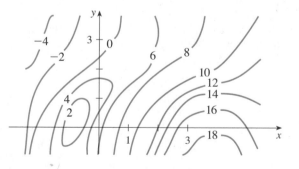

51–56 ■ Find $\partial z/\partial x$ and $\partial z/\partial y$.

51. $z = f(x) + g(y)$

52. $z = f(x)g(y)$

53. $z = f(x + y)$

54. $z = f(xy)$

55. $z = f(x/y)$

56. $z = f(ax + by)$

57–62 ■ Find all the second partial derivatives.

57. $f(x, y) = x^2y + x\sqrt{y}$

58. $f(x, y) = \sin(x + y) + \cos(x - y)$

59. $z = (x^2 + y^2)^{3/2}$

60. $z = \cos^2(5x + 2y)$

61. $z = t\sin^{-1}\sqrt{x}$

62. $z = x^{\ln t}$

63–66 ■ Verify that the conclusion of Clairaut's Theorem holds, that is, $u_{xy} = u_{yx}$.

63. $u = x^5y^4 - 3x^2y^3 + 2x^2$

64. $u = \sin^2x \cos y$

65. $u = \sin^{-1}(xy^2)$

66. $u = x^2y^3z^4$

67–74 ■ Find the indicated partial derivative.

67. $f(x, y) = x^2y^3 - 2x^4y$; f_{xxx}

68. $f(x, y) = e^{xy^2}$; f_{xxy}

69. $f(x, y, z) = x^5 + x^4y^4z^3 + yz^2$; f_{xyz}

70. $f(x, y, z) = e^{xyz}$; f_{yzy}

71. $z = x\sin y$; $\dfrac{\partial^3 z}{\partial y^2\, \partial x}$

72. $z = \ln\sin(x - y)$; $\dfrac{\partial^3 z}{\partial y\, \partial x^2}$

73. $u = \ln(x + 2y^2 + 3z^3)$; $\dfrac{\partial^3 u}{\partial x\, \partial y\, \partial z}$

74. $u = x^ay^bz^c$; $\dfrac{\partial^6 u}{\partial x\, \partial y^2\, \partial z^3}$

75. Verify that the function $u = e^{-\alpha^2k^2t}\sin kx$ is a solution of the heat conduction equation $u_t = \alpha^2 u_{xx}$.

76. Determine whether each of the following functions is a solution of Laplace's equation $u_{xx} + u_{yy} = 0$.
   (a) $u = x^2 + y^2$
   (b) $u = x^2 - y^2$
   (c) $u = x^3 + 3xy^2$
   (d) $u = \ln\sqrt{x^2 + y^2}$
   (e) $u = \sin x \cosh y + \cos x \sinh y$
   (f) $u = e^{-x}\cos y - e^{-y}\cos x$

77. Verify that the function $u = 1/\sqrt{x^2 + y^2 + z^2}$ is a solution of the three-dimensional Laplace equation
$$u_{xx} + u_{yy} + u_{zz} = 0.$$

78. Show that each of the following functions is a solution of the wave equation $u_{tt} = a^2 u_{xx}$.
   (a) $u = \sin(kx)\sin(akt)$
   (b) $u = t/(a^2t^2 - x^2)$
   (c) $u = (x - at)^6 + (x + at)^6$
   (d) $u = \sin(x - at) + \ln(x + at)$

79. If f and g are twice differentiable functions of a single variable, show that the function

$$u(x, t) = f(x + at) + g(x - at)$$

is a solution of the wave equation given in Exercise 78.

80. If f and g are twice differentiable functions of a single variable, show that the function

$$u(x, y) = xf(x + y) + yg(x + y)$$

satisfies the equation $u_{xx} - 2u_{xy} + u_{yy} = 0$.

81. Show that the function $z = xe^y + ye^x$ is a solution of the equation

$$\frac{\partial^3 z}{\partial x^3} + \frac{\partial^3 z}{\partial y^3} = x\frac{\partial^3 z}{\partial x\, \partial y^2} + y\frac{\partial^3 z}{\partial x^2\, \partial y}$$

82. If $u = e^{a_1 x_1 + a_2 x_2 + \cdots + a_n x_n}$, where $a_1^2 + a_2^2 + \cdots + a_n^2 = 1$, show that

$$\frac{\partial^2 u}{\partial x_1^2} + \frac{\partial^2 u}{\partial x_2^2} + \cdots + \frac{\partial^2 u}{\partial x_n^2} = u$$

83. Show that the function

$$f(x_1, \ldots, x_n) = (x_1^2 + \cdots + x_n^2)^{(2-n)/2}$$

satisfies the equation

$$\frac{\partial^2 f}{\partial x_1^2} + \cdots + \frac{\partial^2 f}{\partial x_n^2} = 0$$

84. The temperature at a point (x, y) on a flat metal plate is given by $T(x, y) = 60/(1 + x^2 + y^2)$, where T is measured in °C and x, y in meters. Find the rate of change of temperature with respect to distance at the point $(2, 1)$ in (a) the x-direction and (b) the y-direction.

85. The total resistance R produced by three conductors with resistances R_1, R_2, R_3 connected in a parallel electrical circuit is given by the formula

$$\frac{1}{R} = \frac{1}{R_1} + \frac{1}{R_2} + \frac{1}{R_3}$$

Find $\partial R/\partial R_1$.

86. The gas law for a fixed mass m of an ideal gas at absolute temperature T, pressure P, and volume V is $PV = mRT$, where R is the gas constant. Show that

$$\frac{\partial P}{\partial V}\frac{\partial V}{\partial T}\frac{\partial T}{\partial P} = -1$$

87. The kinetic energy of a body with mass m and velocity v is $K = \frac{1}{2}mv^2$. Show that

$$\frac{\partial K}{\partial m}\frac{\partial^2 K}{\partial v^2} = K$$

88. If a, b, c are the sides of a triangle and A, B, C are the opposite angles, find $\partial A/\partial a$, $\partial A/\partial b$, $\partial A/\partial c$ by implicit differentiation of the Law of Cosines.

89. You are told that there is a function $f(x, y)$ whose partial derivatives are $f_x(x, y) = x + 4y$ and $f_y(x, y) = 3x - y$. Should you believe it?

90. The paraboloid $z = 6 - x - x^2 - 2y^2$ intersects the plane $x = 1$ in a parabola. Find parametric equations for the tangent line to this parabola at the point $(1, 2, -4)$. Use a computer to graph the paraboloid, the parabola, and the tangent line on the same screen.

91. The ellipsoid $4x^2 + 2y^2 + z^2 = 16$ intersects the plane $y = 2$ in an ellipse. Find parametric equations for the tangent line to this ellipse at the point $(1, 2, 2)$.

92. In a study of frost penetration it was found that the temperature T at time t (measured in days) at a depth x (measured in feet) can be modeled by the function

$$T(x, t) = T_0 + T_1 e^{-\lambda x} \sin(\omega t - \lambda x)$$

where $\omega = 2\pi/365$ and λ is a positive constant.
(a) Find $\partial T/\partial x$. What is its physical significance?
(b) Find $\partial T/\partial t$. What is its physical significance?
(c) Show that T satisfies the heat equation $T_t = kT_{xx}$ for a certain constant k.
(d) If $\lambda = 0.2$, $T_0 = 0$, and $T_1 = 10$, use a computer to graph $T(x, t)$.
(e) What is the physical significance of the term $-\lambda x$ in the expression $\sin(\omega t - \lambda x)$?

93. Use Clairaut's Theorem to show that if the third-order partial derivatives of f are continuous, then

$$f_{xyy} = f_{yxy} = f_{yyx}$$

94. (a) How many nth-order partial derivatives does a function of two variables have?
(b) If these partial derivatives are all continuous, how many of them can be distinct?
(c) Answer the question in part (a) for a function of three variables.

95. If $f(x, y) = x(x^2 + y^2)^{-3/2} e^{\sin(x^2 y)}$, find $f_x(1, 0)$.
[*Hint:* Instead of finding $f_x(x, y)$ first, note that it is easier to use Equation 1 or Equation 2.]

96. If $f(x, y) = \sqrt[3]{x^3 + y^3}$, find $f_x(0, 0)$.

97. Let

$$f(x, y) = \begin{cases} \dfrac{x^3 y - xy^3}{x^2 + y^2} & \text{if } (x, y) \neq (0, 0) \\ 0 & \text{if } (x, y) = (0, 0) \end{cases}$$

(a) Use a computer to graph f.
(b) Find $f_x(x, y)$ and $f_y(x, y)$ when $(x, y) \neq (0, 0)$.
(c) Find $f_x(0, 0)$ and $f_y(0, 0)$ using Equations 2 and 3.
(d) Show that $f_{xy}(0, 0) = -1$ and $f_{yx}(0, 0) = 1$.
(e) Does the result of part (d) contradict Clairaut's Theorem? Use graphs of f_{xy} and f_{yx} to illustrate your answer.

12.4 TANGENT PLANES AND DIFFERENTIALS

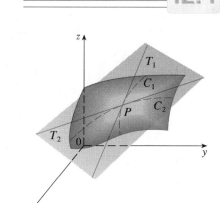

FIGURE 1

Suppose a surface S has equation $z = f(x, y)$, where f has continuous first partial derivatives, and let $P(x_0, y_0, z_0)$ be a point on S. As in the preceding section, let C_1 and C_2 be the curves obtained by intersecting the vertical planes $y = y_0$ and $x = x_0$ with the surface S. Then the point P lies on both C_1 and C_2. Let T_1 and T_2 be the tangent lines to the curves C_1 and C_2 at the point P. Then the **tangent plane** to the surface S at the point P is defined to be the plane that contains both of the tangent lines T_1 and T_2 (see Figure 1).

We will see in Section 12.6 that if C is any other curve that lies on the surface S and passes through P, then its tangent line at P also lies in the tangent plane. Therefore, you can think of the tangent plane to S at P as consisting of all possible tangent lines at P to curves that lie on S and pass through P. The tangent plane at P is the plane that most closely approximates the surface S near the point P.

We know from Equation 11.5.6 that any plane passing through $P(x_0, y_0, z_0)$ has an equation of the form

$$A(x - x_0) + B(y - y_0) + C(z - z_0) = 0$$

By dividing this equation by C and letting $a = -A/C$ and $b = -B/C$, we can write it in the form

(1) $$z - z_0 = a(x - x_0) + b(y - y_0)$$

If Equation 1 represents the tangent plane at P, then its intersection with the plane $y = y_0$ must be the tangent line T_1. Setting $y = y_0$ in Equation 1 gives

$$z - z_0 = a(x - x_0) \qquad y = y_0$$

and we recognize these as the equations (in slope-point form) of a line with slope a. But from Section 12.3 we know that the slope of T_1 is $f_x(x_0, y_0)$. Therefore, we have $a = f_x(x_0, y_0)$.

Similarly, putting $x = x_0$ in Equation 1, we get $z - z_0 = b(y - y_0)$, which must represent the tangent line T_2, so $b = f_y(x_0, y_0)$.

Note the similarity between the equation of a tangent plane and the equation of a tangent line:

$$y - y_0 = f'(x_0)(x - x_0)$$

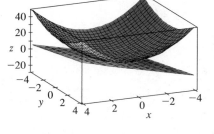

FIGURE 2

Figure 2 shows the elliptic paraboloid and the tangent plane in Example 1.

(2) An equation of the tangent plane to the surface $z = f(x, y)$ at the point $P(x_0, y_0, z_0)$ is

$$z - z_0 = f_x(x_0, y_0)(x - x_0) + f_y(x_0, y_0)(y - y_0)$$

EXAMPLE 1 Find the tangent plane to the elliptic paraboloid $z = 2x^2 + y^2$ at the point $(1, 1, 3)$.

SOLUTION Let $f(x, y) = 2x^2 + y^2$. Then

$$f_x(x, y) = 4x \qquad f_y(x, y) = 2y$$
$$f_x(1, 1) = 4 \qquad f_y(1, 1) = 2$$

Then (2) gives the equation of the tangent plane at $(1, 1, 3)$ as

$$z - 3 = 4(x - 1) + 2(y - 1)$$

or $\qquad\qquad 4x + 2y - z = 3$ ∎

DIFFERENTIALS

Recall that for a function of one variable, $y = f(x)$, we defined the increment of y as

$$\Delta y = f(x + \Delta x) - f(x)$$

and the differential of y as

(3) $$dy = f'(x)\, dx$$

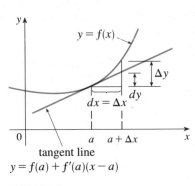

tangent line
$y = f(a) + f'(a)(x - a)$

FIGURE 3

(See Section 2.9.) Figure 3 shows the relationship between Δy and dy: Δy represents the change in height of the curve $y = f(x)$ and dy represents the change in height of the tangent line when x changes by an amount $dx = \Delta x$.

We note that $\Delta y - dy$ approaches 0 faster than Δx because if we let ε denote the ratio of these quantities, then

$$\varepsilon = \frac{\Delta y - dy}{\Delta x} = \frac{f(x + \Delta x) - f(x) - f'(x)\,\Delta x}{\Delta x}$$

$$= \frac{f(x + \Delta x) - f(x)}{\Delta x} - f'(x)$$

$$\to f'(x) - f'(x) = 0 \qquad \text{as } \Delta x \to 0$$

Therefore, we have

(4) $$\Delta y = dy + \varepsilon\,\Delta x \qquad \text{where } \varepsilon \to 0 \text{ as } \Delta x \to 0$$

Now consider a function of two variables, $z = f(x, y)$. If x and y are given increments Δx and Δy, then the corresponding **increment** of z is

(5) $$\Delta z = f(x + \Delta x,\, y + \Delta y) - f(x, y)$$

Thus the increment Δz represents the change in the value of f when (x, y) changes to $(x + \Delta x, y + \Delta y)$.

The **differentials** dx and dy are independent variables; that is, they can be given any values. Then the **differential** dz, also called the **total differential,** is defined by

(6) $$dz = f_x(x, y)\, dx + f_y(x, y)\, dy = \frac{\partial z}{\partial x}\, dx + \frac{\partial z}{\partial y}\, dy$$

(Compare with Equation 3.) Sometimes the notation df is used in place of dz.

If we take

$$dx = \Delta x = x - a \qquad dy = \Delta y = y - b$$

in Equation 6, then the differential of z is

(7) $$dz = f_x(a, b)\,(x - a) + f_y(a, b)\,(y - b)$$

On the other hand, if f_x and f_y are continuous, we see from (2) that the equation of the tangent plane to the surface $z = f(x, y)$ at the point $(a, b, f(a, b))$ is

(8) $$z - f(a, b) = f_x(a, b)\,(x - a) + f_y(a, b)\,(y - b)$$

Comparing Equations 7 and 8, we see that dz represents the change in height of the tangent plane, whereas Δz represents the change in height of the surface $z = f(x, y)$ when

(x, y) changes from (a, b) to $(a + \Delta x, b + \Delta y)$. (See Figure 4 and compare it with Figure 3.)

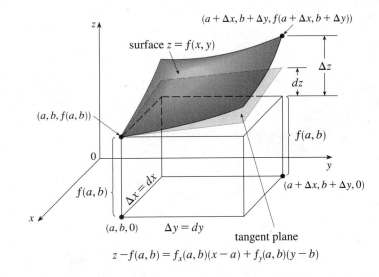

FIGURE 4

$z - f(a, b) = f_x(a, b)(x - a) + f_y(a, b)(y - b)$

It is proved later in this section (Theorem 10) that if f_x and f_y are continuous, then

$$\Delta z - dz = \varepsilon_1 \Delta x + \varepsilon_2 \Delta y$$

where ε_1 and ε_2 are functions of Δx and Δy that approach 0 as Δx and Δy approach 0. This means that $\Delta z - dz \approx 0$ and so

$$\Delta z \approx dz$$

which says that the actual change in z is approximately equal to the differential dz when Δx and Δy are small. This allows us to estimate the value of $f(a + \Delta x, b + \Delta y)$ when $f(a, b)$ is known:

(9) $$f(a + \Delta x, b + \Delta y) \approx f(a, b) + dz$$

When we use the approximation in (9) we are using the tangent plane at $P(a, b, f(a, b))$ as an approximation to the surface $z = f(x, y)$ when (x, y) is close to (a, b).

EXAMPLE 2
(a) If $z = f(x, y) = x^2 + 3xy - y^2$, find the differential dz.
(b) If x changes from 2 to 2.05 and y changes from 3 to 2.96, compare the values of Δz and dz.

SOLUTION
(a) Definition 6 gives

$$dz = \frac{\partial z}{\partial x} \, dx + \frac{\partial z}{\partial y} \, dy = (2x + 3y) \, dx + (3x - 2y) \, dy$$

(b) Putting $x = 2$, $dx = \Delta x = 0.05$, $y = 3$, and $dy = \Delta y = -0.04$, we get

$$dz = [2(2) + 3(3)]0.05 + [3(2) - 2(3)](-0.04)$$

$$= 0.65$$

In Example 2, dz is close to Δz because the tangent plane is a good approximation to the surface $z = x^2 + 3xy - y^2$ near $(2, 3, 13)$. (See Figure 5.)

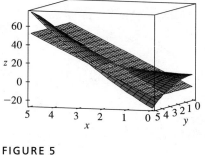

FIGURE 5

The increment of z is

$$\Delta z = f(2.05, 2.96) - f(2, 3)$$

$$= [(2.05)^2 + 3(2.05)(2.96) - (2.96)^2] - [2^2 + 3(2)(3) - 3^2]$$

$$= 0.6449$$

Notice that $\Delta z \approx dz$ but dz is easier to compute. ∎

EXAMPLE 3　Use differentials to find an approximate value for

$$\sqrt{9(1.95)^2 + (8.1)^2}$$

SOLUTION　Consider the function $z = f(x, y) = \sqrt{9x^2 + y^2}$ and observe that we can easily calculate $f(2, 8) = 10$. Therefore, we take $a = 2$, $b = 8$, $dx = \Delta x = -0.05$, and $dy = \Delta y = 0.1$ in (9). Since

$$f_x(x, y) = \frac{9x}{\sqrt{9x^2 + y^2}} \qquad \text{and} \qquad f_y(x, y) = \frac{y}{\sqrt{9x^2 + y^2}}$$

we have

$$\sqrt{9(1.95)^2 + (8.1)^2} = f(1.95, 8.1) \approx f(2, 8) + dz$$

$$= f(2, 8) + f_x(2, 8)\,dx + f_y(2, 8)\,dy$$

You can check with a calculator that this approximation is accurate to two decimal places.

$$= 10 + \tfrac{18}{10}(-0.05) + \tfrac{8}{10}(0.1)$$

$$= 9.99$$
∎

EXAMPLE 4　The base radius and height of a right circular cone are measured as 10 cm and 25 cm, respectively, with a possible error in measurement of as much as 0.1 cm in each. Use differentials to estimate the maximum error in the calculated volume of the cone.

SOLUTION　The volume V of a cone with base radius r and height h is $V = \pi r^2 h / 3$. So the differential of V is

$$dV = \frac{\partial V}{\partial r}\,dr + \frac{\partial V}{\partial h}\,dh = \frac{2\pi r h}{3}\,dr + \frac{\pi r^2}{3}\,dh$$

Since the errors are at most 0.1 cm, we have $|\Delta x| \leq 0.1$, $|\Delta y| \leq 0.1$. To find the largest error in the volume we take the largest error in the measurement of r and h. Therefore, we take $dr = 0.1$ and $dh = 0.1$ along with $r = 10$, $h = 25$. This gives

$$dV = \frac{500\pi}{3}(0.1) + \frac{100\pi}{3}(0.1) = 20\pi$$

Thus the maximum error in the calculated volume is about 20π cm$^3 \approx 63$ cm^3. ∎

The following theorem says that dz is a good approximation to Δz when Δx and Δy are small, provided that f_x and f_y are both continuous.

> **(10) THEOREM** Suppose that f_x and f_y exist on a rectangular region R with sides parallel to the axes and containing the points (a, b) and $(a + \Delta x, b + \Delta y)$. Suppose that f_x and f_y are continuous at the point (a, b) and let
>
> $$\Delta z = f(a + \Delta x, b + \Delta y) - f(a, b)$$
>
> Then $$\Delta z = f_x(a, b)\,\Delta x + f_y(a, b)\,\Delta y + \varepsilon_1\,\Delta x + \varepsilon_2\,\Delta y$$
>
> where ε_1 and ε_2 are functions of Δx and Δy that approach 0 as $(\Delta x, \Delta y) \to (0, 0)$.

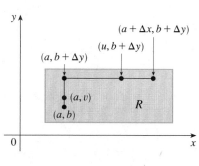

FIGURE 6

PROOF Referring to Figure 6, we write

(11) $\Delta z = [\, f(a + \Delta x, b + \Delta y) - f(a, b + \Delta y)\,] + [\, f(a, b + \Delta y) - f(a, b)\,]$

Observe that the function of a single variable

$$g(x) = f(x, b + \Delta y)$$

is defined on the interval $[a, a + \Delta x]$ and $g'(x) = f_x(x, b + \Delta y)$. If we apply the Mean Value Theorem to g, we get

$$g(a + \Delta x) - g(a) = g'(u)\,\Delta x$$

where u is some number between a and $a + \Delta x$. In terms of f, this equation becomes

$$f(a + \Delta x, b + \Delta y) - f(a, b + \Delta y) = f_x(u, b + \Delta y)\,\Delta x$$

This gives us an expression for the first part of the right side of Equation 11. For the second part we let $h(y) = f(a, y)$. Then h is a function of a single variable defined on the interval $[b, b + \Delta y]$ and $h'(y) = f_y(a, y)$. A second application of the Mean Value Theorem then gives

$$h(b + \Delta y) - h(b) = h'(v)\,\Delta y$$

where v is some number between b and $b + \Delta y$. In terms of f, this becomes

$$f(a, b + \Delta y) - f(a, b) = f_y(a, v)\,\Delta y$$

We now substitute these expressions into Equation 11 and obtain

$$\Delta z = f_x(u, b + \Delta y)\,\Delta x + f_y(a, v)\,\Delta y$$
$$= f_x(a, b)\,\Delta x + [\, f_x(u, b + \Delta y) - f_x(a, b)\,]\Delta x + f_y(a, b)\,\Delta y$$
$$+ [\, f_y(a, v) - f_y(a, b)\,]\Delta y$$
$$= f_x(a, b)\,\Delta x + f_y(a, b)\,\Delta y + \varepsilon_1\,\Delta x + \varepsilon_2\,\Delta y$$

where $$\varepsilon_1 = f_x(u, b + \Delta y) - f_x(a, b)$$

$$\varepsilon_2 = f_y(a, v) - f_y(a, b)$$

Since $(u, b + \Delta y) \to (a, b)$ and $(a, v) \to (a, b)$ as $(\Delta x, \Delta y) \to (0, 0)$ and since f_x and f_y are continuous at (a, b), we see that $\varepsilon_1 \to 0$ and $\varepsilon_2 \to 0$ as $(\Delta x, \Delta y) \to (0, 0)$. □

The conclusion of Theorem 10 can be written as

$$\Delta z = dz + \varepsilon_1 \Delta x + \varepsilon_2 \Delta y$$

where ε_1 and $\varepsilon_2 \to 0$ as $(\Delta x, \Delta y) \to (0, 0)$. This is the two-dimensional version of Equation 4, which is equivalent to the differentiability of a function of a single variable. Therefore, we use the following definition for the differentiability of a function of two variables.

---

(12) DEFINITION   If $z = f(x, y)$, then f is **differentiable** at (a, b) if Δz can be expressed in the form

$$\Delta z = f_x(a, b) \Delta x + f_y(a, b) \Delta y + \varepsilon_1 \Delta x + \varepsilon_2 \Delta y$$

where ε_1 and $\varepsilon_2 \to 0$ as $(\Delta x, \Delta y) \to (0, 0)$.

Definition 12 says that a function f is differentiable at (a, b) if the differential dz is a good approximation to the increment Δz. In other words, the linear function

$$z = f(a, b) + f_x(a, b)(x - a) + f_y(a, b)(y - b)$$

is a good approximation to the function f near (a, b).

Thus Theorem 10 says that if f_x and f_y exist near (a, b) and are continuous at (a, b), then f is differentiable at (a, b) and the tangent plane is a good approximation to the graph of f near (a, b). In particular, polynomials and rational functions are differentiable on their domains because their partial derivatives are continuous.

It can be proved that, just as in single-variable calculus, all differentiable functions are continuous (see Exercise 39). But, unlike the situation in one-dimensional calculus, Exercise 40 gives an example of a function whose partial derivatives exist but that is not differentiable.

FUNCTIONS OF THREE OR MORE VARIABLES

Differentials and differentiability can be defined in a similar manner for functions of more than two variables. For instance, if $w = f(x, y, z)$, then the **increment** of w is

$$\Delta w = f(x + \Delta x, y + \Delta y, z + \Delta z) - f(x, y, z)$$

The **differential** dw is defined in terms of the differentials dx, dy, and dz of the independent variables by

$$dw = \frac{\partial w}{\partial x} \, dx + \frac{\partial w}{\partial y} \, dy + \frac{\partial w}{\partial z} \, dz$$

If $dx = \Delta x$, $dy = \Delta y$, and $dz = \Delta z$ are all small and f has continuous partial derivatives, then dw can be used to approximate Δw.

Differentiability can be defined by an expression similar to the one in Definition 12.

EXAMPLE 5   The dimensions of a rectangular box are measured to be 75 cm, 60 cm, and 40 cm, and each measurement is correct to within 0.2 cm. Use differentials to estimate the largest possible error when the volume of the box is calculated from these measurements.

SOLUTION   If the dimensions of the box are x, y, and z, its volume is $V = xyz$ and so

$$dV = \frac{\partial V}{\partial x} \, dx + \frac{\partial V}{\partial y} \, dy + \frac{\partial V}{\partial z} \, dz = yz \, dx + xz \, dy + xy \, dz$$

We are given that $|\Delta x| \leq 0.2$, $|\Delta y| \leq 0.2$, and $|\Delta z| \leq 0.2$. To find the largest error in the volume, we therefore use $dx = 0.2$, $dy = 0.2$, and $dz = 0.2$ together with $x = 75$, $y = 60$, and $z = 40$:

$$\Delta V \approx dV = (60)(40)(0.2) + (75)(40)(0.2) + (75)(60)(0.2)$$

$$= 1980$$

Thus an error of only 0.2 cm in measuring each dimension could lead to an error of as much as 1980 cm^3 in the calculated volume! ∎

EXERCISES 12.4

1–6 ■ Find an equation of the tangent plane to the given surface at the specified point.

1. $z = x^2 + 4y^2$, $(2, 1, 8)$

2. $z = x^2 - y^2$, $(3, -2, 5)$

3. $z = 5 + (x - 1)^2 + (y + 2)^2$, $(2, 0, 10)$

4. $z = \sin(x + y)$, $(1, -1, 0)$

5. $z = \ln(2x + y)$, $(-1, 3, 0)$

6. $z = e^x \ln y$, $(3, 1, 0)$

7–8 ■ Use a computer to graph the surface and the tangent plane at the given point. Choose the domain and viewpoint so that you get a good view of both the surface and the tangent plane.

7. $z = xy$, $(-1, 2, -2)$ **8.** $z = \sqrt{x - y}$, $(5, 1, 2)$

9–10 ■ Draw the graph of f and its tangent plane at the given point. Use your computer algebra system both to compute the partial derivatives and to graph the surface and its tangent plane.

9. $f(x, y) = e^{-(x^2+y^2)/15}(\sin^2 x + \cos^2 y)$, $(2, 3, f(2, 3))$

10. $f(x, y) = \dfrac{\sqrt{1 + 4x^2 + 4y^2}}{1 + x^4 + y^4}$, $(1, 1, 1)$

11–18 ■ Find the differential of the function.

11. $z = x^2 y^3$ **12.** $z = ye^{xy}$

13. $u = e^x \cos xy$ **14.** $v = \ln(2x - 3y)$

15. $w = x^2 y + y^2 z$ **16.** $w = x \sin yz$

17. $w = \ln \sqrt{x^2 + y^2 + z^2}$ **18.** $w = \dfrac{x + y}{y + z}$

19. If $z = 5x^2 + y^2$ and (x, y) changes from $(1, 2)$ to $(1.05, 2.1)$, compare the values of Δz and dz.

20. If $z = x^2 - xy + 3y^2$ and (x, y) changes from $(3, -1)$ to $(2.96, -0.95)$, compare the values of Δz and dz.

21–24 ■ Use differentials to approximate the value of f at the given point.

21. $f(x, y) = \sqrt{20 - x^2 - 7y^2}$, $(1.95, 1.08)$

22. $f(x, y) = \ln(x - 3y)$, $(6.9, 2.06)$

23. $f(x, y, z) = x^2 y^3 z^4$, $(1.05, 0.9, 3.01)$

24. $f(x, y, z) = xy^2 \sin \pi z$, $(3.99, 4.98, 4.03)$

25–28 ■ Use differentials to approximate the number.

25. $8.94 \sqrt{9.99 - (1.01)^3}$ **26.** $(\sqrt{99} + \sqrt[3]{124})^4$

27. $\sqrt{0.99} \, e^{0.02}$ **28.** $\sqrt{(3.02)^2 + (1.97)^2 + (5.99)^2}$

29. The length and width of a rectangle are measured as 30 cm and 24 cm, respectively, with an error in measurement of at most 0.1 cm in each. Use differentials to estimate the maximum error in the calculated area of the rectangle.

30. The dimensions of a closed rectangular box are measured as 80 cm, 60 cm, and 50 cm, respectively, with a possible error of 0.2 cm in each dimension. Use differentials to estimate the maximum error in calculating the surface area of the box.

31. Use differentials to estimate the amount of tin in a closed tin can with diameter 8 cm and height 12 cm if the tin is 0.04 cm thick.

32. Use differentials to estimate the amount of metal in a closed cylindrical can that is 10 cm high and 4 cm in diameter if the metal in the wall is 0.05 cm thick and the metal in the top and bottom is 0.1 cm thick.

33. A boundary stripe 3 in. wide is painted around a rectangle whose dimensions are 100 ft by 200 ft. Use differentials to approximate the number of square feet of paint in the stripe.

34. The pressure, volume, and temperature of a mole of an ideal gas are related by the equation $PV = 8.31T$, where P is measured in kilopascals, V in liters, and T in kelvins. Use differentials to find the approximate change in the pressure if the volume increases from 12 L to 12.3 L and the temperature decreases from 310 K to 305 K.

35. If R is the total resistance of three resistors, connected in parallel, with resistances R_1, R_2, R_3, then

$$\frac{1}{R} = \frac{1}{R_1} + \frac{1}{R_2} + \frac{1}{R_3}$$

If the resistances are measured as $R_1 = 25$ ohms, $R_2 = 40$ ohms, and $R_3 = 50$ ohms, with possible errors of 0.5% in each case, estimate the maximum error in the calculated value of R.

36. Four positive numbers, each less than 50, are rounded to the first decimal place and then multiplied together. Use differentials to estimate the maximum possible error in the computed product that might result from the rounding.

37–38 ■ Show that the function is differentiable by finding values of ε_1 and ε_2 that satisfy Definition 12.

37. $f(x, y) = x^2 + y^2$ **38.** $f(x, y) = xy - 5y^2$

39. Prove that if f is a function of two variables that is differentiable at (a, b), then f is continuous at (a, b). *Hint:* Show that

$$\lim_{(\Delta x, \Delta y) \to (0, 0)} f(a + \Delta x, b + \Delta y) = f(a, b)$$

40. The function

$$f(x, y) = \begin{cases} \dfrac{xy}{x^2 + y^2} & \text{if } (x, y) \neq (0, 0) \\ 0 & \text{if } (x, y) = (0, 0) \end{cases}$$

was graphed in Figure 8 in Section 12.2. Show that $f_x(0, 0)$ and $f_y(0, 0)$ both exist but f is not differentiable at $(0, 0)$. [*Hint:* Use the result of Exercise 39.]

12.5   THE CHAIN RULE

We recall that the Chain Rule for functions of a single variable gives the rule for differentiating a composite function: If $y = f(x)$ and $x = g(t)$, where f and g are differentiable functions, then y is indirectly a differentiable function of t and

(1)
$$\frac{dy}{dt} = \frac{dy}{dx} \frac{dx}{dt}$$

For functions of more than one variable, the Chain Rule has several versions, each of them giving a rule for differentiating a composite function. The first version (Theorem 2) deals with the case where $z = f(x, y)$ and each of the variables x and y is in turn a function of a variable t. This means that z is indirectly a function of t $[z = f(g(t), h(t))]$ and the Chain Rule gives a formula for differentiating z as a function of t. We assume that f is differentiable (Definition 12.4.12). Recall from Theorem 12.4.10 that this is the case when f_x and f_y are continuous.

---

(2) THE CHAIN RULE (CASE 1)   Suppose that $z = f(x, y)$ is a differentiable function of x and y, where $x = g(t)$ and $y = h(t)$ are both differentiable functions of t. Then z is a differentiable function of t and

$$\frac{dz}{dt} = \frac{\partial f}{\partial x} \frac{dx}{dt} + \frac{\partial f}{\partial y} \frac{dy}{dt}$$

PROOF   A change of Δt in t produces changes of Δx in x and Δy in y. These, in turn, produce a change of Δz in z, and from Definition 12.4.12 we have

$$\Delta z = \frac{\partial f}{\partial x} \Delta x + \frac{\partial f}{\partial y} \Delta y + \varepsilon_1 \Delta x + \varepsilon_2 \Delta y$$

where $\varepsilon_1 \to 0$ and $\varepsilon_2 \to 0$ as $(\Delta x, \Delta y) \to (0, 0)$. [If the functions ε_1 and ε_2 are not defined at $(0, 0)$, we can define them to be 0 there.] Dividing both sides of this equation by Δt, we have

$$\frac{\Delta z}{\Delta t} = \frac{\partial f}{\partial x} \frac{\Delta x}{\Delta t} + \frac{\partial f}{\partial y} \frac{\Delta y}{\Delta t} + \varepsilon_1 \frac{\Delta x}{\Delta t} + \varepsilon_2 \frac{\Delta y}{\Delta t}$$

If we now let $\Delta t \to 0$, then $\Delta x = g(t + \Delta t) - g(t) \to 0$ because g is differentiable and therefore continuous. Similarly, $\Delta y \to 0$. This, in turn, means that $\varepsilon_1 \to 0$ and $\varepsilon_2 \to 0$, so

$$
\begin{aligned}
\frac{dz}{dt} &= \lim_{\Delta t \to 0} \frac{\Delta z}{\Delta t} \\[2mm]
&= \frac{\partial f}{\partial x} \lim_{\Delta t \to 0} \frac{\Delta x}{\Delta t} + \frac{\partial f}{\partial y} \lim_{\Delta t \to 0} \frac{\Delta y}{\Delta t} + \lim_{\Delta t \to 0} \varepsilon_1 \lim_{\Delta t \to 0} \frac{\Delta x}{\Delta t} + \lim_{\Delta t \to 0} \varepsilon_2 \lim_{\Delta t \to 0} \frac{\Delta y}{\Delta t} \\[2mm]
&= \frac{\partial f}{\partial x} \frac{dx}{dt} + \frac{\partial f}{\partial y} \frac{dy}{dt} + 0 \cdot \frac{dx}{dt} + 0 \cdot \frac{dy}{dt} \\[2mm]
&= \frac{\partial f}{\partial x} \frac{dx}{dt} + \frac{\partial f}{\partial y} \frac{dy}{dt}
\end{aligned}
$$

Since we often write $\partial z/\partial x$ in place of $\partial f/\partial x$, we can rewrite the Chain Rule in the form

$$
\frac{dz}{dt} = \frac{\partial z}{\partial x} \frac{dx}{dt} + \frac{\partial z}{\partial y} \frac{dy}{dt}
$$

Notice the similarity to the definition of the differential:

$$
dz = \frac{\partial z}{\partial x} dx + \frac{\partial z}{\partial y} dy
$$

EXAMPLE 1   If $z = x^2 y + 3xy^4$, where $x = e^t$ and $y = \sin t$, find dz/dt.

SOLUTION   The Chain Rule gives

$$
\begin{aligned}
\frac{dz}{dt} &= \frac{\partial z}{\partial x} \frac{dx}{dt} + \frac{\partial z}{\partial y} \frac{dy}{dt} \\[2mm]
&= (2xy + 3y^4)e^t + (x^2 + 12xy^3)\cos t \\[2mm]
&= (2e^t \sin t + 3\sin^4 t)e^t + (e^{2t} + 12e^t \sin^3 t)\cos t
\end{aligned}
$$

NOTE 1:   Although we have expressed the answer to Example 1 totally in terms of t, the expression in terms of x, y, and t is adequate for some purposes. For instance, if we had been asked to find the value of dz/dt when $t = 0$, we could simply observe that $x = 1$ and $y = 0$ when $t = 0$ and so

$$
\frac{dz}{dt}\bigg|_{t=0} = 0e^0 + 1\cos 0 = 1
$$

NOTE 2:   If $T(x, y)$ represents the temperature at a point (x, y) and $x = f(t)$ and $y = g(t)$ are the parametric equations of a curve C, then the composite function $z = T(f(t), g(t))$ represents the temperature at points on C and the derivative dz/dt represents the rate of change of temperature along the curve.

EXAMPLE 2   The pressure P (in kilopascals), volume V (in liters), and temperature T (in kelvins) of a mole of an ideal gas are related by the equation $PV = 8.31T$. Find the rate at which the pressure is changing when the temperature is 300 K and increasing at a rate of 0.1 K/s and the volume is 100 L and increasing at a rate of 0.2 L/s.

SOLUTION   If t represents the time elapsed in seconds, then at the given instant we have $T = 300$, $dT/dt = 0.1$, $V = 100$, $dV/dt = 0.2$. Since

$$
P = 8.31 \frac{T}{V}
$$

the Chain Rule gives

$$\frac{dP}{dt} = \frac{\partial P}{\partial T}\frac{dT}{dt} + \frac{\partial P}{\partial V}\frac{dV}{dt} = \frac{8.31}{V}\frac{dT}{dt} - \frac{8.31T}{V^2}\frac{dV}{dt}$$

$$= \frac{8.31}{100}(0.1) - \frac{8.31(300)}{100^2}(0.2)$$

$$= -0.04155$$

The pressure is decreasing at a rate of about 0.042 kPa/s.   ∎

We now consider the situation where $z = f(x, y)$ but each of x and y is a function of two variables s and t: $x = g(s, t)$, $y = h(s, t)$. Then z is indirectly a function of s and t and we wish to find $\partial z/\partial s$ and $\partial z/\partial t$. Recall that in computing $\partial z/\partial t$ we hold s fixed and compute the ordinary derivative of z with respect to t. Therefore, we can apply Theorem 2 to obtain

$$\frac{\partial z}{\partial t} = \frac{\partial z}{\partial x}\frac{\partial x}{\partial t} + \frac{\partial z}{\partial y}\frac{\partial y}{\partial t}$$

A similar argument holds for $\partial z/\partial s$ and so we have proved the following version of the Chain Rule.

---

(3) THE CHAIN RULE (CASE 2)   Suppose that $z = f(x, y)$ is a differentiable function of x and y, where $x = g(s, t)$, $y = h(s, t)$, and the partial derivatives g_s, g_t, h_s, and h_t exist. Then

$$\frac{\partial z}{\partial s} = \frac{\partial z}{\partial x}\frac{\partial x}{\partial s} + \frac{\partial z}{\partial y}\frac{\partial y}{\partial s}$$

$$\frac{\partial z}{\partial t} = \frac{\partial z}{\partial x}\frac{\partial x}{\partial t} + \frac{\partial z}{\partial y}\frac{\partial y}{\partial t}$$

EXAMPLE 3   If $z = e^x \sin y$, where $x = st^2$ and $y = s^2 t$, find $\partial z/\partial s$ and $\partial z/\partial t$.

SOLUTION   Applying Case 2 of the Chain Rule, we get

$$\frac{\partial z}{\partial s} = \frac{\partial z}{\partial x}\frac{\partial x}{\partial s} + \frac{\partial z}{\partial y}\frac{\partial y}{\partial s} = (e^x \sin y)(t^2) + (e^x \cos y)(2st)$$

$$= t^2 e^{st^2} \sin(s^2 t) + 2st e^{st^2} \cos(s^2 t)$$

$$\frac{\partial z}{\partial t} = \frac{\partial z}{\partial x}\frac{\partial x}{\partial t} + \frac{\partial z}{\partial y}\frac{\partial y}{\partial t} = (e^x \sin y)(2st) + (e^x \cos y)(s^2)$$

$$= 2st e^{st^2} \sin(s^2 t) + s^2 e^{st^2} \cos(s^2 t)$$ ∎

Case 2 of the Chain Rule contains three types of variables: s and t are **independent** variables, x and y are called **intermediate** variables, and z is the **dependent** variable. Notice that Theorem 3 has one term for each intermediate variable and each of these terms resembles the one-dimensional Chain Rule in Equation 1.

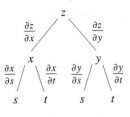

FIGURE 1

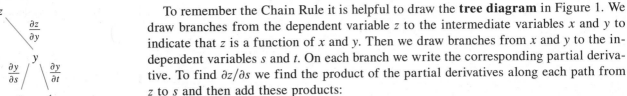

To remember the Chain Rule it is helpful to draw the **tree diagram** in Figure 1. We draw branches from the dependent variable z to the intermediate variables x and y to indicate that z is a function of x and y. Then we draw branches from x and y to the independent variables s and t. On each branch we write the corresponding partial derivative. To find $\partial z/\partial s$ we find the product of the partial derivatives along each path from z to s and then add these products:

$$\frac{\partial z}{\partial s} = \frac{\partial z}{\partial x}\frac{\partial x}{\partial s} + \frac{\partial z}{\partial y}\frac{\partial y}{\partial s}$$

Similarly, we find $\partial z/\partial t$ by using the paths from z to t.

Now we consider the general situation in which a dependent variable u is a function of n intermediate variables $x_1, \ldots, x_n$, each of which is, in turn, a function of m independent variables $t_1, \ldots, t_m$. Notice that there are n terms, one for each intermediate variable. The proof is similar to that of Case 1.

> **(4) THE CHAIN RULE (GENERAL VERSION)**   Suppose that u is a differentiable function of the n variables $x_1, x_2, \ldots, x_n$ and each x_j is a function of the m variables $t_1, t_2, \ldots, t_m$ such that all the partial derivatives $\partial x_j/\partial t_i$ exist $(j = 1, 2, \ldots, n; i = 1, 2, \ldots, m)$. Then u is a function of $t_1, t_2, \ldots, t_m$ and
>
> $$\frac{\partial u}{\partial t_i} = \frac{\partial u}{\partial x_1}\frac{\partial x_1}{\partial t_i} + \frac{\partial u}{\partial x_2}\frac{\partial x_2}{\partial t_i} + \cdots + \frac{\partial u}{\partial x_n}\frac{\partial x_n}{\partial t_i}$$
>
> for each $i = 1, 2, \ldots, m$.

EXAMPLE 4   Write out the Chain Rule for the case where $w = f(x, y, z, t)$ and $x = x(u, v)$, $y = y(u, v)$, $z = z(u, v)$, and $t = t(u, v)$.

SOLUTION   We apply Theorem 4 with $n = 4$ and $m = 2$. Figure 2 shows the tree diagram. Although we have not written the derivatives on the branches, it is understood that if a branch leads from y to u, then the partial derivative for that branch is $\partial y/\partial u$. With the aid of the tree diagram we can now write the required expressions:

$$\frac{\partial w}{\partial u} = \frac{\partial w}{\partial x}\frac{\partial x}{\partial u} + \frac{\partial w}{\partial y}\frac{\partial y}{\partial u} + \frac{\partial w}{\partial z}\frac{\partial z}{\partial u} + \frac{\partial w}{\partial t}\frac{\partial t}{\partial u}$$

$$\frac{\partial w}{\partial v} = \frac{\partial w}{\partial x}\frac{\partial x}{\partial v} + \frac{\partial w}{\partial y}\frac{\partial y}{\partial v} + \frac{\partial w}{\partial z}\frac{\partial z}{\partial v} + \frac{\partial w}{\partial t}\frac{\partial t}{\partial v}$$ ■

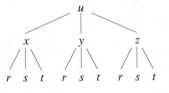

FIGURE 2

EXAMPLE 5   If $u = x^4 y + y^2 z^3$, where $x = rse^t$, $y = rs^2 e^{-t}$, and $z = r^2 s \sin t$, find the value of $\partial u/\partial s$ when $r = 2$, $s = 1$, $t = 0$.

SOLUTION   With the help of the tree diagram in Figure 3, we have

$$\frac{\partial u}{\partial s} = \frac{\partial u}{\partial x}\frac{\partial x}{\partial s} + \frac{\partial u}{\partial y}\frac{\partial y}{\partial s} + \frac{\partial u}{\partial z}\frac{\partial z}{\partial s}$$

$$= (4x^3 y)(re^t) + (x^4 + 2yz^3)(2rse^{-t}) + (3y^2 z^2)(r^2 \sin t)$$

When $r = 2$, $s = 1$, and $t = 0$, we have $x = 2$, $y = 2$, and $z = 0$, so

$$\frac{\partial u}{\partial s} = (64)(2) + (16)(4) + (0)(0) = 192$$ ■

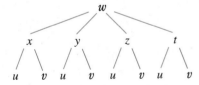

FIGURE 3

EXAMPLE 6   If $g(s, t) = f(s^2 - t^2, t^2 - s^2)$ and f is differentiable, show that g satisfies the equation

$$t \frac{\partial g}{\partial s} + s \frac{\partial g}{\partial t} = 0$$

SOLUTION   Let $x = s^2 - t^2$ and $y = t^2 - s^2$. Then $g(s, t) = f(x, y)$ and the Chain Rule gives

$$\frac{\partial g}{\partial s} = \frac{\partial f}{\partial x} \frac{\partial x}{\partial s} + \frac{\partial f}{\partial y} \frac{\partial y}{\partial s} = \frac{\partial f}{\partial x} (2s) + \frac{\partial f}{\partial y} (-2s)$$

$$\frac{\partial g}{\partial t} = \frac{\partial f}{\partial x} \frac{\partial x}{\partial t} + \frac{\partial f}{\partial y} \frac{\partial y}{\partial t} = \frac{\partial f}{\partial x} (-2t) + \frac{\partial f}{\partial y} (2t)$$

Therefore

$$t \frac{\partial g}{\partial s} + s \frac{\partial g}{\partial t} = \left(2st \frac{\partial f}{\partial x} - 2st \frac{\partial f}{\partial y} \right) + \left(-2st \frac{\partial f}{\partial x} + 2st \frac{\partial f}{\partial y} \right) = 0 \qquad \blacksquare$$

EXAMPLE 7   If $z = f(x, y)$ has continuous second-order partial derivatives and $x = r^2 + s^2$ and $y = 2rs$, find (a) $\partial z/\partial r$ and (b) $\partial^2 z/\partial r^2$.

SOLUTION
(a) The Chain Rule gives

$$\frac{\partial z}{\partial r} = \frac{\partial z}{\partial x} \frac{\partial x}{\partial r} + \frac{\partial z}{\partial y} \frac{\partial y}{\partial r} = \frac{\partial z}{\partial x} (2r) + \frac{\partial z}{\partial y} (2s)$$

(b) Applying the Product Rule to the expression in part (a), we get

(5)
$$\frac{\partial^2 z}{\partial r^2} = \frac{\partial}{\partial r} \left(2r \frac{\partial z}{\partial x} + 2s \frac{\partial z}{\partial y} \right)$$

$$= 2 \frac{\partial z}{\partial x} + 2r \frac{\partial}{\partial r} \left(\frac{\partial z}{\partial x} \right) + 2s \frac{\partial}{\partial r} \left(\frac{\partial z}{\partial y} \right)$$

But, using the Chain Rule again, we have

$$\frac{\partial}{\partial r} \left(\frac{\partial z}{\partial x} \right) = \frac{\partial}{\partial x} \left(\frac{\partial z}{\partial x} \right) \frac{\partial x}{\partial r} + \frac{\partial}{\partial y} \left(\frac{\partial z}{\partial x} \right) \frac{\partial y}{\partial r}$$

$$= \frac{\partial^2 z}{\partial x^2} (2r) + \frac{\partial^2 z}{\partial y \, \partial x} (2s)$$

$$\frac{\partial}{\partial r} \left(\frac{\partial z}{\partial y} \right) = \frac{\partial}{\partial x} \left(\frac{\partial z}{\partial y} \right) \frac{\partial x}{\partial r} + \frac{\partial}{\partial y} \left(\frac{\partial z}{\partial y} \right) \frac{\partial y}{\partial r}$$

$$= \frac{\partial^2 z}{\partial x \, \partial y} (2r) + \frac{\partial^2 z}{\partial y^2} (2s)$$

Putting these expressions into Equation 5 and using the equality of the mixed second-order derivatives, we obtain

$$\frac{\partial^2 z}{\partial r^2} = 2\,\frac{\partial z}{\partial x} + 2r\left(2r\,\frac{\partial^2 z}{\partial x^2} + 2s\,\frac{\partial^2 z}{\partial y\,\partial x}\right) + 2s\left(2r\,\frac{\partial^2 z}{\partial x\,\partial y} + 2s\,\frac{\partial^2 z}{\partial y^2}\right)$$

$$= 2\,\frac{\partial z}{\partial x} + 4r^2\,\frac{\partial^2 z}{\partial x^2} + 8rs\,\frac{\partial^2 z}{\partial x\,\partial y} + 4s^2\,\frac{\partial^2 z}{\partial y^2} \qquad \blacksquare$$

IMPLICIT DIFFERENTIATION

The Chain Rule can be used to give a more complete description of the process of implicit differentiation that was introduced in Sections 2.6 and 12.3. We suppose that an equation of the form $F(x, y) = 0$ defines y implicitly as a differentiable function of x, that is, $y = f(x)$, where $F(x, f(x)) = 0$ for all x in the domain of f. If F is differentiable, we can apply Case 1 of the Chain Rule to differentiate both sides of the equation $F(x, y) = 0$ with respect to x. Since both x and y are functions of x, we obtain

$$\frac{\partial F}{\partial x}\,\frac{dx}{dx} + \frac{\partial F}{\partial y}\,\frac{dy}{dx} = 0$$

But $dx/dx = 1$, so if $\partial F/\partial y \neq 0$ we solve for dy/dx and obtain

(6)
$$\frac{dy}{dx} = -\frac{\dfrac{\partial F}{\partial x}}{\dfrac{\partial F}{\partial y}} = -\frac{F_x}{F_y}$$

To derive this equation we assumed that $F(x, y) = 0$ defines y implicitly as a function of x. The **Implicit Function Theorem,** proved in advanced calculus, gives conditions under which this assumption is valid. It states that if F is defined on an open disk containing (a, b), where $F(a, b) = 0$, $F_y(a, b) \neq 0$, and F_x and F_y are continuous on the disk, then the equation $F(x, y) = 0$ defines y as a function of x near the point (a, b) and the derivative of this function is given by Equation 6.

EXAMPLE 8   Find y' if $x^3 + y^3 = 6xy$.

SOLUTION   The given equation can be written as

$$F(x, y) = x^3 + y^3 - 6xy = 0$$

so Equation 6 gives

The solution to Example 8 should be compared to the one in Example 2 in Section 2.6.

$$\frac{dy}{dx} = -\frac{F_x}{F_y} = -\frac{3x^2 - 6y}{3y^2 - 6x} = -\frac{x^2 - 2y}{y^2 - 2x} \qquad \blacksquare$$

Now we suppose that z is given implicitly as a function $z = f(x, y)$ by an equation of the form $F(x, y, z) = 0$. This means that $F(x, y, f(x, y)) = 0$ for all (x, y) in the domain of f. If F is differentiable and f_x and f_y exist, then we can use the Chain Rule to differentiate the equation $F(x, y, z) = 0$ as follows:

$$\frac{\partial F}{\partial x}\,\frac{\partial x}{\partial x} + \frac{\partial F}{\partial y}\,\frac{\partial y}{\partial x} + \frac{\partial F}{\partial z}\,\frac{\partial z}{\partial x} = 0$$

But $\qquad \dfrac{\partial}{\partial x}(x) = 1 \qquad$ and $\qquad \dfrac{\partial}{\partial x}(y) = 0$

so this equation becomes

$$\frac{\partial F}{\partial x} + \frac{\partial F}{\partial z}\frac{\partial z}{\partial x} = 0$$

If $\partial F / \partial z \neq 0$, we solve for $\partial z / \partial x$ and obtain the first formula in Equations 7. The formula for $\partial z / \partial y$ is obtained in a similar manner.

(7)
$$\frac{\partial z}{\partial x} = -\frac{\dfrac{\partial F}{\partial x}}{\dfrac{\partial F}{\partial z}} \qquad \frac{\partial z}{\partial y} = -\frac{\dfrac{\partial F}{\partial y}}{\dfrac{\partial F}{\partial z}}$$

Again, a version of the **Implicit Function Theorem** gives conditions under which our assumption is valid. If F is defined within a sphere containing (a, b, c), where $F(a, b, c) = 0$, $F_z(a, b, c) \neq 0$, and F_x, F_y, and F_z are continuous inside the sphere, then the equation $F(x, y, z) = 0$ defines z as a function of x and y near the point (a, b, c) and the partial derivatives of this function are given by (7).

EXAMPLE 9 Find $\dfrac{\partial z}{\partial x}$ and $\dfrac{\partial z}{\partial y}$ if $x^3 + y^3 + z^3 + 6xyz = 1$.

SOLUTION   Let $F(x, y, z) = x^3 + y^3 + z^3 + 6xyz - 1$. Then, from Equations 7, we have

$$\frac{\partial z}{\partial x} = -\frac{F_x}{F_z} = -\frac{3x^2 + 6yz}{3z^2 + 6xy} = -\frac{x^2 + 2yz}{z^2 + 2xy}$$

The solution to Example 9 should be compared to the one in Example 4 in Section 12.3.

$$\frac{\partial z}{\partial y} = -\frac{F_y}{F_z} = -\frac{3y^2 + 6xz}{3z^2 + 6xy} = -\frac{y^2 + 2xz}{z^2 + 2xy}$$ ∎

EXERCISES 12.5

1–6 ■ Use the Chain Rule to find dz/dt or dw/dt.

1. $z = x^2 + y^2$, $x = t^3$, $y = 1 + t^2$

2. $z = x^2 y^3$, $x = 1 + \sqrt{t}$, $y = 1 - \sqrt{t}$

3. $z = \ln(x + y^2)$, $x = \sqrt{1 + t}$, $y = 1 + \sqrt{t}$

4. $z = xe^{x/y}$, $x = \cos t$, $y = e^{2t}$

5. $w = xy^2 z^3$, $x = \sin t$, $y = \cos t$, $z = 1 + e^{2t}$

6. $w = \dfrac{x}{y} + \dfrac{y}{z}$, $x = \sqrt{t}$, $y = \cos 2t$, $z = e^{-3t}$

7–12 ■ Use the Chain Rule to find $\partial z/\partial s$ and $\partial z/\partial t$.

7. $z = x^2 \sin y$, $x = s^2 + t^2$, $y = 2st$

8. $z = \sin x \cos y$, $x = (s - t)^2$, $y = s^2 - t^2$

9. $z = x^2 - 3x^2 y^3$, $x = se^t$, $y = se^{-t}$

10. $z = x \tan^{-1}(xy)$, $x = t^2$, $y = se^t$

11. $z = 2^{x-3y}$, $x = s^2 t$, $y = st^2$

12. $z = xe^y + ye^{-x}$, $x = e^t$, $y = st^2$

13–16 ■ Write out the Chain Rule for the given case (as in Example 4).

13. $u = f(x, y)$, where $x = x(r, s, t)$, $y = y(r, s, t)$

14. $w = f(x, y, z)$, where $x = x(t, u)$, $y = y(t, u)$, $z = z(t, u)$

15. $v = f(p, q, r)$, where $p = p(x, y, z)$, $q = q(x, y, z)$, $r = r(x, y, z)$

16. $u = f(s, t)$, where $s = s(w, x, y, z)$, $t = t(w, x, y, z)$

17–22 ■ Use the Chain Rule to find the indicated partial derivatives.

17. $w = x^2 + y^2 + z^2$, $x = st$, $y = s\cos t$, $z = s\sin t$;

$\dfrac{\partial w}{\partial s}, \dfrac{\partial w}{\partial t}$ when $s = 1$, $t = 0$

18. $u = xy + yz + zx$, $x = st$, $y = e^{st}$, $z = t^2$;

$\dfrac{\partial u}{\partial s}, \dfrac{\partial u}{\partial t}$ when $s = 0$, $t = 1$

19. $z = y^2\tan x$, $x = t^2uv$, $y = u + tv^2$;

$\dfrac{\partial z}{\partial t}, \dfrac{\partial z}{\partial u}, \dfrac{\partial z}{\partial v}$ when $t = 2$, $u = 1$, $v = 0$

20. $z = \dfrac{x}{y}$, $x = re^{st}$, $y = rse^t$;

$\dfrac{\partial z}{\partial r}, \dfrac{\partial z}{\partial s}, \dfrac{\partial z}{\partial t}$ when $r = 1$, $s = 2$, $t = 0$

21. $u = \dfrac{x + y}{y + z}$, $x = p + r + t$, $y = p - r + t$,

$z = p + r - t$; $\dfrac{\partial u}{\partial p}, \dfrac{\partial u}{\partial r}, \dfrac{\partial u}{\partial t}$

22. $t = z\sec(xy)$, $x = uv$, $y = vw$, $z = wu$; $\dfrac{\partial t}{\partial u}, \dfrac{\partial t}{\partial v}, \dfrac{\partial t}{\partial w}$

23–26 ■ Use Equation 6 to find dy/dx.

23. $x^2 - xy + y^3 = 8$ **24.** $y^5 + 3x^2y^2 + 5x^4 = 12$

25. $2y^2 + \sqrt[3]{xy} = 3x^2 + 17$ **26.** $x\cos y + y\cos x = 1$

27–32 ■ Use Equations 7 to find $\partial z/\partial x$ and $\partial z/\partial y$.

27. $xy + yz - xz = 0$ **28.** $xyz = \cos(x + y + z)$

29. $x^2 + y^2 - z^2 = 2x(y + z)$

30. $xy^2z^3 + x^3y^2z = x + y + z$

31. $xe^y + yz + ze^x = 0$

32. $y^2ze^{x+y} - \sin(xyz) = 0$

33. The radius of a right circular cylinder is decreasing at a rate of 1.2 cm/s while its height is increasing at a rate of 3 cm/s. At what rate is the volume of the cylinder changing when the radius is 80 cm and the height is 150 cm?

34. The radius of a right circular cone is increasing at a rate of 1.8 in/s while its height is decreasing at a rate of 2.5 in/s. At what rate is the volume of the cone changing when the radius is 120 in. and the height is 140 in.?

35. The length ℓ, width w, and height h of a box change with time. At a certain instant the dimensions are $\ell = 1$ m and $w = h = 2$ m, and ℓ and w are increasing at a rate of 2 m/s while h is decreasing at a rate of 3 m/s. At that instant find the rates at which the following quantities are changing.
 (a) The volume
 (b) The surface area
 (c) The length of a diagonal

36. The voltage V in a simple electrical circuit is slowly decreasing as the battery wears out. The resistance R is slowly increasing as the resistor heats up. Use Ohm's Law, $V = IR$, to find how the current I is changing at the moment when $R = 400\,\Omega$, $I = 0.08$ A, $dV/dt = -0.01$ V/s, and $dR/dt = 0.03\,\Omega$/s.

37. The pressure of 1 mole of an ideal gas is increasing at a rate of 0.05 kPa/s and the temperature is increasing at a rate of 0.15 K/s. Use the equation in Example 2 to find the rate of change of the volume when the pressure is 20 kPa and the temperature is 320 K.

38. Car A is traveling north on Highway 16 at 90 km/h. Car B is traveling west on Highway 83 at 80 km/h. Each car is approaching the intersection of these highways. How fast is the distance between the cars changing when car A is 0.3 km from the intersection and car B is 0.4 km from the intersection?

39–42 ■ Assume that all the given functions are differentiable.

39. If $z = f(x, y)$, where $x = r\cos\theta$ and $y = r\sin\theta$, (a) find $\partial z/\partial r$ and $\partial z/\partial\theta$ and (b) show that

$$\left(\frac{\partial z}{\partial x}\right)^2 + \left(\frac{\partial z}{\partial y}\right)^2 = \left(\frac{\partial z}{\partial r}\right)^2 + \frac{1}{r^2}\left(\frac{\partial z}{\partial\theta}\right)^2$$

40. If $u = f(x, y)$, where $x = e^s\cos t$ and $y = e^s\sin t$, show that

$$\left(\frac{\partial u}{\partial x}\right)^2 + \left(\frac{\partial u}{\partial y}\right)^2 = e^{-2s}\left[\left(\frac{\partial u}{\partial s}\right)^2 + \left(\frac{\partial u}{\partial t}\right)^2\right]$$

41. If $z = f(x - y)$, show that $\dfrac{\partial z}{\partial x} + \dfrac{\partial z}{\partial y} = 0$.

42. If $z = f(x, y)$, where $x = s + t$ and $y = s - t$, show that

$$\left(\frac{\partial z}{\partial x}\right)^2 - \left(\frac{\partial z}{\partial y}\right)^2 = \frac{\partial z}{\partial s}\frac{\partial z}{\partial t}$$

43–48 ■ Assume that all the given functions have continuous second-order partial derivatives.

43. Show that any function of the form

$$z = f(x + at) + g(x - at)$$

is a solution of the wave equation

$$\frac{\partial^2 z}{\partial t^2} = a^2\frac{\partial^2 z}{\partial x^2}$$

[*Hint:* Let $u = x + at$, $v = x - at$.]

44. If $u = f(x, y)$, where $x = e^s\cos t$ and $y = e^s\sin t$, show that

$$\frac{\partial^2 u}{\partial x^2} + \frac{\partial^2 u}{\partial y^2} = e^{-2s}\left[\frac{\partial^2 u}{\partial s^2} + \frac{\partial^2 u}{\partial t^2}\right]$$

45. If $z = f(x, y)$, where $x = r^2 + s^2$, $y = 2rs$, find $\partial^2 z/\partial r\,\partial s$. (Compare with Example 7.)

46. If $z = f(x, y)$, where $x = r\cos\theta$, $y = r\sin\theta$, find (a) $\partial z/\partial r$, (b) $\partial z/\partial\theta$, and (c) $\partial^2 z/\partial r\,\partial\theta$.

47. If $z = f(x, y)$, where $x = r\cos\theta$, $y = r\sin\theta$, show that

$$\frac{\partial^2 z}{\partial x^2} + \frac{\partial^2 z}{\partial y^2} = \frac{\partial^2 z}{\partial r^2} + \frac{1}{r^2}\frac{\partial^2 z}{\partial\theta^2} + \frac{1}{r}\frac{\partial z}{\partial r}$$

48. Suppose $z = f(x, y)$, where $x = g(s, t)$ and $y = h(s, t)$.
    (a) Show that

$$\frac{\partial^2 z}{\partial t^2} = \frac{\partial^2 z}{\partial x^2}\left(\frac{\partial x}{\partial t}\right)^2 + 2\frac{\partial^2 z}{\partial x\,\partial y}\frac{\partial x}{\partial t}\frac{\partial y}{\partial t} + \frac{\partial^2 z}{\partial y^2}\left(\frac{\partial y}{\partial t}\right)^2$$
$$+ \frac{\partial z}{\partial x}\frac{\partial^2 x}{\partial t^2} + \frac{\partial z}{\partial y}\frac{\partial^2 y}{\partial t^2}$$

    (b) Find a similar formula for $\partial^2 z/\partial s\,\partial t$.

49–51 ■ Each function f is **homogeneous of degree n**; that is, f satisfies the equation $f(tx, ty) = t^n f(x, y)$ for all t, where n is a positive integer and f has continuous second-order partial derivatives. Verify that f satisfies the equation.

49. $x\dfrac{\partial f}{\partial x} + y\dfrac{\partial f}{\partial y} = nf(x, y)$

50. $x^2\dfrac{\partial^2 f}{\partial x^2} + 2xy\dfrac{\partial^2 f}{\partial x\,\partial y} + y^2\dfrac{\partial^2 f}{\partial y^2} = n(n - 1)f(x, y)$

51. $f_x(tx, ty) = t^{n-1}f_x(x, y)$

52. Suppose that the equation $F(x, y, z) = 0$ implicitly defines each of the three variables x, y, and z as functions of the other two: $z = f(x, y)$, $y = g(x, z)$, $x = h(y, z)$. If F is differentiable and F_x, F_y, and F_z are all nonzero, show that

$$\frac{\partial z}{\partial x}\frac{\partial x}{\partial y}\frac{\partial y}{\partial z} = -1$$

12.6   DIRECTIONAL DERIVATIVES AND THE GRADIENT VECTOR

Recall that if $z = f(x, y)$, then the partial derivatives f_x and f_y are defined as

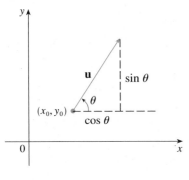

FIGURE 1

(1)
$$f_x(x_0, y_0) = \lim_{h\to 0}\frac{f(x_0 + h, y_0) - f(x_0, y_0)}{h}$$

$$f_y(x_0, y_0) = \lim_{h\to 0}\frac{f(x_0, y_0 + h) - f(x_0, y_0)}{h}$$

and represent the rates of change of z in the x- and y-directions, that is, in the directions of the unit vectors $\mathbf{i}$ and $\mathbf{j}$.

Suppose that we now wish to find the rate of change of z at (x_0, y_0) in the direction of an arbitrary unit vector $\mathbf{u} = \langle a, b\rangle$ (see Figure 1). To do this we consider the surface S with equation $z = f(x, y)$ (the graph of f) and we let $z_0 = f(x_0, y_0)$. Then the point $P(x_0, y_0, z_0)$ lies on S. The vertical plane that passes through P in the direction $\mathbf{u}$ intersects S in a curve C (see Figure 2). The slope of the tangent line T to C at P is the rate of change of z in the direction of $\mathbf{u}$.

If $Q(x, y, z)$ is another point on C and P', Q' are the projections of P, Q on the xy-plane, then the vector $\overrightarrow{P'Q'}$ is parallel to $\mathbf{u}$ and so

$$\overrightarrow{P'Q'} = h\mathbf{u} = \langle ha, hb\rangle$$

for some scalar h. Therefore, $x - x_0 = ha$, $y - y_0 = hb$, so $x = x_0 + ha$, $y = y_0 + hb$, and

$$\frac{\Delta z}{h} = \frac{z - z_0}{h} = \frac{f(x_0 + ha, y_0 + hb) - f(x_0, y_0)}{h}$$

If we take the limit as $h \to 0$, we obtain the rate of change of z (with respect to distance) in the direction of $\mathbf{u}$, which is called the directional derivative of f in the direction of $\mathbf{u}$.

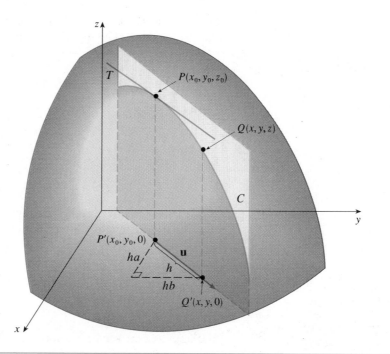

FIGURE 2

---

(2) DEFINITION   The **directional derivative** of f at (x_0, y_0) in the direction of a unit vector $\mathbf{u} = \langle a, b \rangle$ is

$$D_{\mathbf{u}} f(x_0, y_0) = \lim_{h \to 0} \frac{f(x_0 + ha, \, y_0 + hb) - f(x_0, y_0)}{h}$$

if this limit exists.

By comparing Definition 2 with Equations 1, we see that if $\mathbf{u} = \mathbf{i} = \langle 1, 0 \rangle$, then $D_{\mathbf{i}} f = f_x$ and if $\mathbf{u} = \mathbf{j} = \langle 0, 1 \rangle$, then $D_{\mathbf{j}} f = f_y$. In other words, the partial derivatives of f with respect to x and y are just special cases of the directional derivative.

For computational purposes we generally use the formula given by the following theorem.

---

(3) THEOREM   If f is a differentiable function of x and y, then f has a directional derivative in the direction of any unit vector $\mathbf{u} = \langle a, b \rangle$ and

$$D_{\mathbf{u}} f(x, y) = f_x(x, y)a + f_y(x, y)b$$

PROOF   If we define a function g of the single variable h by

$$g(h) = f(x_0 + ha, y_0 + hb)$$

then by the definition of a derivative we have

$$(4) \qquad g'(0) = \lim_{h \to 0} \frac{g(h) - g(0)}{h} = \lim_{h \to 0} \frac{f(x_0 + ha, y_0 + hb) - f(x_0, y_0)}{h}$$

$$= D_{\mathbf{u}} f(x_0, y_0)$$

On the other hand, we can write $g(h) = f(x, y)$, where $x = x_0 + ha$, $y = y_0 + hb$, so the Chain Rule (Theorem 12.5.2) gives

$$g'(h) = \frac{\partial f}{\partial x}\frac{dx}{dh} + \frac{\partial f}{\partial y}\frac{dy}{dh} = f_x(x, y)a + f_y(x, y)b$$

If we now put $h = 0$, then $x = x_0$, $y = y_0$, and

(5)
$$g'(0) = f_x(x_0, y_0)a + f_y(x_0, y_0)b$$

Comparing Equations 4 and 5, we see that

$$D_{\mathbf{u}} f(x_0, y_0) = f_x(x_0, y_0)a + f_y(x_0, y_0)b \qquad \square$$

If the unit vector $\mathbf{u}$ makes an angle θ with the positive x-axis (as in Figure 1), then we can write $\mathbf{u} = \langle \cos\theta, \sin\theta \rangle$ and the formula in Theorem 3 becomes

(6)
$$D_{\mathbf{u}} f(x, y) = f_x(x, y)\cos\theta + f_y(x, y)\sin\theta$$

The directional derivative $D_{\mathbf{u}} f(1, 2)$ in Example 1 represents the rate of change of z in the direction of $\mathbf{u}$. This is the slope of the tangent line to the curve of intersection of the surface $z = x^3 - 3xy + 4y^2$ and the vertical plane through $(1, 2, 0)$ in the direction of $\mathbf{u}$ shown in Figure 3.

EXAMPLE 1 Find the directional derivative $D_{\mathbf{u}} f(x, y)$ if
$$f(x, y) = x^3 - 3xy + 4y^2$$
and $\mathbf{u}$ is the unit vector given by angle $\theta = \pi/6$. What is $D_{\mathbf{u}} f(1, 2)$?

SOLUTION Formula 6 gives
$$D_{\mathbf{u}} f(x, y) = f_x(x, y)\cos\frac{\pi}{6} + f_y(x, y)\sin\frac{\pi}{6}$$
$$= (3x^2 - 3y)\frac{\sqrt{3}}{2} + (-3x + 8y)\tfrac{1}{2}$$
$$= \tfrac{1}{2}[3\sqrt{3}x^2 - 3x + (8 - 3\sqrt{3})y]$$

Therefore
$$D_{\mathbf{u}} f(1, 2) = \tfrac{1}{2}[3\sqrt{3}(1)^2 - 3(1) + (8 - 3\sqrt{3})(2)] = \frac{13 - 3\sqrt{3}}{2} \qquad \blacksquare$$

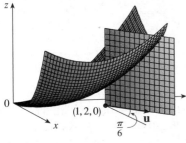
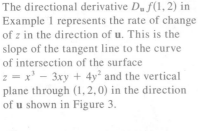

FIGURE 3

Notice from Theorem 3 that the directional derivative can be written as the dot product of two vectors:

(7)
$$D_{\mathbf{u}} f(x, y) = f_x(x, y)a + f_y(x, y)b$$
$$= \langle f_x(x, y), f_y(x, y) \rangle \cdot \langle a, b \rangle$$
$$= \langle f_x(x, y), f_y(x, y) \rangle \cdot \mathbf{u}$$

The first vector in this dot product occurs not only in computing directional derivatives but in many other contexts as well. So we give it a special name (the *gradient* of f) and a special notation (**grad** f or ∇f, which is read "del f").

(8) **DEFINITION** If f is a function of two variables x and y, then the **gradient** of f is the vector function ∇f defined by
$$\nabla f(x, y) = \langle f_x(x, y), f_y(x, y) \rangle = \frac{\partial f}{\partial x}\mathbf{i} + \frac{\partial f}{\partial y}\mathbf{j}$$

EXAMPLE 2 If $f(x, y) = \sin x + e^{xy}$, then

$$\nabla f(x, y) = \langle \cos x + ye^{xy}, xe^{xy} \rangle$$

and

$$\nabla f(0, 1) = \langle 2, 0 \rangle$$

■

With this notation for the gradient vector, we can rewrite the expression (7) for the directional derivative as

(9)

$$D_{\mathbf{u}} f(x, y) = \nabla f(x, y) \cdot \mathbf{u}$$

This expresses the directional derivative in the direction of $\mathbf{u}$ as the scalar projection of the gradient vector onto $\mathbf{u}$.

EXAMPLE 3 Find the directional derivative of the function $f(x, y) = x^2 y^3 - 4y$ at the point $(2, -1)$ in the direction of the vector $\mathbf{v} = 2\mathbf{i} + 5\mathbf{j}$.

SOLUTION We first compute the gradient vector at $(2, -1)$:

$$\nabla f(x, y) = 2xy^3 \mathbf{i} + (3x^2 y^2 - 4)\mathbf{j}$$

$$\nabla f(2, -1) = -4\mathbf{i} + 8\mathbf{j}$$

Note that $\mathbf{v}$ is not a unit vector, but since $|\mathbf{v}| = \sqrt{29}$, the unit vector in the direction of $\mathbf{v}$ is

$$\mathbf{u} = \frac{\mathbf{v}}{|\mathbf{v}|} = \frac{2}{\sqrt{29}} \mathbf{i} + \frac{5}{\sqrt{29}} \mathbf{j}$$

Therefore, by Equation 9, we have

$$D_{\mathbf{u}} f(2, -1) = \nabla f(2, -1) \cdot \mathbf{u} = (-4\mathbf{i} + 8\mathbf{j}) \cdot \left(\frac{2}{\sqrt{29}} \mathbf{i} + \frac{5}{\sqrt{29}} \mathbf{j} \right)$$

$$= \frac{-4 \cdot 2 + 8 \cdot 5}{\sqrt{29}} = \frac{32}{\sqrt{29}}$$

■

The gradient vector $\nabla f(2, -1)$ in Example 3 is shown in Figure 4 with initial point $(2, -1)$. Also shown is the vector $\mathbf{v}$ that gives the direction of the directional derivative. Both of these vectors are superimposed on a contour plot of the graph of f.

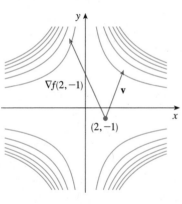

FIGURE 4

For functions of three variables we can define directional derivatives in a similar manner. Again $D_{\mathbf{u}} f(x, y, z)$ can be interpreted as the rate of change of the function in the direction of a unit vector $\mathbf{u}$.

(10) **DEFINITION** The **directional derivative** of f at (x_0, y_0, z_0) in the direction of a unit vector $\mathbf{u} = \langle a, b, c \rangle$ is

$$D_{\mathbf{u}} f(x_0, y_0, z_0) = \lim_{h \to 0} \frac{f(x_0 + ha, y_0 + hb, z_0 + hc) - f(x_0, y_0, z_0)}{h}$$

if this limit exists.

If we use vector notation, then we can write both definitions (2 and 10) of the directional derivative in the compact form

(11)
$$D_{\mathbf{u}} f(\mathbf{x}_0) = \lim_{h \to 0} \frac{f(\mathbf{x}_0 + h\mathbf{u}) - f(\mathbf{x}_0)}{h}$$

where $\mathbf{x}_0 = \langle x_0, y_0 \rangle$ if $n = 2$ and $\mathbf{x}_0 = \langle x_0, y_0, z_0 \rangle$ if $n = 3$. This is reasonable because the vector equation of the line through $\mathbf{x}_0$ in the direction of the vector $\mathbf{u}$ is given by $\mathbf{x} = \mathbf{x}_0 + t\mathbf{u}$ (Equation 11.5.1) and so $f(\mathbf{x}_0 + h\mathbf{u})$ represents the value of f at a point on this line.

If $f(x, y, z)$ is differentiable and $\mathbf{u} = \langle a, b, c \rangle$, then the same method that was used to prove Theorem 3 can be used to show that

(12)
$$D_{\mathbf{u}} f(x, y, z) = f_x(x, y, z)a + f_y(x, y, z)b + f_z(x, y, z)c$$

For a function f of three variables, the **gradient vector,** denoted by ∇f or **grad** f, is

$$\nabla f(x, y, z) = \langle f_x(x, y, z), f_y(x, y, z), f_z(x, y, z) \rangle$$

or, for short,

(13)
$$\nabla f = \langle f_x, f_y, f_z \rangle = \frac{\partial f}{\partial x} \mathbf{i} + \frac{\partial f}{\partial y} \mathbf{j} + \frac{\partial f}{\partial z} \mathbf{k}$$

Then, just as with functions of two variables, Formula 12 for the directional derivative can be rewritten as

(14)
$$D_{\mathbf{u}} f(x, y, z) = \nabla f(x, y, z) \cdot \mathbf{u}$$

EXAMPLE 4   If $f(x, y, z) = x \sin yz$, (a) find the gradient of f and (b) find the directional derivative of f at $(1, 3, 0)$ in the direction of $\mathbf{v} = \mathbf{i} + 2\mathbf{j} - \mathbf{k}$.

SOLUTION
(a) The gradient of f is

$$\nabla f(x, y, z) = \langle f_x(x, y, z), f_y(x, y, z), f_z(x, y, z) \rangle$$
$$= \langle \sin yz, xz \cos yz, xy \cos yz \rangle$$

(b) At $(1, 3, 0)$ we have $\nabla f(1, 3, 0) = \langle 0, 0, 3 \rangle$. The unit vector in the direction of $\mathbf{v} = \mathbf{i} + 2\mathbf{j} - \mathbf{k}$ is

$$\mathbf{u} = \frac{1}{\sqrt{6}} \mathbf{i} + \frac{2}{\sqrt{6}} \mathbf{j} - \frac{1}{\sqrt{6}} \mathbf{k}$$

Therefore, Equation 14 gives

$$D_{\mathbf{u}} f(1, 3, 0) = \nabla f(1, 3, 0) \cdot \mathbf{u}$$
$$= 3\mathbf{k} \cdot \left(\frac{1}{\sqrt{6}} \mathbf{i} + \frac{2}{\sqrt{6}} \mathbf{j} - \frac{1}{\sqrt{6}} \mathbf{k} \right)$$
$$= 3 \left(-\frac{1}{\sqrt{6}} \right) = -\sqrt{\frac{3}{2}}$$

Suppose we have a function f of two or three variables and we consider all possible directional derivatives of f at a given point. These give the rates of change of f in all possible directions. We can then ask the questions: In which of these directions does f change fastest and what is the maximum rate of change? The answers are provided by the following theorem.

> **(15) THEOREM** Suppose f is a differentiable function of two or three variables. The maximum value of the directional derivative $D_{\mathbf{u}} f(\mathbf{x})$ is $|\nabla f(\mathbf{x})|$ and it occurs when $\mathbf{u}$ has the same direction as the gradient vector $\nabla f(\mathbf{x})$.

PROOF From Equation 9 or 14 we have

$$\begin{aligned} D_{\mathbf{u}} f &= \nabla f \cdot \mathbf{u} \\ &= |\nabla f||\mathbf{u}|\cos\theta \qquad \text{(from Theorem 11.3.3)} \\ &= |\nabla f|\cos\theta \end{aligned}$$

where θ is the angle between ∇f and $\mathbf{u}$. The maximum value of $\cos\theta$ is 1 and this occurs when $\theta = 0$. Therefore, the maximum value of $D_{\mathbf{u}} f$ is $|\nabla f|$ and it occurs when $\theta = 0$, that is, when $\mathbf{u}$ has the same direction as ∇f. □

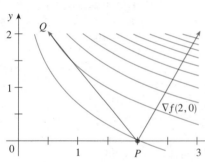

FIGURE 5

At (2, 0) the function in Example 5 increases fastest in the direction of the gradient vector $\nabla f(2,0) = \langle 1, 2\rangle$. Notice from Figure 5 that this vector appears to be perpendicular to the level curve through (2, 0). Figure 6 shows the graph of f and the gradient vector.

EXAMPLE 5

(a) If $f(x, y) = xe^y$, find the rate of change of f at the point $P(2, 0)$ in the direction from P to $Q(\frac{1}{2}, 2)$.

(b) In what direction does f have the maximum rate of change? What is this maximum rate of change?

SOLUTION

(a) We first compute the gradient vector:

$$\nabla f(x, y) = \langle f_x, f_y\rangle = \langle e^y, xe^y\rangle$$

$$\nabla f(2, 0) = \langle 1, 2\rangle$$

The unit vector in the direction of $\overrightarrow{PQ} = \langle -1.5, 2\rangle$ is $\mathbf{u} = \langle -\frac{3}{5}, \frac{4}{5}\rangle$, so the rate of change of f in the direction from P to Q is

$$\begin{aligned} D_{\mathbf{u}} f(2, 0) &= \nabla f(2, 0) \cdot \mathbf{u} = \langle 1, 2\rangle \cdot \langle -\tfrac{3}{5}, \tfrac{4}{5}\rangle \\ &= 1(-\tfrac{3}{5}) + 2(\tfrac{4}{5}) = 1 \end{aligned}$$

(b) According to Theorem 15, f increases fastest in the direction of the gradient vector $\nabla f(2, 0) = \langle 1, 2\rangle$. The maximum rate of change is

$$|\nabla f(2, 0)| = |\langle 1, 2\rangle| = \sqrt{5}$$ ■

FIGURE 6

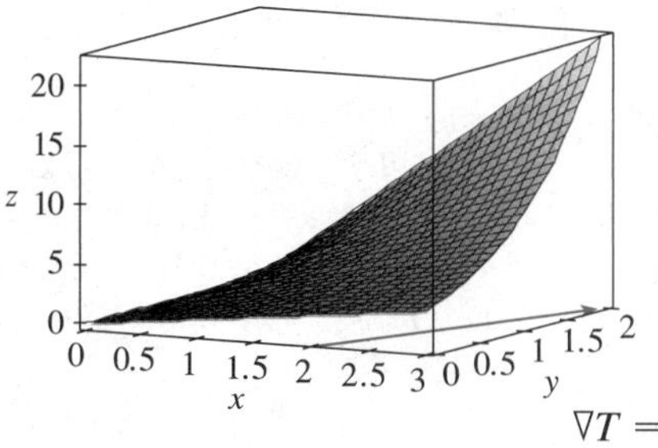

EXAMPLE 6 Suppose that the temperature at a point (x, y, z) in space is given by $T(x, y, z) = 80/(1 + x^2 + 2y^2 + 3z^2)$, where T is measured in °C and x, y, z in meters. In which direction does the temperature increase fastest at the point $(1, 1, -2)$? What is the maximum rate of increase?

SOLUTION The gradient of T is

$$\begin{aligned} \nabla T &= \frac{\partial T}{\partial x}\mathbf{i} + \frac{\partial T}{\partial y}\mathbf{j} + \frac{\partial T}{\partial z}\mathbf{k} \\ &= -\frac{160x}{(1 + x^2 + 2y^2 + 3z^2)^2}\mathbf{i} - \frac{320y}{(1 + x^2 + 2y^2 + 3z^2)^2}\mathbf{j} - \frac{480z}{(1 + x^2 + 2y^2 + 3z^2)^2}\mathbf{k} \end{aligned}$$

$$= \frac{160}{(1 + x^2 + 2y^2 + 3z^2)^2}(-x\,\mathbf{i} - 2y\,\mathbf{j} - 3z\,\mathbf{k})$$

At the point $(1, 1, -2)$ the gradient vector is

$$\nabla T(1, 1, -2) = \tfrac{160}{256}(-\mathbf{i} - 2\,\mathbf{j} + 6\,\mathbf{k})$$

$$= \tfrac{5}{8}(-\mathbf{i} - 2\,\mathbf{j} + 6\,\mathbf{k})$$

By Theorem 15 the temperature increases fastest in the direction of the gradient vector $\nabla T(1, 1, -2) = \tfrac{5}{8}(-\mathbf{i} - 2\,\mathbf{j} + 6\,\mathbf{k})$ or, equivalently, in the direction of $-\mathbf{i} - 2\,\mathbf{j} + 6\,\mathbf{k}$ or the unit vector $(-\mathbf{i} - 2\,\mathbf{j} + 6\,\mathbf{k})/\sqrt{41}$. The maximum rate of increase is the length of the gradient vector:

$$|\nabla T(1, 1, -2)| = \tfrac{5}{8}|-\mathbf{i} - 2\,\mathbf{j} + 6\,\mathbf{k}|$$

$$= \frac{5\sqrt{41}}{8}$$

Therefore, the maximum rate of increase of temperature is $5\sqrt{41}/8 \approx 4\,°C/m$.   ∎

TANGENT PLANES TO LEVEL SURFACES

Suppose S is a surface with equation $F(x, y, z) = k$, that is, it is a level surface of a function F of three variables, and let $P(x_0, y_0, z_0)$ be a point on S. Let C be any curve that lies on the surface S and passes through the point P. Recall from Section 11.7 that C is described by a continuous vector function $\mathbf{r}(t) = \langle x(t), y(t), z(t) \rangle$. Let t_0 be the parameter value corresponding to P; that is, $\mathbf{r}(t_0) = \langle x_0, y_0, z_0 \rangle$. Since C lies on S, any point $(x(t), y(t), z(t))$ must satisfy the equation of S, that is,

(16) $$F(x(t), y(t), z(t)) = k$$

If x, y, and z are differentiable functions of t and F is also differentiable, then we can use the Chain Rule to differentiate both sides of Equation 16 as follows:

(17) $$\frac{\partial F}{\partial x}\frac{dx}{dt} + \frac{\partial F}{\partial y}\frac{dy}{dt} + \frac{\partial F}{\partial z}\frac{dz}{dt} = 0$$

But, since $\nabla F = \langle F_x, F_y, F_z \rangle$ and $\mathbf{r}'(t) = \langle x'(t), y'(t), z'(t) \rangle$, Equation 17 can be written in terms of a dot product as

$$\nabla F \cdot \mathbf{r}'(t) = 0$$

In particular, when $t = t_0$ we have $\mathbf{r}(t_0) = \langle x_0, y_0, z_0 \rangle$, so

(18) $$\nabla F(x_0, y_0, z_0) \cdot \mathbf{r}'(t_0) = 0$$

Equation 18 says that *the gradient vector at P, $\nabla F(x_0, y_0, z_0)$, is perpendicular to the tangent vector $\mathbf{r}'(t_0)$ to any curve C on S that passes through P* (see Figure 7). If $\nabla F(x_0, y_0, z_0) \neq \mathbf{0}$, it is therefore natural to define the **tangent plane to the level surface** $F(x, y, z) = k$ **at** $P(x_0, y_0, z_0)$ as the plane that passes through P and has normal vector $\nabla F(x_0, y_0, z_0)$. Using the standard equation of a plane (11.5.6), we can write the equation of this tangent plane as

(19) $$F_x(x_0, y_0, z_0)(x - x_0) + F_y(x_0, y_0, z_0)(y - y_0) + F_z(x_0, y_0, z_0)(z - z_0) = 0$$

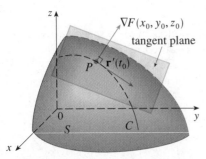

FIGURE 7

The **normal line** to S at P is the line passing through P and perpendicular to the tangent plane. Its direction is therefore given by the gradient vector $\nabla F(x_0, y_0, z_0)$ and so, by Equation 11.5.3, its symmetric equations are

(20)
$$\frac{x - x_0}{F_x(x_0, y_0, z_0)} = \frac{y - y_0}{F_y(x_0, y_0, z_0)} = \frac{z - z_0}{F_z(x_0, y_0, z_0)}$$

In the special case in which the equation of a surface S is of the form $z = f(x, y)$ (that is, S is the graph of a function f of two variables), we can rewrite it as

$$F(x, y, z) = f(x, y) - z = 0$$

and regard S as a level surface (with $k = 0$) of F. Then

$$F_x(x_0, y_0, z_0) = f_x(x_0, y_0) \qquad F_y(x_0, y_0, z_0) = f_y(x_0, y_0) \qquad F_z(x_0, y_0, z_0) = -1$$

so Equation 19 becomes

$$f_x(x_0, y_0)(x - x_0) + f_y(x_0, y_0)(y - y_0) - (z - z_0) = 0$$

which is equivalent to (12.4.2). Thus our new, more general definition of a tangent plane is consistent with the definition that was given for the special case of Section 12.4.

EXAMPLE 7   Find the equations of the tangent plane and normal line at the point $(-2, 1, -3)$ to the ellipsoid

$$\frac{x^2}{4} + y^2 + \frac{z^2}{9} = 3$$

Figure 8 shows the ellipsoid, tangent plane, and normal line in Example 7.

SOLUTION   The ellipsoid is the level surface (with $k = 3$) of the function

$$F(x, y, z) = \frac{x^2}{4} + y^2 + \frac{z^2}{9}$$

Therefore, we have

$$F_x(x, y, z) = \frac{x}{2} \qquad\qquad F_y(x, y, z) = 2y \qquad\qquad F_z(x, y, z) = \frac{2z}{9}$$

$$F_x(-2, 1, -3) = -1 \qquad F_y(-2, 1, -3) = 2 \qquad F_z(-2, 1, -3) = -\tfrac{2}{3}$$

Then Equation 19 gives the equation of the tangent plane at $(-2, 1, -3)$ as

$$-1(x + 2) + 2(y - 1) - \tfrac{2}{3}(z + 3) = 0$$

which simplifies to $3x - 6y + 2z + 18 = 0$.

By Equation 20, symmetric equations of the normal line are

$$\frac{x + 2}{-1} = \frac{y - 1}{2} = \frac{z + 3}{-\tfrac{2}{3}}$$

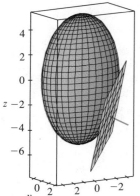

FIGURE 8

SIGNIFICANCE OF THE GRADIENT VECTOR

Let us summarize the ways in which the gradient vector is significant. We first consider a function f of three variables and a point $P(x_0, y_0, z_0)$ in its domain. On the one hand, we know from Theorem 15 that the gradient vector $\nabla f(x_0, y_0, z_0)$ gives the direction of fastest increase of f. On the other hand, we know that $\nabla f(x_0, y_0, z_0)$ is orthogonal to the level surface S of f through P. (Refer to Figure 7.) These two properties are quite compatible intuitively because as we move away from P on the level surface S, the value of f does not change at all. So it seems reasonable that if we move in the perpendicular direction, we get the maximum increase.

In like manner we consider a function f of two variables and a point $P(x_0, y_0)$ in its domain. Again the gradient vector $\nabla f(x_0, y_0)$ gives the direction of fastest increase of f. Also, by considerations similar to our discussion of tangent planes, it can be shown that $\nabla f(x_0, y_0)$ is perpendicular to the level curve $f(x, y) = k$ that passes through P. Again this is intuitively plausible because the values of f remain constant as we move along the curve. (See Figure 9.)

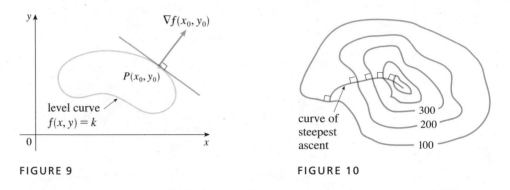

FIGURE 9

FIGURE 10

If we consider a topographical map of a hill and let $f(x, y)$ represent the height above sea level at a point with coordinates (x, y), then a curve of steepest ascent can be drawn as in Figure 10 by making it perpendicular to all of the contour lines. This phenomenon can also be noticed in Figure 10 in Section 12.1, where Lonesome Creek follows a curve of steepest descent.

EXERCISES 12.6

1–4 ■ Find the directional derivative of f at the given point in the direction indicated by the angle θ.

1. $f(x, y) = x^2y^3 + 2x^4y$, $(1, -2)$, $\theta = \pi/3$

2. $f(x, y) = (x^2 - y)^3$, $(3, 1)$, $\theta = 3\pi/4$

3. $f(x, y) = y^x$, $(1, 2)$, $\theta = \pi/2$

4. $f(x, y) = \sin(x + 2y)$, $(4, -2)$, $\theta = -2\pi/3$

5–8 ■
(a) Find the gradient of f.
(b) Evaluate the gradient at the point P.
(c) Find the rate of change of f at P in the direction of the vector $\mathbf{u}$.

5. $f(x, y) = x^3 - 4x^2y + y^2$, $P(0, -1)$, $\mathbf{u} = \langle \frac{3}{5}, \frac{4}{5} \rangle$

6. $f(x, y) = e^x \sin y$, $P(1, \pi/4)$, $\mathbf{u} = \left\langle \dfrac{-1}{\sqrt{5}}, \dfrac{2}{\sqrt{5}} \right\rangle$

7. $f(x, y, z) = xy^2z^3$, $P(1, -2, 1)$, $\mathbf{u} = \left\langle \dfrac{1}{\sqrt{3}}, \dfrac{-1}{\sqrt{3}}, \dfrac{1}{\sqrt{3}} \right\rangle$

8. $f(x, y, z) = xy + yz^2 + xz^3$, $P(2, 0, 3)$, $\mathbf{u} = \langle -\frac{2}{3}, -\frac{1}{3}, \frac{2}{3} \rangle$

9–16 ■ Find the directional derivative of the function at the given point in the direction of the vector $\mathbf{v}$.

9. $f(x, y) = \sqrt{x - y}$, $(5, 1)$, $\mathbf{v} = \langle 12, 5 \rangle$

10. $f(x, y) = x/y$, $(6, -2)$, $\mathbf{v} = \langle -1, 3 \rangle$

11. $g(x, y) = xe^{xy}$, $(-3, 0)$, $\mathbf{v} = 2\mathbf{i} + 3\mathbf{j}$

12. $g(x, y) = e^x \cos y$, $(1, \pi/6)$, $\mathbf{v} = \mathbf{i} - \mathbf{j}$

13. $f(x, y, z) = \sqrt{xyz}$, $(2, 4, 2)$, $\mathbf{v} = \langle 4, 2, -4 \rangle$

14. $f(x, y, z) = z^3 - x^2y$, $(1, 6, 2)$, $\mathbf{v} = \langle 3, 4, 12 \rangle$

15. $g(x, y, z) = x \tan^{-1}(y/z)$, $(1, 2, -2)$, $\mathbf{v} = \mathbf{i} + \mathbf{j} - \mathbf{k}$

16. $g(x, y, z) = xe^{yz} + xye^z$, $(-2, 1, 1)$, $\mathbf{v} = \mathbf{i} - 2\mathbf{j} + 3\mathbf{k}$

17–22 ■ Find the maximum rate of change of f at the given point and the direction in which it occurs.

17. $f(x, y) = xe^{-y} + 3y$, $(1, 0)$

18. $f(x, y) = \ln(x^2 + y^2)$, $(1, 2)$

19. $f(x, y) = \sqrt{x^2 + 2y}$, $(4, 10)$

20. $f(x, y, z) = x + y/z$, $(4, 3, -1)$

21. $f(x, y) = \cos(3x + 2y)$, $(\pi/6, -\pi/8)$

22. $f(x, y, z) = \dfrac{x}{y} + \dfrac{y}{z}$, $(4, 2, 1)$

23. Show that a differentiable function f decreases most rapidly at $\mathbf{x}$ in the direction opposite to the gradient vector, that is, in the direction of $-\nabla f(\mathbf{x})$.

24. Use the result of Exercise 23 to find the direction in which the function $f(x, y) = x^4 y - x^2 y^3$ decreases fastest at the point $(2, -3)$.

25. The temperature T in a metal ball is inversely proportional to the distance from the center of the ball, which we take to be the origin. The temperature at the point $(1, 2, 2)$ is $120°$.
   (a) Find the rate of change of T at $(1, 2, 2)$ in the direction toward the point $(2, 1, 3)$.
   (b) Show that at any point in the ball the direction of greatest increase in temperature is given by a vector that points toward the origin.

26. The temperature at a point (x, y, z) is given by
$$T(x, y, z) = 200 e^{-x^2 - 3y^2 - 9z^2}$$
   where T is measured in °C and x, y, z in meters.
   (a) Find the rate of change of temperature at the point $P(2, -1, 2)$ in the direction toward the point $(3, -3, 3)$.
   (b) In which direction does the temperature increase fastest at P?
   (c) Find the maximum rate of increase at P.

27. Suppose that over a certain region of space the electrical potential V is given by $V(x, y, z) = 5x^2 - 3xy + xyz$.
   (a) Find the rate of change of the potential at $P(3, 4, 5)$ in the direction of the vector $\mathbf{v} = \mathbf{i} + \mathbf{j} - \mathbf{k}$.
   (b) In which direction does V change most rapidly at P?
   (c) What is the maximum rate of change at P?

28. Suppose that you are climbing a hill whose shape is given by the equation $z = 1000 - 0.01x^2 - 0.02y^2$ and you are standing at a point with coordinates $(60, 100, 764)$.
   (a) In which direction should you proceed initially in order to reach the top of the hill fastest?
   (b) If you climb in that direction, at what angle above the horizontal will you be climbing initially?

29. Let f be a function of two variables that has continuous partial derivatives and consider the points $A(1, 3)$, $B(3, 3)$, $C(1, 7)$, and $D(6, 15)$. The directional derivative of f at A in the direction of the vector $\overrightarrow{AB}$ is 3 and the directional derivative at A in the direction of $\overrightarrow{AC}$ is 26. Find the directional derivative of f at A in the direction of the vector $\overrightarrow{AD}$.

30. For the given contour map draw the curves of steepest ascent starting at P and at Q.

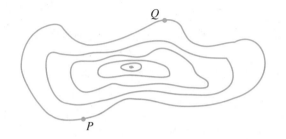

31–34 ■ Show that the operation of taking the gradient of a function has the given property. Assume that u and v are differentiable functions of x and y and a, b are constants.

31. $\nabla(au + bv) = a\nabla u + b\nabla v$ **32.** $\nabla(uv) = u\nabla v + v\nabla u$

33. $\nabla\left(\dfrac{u}{v}\right) = \dfrac{v\nabla u - u\nabla v}{v^2}$ **34.** $\nabla u^n = nu^{n-1}\nabla u$

35–40 ■ Find equations of (a) the tangent plane and (b) the normal line to the given surface at the specified point.

35. $4x^2 + y^2 + z^2 = 24$, $(2, 2, 2)$

36. $x^2 - 2y^2 + z^2 = 3$, $(-1, 1, -2)$

37. $x^2 + y^2 - z^2 - 2xy + 4xz = 4$, $(1, 0, 1)$

38. $x^2 - 2y^2 - 3z^2 + xyz = 4$, $(3, -2, -1)$

39. $z + 1 = xe^y \cos z$, $(1, 0, 0)$

40. $xe^{yz} = 1$, $(1, 0, 5)$

41–42 ■ Use a computer to graph the surface, the tangent plane, and the normal line on the same screen. Choose the domain carefully so that you avoid extraneous vertical planes. Choose the viewpoint so that you get a good view of all three objects.

41. $xy + yz + zx = 3$, $(1, 1, 1)$

42. $xyz = 6$, $(1, 2, 3)$

43. If $f(x, y) = x^2 + 4y^2$, find the gradient vector $\nabla f(2, 1)$ and use it to find the tangent line to the level curve $f(x, y) = 8$ at the point $(2, 1)$. Sketch the level curve, the tangent line, and the gradient vector.

44. If $g(x, y) = x - y^2$, find the gradient vector $\nabla g(3, -1)$ and use it to find the tangent line to the level curve $g(x, y) = 2$ at the point $(3, -1)$. Sketch the level curve, the tangent line, and the gradient vector.

45. Show that the equation of the tangent plane to the ellipsoid $x^2/a^2 + y^2/b^2 + z^2/c^2 = 1$ at the point (x_0, y_0, z_0) can be written as
$$\frac{xx_0}{a^2} + \frac{yy_0}{b^2} + \frac{zz_0}{c^2} = 1$$

46. Find the equation of the tangent plane to the hyperboloid $x^2/a^2 + y^2/b^2 - z^2/c^2 = 1$ at (x_0, y_0, z_0) and express it in a form similar to the one in Exercise 45.

47. Show that the equation of the tangent plane to the elliptic paraboloid $z/c = x^2/a^2 + y^2/b^2$ at the point (x_0, y_0, z_0) can be written as

$$\frac{2xx_0}{a^2} + \frac{2yy_0}{b^2} = \frac{z + z_0}{c}$$

48. Find the points on the ellipsoid $x^2 + 2y^2 + 3z^2 = 1$ where the tangent plane is parallel to the plane $3x - y + 3z = 1$.

49. Find the points on the hyperboloid $x^2 - y^2 + 2z^2 = 1$ where the normal line is parallel to the line that joins the points $(3, -1, 0)$ and $(5, 3, 6)$.

50. Show that the ellipsoid $3x^2 + 2y^2 + z^2 = 9$ and the sphere $x^2 + y^2 + z^2 - 8x - 6y - 8z + 24 = 0$ are tangent to each other at the point $(1, 1, 2)$. (This means that they have a common tangent plane at the point.)

51. Show that every plane that is tangent to the cone $x^2 + y^2 = z^2$ passes through the origin.

52. Show that every normal line to the sphere $x^2 + y^2 + z^2 = r^2$ passes through the center of the sphere.

53. Show that the sum of the x-, y-, and z-intercepts of any tangent plane to the surface $\sqrt{x} + \sqrt{y} + \sqrt{z} = \sqrt{c}$ is a constant.

54. Show that the product of the x-, y-, and z-intercepts of any tangent plane to the surface $xyz = c^3$ is a constant.

55. Find parametric equations for the tangent line to the curve of intersection of the paraboloid $z = x^2 + y^2$ and the ellipsoid $4x^2 + y^2 + z^2 = 9$ at the point $(-1, 1, 2)$.

56. (a) The plane $y + z = 3$ intersects the cylinder $x^2 + y^2 = 5$ in an ellipse. Find parametric equations for the tangent line to this ellipse at the point $(1, 2, 1)$.
   (b) Use a computer to graph the cylinder, the plane, and the tangent line on the same screen.

57. (a) Two surfaces are called **orthogonal** at a point of intersection if their normal lines are perpendicular at that point. Show that surfaces with equations $F(x, y, z) = 0$ and $G(x, y, z) = 0$ are orthogonal at a point P where $\nabla F \neq \mathbf{0}$ and $\nabla G \neq \mathbf{0}$ if and only if

$$F_x G_x + F_y G_y + F_z G_z = 0$$

   at P.
   (b) Use part (a) to show that the surfaces $z^2 = x^2 + y^2$ and $x^2 + y^2 + z^2 = r^2$ are orthogonal at every point of intersection. Can you see why this is true without using calculus?

58. (a) Show that the function $f(x, y) = \sqrt[3]{xy}$ is continuous and the partial derivatives f_x and f_y exist at the origin but the directional derivatives in all other directions do not exist.
   (b) Use a computer to graph f near the origin and comment on how the graph confirms part (a).

59. Suppose that the directional derivatives of $f(x, y)$ are known at a given point in two nonparallel directions given by unit vectors $\mathbf{u}$ and $\mathbf{v}$. Is it possible to find ∇f at this point? If so, how would you do it?

60. Show that if $z = f(x, y)$ is differentiable at $\mathbf{x}_0 = \langle x_0, y_0 \rangle$, then

$$\lim_{\mathbf{x} \to \mathbf{x}_0} \frac{f(\mathbf{x}) - f(\mathbf{x}_0) - \nabla f(\mathbf{x}_0) \cdot (\mathbf{x} - \mathbf{x}_0)}{|\mathbf{x} - \mathbf{x}_0|} = 0$$

[*Hint:* Use Definition 12.4.12 directly.]

12.7   MAXIMUM AND MINIMUM VALUES

As we saw in Chapter 3, one of the main uses of ordinary derivatives is in finding maximum and minimum values. In this section we see how to use partial derivatives to locate maxima and minima of functions of two variables.

> **(1) DEFINITION**   A function of two variables has a **local maximum** at (a, b) if $f(x, y) \leq f(a, b)$ for all points (x, y) in some disk with center (a, b). The number $f(a, b)$ is called a **local maximum value**. If $f(x, y) \geq f(a, b)$ for all (x, y) in such a disk, $f(a, b)$ is a **local minimum value**.

If the inequalities in Definition 1 hold for *all* points (x, y) in the domain of f, then f has an **absolute maximum** (or **absolute minimum**) at (a, b).

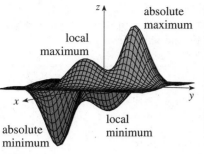

FIGURE 1

The graph of a function with several maxima and minima is shown in Figure 1. You can think of the local maxima as mountain peaks and the local minima as valley bottoms.

> **(2) THEOREM** If f has a local extremum (that is, a local maximum or minimum) at (a, b) and the first-order partial derivatives of f exist there, then $f_x(a, b) = 0$ and $f_y(a, b) = 0$.

PROOF Let $g(x) = f(x, b)$. If f has a local extremum at (a, b), then g has a local extremum at a, so $g'(a) = 0$ by Fermat's Theorem (3.1.4). But, by Equation 12.3.1, $g'(a) = f_x(a, b)$ and so $f_x(a, b) = 0$. Similarly, by applying Fermat's Theorem to the function $G(y) = f(a, y)$, we obtain $f_y(a, b) = 0$. ☐

If we put $f_x(a, b) = 0$ and $f_y(a, b) = 0$ in the equation of a tangent plane (12.4.2), we get $z = z_0$. Thus the geometric interpretation of Theorem 2 is that if the graph of f has a tangent plane at a local extremum, then the tangent plane must be horizontal.

A point (a, b) such that $f_x(a, b) = 0$ and $f_y(a, b) = 0$, or one of these partial derivatives does not exist, is called a **critical point** (or *stationary point*) of f. Theorem 2 says that if f has a local extremum at (a, b), then (a, b) is a critical point of f. However, as in single-variable calculus, not all critical points give rise to extrema. At a critical point, a function could have a local maximum or a local minimum or neither.

EXAMPLE 1 Let $f(x, y) = x^2 + y^2 - 2x - 6y + 14$. Then

$$f_x(x, y) = 2x - 2 \qquad f_y(x, y) = 2y - 6$$

These partial derivatives are equal to 0 when $x = 1$ and $y = 3$, so the only critical point is $(1, 3)$. By completing the square, we find that

$$f(x, y) = 4 + (x - 1)^2 + (y - 3)^2$$

Since $(x - 1)^2 \geq 0$ and $(y - 3)^2 \geq 0$, we have $f(x, y) \geq 4$ for all values of x and y. Therefore, $f(1, 3) = 4$ is a local minimum, and in fact it is the absolute minimum of f. This can be confirmed geometrically from the graph of f, which is the elliptic paraboloid with vertex $(1, 3, 4)$ shown in Figure 2. ■

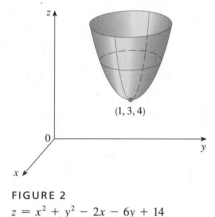

FIGURE 2
$z = x^2 + y^2 - 2x - 6y + 14$

EXAMPLE 2 Find the extreme values of $f(x, y) = y^2 - x^2$.

SOLUTION Since $f_x = -2x$ and $f_y = 2y$, the only critical point is $(0, 0)$. Notice that for points on the x-axis we have $y = 0$, so $f(x, y) = -x^2 < 0$ (if $x \neq 0$). However, for points on the y-axis we have $x = 0$, so $f(x, y) = y^2 > 0$ (if $y \neq 0$). Thus every disk with center $(0, 0)$ contains points where f takes positive values as well as points where f takes negative values. Therefore, $f(0, 0) = 0$ cannot be an extreme value for f, so f has no extreme values. ■

Example 2 illustrates the fact that a function need not have a maximum or minimum value at a critical point. Figure 3 shows how this is possible. The graph of f is the hyperbolic paraboloid $z = y^2 - x^2$, which has a horizontal tangent plane ($z = 0$) at the origin. You can see that $f(0, 0) = 0$ is a maximum in the direction of the x-axis but a minimum in the direction of the y-axis. Near the origin the graph has the shape of a saddle and so $(0, 0)$ is called a *saddle point* of f.

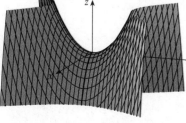

FIGURE 3
$z = y^2 - x^2$

We need to be able to determine whether or not a function has an extreme value at a critical point. The following test, which is proved at the end of this section, is analogous to the Second Derivative Test for functions of one variable.

---

(3) SECOND DERIVATIVES TEST   Suppose the second partial derivatives of f are continuous in a disk with center (a, b), and suppose that $f_x(a, b) = 0$ and $f_y(a, b) = 0$ [that is, (a, b) is a critical point of f]. Let

$$D = D(a, b) = f_{xx}(a, b) f_{yy}(a, b) - [f_{xy}(a, b)]^2$$

(a) If $D > 0$ and $f_{xx}(a, b) > 0$, then $f(a, b)$ is a local minimum.
(b) If $D > 0$ and $f_{xx}(a, b) < 0$, then $f(a, b)$ is a local maximum.
(c) If $D < 0$, then $f(a, b)$ is not a local extremum.

NOTE 1:   In case (c) the point (a, b) is called a **saddle point** of f and the graph of f crosses its tangent plane at (a, b).

NOTE 2:   If $D = 0$, the test gives no information: f could have a local maximum or local minimum at (a, b), or (a, b) could be a saddle point of f.

NOTE 3:   To remember the formula for D it is helpful to write it as a determinant:

$$D = \begin{vmatrix} f_{xx} & f_{xy} \\ f_{yx} & f_{yy} \end{vmatrix} = f_{xx}f_{yy} - (f_{xy})^2$$

EXAMPLE 3   Find the local extrema of $f(x, y) = x^4 + y^4 - 4xy + 1$.

SOLUTION   We first locate the critical points:

$$f_x = 4x^3 - 4y \qquad f_y = 4y^3 - 4x$$

Setting these partial derivatives equal to 0, we obtain the equations

$$x^3 - y = 0 \qquad \text{and} \qquad y^3 - x = 0$$

To solve these equations we substitute $y = x^3$ from the first equation into the second one. This gives

$$0 = x^9 - x = x(x^8 - 1) = x(x^4 - 1)(x^4 + 1) = x(x^2 - 1)(x^2 + 1)(x^4 + 1)$$

so there are three real roots: $x = 0, 1, -1$. The three critical points are $(0, 0)$, $(1, 1)$, and $(-1, -1)$.

Next we calculate the second partial derivatives and $D(x, y)$:

$$f_{xx} = 12x^2 \qquad f_{xy} = -4 \qquad f_{yy} = 12y^2$$

$$D(x, y) = f_{xx}f_{yy} - (f_{xy})^2 = 144x^2y^2 - 16$$

Since $D(0, 0) = -16 < 0$, it follows from case (c) of the Second Derivatives Test that the origin is a saddle point; that is, f has no local extremum at $(0, 0)$. Since $D(1, 1) = 128 > 0$ and $f_{xx}(1, 1) = 12 > 0$, we see from case (a) of the test that $f(1, 1) = -1$ is a local minimum. Similarly, we have $D(-1, -1) = 128 > 0$ and $f_{xx}(-1, -1) = 12 > 0$, so $f(-1, -1) = -1$ is also a local minimum.

The graph of f and its contour map are shown in Figures 4 and 5.   ∎

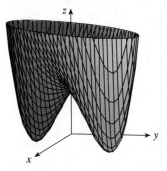

FIGURE 4
$z = x^4 + y^4 - 4xy + 1$

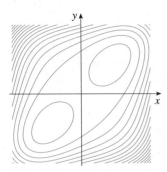

FIGURE 5
Contour map of
$z = x^4 + y^4 - 4xy + 1$

EXAMPLE 4 Find and classify the critical points of the function

$$f(x, y) = 10x^2y - 5x^2 - 4y^2 - x^4 - 2y^4$$

Also find the highest point on the graph of f.

SOLUTION   The first-order partial derivatives are

$$f_x = 20xy - 10x - 4x^3 \qquad f_y = 10x^2 - 8y - 8y^3$$

So to find the critical points we need to solve the equations

(4) $$2x(10y - 5 - 2x^2) = 0$$

(5) $$5x^2 - 4y - 4y^3 = 0$$

From Equation 4 we see that either

$$x = 0 \qquad \text{or} \qquad 10y - 5 - 2x^2 = 0$$

In the first case ($x = 0$), Equation 5 becomes $-4y(1 + y^2) = 0$, so $y = 0$ and we have the critical point $(0, 0)$.

In the second case ($10y - 5 - 2x^2 = 0$), we get

(6) $$x^2 = 5y - 2.5$$

and, putting this in Equation 5, we have $25y - 12.5 - 4y - 4y^3 = 0$. So we have to solve the cubic equation

(7) $$4y^3 - 21y + 12.5 = 0$$

Using a graphing calculator or computer to graph the function

$$g(y) = 4y^3 - 21y + 12.5$$

as in Figure 6, we see that Equation 7 has three real roots. By zooming in, we can find the roots to four decimal places:

$$y \approx -2.5452 \qquad y \approx 0.6468 \qquad y \approx 1.8984$$

(Alternatively, we could have used Newton's method or a computer algebra system to locate these roots.) From Equation 6, the corresponding x-values are given by

$$x = \pm\sqrt{5y - 2.5}$$

If $y \approx -2.5452$, then there is no corresponding real value of x. If $y \approx 0.6468$, then $x \approx \pm 0.8567$. If $y \approx 1.8984$, then $x \approx \pm 2.6442$. So we have a total of five critical points, which are analyzed in the following chart. All quantities are rounded to two decimal places.

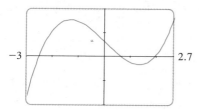

FIGURE 6

Critical point	Value of f	f_{xx}	D	Conclusion
$(0, 0)$	0.00	-10.00	80.00	local maximum
$(\pm 2.64, 1.90)$	8.50	-55.93	2488.71	local maximum
$(\pm 0.86, 0.65)$	-1.48	-5.87	-187.64	saddle point

-3 2.7

The five critical points are shown in blue in Figure 7. Notice that the level curves have shapes similar to ellipses near the maximum points and similar to hyperbolas near the saddle points.

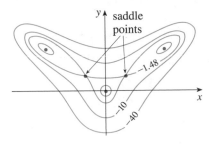

FIGURE 7

Figure 7 shows the location of the critical points on a contour map of f. Figures 8 and 9 give two views of the graph of f and we see that the surface opens downward. [This can also be seen from the expression for $f(x, y)$: the dominant terms are $-x^2 - 2y^4$ when $|x|$ and $|y|$ are large.] Comparing the values of f at its local maximum points, we see that the absolute maximum value of f is $f(\pm 2.64, 1.90) \approx 8.50$. In other words, the highest points on the graph of f are $(\pm 2.64, 1.90, 8.50)$.

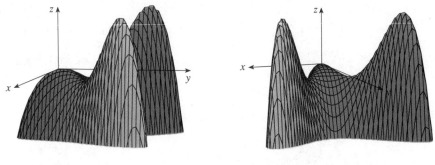

FIGURE 8   **FIGURE 9**   ■

EXAMPLE 5 Find the shortest distance from the point $(1, 0, -2)$ to the plane $x + 2y + z = 4$.

SOLUTION The distance from any point (x, y, z) to the point $(1, 0, -2)$ is

$$d = \sqrt{(x - 1)^2 + y^2 + (z + 2)^2}$$

but if (x, y, z) lies on the plane $x + 2y + z = 4$, then $z = 4 - x - 2y$ and so we have $d = \sqrt{(x - 1)^2 + y^2 + (6 - x - 2y)^2}$. We can minimize d by minimizing the simpler expression

$$d^2 = f(x, y) = (x - 1)^2 + y^2 + (6 - x - 2y)^2$$

By solving the equations

$$f_x = 2(x - 1) - 2(6 - x - 2y) = 4x + 4y - 14 = 0$$

$$f_y = 2y - 4(6 - x - 2y) = 4x + 10y - 24 = 0$$

we find that the only critical point is $\left(\frac{11}{6}, \frac{5}{3}\right)$. Since $f_{xx} = 4$, $f_{xy} = 4$, and $f_{yy} = 10$, we have $D(x, y) = f_{xx} f_{yy} - (f_{xy})^2 = 24 > 0$ and $f_{xx} > 0$, so by the Second Derivatives Test f has a local minimum at $\left(\frac{11}{6}, \frac{5}{3}\right)$. Intuitively, we can see that this local minimum is actually an absolute minimum because there must be a point on the given plane that is closest to $(1, 0, -2)$. If $x = \frac{11}{6}$ and $y = \frac{5}{3}$, then

$$d = \sqrt{(x - 1)^2 + y^2 + (6 - x - 2y)^2} = \sqrt{\left(\tfrac{5}{6}\right)^2 + \left(\tfrac{5}{3}\right)^2 + \left(\tfrac{5}{6}\right)^2} = \frac{5\sqrt{6}}{6}$$

Example 5 could also be solved using vectors. Compare with the methods of Section 11.5.

The shortest distance from $(1, 0, -2)$ to the plane $x + 2y + z = 4$ is $5\sqrt{6}/6$.   ■

EXAMPLE 6 A rectangular box without a lid is to be made from 12 m^2 of cardboard. Find the maximum volume of such a box.

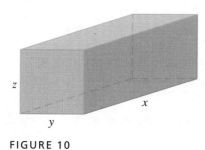

FIGURE 10

SOLUTION   Let the length, width, and height of the box (in meters) be x, y, and z, as shown in Figure 10. Then the volume of the box is

$$V = xyz$$

We can express V as a function of just two variables x and y by using the fact that the area of the four sides and the bottom of the box is

$$2xz + 2yz + xy = 12$$

Solving this equation for z, we get $z = (12 - xy)/[2(x + y)]$, so the expression for V becomes

$$V = xy\frac{12 - xy}{2(x + y)} = \frac{12xy - x^2y^2}{2(x + y)}$$

We compute the partial derivatives:

$$\frac{\partial V}{\partial x} = \frac{y^2(12 - 2xy - x^2)}{2(x + y)^2} \qquad \frac{\partial V}{\partial y} = \frac{x^2(12 - 2xy - y^2)}{2(x + y)^2}$$

If V is a maximum, then $\partial V/\partial x = \partial V/\partial y = 0$, but $x = 0$ or $y = 0$ gives $V = 0$, so we must solve the equations

$$12 - 2xy - x^2 = 0 \qquad 12 - 2xy - y^2 = 0$$

These imply that $x^2 = y^2$ and so $x = y$. (Note that x and y must both be positive in this problem.) If we put $x = y$ in either equation we get $12 - 3x^2 = 0$, which gives $x = 2$, $y = 2$, and $z = (12 - 2 \cdot 2)/[2(2 + 2)] = 1$.

We could use the Second Derivatives Test to show that this gives a local maximum of V, or we could simply argue from the physical nature of this problem that there must be an absolute maximum volume, which has to occur at a critical point of V, so it must occur when $x = 2$, $y = 2$, $z = 1$. Then $V = 2 \cdot 2 \cdot 1 = 4$, so the maximum volume of the box is 4 m³.   ∎

ABSOLUTE MAXIMUM AND MINIMUM VALUES

For a function f of one variable the Extreme Value Theorem says that if f is continuous on a closed interval $[a, b]$, then f has an absolute minimum value and an absolute maximum value. According to the procedure in (3.1.8) we found these by evaluating f not only at the critical numbers but also at the endpoints a and b.

There is a similar situation for functions of two variables. Just as a closed interval contains its endpoints, a **closed set** in $\mathbb{R}^2$ is one that contains all its boundary points. (See the discussion following Definition 12.2.3.) For instance, the disk

$$D = \{(x, y) \mid x^2 + y^2 \le 1\}$$

consisting of all points on and inside the circle $x^2 + y^2 = 1$ is a closed set because it contains all of its boundary points (which are the points on the circle $x^2 + y^2 = 1$). But if even one point on the boundary curve were omitted, the set would not be closed. (See Figure 11.)

A **bounded set** in $\mathbb{R}^2$ is one that is contained within some disk. In other words, it is finite in extent. Then, in terms of closed and bounded sets, we can state the following analogue of the Extreme Value Theorem in two dimensions.

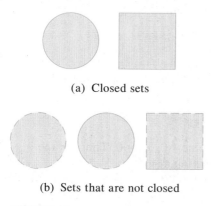

(a) Closed sets

(b) Sets that are not closed

FIGURE 11

> **(8) EXTREME VALUE THEOREM FOR FUNCTIONS OF TWO VARIABLES**
> If f is continuous on a closed, bounded set D in $\mathbb{R}^2$, then f attains an absolute maximum value $f(x_1, y_1)$ and an absolute minimum value $f(x_2, y_2)$ at some points (x_1, y_1) and (x_2, y_2) in D.

To find the extreme values guaranteed by Theorem 8, we note that, by Theorem 2, if f has an extreme value at (x_1, y_1), then (x_1, y_1) is either a critical point of f or a boundary point of D. Thus we have the following analogue of (3.1.8):

> **(9)** To find the absolute maximum and minimum values of a continuous function f on a closed, bounded set D:
> **1.** Find the values of f at the critical points of f in D.
> **2.** Find the extreme values of f on the boundary of D.
> **3.** The largest of the values from steps 1 and 2 is the absolute maximum value; the smallest of these values is the absolute minimum value.

EXAMPLE 7 Find the absolute maximum and minimum values of the function $f(x, y) = x^2 - 2xy + 2y$ on the rectangle $D = \{(x, y) \mid 0 \le x \le 3, 0 \le y \le 2\}$.

SOLUTION Since f is a polynomial, it is continuous on the closed, bounded rectangle D, so Theorem 8 tells us there is both an absolute maximum and an absolute minimum. According to step 1 we first find the critical points. These occur when

$$f_x = 2x - 2y = 0 \qquad f_y = -2x + 2 = 0$$

so the only critical point is $(1, 1)$, and the value of f there is $f(1, 1) = 1$.

In step 2 we look at the values of f on the boundary of D, which consists of the four line segments L_1, L_2, L_3, L_4 shown in Figure 12. On L_1 we have $y = 0$ and

$$f(x, 0) = x^2 \qquad 0 \le x \le 3$$

This is an increasing function of x, so its minimum value is $f(0, 0) = 0$ and its maximum value is $f(3, 0) = 9$. On L_2 we have $x = 3$ and

$$f(3, y) = 9 - 4y \qquad 0 \le y \le 2$$

This is a decreasing function of y, so its maximum value is $f(3, 0) = 9$ and its minimum value is $f(3, 2) = 1$. On L_3 we have $y = 2$ and

$$f(x, 2) = x^2 - 4x + 4 \qquad 0 \le x \le 3$$

By the methods of Chapter 3, or simply by observing that $f(x, 2) = (x - 2)^2$, we see that the minimum value of this function is $f(2, 2) = 0$ and the maximum value is $f(0, 2) = 4$. Finally, on L_4 we have $x = 0$ and

$$f(0, y) = 2y \qquad 0 \le y \le 2$$

with maximum value $f(0, 2) = 4$ and minimum value $f(0, 0) = 0$. Thus, on the boundary, the minimum value of f is 0 and the maximum is 9.

In step 3 we compare these values with the value $f(1, 1) = 1$ at the critical point and conclude that the absolute maximum value of f on D is $f(3, 0) = 9$ and the absolute minimum value is $f(0, 0) = f(2, 2) = 0$. Figure 13 shows the graph of f. ∎

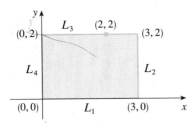

FIGURE 12

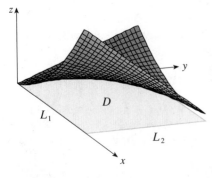

FIGURE 13
$f(x, y) = x^2 - 2xy + 2y$

We close this section by giving a proof of the first part of the Second Derivatives Test. Notice the similarity to the proof of the Second Derivative Test for functions of one variable. Parts (b) and (c) have similar proofs.

PROOF OF THEOREM 3, PART (a) We compute the second-order directional derivative of f in the direction of $\mathbf{u} = \langle h, k \rangle$. The first-order derivative is given by Theorem 12.6.3:

$$D_{\mathbf{u}} f = f_x h + f_y k$$

Applying this theorem a second time, we have

$$D_{\mathbf{u}}^2 f = D_{\mathbf{u}}(D_{\mathbf{u}} f) = \frac{\partial}{\partial x}(D_{\mathbf{u}} f)h + \frac{\partial}{\partial y}(D_{\mathbf{u}} f)k$$
$$= (f_{xx} h + f_{yx} k)h + (f_{xy} h + f_{yy} k)k$$
$$= f_{xx} h^2 + 2f_{xy} hk + f_{yy} k^2 \qquad \text{(by Clairaut's Theorem)}$$

If we complete the square in this expression, we obtain

$$(10) \qquad D_{\mathbf{u}}^2 f = f_{xx}\left(h + \frac{f_{xy}}{f_{xx}}k\right)^2 + \frac{k^2}{f_{xx}}(f_{xx} f_{yy} - f_{xy}^2)$$

We are given that $f_{xx}(a, b) > 0$ and $D(a, b) > 0$. But f_{xx} and $D = f_{xx} f_{yy} - f_{xy}^2$ are continuous functions, so there is a disk B with center (a, b) and radius $\delta > 0$ such that $f_{xx}(x, y) > 0$ and $D(x, y) > 0$ whenever (x, y) is in B. Therefore, by looking at Equation 10, we see that $D_{\mathbf{u}}^2 f(x, y) > 0$ whenever (x, y) is in B. This means that if C is the curve obtained by intersecting the graph of f with the vertical plane through $P(a, b, f(a, b))$ in the direction of $\mathbf{u}$, then C is concave upward on an interval of length 2δ. This is true in the direction of every vector $\mathbf{u}$, so if we restrict (x, y) to lie in B, the graph of f lies above its horizontal tangent plane at P. Thus $f(x, y) \geq f(a, b)$ whenever (x, y) is in B. This shows that $f(a, b)$ is a local minimum. $\quad\square$

EXERCISES 12.7

1–16 ■ Find the local maximum and minimum values and saddle points of the function. If you have three-dimensional graphing software, use a computer to graph the function with a domain and viewpoint that reveal all the important aspects of the function.

1. $f(x, y) = x^2 + y^2 + 4x - 6y$

2. $f(x, y) = 4x^2 + y^2 - 4x + 2y$

3. $f(x, y) = 2x^2 + y^2 + 2xy + 2x + 2y$

4. $f(x, y) = 1 + 2xy - x^2 - y^2$

5. $f(x, y) = x^2 + y^2 + x^2 y + 4$

6. $f(x, y) = 2x^3 + xy^2 + 5x^2 + y^2$

7. $f(x, y) = x^3 - 3xy + y^3$

8. $f(x, y) = y\sqrt{x} - y^2 - x + 6y$

9. $f(x, y) = xy - 2x - y$

10. $f(x, y) = xy(1 - x - y)$

11. $f(x, y) = \dfrac{x^2 y^2 - 8x + y}{xy}$

12. $f(x, y) = x^2 + y^2 + \dfrac{1}{x^2 y^2}$

13. $f(x, y) = e^x \cos y$

14. $f(x, y) = (2x - x^2)(2y - y^2)$

15. $f(x, y) = x \sin y$

16. $f(x, y) = \dfrac{(x + y + 1)^2}{x^2 + y^2 + 1}$

17–20 ■ Use a graph and/or level curves to estimate the local maximum and minimum values and saddle points of the function. Then use calculus to find these values precisely.

17. $f(x, y) = 3x^2 y + y^3 - 3x^2 - 3y^2 + 2$

18. $f(x, y) = xye^{-x^2 - y^2}$

19. $f(x, y) = \sin x + \sin y + \sin(x + y)$,
  $0 \leqslant x \leqslant 2\pi, 0 \leqslant y \leqslant 2\pi$

20. $f(x, y) = \sin x + \sin y + \cos(x + y)$,
  $0 \leqslant x \leqslant \pi/4, 0 \leqslant y \leqslant \pi/4$

21–24 ■ Use a graphing calculator or computer as in Example 4 (or Newton's method or a CAS) to find the critical points of f correct to three decimal places. Then classify the critical points and find the highest or lowest points on the graph.

21. $f(x, y) = x^4 - 5x^2 + y^2 + 3x + 2$

22. $f(x, y) = 5 - 10xy - 4x^2 + 3y - y^4$

23. $f(x, y) = 2x + 4x^2 - y^2 + 2xy^2 - x^4 - y^4$

24. $f(x, y) = e^x + y^4 - x^3 + 4 \cos y$

25–32 ■ Find the absolute maximum and minimum values of f on the set D.

25. $f(x, y) = 5 - 3x + 4y$, D is the closed triangular region with vertices $(0, 0)$, $(4, 0)$, and $(4, 5)$

26. $f(x, y) = x^2 + 2xy + 3y^2$, D is the closed triangular region with vertices $(-1, 1)$, $(2, 1)$, and $(-1, -2)$

27. $f(x, y) = x^2 + y^2 + x^2y + 4$, $D = \{(x, y) \,|\, |x| \leqslant 1, |y| \leqslant 1\}$

28. $f(x, y) = y\sqrt{x} - y^2 - x + 6y$,
  $D = \{(x, y) \,|\, 0 \leqslant x \leqslant 9, 0 \leqslant y \leqslant 5\}$

29. $f(x, y) = 1 + xy - x - y$, D is the region bounded by the parabola $y = x^2$ and the line $y = 4$

30. $f(x, y) = 2x^2 + x + y^2 - 2$, $D = \{(x, y) \,|\, x^2 + y^2 \leqslant 4\}$

31. $f(x, y) = 2x^3 + y^4$, $D = \{(x, y) \,|\, x^2 + y^2 \leqslant 1\}$

32. $f(x, y) = x^3 - 3x - y^3 + 12y$, D is the quadrilateral whose vertices are $(-2, 3)$, $(2, 3)$, $(2, 2)$, and $(-2, -2)$.

33. It is impossible for a continuous function of one variable to have two local maxima and no local minimum. But there do exist functions of *two* variables with this property. Show that the function

$$f(x, y) = -(x^2 - 1)^2 - (x^2y - x - 1)^2$$

has only two critical points, but has local maxima at both of them. Then use a computer to produce a graph with a carefully chosen domain and viewpoint to see how this is possible.

34. If a function of one variable is continuous on an interval and has only one local maximum, then the maximum has to be an absolute maximum. But this is not true for functions of two variables. Show that the function

$$f(x, y) = 3xe^y - x^3 - e^{3y}$$

has exactly one critical point, and that f has a local maximum there that is not an absolute maximum. Then use a computer to produce a graph with a carefully chosen domain and viewpoint to see how this is possible.

35. Find the point on the plane $x + 2y + 3z = 4$ that is closest to the origin.

36. Find the point on the plane $2x - y + z = 1$ that is closest to the point $(-4, 1, 3)$.

37. Find the shortest distance from the point $(2, -2, 3)$ to the plane $6x + 4y - 3z = 2$.

38. Find the shortest distance from the point (x_0, y_0, z_0) to the plane $Ax + By + Cz + D = 0$.

39. Find the points on the surface $z^2 = xy + 1$ that are closest to the origin.

40. Find the points on the surface $x^2y^2z = 1$ that are closest to the origin.

41. Find three positive numbers whose sum is 100 and whose product is a maximum.

42. Find three positive numbers x, y, and z whose sum is 100 such that $x^ay^bz^c$ is a maximum.

43. Find the volume of the largest rectangular box with edges parallel to the axes that can be inscribed in the ellipsoid $9x^2 + 36y^2 + 4z^2 = 36$.

44. Solve the problem in Exercise 43 for a general ellipsoid $x^2/a^2 + y^2/b^2 + z^2/c^2 = 1$.

45. Find the volume of the largest rectangular box in the first octant with three faces in the coordinate planes and one vertex in the plane $x + 2y + 3z = 6$.

46. Solve the problem in Exercise 45 for a general plane $x/a + y/b + z/c = 1$, where a, b, c are positive numbers.

47. Find the dimensions of a rectangular box of maximum volume such that the sum of the lengths of its 12 edges is a constant c.

48. Find the dimensions of the rectangular box with largest volume if the total surface area is given as 64 cm^2.

49. A cardboard box without a lid is to have a volume of 32,000 cm^3. Find the dimensions that minimize the amount of cardboard used.

50. The base of an aquarium with given volume V is made of slate and the sides are made of glass. If slate costs five times as much (per unit area) as glass, find the dimensions of the aquarium that minimize the cost of the materials.

51. Suppose that a scientist has reason to believe that two quantities x and y are related linearly, that is, $y = mx + b$, at least approximately, for some values of m and b. The

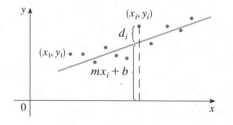

scientist performs an experiment and collects data in the form of points (x_1, y_1), (x_2, y_2), ..., (x_n, y_n), and then plots these points. The points do not lie exactly on a straight line, so the scientist wants to find constants m and b so that the line $y = mx + b$ "fits" the points as well as possible. (See the figure.) Let $d_i = y_i - (mx_i + b)$ be the vertical deviation of the point (x_i, y_i) from the line. The **method of least squares** determines m and b so as to minimize $\sum_{i=1}^{n} d_i^2$, the sum of the squares of these deviations. Show that, according to this method, the line of best fit is obtained when

$$m \sum_{i=1}^{n} x_i + bn = \sum_{i=1}^{n} y_i$$

$$m \sum_{i=1}^{n} x_i^2 + b \sum_{i=1}^{n} x_i = \sum_{i=1}^{n} x_i y_i$$

Thus the line is found by solving these two equations in the two unknowns m and b.

52. (a) Use the method of least squares (Exercise 51) to find the straight line $y = mx + b$ that best fits the data points $(1, 5.1)$, $(2, 6.8)$, $(3, 9.4)$, and $(4, 10.5)$.
    (b) As a check on your calculations in part (a), plot the four points and graph the line.

53. The following data give the height x (in inches) and weight y (in pounds) of ten 18-year-old boys. Use the method of least squares (Exercise 51) to fit these data to a straight line. Then use it to predict the weight of an 18-year-old boy who is 6 ft tall.

x	69	65	71	73	68	63	70	67	69	70
y	138	127	178	185	141	122	158	135	145	162

54. Find an equation of the plane that passes through the point $(1, 2, 3)$ and cuts off the smallest volume in the first octant.

12.8 LAGRANGE MULTIPLIERS

In Example 6 in Section 12.7 we maximized a volume function $V = xyz$ subject to the constraint $2xz + 2yz + xy = 12$, which expressed the side condition that the surface area was 12 m². In this section we present Lagrange's method for maximizing or minimizing a general function $f(x, y, z)$ subject to a constraint (or side condition) of the form $g(x, y, z) = k$.

It is easier to explain the geometric basis of Lagrange's method for functions of two variables. So we start by trying to find the extreme values of $f(x, y)$ subject to a constraint of the form $g(x, y) = k$. In other words, we seek the extreme values of $f(x, y)$ when the point (x, y) is restricted to lie on the level curve $g(x, y) = k$. Figure 1 shows this curve together with several level curves of f. These have the equations $f(x, y) = c$, where $c = 7, 8, 9, 10, 11$. To maximize $f(x, y)$ subject to $g(x, y) = k$ is to find the largest value of c such that the level curve $f(x, y) = c$ intersects $g(x, y) = k$. It appears from Figure 1 that this happens when these curves just touch each other, that is, when they have a common tangent line. This means that the normal lines at the point (x_0, y_0) where they touch are identical. Therefore, the gradient vectors are parallel; that is, $\nabla f(x_0, y_0) = \lambda \nabla g(x_0, y_0)$ for some scalar λ.

This kind of argument also applies to the problem of finding the extreme values of $f(x, y, z)$ subject to the constraint $g(x, y, z) = k$. Thus the point (x, y, z) is restricted to lie on the level surface S with equation $g(x, y, z) = k$. Instead of the level curves in Figure 1, we consider the level surfaces $f(x, y, z) = c$ and argue that if the maximum value of f is $f(x_0, y_0, z_0) = c$, then the level surface $f(x, y, z) = c$ is tangent to the level surface $g(x, y, z) = k$ and so the corresponding gradient vectors are parallel.

This intuitive argument can be made precise as follows. Suppose that f has an extreme value at a point $P(x_0, y_0, z_0)$ on S and let C be a curve with vector equation $\mathbf{r}(t) = \langle x(t), y(t), z(t) \rangle$ that lies on S and passes through P. If t_0 is the parameter value corresponding to the point P, then $\mathbf{r}(t_0) = \langle x_0, y_0, z_0 \rangle$. The composite function $h(t) = f(x(t), y(t), z(t))$ represents the values that f takes on the curve C. Since f has an extreme value at (x_0, y_0, z_0), it follows that h has an extreme value at t_0, so $h'(t_0) = 0$.

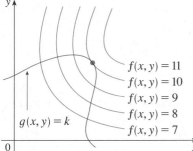

$f(x, y) = 11$
$f(x, y) = 10$
$f(x, y) = 9$
$f(x, y) = 8$
$f(x, y) = 7$
$g(x, y) = k$

FIGURE 1

But if f is differentiable, we can use the Chain Rule to write

$$0 = h'(t_0) = f_x(x_0, y_0, z_0)x'(t_0) + f_y(x_0, y_0, z_0)y'(t_0) + f_z(x_0, y_0, z_0)z'(t_0)$$
$$= \nabla f(x_0, y_0, z_0) \cdot \mathbf{r}'(t_0)$$

This shows that the gradient vector $\nabla f(x_0, y_0, z_0)$ is orthogonal to the tangent vector $\mathbf{r}'(t_0)$ to every such curve C. But we already know from Section 12.6 that the gradient vector of g, $\nabla g(x_0, y_0, z_0)$, is also orthogonal to $\mathbf{r}'(t_0)$ (see Equation 12.6.18). This means that the gradient vectors $\nabla f(x_0, y_0, z_0)$ and $\nabla g(x_0, y_0, z_0)$ must be parallel. Therefore, if $\nabla g(x_0, y_0, z_0) \neq \mathbf{0}$, there is a number λ such that

(1)
$$\nabla f(x_0, y_0, z_0) = \lambda \nabla g(x_0, y_0, z_0)$$

The number λ in Equation 1 is called a **Lagrange multiplier.** The procedure based on Equation 1 is called the **method of Lagrange multipliers** and is as follows.

Lagrange multipliers are named after the French-Italian mathematician Joseph-Louis Lagrange (1736–1813). See page 193 for a biographical sketch of Lagrange.

To find the maximum and minimum values of $f(x, y, z)$ subject to the constraint $g(x, y, z) = k$ (assuming that these extreme values exist):

(a) Find all values of x, y, z, and λ such that

$$\nabla f(x, y, z) = \lambda \nabla g(x, y, z)$$

and
$$g(x, y, z) = k$$

(b) Evaluate f at all the points (x, y, z) that arise from step (a). The largest of these values is the maximum value of f; the smallest is the minimum value of f.

If we write the vector equation $\nabla f = \lambda \nabla g$ in terms of its components, then the equations in step (a) become

$$f_x = \lambda g_x \qquad f_y = \lambda g_y \qquad f_z = \lambda g_z \qquad g(x, y, z) = k$$

This is a system of four equations in the four unknowns x, y, z, and λ, but it is not necessary to find explicit values for λ.

For functions of two variables the method of Lagrange multipliers is similar to the method just described. To find the extreme values of $f(x, y)$ subject to the constraint $g(x, y) = k$, we look for values of x, y, and λ such that

$$\nabla f(x, y) = \lambda \nabla g(x, y) \qquad \text{and} \qquad g(x, y) = k$$

This amounts to solving three equations in three unknowns:

$$f_x = \lambda g_x \qquad f_y = \lambda g_y \qquad g(x, y) = k$$

Our first illustration of Lagrange's method is to reconsider the problem given in Example 6 in Section 12.7.

EXAMPLE 1   A rectangular box without a lid is to be made from 12 m^2 of cardboard. Find the maximum volume of such a box.

SOLUTION   As in Example 6 in Section 12.7 we let x, y, and z be the length, width, and height, respectively, of the box in meters. Then we wish to maximize

$$V = xyz$$

subject to the constraint

$$g(x, y, z) = 2xz + 2yz + xy = 12$$

Using the method of Lagrange multipliers, we look for values of x, y, z, and λ such that $\nabla V = \lambda \nabla g$ and $g(x, y, z) = 12$. This gives the equations

$$V_x = \lambda g_x \qquad V_y = \lambda g_y \qquad V_z = \lambda g_z \qquad 2xz + 2yz + xy = 12$$

which become

(2) $$yz = \lambda(2z + y)$$

(3) $$xz = \lambda(2z + x)$$

(4) $$xy = \lambda(2x + 2y)$$

(5) $$2xz + 2yz + xy = 12$$

There are no general rules for solving systems of equations. Sometimes some ingenuity is required. In the present example you might notice that if we multiply (2) by x, (3) by y, and (4) by z, then the left sides of these equations will be identical. Doing this, we have

(6) $$xyz = \lambda(2xz + xy)$$

(7) $$xyz = \lambda(2yz + xy)$$

(8) $$xyz = \lambda(2xz + 2yz)$$

We observe that $\lambda \neq 0$ because $\lambda = 0$ would imply $yz = xz = xy = 0$ from (2), (3), and (4) and this would contradict (5). Therefore, from (6) and (7) we have

$$2xz + xy = 2yz + xy$$

which gives $xz = yz$. But $z \neq 0$ (since $z = 0$ would give $V = 0$), so $x = y$. From (7) and (8) we have

$$2yz + xy = 2xz + 2yz$$

which gives $2xz = xy$ and so (since $x \neq 0$) $y = 2z$. If we now put $x = y = 2z$ in (5), we get

$$4z^2 + 4z^2 + 4z^2 = 12$$

Since x, y, and z are all positive, we therefore have $z = 1$, $x = 2$, and $y = 2$ as before. ∎

EXAMPLE 2 Find the extreme values of the function $f(x, y) = x^2 + 2y^2$ on the circle $x^2 + y^2 = 1$.

SOLUTION We are asked for the extreme values of f subject to the constraint $g(x, y) = x^2 + y^2 = 1$. Using Lagrange multipliers, we solve the equations $\nabla f = \lambda \nabla g$, $g(x, y) = 1$, which can be written as

$$f_x = \lambda g_x \qquad f_y = \lambda g_y \qquad g(x, y) = 1$$

In geometric terms, Example 2 asks for the highest and lowest points on the curve C in Figure 2 that lies on the paraboloid $z = x^2 + 2y^2$ and directly above the constraint circle $x^2 + y^2 = 1$.

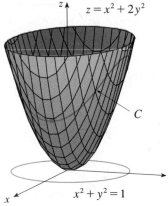

FIGURE 2

The geometry behind the use of Lagrange multipliers in Example 2 is shown in Figure 3. The extreme values of $f(x, y) = x^2 + 2y^2$ correspond to the level curves that touch the circle $x^2 + y^2 = 1$.

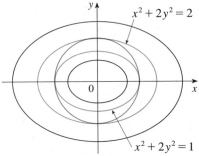

FIGURE 3

or as

$$(9) \qquad 2x = 2x\lambda$$

$$(10) \qquad 4y = 2y\lambda$$

$$(11) \qquad x^2 + y^2 = 1$$

From (9) we have $x = 0$ or $\lambda = 1$. If $x = 0$, then (11) gives $y = \pm 1$. If $\lambda = 1$, then $y = 0$ from (10), so then (11) gives $x = \pm 1$. Therefore, f has possible extreme values at the points $(0, 1)$, $(0, -1)$, $(1, 0)$, and $(-1, 0)$. Evaluating f at these four points, we find that

$$f(0, 1) = 2 \qquad f(0, -1) = 2 \qquad f(1, 0) = 1 \qquad f(-1, 0) = 1$$

Therefore, the maximum value of f on the circle $x^2 + y^2 = 1$ is $f(0, \pm 1) = 2$ and the minimum value is $f(\pm 1, 0) = 1$. Checking with Figure 2, we see that these values look reasonable. ∎

EXAMPLE 3   Find the extreme values of $f(x, y) = x^2 + 2y^2$ on the disk $x^2 + y^2 \le 1$.

SOLUTION   According to the procedure in (12.7.9), we compare the values of f at the critical points with those at the points on the boundary. Since $f_x = 2x$ and $f_y = 4y$, the only critical point is $(0, 0)$. We compare the value of f at that point with the extreme values on the boundary from Example 2:

$$f(0, 0) = 0 \qquad f(\pm 1, 0) = 1 \qquad f(0, \pm 1) = 2$$

Therefore, the maximum value of f on the disk $x^2 + y^2 \le 1$ is $f(0, \pm 1) = 2$ and the minimum value is $f(0, 0) = 0$. ∎

EXAMPLE 4   Find the points on the sphere $x^2 + y^2 + z^2 = 4$ that are closest to and farthest from the point $(3, 1, -1)$.

SOLUTION   The distance from a point (x, y, z) to the point $(3, 1, -1)$ is

$$d = \sqrt{(x - 3)^2 + (y - 1)^2 + (z + 1)^2}$$

but the algebra is simpler if we instead maximize and minimize the square of the distance:

$$d^2 = f(x, y, z) = (x - 3)^2 + (y - 1)^2 + (z + 1)^2$$

The constraint is that the point (x, y, z) lies on the sphere, that is,

$$g(x, y, z) = x^2 + y^2 + z^2 = 4$$

According to the method of Lagrange multipliers, we solve $\nabla f = \lambda \nabla g$, $g = 4$. This gives

$$(12) \qquad 2(x - 3) = 2x\lambda$$

$$(13) \qquad 2(y - 1) = 2y\lambda$$

$$(14) \qquad 2(z + 1) = 2z\lambda$$

$$(15) \qquad x^2 + y^2 + z^2 = 4$$

The simplest way to solve these equations is to solve for x, y, and z in terms of λ from (12), (13), and (14), and then substitute these values into (15). From (12) we have

$$x - 3 = x\lambda \quad \text{or} \quad x(1 - \lambda) = 3 \quad \text{or} \quad x = \frac{3}{1 - \lambda}$$

[Note that $1 - \lambda \neq 0$ because $\lambda = 1$ is impossible from (1).] Similarly, (13) and (14) give

$$y = \frac{1}{1 - \lambda} \qquad z = -\frac{1}{1 - \lambda}$$

Therefore, from (15) we have

$$\frac{3^2}{(1 - \lambda)^2} + \frac{1^2}{(1 - \lambda)^2} + \frac{(-1)^2}{(1 - \lambda)^2} = 4$$

which gives $(1 - \lambda)^2 = \frac{11}{4}$, $1 - \lambda = \pm\sqrt{11}/2$, so

$$\lambda = 1 \pm \frac{\sqrt{11}}{2}$$

These values of λ then give the corresponding points (x, y, z):

$$\left(\frac{6}{\sqrt{11}}, \frac{2}{\sqrt{11}}, -\frac{2}{\sqrt{11}} \right) \quad \text{and} \quad \left(-\frac{6}{\sqrt{11}}, -\frac{2}{\sqrt{11}}, \frac{2}{\sqrt{11}} \right)$$

It is easy to see that f has a smaller value at the first of these points, so the closest point is $(6/\sqrt{11}, 2/\sqrt{11}, -2/\sqrt{11})$ and the farthest is $(-6/\sqrt{11}, -2/\sqrt{11}, 2/\sqrt{11})$. ∎

Figure 4 shows the sphere and the nearest point P in Example 4. Can you see how to find the coordinates of P without using calculus?

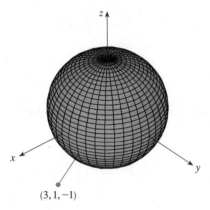

$(3, 1, -1)$

FIGURE 4

TWO CONSTRAINTS

Suppose now that we want to find the maximum and minimum values of $f(x, y, z)$ subject to two constraints (side conditions) of the form $g(x, y, z) = k$ and $h(x, y, z) = c$. Geometrically, this means that we are looking for the extreme values of f when (x, y, z) is restricted to lie on the curve of intersection of the level surfaces $g(x, y, z) = k$ and $h(x, y, z) = c$. It can be shown that if an extreme value occurs at (x_0, y_0, z_0), then the gradient vector $\nabla f(x_0, y_0, z_0)$ is in the plane determined by $\nabla g(x_0, y_0, z_0)$ and $\nabla h(x_0, y_0, z_0)$. (We assume that these gradient vectors are not parallel.) So there are numbers λ and μ (called Lagrange multipliers) such that

(16)
$$\nabla f(x_0, y_0, z_0) = \lambda \nabla g(x_0, y_0, z_0) + \mu \nabla h(x_0, y_0, z_0)$$

In this case Lagrange's method is to look for extreme values by solving five equations in the five unknowns x, y, z, λ, and μ. These equations are obtained by writing Equation 16 in terms of its components and using the constraint equations:

$$f_x = \lambda g_x + \mu h_x$$
$$f_y = \lambda g_y + \mu h_y$$
$$f_z = \lambda g_z + \mu h_z$$
$$g(x, y, z) = k$$
$$h(x, y, z) = c$$

The cylinder $x^2 + y^2 = 1$ intersects the plane $x - y + z = 1$ in an ellipse (Figure 5). Example 5 asks for the maximum value of f when (x, y, z) is restricted to lie on the ellipse.

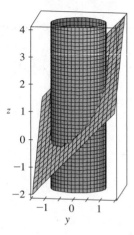

FIGURE 5

EXAMPLE 5   Find the maximum value of the function $f(x, y, z) = x + 2y + 3z$ on the curve of intersection of the plane $x - y + z = 1$ and the cylinder $x^2 + y^2 = 1$.

SOLUTION   We maximize the function $f(x, y, z) = x + 2y + 3z$ subject to the constraints $g(x, y, z) = x - y + z = 1$ and $h(x, y, z) = x^2 + y^2 = 1$. The Lagrange condition is $\nabla f = \lambda \nabla g + \mu \nabla h$, so we solve the equations

(17) $1 = \lambda + 2x\mu$

(18) $2 = -\lambda + 2y\mu$

(19) $3 = \lambda$

(20) $x - y + z = 1$

(21) $x^2 + y^2 = 1$

Putting $\lambda = 3$ [from (19)] in (17), we get $2x\mu = -2$, so $x = -1/\mu$. Similarly, (18) gives $y = 5/(2\mu)$. Substitution in (21) then gives

$$\frac{1}{\mu^2} + \frac{25}{4\mu^2} = 1$$

and so $\mu^2 = \frac{29}{4}$, $\mu = \pm\sqrt{29}/2$. Then $x = \mp 2/\sqrt{29}$, $y = \pm 5/\sqrt{29}$, and, from (20), $z = 1 - x + y = 1 \pm 7/\sqrt{29}$. The corresponding values of f are

$$\mp \frac{2}{\sqrt{29}} + 2\left(\pm\frac{5}{\sqrt{29}}\right) + 3\left(1 \pm \frac{7}{\sqrt{29}}\right) = 3 \pm \sqrt{29}$$

Therefore, the maximum value of f on the given curve is $3 + \sqrt{29}$.   ∎

EXERCISES 12.8

1–15 ■ Use Lagrange multipliers to find the maximum and minimum values of the function subject to the given constraint or constraints.

1. $f(x, y) = x^2 - y^2$; $x^2 + y^2 = 1$

2. $f(x, y) = 2x + y$; $x^2 + 4y^2 = 1$

3. $f(x, y) = xy$; $9x^2 + y^2 = 4$

4. $f(x, y) = x^2 + y^2$; $x^4 + y^4 = 1$

5. $f(x, y, z) = x + 3y + 5z$; $x^2 + y^2 + z^2 = 1$

6. $f(x, y, z) = x - y + 3z$; $x^2 + y^2 + 4z^2 = 4$

7. $f(x, y, z) = xyz$; $x^2 + 2y^2 + 3z^2 = 6$

8. $f(x, y, z) = x^2 y^2 z^2$; $x^2 + y^2 + z^2 = 1$

9. $f(x, y, z) = x^2 + y^2 + z^2$; $x^4 + y^4 + z^4 = 1$

10. $f(x, y, z) = x^4 + y^4 + z^4$; $x^2 + y^2 + z^2 = 1$

11. $f(x, y, z, t) = x + y + z + t$; $x^2 + y^2 + z^2 + t^2 = 1$

12. $f(x_1, x_2, \ldots, x_n) = x_1 + x_2 + \cdots + x_n$;
$x_1^2 + x_2^2 + \cdots + x_n^2 = 1$

13. $f(x, y, z) = x + 2y$; $x + y + z = 1$, $y^2 + z^2 = 4$

14. $f(x, y, z) = 3x - y - 3z$; $x + y - z = 0$, $x^2 + 2z^2 = 1$

15. $f(x, y, z) = yz + xy$; $xy = 1$, $y^2 + z^2 = 1$

16–17 ■ Find the extreme values of f on the region described by the inequality.

16. $f(x, y) = 2x^2 + 3y^2 - 4x - 5$, $x^2 + y^2 \leq 16$

17. $f(x, y) = e^{-xy}$, $x^2 + 4y^2 \leq 1$

18. (a) Use a graphing calculator or computer to graph the circle $x^2 + y^2 = 1$. On the same screen graph several curves of the form $x^2 + y = c$ until you find two that just touch the circle. What is the significance of the values of c for these two curves?

(b) Use Lagrange multipliers to find the extreme values of $f(x, y) = x^2 + y$ subject to the constraint $x^2 + y^2 = 1$. Compare your answers with those in part (a).

19. (a) If your computer algebra system plots implicitly defined curves, use it to estimate the minimum and maximum values of $f(x, y) = x^3 + y^3 + 3xy$ subject to the constraint $(x - 3)^2 + (y - 3)^2 = 9$ by graphical methods.

(b) Solve the problem in part (a) with the aid of Lagrange multipliers. Use your CAS to solve the equations numerically. Compare your answers with those in part (a).

20. Find the maximum and minimum volumes of a rectangular box whose surface area is 1500 cm^2 and whose total edge length is 200 cm.

21. A manufacturer produces a quantity Q of a certain product and Q depends on the amount x of labor used and the amount y of capital. A simple model that is sometimes used to express Q explicitly as a function of x and y is the Cobb-Douglas production function: $Q = Kx^\alpha y^{1-\alpha}$, where K and α are constants with $K > 0$ and $0 < \alpha < 1$. If the cost of a unit of labor is m and the cost of a unit of capital is n, and the company can spend only p dollars as its total budget, then maximizing the production Q is subject to the constraint $mx + ny = p$. Show that the maximum production occurs when

$$x = \frac{\alpha p}{m} \quad \text{and} \quad y = \frac{(1-\alpha)p}{n}$$

22. Referring to Exercise 21, we now suppose that the production is fixed at $Kx^\alpha y^{1-\alpha} = Q$, where Q is a constant. What values of x and y minimize the cost function $C(x, y) = mx + ny$?

23. Use Lagrange multipliers to prove that the rectangle with maximum area that has a given perimeter p is a square.

24. Use Lagrange multipliers to prove that the triangle with maximum area that has a given perimeter p is equilateral. [*Hint:* Use Heron's formula for the area:

$A = \sqrt{s(s-x)(s-y)(s-z)}$, where $s = p/2$ and x, y, z are the lengths of the sides.]

25–40 ■ Use Lagrange multipliers to give an alternate solution to the indicated exercise in Section 12.7.

25. Exercise 35   26. Exercise 36

27. Exercise 37   28. Exercise 38

29. Exercise 39   30. Exercise 40

31. Exercise 41   32. Exercise 42

33. Exercise 43   34. Exercise 44

35. Exercise 45   36. Exercise 46

37. Exercise 47   38. Exercise 48

39. Exercise 49   40. Exercise 50

41. The plane $x + y + 2z = 2$ intersects the paraboloid $z = x^2 + y^2$ in an ellipse. Find the points on this ellipse that are nearest to and farthest from the origin.

42. (a) Maximize $\sum_{i=1}^{n} x_i y_i$ subject to the constraints $\sum_{i=1}^{n} x_i^2 = 1$ and $\sum_{i=1}^{n} y_i^2 = 1$.
    (b) Put

$$x_i = \frac{a_i}{\sqrt{\sum a_i^2}} \quad \text{and} \quad y_i = \frac{b_i}{\sqrt{\sum b_i^2}}$$

to show that

$$\sum a_i b_i \leq \sqrt{\sum a_i^2} \sqrt{\sum b_i^2}$$

for any numbers $a_1, \ldots, a_n, b_1, \ldots, b_n$. This inequality is known as the Cauchy-Schwarz Inequality.

---

12   REVIEW

KEY TOPICS ■ Define, state, or discuss the following.

1. Functions of two variables and their graphs

2. Level curves

3. Functions of three variables

4. Level surfaces

5. Limit of a function of two or three variables

6. Continuity of a function of two or three variables

7. Partial derivatives

8. Clairaut's Theorem

9. Equation of a tangent plane to the surface $z = f(x, y)$

10. Equation of a tangent plane to the surface $F(x, y, z) = k$

11. Differential

12. Differentiable function

13. The Chain Rule

14. Implicit differentiation

15. Directional derivative

16. Gradient and its significance

17. Local and absolute maximum and minimum values

18. Critical point

19. Saddle point

20. Second Derivatives Test

21. Extreme Value Theorem for functions of two variables

22. Lagrange multipliers

EXERCISES

1–12 ■ Determine whether the statement is true or false.

1. $f_y(a, b) = \lim\limits_{y \to b} \dfrac{f(a, y) - f(a, b)}{y - b}$

2. There exists a function f whose partial derivatives are $f_x(x, y) = x + y^2$ and $f_y(x, y) = x - y^2$.

3. $f_{xy} = \dfrac{\partial^2 f}{\partial x \, \partial y}$

4. $D_{\mathbf{k}} f(x, y, z) = f_z(x, y, z)$

5. If $f(x, y) \to L$ as $(x, y) \to (a, b)$ along every straight line through (a, b), then $\lim_{(x, y) \to (a, b)} f(x, y) = L$.

6. If $f_x(a, b)$ and $f_y(a, b)$ both exist, then f is differentiable at (a, b).

7. If f has a local minimum at (a, b) and f is differentiable at (a, b), then $\nabla f(a, b) = 0$.

8. $\lim\limits_{(x, y) \to (1, 1)} \dfrac{x - y}{x^2 - y^2} = \lim\limits_{(x, y) \to (1, 1)} \dfrac{1}{x + y} = \dfrac{1}{2}$

9. If $f(x, y) = \ln y$, then $\nabla f(x, y) = 1/y$.

10. If (a, b) is a critical point of f and $f_{xx}(a, b)f_{yy}(a, b) < [\, f_{xy}(a, b)]^2$, then f has a saddle point at (a, b).

11. If $f(x, y) = \sin x + \sin y$, then $-\sqrt{2} \leqslant D_{\mathbf{u}} f(x, y) \leqslant \sqrt{2}$.

12. If $f(x, y)$ has two local maxima, then f must have a local minimum.

13–16 ■ Find and sketch the domain of the function.

13. $f(x, y) = \dfrac{\ln(x + y + 1)}{x - 1}$ **14.** $f(x, y) = \sqrt{\sin \pi(x^2 + y^2)}$

15. $f(x, y) = \cos^{-1}x + \tan^{-1}y$ **16.** $f(x, y, z) = \sqrt{z - x^2 - y^2}$

17–18 ■ Sketch the graph of the function.

17. $f(x, y) = 1 - x^2 - y^2$ **18.** $f(x, y) = \sqrt{x^2 + y^2 - 1}$

19–20 ■ Sketch several level curves of the function.

19. $f(x, y) = e^{-(x^2 + y^2)}$ **20.** $f(x, y) = x^2 + 4y$

21–22 ■ Evaluate the limit or show that it does not exist.

21. $\lim\limits_{(x, y) \to (0, 0)} \dfrac{x^2 y^2}{x^2 + 2y^2}$ **22.** $\lim\limits_{(x, y) \to (0, 0)} \dfrac{2xy}{x^2 + 2y^2}$

23–28 ■ Find the first partial derivatives.

23. $f(x, y) = 3x^4 - x\sqrt{y}$ **24.** $g(x, y) = \dfrac{x}{\sqrt{x + 2y}}$

25. $f(s, t) = e^{2s} \cos \pi t$ **26.** $g(r, s) = r \sin \sqrt{r^2 + s^2}$

27. $f(x, y, z) = xy^z$ **28.** $g(u, v, w) = w^2 e^{u/v}$

29–32 ■ Find all second partial derivatives of f.

29. $f(x, y) = x^2 y^3 - 2x^4 + y^2$ **30.** $f(x, y) = x^3 \ln(x - y)$

31. $f(x, y, z) = xy^2 z^3$ **32.** $f(x, y, z) = xe^y \cos z$

33. If $u = x^y$, show that $\dfrac{x}{y} \dfrac{\partial u}{\partial x} + \dfrac{1}{\ln x} \dfrac{\partial u}{\partial y} = 2u$.

34. If $\rho = \sqrt{x^2 + y^2 + z^2}$, show that

$$\frac{\partial^2 \rho}{\partial x^2} + \frac{\partial^2 \rho}{\partial y^2} + \frac{\partial^2 \rho}{\partial z^2} = \frac{2}{\rho}$$

35–39 ■ Find an equation of the tangent plane to the given surface at the specified point.

35. $z = x^2 + y^2 + 4y, \quad (0, 1, 5)$

36. $z = xe^y, \quad (1, 0, 1)$

37. $xy^2 z^3 = 12, \quad (3, 2, 1)$

38. $x^2 + y^2 + z^2 = 29, \quad (2, 3, 4)$

39. $x^2 + 2y^2 - 3z^2 = 14, \quad (3, 2, -1)$

40. Use a computer to graph the surface $z = x^3 + 2xy$ and its tangent plane and normal line at $(1, 2, 5)$ on the same screen. Choose the domain and viewpoint so that you get a good view of all three objects.

41. Find the points on the sphere $x^2 + y^2 + z^2 = 1$ where the tangent plane is parallel to the plane $2x + y - 3z = 2$.

42. Find dz if $z = x^2 \tan^{-1}y$.

43. Use differentials to approximate the number

$$(1.98)^3 \sqrt{(3.01)^2 + (3.97)^2}$$

44. The two legs of a right triangle are measured as 5 m and 12 m, respectively, with a possible error in measurement of at most 0.2 cm in each. Use differentials to estimate the maximum error in the calculated value of (a) the area of the triangle and (b) the length of the hypotenuse.

45. If $w = \sqrt{x} + y^2/z$, where $x = e^{2t}$, $y = t^3 + 4t$, and $z = t^2 - 4$, use the Chain Rule to find dw/dt.

46. If $z = \cos xy + y \cos x$, where $x = u^2 + v$ and $y = u - v^2$, use the Chain Rule to find $\partial z/\partial u$ and $\partial z/\partial v$.

47. If $z = y + f(x^2 - y^2)$, where f is differentiable, show that

$$y \frac{\partial z}{\partial x} + x \frac{\partial z}{\partial y} = x$$

48. The length x of a side of a triangle is increasing at a rate of 3 in/s, the length y of another side is decreasing at a rate of 2 in/s, and the contained angle θ is increasing at a rate of 0.05 radian/s. How fast is the area of the triangle changing when $x = 40$ in., $y = 50$ in., and $\theta = \pi/6$?

49. If $z = f(u, v)$, where $u = xy$, $v = y/x$, and f has continuous second partial derivatives, show that

$$x^2 \frac{\partial^2 z}{\partial x^2} - y^2 \frac{\partial^2 z}{\partial y^2} = -4uv \frac{\partial^2 z}{\partial u \, \partial v} + 2v \frac{\partial z}{\partial v}$$

50. If $yz^4 + x^2 z^3 = e^{xyz}$, find $\dfrac{\partial z}{\partial x}$ and $\dfrac{\partial z}{\partial y}$.

51. Find the gradient of the function $f(x, y, z) = z^2 e^{x\sqrt{y}}$.

52. (a) How are the directional derivative and the gradient of $f(x, y)$ related?
(b) When is the directional derivative of f a maximum?
(c) When is it a minimum?
(d) When is it 0?
(e) When is it half of its maximum value?

53–54 ■ Find the directional derivative of f at the given point in the indicated direction.

53. $f(x, y) = 2\sqrt{x} - y^2$, $(1, 5)$, in the direction toward the point $(4, 1)$

54. $f(x, y, z) = x^2 y + x\sqrt{1 + z}$, $(1, 2, 3)$, in the direction of $\mathbf{v} = 2\mathbf{i} + \mathbf{j} - 2\mathbf{k}$

55. Find the maximum rate of change of $f(x, y) = x^2 y + \sqrt{y}$ at the point $(2, 1)$. In which direction does it occur?

56. Find the direction in which $f(x, y, z) = ze^{xy}$ increases most rapidly at the point $(0, 1, 2)$. What is the maximum rate of increase?

57. Evaluate $\displaystyle \lim_{(x, y) \to (1, 0)} \tan^{-1}\left(\frac{x^2 + y^2}{(x - 1)^2 + y^2}\right)$.

58. Find parametric equations of the tangent line at the point $(-2, 2, 4)$ to the curve of intersection of the surface $z = 2x^2 - y^2$ and the plane $z = 4$.

59–62 ■ Find the local maximum and minimum values and saddle points of the function. If you have three-dimensional graphing software, use a computer to graph the function with a domain and viewpoint that reveal all the important aspects of the function.

59. $f(x, y) = x^2 - xy + y^2 + 9x - 6y + 10$

60. $f(x, y) = x^3 - 6xy + 8y^3$

61. $f(x, y) = 3xy - x^2 y - xy^2$

62. $f(x, y) = (x^2 + y)e^{y/2}$

63–64 ■ Find the absolute maximum and minimum values of f on the set D.

63. $f(x, y) = 4xy^2 - x^2 y^2 - xy^3$; D is the closed triangular region in the xy-plane with vertices $(0, 0)$, $(0, 6)$, and $(6, 0)$

64. $f(x, y) = e^{-x^2 - y^2}(x^2 + 2y^2)$; D is the disk $x^2 + y^2 \le 4$

65. Use a graph and/or level curves to estimate the local maximum and minimum values and saddle points of $f(x, y) = x^3 - 3x + y^4 - 2y^2$. Then use calculus to find these values precisely.

66. Use a graphing calculator or computer (or Newton's method or a computer algebra system) to find the critical points of $f(x, y) = 12 + 10y - 2x^2 - 8xy - y^4$ correct to three decimal places. Then classify the critical points and find the highest point on the graph.

67–70 ■ Use Lagrange multipliers to find the maximum and minimum values of f subject to the given constraints.

67. $f(x, y) = x^2 y$; $x^2 + y^2 = 1$

68. $f(x, y) = \dfrac{1}{x} + \dfrac{1}{y}$; $\dfrac{1}{x^2} + \dfrac{1}{y^2} = 1$

69. $f(x, y, z) = x + y + z$; $\dfrac{1}{x} + \dfrac{1}{y} + \dfrac{1}{z} = 1$

70. $f(x, y, z) = x^2 + 2y^2 + 3z^2$; $x + y + z = 1$, $x - y + 2z = 2$

71. Find the points on the surface $xy^2 z^3 = 2$ that are closest to the origin.

72. A package in the shape of a rectangular box can be mailed parcel post if the sum of its length and girth (the perimeter of a cross-section perpendicular to the length) is at most 84 in. Find the dimensions of the package with largest volume that can be mailed parcel post.

73. A pentagon is formed by placing an isosceles triangle on a rectangle, as shown in the figure. If the pentagon has fixed perimeter P, find the lengths of the sides of the pentagon that maximize the area of the pentagon.

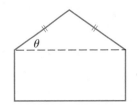

74. A particle of mass m moves on the surface $z = f(x, y)$. Let $x = x(t)$, $y = y(t)$ be the x- and y-coordinates of the particle at time t.
(a) Find the velocity vector $\mathbf{v}$ and the kinetic energy $K = \frac{1}{2} m |\mathbf{v}|^2$ of the particle.
(b) Determine the acceleration vector $\mathbf{a}$.
(c) Let $z = x^2 + y^2$ and $x(t) = t\cos t$, $y(t) = t\sin t$. Find the velocity vector, the kinetic energy, and the acceleration vector.

PROBLEMS PLUS

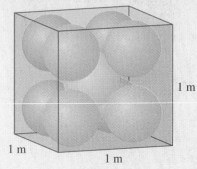

FIGURE FOR PROBLEM 1

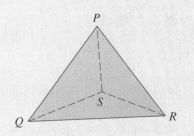

FIGURE FOR PROBLEM 2

1. Each edge of a cubical box has length 1 m. The box contains nine spherical balls with the same radius r. The center of one ball is at the center of the cube and it touches the other eight balls. Each of the other eight balls touches three sides of the box. Thus the balls are tightly packed in the box. Find r. (If you have trouble with this problem, read about the problem-solving strategy entitled *Use analogy* in Section 4 of Review and Preview.)

2. A tetrahedron is a solid with four vertices, P, Q, R, and S, and four triangular faces. (See the figure.)
   (a) Let $\mathbf{v}_1$, $\mathbf{v}_2$, $\mathbf{v}_3$, and $\mathbf{v}_4$ be vectors with magnitudes equal to the areas of the faces opposite the vertices P, Q, R, and S, respectively, and directions parallel to the outward normals of the respective faces. Show that

$$\mathbf{v}_1 + \mathbf{v}_2 + \mathbf{v}_3 + \mathbf{v}_4 = \mathbf{0}$$

   (b) The volume V of a tetrahedron is one-third the distance from a vertex to the opposite face, times the area of that face. Find a formula for the volume of the tetrahedron with vertices P, Q, R, and S.
   (c) Determine the volume of the tetrahedron whose vertices are $P(1, 1, 1)$, $Q(1, 2, 3)$, $R(1, 1, 2)$, and $S(3, -1, 2)$.

3. Suppose the tetrahedron in Problem 2 has a trirectangular vertex S. (This means that the three angles at S are all right angles.) Let A, B, and C be the areas of the three faces that meet at S and let D be the area of the opposite face PQR. Show that

$$D^2 = A^2 + B^2 + C^2$$

   (This is a three-dimensional version of the Pythagorean Theorem.)

4. Let B be a solid box with length L, width W, and height H. Let S be the set of all points that are a distance at most 1 from some point of B. Express the volume of S in terms of L, W, and H.

5. A rectangle with length L and width W is cut into four smaller rectangles by two lines parallel to the sides. Find the maximum and minimum values of the sum of the squares of the areas of the smaller rectangles.

6. Find the curvature of the curve with parametric equations

$$x = \int_0^t \sin \tfrac{1}{2}\pi\theta^2 \, d\theta \qquad y = \int_0^t \cos \tfrac{1}{2}\pi\theta^2 \, d\theta$$

7. Suppose f is a differentiable function of one variable. Show that all tangent planes to the surface $z = xf(y/x)$ intersect in a common point.

8. The plane $x + y + z = 24$ intersects the paraboloid $z = x^2 + y^2$ in an ellipse. Find the highest and lowest points on the ellipse.

9. (a) Show that when Laplace's equation

$$\frac{\partial^2 u}{\partial x^2} + \frac{\partial^2 u}{\partial y^2} + \frac{\partial^2 u}{\partial z^2} = 0$$

   is written in cylindrical coordinates, it becomes

$$\frac{\partial^2 u}{\partial r^2} + \frac{1}{r}\frac{\partial u}{\partial r} + \frac{1}{r^2}\frac{\partial^2 u}{\partial \theta^2} + \frac{\partial^2 u}{\partial z^2} = 0$$

(b) Show that when Laplace's equation is written in spherical coordinates, it becomes

$$\frac{\partial^2 u}{\partial \rho^2} + \frac{2}{\rho}\frac{\partial u}{\partial \rho} + \frac{\cot\phi}{\rho^2}\frac{\partial u}{\partial \phi} + \frac{1}{\rho^2}\frac{\partial^2 u}{\partial \phi^2} + \frac{1}{\rho^2 \sin^2\phi}\frac{\partial^2 u}{\partial \theta^2} = 0$$

10. The temperature at a point (x, y) on a metal plate is given by $T(x, y) = 100 - x^2 - 2y^2$. A heat-seeking particle starts at the point $(4, 2)$ and moves at each instant in the direction of maximum temperature increase.
    (a) In what direction does the particle move initially?
    (b) Draw a contour map of T and sketch the path followed by the particle.
    (c) Find an equation of this path by setting up and solving two differential equations.

11. Newton's method for approximating a root of an equation $f(x) = 0$ (see Section 2.10) can be adapted to approximating a solution of a system of equations $f(x, y) = 0$ and $g(x, y) = 0$. The surfaces $z = f(x, y)$ and $z = g(x, y)$ intersect in a curve that intersects the xy-plane at the point (r, s), which is the solution of the system. If an initial approximation (x_1, y_1) is close to this point, then the tangent planes to the surfaces at (x_1, y_1) intersect in a straight line that intersects the xy-plane in a point (x_2, y_2), which should be closer to (r, s). (Compare with Figure 1 in Section 2.10.) Show that

$$x_2 = x_1 - \frac{fg_y - f_y g}{f_x g_y - f_y g_x} \quad \text{and} \quad y_2 = y_1 - \frac{f_x g - f g_x}{f_x g_y - f_y g_x}$$

    where f, g, and their partial derivatives are evaluated at (x_1, y_1). If we continue this procedure, we obtain successive approximations (x_n, y_n).

12. Use Newton's method (see Problem 11) to find the points of intersection of the curves $x^2 + y^2 = 4$ and $\sin y = x$ correct to four decimal places.

13. If the ellipse $x^2/a^2 + y^2/b^2 = 1$ is to contain the circle $x^2 + y^2 = 2y$ in its interior, what values of a and b minimize the area of the ellipse?

14. A cable has radius r and length L and is wound around a spool with radius R without overlapping. What is the shortest length along the spool that is covered by the cable?

15. Let L be the line of intersection of the planes $cx + y + z = c$ and $x - cy + cz = -1$, where c is a real number.
    (a) Find symmetric equations for L.
    (b) As the number c varies, the line L sweeps out a surface S. Find an equation for the curve of intersection of S with the horizontal plane $z = t$ (the trace of S in the plane $z = t$).
    (c) Find the volume of the solid bounded by S and the planes $z = 0$ and $z = 1$.

13

MULTIPLE INTEGRALS

In this chapter we extend the idea of a definite integral to double and triple integrals of functions of two or three variables. These ideas are then used to compute volumes, masses, and centroids of more general regions than we were able to consider in Chapters 5 and 8.

13.1   DOUBLE INTEGRALS OVER RECTANGLES

As a guide to defining the double integral of a function of two variables, let us first recall the basic facts concerning definite integrals of functions of a single variable. If f is defined on a closed interval $[a, b]$, we start by taking a partition P of $[a, b]$ into subintervals $[x_{i-1}, x_i]$ where

$$a = x_0 < x_1 < \cdots < x_{n-1} < x_n = b$$

We choose points x_i^* in $[x_{i-1}, x_i]$ and let $\Delta x_i = x_i - x_{i-1}$ and $\|P\| = \max\{\Delta x_i\}$. Then we form the Riemann sum

(1)
$$\sum_{i=1}^{n} f(x_i^*)\, \Delta x_i$$

and take the limit of such sums as $\|P\| \to 0$ to obtain the definite integral of f from a to b:

(2)
$$\int_a^b f(x)\, dx = \lim_{\|P\| \to 0} \sum_{i=1}^{n} f(x_i^*)\, \Delta x_i$$

In the special case where $f(x) \geqslant 0$, the Riemann sum can be interpreted as the sum of the areas of the approximating rectangles in Figure 1, and $\int_a^b f(x)\, dx$ represents the area under the curve $y = f(x)$ from a to b.

In a similar manner we define the double integral of a function f of two variables that is defined on a closed rectangle

$$R = [a, b] \times [c, d] = \{(x, y) \in \mathbb{R}^2 \mid a \leqslant x \leqslant b,\ c \leqslant y \leqslant d\}$$

The first step is to take a partition P of R into subrectangles. This is accomplished by partitioning the intervals $[a, b]$ and $[c, d]$ as follows:

$$a = x_0 < x_1 < \cdots < x_{i-1} < x_i < \cdots < x_m = b$$

$$c = y_0 < y_1 < \cdots < y_{j-1} < y_j < \cdots < y_n = d$$

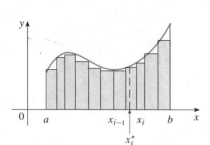

FIGURE 1

By drawing lines parallel to the coordinate axes through these partition points as in Figure 2, we form the subrectangles

$$R_{ij} = [x_{i-1}, x_i] \times [y_{j-1}, y_j] = \{(x, y) \mid x_{i-1} \le x \le x_i, y_{j-1} \le y \le y_j\}$$

for $i = 1, \ldots, m$ and $j = 1, \ldots, n$. There are mn of these subrectangles, and they cover R. If we let

$$\Delta x_i = x_i - x_{i-1} \qquad \Delta y_j = y_j - y_{j-1}$$

then the area of R_{ij} is $\qquad \Delta A_{ij} = \Delta x_i \, \Delta y_j$

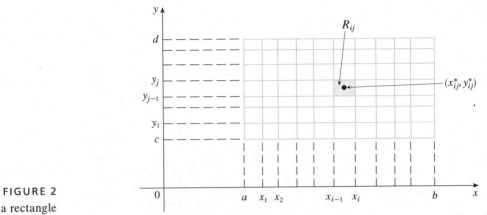

FIGURE 2
Partition of a rectangle

Next we choose a point (x_{ij}^*, y_{ij}^*) in R_{ij} and, by analogy with the Riemann sum (1), we form the **double Riemann sum**

(3)
$$\sum_{i=1}^{m} \sum_{j=1}^{n} f(x_{ij}^*, y_{ij}^*) \, \Delta A_{ij}$$

This double sum means that for each subrectangle we evaluate f at the chosen point and multiply by the area of the subrectangle, and then we add the results. When written out in full, the double Riemann sum (3) becomes

$$f(x_{11}^*, y_{11}^*) \Delta A_{11} + f(x_{12}^*, y_{12}^*) \Delta A_{12} + \cdots + f(x_{1n}^*, y_{1n}^*) \Delta A_{1n}$$
$$+ f(x_{21}^*, y_{21}^*) \Delta A_{21} + f(x_{22}^*, y_{22}^*) \Delta A_{22} + \cdots + f(x_{2n}^*, y_{2n}^*) \Delta A_{2n}$$
$$+ \cdots$$
$$+ f(x_{m1}^*, y_{m1}^*) \Delta A_{m1} + f(x_{m2}^*, y_{m2}^*) \Delta A_{m2} + \cdots + f(x_{mn}^*, y_{mn}^*) \Delta A_{mn}$$

The final ingredient that we need for the definition of a double integral is the **norm** of the partition, which is the length of the longest diagonal of all the subrectangles R_{ij} and is denoted by $\|P\|$. Note that if we let $\|P\| \to 0$, then the partition becomes finer. By analogy with the defining equation for a single integral (2), we make the following definition.

(4) **DEFINITION**   The **double integral** of f over the rectangle R is

$$\iint\limits_R f(x, y) \, dA = \lim_{\|P\| \to 0} \sum_{i=1}^{m} \sum_{j=1}^{n} f(x_{ij}^*, y_{ij}^*) \, \Delta A_{ij}$$

if this limit exists.

NOTE 1:   The precise meaning of the limit that defines the double integral is as follows:

$$\iint\limits_{R} f(x, y)\, dA = I$$

means that for every $\varepsilon > 0$ there is a corresponding number $\delta > 0$ such that

$$\left| I - \sum_{i=1}^{m} \sum_{j=1}^{n} f(x_{ij}^*, y_{ij}^*)\, \Delta A_{ij} \right| < \varepsilon$$

for all partitions P of R with $\|P\| < \delta$ and for all possible choices of (x_{ij}^*, y_{ij}^*) in R_{ij}. In other words, Riemann sums can be made arbitrarily close to I by taking sufficiently fine partitions.

NOTE 2:   In view of the fact that $\Delta A_{ij} = \Delta x_i\, \Delta y_j$, another notation that is sometimes used for the double integral is

$$\iint\limits_{R} f(x, y)\, dA = \iint\limits_{R} f(x, y)\, dx\, dy$$

NOTE 3:   A function f is called **integrable** if the limit in Definition 4 exists. It is shown in courses on advanced calculus that all continuous functions are integrable. In fact, the double integral of f exists provided that f is "not too discontinuous." In particular, if f is bounded on R and is continuous there, except on a finite number of smooth curves, then f is integrable over R.

EXAMPLE 1   Find an approximate value for the integral $\iint_R (x - 3y^2)\, dA$, where $R = \{(x, y) \mid 0 \le x \le 2,\ 1 \le y \le 2\}$, by computing the double Riemann sum with partition lines $x = 1$ and $y = \frac{3}{2}$ and taking (x_{ij}^*, y_{ij}^*) to be the center of each rectangle.

SOLUTION   The partition is shown in Figure 3. The area of each subrectangle is $\Delta A_{ij} = \frac{1}{2}$, (x_{ij}^*, y_{ij}^*) is the center of R_{ij}, and $f(x, y) = x - 3y^2$. So the corresponding Riemann sum is

$$\sum_{i=1}^{2} \sum_{j=1}^{2} f(x_{ij}^*, y_{ij}^*)\, \Delta A_{ij}$$

$$= f(x_{11}^*, y_{11}^*)\, \Delta A_{11} + f(x_{12}^*, y_{12}^*)\, \Delta A_{12} + f(x_{21}^*, y_{21}^*)\, \Delta A_{21} + f(x_{22}^*, y_{22}^*)\, \Delta A_{22}$$

$$= f\left(\tfrac{1}{2}, \tfrac{5}{4}\right) \Delta A_{11} + f\left(\tfrac{1}{2}, \tfrac{7}{4}\right) \Delta A_{12} + f\left(\tfrac{3}{2}, \tfrac{5}{4}\right) \Delta A_{21} + f\left(\tfrac{3}{2}, \tfrac{7}{4}\right) \Delta A_{22}$$

$$= \left(-\tfrac{67}{16}\right)\tfrac{1}{2} + \left(-\tfrac{139}{16}\right)\tfrac{1}{2} + \left(-\tfrac{51}{16}\right)\tfrac{1}{2} + \left(-\tfrac{123}{16}\right)\tfrac{1}{2}$$

$$= -\tfrac{95}{8} = -11.875$$

Thus we have

$$\iint\limits_{R} (x - 3y^2)\, dA \approx -11.875$$

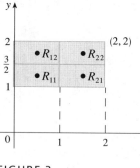

FIGURE 3

NOTE 4:   It is very difficult to calculate double integrals directly from the definition. In the next section we will develop an efficient method for computing double integrals and then we will see that the exact value of the double integral in Example 1 is -12.

Number of subrectangles	Midpoint Rule approximations
1	−11.5000
4	−11.8750
16	−11.9687
64	−11.9922
256	−11.9980
1024	−11.9995

NOTE 5: The methods that we used for approximating single integrals (the Midpoint Rule, the Trapezoidal Rule, Simpson's Rule) all have analogues for double integrals. In the **Midpoint Rule for double integrals** we use a double Riemann sum to approximate the double integral, where the point (x_{ij}^*, y_{ij}^*) in R_{ij} is chosen to be the center $(\bar{x}_i, \bar{y}_j)$ of R_{ij}. In other words, $\bar{x}_i$ is the midpoint of $[x_{i-1}, x_i]$ and $\bar{y}_j$ is the midpoint of $[y_{j-1}, y_j]$. In effect, the approximation we got in Example 1 is the result of using the Midpoint Rule with four subrectangles. If we keep dividing each subrectangle into four smaller ones with similar shape, we get the Midpoint Rule approximations displayed in the chart in the margin. Notice how these approximations approach the exact value of the double integral, -12.

INTERPRETATION OF DOUBLE INTEGRALS AS VOLUMES

Just as single integrals of positive functions can be interpreted as areas, double integrals of positive functions can be interpreted as volumes in the following way. Suppose that $f(x, y) \geq 0$ and f is defined on the rectangle $R = [a, b] \times [c, d]$. The graph of f is a surface with equation $z = f(x, y)$. Let S be the solid that lies above R and under the graph of f, that is,

$$S = \{(x, y, z) \in \mathbb{R}^3 \mid 0 \leq z \leq f(x, y), \ (x, y) \in R\}$$

If we partition R into subrectangles R_{ij} and choose (x_{ij}^*, y_{ij}^*) in R_{ij}, then we can approximate the part of S that lies above R_{ij} by a thin rectangular box (or "column") with base R_{ij} and height $f(x_{ij}^*, y_{ij}^*)$ as shown in Figure 4. (Compare with Figure 1.) The volume of this box is

$$V_{ij} = f(x_{ij}^*, y_{ij}^*) \, \Delta A_{ij}$$

If we follow this procedure for all the rectangles and add the volumes of the corresponding boxes, we get an approximation to the total volume of S:

(5)
$$V \approx \sum_{i=1}^{m} \sum_{j=1}^{n} f(x_{ij}^*, y_{ij}^*) \, \Delta A_{ij}$$

(See Figure 5.) Notice that this double sum is just the Riemann sum (3). Our intuition tells us that the approximation given in (5) becomes better if we use a finer partition P

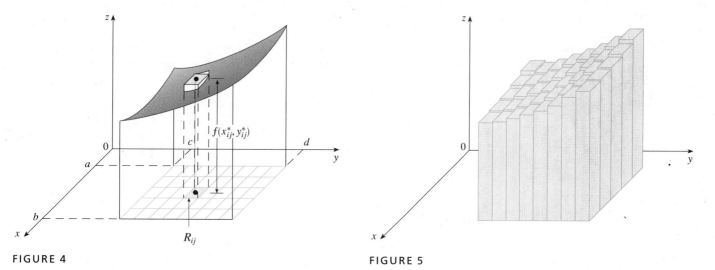

FIGURE 4   FIGURE 5

and so we would expect that

$$V = \lim_{\|P\| \to 0} \sum_{i=1}^{m} \sum_{j=1}^{n} f(x_{ij}^*, y_{ij}^*) \, \Delta A_{ij} = \iint_R f(x, y) \, dA$$

(the latter equality being true by the definition of a double integral). It can be shown, using our previous definition of volume (5.2.1), that this expectation is correct.

---

(6) THEOREM If $f(x, y) \geq 0$ and f is continuous on the rectangle R, then the volume V of the solid that lies above R and under the surface $z = f(x, y)$ is

$$V = \iint_R f(x, y) \, dA$$

It should be remembered that the interpretation of a double integral as a volume is valid only when the integrand f is a *positive* function. The integrand in Example 1 is not a positive function, so its integral is not a volume.

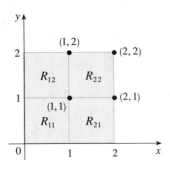

FIGURE 6

EXAMPLE 2 Estimate the volume of the solid that lies above the square $R = [0, 2] \times [0, 2]$ and below the elliptic paraboloid $z = 16 - x^2 - 2y^2$. Use the partition of R into four equal squares and choose (x_{ij}^*, y_{ij}^*) to be the upper right corner of R_{ij}. Sketch the solid and the approximating rectangular boxes.

SOLUTION The partition is shown in Figure 6. The paraboloid is the graph of $f(x, y) = 16 - x^2 - 2y^2$ and the area of each square is 1. Approximating the volume by the Riemann sum, we have

$$V \approx f(1, 1) \, \Delta A_{11} + f(1, 2) \, \Delta A_{12} + f(2, 1) \, \Delta A_{21} + f(2, 2) \, \Delta A_{22}$$

$$= 13(1) + 7(1) + 10(1) + 4(1) = 34$$

This is the volume of the approximating rectangular boxes shown in Figure 7. ∎

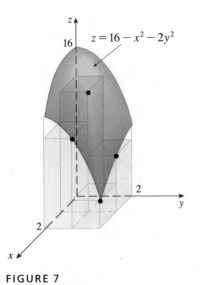

FIGURE 7

We list here three properties of double integrals that can be proved in the same manner as in Section 4.3. We assume that all of the integrals exist. Properties 7 and 8 are referred to as the *linearity* of the integral.

$$(7) \qquad \iint_R [f(x, y) + g(x, y)] \, dA = \iint_R f(x, y) \, dA + \iint_R g(x, y) \, dA$$

$$(8) \qquad \iint_R c f(x, y) \, dA = c \iint_R f(x, y) \, dA \qquad \text{where } c \text{ is a constant}$$

If $f(x, y) \geq g(x, y)$ for all (x, y) in R, then

$$(9) \qquad \iint_R f(x, y) \, dA \geq \iint_R g(x, y) \, dA$$

EXERCISES 13.1

1. Find approximations to $\iint_R (x - 3y^2)\, dA$ using the same partition as in Example 1 but choosing (x_{ij}^*, y_{ij}^*) to be the (a) upper left corner, (b) upper right corner, (c) lower left corner, (d) lower right corner of R_{ij}.

2. Find the approximation to the volume in Example 2 if (x_{ij}^*, y_{ij}^*) is chosen to be the center of each square.

3–8 ■ Calculate the double Riemann sum of f for the partition of R given by the indicated lines and the given choice of (x_{ij}^*, y_{ij}^*). Also calculate the norm of the partition.

3. $f(x, y) = x^2 + 4y$, $\quad R = \{(x, y) \mid 0 \leqslant x \leqslant 2, 0 \leqslant y \leqslant 3\}$, $x = 1, y = 1, y = 2$; $\quad (x_{ij}^*, y_{ij}^*) = $ upper right corner of R_{ij}

4. $f(x, y) = x^2 + 4y$, $\quad R = \{(x, y) \mid 0 \leqslant x \leqslant 2, 0 \leqslant y \leqslant 3\}$, $x = 1, y = 1, y = 2$; $\quad (x_{ij}^*, y_{ij}^*) = $ center of R_{ij}

5. $f(x, y) = xy - y^2$, $\quad R = \{(x, y) \mid 0 \leqslant x \leqslant 5, 0 \leqslant y \leqslant 4\}$, $x = 1, x = 2, x = 3, x = 4, y = 2$; $(x_{ij}^*, y_{ij}^*) = $ center of R_{ij}

6. $f(x, y) = 2x + x^2 y$, $R = \{(x, y) \mid -2 \leqslant x \leqslant 2, -1 \leqslant y \leqslant 1\}$, $x = -1, x = 0, x = 1, y = -\frac{1}{2}, y = 0, y = \frac{1}{2}$; $(x_{ij}^*, y_{ij}^*) = $ lower left corner of R_{ij}

7. $f(x, y) = x^2 - y^2$, $\quad R = [0, 5] \times [0, 2]$, $\quad x = 1, x = 3$, $x = 4, y = \frac{1}{2}, y = 1$; $\quad (x_{ij}^*, y_{ij}^*) = $ upper left corner of R_{ij}

8. $f(x, y) = 5xy^2$, $\quad R = [1, 3] \times [1, 4]$, $\quad x = 1.8, x = 2.5$, $y = 2, y = 3$; $\quad (x_{ij}^*, y_{ij}^*) = $ lower right corner of R_{ij}

9. Let V be the volume of the solid that lies under the graph of $f(x, y) = \sqrt{52 - x^2 - y^2}$ and above the rectangle given by $2 \leqslant x \leqslant 4, 2 \leqslant y \leqslant 6$. For the partition of R defined by the lines $x = 3$ and $y = 4$, let L and U be the Riemann sums computed using lower left corners and upper right corners, respectively. Without calculating the numbers V, L, and U, arrange them in increasing order and explain your reasoning.

10. The figure shows level curves of a function f in the square $R = [0, 1] \times [0, 1]$. Use them to estimate $\iint_R f(x, y)\, dA$ to the nearest integer.

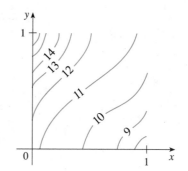

11. Use a programmable calculator or computer (or the sum command on a CAS) to estimate $\iint_R e^{-x^2 - y^2}\, dA$, where $R = [0, 1] \times [0, 1]$. Use the Midpoint Rule with the following numbers of squares of equal size: 1, 4, 16, 64, 256, and 1024.

12. Repeat Exercise 11 for the integral $\iint_R \cos(x^4 + y^4)\, dA$.

13. Evaluate the integral $\iint_R (4 - 2y)\, dA$, where $R = [0, 1] \times [0, 1]$, by first identifying it as the volume of a solid.

14. The integral $\iint_R \sqrt{9 - y^2}\, dA$, where $R = [0, 4] \times [0, 2]$, represents the volume of a solid. Sketch the solid.

15. If f is a constant function, $f(x, y) = k$, and $R = [a, b] \times [c, d]$, show that

$$\iint_R k\, dA = k(b - a)(d - c)$$

16. If $R = [0, 1] \times [0, 1]$, show that

$$0 \leqslant \iint_R \sin(x + y)\, dA \leqslant 1$$

13.2 ITERATED INTEGRALS

Recall that it is usually difficult to evaluate single integrals from first principles, but the Fundamental Theorem of Calculus provides a much easier method. The evaluation of double integrals from first principles is even more difficult, but in this section we see how to express a double integral as an iterated integral, which can then be evaluated by calculating two single integrals.

Suppose f is a function of two variables that is integrable over the rectangle $R = [a, b] \times [c, d]$. We use the notation $\int_c^d f(x, y)\, dy$ to mean that x is held fixed and $f(x, y)$ is integrated with respect to y from $y = c$ to $y = d$. This procedure is called *partial integration with respect to y*. (Notice its similarity to partial differentiation.)

Now $\int_c^d f(x, y)\, dy$ is a number that depends on the value of x, so it defines a function of x:

$$A(x) = \int_c^d f(x, y)\, dy$$

If we now integrate the function A with respect to x from $x = a$ to $x = b$, we get

(1)
$$\int_a^b A(x)\, dx = \int_a^b \left[\int_c^d f(x, y)\, dy \right] dx$$

The integral on the right side of Equation 1 is called an **iterated integral.** Usually the brackets are omitted. Thus

(2)
$$\int_a^b \int_c^d f(x, y)\, dy\, dx = \int_a^b \left[\int_c^d f(x, y)\, dy \right] dx$$

means that we first integrate with respect to y from c to d and then with respect to x from a to b.

Similarly, the iterated integral

(3)
$$\int_c^d \int_a^b f(x, y)\, dx\, dy = \int_c^d \left[\int_a^b f(x, y)\, dx \right] dy$$

means that we first integrate with respect to x (holding y fixed) from $x = a$ to $x = b$ and then we integrate the resulting function of y with respect to y from $y = c$ to $y = d$. Notice that in both Equations 2 and 3 we work from the inside out.

EXAMPLE 1   Evaluate the iterated integrals:

(a) $\displaystyle\int_0^3 \int_1^2 x^2 y\, dy\, dx$ (b) $\displaystyle\int_1^2 \int_0^3 x^2 y\, dx\, dy$

SOLUTION
(a) Regarding x as a constant, we obtain

$$\int_1^2 x^2 y\, dy = \left[x^2 \frac{y^2}{2} \right]_{y=1}^{y=2}$$

$$= x^2 \left(\frac{2^2}{2} \right) - x^2 \left(\frac{1^2}{2} \right) = \tfrac{3}{2} x^2$$

Thus the function A in the preceding discussion is given by $A(x) = \frac{3}{2} x^2$ in this example. We now integrate this function of x from 0 to 3:

$$\int_0^3 \int_1^2 x^2 y\, dy\, dx = \int_0^3 \left[\int_1^2 x^2 y\, dy \right] dx$$

$$= \int_0^3 \tfrac{3}{2} x^2\, dx = \frac{x^3}{2} \bigg]_0^3 = \frac{27}{2}$$

(b) Here we first integrate with respect to x:

$$\int_1^2 \int_0^3 x^2 y\, dx\, dy = \int_1^2 \left[\int_0^3 x^2 y\, dx \right] dy = \int_1^2 \left[\frac{x^3}{3} y \right]_{x=0}^{x=3} dy$$

$$= \int_1^2 9y\, dy = 9 \frac{y^2}{2} \bigg]_1^2 = \frac{27}{2}$$ ∎

Notice that in Example 1 we obtained the same answer whether we integrated with respect to y or x first. In general, it turns out (see Theorem 4) that the two iterated integrals in Equations 2 and 3 are always equal; that is, the order of integration does not matter. (This is similar to Clairaut's Theorem on the equality of the mixed partial derivatives.)

The following theorem gives a practical method for evaluating a double integral by expressing it as an iterated integral (in either order).

Theorem 4 is named after the Italian mathematician Guido Fubini (1879–1943), who proved a very general version of this theorem in 1907. But the version for continuous functions was known to the French mathematician Augustin-Louis Cauchy almost a century earlier.

---

(4) FUBINI'S THEOREM   If f is continuous on the rectangle $R = \{(x, y) \mid a \leq x \leq b, c \leq y \leq d\}$, then

$$\iint\limits_{R} f(x, y)\, dA = \int_a^b \int_c^d f(x, y)\, dy\, dx = \int_c^d \int_a^b f(x, y)\, dx\, dy$$

More generally, this is true if we assume that f is bounded on R, f is discontinuous only on a finite number of smooth curves, and the iterated integrals exist.

The proof of Fubini's Theorem is too difficult to include in this book, but we can at least give an intuitive indication of why it is true for the case where $f(x, y) \geq 0$. Recall that if f is positive, then we can interpret the double integral $\iint_R f(x, y)\, dA$ as the volume V of the solid S that lies above R and under the surface $z = f(x, y)$. But we have another formula that we used for volume in Chapter 5, namely,

$$V = \int_a^b A(x)\, dx$$

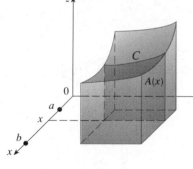

where $A(x)$ is the area of a cross-section of S in the plane through x perpendicular to the x-axis. From Figure 1 you can see that $A(x)$ is the area under the curve C whose equation is $z = f(x, y)$, where x is held constant and $c \leq y \leq d$. Therefore

FIGURE 1

$$A(x) = \int_c^d f(x, y)\, dy$$

and we have

$$\iint\limits_{R} f(x, y)\, dA = V = \int_a^b A(x)\, dx = \int_a^b \int_c^d f(x, y)\, dy\, dx$$

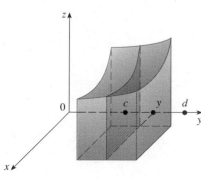

A similar argument, using cross-sections perpendicular to the y-axis as in Figure 2, shows that

$$\iint\limits_{R} f(x, y)\, dA = \int_c^d \int_a^b f(x, y)\, dx\, dy$$

FIGURE 2

EXAMPLE 2   Evaluate the double integral $\iint_R (x - 3y^2)\, dA$, where $R = \{(x, y) \mid 0 \leq x \leq 2,\ 1 \leq y \leq 2\}$. (Compare with Example 1 in Section 13.1.)

SOLUTION 1   Fubini's Theorem gives

$$\iint_R (x - 3y^2)\, dA = \int_0^2 \int_1^2 (x - 3y^2)\, dy\, dx$$

$$= \int_0^2 \left[xy - y^3 \right]_{y=1}^{y=2} dx$$

$$= \int_0^2 (x - 7)\, dx = \frac{x^2}{2} - 7x \Big]_0^2 = -12$$

Notice the negative answer in Example 2; nothing is wrong with that. The function f in that example is not a positive function, so its integral does not represent a volume. From Figure 3 we see that f is always negative on R, so the value of the integral is the *negative* of the volume that lies *above* the graph of f and *below* R.

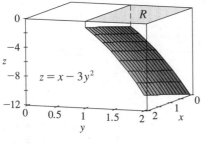

FIGURE 3

SOLUTION 2   Again applying Fubini's Theorem, but this time integrating with respect to x first, we have

$$\iint_R (x - 3y^2)\, dA = \int_1^2 \int_0^2 (x - 3y^2)\, dx\, dy$$

$$= \int_1^2 \left[\frac{x^2}{2} - 3xy^2 \right]_{x=0}^{x=2} dy$$

$$= \int_1^2 (2 - 6y^2)\, dy = 2y - 2y^3 \Big]_1^2 = -12$$

■

EXAMPLE 3   Evaluate $\iint_R y \sin(xy)\, dA$, where $R = [1, 2] \times [0, \pi]$.

SOLUTION 1   If we first integrate with respect to x, we get

$$\iint_R y \sin(xy)\, dA = \int_0^\pi \int_1^2 y \sin(xy)\, dx\, dy$$

$$= \int_0^\pi \left[-\cos(xy) \right]_{x=1}^{x=2} dy$$

$$= \int_0^\pi (-\cos 2y + \cos y)\, dy$$

$$= -\tfrac{1}{2} \sin 2y + \sin y \Big]_0^\pi = 0$$

For a function f that takes on both positive and negative values, $\iint_R f(x, y)\, dA$ is a difference of volumes: $V_1 - V_2$, where V_1 is the volume above R and below the graph of f and V_2 is the volume below R and above the graph. The fact that the integral in Example 3 is 0 means that these two volumes V_1 and V_2 are equal. (See Figure 4.)

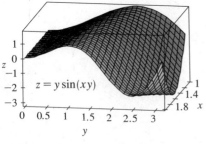

FIGURE 4

SOLUTION 2   If we reverse the order of integration, we get

$$\iint_R y \sin(xy)\, dA = \int_1^2 \int_0^\pi y \sin(xy)\, dy\, dx$$

To evaluate the inner integral we use integration by parts with

$$u = y \qquad\qquad dv = \sin(xy)\, dy$$

$$du = dy \qquad\qquad v = -\frac{\cos(xy)}{x}$$

and so
$$\int_0^\pi y \sin(xy)\, dy = -\frac{y\cos(xy)}{x}\Big]_{y=0}^{y=\pi} + \frac{1}{x}\int_0^\pi \cos(xy)\, dy$$

$$= -\frac{\pi\cos\pi x}{x} + \frac{1}{x^2}\big[\sin(xy)\big]_{y=0}^{y=\pi}$$

$$= -\frac{\pi\cos\pi x}{x} + \frac{\sin\pi x}{x^2}$$

If we now integrate the first term by parts with $u = -1/x$ and $dv = \pi\cos\pi x\, dx$, we get $du = dx/x^2$, $v = \sin\pi x$, and

$$\int\left(-\frac{\pi\cos\pi x}{x}\right) dx = -\frac{\sin\pi x}{x} - \int\frac{\sin\pi x}{x^2}\, dx$$

Therefore
$$\int\left(-\frac{\pi\cos\pi x}{x} + \frac{\sin\pi x}{x^2}\right) dx = -\frac{\sin\pi x}{x}$$

and so
$$\int_1^2\int_0^\pi y\sin(xy)\, dy\, dx = \left[-\frac{\sin\pi x}{x}\right]_1^2$$

$$= -\frac{\sin 2\pi}{2} + \sin\pi = 0 \qquad\blacksquare$$

In Example 2, Solutions 1 and 2 are equally straightforward, but in Example 3 the first solution is much easier than the second one. Therefore, when we evaluate double integrals it is wise to choose the order of integration that gives simpler integrals.

EXAMPLE 4   Find the volume of the solid S that is bounded by the elliptic paraboloid $x^2 + 2y^2 + z = 16$, the planes $x = 2$ and $y = 2$, and the three coordinate planes.

SOLUTION   We first observe that S is the solid that lies under the surface $z = 16 - x^2 - 2y^2$ and above the square $R = [0,2]\times[0,2]$ (see Figure 5). This solid was considered in Example 2 in Section 13.1, but we are now in a position to evaluate the double integral using Fubini's Theorem. Therefore

$$V = \iint\limits_R (16 - x^2 - 2y^2)\, dA = \int_0^2\int_0^2 (16 - x^2 - 2y^2)\, dx\, dy$$

$$= \int_0^2\left[16x - \frac{x^3}{3} - 2y^2x\right]_{x=0}^{x=2} dy$$

$$= \int_0^2 \left(\tfrac{88}{3} - 4y^2\right) dy = \left[\tfrac{88}{3}y - \tfrac{4}{3}y^3\right]_0^2 = 48 \qquad\blacksquare$$

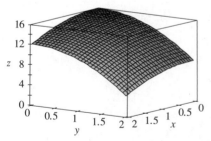

FIGURE 5

In the special case where $f(x, y)$ can be factored as the product of a function of x only and a function of y only, the double integral of f can be written in a particularly simple form. To be specific, suppose that $f(x, y) = g(x)h(y)$ and $R = [a, b]\times[c, d]$. Then Fubini's Theorem gives

$$\iint\limits_R f(x, y)\, dA = \int_c^d\int_a^b g(x)h(y)\, dx\, dy = \int_c^d\left[\int_a^b g(x)h(y)\, dx\right] dy$$

In the inner integral y is a constant, so $h(y)$ is a constant and we can write

$$\int_c^d \left[\int_a^b g(x)h(y)\,dx \right] dy = \int_c^d \left[h(y) \left(\int_a^b g(x)\,dx \right) \right] dy$$

$$= \int_a^b g(x)\,dx \int_c^d h(y)\,dy$$

since $\int_a^b g(x)\,dx$ is a constant. Therefore, in this case, the double integral of f can be written as the product of two single integrals:

$$\iint_R g(x)h(y)\,dA = \int_a^b g(x)\,dx \int_c^d h(y)\,dy \qquad \text{where } R = [a,b] \times [c,d]$$

The function $f(x,y) = \sin x \cos y$ in Example 5 is positive on R, so the integral represents the volume of the solid that lies above R and below the graph of f shown in Figure 6.

EXAMPLE 5   If $R = [0, \pi/2] \times [0, \pi/2]$, then

$$\iint_R \sin x \cos y\,dA = \int_0^{\pi/2} \sin x\,dx \int_0^{\pi/2} \cos y\,dy$$

$$= \big[-\cos x \big]_0^{\pi/2} \big[\sin y \big]_0^{\pi/2}$$

$$= 1 \cdot 1 = 1 \qquad\blacksquare$$

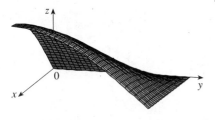

FIGURE 6

EXERCISES 13.2

1–4 ■ Find $\int_0^2 f(x,y)\,dy$ and $\int_0^1 f(x,y)\,dx$.

1. $f(x,y) = x^2 y^3$

2. $f(x,y) = 2xy - 3x^2$

3. $f(x,y) = xe^{x+y}$

4. $f(x,y) = \dfrac{x}{y^2 + 1}$

5–14 ■ Calculate the iterated integral.

5. $\displaystyle\int_0^4 \int_0^2 x\sqrt{y}\,dx\,dy$

6. $\displaystyle\int_0^2 \int_0^3 e^{x-y}\,dy\,dx$

7. $\displaystyle\int_{-1}^1 \int_0^1 (x^3 y^3 + 3xy^2)\,dy\,dx$ **8.** $\displaystyle\int_0^1 \int_1^2 (x^4 - y^2)\,dx\,dy$

9. $\displaystyle\int_0^3 \int_0^1 \sqrt{x+y}\,dx\,dy$

10. $\displaystyle\int_0^{\pi/2} \int_0^{\pi/2} \sin(x+y)\,dy\,dx$

11. $\displaystyle\int_0^{\pi/4} \int_0^3 \sin x\,dy\,dx$

12. $\displaystyle\int_1^2 \int_0^1 (x+y)^{-2}\,dx\,dy$

13. $\displaystyle\int_0^{\ln 2} \int_0^{\ln 5} e^{2x-y}\,dx\,dy$

14. $\displaystyle\int_0^1 \int_0^1 \dfrac{xy}{\sqrt{x^2+y^2+1}}\,dy\,dx$

15–22 ■ Calculate the double integral.

15. $\displaystyle\iint_R (2y^2 - 3xy^3)\,dA$, $R = \{(x,y) \mid 1 \le x \le 2,\, 0 \le y \le 3\}$

16. $\displaystyle\iint_R \left(xy^2 + \dfrac{y}{x} \right) dA$, $R = \{(x,y) \mid 2 \le x \le 3,\, -1 \le y \le 0\}$

17. $\displaystyle\iint_R x\sin y\,dA$, $R = \{(x,y) \mid 1 \le x \le 4,\, 0 \le y \le \pi/6\}$

18. $\displaystyle\iint_R \dfrac{1+x}{1+y}\,dA$, $R = \{(x,y) \mid -1 \le x \le 2,\, 0 \le y \le 1\}$

19. $\displaystyle\iint_R x\sin(x+y)\,dA$, $R = [0, \pi/6] \times [0, \pi/3]$

20. $\displaystyle\iint_R xe^{xy}\,dA$, $R = [0,1] \times [0,1]$

21. $\displaystyle\iint_R \dfrac{1}{x+y}\,dA$, $R = [1,2] \times [0,1]$

22. $\displaystyle\iint_R xye^{xy^2}\,dA$, $R = [0,1] \times [0,1]$

23–24 ■ Sketch the solid whose volume is given by the iterated integral.

23. $\displaystyle\int_0^1 \int_0^1 (4 - x - 2y)\,dx\,dy$ **24.** $\displaystyle\int_0^1 \int_0^1 (2 - x^2 - y^2)\,dy\,dx$

25. Find the volume of the solid lying under the plane $z = 2x + 5y + 1$ and above the rectangle $\{(x,y) \mid -1 \le x \le 0,\, 1 \le y \le 4\}$.

26. Find the volume of the solid lying under the circular paraboloid $z = x^2 + y^2$ and above the rectangle $R = [-2,2] \times [-3,3]$.

27. Find the volume of the solid lying under the elliptic paraboloid $x^2/4 + y^2/9 + z = 1$ and above the square $R = [-1, 1] \times [-2, 2]$.

28. Find the volume of the solid lying under the hyperbolic paraboloid $z = y^2 - x^2$ and above the square $R = [-1, 1] \times [1, 3]$.

29. Find the volume of the solid bounded by the surface $z = x\sqrt{x^2 + y}$ and the planes $x = 0$, $x = 1$, $y = 0$, $y = 1$, and $z = 0$.

30. Find the volume of the solid bounded by the elliptic paraboloid $z = 1 + (x - 1)^2 + 4y^2$, the planes $x = 3$ and $y = 2$, and the coordinate planes.

31. Find the volume of the solid in the first octant bounded by the cylinder $z = 9 - y^2$ and the plane $x = 2$.

32. (a) Find the volume of the solid bounded by the surface $z = 6 - xy$ and the planes $x = 2$, $x = -2$, $y = 0$, $y = 3$, and $z = 0$.
   (b) Use a computer to draw the solid.

33. Use a computer algebra system to find the exact value of the integral $\iint_R x^5 y^3 e^{xy} \, dA$, where $R = [0, 1] \times [0, 1]$. Then use the CAS to draw the solid whose volume is given by the integral.

34. Graph the solid that lies between the surfaces $z = e^{-x^2} \cos(x^2 + y^2)$ and $z = 2 - x^2 - y^2$ for $|x| \leq 1$,

$|y| \leq 1$. Use a computer algebra system to approximate the volume of this solid correct to four decimal places.

35–36 ■ The **average value** of a function $f(x, y)$ over a rectangle R is defined to be

$$f_{\text{ave}} = \frac{1}{A(R)} \iint_R f(x, y) \, dA$$

where $A(R)$ is the area of R. (Compare with the definition for functions of one variable in Equation 5.5.1.) Find the average value of f over the given rectangle.

35. $f(x, y) = x^2 y$, R has vertices $(-1, 0)$, $(-1, 5)$, $(1, 5)$, $(1, 0)$

36. $f(x, y) = x \sin xy$, $R = [0, \pi/2] \times [0, 1]$

37. Use your CAS to compute the iterated integrals

$$\int_0^1 \int_0^1 \frac{x - y}{(x + y)^3} \, dy \, dx \quad \text{and} \quad \int_0^1 \int_0^1 \frac{x - y}{(x + y)^3} \, dx \, dy$$

Do the answers contradict Fubini's Theorem? Explain what is happening.

38. (a) In what way are the theorems of Fubini and Clairaut similar?
   (b) If $f(x, y)$ is continuous on $[a, b] \times [c, d]$ and
$$g(x, y) = \int_a^x \int_c^y f(s, t) \, dt \, ds \quad \text{for } a < x < b, \, c < y < d$$
   show that $g_{xy} = g_{yx} = f(x, y)$.

13.3 DOUBLE INTEGRALS OVER GENERAL REGIONS

For single integrals, the region over which we integrate is always an interval. But, for double integrals, we want to be able to integrate not just over rectangles but also over regions D of more general shape, such as the one illustrated in Figure 1. We suppose that D is a bounded region, which means that D can be enclosed in a rectangular region R as in Figure 2. Then we define a new function F with domain R by

$$(1) \qquad F(x, y) = \begin{cases} f(x, y) & \text{if } (x, y) \text{ is in } D \\ 0 & \text{if } (x, y) \text{ is in } R \text{ but not in } D \end{cases}$$

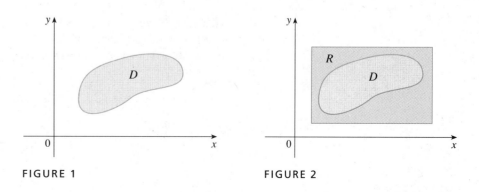

FIGURE 1   FIGURE 2

If F is integrable over R, then we say f is **integrable** over D and we define the **double integral of f over D** by

$$\text{(2)} \qquad \iint_D f(x, y)\, dA = \iint_R F(x, y)\, dA \qquad \text{where } F \text{ is given by Equation 1.}$$

Definition 2 makes sense because R is a rectangle and so $\iint_R F(x, y)\, dA$ has been previously defined in Section 13.1. The procedure that we have used is reasonable because the values of $F(x, y)$ are 0 when (x, y) lies outside D and so they contribute nothing to the integral. This means that it does not matter what rectangle R we use as long as it contains D.

In the case where $f(x, y) \geq 0$ we can still interpret $\iint_D f(x, y)\, dA$ as the volume of the solid that lies above D and under the surface $z = f(x, y)$ (the graph of f). You can see that this is reasonable by comparing the graphs of f and F in Figures 3 and 4 and remembering that $\iint_R F(x, y)\, dA$ is the volume under the graph of F.

Figure 4 also shows that F is likely to have discontinuities at the boundary points of D. Nonetheless if f is continuous on D and the boundary curve of D is "well-behaved" (in a sense outside the scope of this book), then it can be shown that F is integrable over R and therefore f is integrable over D. In particular, this is the case for the following types of regions.

A plane region D is said to be of **type I** if it lies between the graphs of two continuous functions of x, that is,

$$D = \{(x, y) \mid a \leq x \leq b, \ g_1(x) \leq y \leq g_2(x)\}$$

where g_1 and g_2 are continuous on $[a, b]$. Some examples of type I regions are shown in Figure 5.

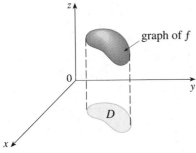

FIGURE 3

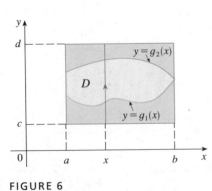

FIGURE 4

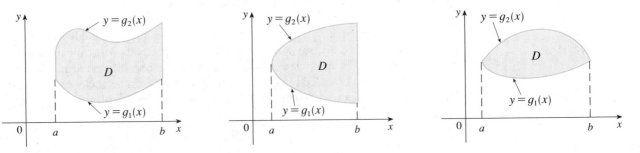

FIGURE 5   Some type I regions

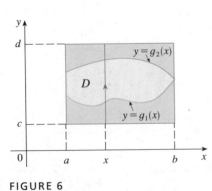

FIGURE 6

In order to evaluate $\iint_D f(x, y)\, dA$ when D is a region of type I, we choose a rectangle $R = [a, b] \times [c, d]$ that contains D, as in Figure 6, and we let F be the function given by Equation 1; that is, F agrees with f on D and F is 0 outside D. Then, by Fubini's Theorem,

$$\iint_D f(x, y)\, dA = \iint_R F(x, y)\, dA = \int_a^b \int_c^d F(x, y)\, dy\, dx$$

Observe that $F(x, y) = 0$ if $y < g_1(x)$ or $y > g_2(x)$ since (x, y) then lies outside D. Therefore

$$\int_c^d F(x, y)\, dy = \int_{g_1(x)}^{g_2(x)} F(x, y)\, dy = \int_{g_1(x)}^{g_2(x)} f(x, y)\, dy$$

because $F(x, y) = f(x, y)$ when $g_1(x) \leq y \leq g_2(x)$. Thus we have the following formula that enables us to evaluate the double integral as an iterated integral.

---

(3) If f is continuous on a type I region D such that

$$D = \{(x, y) \,|\, a \leq x \leq b, g_1(x) \leq y \leq g_2(x)\}$$

then

$$\iint_D f(x, y) \, dA = \int_a^b \int_{g_1(x)}^{g_2(x)} f(x, y) \, dy \, dx$$

The integral on the right side of (3) is an iterated integral that is similar to the ones we considered in the preceding section, except that in the inner integral we regard x as being constant not only in $f(x, y)$ but also in the limits of integration, $g_1(x)$ and $g_2(x)$.

We also consider plane regions of **type II,** which can be expressed as

(4) $$D = \{(x, y) \,|\, c \leq y \leq d, h_1(y) \leq x \leq h_2(y)\}$$

where h_1 and h_2 are continuous. Two such regions are illustrated in Figure 7.

Using the same methods that were used in establishing (4), we can show that

---

(5) $$\iint_D f(x, y) \, dA = \int_c^d \int_{h_1(y)}^{h_2(y)} f(x, y) \, dx \, dy$$

where D is a type II region given by Equation 4.

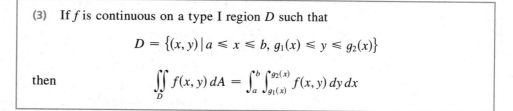

FIGURE 7
Some type II regions

EXAMPLE 1 Evaluate $\iint_D (x + 2y) \, dA$, where D is the region bounded by the parabolas $y = 2x^2$ and $y = 1 + x^2$.

SOLUTION The parabolas intersect when $2x^2 = 1 + x^2$, that is, $x^2 = 1$, so $x = \pm 1$. We note that the region D, sketched in Figure 8, is a type I region but not a type II region and we can write

$$D = \{(x, y) \,|\, -1 \leq x \leq 1, 2x^2 \leq y \leq 1 + x^2\}$$

Since the lower boundary is $y = 2x^2$ and the upper boundary is $y = 1 + x^2$, Equation 3 gives

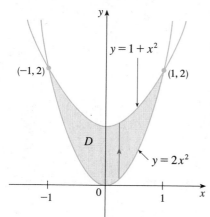

FIGURE 8

$$\iint_D (x + 2y) \, dA = \int_{-1}^1 \int_{2x^2}^{1+x^2} (x + 2y) \, dy \, dx$$

$$= \int_{-1}^1 \left[xy + y^2 \right]_{y=2x^2}^{y=1+x^2} dx$$

$$= \int_{-1}^1 \left[x(1 + x^2) + (1 + x^2)^2 - x(2x^2) - (2x^2)^2 \right] dx$$

$$= \int_{-1}^1 (-3x^4 - x^3 + 2x^2 + x + 1) \, dx$$

$$= -3\frac{x^5}{5} - \frac{x^4}{4} + 2\frac{x^3}{3} + \frac{x^2}{2} + x \Big]_{-1}^1 = \frac{32}{15}$$

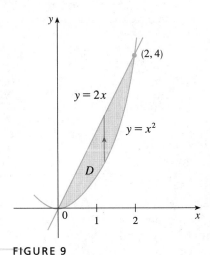

FIGURE 9
D as a type I region

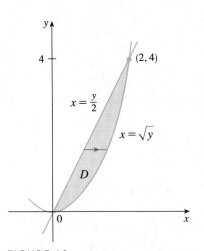

FIGURE 10
D as a type II region

Figure 11 shows the solid whose volume is calculated in Example 2. It lies above the *xy*-plane, below the paraboloid $z = x^2 + y^2$, and between the plane $y = 2x$ and the parabolic cylinder $y = x^2$.

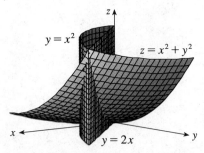

FIGURE 11

NOTE: When we set up a double integral as in Example 1, it is essential to draw a diagram. Often it is helpful to draw a vertical arrow as in Figure 8. Then the limits of integration for the *inner* integral can be read from the diagram as follows: The arrow starts at the lower boundary $y = g_1(x)$, which gives the lower limit in the integral, and the arrow ends at the upper boundary $y = g_2(x)$, which gives the upper limit of integration. For a type II region the arrow is horizontal and goes from the left boundary to the right boundary.

EXAMPLE 2   Find the volume of the solid that lies under the paraboloid $z = x^2 + y^2$ and above the region *D* in the *xy*-plane bounded by the line $y = 2x$ and the parabola $y = x^2$.

SOLUTION 1   From Figure 9 we see that *D* is a type I region and

$$D = \{(x, y) \mid 0 \leq x \leq 2, \ x^2 \leq y \leq 2x\}$$

Therefore, the volume under $z = x^2 + y^2$ and above *D* is

$$V = \iint_D (x^2 + y^2)\, dA = \int_0^2 \int_{x^2}^{2x} (x^2 + y^2)\, dy\, dx$$

$$= \int_0^2 \left[x^2 y + \frac{y^3}{3} \right]_{y=x^2}^{y=2x} dx = \int_0^2 \left[x^2(2x) + \frac{(2x)^3}{3} - x^2 x^2 - \frac{(x^2)^3}{3} \right] dx$$

$$= \int_0^2 \left(-\frac{x^6}{3} - x^4 + \frac{14x^3}{3} \right) dx = -\frac{x^7}{21} - \frac{x^5}{5} + \frac{7x^4}{6} \bigg]_0^2 = \frac{216}{35}$$

SOLUTION 2   From Figure 10 we see that *D* can also be written as a type II region:

$$D = \left\{ (x, y) \ \middle| \ 0 \leq y \leq 4, \ \frac{y}{2} \leq x \leq \sqrt{y} \right\}$$

Therefore, another expression for *V* is

$$V = \iint_D (x^2 + y^2)\, dA = \int_0^4 \int_{y/2}^{\sqrt{y}} (x^2 + y^2)\, dx\, dy$$

$$= \int_0^4 \left[\frac{x^3}{3} + y^2 x \right]_{x=y/2}^{x=\sqrt{y}} dy = \int_0^4 \left(\frac{y^{3/2}}{3} + y^{5/2} - \frac{y^3}{24} - \frac{y^3}{2} \right) dy$$

$$= \frac{2}{15} y^{5/2} + \frac{2}{7} y^{7/2} - \frac{13}{96} y^4 \bigg]_0^4 = \frac{216}{35} \qquad \blacksquare$$

EXAMPLE 3   Evaluate $\iint_D xy\, dA$ where *D* is the region bounded by the line $y = x - 1$ and the parabola $y^2 = 2x + 6$.

SOLUTION   The region *D* is shown in Figure 12. Again *D* is both type I and type II, but the description of *D* as a type I region is more complicated because the lower boundary consists of two parts. Therefore, we prefer to express *D* as a type II region:

$$D = \left\{ (x, y) \ \middle| \ -2 \leq y \leq 4, \ \frac{y^2}{2} - 3 \leq x \leq y + 1 \right\}$$

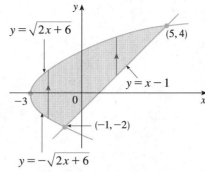

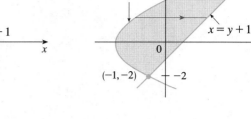

FIGURE 12   (a) D as a type I region   (b) D as a type II region

Then (5) gives

$$\iint_D xy\, dA = \int_{-2}^{4}\int_{y^2/2-3}^{y+1} xy\, dx\, dy = \int_{-2}^{4}\left[\frac{x^2}{2}y\right]_{x=y^2/2-3}^{x=y+1} dy$$

$$= \frac{1}{2}\int_{-2}^{4} y\left[(y+1)^2 - \left(\frac{y^2}{2}-3\right)^2\right] dy$$

$$= \frac{1}{2}\int_{-2}^{4}\left(-\frac{y^5}{4} + 4y^3 + 2y^2 - 8y\right) dy$$

$$= \frac{1}{2}\left[-\frac{y^6}{24} + y^4 + 2\frac{y^3}{3} - 4y^2\right]_{-2}^{4} = 36$$

If we had expressed D as a type I region using Figure 12(a), then we would have obtained

$$\iint_D xy\, dA = \int_{-3}^{-1}\int_{-\sqrt{2x+6}}^{\sqrt{2x+6}} xy\, dy\, dx + \int_{-1}^{5}\int_{x-1}^{\sqrt{2x+6}} xy\, dy\, dx$$

but this would have involved more work than the above method.   ∎

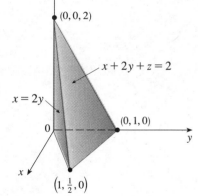

FIGURE 13

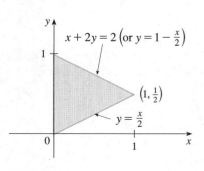

FIGURE 14

EXAMPLE 4   Find the volume of the tetrahedron bounded by the planes $x + 2y + z = 2$, $x = 2y$, $x = 0$, and $z = 0$.

SOLUTION   In a question such as this, it is wise to draw two diagrams: one of the three-dimensional solid and another of the plane region D over which it lies. Figure 13 shows the tetrahedron T bounded by the coordinate planes $x = 0$, $z = 0$, the vertical plane $x = 2y$, and the plane $x + 2y + z = 2$. Since the plane $x + 2y + z = 2$ intersects the xy-plane (whose equation is $z = 0$) in the line $x + 2y = 2$, we see that T lies above the triangular region D in the xy-plane bounded by the lines $x = 2y$, $x + 2y = 2$, and $x = 0$ (see Figure 14).

    The plane $x + 2y + z = 2$ can be written as $z = 2 - x - 2y$, so the required volume lies under the graph of the function $z = 2 - x - 2y$ and above

$$D = \left\{(x, y) \,\middle|\, 0 \le x \le 1,\ \frac{x}{2} \le y \le 1 - \frac{x}{2}\right\}$$

Therefore

$$V = \iint\limits_{D} (2 - x - 2y)\, dA = \int_0^1 \int_{x/2}^{1-x/2} (2 - x - 2y)\, dy\, dx$$

$$= \int_0^1 \left[2y - xy - y^2 \right]_{y=x/2}^{y=1-x/2}\, dy$$

$$= \int_0^1 \left[2 - x - x\left(1 - \frac{x}{2}\right) - \left(1 - \frac{x}{2}\right)^2 - x + \frac{x^2}{2} + \frac{x^2}{4} \right] dx$$

$$= \int_0^1 (x^2 - 2x + 1)\, dx = \frac{x^3}{3} - x^2 + x \Big]_0^1 = \frac{1}{3}$$
∎

EXAMPLE 5   Evaluate the iterated integral $\int_0^1 \int_x^1 \sin(y^2)\, dy\, dx$.

SOLUTION   If we try to evaluate the integral as it stands, we are faced with the task of first evaluating $\int \sin(y^2)\, dy$. But it is impossible to do so in finite terms since $\int \sin(y^2)\, dy$ is not an elementary function. (See the end of Section 7.6.) So we must change the order of integration. This is accomplished by first expressing the given iterated integral as a double integral. Using (3) backward, we have

$$\int_0^1 \int_x^1 \sin(y^2)\, dy\, dx = \iint\limits_{D} \sin(y^2)\, dA$$

where
$$D = \{ (x, y) \mid 0 \le x \le 1,\ x \le y \le 1 \}$$

We sketch this region D in Figure 15. Then from Figure 16 we see that an alternative description of D is

$$D = \{ (x, y) \mid 0 \le y \le 1,\ 0 \le x \le y \}$$

This enables us to use (5) to express the double integral as an iterated integral in the reverse order:

$$\int_0^1 \int_x^1 \sin(y^2)\, dy\, dx = \iint\limits_{D} \sin(y^2)\, dA$$

$$= \int_0^1 \int_0^y \sin(y^2)\, dx\, dy = \int_0^1 \left[x \sin(y^2) \right]_{x=0}^{x=y}\, dy$$

$$= \int_0^1 y \sin(y^2)\, dy = -\tfrac{1}{2} \cos(y^2) \Big]_0^1$$

$$= \tfrac{1}{2}(1 - \cos 1)$$
∎

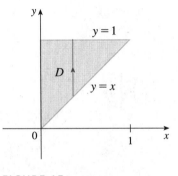

FIGURE 15
D as a type I region

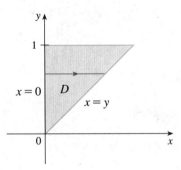

FIGURE 16
D as a type II region

PROPERTIES OF DOUBLE INTEGRALS

We assume that all of the following integrals exist. The first three properties of double integrals over a region D follow immediately from Definition 2 and Properties 13.1.7, 13.1.8, and 13.1.9.

(6)
$$\iint\limits_{D} [f(x, y) + g(x, y)]\, dA = \iint\limits_{D} f(x, y)\, dA + \iint\limits_{D} g(x, y)\, dA$$

(7)
$$\iint\limits_{D} cf(x, y)\, dA = c \iint\limits_{D} f(x, y)\, dA$$

If $f(x, y) \geqslant g(x, y)$ for all (x, y) in D, then

$$(8) \qquad \iint_D f(x, y) \, dA \geqslant \iint_D g(x, y) \, dA$$

The next property of double integrals is the analogue of the property of single integrals given by the equation $\int_a^b f(x) \, dx = \int_a^c f(x) \, dx + \int_c^b f(x) \, dx$.

If $D = D_1 \cup D_2$, where D_1 and D_2 do not overlap except perhaps on their boundaries (see Figure 17), then

$$(9) \qquad \boxed{\iint_D f(x, y) \, dA = \iint_{D_1} f(x, y) \, dA + \iint_{D_2} f(x, y) \, dA}$$

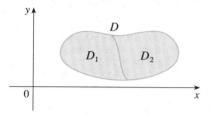

D D_1 D_2

FIGURE 17

Property 9 can be used to evaluate double integrals over regions D that are neither type I nor type II but can be expressed as a union of type I or II regions. Figure 18 illustrates this procedure. (See Exercises 45 and 46.)

The next property of integrals says that if we integrate the constant function $f(x, y) = 1$ over a region D, we get the area of D:

$$(10) \qquad \boxed{\iint_D 1 \, dA = A(D)}$$

For instance, if D is a type I region and we put $f(x, y) = 1$ in Formula 3, we get

$$\iint_D 1 \, dA = \int_a^b \int_{g_1(x)}^{g_2(x)} 1 \, dy \, dx = \int_a^b [g_2(x) - g_1(x)] \, dx = A(D)$$

by Equation 5.1.2.

Finally, we obtain an analogue of Property 8 of single integrals (see Section 4.3) by combining Properties 7, 8, and 10. (See Exercise 49.)

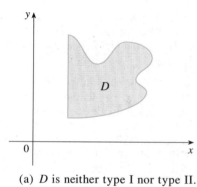

(a) D is neither type I nor type II.

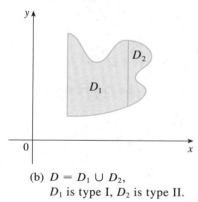

D_2 D_1

(b) $D = D_1 \cup D_2$,
D_1 is type I, D_2 is type II.

FIGURE 18

$$(11) \quad \boxed{\text{If } m \leqslant f(x, y) \leqslant M \text{ for all } (x, y) \text{ in } D, \text{ then} \\[2mm] mA(D) \leqslant \iint_D f(x, y) \, dA \leqslant MA(D)}$$

EXAMPLE 6   Use Property 11 to estimate the integral $\iint_D e^{\sin x \cos y} \, dA$, where D is the disk with center the origin and radius 2.

SOLUTION   Since $-1 \leqslant \sin x \leqslant 1$ and $-1 \leqslant \cos y \leqslant 1$, we have $-1 \leqslant \sin x \cos y \leqslant 1$ and therefore

$$e^{-1} \leqslant e^{\sin x \cos y} \leqslant e^1 = e$$

Thus, using $m = e^{-1} = 1/e$, $M = e$, and $A(D) = \pi(2)^2$ in Property 11, we obtain

$$\frac{4\pi}{e} \leqslant \iint_D e^{\sin x \cos y} \, dA \leqslant 4\pi e$$

∎

EXERCISES 13.3

1–6 ■ Evaluate the iterated integral.

1. $\displaystyle\int_0^1 \int_0^y x \, dx \, dy$

2. $\displaystyle\int_0^1 \int_0^y y \, dx \, dy$

3. $\displaystyle\int_0^2 \int_{\sqrt{x}}^3 (x^2 + y) \, dy \, dx$

4. $\displaystyle\int_0^1 \int_{1-x}^{1+x} (2x - 3y^2) \, dy \, dx$

5. $\displaystyle\int_0^1 \int_0^x \sin(x^2) \, dy \, dx$

6. $\displaystyle\int_0^1 \int_{x-1}^0 \frac{2y}{x+1} \, dy \, dx$

7–18 ■ Evaluate the double integral.

7. $\displaystyle\iint_D xy \, dA, \quad D = \{(x, y) \mid 0 \le x \le 1, \ x^2 \le y \le \sqrt{x}\}$

8. $\displaystyle\iint_D (x - 2y) \, dA,$
$D = \{(x, y) \mid 1 \le x \le 3, \ 1 + x \le y \le 2x\}$

9. $\displaystyle\iint_D (x^2 - 2xy) \, dA,$
$D = \{(x, y) \mid 0 \le x \le 1, \ \sqrt{x} \le y \le 2 - x\}$

10. $\displaystyle\iint_D x \sin y \, dA, \quad D = \{(x, y) \mid 0 \le y \le \pi/2, \ 0 \le x \le \cos y\}$

11. $\displaystyle\iint_D e^{x/y} \, dA, \quad D = \{(x, y) \mid 1 \le y \le 2, \ y \le x \le y^3\}$

12. $\displaystyle\iint_D \frac{1}{x} \, dA, \quad D = \{(x, y) \mid 1 \le y \le e, \ y^2 \le x \le y^4\}$

13. $\displaystyle\iint_D x \cos y \, dA, \quad D$ is bounded by $y = 0$, $y = x^2$, $x = 1$

14. $\displaystyle\iint_D e^{x+y} \, dA, \quad D$ is bounded by $y = 0$, $y = x$, $x = 1$

15. $\displaystyle\iint_D 4y^3 \, dA, \quad D$ is bounded by $y = x - 6$, $y^2 = x$

16. $\displaystyle\iint_D (y^2 - x) \, dA, \quad D$ is bounded by $x = y^2$, $x = 3 - 2y^2$

17. $\displaystyle\iint_D xy \, dA, \quad D$ is the first-quadrant part of the disk with center $(0, 0)$ and radius 1

18. $\displaystyle\iint_D ye^x \, dA, \quad D$ is the triangular region with vertices $(0, 0)$, $(2, 4)$, and $(6, 0)$

19–28 ■ Find the volume of the given solid.

19. Under the paraboloid $z = x^2 + y^2$ and above the region bounded by $y = x^2$ and $x = y^2$

20. Under the paraboloid $z = 3x^2 + y^2$ and above the region bounded by $y = x$ and $x = y^2 - y$

21. Under the surface $z = xy$ and above the triangle with vertices $(1, 1)$, $(4, 1)$, and $(1, 2)$

22. Bounded by the paraboloid $z = x^2 + y^2 + 4$ and the planes $x = 0$, $y = 0$, $z = 0$, $x + y = 1$

23. Bounded by the cylinder $x^2 + z^2 = 9$ and the planes $x = 0$, $y = 0$, $z = 0$, $x + 2y = 2$ in the first octant

24. Bounded by the cylinder $y^2 + z^2 = 4$ and the planes $x = 2y$, $x = 0$, $z = 0$ in the first octant

25. Bounded by the planes $x = 0$, $y = 0$, $z = 0$, and $x + y + z = 1$

26. Bounded by the planes $y = 0$, $z = 0$, $y = x$, and $6x + 2y + 3z = 6$

27. Bounded by the cylinder $x^2 + y^2 = 1$ and the planes $y = z$, $x = 0$, $z = 0$ in the first octant

28. Bounded by the cylinders $x^2 + y^2 = r^2$ and $y^2 + z^2 = r^2$

29. Use a graphing calculator or computer to estimate the x-coordinates of the points of intersection of the curves $y = x^4$ and $y = 3x - x^2$. If D is the region bounded by these curves, estimate $\iint_D x \, dA$.

30. Find the approximate volume of the solid in the first octant that is bounded by the planes $y = x$, $z = 0$, and $z = x$ and the cylinder $y = \cos x$. (Use a graphing device to estimate the points of intersection.)

CAS **31–32** ■ Use a computer algebra system to find the exact volume of the solid.

31. Under the surface $z = x^3 y^4 + xy^2$ and above the region bounded by the curves $y = x^3 - x$ and $y = x^2 + x$ for $x \ge 0$

32. Between the paraboloids $z = 2x^2 + y^2$ and $z = 8 - x^2 - 2y^2$ and inside the cylinder $x^2 + y^2 = 1$

33–38 ■ Sketch the region of integration and change the order of integration.

33. $\displaystyle\int_0^1 \int_0^x f(x, y) \, dy \, dx$

34. $\displaystyle\int_0^{\pi/2} \int_0^{\sin x} f(x, y) \, dy \, dx$

35. $\displaystyle\int_1^2 \int_0^{\ln x} f(x, y) \, dy \, dx$

36. $\displaystyle\int_0^1 \int_{y^2}^{2-y} f(x, y) \, dx \, dy$

37. $\displaystyle\int_0^4 \int_{y/2}^2 f(x, y) \, dx \, dy$

38. $\displaystyle\int_0^1 \int_{\arctan x}^{\pi/4} f(x, y) \, dy \, dx$

39–44 ■ Evaluate the integral by reversing the order of integration.

39. $\displaystyle\int_0^1 \int_{3y}^3 e^{x^2} \, dx \, dy$

40. $\displaystyle\int_0^1 \int_{\sqrt{y}}^1 \sqrt{x^3 + 1} \, dx \, dy$

41. $\displaystyle\int_0^3 \int_{y^2}^9 y \cos(x^2) \, dx \, dy$

42. $\displaystyle\int_0^1 \int_{x^2}^1 x^3 \sin(y^3) \, dy \, dx$

43. $\displaystyle\int_0^1 \int_{\arcsin y}^{\pi/2} \cos x \sqrt{1 + \cos^2 x} \, dx \, dy$

44. $\int_0^8 \int_{\sqrt[3]{y}}^2 e^{x^4} \, dx \, dy$

45–46 ■ Express D as a union of regions of type I or type II and evaluate the integral.

45. $\iint_D x^2 \, dA$

46. $\iint_D xy \, dA$

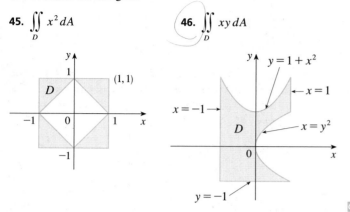

47–48 ■ Use Property 11 to estimate the value of the integral.

47. $\iint_D \sqrt{x^3 + y^3} \, dA, \quad D = [0, 1] \times [0, 1]$

48. $\iint_D e^{x^2+y^2} \, dA, \quad D$ is the disk with center the origin and radius $\frac{1}{2}$

49. Prove Property 11.

50. In evaluating a double integral over a region D, a sum of iterated integrals was obtained as follows:

$$\iint_D f(x, y) \, dA = \int_0^1 \int_0^{2y} f(x, y) \, dx \, dy + \int_1^3 \int_0^{3-y} f(x, y) \, dx \, dy$$

Sketch the region D and express the double integral as an iterated integral with reversed order of integration.

51. Evaluate $\iint_D (x^2 \tan x + y^3 + 4) \, dA$, where $D = \{(x, y) \,|\, x^2 + y^2 \leqslant 2\}$.
[*Hint:* Exploit the fact that D is symmetric with respect to both axes.]

CAS 52. Graph the solid bounded by the plane $x + y + z = 1$ and the paraboloid $z = 4 - x^2 - y^2$ and find its exact volume. (Use your CAS to do the graphing, to find the equations of the boundary curves of the region of integration, and to evaluate the double integral.)

13.4 DOUBLE INTEGRALS IN POLAR COORDINATES

Suppose that we want to evaluate a double integral $\iint_R f(x, y) \, dA$, where R is one of the regions shown in Figure 1. In either case the description of R in terms of rectangular coordinates is rather complicated but R is easily described using polar coordinates.

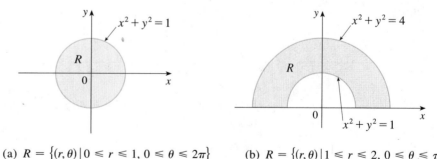

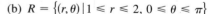

FIGURE 1   (a) $R = \{(r, \theta) \,|\, 0 \leqslant r \leqslant 1, 0 \leqslant \theta \leqslant 2\pi\}$ (b) $R = \{(r, \theta) \,|\, 1 \leqslant r \leqslant 2, 0 \leqslant \theta \leqslant \pi\}$

Recall from Section 9.4 that the polar coordinates (r, θ) of a point are related to the rectangular coordinates by the equations

$$r^2 = x^2 + y^2 \qquad x = r\cos\theta \qquad y = r\sin\theta$$

The regions in Figure 1 are special cases of a **polar rectangle**

$$R = \{(r, \theta) \,|\, a \leqslant r \leqslant b, \alpha \leqslant \theta \leqslant \beta\}$$

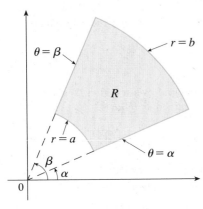

FIGURE 2
Polar rectangle

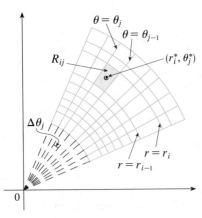

FIGURE 3
Polar partition

which is shown in Figure 2. In order to compute $\iint_R f(x, y)\, dA$, where R is a polar rectangle, we start with a partition of $[a, b]$ into m subintervals:

$$a = r_0 < r_1 < r_2 < \cdots < r_{i-1} < r_i < \cdots < r_m = b$$

and a partition of $[\alpha, \beta]$ into n subintervals:

$$\alpha = \theta_0 < \theta_1 < \theta_2 < \cdots < \theta_{j-1} < \theta_j < \cdots < \theta_n = \beta$$

Then the circles $r = r_i$ and the rays $\theta = \theta_j$ determine a **polar partition** P of R into the small polar rectangles shown in Figure 3. The norm $\|P\|$ of the polar partition is the length of the longest diagonal of all the polar subrectangles.

The "center" of the polar subrectangle

$$R_{ij} = \{(r, \theta) \mid r_{i-1} \le r \le r_i,\ \theta_{j-1} \le \theta \le \theta_j\}$$

has polar coordinates

$$r_i^* = \tfrac{1}{2}(r_{i-1} + r_i) \qquad \theta_j^* = \tfrac{1}{2}(\theta_{j-1} + \theta_j)$$

We compute the area of R_{ij} using the fact that the area of a sector of a circle with radius r and central angle θ is $\tfrac{1}{2}r^2\theta$. Subtracting the areas of two such sectors, each of which has central angle $\Delta\theta_j = \theta_j - \theta_{j-1}$, we find that the area of R_{ij} is

$$\begin{aligned}
\Delta A_{ij} &= \tfrac{1}{2}r_i^2\,\Delta\theta_j - \tfrac{1}{2}r_{i-1}^2\,\Delta\theta_j \\
&= \tfrac{1}{2}(r_i^2 - r_{i-1}^2)\,\Delta\theta_j \\
&= \tfrac{1}{2}(r_i + r_{i-1})(r_i - r_{i-1})\,\Delta\theta_j \\
&= r_i^*\,\Delta r_i\,\Delta\theta_j
\end{aligned}$$

where $\Delta r_i = r_i - r_{i-1}$.

Although we have defined the double integral $\iint_R f(x, y)\, dA$ in terms of rectangular partitions, it can be shown that, for continuous functions f, we always obtain the same answer using polar partitions. The rectangular coordinates of the center of R_{ij} are $(r_i^* \cos\theta_j^*, r_i^* \sin\theta_j^*)$, so a typical Riemann sum is

$$(1) \quad \sum_{i=1}^{m}\sum_{j=1}^{n} f(r_i^* \cos\theta_j^*, r_i^* \sin\theta_j^*)\,\Delta A_{ij} = \sum_{i=1}^{m}\sum_{j=1}^{n} f(r_i^* \cos\theta_j^*, r_i^* \sin\theta_j^*)\, r_i^*\,\Delta r_i\,\Delta\theta_j$$

If we write $g(r, \theta) = rf(r\cos\theta, r\sin\theta)$, then the Riemann sum in Equation 1 can be written as

$$\sum_{i=1}^{m}\sum_{j=1}^{n} g(r_i^*, \theta_j^*)\,\Delta r_i\,\Delta\theta_j$$

which is a Riemann sum for the double integral

$$\int_{\alpha}^{\beta}\int_{a}^{b} g(r, \theta)\, dr\, d\theta$$

Therefore, we have

$$\iint_R f(x, y)\, dA = \lim_{\|P\|\to 0} \sum_{i=1}^{m}\sum_{j=1}^{n} f(r_i^* \cos\theta_j^*, r_i^* \sin\theta_j^*)\,\Delta A_{ij}$$

$$= \lim_{\|P\|\to 0} \sum_{i=1}^{m} \sum_{j=1}^{n} g(r_i^*, \theta_j^*) \,\Delta r_i \,\Delta\theta_j$$

$$= \int_{\alpha}^{\beta} \int_{a}^{b} g(r, \theta) \, dr \, d\theta$$

$$= \int_{\alpha}^{\beta} \int_{a}^{b} f(r\cos\theta, r\sin\theta) \, r \, dr \, d\theta$$

(2) CHANGE TO POLAR COORDINATES IN A DOUBLE INTEGRAL If f is continuous on a polar rectangle R given by $0 \le a \le r \le b$, $\alpha \le \theta \le \beta$, where $0 \le \beta - \alpha \le 2\pi$, then

$$\iint\limits_{R} f(x, y) \, dA = \int_{\alpha}^{\beta} \int_{a}^{b} f(r\cos\theta, r\sin\theta) \, r \, dr \, d\theta$$

The formula in (2) says that we convert from rectangular to polar coordinates in a double integral by writing $x = r\cos\theta$ and $y = r\sin\theta$, using the appropriate limits of integration for r and θ, and replacing dA by $r\,dr\,d\theta$. Be careful not to forget the additional factor r on the right side of Formula 2. A classical method for remembering this is shown in Figure 4, where the "infinitesimal" polar rectangle can be thought of as an ordinary rectangle with dimensions $r\,d\theta$ and dr and therefore has "area" $dA = r\,dr\,d\theta$.

EXAMPLE 1 Evaluate $\iint_{R} (3x + 4y^2)\,dA$, where R is the region in the upper half-plane bounded by the circles $x^2 + y^2 = 1$ and $x^2 + y^2 = 4$.

SOLUTION The region R can be described as

$$R = \{(x, y) \mid y \ge 0, \, 1 \le x^2 + y^2 \le 4\}$$

It is the half-ring shown in Figure 1(b), and in polar coordinates it is given by $1 \le r \le 2$, $0 \le \theta \le \pi$. Therefore, by Formula 2,

$$\iint\limits_{R} (3x + 4y^2)\,dA = \int_{0}^{\pi} \int_{1}^{2} (3r\cos\theta + 4r^2\sin^2\theta) \, r \, dr \, d\theta$$

$$= \int_{0}^{\pi} \int_{1}^{2} (3r^2\cos\theta + 4r^3\sin^2\theta) \, dr \, d\theta$$

$$= \int_{0}^{\pi} \left[r^3\cos\theta + r^4\sin^2\theta \right]_{r=1}^{r=2} d\theta$$

$$= \int_{0}^{\pi} (7\cos\theta + 15\sin^2\theta) \, d\theta$$

$$= \int_{0}^{\pi} \left[7\cos\theta + \tfrac{15}{2}(1 - \cos 2\theta) \right] d\theta$$

$$= 7\sin\theta + \frac{15\theta}{2} - \frac{15}{4}\sin 2\theta \Bigg]_{0}^{\pi} = \frac{15\pi}{2}$$

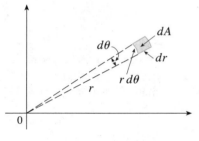

FIGURE 4

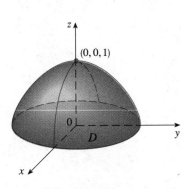

FIGURE 5

EXAMPLE 2   Find the volume of the solid bounded by the plane $z = 0$ and the paraboloid $z = 1 - x^2 - y^2$.

SOLUTION   If we put $z = 0$ in the equation of the paraboloid, we get $x^2 + y^2 = 1$. This means that the plane intersects the paraboloid in the circle $x^2 + y^2 = 1$, so the solid lies under the paraboloid and above the circular disk D given by $x^2 + y^2 \leqslant 1$ [see Figures 5 and 1(a)]. In polar coordinates D is given by $0 \leqslant r \leqslant 1$, $0 \leqslant \theta \leqslant 2\pi$. Since $1 - x^2 - y^2 = 1 - r^2$, the volume is

$$V = \iint_D (1 - x^2 - y^2)\, dA = \int_0^{2\pi} \int_0^1 (1 - r^2)\, r\, dr\, d\theta$$

$$= \int_0^{2\pi} d\theta \int_0^1 (r - r^3)\, dr = 2\pi \left[\frac{r^2}{2} - \frac{r^4}{4} \right]_0^1 = \frac{\pi}{2}$$

If we had used rectangular coordinates instead of polar coordinates, then we would have obtained

$$V = \iint_D (1 - x^2 - y^2)\, dA = \int_{-1}^1 \int_{-\sqrt{1-x^2}}^{\sqrt{1-x^2}} (1 - x^2 - y^2)\, dy\, dx$$

which is not easy to evaluate because it involves finding the following integrals:

$$\int \sqrt{1 - x^2}\, dx \qquad \int x^2 \sqrt{1 - x^2}\, dx \qquad \int (1 - x^2)^{3/2}\, dx \qquad \blacksquare$$

What we have done so far can be extended to the more complicated type of region shown in Figure 6. We could call it a type II polar region because it is analogous to the type II rectangular regions considered in Section 13.3. In fact, by combining Formulas 2 and 13.3.5, we obtain the following formula:

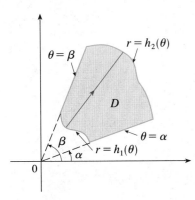

FIGURE 6

Type II polar region,
$D = \{(r, \theta) \mid \alpha \leqslant \theta \leqslant \beta,$
$h_1(\theta) \leqslant r \leqslant h_2(\theta)\}$

---

(3)   If f is continuous on a polar region of the form

$$D = \{(r, \theta) \mid \alpha \leqslant \theta \leqslant \beta,\ h_1(\theta) \leqslant r \leqslant h_2(\theta)\}$$

then

$$\iint_D f(x, y)\, dA = \int_\alpha^\beta \int_{h_1(\theta)}^{h_2(\theta)} f(r\cos\theta, r\sin\theta)\, r\, dr\, d\theta$$

In particular, taking $f(x, y) = 1$, $h_1(\theta) = 0$, and $h_2(\theta) = h(\theta)$ in this formula, we see that the area of the region D bounded by $\theta = \alpha$, $\theta = \beta$, and $r = h(\theta)$ is

$$A(D) = \iint_D 1\, dA = \int_\alpha^\beta \int_0^{h(\theta)} r\, dr\, d\theta$$

$$= \int_\alpha^\beta \left[\frac{r^2}{2} \right]_0^{h(\theta)} d\theta = \int_\alpha^\beta \tfrac{1}{2}[h(\theta)]^2\, d\theta$$

and this agrees with Formula 9.5.3.

Similarly, the analogue of a type I rectangular region is a polar region of the form

$$D = \{(r, \theta) \mid a \leqslant r \leqslant b,\ g_1(r) \leqslant \theta \leqslant g_2(r)\}$$

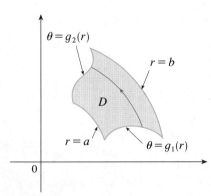

FIGURE 7

Type I polar region,
$D = \{(r, \theta) \mid a \leq r \leq b,$
$g_1(r) \leq \theta \leq g_2(r)\}$

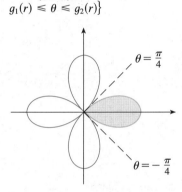

FIGURE 8

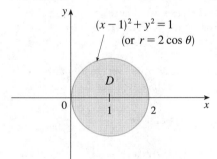

FIGURE 9

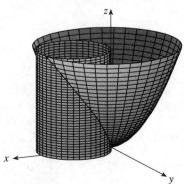

FIGURE 10

(see Figure 7), and Formula 13.3.3, together with Formula 2, shows that

$$(4) \qquad \iint_D f(x, y)\, dA = \int_a^b \int_{g_1(r)}^{g_2(r)} f(r\cos\theta, r\sin\theta)\, r\, d\theta\, dr$$

EXAMPLE 3   Use a double integral to find the area enclosed by one loop of the four-leaved rose $r = \cos 2\theta$.

SOLUTION   From the sketch of the curve in Figure 8 we see that a loop is given by the region

$$D = \left\{(r, \theta) \ \middle| \ -\frac{\pi}{4} \leq \theta \leq \frac{\pi}{4},\ 0 \leq r \leq \cos 2\theta\right\}$$

So the area is

$$A(D) = \iint_D dA = \int_{-\pi/4}^{\pi/4} \int_0^{\cos 2\theta} r\, dr\, d\theta$$

$$= \int_{-\pi/4}^{\pi/4} \left[\tfrac{1}{2} r^2\right]_0^{\cos 2\theta} d\theta = \tfrac{1}{2} \int_{-\pi/4}^{\pi/4} \cos^2 2\theta\, d\theta$$

$$= \tfrac{1}{4} \int_{-\pi/4}^{\pi/4} (1 + \cos 4\theta)\, d\theta = \tfrac{1}{4}\left[\theta + \tfrac{1}{4}\sin 4\theta\right]_{-\pi/4}^{\pi/4} = \frac{\pi}{8}$$

■

EXAMPLE 4   Find the volume of the solid that lies under the paraboloid $z = x^2 + y^2$, above the xy-plane, and inside the cylinder $x^2 + y^2 = 2x$.

SOLUTION   The solid lies above the disk D whose boundary circle has equation $x^2 + y^2 = 2x$ or

$$(x - 1)^2 + y^2 = 1$$

(see Figures 9 and 10). In polar coordinates we have $x^2 + y^2 = r^2$ and $x = r\cos\theta$, so the boundary circle becomes $r^2 = 2r\cos\theta$ or $r = 2\cos\theta$. Thus the disk D is given by

$$D = \left\{(r, \theta) \ \middle| \ -\frac{\pi}{2} \leq \theta \leq \frac{\pi}{2},\ 0 \leq r \leq 2\cos\theta\right\}$$

and, by Formula 3, we have

$$V = \iint_D (x^2 + y^2)\, dA = \int_{-\pi/2}^{\pi/2} \int_0^{2\cos\theta} r^2\, r\, dr\, d\theta$$

$$= \int_{-\pi/2}^{\pi/2} \left[\frac{r^4}{4}\right]_0^{2\cos\theta} d\theta = 4\int_{-\pi/2}^{\pi/2} \cos^4\theta\, d\theta$$

$$= 8\int_0^{\pi/2} \cos^4\theta\, d\theta = 8\int_0^{\pi/2} \left(\frac{1 + \cos 2\theta}{2}\right)^2 d\theta$$

$$= 2\int_0^{\pi/2} \left[1 + 2\cos 2\theta + \tfrac{1}{2}(1 + \cos 4\theta)\right] d\theta$$

$$= 2\left[\tfrac{3}{2}\theta + \sin 2\theta + \tfrac{1}{8}\sin 4\theta\right]_0^{\pi/2} = 2\left(\frac{3}{2}\right)\left(\frac{\pi}{2}\right) = \frac{3\pi}{2}$$

■

EXERCISES 13.4

1–8 ■ Evaluate the given integral by changing to polar coordinates.

1. $\iint_R x\,dA$, where R is the disk with center the origin and radius 5

2. $\iint_R y\,dA$, where R is the region in the first quadrant bounded by the circle $x^2 + y^2 = 9$ and the lines $y = x$ and $y = 0$

3. $\iint_R xy\,dA$, where R is the region in the first quadrant that lies between the circles $x^2 + y^2 = 4$ and $x^2 + y^2 = 25$

4. $\iint_R \sin(x^2 + y^2)\,dA$, where R is the annular region $1 \leqslant x^2 + y^2 \leqslant 16$

5. $\iint_D 1/\sqrt{x^2 + y^2}\,dA$, where D is the region that lies inside the cardioid $r = 1 + \sin\theta$ and outside the circle $r = 1$

6. $\iint_D \sqrt{x^2 + y^2}\,dA$, where D is the region bounded by the cardioid $r = 1 + \cos\theta$

7. $\iint_D (x^2 + y^2)\,dA$, where D is the region bounded by the spirals $r = \theta$ and $r = 2\theta$ for $0 \leqslant \theta \leqslant 2\pi$

8. $\iint_D x\,dA$, where D is the region in the first quadrant that lies between the circles $x^2 + y^2 = 4$ and $x^2 + y^2 = 2x$

9–14 ■ Use a double integral to find the area of the region.

9. One loop of the rose $r = \cos 3\theta$

10. The region enclosed by the cardioid $r = 1 - \sin\theta$

11. The region enclosed by the lemniscate $r^2 = 4\cos 2\theta$

12. The region inside the circle $r = 4\sin\theta$ and outside the circle $r = 2$

13. The region inside the circle $r = 3\cos\theta$ and outside the cardioid $r = 1 + \cos\theta$

14. The smaller region bounded by the spiral $r\theta = 1$, the circles $r = 1$ and $r = 3$, and the polar axis

15–23 ■ Use polar coordinates to find the volume of the given solid.

15. Under the paraboloid $z = x^2 + y^2$ and above the disk $x^2 + y^2 \leqslant 9$

16. Under the cone $z = \sqrt{x^2 + y^2}$ and above the ring $4 \leqslant x^2 + y^2 \leqslant 25$

17. Under the plane $6x + 4y + z = 12$ and above the disk with boundary circle $x^2 + y^2 = y$

18. Bounded by the paraboloid $z = 10 - 3x^2 - 3y^2$ and the plane $z = 4$

19. Above the cone $z = \sqrt{x^2 + y^2}$ and below the sphere $x^2 + y^2 + z^2 = 1$

20. Bounded by the paraboloids $z = 3x^2 + 3y^2$ and $z = 4 - x^2 - y^2$

21. Inside both the cylinder $x^2 + y^2 = 4$ and the ellipsoid $4x^2 + 4y^2 + z^2 = 64$

22. Inside the sphere $x^2 + y^2 + z^2 = 4a^2$ and outside the cylinder $x^2 + y^2 = 2ax$

23. A sphere of radius a

24. (a) A cylindrical drill with radius r_1 is used to bore a hole through the center of a sphere of radius r_2. Find the volume of the ring-shaped solid that remains.
(b) Express the volume in part (a) in terms of the height h of the ring. Notice that the volume depends only on h, not on r_1 or r_2.

25–28 ■ Evaluate the given iterated integral by converting to polar coordinates.

25. $\int_0^1 \int_0^{\sqrt{1-x^2}} e^{x^2+y^2}\,dy\,dx$

26. $\int_{-a}^{a} \int_0^{\sqrt{a^2-y^2}} (x^2 + y^2)^{3/2}\,dx\,dy$

27. $\int_0^2 \int_{-\sqrt{4-y^2}}^{\sqrt{4-y^2}} x^2 y^2\,dx\,dy$

28. $\int_0^2 \int_0^{\sqrt{2x-x^2}} \sqrt{x^2 + y^2}\,dy\,dx$

29. Convert the integral $\int_0^1 \int_0^{\sqrt{1-x^2}} e^{-(x^2+y^2)^2}\,dy\,dx$ to polar coordinates. If your calculator or CAS evaluates definite integrals numerically, use it to find the integral correct to three decimal places. If not, use Simpson's Rule.

30. An agricultural sprinkler distributes water in a circular pattern of radius 100 ft. It supplies water to a depth of e^{-r} feet per hour at a distance of r feet from the sprinkler.
(a) What is the total amount of water supplied per hour to the region inside the circle of radius R centered at the sprinkler?
(b) Determine an expression for the average amount of water per hour per square foot supplied to the region inside the circle of radius R.

31. Use polar coordinates to combine the sum
$$\int_{1/\sqrt{2}}^1 \int_{\sqrt{1-x^2}}^x xy\,dy\,dx + \int_1^{\sqrt{2}} \int_0^x xy\,dy\,dx + \int_{\sqrt{2}}^2 \int_0^{\sqrt{4-x^2}} xy\,dy\,dx$$
into one double integral. Then evaluate the double integral.

32. (a) We define the improper integral (over the entire plane $\mathbb{R}^2$)
$$I = \iint_{\mathbb{R}^2} e^{-(x^2+y^2)}\,dA = \int_{-\infty}^{\infty} \int_{-\infty}^{\infty} e^{-(x^2+y^2)}\,dy\,dx$$
$$= \lim_{a \to \infty} \iint_{D_a} e^{-(x^2+y^2)}\,dA$$
where D_a is the disk with radius a and center the origin. Show that
$$\int_{-\infty}^{\infty} \int_{-\infty}^{\infty} e^{-(x^2+y^2)}\,dA = \pi$$

(b) An equivalent definition of the improper integral in part (a) is

$$\iint\limits_{\mathbb{R}^2} e^{-(x^2+y^2)} dA = \lim_{a \to \infty} \iint\limits_{S_a} e^{-(x^2+y^2)} dA$$

where S_a is the square with vertices $(\pm a, \pm a)$. Use this to show that

$$\int_{-\infty}^{\infty} e^{-x^2} dx \int_{-\infty}^{\infty} e^{-y^2} dy = \pi$$

(c) Deduce that

$$\int_{-\infty}^{\infty} e^{-x^2} dx = \sqrt{\pi}$$

(d) By making the change of variable $t = \sqrt{2}\,x$, show that

$$\int_{-\infty}^{\infty} e^{-x^2/2} dx = \sqrt{2\pi}$$

(This is a fundamental result for probability and statistics.)

33. Use the result of Exercise 32 part (c) to evaluate the following integrals.

(a) $\int_0^{\infty} x^2 e^{-x^2} dx$ (b) $\int_0^{\infty} \sqrt{x}\, e^{-x} dx$

13.5   APPLICATIONS OF DOUBLE INTEGRALS

We have already seen one application of double integrals: computing volumes. Another geometric application is finding areas of surfaces and this will be done in the next section. In this section we explore physical applications such as computing mass, electric charge, center of mass, and moment of inertia.

In Chapter 8 we were able to use single integrals to compute moments and the center of mass of a thin plate or lamina with constant density. But now, equipped with the double integral, we can consider a lamina with variable density. Suppose the lamina occupies a region D of the xy-plane and its density (in units of mass per unit area) at a point (x, y) in D is given by $\rho(x, y)$, where ρ is a continuous function on D. This means that

$$\rho(x, y) = \lim \frac{\Delta m}{\Delta A}$$

where Δm and ΔA are the mass and area of a small rectangle that contains (x, y) and the limit is taken as the dimensions of the rectangle approach 0 (see Figure 1).

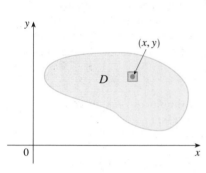

FIGURE 1

To find the total mass m of the lamina we partition a rectangle R containing D into subrectangles R_{ij} (as in Figure 2) and consider $\rho(x, y)$ to be 0 outside D. If we choose a point (x_{ij}^*, y_{ij}^*) in R_{ij}, then the mass of the part of the lamina that occupies R_{ij} is approximately $\rho(x_{ij}^*, y_{ij}^*) \Delta A_{ij}$, where ΔA_{ij} is the area of R_{ij}. If we add all such masses, we get an approximation to the total mass:

$$m \approx \sum_{i=1}^{m} \sum_{j=1}^{n} \rho(x_{ij}^*, y_{ij}^*) \Delta A_{ij}$$

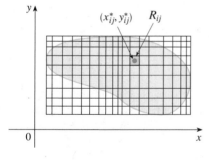

FIGURE 2

If we now take finer partitions, we obtain the total mass m of the lamina as the limiting value of the approximations:

(1)
$$m = \lim_{\|P\| \to 0} \sum_{i=1}^{m} \sum_{j=1}^{n} \rho(x_{ij}^*, y_{ij}^*) \Delta A_{ij} = \iint\limits_{D} \rho(x, y)\, dA$$

Physicists also consider other types of density that can be treated in the same manner. For example, if an electric charge is distributed over a region D and the charge density (in units of charge per unit area) is given by $\sigma(x, y)$ at a point (x, y) in D, then the total charge Q is given by

(2)
$$Q = \iint\limits_{D} \sigma(x, y)\, dA$$

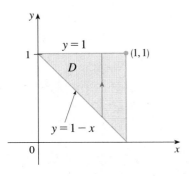

FIGURE 3

EXAMPLE 1   Charge is distributed over the triangular region D in Figure 3 so that the charge density at (x, y) is $\sigma(x, y) = xy$, measured in coulombs per square meter (C/m²). Find the total charge.

SOLUTION   From Equation 2 and Figure 3 we have

$$Q = \iint_D \sigma(x, y)\, dA = \int_0^1 \int_{1-x}^1 xy\, dy\, dx$$

$$= \int_0^1 \left[x\frac{y^2}{2} \right]_{y=1-x}^{y=1} dx = \int_0^1 \frac{x}{2}[1^2 - (1 - x)^2]\, dx$$

$$= \frac{1}{2} \int_0^1 (2x^2 - x^3)\, dx = \frac{1}{2}\left[\frac{2x^3}{3} - \frac{x^4}{4} \right]_0^1 = \frac{5}{24}$$

Thus the total charge is $\frac{5}{24}$ C.   ∎

Now let us find the center of mass of a lamina with density function $\rho(x, y)$ that occupies a region D. Recall that in Section 8.4 we defined the moment of a particle about an axis as the product of its mass and its directed distance from the axis. We partition D into small rectangles as in Figure 2. Then the mass of R_{ij} is approximately $\rho(x_{ij}^*, y_{ij}^*)\, \Delta A_{ij}$, so we can approximate the moment of R_{ij} with respect to the x-axis by

$$[\rho(x_{ij}^*, y_{ij}^*)\, \Delta A_{ij}]\, y_{ij}^*$$

If we now add these quantities and take the limit as the norm $\|P\|$ approaches 0, we obtain the **moment** of the entire lamina **about the x-axis:**

(3)
$$M_x = \lim_{\|P\|\to 0} \sum_{i=1}^m \sum_{j=1}^n y_{ij}^* \rho(x_{ij}^*, y_{ij}^*)\, \Delta A_{ij} = \iint_D y\rho(x, y)\, dA$$

Similarly, the **moment about the y-axis is**

(4)
$$M_y = \lim_{\|P\|\to 0} \sum_{i=1}^m \sum_{j=1}^n x_{ij}^* \rho(x_{ij}^*, y_{ij}^*)\, \Delta A_{ij} = \iint_D x\rho(x, y)\, dA$$

As before, we define the center of mass $(\bar{x}, \bar{y})$ so that $m\bar{x} = M_y$ and $m\bar{y} = M_x$. The physical significance is that the lamina behaves as if its entire mass is concentrated at its center of mass. Thus the lamina balances horizontally when supported at its center of mass. (See Figure 1 in Section 8.4.)

(5)   The coordinates $(\bar{x}, \bar{y})$ of the center of mass of a lamina occupying the region D and having density function $\rho(x, y)$ are

$$\bar{x} = \frac{M_y}{m} = \frac{1}{m} \iint_D x\rho(x, y)\, dA \qquad \bar{y} = \frac{M_x}{m} = \frac{1}{m} \iint_D y\rho(x, y)\, dA$$

where the mass m is given by

$$m = \iint_D \rho(x, y)\, dA$$

EXAMPLE 2 Find the mass and center of mass of a triangular lamina with vertices $(0, 0)$, $(1, 0)$, and $(0, 2)$ if the density function is $\rho(x, y) = 1 + 3x + y$.

SOLUTION The triangle is shown in Figure 4. (Note that the equation of the upper boundary is $y = 2 - 2x$.) The mass of the lamina is

$$m = \iint_D \rho(x, y)\, dA = \int_0^1 \int_0^{2-2x} (1 + 3x + y)\, dy\, dx$$

$$= \int_0^1 \left[y + 3xy + \frac{y^2}{2} \right]_{y=0}^{y=2-2x} dx$$

$$= 4 \int_0^1 (1 - x^2)\, dx = 4\left[x - \frac{x^3}{3} \right]_0^1 = \frac{8}{3}$$

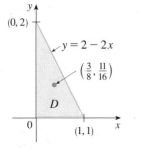

FIGURE 4

Then the formulas in (5) give

$$\bar{x} = \frac{1}{m} \iint_D x\rho(x, y)\, dA = \frac{3}{8} \int_0^1 \int_0^{2-2x} (x + 3x^2 + xy)\, dy\, dx$$

$$= \frac{3}{8} \int_0^1 \left[xy + 3x^2 y + x\frac{y^2}{2} \right]_{y=0}^{y=2-2x} dx$$

$$= \frac{3}{2} \int_0^1 (x - x^3)\, dx = \left[\frac{x^2}{2} - \frac{x^4}{4} \right]_0^1 = \frac{3}{8}$$

$$\bar{y} = \frac{1}{m} \iint_D y\rho(x, y)\, dA = \frac{3}{8} \int_0^1 \int_0^{2-2x} (y + 3xy + y^2)\, dy\, dx$$

$$= \frac{3}{8} \int_0^1 \left[\frac{y^2}{2} + 3x\frac{y^2}{2} + \frac{y^3}{3} \right]_{y=0}^{y=2-2x} dx$$

$$= \frac{9}{4} \int_0^1 (7 - 9x - 3x^2 + 5x^3)\, dx$$

$$= \frac{1}{4} \left[7x - 9\frac{x^2}{2} - x^3 + 5\frac{x^4}{4} \right]_0^1 = \frac{11}{16}$$

The center of mass is at the point $\left(\frac{3}{8}, \frac{11}{16} \right)$. ∎

EXAMPLE 3 The density at any point on a semicircular lamina is proportional to the distance from the center of the circle. Find the center of mass of the lamina.

SOLUTION Let us place the lamina as the upper half of the circle $x^2 + y^2 = a^2$ (see Figure 5). Then the distance from a point (x, y) to the center of the circle (the origin) is $\sqrt{x^2 + y^2}$. Therefore, the density function is

$$\rho(x, y) = K\sqrt{x^2 + y^2}$$

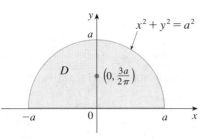

FIGURE 5

where K is some constant. Both the density function and the shape of the lamina suggest that we convert to polar coordinates. Then $\sqrt{x^2 + y^2} = r$ and the region D is given by $0 \leq r \leq a$, $0 \leq \theta \leq \pi$. Thus the mass of the lamina is

$$m = \iint_D \rho(x, y)\, dA = \iint_D K\sqrt{x^2 + y^2}\, dA$$

$$= \int_0^\pi \int_0^a (Kr)r\,dr\,d\theta = K \int_0^\pi d\theta \int_0^a r^2\,dr$$

$$= K\pi \frac{r^3}{3}\bigg]_0^a = \frac{K\pi a^3}{3}$$

Both the lamina and the density function are symmetric with respect to the y-axis, so the center of mass must lie on the y-axis, that is, $\bar{x} = 0$. The y-coordinate is given by

$$\bar{y} = \frac{1}{m} \iint_D y\rho(x, y)\,dA$$

$$= \frac{3}{K\pi a^3} \int_0^\pi \int_0^a r\sin\theta\,(Kr)r\,dr\,d\theta$$

$$= \frac{3}{\pi a^3} \int_0^\pi \sin\theta\,d\theta \int_0^a r^3\,dr$$

$$= \frac{3}{\pi a^3} \big[-\cos\theta\big]_0^\pi \left[\frac{r^4}{4}\right]_0^a$$

$$= \frac{3}{\pi a^3} \frac{2a^4}{4} = \frac{3a}{2\pi}$$

Therefore, the center of mass is at the point $(0, 3a/(2\pi))$. [Compare this with Example 3 in Section 8.4 where we found that the center of mass of a lamina with the same shape but uniform density is at the point $(0, 4a/(3\pi))$.]   ■

MOMENT OF INERTIA

The **moment of inertia** (also called the **second moment**) of a particle of mass m about an axis is defined to be mr^2, where r is the distance from the particle to the axis. We extend this concept to a lamina with density function $\rho(x, y)$ and occupying a region D by proceeding as we did for ordinary moments. We partition D into small rectangles, approximate the moment of inertia of each subrectangle about the x-axis, and take the limit of the sum as $\|P\|$ approaches 0. The result is the **moment of inertia** of the lamina **about the x-axis:**

$$(6) \qquad I_x = \lim_{\|P\| \to 0} \sum_{i=1}^m \sum_{j=1}^n (y_{ij}^*)^2 \rho(x_{ij}^*, y_{ij}^*)\,\Delta A_{ij} = \iint_D y^2 \rho(x, y)\,dA$$

Similarly, the **moment of inertia about the y-axis** is

$$(7) \qquad I_y = \lim_{\|P\| \to 0} \sum_{i=1}^m \sum_{j=1}^n (x_{ij}^*)^2 \rho(x_{ij}^*, y_{ij}^*)\,\Delta A_{ij} = \iint_D x^2 \rho(x, y)\,dA$$

It is also of interest to consider the **moment of inertia about the origin,** also called the **polar moment of inertia:**

$$(8) \quad I_0 = \lim_{\|P\| \to 0} \sum_{i=1}^m \sum_{j=1}^n \big[(x_{ij}^*)^2 + (y_{ij}^*)^2\big]\rho(x_{ij}^*, y_{ij}^*)\,\Delta A_{ij} = \iint_D (x^2 + y^2)\rho(x, y)\,dA$$

Note that $I_0 = I_x + I_y$.

EXAMPLE 4   Find the moments of inertia I_x, I_y, and I_0 of a homogeneous disk D with density $\rho(x, y) = \rho$, center the origin, and radius a.

SOLUTION   The boundary of D is the circle $x^2 + y^2 = a^2$ and in polar coordinates D is described by $0 \le \theta \le 2\pi$, $0 \le r \le a$. Let us compute I_0 first:

$$I_0 = \iint\limits_{D} (x^2 + y^2)\rho \, dA = \rho \int_0^{2\pi} \int_0^a r^2 r \, dr \, d\theta$$

$$= \rho \int_0^{2\pi} d\theta \int_0^a r^3 \, dr = 2\pi\rho \left[\frac{r^4}{4} \right]_0^a = \frac{\pi\rho a^4}{2}$$

Instead of computing I_x and I_y directly, we use the facts that $I_x + I_y = I_0$ and $I_x = I_y$ (from the symmetry of the problem). Thus

$$I_x = I_y = \frac{I_0}{2} = \frac{\pi\rho a^4}{4}$$ ∎

In Example 4 notice that the mass of the disk is

$$m = \text{density} \times \text{area} = \rho(\pi a^2)$$

so the moment of inertia of the disk about the origin (like a wheel about its axle) can be written as

$$I_0 = \tfrac{1}{2}ma^2$$

Thus if we increase the mass or the radius of the disk, we thereby increase the moment of inertia. In general, the moment of inertia plays much the same role in rotational motion that mass plays in linear motion. The moment of inertia of a wheel is what makes it difficult to start or stop the rotation of the wheel, just as the mass of a car is what makes it difficult to start or stop the motion of the car.

The **radius of gyration of a lamina about an axis** is the number R such that

(9) $$mR^2 = I$$

where m is the mass of the lamina and I is the moment of inertia about the given axis. Equation 9 says that if the mass of the lamina were concentrated at a distance R from the axis, then the moment of inertia of this "point mass" would be the same as the moment of inertia of the lamina.

In particular, the radius of gyration $\bar{\bar{y}}$ with respect to the x-axis and the radius of gyration $\bar{\bar{x}}$ with respect to the y-axis are given by the equations

(10) $$m\bar{\bar{y}}^2 = I_x \qquad m\bar{\bar{x}}^2 = I_y$$

Thus $(\bar{\bar{x}}, \bar{\bar{y}})$ is the point at which the mass of the lamina can be concentrated without changing the moments of inertia with respect to the coordinate axes. (Note the analogy with the center of mass.)

EXAMPLE 5   Find the radius of gyration about the x-axis of the disk in Example 4.

SOLUTION   As noted, the mass of the disk is $m = \rho\pi a^2$, so from Equations 10 we have

$$\bar{\bar{y}}^2 = \frac{I_x}{m} = \frac{\tfrac{1}{4}\pi\rho a^4}{\rho\pi a^2} = \frac{a^2}{4}$$

Therefore, the radius of gyration about the x-axis is

$$\bar{\bar{y}} = \frac{a}{2}$$

which is half the radius of the disk.   ∎

EXERCISES 13.5

1. Electric charge is distributed over the rectangle $0 \leqslant x \leqslant 2$, $1 \leqslant y \leqslant 2$ so that the charge density at (x, y) is $\sigma(x, y) = x^2 + 3y^2$ (measured in coulombs per square meter). Find the total charge on the rectangle.

2. Electric charge is distributed over the unit disk $x^2 + y^2 \leqslant 1$ so that the charge density at (x, y) is $\sigma(x, y) = 1 + x^2 + y^2$ (measured in coulombs per square meter). Find the total charge on the disk.

3–12 ■ Find the mass and center of mass of the lamina that occupies the region D and has the given density function ρ.

3. $D = \{(x, y) \mid -1 \leqslant x \leqslant 1, 0 \leqslant y \leqslant 1\}$; $\rho(x, y) = x^2$

4. $D = \{(x, y) \mid 0 \leqslant x \leqslant 2, 0 \leqslant y \leqslant 3\}$; $\rho(x, y) = y$

5. D is the triangular region with vertices $(0, 0)$, $(2, 1)$, $(0, 3)$; $\rho(x, y) = x + y$

6. D is the triangular region with vertices $(0, 0)$, $(1, 1)$, $(4, 0)$; $\rho(x, y) = x$

7. D is the region in the first quadrant bounded by the parabola $y = x^2$ and the line $y = 1$; $\rho(x, y) = xy$

8. D is bounded by the parabola $y = 9 - x^2$ and the x-axis; $\rho(x, y) = y$

9. D is bounded by the parabola $x = y^2$ and the line $y = x - 2$; $\rho(x, y) = 3$

CAS 10. D is bounded by the cardioid $r = 1 + \sin\theta$; $\rho(x, y) = 2$

11. $D = \{(x, y) \mid 0 \leqslant y \leqslant \sin x, 0 \leqslant x \leqslant \pi\}$; $\rho(x, y) = y$

12. $D = \{(x, y) \mid 0 \leqslant y \leqslant \cos x, 0 \leqslant x \leqslant \pi/2\}$; $\rho(x, y) = x$

13. A lamina occupies the part of the disk $x^2 + y^2 \leqslant 1$ in the first quadrant. Find its center of mass if the density at any point is proportional to its distance from the x-axis.

14. Find the center of mass of the lamina in Exercise 13 if the density at any point is proportional to the square of its distance from the origin.

15. Find the center of mass of a lamina in the shape of an isosceles right triangle with equal sides of length a if the density at any point is proportional to the square of the distance from the vertex opposite the hypotenuse.

16. A lamina occupies the region inside the circle $x^2 + y^2 = 2y$ but outside the circle $x^2 + y^2 = 1$. Find the center of mass

if the density at any point is inversely proportional to its distance from the origin.

17. Find the moments of inertia I_x, I_y, I_0 for the lamina of Exercise 7.

18. Find the moments of inertia I_x, I_y, I_0 for the lamina of Exercise 4.

19. Find the moments of inertia I_x, I_y, I_0 for the lamina of Exercise 9.

20. Find the moments of inertia I_x, I_y, I_0 for the lamina of Exercise 14.

21. A lamina with constant density $\rho(x, y) = \rho$ occupies a square with vertices $(0, 0)$, $(a, 0)$, (a, a), and $(0, a)$. Find the moments of inertia I_x and I_y and the radii of gyration $\bar{\bar{x}}$ and $\bar{\bar{y}}$.

22. A lamina with constant density $\rho(x, y) = \rho$ occupies the region under the curve $y = \sin x$ from $x = 0$ to $x = \pi$. Find the moments of inertia I_x and I_y and the radii of gyration $\bar{\bar{x}}$ and $\bar{\bar{y}}$.

23. Show that the formulas for $\bar{x}$ and $\bar{y}$ in (5) become Equations 8.4.11 when $\rho(x, y) = \rho$ (the density is constant).

24. (a) A lamina has constant density ρ and takes the shape of a disk with center the origin and radius R. Use Newton's Law of Gravitation (see Section 11.9) to show that the magnitude of the force of attraction that the lamina exerts on a body with mass m located at the point $(0, 0, d)$ on the positive z-axis is

$$F = 2\pi G m \rho d \left(\frac{1}{d} - \frac{1}{\sqrt{R^2 + d^2}} \right)$$

[*Hint:* Consider a polar partition as in Figure 3 in Section 13.4 and first compute the vertical component of the force exerted by the polar subrectangle R_{ij}.]

   (b) Show that the magnitude of the force of attraction of a lamina with density ρ that occupies an entire plane on an object with mass m located at a distance d from the plane is

$$F = 2\pi G m \rho$$

Notice that this expression does not depend on d.

13.6 SURFACE AREA

In Section 14.6 we will deal with areas of more general surfaces, called parametric surfaces, and so it is possible to omit this section if that later section will be covered.

In this section we apply double integrals to the problem of computing the area of a surface. In Section 8.3 we found the area of a very special type of surface—a surface of revolution—by the methods of single-variable calculus. Here we compute the area of a surface with equation $z = f(x, y)$, the graph of a function of two variables.

Let S be a surface with equation $z = f(x, y)$, where f has continuous partial derivatives. For simplicity in deriving the surface area formula, we assume that $f(x, y) \geq 0$ and the domain D of f is a rectangle. We consider a partition P of D into small rectangles R_{ij} with areas $\Delta A_{ij} = \Delta x_i \Delta y_j$. If (x_i, y_j) is the corner of R_{ij} closest to the origin, let $P_{ij}(x_i, y_j, f(x_i, y_j))$ be the point on S directly above it (see Figure 1). The tangent plane to S at P_{ij} is an approximation to S near P_{ij}. So the area ΔT_{ij} of the part of this tangent plane (a parallelogram) that lies directly above R_{ij} is an approximation to the area ΔS_{ij} of the part of S that lies directly above R_{ij}. Thus the sum $\Sigma\Sigma \, \Delta T_{ij}$ is an approximation to the total area of S, and this approximation appears to improve as $\|P\| \to 0$. Therefore, we define the **surface area** of S to be

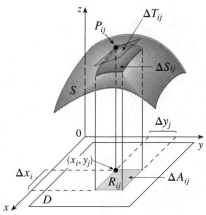

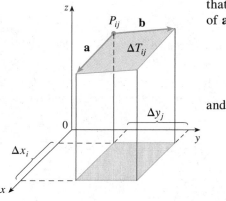

FIGURE 1

(1)
$$A(S) = \lim_{\|P\| \to 0} \sum_{i=1}^{m} \sum_{j=1}^{n} \Delta T_{ij}$$

To find a formula that is more convenient than Equation 1 for computational purposes, we let $\mathbf{a}$ and $\mathbf{b}$ be the vectors that start at P_{ij} and lie along the sides of the parallelogram with area ΔT_{ij} (see Figure 2). Then $\Delta T_{ij} = |\mathbf{a} \times \mathbf{b}|$. Recall from Section 12.3 that $f_x(x_i, y_j)$ and $f_y(x_i, y_j)$ are the slopes of the tangent lines through P_{ij} in the directions of $\mathbf{a}$ and $\mathbf{b}$. Therefore

$$\mathbf{a} = \Delta x_i \, \mathbf{i} + f_x(x_i, y_j) \, \Delta x_i \, \mathbf{k}$$

$$\mathbf{b} = \Delta y_j \, \mathbf{j} + f_y(x_i, y_j) \, \Delta y_j \, \mathbf{k}$$

and

$$\mathbf{a} \times \mathbf{b} = \begin{vmatrix} \mathbf{i} & \mathbf{j} & \mathbf{k} \\ \Delta x_i & 0 & f_x(x_i, y_j) \, \Delta x_i \\ 0 & \Delta y_j & f_y(x_i, y_j) \, \Delta y_j \end{vmatrix}$$

$$= -f_x(x_i, y_j) \, \Delta x_i \Delta y_j \, \mathbf{i} - f_y(x_i, y_j) \, \Delta x_i \Delta y_j \, \mathbf{j} + \Delta x_i \Delta y_j \, \mathbf{k}$$

$$= [-f_x(x_i, y_j) \, \mathbf{i} - f_y(x_i, y_j) \, \mathbf{j} + \mathbf{k}] \Delta A_{ij}$$

Thus
$$\Delta T_{ij} = |\mathbf{a} \times \mathbf{b}| = \sqrt{[f_x(x_i, y_j)]^2 + [f_y(x_i, y_j)]^2 + 1} \, \Delta A_{ij}$$

FIGURE 2

From Definition 1 we then have

$$A(S) = \lim_{\|P\| \to 0} \sum_{i=1}^{m} \sum_{j=1}^{n} \Delta T_{ij}$$

$$= \lim_{\|P\| \to 0} \sum_{i=1}^{m} \sum_{j=1}^{n} \sqrt{[f_x(x_i, y_j)]^2 + [f_y(x_i, y_j)]^2 + 1} \, \Delta A_{ij}$$

and by the definition of a double integral we get the following formula.

(2) The area of the surface with equation $z = f(x, y)$, $(x, y) \in D$, where f_x and f_y are continuous, is

$$A(S) = \iint_D \sqrt{[f_x(x, y)]^2 + [f_y(x, y)]^2 + 1} \ dA$$

We will verify in Section 14.6 that this formula is consistent with our previous formula for the area of a surface of revolution. Notice the similarity of Formula 2 to the arc length formula

$$L = \int_a^b \sqrt{[f'(x)]^2 + 1} \ dx$$

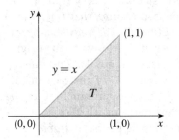

FIGURE 3

EXAMPLE 1   Find the surface area of the part of the surface $z = x^2 + 2y$ that lies above the triangular region T in the xy-plane with vertices $(0, 0)$, $(1, 0)$, and $(1, 1)$.

SOLUTION   The region T is shown in Figure 3 and is described by

$$T = \{(x, y) \mid 0 \le x \le 1, 0 \le y \le x\}$$

Using Formula 2 with $f(x, y) = x^2 + 2y$, we get

$$A = \iint_T \sqrt{(2x)^2 + (2)^2 + 1} \ dA = \int_0^1 \int_0^x \sqrt{4x^2 + 5} \ dy \, dx$$

$$= \int_0^1 x \sqrt{4x^2 + 5} \ dx = \tfrac{1}{8} \cdot \tfrac{2}{3} (4x^2 + 5)^{3/2} \big]_0^1 = \tfrac{1}{12} (27 - 5\sqrt{5})$$

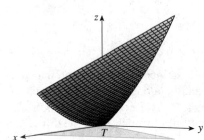

FIGURE 4

Figure 4 shows the portion of the surface whose area we have just computed. ∎

EXAMPLE 2   Find the area of the part of the paraboloid $z = x^2 + y^2$ that lies under the plane $z = 9$.

SOLUTION   The plane intersects the paraboloid in the circle $x^2 + y^2 = 9$, $z = 9$. Therefore, the given surface lies above the disk D with center the origin and radius 3 (see Figure 5). Using Formula 2, we have

$$A = \iint_D \sqrt{(2x)^2 + (2y)^2 + 1} \ dA = \iint_D \sqrt{4(x^2 + y^2) + 1} \ dA$$

Converting to polar coordinates, we obtain

$$A = \int_0^{2\pi} \int_0^3 \sqrt{4r^2 + 1} \ r \, dr \, d\theta = \int_0^{2\pi} d\theta \int_0^3 \tfrac{1}{8} \sqrt{4r^2 + 1} \ (8r) \, dr$$

$$= 2\pi(\tfrac{1}{8}) \tfrac{2}{3} (4r^2 + 1)^{3/2} \big]_0^3 = \frac{\pi}{6} (37\sqrt{37} - 1)$$

∎

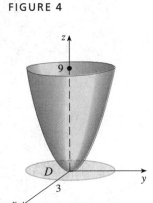

FIGURE 5

EXERCISES 13.6

1–10 ■ Find the area of the surface.

1. The part of the plane $x + 2y + z = 4$ that lies inside the cylinder $x^2 + y^2 = 4$

2. The part of the plane $2x + 3y - z + 1 = 0$ that lies above the rectangle $[1, 4] \times [2, 4]$

3. The part of the cylinder $y^2 + z^2 = 9$ that lies above the rectangle with vertices $(0, 0)$, $(4, 0)$, $(0, 2)$, and $(4, 2)$

4. The part of the surface $z = x + y^2$ that lies above the triangle with vertices $(0, 0)$, $(1, 1)$, and $(0, 1)$

5. The part of the hyperbolic paraboloid $z = y^2 - x^2$ that lies between the cylinders $x^2 + y^2 = 1$ and $x^2 + y^2 = 4$

6. The part of the paraboloid $z = 4 - x^2 - y^2$ that lies above the xy-plane

7. The part of the surface $z = xy$ that lies within the cylinder $x^2 + y^2 = 1$

8. The part of the sphere $x^2 + y^2 + z^2 = 4$ that lies above the plane $z = 1$

9. The part of the sphere $x^2 + y^2 + z^2 = a^2$ that lies within the cylinder $x^2 + y^2 = ax$ and above the xy-plane

10. The part of the sphere $x^2 + y^2 + z^2 = 4z$ that lies inside the paraboloid $z = x^2 + y^2$

11. Use the Midpoint Rule for double integrals (see Section 13.1) with four squares to estimate the surface area of the portion of the paraboloid $z = x^2 + y^2$ that lies above the square $[0, 1] \times [0, 1]$.

CAS 12. Use a computer algebra system to approximate the surface area in Exercise 11 to four decimal places. Compare with the answer to that exercise.

CAS 13. Find the exact area of the surface $z = x^2 + 2y$, $0 \le x \le 1$, $0 \le y \le 1$.

CAS 14. Find the exact area of the surface

$$z = 1 + x + y + x^2, \quad -2 \le x \le 1, -1 \le y \le 1$$

Illustrate by graphing the surface.

CAS 15. Find, to four decimal places, the area of the part of the surface $z = 1 + x^2y^2$ that lies above the disk $x^2 + y^2 \le 1$.

CAS 16. Find, to four decimal places, the area of the part of the surface $z = (1 + x^2)/(1 + y^2)$ that lies above the square $|x| + |y| \le 1$. Illustrate by graphing this part of the surface.

17. Show that the area of the part of the plane $z = ax + by + c$ that projects onto a region D in the xy-plane with area $A(D)$ is $\sqrt{a^2 + b^2 + 1}\ A(D)$.

18. If you attempt to use Formula 2 to find the area of the top half of the sphere $x^2 + y^2 + z^2 = a^2$, you have a slight problem because the double integral is improper. In fact, the integrand has an infinite discontinuity at every point of the boundary circle $x^2 + y^2 = a^2$. However, the integral can be computed as the limit of the integral over the disk $x^2 + y^2 \le t^2$ as $t \to a^-$. Use this method to show that the area of a sphere of radius a is $4\pi a^2$.

13.7 TRIPLE INTEGRALS

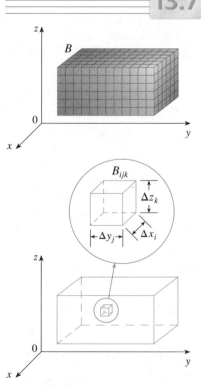

FIGURE 1

Just as we defined single integrals for functions of one variable and double integrals for functions of two variables, so we can define triple integrals for functions of three variables. Let us first deal with the simplest case where f is defined on a rectangular box:

(1) $$B = \{(x, y, z) \mid a \le x \le b, c \le y \le d, r \le z \le s\}$$

The first step is to partition the intervals $[a, b]$, $[c, d]$, and $[r, s]$ as follows:

$$a = x_0 < x_1 < \cdots < x_{i-1} < x_i < \cdots < x_l = b$$

$$c = y_0 < y_1 < \cdots < y_{j-1} < y_j < \cdots < y_m = d$$

$$r = z_0 < z_1 < \cdots < z_{k-1} < z_k < \cdots < z_n = s$$

The planes through these partition points parallel to the coordinate planes divide the box B into lmn sub-boxes

$$B_{ijk} = [x_{i-1}, x_i] \times [y_{j-1}, y_j] \times [z_{k-1}, z_k]$$

which are shown in Figure 1.

The volume of B_{ijk} is

$$\Delta V_{ijk} = \Delta x_i\, \Delta y_j\, \Delta z_k$$

where $\quad \Delta x_i = x_i - x_{i-1} \qquad \Delta y_j = y_j - y_{j-1} \qquad \Delta z_k = z_k - z_{k-1}$

Then we form the **triple Riemann sum**

(2)
$$\sum_{i=1}^{l} \sum_{j=1}^{m} \sum_{k=1}^{n} f(x_{ijk}^*, y_{ijk}^*, z_{ijk}^*) \, \Delta V_{ijk}$$

where $(x_{ijk}^*, y_{ijk}^*, z_{ijk}^*)$ is in B_{ijk}. We define the **norm** of the partition P to be the length of the longest diagonal of all the boxes B_{ijk} and we denote the norm by $\|P\|$. Then, by analogy with the definition of a double integral (13.1.4), we define the triple integral as the limit of the triple Riemann sums in (2).

---

(3) DEFINITION   The **triple integral** of f over the box B is

$$\iiint_{B} f(x, y, z) \, dV = \lim_{\|P\| \to 0} \sum_{i=1}^{l} \sum_{j=1}^{m} \sum_{k=1}^{n} f(x_{ijk}^*, y_{ijk}^*, z_{ijk}^*) \, \Delta V_{ijk}$$

if this limit exists.

The precise meaning of the limit in Definition 3 is similar to that given in Note 1 after Definition 13.1.4. Again, the triple integral always exists if f is continuous. An alternative notation is

$$\iiint_{B} f(x, y, z) \, dV = \iiint_{B} f(x, y, z) \, dx \, dy \, dz$$

Just as for double integrals, the practical method for evaluating triple integrals is to express them as iterated integrals as follows:

---

(4) FUBINI'S THEOREM FOR TRIPLE INTEGRALS   If f is continuous on the rectangular box $B = [a, b] \times [c, d] \times [r, s]$, then

$$\iiint_{B} f(x, y, z) \, dV = \int_{r}^{s} \int_{c}^{d} \int_{a}^{b} f(x, y, z) \, dx \, dy \, dz$$

The iterated integral on the right side of Fubini's Theorem means that we integrate first with respect to x (keeping y and z fixed), then we integrate with respect to y (keeping z fixed), and finally we integrate with respect to z. There are five other possible orders in which we can integrate. For instance, if we integrate with respect to y, then z, and then x, we have

$$\iiint_{B} f(x, y, z) \, dV = \int_{a}^{b} \int_{r}^{s} \int_{c}^{d} f(x, y, z) \, dy \, dz \, dx$$

EXAMPLE 1   Evaluate the triple integral $\iiint_{B} xyz^2 \, dV$, where B is the rectangular box given by

$$B = \{(x, y, z) \mid 0 \leqslant x \leqslant 1, \, -1 \leqslant y \leqslant 2, \, 0 \leqslant z \leqslant 3\}$$

SOLUTION   We could use any of the six possible orders of integration. If we choose to integrate with respect to x, then y, and then z, we obtain

$$\iiint_B xyz^2 \, dV = \int_0^3 \int_{-1}^2 \int_0^1 xyz^2 \, dx \, dy \, dz = \int_0^3 \int_{-1}^2 \left[\frac{x^2yz^2}{2} \right]_{x=0}^{x=1} dy \, dz$$

$$= \int_0^3 \int_{-1}^2 \frac{yz^2}{2} \, dy \, dz = \int_0^3 \left[\frac{y^2z^2}{4} \right]_{y=-1}^{y=2} dz$$

$$= \int_0^3 \frac{3z^2}{4} \, dz = \frac{z^3}{4} \Big]_0^3 = \frac{27}{4}$$

Now we define the **triple integral over a general bounded region E** in three-dimensional space (a solid) by the same procedure that we used for double integrals (13.3.2). We enclose E in a box B of the type given by Equation 1. Then we define F so that it agrees with f on E but is 0 for points in B that are outside E. By definition,

$$\iiint_E f(x, y, z) \, dV = \iiint_B F(x, y, z) \, dV$$

This integral exists if f is continuous and the boundary of E is "reasonably smooth." The triple integral has essentially the same properties as the double integral (13.3.6–13.3.9).

We restrict our attention to continuous functions f and to certain simple types of regions. A solid region E is said to be of **type 1** if it lies between the graphs of two continuous functions of x and y, that is,

(5)
$$E = \{(x, y, z) \mid (x, y) \in D, \, \phi_1(x, y) \le z \le \phi_2(x, y)\}$$

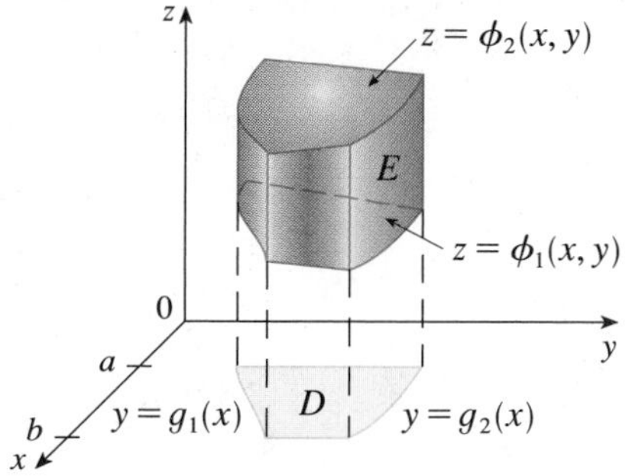

where D is the projection of E onto the xy-plane as shown in Figure 2. Notice that the upper boundary of the solid E is the surface with equation $z = \phi_2(x, y)$, while the lower boundary is the surface $z = \phi_1(x, y)$.

By the same sort of argument that led to (13.3.3), it can be shown that, if E is a type 1 region given by Equation 5, then

(6)
$$\iiint_E f(x, y, z) \, dV = \iint_D \left[\int_{\phi_1(x,y)}^{\phi_2(x,y)} f(x, y, z) \, dz \right] dA$$

The meaning of the inner integral on the right side of Equation 6 is that x and y are held fixed, and therefore $\phi_1(x, y)$ and $\phi_2(x, y)$ are regarded as constants, while $f(x, y, z)$ is integrated with respect to z.

In particular, if the projection D of E onto the xy-plane is a type I plane region (as in Figure 3), then

$$E = \{(x, y, z) \mid a \le x \le b, \, g_1(x) \le y \le g_2(x), \, \phi_1(x, y) \le z \le \phi_2(x, y)\}$$

and Equation 6 becomes

(7)
$$\iiint_E f(x, y, z) \, dV = \int_a^b \int_{g_1(x)}^{g_2(x)} \int_{\phi_1(x,y)}^{\phi_2(x,y)} f(x, y, z) \, dz \, dy \, dx$$

If, on the other hand, D is a type II plane region (as in Figure 4), then

$$E = \{(x, y, z) \mid c \le y \le d, \, h_1(y) \le x \le h_2(y), \, \phi_1(x, y) \le z \le \phi_2(x, y)\}$$

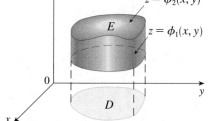

FIGURE 2
A type 1 solid region

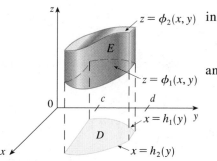

FIGURE 3
A type 1 solid region

FIGURE 4
Another type 1 solid region

and Equation 6 becomes

(8)
$$\iiint_E f(x, y, z)\, dV = \int_c^d \int_{h_1(y)}^{h_2(y)} \int_{\phi_1(x, y)}^{\phi_2(x, y)} f(x, y, z)\, dz\, dx\, dy$$

EXAMPLE 2   Evaluate $\iiint_E z\, dV$, where E is the solid tetrahedron bounded by the four planes $x = 0$, $y = 0$, $z = 0$, and $x + y + z = 1$.

SOLUTION   When we set up a triple integral it is wise to draw *two* diagrams: one of the solid region E (see Figure 5) and one of its projection D on the xy-plane (see Figure 6). The lower boundary of the tetrahedron is the plane $z = 0$ and the upper boundary is the plane $x + y + z = 1$ (or $z = 1 - x - y$), so we use $\phi_1(x, y) = 0$ and $\phi_2(x, y) = 1 - x - y$ in Formula 7. Notice that the planes $x + y + z = 1$ and $z = 0$ intersect in the line $x + y = 1$ (or $y = 1 - x$) in the xy-plane. So the projection of E is the triangular region shown in Figure 6, and we have

(9) $E = \{(x, y, z) \mid 0 \leqslant x \leqslant 1,\ 0 \leqslant y \leqslant 1 - x,\ 0 \leqslant z \leqslant 1 - x - y\}$

This description of E as a type 1 region enables us to evaluate the integral as follows:

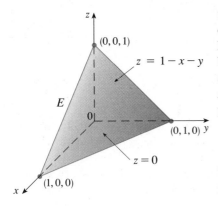

(0, 0, 1)

$z = 1 - x - y$

E

0

(0, 1, 0) y

$z = 0$

(1, 0, 0)

x

FIGURE 5

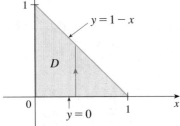

y

1

$y = 1 - x$

D

0

$y = 0$

1

x

FIGURE 6

$$\iiint_E z\, dV = \int_0^1 \int_0^{1-x} \int_0^{1-x-y} z\, dz\, dy\, dx$$

$$= \int_0^1 \int_0^{1-x} \left[\frac{z^2}{2} \right]_{z=0}^{z=1-x-y} dy\, dx$$

$$= \frac{1}{2} \int_0^1 \int_0^{1-x} (1 - x - y)^2\, dy\, dx$$

$$= \frac{1}{2} \int_0^1 \left[-\frac{(1 - x - y)^3}{3} \right]_{y=0}^{y=1-x} dx$$

$$= \frac{1}{6} \int_0^1 (1 - x)^3\, dx = \frac{1}{6} \left[-\frac{(1 - x)^4}{4} \right]_0^1 = \frac{1}{24}$$

A solid region E is of **type 2** if it is of the form

$$E = \{(x, y, z) \mid (y, z) \in D,\ \phi_1(y, z) \leqslant x \leqslant \phi_2(y, z)\}$$

where, this time, D is the projection of E onto the yz-plane (see Figure 7). The back surface is $x = \phi_1(y, z)$, the front surface is $x = \phi_2(y, z)$, and we have

(10) $$\iiint_E f(x, y, z)\, dV = \iint_D \left[\int_{\phi_1(y, z)}^{\phi_2(y, z)} f(x, y, z)\, dx \right] dA$$

Finally, a **type 3** region is of the form

$$E = \{(x, y, z) \mid (x, z) \in D,\ \phi_1(x, z) \leqslant y \leqslant \phi_2(x, z)\}$$

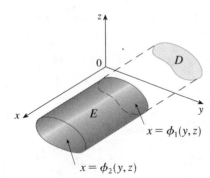

0

D

x

y

E

$x = \phi_1(y, z)$

$x = \phi_2(y, z)$

FIGURE 7
A type 2 region

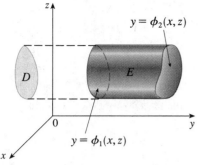

FIGURE 8
A type 3 region

where D is the projection of E onto the xz-plane, $y = \phi_1(x, z)$ is the left surface, and $y = \phi_2(x, z)$ is the right surface (see Figure 8). For this type of region we have

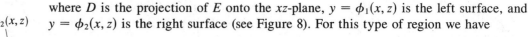

(11)
$$\iiint_E f(x, y, z) \, dV = \iint_D \left[\int_{\phi_1(x,z)}^{\phi_2(x,z)} f(x, y, z) \, dy \right] dA$$

In each of Equations 10 and 11 there may be two possible expressions for the integral depending on whether D is a type I or II plane region (and corresponding to Equations 7 and 8).

EXAMPLE 3   Evaluate $\iiint_E \sqrt{x^2 + z^2} \, dV$, where E is the region bounded by the paraboloid $y = x^2 + z^2$ and the plane $y = 4$.

SOLUTION   The solid E is shown in Figure 9. If we regard it as a type 1 region, then we need to consider its projection D_1 onto the xy-plane, which is the parabolic region in Figure 10. (The trace of $y = x^2 + z^2$ in the plane $z = 0$ is the parabola $y = x^2$.)

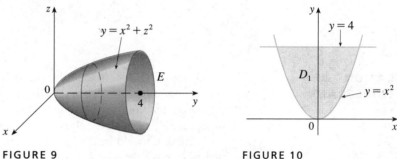

FIGURE 9
Region of integration

FIGURE 10
Projection on xy-plane

From $y = x^2 + z^2$ we obtain $z = \pm\sqrt{y - x^2}$, so the lower boundary surface of E is $z = -\sqrt{y - x^2}$ and the upper surface is $z = \sqrt{y - x^2}$. Therefore, the description of E as a type 1 region is

$$E = \left\{ (x, y, z) \mid -2 \le x \le 2, \ x^2 \le y \le 4, \ -\sqrt{y - x^2} \le z \le \sqrt{y - x^2} \right\}$$

and so we obtain

$$\iiint_E \sqrt{x^2 + z^2} \, dV = \int_{-2}^{2} \int_{x^2}^{4} \int_{-\sqrt{y-x^2}}^{\sqrt{y-x^2}} \sqrt{x^2 + z^2} \, dz \, dy \, dx$$

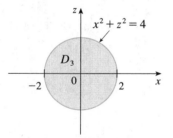

FIGURE 11
Projection on xz-plane

⊘ The most difficult step in evaluating a triple integral is setting up an expression for the region of integration (such as Equation 9 in Example 2). Remember that the limits of integration in the inner integral contain at most two variables, the limits of integration in the middle integral contain at most one variable, and the limits of integration in the outer integral must be constants.

Although this expression is correct, it is extremely difficult to evaluate. So let us instead consider E as a type 3 region. As such, its projection D_3 onto the xz-plane is the disk $x^2 + z^2 \le 4$ shown in Figure 11.

Then the left boundary of E is the paraboloid $y = x^2 + z^2$ and the right boundary is the plane $y = 4$, so taking $\phi_1(x, z) = x^2 + z^2$ and $\phi_2(x, z) = 4$ in Equation 11, we have

$$\iiint_E \sqrt{x^2 + z^2} \, dV = \iint_{D_3} \left[\int_{x^2+z^2}^{4} \sqrt{x^2 + z^2} \, dy \right] dA$$

$$= \iint_{D_3} (4 - x^2 - z^2) \sqrt{x^2 + z^2} \, dA$$

Although this integral could be written as

$$\int_{-2}^{2} \int_{-\sqrt{4-x^2}}^{\sqrt{4-x^2}} (4 - x^2 - z^2)\sqrt{x^2 + z^2}\ dz\ dx$$

it is easier to convert to polar coordinates in the xz-plane: $x = r\cos\theta$, $z = r\sin\theta$. This gives

$$\iiint_{E} \sqrt{x^2 + z^2}\ dV = \iint_{D_3} (4 - x^2 - z^2)\sqrt{x^2 + z^2}\ dA$$

$$= \int_{0}^{2\pi} \int_{0}^{2} (4 - r^2)r\,r\,dr\,d\theta = \int_{0}^{2\pi} d\theta \int_{0}^{2} (4r^2 - r^4)\,dr$$

$$= 2\pi \left[\frac{4r^3}{3} - \frac{r^5}{5} \right]_{0}^{2} = \frac{128\pi}{15} \qquad \blacksquare$$

APPLICATIONS OF TRIPLE INTEGRALS

Recall that if $f(x) \ge 0$, then the single integral $\int_{a}^{b} f(x)\,dx$ represents the area under the curve $y = f(x)$ from a to b, and if $f(x, y) \ge 0$, then the double integral $\iint_{D} f(x, y)\,dA$ represents the volume under the surface $z = f(x, y)$ and above D. The corresponding interpretation of a triple integral $\iiint_{E} f(x, y, z)\,dV$, where $f(x, y, z) \ge 0$, is not very useful because it would be the "hypervolume" of a four-dimensional object and, of course, that is impossible to visualize. (Remember that E is just the *domain* of the function f; the graph of f lies in four-dimensional space.) Nonetheless, the triple integral $\iiint_{E} f(x, y, z)\,dV$ can be interpreted in different ways in different physical situations, depending on the physical interpretations of x, y, z and $f(x, y, z)$.

Let us begin with the special case where $f(x, y, z) = 1$ for all points in E. Then the triple integral does represent the volume of E:

(12)

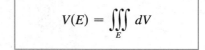

$$V(E) = \iiint_{E} dV$$

For example, you can see this in the case of a type 1 region by putting $f(x, y, z) = 1$ in Formula 6:

$$\iiint_{E} 1\,dV = \iint_{D} \left[\int_{\phi_1(x, y)}^{\phi_2(x, y)} dz \right] dA = \iint_{D} [\phi_2(x, y) - \phi_1(x, y)]\,dA$$

and from Section 13.3 we know this represents the volume that lies between the surfaces $z = \phi_1(x, y)$ and $z = \phi_2(x, y)$.

EXAMPLE 4 Use a triple integral to find the volume of the tetrahedron T bounded by the planes $x + 2y + z = 2$, $x = 2y$, $x = 0$, and $z = 0$.

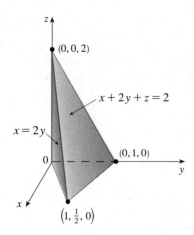

$(0, 0, 2)$

$x + 2y + z = 2$

$x = 2y$

0

$(0, 1, 0)$

$\left(1, \frac{1}{2}, 0\right)$

FIGURE 12

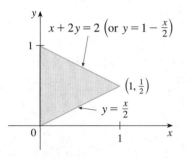

$x + 2y = 2 \left(\text{or } y = 1 - \frac{x}{2}\right)$

$\left(1, \frac{1}{2}\right)$

$y = \frac{x}{2}$

FIGURE 13

SOLUTION   The tetrahedron T and its projection D on the xy-plane are shown in Figures 12 and 13. The lower boundary of T is the plane $z = 0$ and the upper boundary is the plane $x + 2y + z = 2$, that is, $z = 2 - x - 2y$. Therefore, we have

$$V(T) = \iiint_T dV = \int_0^1 \int_{x/2}^{1-x/2} \int_0^{2-x-2y} dz\, dy\, dx$$

$$= \int_0^1 \int_{x/2}^{1-x/2} (2 - x - 2y)\, dy\, dx = \tfrac{1}{3}$$

by the same calculation as in Example 4 in Section 13.3.

(Notice that it is not necessary to use triple integrals to compute volumes. They simply give an alternative method for setting up the calculation.) ∎

All the applications of double integrals in Section 13.5 can be immediately extended to triple integrals. For example, if the density function of a solid object that occupies the region E is $\rho(x, y, z)$, in units of mass per unit volume, at any given point (x, y, z), then its **mass** is

$$(13) \qquad m = \iiint_E \rho(x, y, z)\, dV$$

and its **moments** about the three coordinate planes are

$$(14) \qquad M_{yz} = \iiint_E x\rho(x, y, z)\, dV \qquad M_{xz} = \iiint_E y\rho(x, y, z)\, dV$$

$$M_{xy} = \iiint_E z\rho(x, y, z)\, dV$$

The **center of mass** is located at the point $(\bar{x}, \bar{y}, \bar{z})$ where

$$(15) \qquad \bar{x} = \frac{M_{yz}}{m} \qquad \bar{y} = \frac{M_{xz}}{m} \qquad \bar{z} = \frac{M_{xy}}{m}$$

If the density is constant, the center of mass of the solid is called the **centroid** of E. The **moments of inertia** about the three coordinate axes are

$$(16) \quad I_x = \iiint_E (y^2 + z^2)\rho(x, y, z)\, dV \qquad I_y = \iiint_E (x^2 + z^2)\rho(x, y, z)\, dV$$

$$I_z = \iiint_E (x^2 + y^2)\rho(x, y, z)\, dV$$

Also, as in Section 13.5, the total *electric charge* on a solid object occupying a region E and having charge density $\sigma(x, y, z)$ is

$$Q = \iiint_E \sigma(x, y, z)\, dV$$

EXAMPLE 5 Find the center of mass of a solid of constant density that is bounded by the parabolic cylinder $x = y^2$ and the planes $x = z$, $z = 0$, and $x = 1$.

SOLUTION The solid E and its projection onto the xy-plane are shown in Figure 14. The lower and upper surfaces of E are the planes $z = 0$ and $z = x$, so we describe E as a type 1 region:

$$E = \{(x, y, z) \,|\, -1 \leq y \leq 1, y^2 \leq x \leq 1, 0 \leq z \leq x\}$$

Then, if the density is $\rho(x, y, z) = \rho$, the mass is

$$m = \iiint_E \rho \, dV = \int_{-1}^1 \int_{y^2}^1 \int_0^x \rho \, dz \, dx \, dy$$

$$= \rho \int_{-1}^1 \int_{y^2}^1 x \, dx \, dy = \rho \int_{-1}^1 \left[\frac{x^2}{2} \right]_{x=y^2}^{x=1} dy$$

$$= \frac{\rho}{2} \int_{-1}^1 (1 - y^4) \, dy = \rho \int_0^1 (1 - y^4) \, dy$$

$$= \rho \left[y - \frac{y^5}{5} \right]_0^1 = \frac{4\rho}{5}$$

Because of the symmetry of E and ρ about the xz-plane, we can immediately say that $M_{xz} = 0$ and, therefore, $\bar{y} = 0$. The other moments are

$$M_{yz} = \iiint_E x\rho \, dV = \int_{-1}^1 \int_{y^2}^1 \int_0^x x\rho \, dz \, dx \, dy$$

$$= \rho \int_{-1}^1 \int_{y^2}^1 x^2 \, dx \, dy = \rho \int_{-1}^1 \left[\frac{x^3}{3} \right]_{x=y^2}^{x=1} dy$$

$$= \frac{2\rho}{3} \int_0^1 (1 - y^6) \, dy = \frac{2\rho}{3} \left[y - \frac{y^7}{7} \right]_0^1 = \frac{4\rho}{7}$$

$$M_{xy} = \iiint_E z\rho \, dV = \int_{-1}^1 \int_{y^2}^1 \int_0^x z\rho \, dz \, dx \, dy$$

$$= \rho \int_{-1}^1 \int_{y^2}^1 \left[\frac{z^2}{2} \right]_{z=0}^{z=x} dx \, dy = \frac{\rho}{2} \int_{-1}^1 \int_{y^2}^1 x^2 \, dx \, dy$$

$$= \frac{\rho}{3} \int_0^1 (1 - y^6) \, dy = \frac{2\rho}{7}$$

Therefore, the center of mass is

$$(\bar{x}, \bar{y}, \bar{z}) = \left(\frac{M_{yz}}{m}, \frac{M_{xz}}{m}, \frac{M_{xy}}{m} \right) = \left(\tfrac{5}{7}, 0, \tfrac{5}{14} \right) \qquad \blacksquare$$

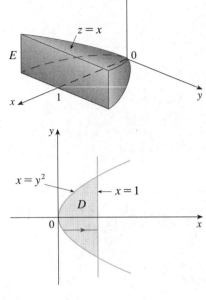

FIGURE 14

EXERCISES 13.7

1. Evaluate the integral in Example 1, integrating first with respect to z, then x, and then y.

2. Evaluate the integral $\iiint_E (x^2 + yz) \, dV$, where

$$E = \{(x, y, z) \,|\, 0 \leq x \leq 2, -3 \leq y \leq 0, -1 \leq z \leq 1\}$$

using three different orders of integration.

3–6 ■ Evaluate the iterated integral.

3. $\int_0^1 \int_0^z \int_0^y xyz \, dx \, dy \, dz$

4. $\int_0^1 \int_x^{2x} \int_0^{x+y} 2xy \, dz \, dy \, dx$

5. $\int_0^\pi \int_0^2 \int_0^{\sqrt{4-z^2}} z \sin y \, dx \, dz \, dy$

6. $\int_0^3 \int_0^{\sqrt{9-x^2}} \int_0^x yz \, dy \, dz \, dx$

7–16 ■ Evaluate the triple integral.

7. $\iiint_E yz\, dV$, where
$E = \{(x, y, z) \mid 0 \le z \le 1,\ 0 \le y \le 2z,\ 0 \le x \le z + 2\}$

8. $\iiint_E e^x\, dV$, where
$E = \{(x, y, z) \mid 0 \le y \le 1,\ 0 \le x \le y,\ 0 \le z \le x + y\}$

9. $\iiint_E y\, dV$, where E lies under the plane $z = x + 2y$ and above the region in the xy-plane bounded by the curves $y = x^2$, $y = 0$, and $x = 1$

10. $\iiint_E x\, dV$, where E is bounded by the planes $x = 0$, $y = 0$, $z = 0$, and $3x + 2y + z = 6$

11. $\iiint_E xy\, dV$, where E is the solid tetrahedron with vertices $(0, 0, 0)$, $(1, 0, 0)$, $(0, 2, 0)$, and $(0, 0, 3)$

12. $\iiint_E xz\, dV$, where E is the solid tetrahedron with vertices $(0, 0, 0)$, $(0, 1, 0)$, $(1, 1, 0)$, and $(0, 1, 1)$

13. $\iiint_E z\, dV$, where E is bounded by the planes $x = 0$, $y = 0$, $z = 0$, $y + z = 1$, and $x + z = 1$

14. $\iiint_E (x + 2y)\, dV$, where E is bounded by the parabolic cylinder $y = x^2$ and the planes $x = z$, $x = y$, and $z = 0$

15. $\iiint_E x\, dV$, where E is bounded by the paraboloid $x = 4y^2 + 4z^2$ and the plane $x = 4$

16. $\iiint_E z\, dV$, where E is bounded by the cylinder $y^2 + z^2 = 9$ and the planes $x = 0$, $y = 3x$, and $z = 0$ in the first octant

17–20 ■ Use a triple integral to find the volume of the given solid.

17. The tetrahedron bounded by the coordinate planes and the plane $2x + 3y + 6z = 12$

18. The solid bounded by the elliptic cylinder $4x^2 + z^2 = 4$ and the planes $y = 0$ and $y = z + 2$

19. The solid bounded by the cylinder $x = y^2$ and the planes $z = 0$ and $x + z = 1$

20. The solid enclosed by the paraboloids $z = x^2 + y^2$ and $z = 18 - x^2 - y^2$

21. (a) Express the volume of the wedge in the first octant that is cut from the cylinder $y^2 + z^2 = 1$ by the planes $y = x$ and $x = 1$ as a triple integral.
    (b) Use either the Table of Integrals (on the back endpapers) or a computer algebra system to find the exact value of the triple integral in part (a).

22. Evaluate the triple integral exactly:
$$\int_0^2 \int_{-1}^{\sin x} \int_{z-x}^{z+x} e^{3x}(5y + 2z)\, dy\, dz\, dx$$

23. In the **Midpoint Rule for triple integrals** we use a triple Riemann sum to approximate a triple integral over a box B, where $f(x, y, z)$ is evaluated at the center $(\bar{x}_i, \bar{y}_j, \bar{z}_k)$ of the box B_{ijk}. Use the Midpoint Rule to estimate $\iiint_B e^{-x^2 - y^2 - z^2}\, dV$, where B is the cube defined by $0 \le x \le 1$, $0 \le y \le 1$, $0 \le z \le 1$. Use the partition of B into eight cubes of equal size. What is the norm of this partition?

24. Use a computer algebra system to approximate the integral in Exercise 23 to two decimal places. Compare with the answer to that exercise.

25–26 ■ Sketch the solid whose volume is given by the iterated integral.

25. $\int_0^1 \int_0^{1-x} \int_0^{2-2z} dy\, dz\, dx$ **26.** $\int_0^2 \int_0^{2-y} \int_0^{4-y^2} dx\, dz\, dy$

27–30 ■ Express the integral $\iiint_E f(x, y, z)\, dV$ as an iterated integral in six different ways, where E is the solid bounded by the given surfaces.

27. $x^2 + z^2 = 4$, $y = 0$, $y = 6$

28. $z = 0$, $x = 0$, $y = 2$, $z = y - 2x$

29. $z = 0$, $z = y$, $x^2 = 1 - y$

30. $9x^2 + 4y^2 + z^2 = 1$

31. The figure shows the region of integration for the integral
$$\int_0^1 \int_{\sqrt{x}}^1 \int_0^{1-y} f(x, y, z)\, dz\, dy\, dx$$
Rewrite this integral as an equivalent iterated integral in the five other orders.

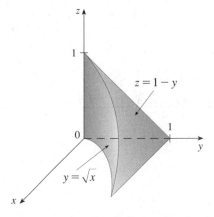

32. The figure shows the region of integration for the integral
$$\int_0^1 \int_0^{1-x^2} \int_0^{1-x} f(x, y, z)\, dy\, dz\, dx$$
Rewrite this integral as an equivalent iterated integral in the five other orders.

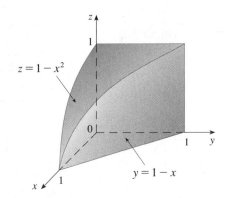

33–34 ■ Write five other iterated integrals that are equal to the given iterated integral.

33. $\int_0^1 \int_y^1 \int_0^y f(x, y, z)\, dz\, dx\, dy$

34. $\int_0^1 \int_0^{x^2} \int_0^y f(x, y, z)\, dz\, dy\, dx$

35–38 ■ Find the mass and center of mass of the given solid E with the given density function ρ.

35. E is the solid of Exercise 9; $\rho(x, y, z) = 2$

36. E is bounded by the parabolic cylinder $z = 1 - y^2$ and the planes $x + z = 1$, $x = 0$, and $z = 0$; $\rho(x, y, z) = 4$

37. E is the cube given by $0 \le x \le a$, $0 \le y \le a$, $0 \le z \le a$; $\rho(x, y, z) = x^2 + y^2 + z^2$

38. E is the tetrahedron bounded by the planes $x = 0$, $y = 0$, $z = 0$, $x + y + z = 1$; $\rho(x, y, z) = y$

39–42 ■ Set up, but do not evaluate, integral expressions for (a) the mass, (b) the center of mass, and (c) the moment of inertia about the z-axis.

39. The solid in the first octant bounded by the cylinder $x^2 + y^2 = 1$ and the planes $y = z$, $x = 0$, and $z = 0$; $\rho(x, y, z) = 1 + x + y + z$

40. The hemisphere $x^2 + y^2 + z^2 = 1$, $z \ge 0$; $\rho(x, y, z) = \sqrt{x^2 + y^2 + z^2}$

41. The solid of Exercise 15; $\rho(x, y, z) = x^2 + y^2 + z^2$

42. The solid of Exercise 16; $\rho(x, y, z) = x^2 + y^2$

CAS **43.** Evaluate the quantities in Exercise 39 exactly.

CAS **44.** Evaluate the quantities in Exercise 42 correct to three decimal places.

45. Find the moments of inertia for a cube of constant density k and side length L if one vertex is at the origin and three edges lie along the coordinate axes.

46. Find the moments of inertia for a rectangular brick with dimensions a, b, and c, mass M, and constant density if the center of the brick is at the origin and the edges are parallel to the coordinate axes.

47–48 ■ The average value of a function $f(x, y, z)$ over a solid region E is defined to be

$$f_{\text{ave}} = \frac{1}{V(E)} \iiint_E f(x, y, z)\, dV$$

where $V(E)$ is the volume of E. For instance, if ρ is a density function, then ρ_{ave} is the average density of E.

47. Find the average value of the function $f(x, y, z) = xyz$ over the cube with side length L that lies in the first octant with one vertex at the origin and edges parallel to the coordinate axes.

48. Find the average value of the function $f(x, y, z) = x + y + z$ over the tetrahedron with vertices $(0, 0, 0)$, $(1, 0, 0)$, $(0, 1, 0)$, and $(0, 0, 1)$.

49. Find the region E for which the triple integral

$$\iiint_E (1 - x^2 - 2y^2 - 3z^2)\, dV$$

is a maximum.

13.8   TRIPLE INTEGRALS IN CYLINDRICAL AND SPHERICAL COORDINATES

We saw in Section 13.4 that some double integrals are easier to evaluate using polar coordinates. In this section we see that some triple integrals are easier to evaluate using cylindrical or spherical coordinates.

CYLINDRICAL COORDINATES

Recall from Section 11.10 that the cylindrical coordinates of a point P are (r, θ, z), where r, θ, and z are shown in Figure 1. Suppose that E is a type 1 region whose projection D on the xy-plane is conveniently described in polar coordinates (see Figure 2). In particular, suppose that f is continuous and

$$E = \{(x, y, z) \mid (x, y) \in D, \phi_1(x, y) \le z \le \phi_2(x, y)\}$$

where D is given in polar coordinates by

$$D = \{(r, \theta) \mid \alpha \le \theta \le \beta, h_1(\theta) \le r \le h_2(\theta)\}$$

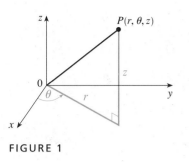

FIGURE 1

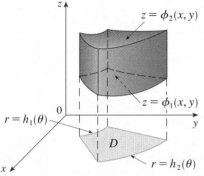

FIGURE 2

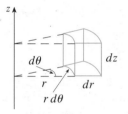

FIGURE 3
Volume element in cylindrical coordinates: $dV = r\,dz\,dr\,d\theta$

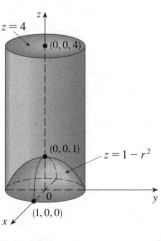

FIGURE 4

We know from Equations 13.7.6 that

(1)
$$\iiint_E f(x, y, z)\, dV = \iint_D \left[\int_{\phi_1(x, y)}^{\phi_2(x, y)} f(x, y, z)\, dz \right] dA$$

But we also know how to evaluate double integrals in polar coordinates. In fact, combining Equation 1 with (13.4.3), we obtain

(2)
$$\iiint_E f(x, y, z)\, dV = \int_\alpha^\beta \int_{h_1(\theta)}^{h_2(\theta)} \int_{\phi_1(r\cos\theta,\, r\sin\theta)}^{\phi_2(r\cos\theta,\, r\sin\theta)} f(r\cos\theta, r\sin\theta, z)\, r\,dz\,dr\,d\theta$$

Formula 2 is the **formula for triple integration in cylindrical coordinates.** It says that we convert a triple integral from rectangular to cylindrical coordinates by writing $x = r\cos\theta$, $y = r\sin\theta$, leaving z as it is, using the appropriate limits of integration for z, r, and θ, and replacing dV by $r\,dz\,dr\,d\theta$. (Figure 3 shows how to remember this.) It is worthwhile to use this formula when E is a solid region easily described in cylindrical coordinates, and especially when the function $f(x, y, z)$ involves the expression $x^2 + y^2$.

EXAMPLE 1 A solid E lies within the cylinder $x^2 + y^2 = 1$, below the plane $z = 4$, and above the paraboloid $z = 1 - x^2 - y^2$ (see Figure 4). The density at any point is proportional to its distance from the axis of the cylinder. Find the mass of E.

SOLUTION In cylindrical coordinates the cylinder is $r = 1$ and the paraboloid is $z = 1 - r^2$, so we can write

$$E = \{(r, \theta, z)\,|\, 0 \le \theta \le 2\pi,\ 0 \le r \le 1,\ 1 - r^2 \le z \le 4\}$$

Since the density at (x, y, z) is proportional to the distance from the z-axis, the density function is

$$f(x, y, z) = K\sqrt{x^2 + y^2} = Kr$$

where K is the proportionality constant. Therefore, from Formula 13.7.13, the mass of E is

$$m = \iiint_E K\sqrt{x^2 + y^2}\, dV = \int_0^{2\pi} \int_0^1 \int_{1-r^2}^4 (Kr)\, r\,dz\,dr\,d\theta$$

$$= \int_0^{2\pi} d\theta \int_0^1 Kr^2[4 - (1 - r^2)]\, dr = 2\pi K \int_0^1 (3r^2 + r^4)\, dr$$

$$= 2\pi K \left[r^3 + \frac{r^5}{5} \right]_0^1 = \frac{12\pi K}{5} \qquad \blacksquare$$

EXAMPLE 2 Evaluate $\displaystyle\int_{-2}^2 \int_{-\sqrt{4-x^2}}^{\sqrt{4-x^2}} \int_{\sqrt{x^2+y^2}}^2 (x^2 + y^2)\, dz\,dy\,dx$.

SOLUTION This iterated integral is a triple integral over the solid region

$$E = \{(x, y, z)\,|\, -2 \le x \le 2,\ -\sqrt{4 - x^2} \le y \le \sqrt{4 - x^2},\ \sqrt{x^2 + y^2} \le z \le 2\}$$

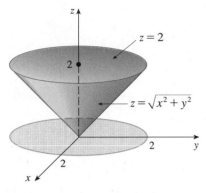

FIGURE 5

and the projection of E in the xy-plane is the disk $x^2 + y^2 \le 4$. The lower surface of E is the cone $z = \sqrt{x^2 + y^2}$ and its upper surface is the plane $z = 2$ (see Figure 5). This region has a much simpler description in cylindrical coordinates:

$$E = \{(r, \theta, z) \,|\, 0 \le \theta \le 2\pi,\ 0 \le r \le 2,\ r \le z \le 2\}$$

Therefore, we have

$$\int_{-2}^{2} \int_{-\sqrt{4-x^2}}^{\sqrt{4-x^2}} \int_{\sqrt{x^2+y^2}}^{2} (x^2 + y^2)\, dz\, dy\, dx = \iiint_E (x^2 + y^2)\, dV$$

$$= \int_0^{2\pi} \int_0^2 \int_r^2 r^2\, r\, dz\, dr\, d\theta$$

$$= \int_0^{2\pi} d\theta \int_0^2 r^3 (2 - r)\, dr$$

$$= 2\pi \left[\tfrac{1}{2} r^4 - \tfrac{1}{5} r^5 \right]_0^2 = \frac{16\pi}{5}$$

SPHERICAL COORDINATES

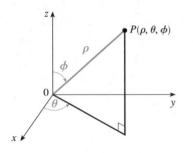

FIGURE 6

Spherical coordinates of P

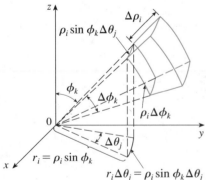

FIGURE 7

In Section 11.10 we defined the spherical coordinates (ρ, θ, ϕ) of a point (see Figure 6) and we demonstrated the following relationships between rectangular coordinates and spherical coordinates:

(3) $x = \rho \sin\phi \cos\theta \qquad y = \rho \sin\phi \sin\theta \qquad z = \rho \cos\phi$

In this coordinate system the analogue of a rectangular box is a **spherical wedge**

$$E = \{(\rho, \theta, \phi) \,|\, a \le \rho \le b,\ \alpha \le \theta \le \beta,\ c \le \phi \le d\}$$

where $a \ge 0$, $\beta - \alpha \le 2\pi$, and $d - c \le \pi$. To integrate over such a region we consider a spherical partition P of E into smaller spherical wedges E_{ijk} by means of spheres $\rho = \rho_i$, half-planes $\theta = \theta_j$, and half-cones $\phi = \phi_k$. The norm of P, $\|P\|$, is the length of the longest diagonal of these wedges. If $\|P\|$ is small, then Figure 7 shows that E_{ijk} is approximately a rectangular box with dimensions $\Delta\rho_i$, $\rho_i \Delta\phi_k$ (arc of a circle with radius ρ_i, angle $\Delta\phi_k$), and $\rho_i \sin\phi_k \Delta\theta_j$ (arc of a circle with radius $\rho_i \sin\phi_k$, angle $\Delta\theta_j$). So an approximation to the volume of E_{ijk} is given by

$$\Delta V_{ijk} \approx \rho_i^2 \sin\phi_k \, \Delta\rho_i \, \Delta\theta_j \, \Delta\phi_k$$

In fact, it can be shown, with the aid of the Mean Value Theorem (Exercise 37), that the volume of E_{ijk} is given exactly by

$$\Delta V_{ijk} = \tilde{\rho}_i^2 \sin\tilde{\phi}_k \, \Delta\rho_i \, \Delta\theta_j \, \Delta\phi_k$$

where $(\tilde{\rho}_i, \tilde{\theta}_j, \tilde{\phi}_k)$ is some point in E_{ijk}. Let $(x_{ijk}^*, y_{ijk}^*, z_{ijk}^*)$ be the rectangular coordinates of this point. Although triple integrals were defined using rectangular partitions, it can be shown that spherical partitions could be used instead. Therefore

$$\iiint_E f(x, y, z)\, dV = \lim_{\|P\| \to 0} \sum_{i=1}^{l} \sum_{j=1}^{m} \sum_{k=1}^{n} f(x_{ijk}^*, y_{ijk}^*, z_{ijk}^*)\, \Delta V_{ijk}$$

$$= \lim_{\|P\| \to 0} \sum_{i=1}^{l} \sum_{j=1}^{m} \sum_{k=1}^{n} f(\tilde{\rho}_i \sin\tilde{\phi}_k \cos\tilde{\theta}_j,\ \tilde{\rho}_i \sin\tilde{\phi}_k \sin\tilde{\theta}_j,\ \tilde{\rho}_i \cos\tilde{\phi}_k)\, \tilde{\rho}_i^2 \sin\tilde{\phi}_k \, \Delta\rho_i \, \Delta\theta_j \, \Delta\phi_k$$

But this sum is a Riemann sum for the function

$$F(\rho, \theta, \phi) = \rho^2 \sin \phi \, f(\rho \sin \phi \cos \theta, \rho \sin \phi \sin \theta, \rho \cos \phi)$$

Consequently, we have the following **formula for triple integration in spherical coordinates:**

---

(4) $$\iiint_E f(x, y, z) \, dV$$

$$= \int_c^d \int_\alpha^\beta \int_a^b f(\rho \sin \phi \cos \theta, \rho \sin \phi \sin \theta, \rho \cos \phi) \, \rho^2 \sin \phi \, d\rho \, d\theta \, d\phi$$

where E is a spherical wedge given by

$$E = \{(\rho, \theta, \phi) \mid a \leqslant \rho \leqslant b, \alpha \leqslant \theta \leqslant \beta, c \leqslant \phi \leqslant d\}$$

Formula 4 says that we convert a triple integral from rectangular coordinates to spherical coordinates by writing

$$x = \rho \sin \phi \cos \theta \qquad y = \rho \sin \phi \sin \theta \qquad z = \rho \cos \phi$$

using the appropriate limits of integration, and replacing dV by $\rho^2 \sin \phi \, d\rho \, d\theta \, d\phi$. This is illustrated in Figure 8.

This formula can be extended to include more general spherical regions such as

$$E = \{(\rho, \theta, \phi) \mid \alpha \leqslant \theta \leqslant \beta, c \leqslant \phi \leqslant d, g_1(\theta, \phi) \leqslant \rho \leqslant g_2(\theta, \phi)\}$$

In this case the formula is the same as in (4) except that the limits of integration for ρ are $g_1(\theta, \phi)$ and $g_2(\theta, \phi)$.

Usually spherical coordinates are used in triple integrals when surfaces such as cones and spheres form the boundary of the region of integration.

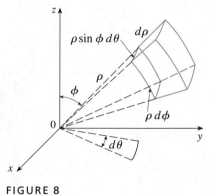

FIGURE 8

Volume element in spherical coordinates: $dV = \rho^2 \sin \phi \, d\rho \, d\theta \, d\phi$

EXAMPLE 3   Evaluate $\iiint_B e^{(x^2+y^2+z^2)^{3/2}} \, dV$, where B is the unit ball:

$$B = \{(x, y, z) \mid x^2 + y^2 + z^2 \leqslant 1\}$$

SOLUTION   Since the boundary of B is a sphere, we use spherical coordinates:

$$B = \{(\rho, \theta, \phi) \mid 0 \leqslant \rho \leqslant 1, 0 \leqslant \theta \leqslant 2\pi, 0 \leqslant \phi \leqslant \pi\}$$

In addition, spherical coordinates are appropriate because

$$x^2 + y^2 + z^2 = \rho^2$$

Thus (4) gives

$$\iiint_B e^{(x^2+y^2+z^2)^{3/2}} \, dV = \int_0^\pi \int_0^{2\pi} \int_0^1 e^{(\rho^2)^{3/2}} \rho^2 \sin \phi \, d\rho \, d\theta \, d\phi$$

$$= \int_0^\pi \sin \phi \, d\phi \int_0^{2\pi} d\theta \int_0^1 \rho^2 e^{\rho^3} \, d\rho$$

$$= \left[-\cos \phi \right]_0^\pi (2\pi) \left[\tfrac{1}{3} e^{\rho^3} \right]_0^1 = \frac{4\pi}{3}(e - 1)$$

NOTE: It would have been extremely awkward to evaluate the integral in Example 3 without spherical coordinates. In rectangular coordinates the iterated integral would have been

$$\int_{-1}^{1}\int_{-\sqrt{1-x^2}}^{\sqrt{1-x^2}}\int_{-\sqrt{1-x^2-y^2}}^{\sqrt{1-x^2-y^2}} e^{(x^2+y^2+z^2)^{3/2}}\, dz\, dy\, dx$$

EXAMPLE 4   Use spherical coordinates to find the volume of the solid that lies above the cone $z = \sqrt{x^2 + y^2}$ and below the sphere $x^2 + y^2 + z^2 = z$ (see Figure 9).

SOLUTION   Notice that the sphere passes through the origin and has center $(0, 0, \frac{1}{2})$. We write the equation of the sphere in spherical coordinates as

$$\rho^2 = \rho \cos \phi \qquad \text{or} \qquad \rho = \cos \phi$$

The cone can be written as

$$\rho \cos \phi = \sqrt{\rho^2 \sin^2\phi \cos^2\theta + \rho^2 \sin^2\phi \sin^2\theta} = \rho \sin \phi$$

This gives $\sin \phi = \cos \phi$ or $\phi = \pi/4$. Therefore, the description of the solid E in spherical coordinates is

$$E = \{(\rho, \theta, \phi)\,|\,0 \leqslant \theta \leqslant 2\pi,\ 0 \leqslant \phi \leqslant \pi/4,\ 0 \leqslant \rho \leqslant \cos \phi\}$$

and its volume is

$$V(E) = \iiint_E dV = \int_0^{2\pi}\int_0^{\pi/4}\int_0^{\cos\phi} \rho^2 \sin\phi\, d\rho\, d\phi\, d\theta$$

$$= \int_0^{2\pi} d\theta \int_0^{\pi/4} \sin\phi \left[\frac{\rho^3}{3}\right]_{\rho=0}^{\rho=\cos\phi} d\phi$$

$$= \frac{2\pi}{3} \int_0^{\pi/4} \sin\phi \cos^3\phi\, d\phi = \frac{2\pi}{3}\left[-\frac{\cos^4\phi}{4}\right]_0^{\pi/4} = \frac{\pi}{8} \qquad \blacksquare$$

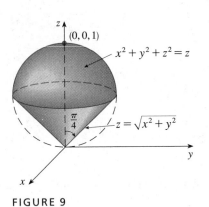

$(0, 0, 1)$

$x^2 + y^2 + z^2 = z$

$\dfrac{\pi}{4}$

$z = \sqrt{x^2 + y^2}$

FIGURE 9

Figure 10 gives another look (this time drawn by Maple) at the solid of Example 4.

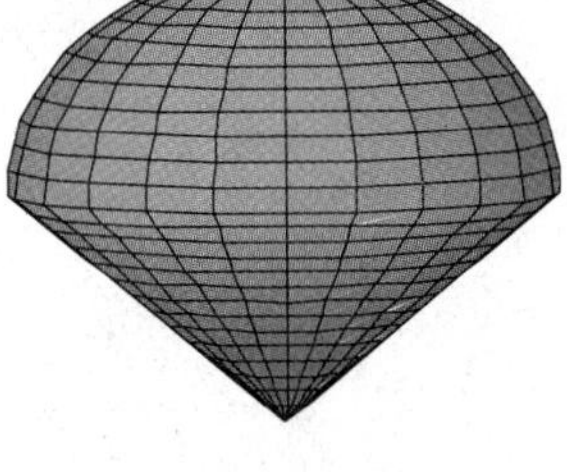

FIGURE 10

EXERCISES 13.8

1–4 ■ Sketch the solid whose volume is given by the integral and evaluate the integral.

1. $\displaystyle\int_0^{2\pi}\int_0^2\int_0^{4-r^2} r\, dz\, dr\, d\theta$

2. $\displaystyle\int_1^3\int_0^{\pi/2}\int_r^3 r\, dz\, d\theta\, dr$

3. $\displaystyle\int_0^{\pi/2}\int_0^{\pi/2}\int_0^1 \rho^2 \sin\phi\, d\rho\, d\theta\, d\phi$

4. $\displaystyle\int_0^{\pi/3}\int_0^{2\pi}\int_0^{\sec\phi} \rho^2 \sin\phi\, d\rho\, d\theta\, d\phi$

5–14 ■ Use cylindrical coordinates.

5. Evaluate $\iiint_E (x^2 + y^2)\, dV$, where E is the region bounded by the cylinder $x^2 + y^2 = 4$ and the planes $z = -1$ and $z = 2$.

6. Evaluate $\iiint_E \sqrt{x^2 + y^2}\, dV$, where E is the solid bounded by the paraboloid $z = 9 - x^2 - y^2$ and the xy-plane.

7. Evaluate $\iiint_E y\, dV$, where E is the solid that lies between the cylinders $x^2 + y^2 = 1$ and $x^2 + y^2 = 4$, above the xy-plane, and below the plane $z = x + 2$.

8. Evaluate $\iiint_E xz\, dV$, where E is bounded by the planes $z = 0$, $z = y$, and the cylinder $x^2 + y^2 = 1$ in the half-space $y \geqslant 0$.

9. Evaluate $\iiint_E x^2\, dV$, where E is the solid that lies within the cylinder $x^2 + y^2 = 1$, above the plane $z = 0$, and below the cone $z^2 = 4x^2 + 4y^2$.

10. (a) Find the volume of the solid that the cylinder $r = a\cos\theta$ cuts out of the sphere of radius a centered at the origin.
    (b) Illustrate the solid of part (a) by graphing the sphere and the cylinder on the same screen.

11. Find the volume of the region E bounded by the paraboloids $z = x^2 + y^2$ and $z = 36 - 3x^2 - 3y^2$.

12. Find the centroid of the region E of Exercise 11.

13. Find the mass and center of mass of the solid S bounded by the paraboloid $z = 4x^2 + 4y^2$ and the plane $z = a$ $(a > 0)$ if S has constant density K.

14. Find the mass of a ball B given by $x^2 + y^2 + z^2 \le a^2$ if the density at any point is proportional to its distance from the z-axis.

15–28 ■ Use spherical coordinates.

15. Evaluate $\iiint_B (x^2 + y^2 + z^2)\, dV$, where B is the unit ball $x^2 + y^2 + z^2 \le 1$.

16. Evaluate $\iiint_H (x^2 + y^2)\, dV$, where H is the hemispherical region that lies above the xy-plane and below the sphere $x^2 + y^2 + z^2 = 1$.

17. Evaluate $\iiint_E y^2\, dV$, where E is the part of the unit ball $x^2 + y^2 + z^2 \le 1$ that lies in the first octant.

18. Evaluate $\iiint_E xe^{(x^2+y^2+z^2)^2}\, dV$, where E is the solid that lies between the spheres $x^2 + y^2 + z^2 = 1$ and $x^2 + y^2 + z^2 = 4$ in the first octant.

19. Evaluate $\iiint_E \sqrt{x^2 + y^2 + z^2}\, dV$, where E is bounded below by the cone $\phi = \pi/6$ and above by the sphere $\rho = 2$.

20. Evaluate $\iiint_E x^2\, dV$, where E lies between the spheres $\rho = 1$ and $\rho = 3$ and above the cone $\phi = \pi/4$.

21. Find the volume of the solid that lies above the cone $\phi = \pi/3$ and below the sphere $\rho = 4\cos\phi$.

22. Find the centroid of the solid in Exercise 21.

23. Find the mass of a solid hemisphere H of radius a if the density at any point is proportional to its distance from the center of the base.

24. Find the center of mass of the solid H in Exercise 23.

25. Find the moment of inertia of the solid H in Exercise 23 about its axis.

26. Find the centroid of a solid homogeneous hemisphere of radius a.

27. Find the moment of inertia about a diameter of the base of a solid homogeneous hemisphere of radius a.

28. Find the mass and center of mass of a solid hemisphere of radius a if the density at any point is proportional to its distance from the base.

29–32 ■ Use cylindrical or spherical coordinates, whichever seems more appropriate.

29. Find the volume and centroid of the solid E that lies above the cone $z = \sqrt{x^2 + y^2}$ and below the sphere $x^2 + y^2 + z^2 = 1$.

30. Find the volume of the smaller wedge cut from a sphere of radius a by two planes that intersect along a diameter at an angle of $\pi/6$.

CAS **31.** Evaluate $\iiint_E z\, dV$, where E lies above the paraboloid $z = x^2 + y^2$ and below the plane $z = 2y$. Use either the Table of Integrals (on the back endpapers) or a computer algebra system to evaluate the integral.

32. (a) Find the volume enclosed by the torus $\rho = \sin\phi$.
(b) Use a computer to draw the torus.

33–34 ■ Evaluate the integral by changing to cylindrical coordinates.

33. $\int_{-1}^{1} \int_{-\sqrt{1-x^2}}^{\sqrt{1-x^2}} \int_{x^2+y^2}^{2-x^2-y^2} (x^2 + y^2)^{3/2}\, dz\, dy\, dx$

34. $\int_0^1 \int_0^{\sqrt{1-y^2}} \int_{x^2+y^2}^{\sqrt{x^2+y^2}} xyz\, dz\, dx\, dy$

35–36 ■ Evaluate the integral by changing to spherical coordinates.

35. $\int_{-3}^{3} \int_{-\sqrt{9-x^2}}^{\sqrt{9-x^2}} \int_0^{\sqrt{9-x^2-y^2}} z\sqrt{x^2 + y^2 + z^2}\, dz\, dy\, dx$

36. $\int_0^3 \int_0^{\sqrt{9-y^2}} \int_{\sqrt{x^2+y^2}}^{\sqrt{18-x^2-y^2}} (x^2 + y^2 + z^2)\, dz\, dx\, dy$

37. (a) Use cylindrical coordinates to show that the volume of the solid bounded above by the sphere $r^2 + z^2 = a^2$ and below by the cone $z = r\cot\phi_0$ (or $\phi = \phi_0$), where $0 < \phi_0 < \pi/2$, is
$$V = \frac{2\pi a^3}{3}(1 - \cos\phi_0)$$

(b) Deduce that the volume of the spherical wedge given by $\rho_1 \le \rho \le \rho_2$, $\theta_1 \le \theta \le \theta_2$, $\phi_1 \le \phi \le \phi_2$ is
$$\Delta V = \frac{\rho_2^3 - \rho_1^3}{3}(\cos\phi_1 - \cos\phi_2)(\theta_2 - \theta_1)$$

(c) Use the Mean Value Theorem to show that the volume in part (b) can be written as
$$\Delta V = \tilde{\rho}^2 \sin\tilde{\phi}\, \Delta\rho\, \Delta\theta\, \Delta\phi$$
where $\tilde{\rho}$ lies between ρ_1 and ρ_2, $\tilde{\phi}$ lies between ϕ_1 and ϕ_2, $\Delta\rho = \rho_2 - \rho_1$, $\Delta\theta = \theta_2 - \theta_1$, and $\Delta\phi = \phi_2 - \phi_1$.

38. Show that
$$\int_{-\infty}^{\infty} \int_{-\infty}^{\infty} \int_{-\infty}^{\infty} \sqrt{x^2 + y^2 + z^2}\; e^{-(x^2+y^2+z^2)}\, dx\, dy\, dz = 2\pi$$
(The improper triple integral is defined as the limit of a triple integral over a solid sphere as the radius of the sphere increases indefinitely.)

13.9   CHANGE OF VARIABLES IN MULTIPLE INTEGRALS

In one-dimensional calculus we often use a change of variable (a substitution) to simplify an integral. By reversing the roles of x and u, we can write the Substitution Rule (4.5.5) as

(1)
$$\int_a^b f(x)\, dx = \int_c^d f(g(u))g'(u)\, du$$

where $x = g(u)$ and $a = g(c)$, $b = g(d)$. Another way of writing Formula 1 is as follows:

(2)
$$\int_a^b f(x)\, dx = \int_c^d f(x(u))\, \frac{dx}{du}\, du$$

A change of variables can also be useful in double integrals. We have already seen one example of this: conversion to polar coordinates. The new variables r and θ are related to the old variables x and y by the equations

$$x = r\cos\theta \qquad y = r\sin\theta$$

and the change of variables formula (13.4.2) can be written as

$$\iint_R f(x, y)\, dA = \iint_S f(r\cos\theta, r\sin\theta)\, r\, dA$$

where S is the region in the $r\theta$-plane that corresponds to the region R in the xy-plane.

More generally, we consider a change of variables that is given by a **transformation** T from the uv-plane to the xy-plane:

$$T(u, v) = (x, y)$$

where x and y are related to u and v by the equations

(3)
$$x = g(u, v) \qquad y = h(u, v)$$

or, as we sometimes write,

$$x = x(u, v) \qquad y = y(u, v)$$

We usually assume that T is a C^1 **transformation,** which means that g and h have continuous first-order partial derivatives.

A transformation T is really just a function whose domain and range are both subsets of $\mathbb{R}^2$. If $T(u_1, v_1) = (x_1, y_1)$, then the point (x_1, y_1) is called the **image** of the point (u_1, v_1). If no two points have the same image, T is called **one-to-one.** Figure 1 shows

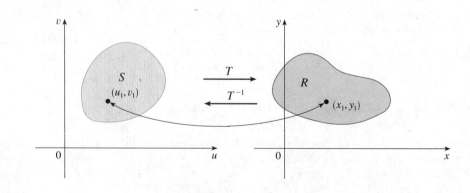

FIGURE 1

the effect of a transformation T on a region S in the uv-plane. T transforms S into a region R in the xy-plane called the **image of S,** consisting of the images of all points in S.

If T is a one-to-one transformation, then it has an **inverse transformation** T^{-1} from the xy-plane to the uv-plane and it may be possible to solve Equations 3 for u and v in terms of x and y:

$$u = G(x, y) \qquad v = H(x, y)$$

EXAMPLE 1 A transformation is defined by the equations

$$x = u^2 - v^2 \qquad y = 2uv$$

Find the image of the square $S = \{(u, v) \mid 0 \le u \le 1, 0 \le v \le 1\}$.

SOLUTION   First let us find the images of the sides of S. The first side, S_1, is given by $v = 0$ ($0 \le u \le 1$) (see Figure 2). From the given equations we have $x = u^2$, $y = 0$, and so $0 \le x \le 1$. Thus S_1 is mapped into the line segment from $(0, 0)$ to $(1, 0)$ in the xy-plane. The second side, S_2, is $u = 1$ ($0 \le v \le 1$) and, putting $u = 1$ in the given equations, we get

$$x = 1 - v^2 \qquad y = 2v$$

Eliminating v, we obtain

(4) $$x = 1 - \frac{y^2}{4} \qquad 0 \le x \le 1$$

which is part of a parabola. Similarly, S_3 is given by $v = 1$ ($0 \le u \le 1$), whose image is the parabolic arc

(5) $$x = \frac{y^2}{4} - 1 \qquad -1 \le x \le 0$$

Finally, S_4 is given by $u = 0$ ($0 \le v \le 1$) whose image is $x = -v^2$, $y = 0$, that is, $-1 \le x \le 0$. (Notice that as we move around the square in the counterclockwise direction, we also move around the parabolic region in the counterclockwise direction.) The image of S is the region R (shown in Figure 2) bounded by the x-axis and the parabolas given by Equations 4 and 5. ∎

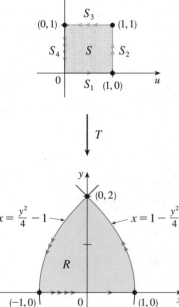

FIGURE 2

Now let us see how a change of variables affects a double integral. We start with a small rectangle S in the uv-plane whose lower left corner is the point (u_0, v_0) and whose dimensions are Δu and Δv (see Figure 3). Its image is a region R in the xy-plane, one of

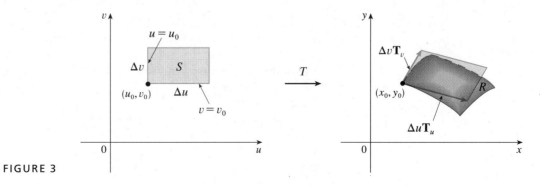

FIGURE 3

whose boundary points is $(x_0, y_0) = T(u_0, v_0)$. The equation of the lower side of S is $v = v_0$, whose image curve is given by $x = g(u, v_0)$, $y = h(u, v_0)$, or, in vector form,

$$g(u, v_0)\, \mathbf{i} + h(u, v_0)\, \mathbf{j}$$

The tangent vector at (x_0, y_0) to this image curve is

$$\mathbf{T}_u = g_u(u_0, v_0)\, \mathbf{i} + h_u(u_0, v_0)\, \mathbf{j} = \frac{\partial x}{\partial u}\, \mathbf{i} + \frac{\partial y}{\partial u}\, \mathbf{j}$$

Similarly, the tangent vector at (x_0, y_0) to the image curve of the left side of S (namely, $u = u_0$) is

$$\mathbf{T}_v = g_v(u_0, v_0)\, \mathbf{i} + h_v(u_0, v_0)\, \mathbf{j} = \frac{\partial x}{\partial v}\, \mathbf{i} + \frac{\partial y}{\partial v}\, \mathbf{j}$$

This means that we can approximate the image region $R = T(S)$ by a parallelogram determined by the vectors $\Delta u\, \mathbf{T}_u$ and $\Delta v\, \mathbf{T}_v$ (see Figure 3). Therefore, we can approximate the area of R by the area of this parallelogram, which, from Section 11.4, is

$$(6) \qquad \left| (\Delta u\, \mathbf{T}_u) \times (\Delta v\, \mathbf{T}_v) \right| = \left| \mathbf{T}_u \times \mathbf{T}_v \right| \Delta u\, \Delta v$$

Computing the cross product, we obtain

$$\mathbf{T}_u \times \mathbf{T}_v = \begin{vmatrix} \mathbf{i} & \mathbf{j} & \mathbf{k} \\ \dfrac{\partial x}{\partial u} & \dfrac{\partial y}{\partial u} & 0 \\ \dfrac{\partial x}{\partial v} & \dfrac{\partial y}{\partial v} & 0 \end{vmatrix} = \begin{vmatrix} \dfrac{\partial x}{\partial u} & \dfrac{\partial y}{\partial u} \\ \dfrac{\partial x}{\partial v} & \dfrac{\partial y}{\partial v} \end{vmatrix} \mathbf{k} = \begin{vmatrix} \dfrac{\partial x}{\partial u} & \dfrac{\partial x}{\partial v} \\ \dfrac{\partial y}{\partial u} & \dfrac{\partial y}{\partial v} \end{vmatrix} \mathbf{k}$$

The Jacobian is named after the German mathematician Carl Gustav Jacob Jacobi (1804–1851). Although the French mathematician Cauchy first used these special determinants involving partial derivatives, Jacobi developed them into a method for evaluating multiple integrals.

The determinant that arises in this calculation is called the *Jacobian* of the transformation and is given a special notation.

---

(7) DEFINITION   The **Jacobian** of the transformation T given by $x = g(u, v)$ and $y = h(u, v)$ is

$$\frac{\partial(x, y)}{\partial(u, v)} = \begin{vmatrix} \dfrac{\partial x}{\partial u} & \dfrac{\partial x}{\partial v} \\ \dfrac{\partial y}{\partial u} & \dfrac{\partial y}{\partial v} \end{vmatrix} = \frac{\partial x}{\partial u}\frac{\partial y}{\partial v} - \frac{\partial x}{\partial v}\frac{\partial y}{\partial u}$$

With this notation we can use Equation 6 to give an approximation to the area ΔA of R:

$$(8) \qquad \Delta A \approx \left| \frac{\partial(x, y)}{\partial(u, v)} \right| \Delta u\, \Delta v$$

where the Jacobian is evaluated at (u_0, v_0).

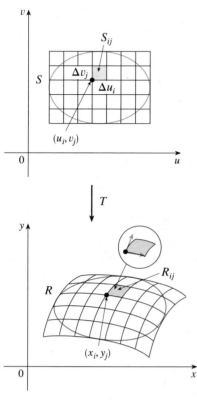

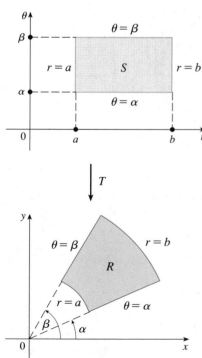

FIGURE 4

FIGURE 5

The polar coordinate transformation

Next we partition a region S in the uv-plane into rectangles S_{ij} and call their images in the xy-plane R_{ij} (see Figure 4). Applying the approximation (8) to each R_{ij}, we approximate the double integral of f over R as follows:

$$\iint_R f(x, y)\, dA \approx \sum_{i=1}^{m} \sum_{j=1}^{n} f(x_i, y_j)\, \Delta A_{ij}$$

$$\approx \sum_{i=1}^{m} \sum_{j=1}^{n} f(g(u_i, v_j),\, h(u_i, v_j)) \left| \frac{\partial(x, y)}{\partial(u, v)} \right| \Delta u_i\, \Delta v_j$$

where the Jacobian is evaluated at (u_i, v_j). Notice that this double sum is a Riemann sum for the integral

$$\iint_S f(g(u, v),\, h(u, v)) \left| \frac{\partial(x, y)}{\partial(u, v)} \right| du\, dv$$

The foregoing argument suggests that the following theorem is true. (A full proof is given in books on advanced calculus.)

---

(9) CHANGE OF VARIABLES IN A DOUBLE INTEGRAL   Suppose that T is a one-to-one C^1 transformation whose Jacobian is nonzero and that maps a region S in the uv-plane onto a region R in the xy-plane. Suppose that f is continuous on R and that R and S are type I or type II plane regions. Then

$$\iint_R f(x, y)\, dx\, dy = \iint_S f(x(u, v),\, y(u, v)) \left| \frac{\partial(x, y)}{\partial(u, v)} \right| du\, dv$$

Theorem 9 says that we change from an integral in x and y to an integral in u and v by expressing x and y in terms of u and v and writing

$$dA = \left| \frac{\partial(x, y)}{\partial(u, v)} \right| du\, dv$$

Notice the similarity between Theorem 9 and the one-dimensional formula in Equation 2. Instead of the derivative dx/du, we have the absolute value of the Jacobian: $|\partial(x, y)/\partial(u, v)|$.

As a first illustration of Theorem 9, let us show that the formula for integration in polar coordinates is just a special case. Here the transformation T from the $r\theta$-plane to the xy-plane is given by

$$x = g(r, \theta) = r\cos\theta \qquad y = h(r, \theta) = r\sin\theta$$

and the geometry of the transformation is shown in Figure 5. T maps an ordinary rectangle in the $r\theta$-plane to a polar rectangle in the xy-plane. The Jacobian of T is

$$\frac{\partial(x, y)}{\partial(r, \theta)} = \begin{vmatrix} \dfrac{\partial x}{\partial r} & \dfrac{\partial x}{\partial \theta} \\[2mm] \dfrac{\partial y}{\partial r} & \dfrac{\partial y}{\partial \theta} \end{vmatrix} = \begin{vmatrix} \cos\theta & -r\sin\theta \\ \sin\theta & r\cos\theta \end{vmatrix} = r\cos^2\theta + r\sin^2\theta = r > 0$$

Thus Theorem 9 gives

$$\iint\limits_{R} f(x, y)\, dx\, dy = \iint\limits_{S} f(r\cos\theta, r\sin\theta)\left|\frac{\partial(x, y)}{\partial(r, \theta)}\right| dr\, d\theta$$

$$= \int_{\alpha}^{\beta}\int_{a}^{b} f(r\cos\theta, r\sin\theta)\, r\, dr\, d\theta$$

which is the same as (13.4.2).

EXAMPLE 2   Use the change of variables $x = u^2 - v^2$, $y = 2uv$ to evaluate the integral $\iint_{R} y\, dA$, where R is the region bounded by the x-axis and the parabolas $y^2 = 4 - 4x$ and $y^2 = 4 + 4x$.

SOLUTION   The region R is pictured in Figure 2. In Example 1 we discovered that $T(S) = R$, where S is the square $[0, 1] \times [0, 1]$. Indeed, the reason for making the change of variables to evaluate the integral is that S is a much simpler region than R. First we need to compute the Jacobian:

$$\frac{\partial(x, y)}{\partial(u, v)} = \begin{vmatrix} \dfrac{\partial x}{\partial u} & \dfrac{\partial x}{\partial v} \\[2mm] \dfrac{\partial y}{\partial u} & \dfrac{\partial y}{\partial v} \end{vmatrix} = \begin{vmatrix} 2u & -2v \\[1mm] 2v & 2u \end{vmatrix} = 4u^2 + 4v^2 > 0$$

Therefore, by Theorem 9,

$$\iint\limits_{R} y\, dA = \iint\limits_{S} 2uv \left|\frac{\partial(x, y)}{\partial(u, v)}\right| dA$$

$$= \int_{0}^{1}\int_{0}^{1} (2uv)4(u^2 + v^2)\, du\, dv$$

$$= 8\int_{0}^{1}\int_{0}^{1} (u^3 v + uv^3)\, du\, dv = 8\int_{0}^{1}\left[\frac{u^4}{4}v + \frac{u^2}{2}v^3\right]_{u=0}^{u=1} dv$$

$$= \int_{0}^{1} (2v + 4v^3)\, dv = \left[v^2 + v^4\right]_{0}^{1} = 2$$
∎

   NOTE: Example 2 was not a very difficult problem to solve because we were given a suitable change of variables. If we are not supplied with a transformation, then the first step is to think of an appropriate change of variables. If $f(x, y)$ is difficult to integrate, then the form of $f(x, y)$ may suggest a transformation. If the region of integration R is awkward, then the transformation should be chosen so that the corresponding region S in the uv-plane has a convenient description.

EXAMPLE 3   Evaluate the integral $\iint_{R} e^{(x+y)/(x-y)}\, dA$, where R is the trapezoidal region with vertices $(1, 0)$, $(2, 0)$, $(0, -2)$, and $(0, -1)$.

SOLUTION   Since it is not easy to integrate $e^{(x+y)/(x-y)}$, we make a change of variables suggested by the form of this function:

(10) $u = x + y$ $v = x - y$

These equations define a transformation T^{-1} from the xy-plane to the uv-plane. Theorem 9 talks about a transformation T from the uv-plane to the xy-plane. It is obtained by solving Equations 10 for x and y:

(11) $$x = \tfrac{1}{2}(u + v) \qquad y = \tfrac{1}{2}(u - v)$$

The Jacobian of T is

$$\frac{\partial(x, y)}{\partial(u, v)} = \begin{vmatrix} \dfrac{\partial x}{\partial u} & \dfrac{\partial x}{\partial v} \\[2mm] \dfrac{\partial y}{\partial u} & \dfrac{\partial y}{\partial v} \end{vmatrix} = \begin{vmatrix} \tfrac{1}{2} & \tfrac{1}{2} \\[1mm] \tfrac{1}{2} & -\tfrac{1}{2} \end{vmatrix} = -\tfrac{1}{2}$$

To find the region S in the uv-plane corresponding to R, we note that the sides of R lie on the lines

$$y = 0 \qquad x - y = 2 \qquad x = 0 \qquad x - y = 1$$

and, from either Equations 10 or Equations 11, the image lines in the uv-plane are

$$u = v \qquad v = 2 \qquad u = -v \qquad v = 1$$

Thus the region S is the trapezoidal region with vertices $(1, 1)$, $(2, 2)$, $(-2, 2)$, and $(-1, 1)$ shown in Figure 6. Since

$$S = \{(u, v) \mid 1 \leqslant v \leqslant 2, -v \leqslant u \leqslant v\}$$

Theorem 9 gives

$$\iint_R e^{(x+y)/(x-y)} \, dA = \iint_S e^{u/v} \left| \frac{\partial(x, y)}{\partial(u, v)} \right| du \, dv$$

$$= \int_1^2 \int_{-v}^v e^{u/v} \left(\tfrac{1}{2}\right) du \, dv = \tfrac{1}{2} \int_1^2 \left[v e^{u/v} \right]_{u=-v}^{u=v} dv$$

$$= \tfrac{1}{2} \int_1^2 (e - e^{-1}) v \, dv = \tfrac{3}{4}(e - e^{-1}) \qquad \blacksquare$$

There is a similar change of variables formula for triple integrals. Let T be a transformation that maps a region S in uvw-space onto a region R in xyz-space by means of the equations

$$x = g(u, v, w) \qquad y = h(u, v, w) \qquad z = k(u, v, w)$$

The **Jacobian** of T is the following 3×3 determinant:

(12) $$\frac{\partial(x, y, z)}{\partial(u, v, w)} = \begin{vmatrix} \dfrac{\partial x}{\partial u} & \dfrac{\partial x}{\partial v} & \dfrac{\partial x}{\partial w} \\[2mm] \dfrac{\partial y}{\partial u} & \dfrac{\partial y}{\partial v} & \dfrac{\partial y}{\partial w} \\[2mm] \dfrac{\partial z}{\partial u} & \dfrac{\partial z}{\partial v} & \dfrac{\partial z}{\partial w} \end{vmatrix}$$

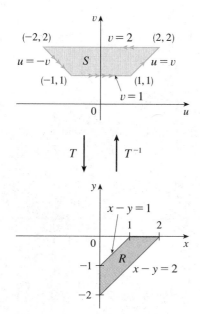

FIGURE 6

Under hypotheses similar to those in Theorem 9, we have the following formula for triple integrals:

$$(13) \quad \iiint_R f(x, y, z)\, dx\, dy\, dz$$

$$= \iiint_S f(x(u, v, w), y(u, v, w), z(u, v, w)) \left| \frac{\partial(x, y, z)}{\partial(u, v, w)} \right| du\, dv\, dw$$

EXAMPLE 4　Use Formula 13 to derive the formula for triple integration in spherical coordinates.

SOLUTION　Here the change of variables is given by

$$x = \rho \sin \phi \cos \theta \qquad y = \rho \sin \phi \sin \theta \qquad z = \rho \cos \phi$$

We compute the Jacobian as follows:

$$\frac{\partial(x, y, z)}{\partial(\rho, \theta, \phi)} = \begin{vmatrix} \sin \phi \cos \theta & -\rho \sin \phi \sin \theta & \rho \cos \phi \cos \theta \\ \sin \phi \sin \theta & \rho \sin \phi \cos \theta & \rho \cos \phi \sin \theta \\ \cos \phi & 0 & -\rho \sin \phi \end{vmatrix}$$

$$= \cos \phi \begin{vmatrix} -\rho \sin \phi \sin \theta & \rho \cos \phi \cos \theta \\ \rho \sin \phi \cos \theta & \rho \cos \phi \sin \theta \end{vmatrix} - \rho \sin \phi \begin{vmatrix} \sin \phi \cos \theta & -\rho \sin \phi \sin \theta \\ \sin \phi \sin \theta & \rho \sin \phi \cos \theta \end{vmatrix}$$

$$= \cos \phi \, (-\rho^2 \sin \phi \cos \phi \sin^2 \theta - \rho^2 \sin \phi \cos \phi \cos^2 \theta)$$

$$\quad - \rho \sin \phi \, (\rho \sin^2 \phi \cos^2 \theta + \rho \sin^2 \phi \sin^2 \theta)$$

$$= -\rho^2 \sin \phi \cos^2 \phi - \rho^2 \sin \phi \sin^2 \phi = -\rho^2 \sin \phi$$

Since $0 \leqslant \phi \leqslant \pi$, we have $\sin \phi \geqslant 0$. Therefore

$$\left| \frac{\partial(x, y, z)}{\partial(\rho, \theta, \phi)} \right| = |-\rho^2 \sin \phi| = \rho^2 \sin \phi$$

and Formula 13 gives

$$\iiint_R f(x, y, z)\, dx\, dy\, dz = \iiint_S f(\rho \sin \phi \cos \theta, \rho \sin \phi \sin \theta, \rho \cos \phi)\rho^2 \sin \phi \, d\rho \, d\theta \, d\phi$$

which is equivalent to Formula 13.8.4.　∎

EXERCISES 13.9

1–6　■　Find the Jacobian of the transformation.

1. $x = u - 2v, \quad y = 2u - v$

2. $x = u - v^2, \quad y = u + v^2$

3. $x = e^{2u} \cos v, \quad y = e^{2u} \sin v$

4. $x = se^t, \quad y = se^{-t}$

5. $x = u + v + w, \quad y = u + v - w, \quad z = u - v + w$

6. $x = 2u, \quad y = 3v^2, \quad z = 4w^3$

7–10　■　Find the image of the set S under the given transformation.

7. $S = \{(u, v) \mid 0 \leqslant u \leqslant 2, 0 \leqslant v \leqslant 1\}$;
$x = u - 2v, y = 2u - v$

8. $S = \{(u, v) \mid 0 \leqslant u \leqslant 1, u \leqslant v \leqslant 1\}; \quad x = u^2, y = v$

9. S is the triangular region with vertices $(0, 0)$, $(1, 0)$, $(0, 1)$; $x = 4u + 3v$, $y = 4v$.

10. S is the disk given by $u^2 + v^2 \leq 1$; $x = au$, $y = bv$.

11–18 ∎ Use the given transformation to evaluate the integral.

11. $\iint_R (3x + 4y) \, dA$, where R is the region bounded by the lines $y = x$, $y = x - 2$, $y = -2x$, and $y = 3 - 2x$; $x = \frac{1}{3}(u + v)$, $y = \frac{1}{3}(v - 2u)$

12. $\iint_R (x + y) \, dA$, where R is the square with vertices $(0, 0)$, $(2, 3)$, $(5, 1)$, and $(3, -2)$; $x = 2u + 3v$, $y = 3u - 2v$

13. $\iint_R x^2 \, dA$, where R is the region bounded by the ellipse $9x^2 + 4y^2 = 36$; $x = 2u$, $y = 3v$

14. $\iint_R (x^2 - xy + y^2) \, dA$, where R is the region bounded by the ellipse $x^2 - xy + y^2 = 2$; $x = \sqrt{2}\, u - \sqrt{2/3}\, v$, $y = \sqrt{2}\, u + \sqrt{2/3}\, v$

15. $\iint_R xy \, dA$, where R is the region in the first quadrant bounded by the lines $y = x$ and $y = 3x$ and the hyperbolas $xy = 1$, $xy = 3$; $x = u/v$, $y = v$

16. $\iint_R y^2 \, dA$, where R is the region bounded by the curves $xy = 1$, $xy = 2$, $xy^2 = 1$, $xy^2 = 2$; $u = xy$, $v = xy^2$. Illustrate by using a graphing calculator or computer to draw R.

17. $\iiint_E dV$, where E is the solid enclosed by the ellipsoid $x^2/a^2 + y^2/b^2 + z^2/c^2 = 1$; $x = au$, $y = bv$, $z = cw$ (This integral gives the volume of the ellipsoid.)

18. $\iiint_E x^2 y \, dV$, where E is the solid of Exercise 17

19–23 ∎ Evaluate the integral by making an appropriate change of variables.

19. $\iint_R xy \, dA$, where R is the region bounded by the lines $2x - y = 1$, $2x - y = -3$, $3x + y = 1$, and $3x + y = -2$

20. $\iint_R \dfrac{x + 2y}{\cos(x - y)} \, dA$, where R is the parallelogram bounded by the lines $y = x$, $y = x - 1$, $x + 2y = 0$, and $x + 2y = 2$

21. $\iint_R \cos\left(\dfrac{y - x}{y + x}\right) dA$, where R is the trapezoidal region with vertices $(1, 0)$, $(2, 0)$, $(0, 2)$, and $(0, 1)$

22. $\iint_R \sin(9x^2 + 4y^2) \, dA$, where R is the region in the first quadrant bounded by the ellipse $9x^2 + 4y^2 = 1$

23. $\iint_R e^{x+y} \, dA$, where R is given by the inequality $|x| + |y| \leq 1$

24. Let f be continuous on $[0, 1]$ and let R be the triangular region with vertices $(0, 0)$, $(1, 0)$, and $(0, 1)$. Show that

$$\iint_R f(x + y) \, dA = \int_0^1 u f(u) \, du$$

---

13 REVIEW

KEY TOPICS ∎ Define, state, or discuss the following.

1. Double Riemann sum

2. Double integral over a rectangle

3. Iterated integral

4. Fubini's Theorem

5. Double integral over a general region

6. Regions of type I and type II

7. Evaluating double integrals

8. Volume

9. Properties of double integrals

10. Conversion from rectangular to polar coordinates in a double integral

11. Electric charge

12. Mass and center of mass of a lamina

13. Moment of inertia

14. Radius of gyration

15. Triple integral

16. Evaluation of triple integrals

17. Mass, center of mass, and moments of inertia of a solid region

18. Surface area

19. Triple integrals in cylindrical coordinates

20. Triple integrals in spherical coordinates

21. Jacobian of a transformation

22. Formula for change of variables in a double integral

23. Formula for change of variables in a triple integral

EXERCISES

1–6 ■ Determine whether the statement is true or false.

1. $\int_{-1}^{2} \int_{0}^{6} x^2 \sin(x - y) \, dx \, dy = \int_{0}^{6} \int_{-1}^{2} x^2 \sin(x - y) \, dy \, dx$

2. $\int_{-1}^{1} \int_{0}^{1} e^{x^2+y^2} \sin y \, dx \, dy = 0$

3. If D is the disk given by $x^2 + y^2 \leqslant 4$, then

$$\iint_{D} \sqrt{4 - x^2 - y^2} \, dA = \frac{16\pi}{3}$$

4. $\int_{1}^{4} \int_{0}^{1} (x^2 + \sqrt{y}) \sin(x^2 y^2) \, dx \, dy \leqslant 9$

5. The integral

$$\int_{0}^{2\pi} \int_{0}^{2} \int_{r}^{2} dz \, dr \, d\theta$$

represents the volume enclosed by the cone $z = \sqrt{x^2 + y^2}$ and the plane $z = 2$.

6. The integral $\iiint_{E} kr^3 \, dz \, dr \, d\theta$ represents the moment of inertia about the z-axis of a solid E with constant density k.

7–12 ■ Calculate the iterated integral.

7. $\int_{-2}^{2} \int_{0}^{4} (4x^3 + 3xy^2) \, dx \, dy$ **8.** $\int_{0}^{\pi} \int_{0}^{1} x \cos(xy) \, dy \, dx$

9. $\int_{1}^{2} \int_{0}^{x^2} \frac{1}{x + y} \, dy \, dx$ **10.** $\int_{0}^{1} \int_{0}^{y^2} e^{y^3} \, dx \, dy$

11. $\int_{0}^{1} \int_{0}^{x^2} \int_{0}^{y} y^2 z \, dz \, dy \, dx$ **12.** $\int_{0}^{1} \int_{\sqrt{y}}^{1} \int_{0}^{y} xy \, dz \, dx \, dy$

13. Describe the region whose area is given by the integral

$$\int_{0}^{\pi} \int_{1}^{1+\sin\theta} r \, dr \, d\theta$$

14. Describe the solid whose volume is given by the integral

$$\int_{0}^{2\pi} \int_{0}^{\pi/6} \int_{1}^{3} \rho^2 \sin\phi \, d\rho \, d\phi \, d\theta$$

and evaluate the integral.

15–16 ■ Calculate the iterated integral by first reversing the order of integration.

15. $\int_{0}^{1} \int_{x}^{1} e^{x/y} \, dy \, dx$ **16.** $\int_{0}^{1} \int_{y^2}^{1} y \sin(x^2) \, dx \, dy$

17–30 ■ Calculate the value of the multiple integral.

17. $\iint_{R} \frac{1}{(x - y)^2} \, dA$, where $R = \{(x, y) \,|\, 0 \leqslant x \leqslant 1, 2 \leqslant y \leqslant 4\}$

18. $\iint_{D} x^3 \, dA$, where $D = \{(x, y) \,|\, -1 \leqslant x \leqslant 1, x^2 - 1 \leqslant y \leqslant x + 1\}$

19. $\iint_{D} xy \, dA$, where D is bounded by $y^2 = x^3$ and $y = x$

20. $\iint_{D} xe^y \, dA$, where D is bounded by $y = 0$, $y = x^2$, $x = 1$

21. $\iint_{D} (xy + 2x + 3y) \, dA$, where D is the region in the first quadrant bounded by $x = 1 - y^2$, $y = 0$, $x = 0$

22. $\iint_{D} y \, dA$, where D is the region in the first quadrant that lies above the hyperbola $xy = 1$ and the line $y = x$ and below the line $y = 2$

23. $\iint_{D} (x^2 + y^2)^{3/2} \, dA$, where D is the region in the first quadrant bounded by the lines $y = 0$ and $y = \sqrt{3} \, x$ and the circle $x^2 + y^2 = 9$

24. $\iint_{D} \sqrt{x^2 + y^2} \, dA$, where D is the closed disk with radius 1 and center $(0, 1)$

25. $\iiint_{E} x^2 z \, dV$, where $E = \{(x, y, z) \,|\, 0 \leqslant x \leqslant 2, 0 \leqslant y \leqslant 2x, 0 \leqslant z \leqslant x\}$

26. $\iiint_{T} y \, dV$, where T is the tetrahedron bounded by the planes $x = 0$, $y = 0$, $z = 0$, and $2x + y + z = 2$

27. $\iiint_{E} y^2 z^2 \, dV$, where E is bounded by the paraboloid $x = 1 - y^2 - z^2$ and the plane $x = 0$

28. $\iiint_{E} z \, dV$, where E is bounded by the planes $y = 0$, $z = 0$, $x + y = 2$ and the cylinder $y^2 + z^2 = 1$ in the first octant

29. $\iiint_{E} yz \, dV$, where E lies above the plane $z = 0$, below the plane $z = y$, and inside the cylinder $x^2 + y^2 = 4$

30. $\iiint_{H} z^3 \sqrt{x^2 + y^2 + z^2} \, dV$, where H is the solid hemisphere with center the origin, radius 1, that lies above the xy-plane

31–36 ■ Find the volume of the given solid.

31. Under the paraboloid $z = x^2 + 4y^2$ and above the rectangle $R = [0, 2] \times [1, 4]$

32. Under the surface $z = x^2 y$ and above the triangle in the xy-plane with vertices $(1, 0)$, $(2, 1)$, and $(4, 0)$

33. The solid tetrahedron with vertices $(0, 0, 0)$, $(0, 0, 1)$, $(0, 2, 0)$, and $(2, 2, 0)$

34. Bounded by the cylinder $x^2 + y^2 = 4$ and the planes $z = 0$ and $y + z = 3$

35. One of the wedges cut from the cylinder $x^2 + 9y^2 = a^2$ by the planes $z = 0$ and $z = mx$

36. Above the paraboloid $z = x^2 + y^2$ and below the half-cone $z = \sqrt{x^2 + y^2}$

37. Find the mass and center of mass of a lamina that occupies the region D bounded by the parabola $x = 1 - y^2$ and the coordinate axes in the first quadrant if the density function is $\rho(x, y) = y$.

38. Find the moments of inertia and radii of gyration about the x- and y-axes for the lamina of Exercise 37.

39. A lamina occupies the part of the disk $x^2 + y^2 \leq a^2$ that lies in the first quadrant. Find the centroid of the lamina.

40. Find the center of mass of the lamina in Exercise 39 if the density function is $\rho(x, y) = xy^2$.

41. Find the centroid of a right circular cone with height h and base radius a. (Place the cone so that its base is in the xy-plane with center the origin and its axis along the positive z-axis.)

42. Find the moment of inertia of the cone of Exercise 41 about its axis (the z-axis).

43. Find the area of the part of the plane $3x + 2y + 6z = 6$ in the first octant.

44. Find the area of the part of the cone $z^2 = a^2(x^2 + y^2)$ between the planes $z = 1$ and $z = 2$.

45. Use polar coordinates to evaluate

$$\int_0^{\sqrt{2}} \int_y^{\sqrt{4-y^2}} \frac{1}{1 + x^2 + y^2}\, dx\, dy$$

46. Use spherical coordinates to evaluate

$$\int_0^1 \int_0^{\sqrt{1-x^2}} \int_0^{\sqrt{1-x^2-y^2}} (x^2 + y^2 + z^2)^2\, dz\, dy\, dx$$

47. If D is the region bounded by the curves $y = 1 - x^2$ and $y = e^x$, find the approximate value of the integral $\iint_D y^2\, dA$. (Use a graphing device to estimate the points of intersection of the curves.)

48. Graph the surface $z = x \sin y$, $-3 \leq x \leq 3$, $-\pi \leq y \leq \pi$, and find its surface area correct to four decimal places.

49. Find the center of mass of the solid tetrahedron with vertices $(0, 0, 0)$, $(1, 0, 0)$, $(0, 2, 0)$, $(0, 0, 3)$ and density function $\rho(x, y, z) = x^2 + y^2 + z^2$.

50. Find the exact value of the triple integral:

$$\int_0^1 \int_0^{e^y} \int_x^{y^2} (x^2 y + z \sin 4y)\, dz\, dx\, dy$$

51. Rewrite the integral

$$\int_{-1}^1 \int_{x^2}^1 \int_0^{1-y} f(x, y, z)\, dz\, dy\, dx$$

as an iterated integral in the order $dx\, dy\, dz$.

52. Give five other iterated integrals that are equal to

$$\int_0^2 \int_0^{y^3} \int_0^{y^2} f(x, y, z)\, dz\, dx\, dy$$

53. Use the transformation $u = x - y$, $v = x + y$ to evaluate $\iint_R (x - y)/(x + y)\, dA$, where R is the square with vertices $(0, 2)$, $(1, 1)$, $(2, 2)$, and $(1, 3)$.

54. Use the transformation $x = u^2$, $y = v^2$, $z = w^2$ to find the volume of the region bounded by the surface $\sqrt{x} + \sqrt{y} + \sqrt{z} = 1$ and the coordinate planes.

55. Use the change of variables formula and an appropriate transformation to evaluate $\iint_R xy\, dA$, where R is the square with vertices $(0, 0)$, $(1, 1)$, $(2, 0)$, and $(1, -1)$.

56. The **Mean Value Theorem for double integrals** says that if f is a continuous function on a plane region D that is of type I or II, then there exists a point (x_0, y_0) in D such that

$$\iint_D f(x, y)\, dA = f(x_0, y_0)\, A(D)$$

Use the Extreme Value Theorem (12.7.8) and Property 13.3.11 of integrals to prove this theorem. (Use the proof of the single-variable version in Section 5.5 as a guide.)

57. Suppose that f is continuous on a domain D in the plane and let (a, b) be an interior point of D. Let D_r be the closed disk with center (a, b) and radius r. Use the Mean Value Theorem for double integrals (see Exercise 56) to show that

$$\lim_{r \to 0} \frac{1}{\pi r^2} \iint_{D_r} f(x, y)\, dA = f(a, b)$$

58. (a) Evaluate $\iint_D \dfrac{1}{(x^2 + y^2)^{n/2}}\, dA$, where n is an integer and D is the region bounded by the circles with center the origin and radii r and R, $0 < r < R$.

(b) For what values of n does the integral in part (a) have a limit as $r \to 0^+$?

(c) Find $\iiint_E \dfrac{1}{(x^2 + y^2 + z^2)^{n/2}}\, dV$, where E is the region bounded by the spheres with center the origin and radii r and R, $0 < r < R$.

(d) For what values of n does the integral in part (c) have a limit as $r \to 0^+$?

APPLICATIONS PLUS

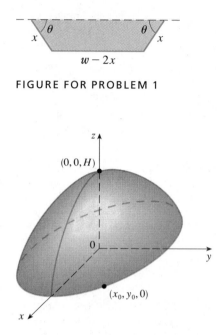

FIGURE FOR PROBLEM 1

$(0,0,H)$

$(x_0, y_0, 0)$

FIGURE FOR PROBLEM 2

1. A long piece of galvanized sheet metal w inches wide is to be bent into a symmetric form with three straight sides to make a rain gutter. The cross-section is shown in the figure.
   (a) Determine the dimensions that allow the maximum possible flow; that is, find the dimensions that give the maximum possible cross-sectional area.
   (b) Would it be better to bend the metal into a gutter with a semicircular cross-section than a three-sided cross-section?

2. A "sugarloaf" mountain has the equation
$$z = f(x, y) = H - \alpha(x^2 + \mu y^2) \qquad \alpha > 0, \ \mu > 1$$
Starting at a point $(x_0, y_0, 0)$ at the mountain base, a climber wants to ascend the mountain by means of the steepest ascent curve.
   (a) Identify the level curves of the surface $z = f(x, y)$ and sketch a few members of the family.
   (b) Calculate the gradient vector for the surface at an arbitrary point (x, y) in the domain of f and determine the differential equation for the steepest ascent curve.
   (c) Find the solution of the differential equation in part (b) that satisfies the initial condition $y(x_0) = y_0$.
   (d) Find parametric equations $x = x(t)$, $y = y(t)$, $z = z(t)$ for the steepest ascent curve on the mountain surface.

3. When studying the spread of an epidemic, we assume that the probability that an infected individual will spread the disease to an uninfected individual is a function of the distance between them. Consider a circular city of radius 10 mi in which the population is uniformly distributed. For an uninfected individual at a fixed point $A(x_0, y_0)$, assume that the probability function is given by
$$f(P) = \tfrac{1}{20}[20 - d(P, A)]$$
where $d(P, A)$ denotes the distance between P and A.
   (a) Suppose the exposure of a person to the disease is the sum of the probabilities of catching the disease from all members of the population. Assume that the infected people are uniformly distributed throughout the city, with k infected individuals per square mile. Find a double integral that represents the exposure of a person residing at A.
   (b) Evaluate the integral for the case in which A is the center of the city and for the case in which A is located on the edge of the city. Where would you prefer to live?

4. Consider a one-parameter family of curves $F(x, y, C) = 0$. A curve Γ is an *envelope* of the given family if for each curve $F(x, y, C) = 0$ in the family, Γ intersects and is tangent to $F(x, y, C) = 0$ at their point of intersection. It can be shown that the equation of an envelope can be obtained by eliminating the parameter α between the two equations
$$F(x, y, \alpha) = 0 \quad \text{and} \quad F_\alpha(x, y, \alpha) = 0$$
   (a) Identify the family of curves defined by
$$\frac{x}{\cos C} + \frac{y}{\sin C} = 1$$
   and determine the envelope for this family.
   (b) Sketch the members of the family and the envelope in the same coordinate system.
   (c) Neglecting air resistance, the path of a projectile fired from the origin at an angle of elevation α and initial velocity v_0 is the parabola whose equation is
$$y = (\tan \alpha)x - \left(\frac{g}{2v_0^2 \cos^2\alpha}\right)x^2$$
   Fix the initial velocity v_0 and regard this equation as a one-parameter family of

FIGURE FOR PROBLEM 5

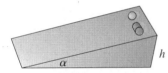

FIGURE FOR PROBLEM 7

curves determined by the angle of elevation α. Find the envelope of this family. (See Applications Plus Problem 3, page 757.)

5. When studying the formation of mountain ranges, geologists estimate the amount of work required to lift a mountain from sea level. Consider a mountain that is essentially in the shape of a right circular cone. Suppose that the weight density of the material in the vicinity of a point P is $g(P)$ and the height is $h(P)$.
   (a) Find a definite integral that represents the total work done in forming the mountain.
   (b) Assume that Mount Fuji in Japan is in the shape of a right circular cone with radius 62,000 ft, height 12,400 ft, and density a constant 200 lb/ft³. How much work was done in forming Mount Fuji if the land was initially at sea level?

6. Marine biologists have determined that when a shark detects the presence of blood in the water, it will swim in the direction in which the concentration of the blood increases most rapidly. Based on certain tests in seawater, the concentration of blood (in parts per million) at a point $P(x, y)$ on the surface is approximated by
$$C(x, y) = e^{-(x^2+2y^2)/10^4}$$
where x and y are measured in meters in a rectangular coordinate system with the blood source at the origin.
   (a) Identify the level curves of the concentration function and sketch several members of this family.
   (b) Suppose a shark is at the point (x_0, y_0) when it first detects the presence of blood in the water. Determine the path that the shark will follow to the source (the origin).

7. Suppose that a solid ball (a marble), a hollow ball (a squash ball), a solid cylinder (a coin), and a hollow cylinder (a ring) roll down a slope. Which of these objects reaches the bottom first? (Make a guess before proceeding.)
   To answer this question we consider a ball or cylinder with mass m, radius r, and moment of inertia I (about the axis of rotation). If the vertical drop is h, then the potential energy at the top is mgh. Suppose the object reaches the bottom with velocity v and angular velocity ω, so $v = \omega r$. The kinetic energy at the bottom consists of two parts: $\frac{1}{2}mv^2$ from translation (moving down the slope) and $\frac{1}{2}I\omega^2$ from rotation. If we assume that energy loss from rolling friction is negligible, then conservation of energy gives
$$mgh = \tfrac{1}{2}mv^2 + \tfrac{1}{2}I\omega^2$$
   (a) Show that
$$v^2 = \frac{2gh}{1 + I^*} \qquad \text{where } I^* = \frac{I}{mr^2}$$
   (b) If $y(t)$ is the vertical distance traveled at time t, then the same reasoning as used in part (a) shows that $v^2 = 2gy/(1 + I^*)$ at any time t. Use this result to show that y satisfies the differential equation
$$\frac{dy}{dt} = \sqrt{\frac{2g}{1 + I^*}}\,(\sin \alpha)\sqrt{y}$$
      where α is the angle of inclination of the plane.
   (c) By solving the differential equation in part (b), show that the total travel time is
$$T = \sqrt{\frac{2h(1 + I^*)}{g \sin^2\alpha}}$$
      This shows that the object with the smallest value of I^* wins the race.
   (d) Show that $I^* = \frac{1}{2}$ for a solid cylinder and $I^* = 1$ for a hollow cylinder.
   (e) Show that $I^* = \frac{2}{5}$ for a solid ball and $I^* = \frac{2}{3}$ for a hollow ball. Thus, the objects finish in the following order: solid ball, solid cylinder, hollow ball, hollow cylinder.
   (f) Calculate I^* for a partly hollow ball with inner radius a and outer radius r. Express your answer in terms of $b = a/r$. What happens as $a \to 0$ and as $a \to r$?

VECTOR CALCULUS

■ It is mathematics that offers the exact mathematical sciences a certain measure of security which, without mathematics, they could not obtain.

ALBERT EINSTEIN

In this chapter we study the calculus of vector fields. (These are functions that assign vectors to points in space.) In particular we define line integrals (which can be used to find the work done by a force field in moving an object along a curve). Then we define surface integrals (which can be used to find the rate of fluid flow across a surface). The connections between these new types of integrals and the single, double, and triple integrals that we have already met are given by the higher-dimensional analogues of the Fundamental Theorem of Calculus: Green's Theorem, Stokes' Theorem, and the Divergence Theorem.

14.1 VECTOR FIELDS

In Section 11.7 we studied vector functions whose domains were sets of real numbers and whose ranges were sets of vectors. Now we look at a type of function, called a vector field, whose domain is a set of points in $\mathbb{R}^2$ (or $\mathbb{R}^3$) and whose range is a set of vectors in V_2 (or V_3). We also see how vector fields arise in various areas of physics.

> **(1) DEFINITION** Let D be a set in $\mathbb{R}^2$ (a plane region). A **vector field on** $\mathbb{R}^2$ is a function $\mathbf{F}$ that assigns to each point (x, y) in D a two-dimensional vector $\mathbf{F}(x, y)$.

The best way to picture a vector field is to draw the arrow representing the vector $\mathbf{F}(x, y)$ starting at the point (x, y). Of course, it is impossible to do this for all points (x, y), but we can gain a reasonable impression of $\mathbf{F}$ by doing it for a few representative points in D as in Figure 1. Since $\mathbf{F}(x, y)$ is a two-dimensional vector, we can write it in terms of its **component functions** P and Q as follows:

$$\mathbf{F}(x, y) = P(x, y)\,\mathbf{i} + Q(x, y)\,\mathbf{j} = \langle P(x, y), Q(x, y)\rangle$$

or, for short,
$$\mathbf{F} = P\mathbf{i} + Q\mathbf{j}$$

Notice that P and Q are scalar functions of two variables and are sometimes called **scalar fields** to distinguish them from vector fields.

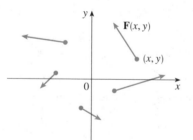

FIGURE 1
Vector field on $\mathbb{R}^2$

> **(2) DEFINITION** Let E be a subset of $\mathbb{R}^3$. A **vector field on** $\mathbb{R}^3$ is a function $\mathbf{F}$ that assigns to each point (x, y, z) in E a three-dimensional vector $\mathbf{F}(x, y, z)$.

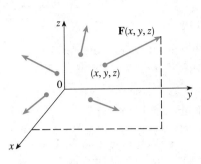

FIGURE 2
Vector field on $\mathbb{R}^3$

A vector field $\mathbf{F}$ on $\mathbb{R}^3$ is pictured in Figure 2. We can express it in terms of its component functions P, Q, and R as

$$\mathbf{F}(x, y, z) = P(x, y, z)\,\mathbf{i} + Q(x, y, z)\,\mathbf{j} + R(x, y, z)\,\mathbf{k}$$

As with the vector functions in Section 11.7, we can define continuity of vector fields and show that $\mathbf{F}$ is continuous if and only if its component functions P, Q, and R are continuous.

We sometimes identify a point (x, y, z) with its position vector $\mathbf{x} = \langle x, y, z \rangle$ and write $\mathbf{F}(\mathbf{x})$ instead of $\mathbf{F}(x, y, z)$. Then $\mathbf{F}$ becomes a function that assigns a vector $\mathbf{F}(\mathbf{x})$ to a vector $\mathbf{x}$.

EXAMPLE 1   A vector field on $\mathbb{R}^2$ is defined by

$$\mathbf{F}(x, y) = -y\,\mathbf{i} + x\,\mathbf{j}$$

Describe F by sketching some of the vectors $\mathbf{F}(x, y)$ as in Figure 1.

SOLUTION   Since $\mathbf{F}(1, 0) = \mathbf{j}$, we draw the vector $\mathbf{j} = \langle 0, 1 \rangle$ starting at the point $(1, 0)$ in Figure 3. Since $\mathbf{F}(0, 1) = -\mathbf{i}$, we draw the vector $\langle -1, 0 \rangle$ with starting point $(0, 1)$. Continuing in this way, we draw a number of representative vectors to represent the vector field in Figure 3.

It appears that each arrow is tangent to a circle with center the origin. To confirm this, we take the dot product of the position vector $\mathbf{x} = x\,\mathbf{i} + y\,\mathbf{j}$ with the vector $\mathbf{F}(\mathbf{x}) = \mathbf{F}(x, y)$:

$$\mathbf{x} \cdot \mathbf{F}(\mathbf{x}) = (x\,\mathbf{i} + y\,\mathbf{j}) \cdot (-y\,\mathbf{i} + x\,\mathbf{j})$$
$$= -xy + yx = 0$$

This shows that $\mathbf{F}(x, y)$ is perpendicular to the position vector $\langle x, y \rangle$ and is therefore tangent to a circle with center the origin and radius $|\mathbf{x}| = \sqrt{x^2 + y^2}$. Notice also that

$$|\mathbf{F}(x, y)| = \sqrt{(-y)^2 + x^2} = \sqrt{x^2 + y^2} = |\mathbf{x}|$$

so the magnitude of the vector $\mathbf{F}(x, y)$ is equal to the radius of the circle.   ∎

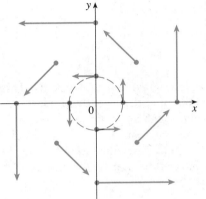

FIGURE 3
$\mathbf{F}(x, y) = -y\,\mathbf{i} + x\,\mathbf{j}$

CAS   Some computer algebra systems are capable of plotting vector fields in two or three dimensions. They give a better impression of the vector field than is possible by hand because the computer can plot a large number of representative vectors. Figure 4 shows a computer plot of the vector field in Example 1; Figures 5 and 6 show two other vector fields. Notice that the computer scales the lengths of the vectors so they are not too long and yet are proportional to their true lengths.

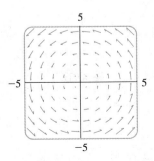

FIGURE 4
$\mathbf{F}(x, y) = \langle -y, x \rangle$

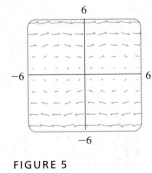

FIGURE 5
$\mathbf{F}(x, y) = \langle y, \sin x \rangle$

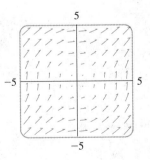

FIGURE 6
$\mathbf{F}(x, y) = \langle \ln(1 + x^2), \ln(1 + y^2) \rangle$

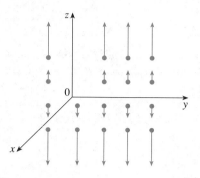

FIGURE 7

$\mathbf{F}(x, y, z) = z\,\mathbf{k}$

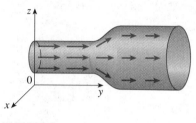

FIGURE 8

Velocity field in fluid flow

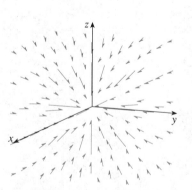

FIGURE 9

Gravitational force field

EXAMPLE 2   Sketch the vector field on $\mathbb{R}^3$ given by $\mathbf{F}(x, y, z) = z\,\mathbf{k}$.

SOLUTION   The sketch is shown in Figure 7. Notice that all vectors are vertical and point upward above the xy-plane and downward below it. The magnitude increases with the distance from the xy-plane. ∎

EXAMPLE 3   Imagine a fluid flowing steadily along a pipe and let $\mathbf{V}(x, y, z)$ be the velocity vector at a point (x, y, z). Then $\mathbf{V}$ assigns a vector to each point (x, y, z) in a certain domain E (the interior of the pipe) and so $\mathbf{V}$ is a vector field on $\mathbb{R}^3$ called a **velocity field.** A possible velocity field is illustrated in Figure 8. The speed at any given point is indicated by the length of the arrow.

Velocity fields also occur in other areas of physics. For instance, the vector field in Example 1 could be used as the velocity field describing the counterclockwise rotation of a wheel. ∎

EXAMPLE 4   Newton's Law of Gravitation states that the magnitude of the gravitational force between two objects with masses m and M is

$$|\mathbf{F}| = \frac{mMG}{r^2}$$

where r is the distance between the objects and G is the gravitational constant. (This is an example of an inverse square law.) Let us assume that the object with mass M is located at the origin in $\mathbb{R}^3$. (For instance, M could be the mass of the earth and the origin would be at the center of the earth.) Let the position vector of the object with mass m be $\mathbf{x} = \langle x, y, z \rangle$. Then $r = |\mathbf{x}|$, so $r^2 = |\mathbf{x}|^2$. The gravitational force exerted on this second object acts toward the origin, and the unit vector in this direction is

$$-\frac{\mathbf{x}}{|\mathbf{x}|}$$

Therefore the gravitational force acting on the object at $\mathbf{x} = \langle x, y, z \rangle$ is

$$(3) \qquad\qquad \mathbf{F}(\mathbf{x}) = -\frac{mMG}{|\mathbf{x}|^3}\,\mathbf{x}$$

[Physicists often use the notation $\mathbf{r}$ instead of $\mathbf{x}$ for the position vector, so you may see Formula 3 written in the form $\mathbf{F} = -(mMG/r^3)\mathbf{r}$.] The function given by Equation 3 is an example of a vector field, called the **gravitational field,** because it associates a vector [the force $\mathbf{F}(\mathbf{x})$] with every point $\mathbf{x}$ in space.

Formula 3 is a compact way of writing the gravitational field, but we can also write it in terms of its component functions by using the facts that $\mathbf{x} = x\,\mathbf{i} + y\,\mathbf{j} + z\,\mathbf{k}$ and $|\mathbf{x}| = \sqrt{x^2 + y^2 + z^2}$:

$$\mathbf{F}(x, y, z) = \frac{-mMGx}{(x^2 + y^2 + z^2)^{3/2}}\,\mathbf{i} + \frac{-mMGy}{(x^2 + y^2 + z^2)^{3/2}}\,\mathbf{j} + \frac{-mMGz}{(x^2 + y^2 + z^2)^{3/2}}\,\mathbf{k}$$

The gravitational field $\mathbf{F}$ is pictured in Figure 9. ∎

EXAMPLE 5   Suppose an electric charge Q is located at the origin. According to Coulomb's Law, the electric force $\mathbf{F}(\mathbf{x})$ exerted by this charge on a charge q located at a point (x, y, z) with position vector $\mathbf{x} = \langle x, y, z \rangle$ is

(4)
$$\mathbf{F}(\mathbf{x}) = \frac{\varepsilon q Q}{|\mathbf{x}|^3} \mathbf{x}$$

where ε is a constant (which depends on the units that are used). For like charges, we have $qQ > 0$ and the force is repulsive; for unlike charges, we have $qQ < 0$ and the force is attractive. Notice the similarity between Formulas 3 and 4. Both vector fields are examples of **force fields.**

Instead of considering the electric force $\mathbf{F}$, physicists often consider the force per unit charge:

$$\mathbf{E}(\mathbf{x}) = \frac{1}{q} \mathbf{F}(\mathbf{x}) = \frac{\varepsilon Q}{|\mathbf{x}|^3} \mathbf{x}$$

Then E is a vector field on $\mathbb{R}^3$ called the **electric field** of Q. ∎

EXAMPLE 6 If f is a scalar function of two variables, recall from Section 12.6 that its gradient ∇f (or grad f) is defined by

$$\nabla f(x, y) = f_x(x, y)\,\mathbf{i} + f_y(x, y)\,\mathbf{j}$$

Figure 10 shows the gradient vector field

$$\nabla f(x, y) = 2xy\,\mathbf{i} + (x^2 - 3y^2)\,\mathbf{j}$$

of the function $f(x, y) = x^2y - y^3$ in Example 6.

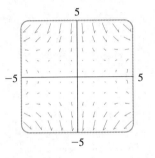

FIGURE 10

Therefore, ∇f is really a vector field on $\mathbb{R}^2$ and is called a **gradient vector field.** For instance, if

$$f(x, y) = x^2y - y^3$$

then its gradient vector field is given by

$$\nabla f(x, y) = 2xy\,\mathbf{i} + (x^2 - 3y^2)\,\mathbf{j}$$

Likewise, if f is a scalar function of three variables, its gradient is a vector field on $\mathbb{R}^3$ given by

$$\nabla f(x, y, z) = f_x(x, y, z)\,\mathbf{i} + f_y(x, y, z)\,\mathbf{j} + f_z(x, y, z)\,\mathbf{k}$$ ∎

A vector field $\mathbf{F}$ is called a **conservative vector field** if it is the gradient of some scalar function, that is, if there exists a function f such that $\mathbf{F} = \nabla f$. In this situation f is called a **potential function** for $\mathbf{F}$.

Not all vector fields are conservative, but such fields do arise frequently in physics. For example, the gravitational field $\mathbf{F}$ in Example 4 is conservative because if we define

$$f(x, y, z) = \frac{mMG}{\sqrt{x^2 + y^2 + z^2}}$$

then

$$\begin{aligned}
\nabla f(x, y, z) &= \frac{\partial f}{\partial x}\,\mathbf{i} + \frac{\partial f}{\partial y}\,\mathbf{j} + \frac{\partial f}{\partial z}\,\mathbf{k} \\
&= \frac{-mMGx}{(x^2 + y^2 + z^2)^{3/2}}\,\mathbf{i} + \frac{-mMGy}{(x^2 + y^2 + z^2)^{3/2}}\,\mathbf{j} + \frac{-mMGz}{(x^2 + y^2 + z^2)^{3/2}}\,\mathbf{k} \\
&= \mathbf{F}(x, y, z)
\end{aligned}$$

Later in this chapter (see Theorems 14.3.6 and 14.5.4) we learn how to tell whether or not a given vector field is conservative.

EXERCISES 14.1

1–10 ■ Sketch the vector field **F** by drawing a diagram like Figure 3 or Figure 7.

1. $\mathbf{F}(x, y) = x\,\mathbf{i} + y\,\mathbf{j}$

2. $\mathbf{F}(x, y) = x\,\mathbf{i} - y\,\mathbf{j}$

3. $\mathbf{F}(x, y) = y\,\mathbf{i} + \mathbf{j}$

4. $\mathbf{F}(x, y) = -x\,\mathbf{i} + 2y\,\mathbf{j}$

5. $\mathbf{F}(x, y) = \dfrac{y\,\mathbf{i} + x\,\mathbf{j}}{\sqrt{x^2 + y^2}}$

6. $\mathbf{F}(x, y) = \dfrac{y\,\mathbf{i} - x\,\mathbf{j}}{\sqrt{x^2 + y^2}}$

7. $\mathbf{F}(x, y, z) = \mathbf{j}$

8. $\mathbf{F}(x, y, z) = z\,\mathbf{j}$

9. $\mathbf{F}(x, y, z) = y\,\mathbf{j}$

10. $\mathbf{F}(x, y, z) = \mathbf{j} + \mathbf{k}$

11–14 ■ Match the vector fields with the plots (labeled I–IV). Give reasons for your choices.

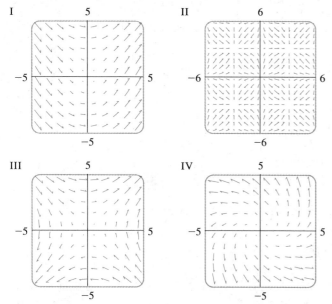

I

II

III

IV

11. $\mathbf{F}(x, y) = \langle y, x \rangle$

12. $\mathbf{F}(x, y) = \langle 2x - 3y, 2x + 3y \rangle$

13. $\mathbf{F}(x, y) = \langle \sin x, \sin y \rangle$

14. $\mathbf{F}(x, y) = \langle \ln(1 + x^2 + y^2), x \rangle$

CAS **15.** If you have a CAS that plots vector fields (the command is fieldplot in Maple and PlotVectorField in Mathematica), use it to plot $\mathbf{F}(x, y) = (y^2 - 2xy)\,\mathbf{i} + (3xy - 6x^2)\,\mathbf{j}$. Explain the appearance by finding the set of points (x, y) such that $\mathbf{F}(x, y) = \mathbf{0}$.

CAS **16.** Let $\mathbf{F}(\mathbf{x}) = (r^2 - 2r)\mathbf{x}$, where $\mathbf{x} = \langle x, y \rangle$ and $r = |\mathbf{x}|$. Use a CAS to plot this vector field in various domains until you can see what is happening. Describe the appearance of the plot and explain it by finding the points where $\mathbf{F}(\mathbf{x}) = \mathbf{0}$.

17–22 ■ Find the gradient vector field of f.

17. $f(x, y) = x^5 - 4x^2y^3$

18. $f(x, y) = \sin(2x + 3y)$

19. $f(x, y) = e^{3x}\cos 4y$

20. $f(x, y, z) = xyz$

21. $f(x, y, z) = xy^2 - yz^3$

22. $f(x, y, z) = x\ln(y - z)$

23–24 ■ Find the gradient vector field ∇f of f and sketch it.

23. $f(x, y) = x^2 - \frac{1}{2}y^2$

24. $f(x, y) = \ln\sqrt{x^2 + y^2}$

25. The **flow lines** (or **streamlines**) of a vector field are the paths followed by a particle whose velocity field is the given vector field. Thus the vectors in a vector field are tangent to the flow lines.
(a) Sketch some flow lines for the vector field $\mathbf{F}(x, y) = x\,\mathbf{i} - y\,\mathbf{j}$. From your sketches, can you guess the equations of the flow lines?
(b) Find an equation of the flow line that passes through the point $(1, 1)$ by solving the differential equations $dx/dt = x$ and $dy/dt = -y$.

14.2 **LINE INTEGRALS**

In this section we define an integral that is similar to a single integral except that instead of integrating over an interval $[a, b]$, we integrate over a curve C. Such integrals are called *line integrals,* although "curve integrals" would be better terminology.

We start with a plane curve C given by the parametric equations

(1) $x = x(t) \qquad y = y(t) \qquad a \leq t \leq b$

or, equivalently, by the vector equation $\mathbf{r}(t) = x(t)\,\mathbf{i} + y(t)\,\mathbf{j}$, and we assume that C is a smooth curve. [This means that $\mathbf{r}'$ is continuous and $\mathbf{r}'(t) \neq \mathbf{0}$. See Section 11.8.] A partition of the parameter interval $[a, b]$ by points t_i with

$$a = t_0 < t_1 < t_2 < \cdots < t_n = b$$

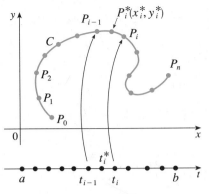

FIGURE 1

A partition of $[a, b]$ determines a partition of C

determines a partition P of the curve by points $P_i(x_i, y_i)$, where $x_i = x(t_i)$ and $y_i = y(t_i)$ (see Figure 1). These points P_i divide C into n subarcs with lengths $\Delta s_1, \Delta s_2, \ldots, \Delta s_n$. The **norm** $\|P\|$ of the partition is the longest of these lengths. We choose any point $P_i^*(x_i^*, y_i^*)$ in the ith subarc. (This corresponds to a point t_i^* in $[t_{i-1}, t_i]$.) Now if f is any function of two variables whose domain includes the curve C, we evaluate f at the point (x_i^*, y_i^*), multiply by the length Δs_i of the subarc, and form the sum

$$\sum_{i=1}^{n} f(x_i^*, y_i^*) \, \Delta s_i$$

which is similar to a Riemann sum. Then we take the limit of these sums and make the following definition by analogy with a single integral.

---

(2) DEFINITION   If f is defined on a smooth curve C given by Equations 1, then the **line integral of f along C** is

$$\int_C f(x, y) \, ds = \lim_{\|P\| \to 0} \sum_{i=1}^{n} f(x_i^*, y_i^*) \, \Delta s_i$$

if this limit exists.

It can be shown that if f is a continuous function, then the limit in Definition 2 always exists and the following formula can be used to evaluate the line integral:

(3)
$$\int_C f(x, y) \, ds = \int_a^b f(x(t), y(t)) \sqrt{\left(\frac{dx}{dt}\right)^2 + \left(\frac{dy}{dt}\right)^2} \, dt$$

The value of the line integral does not depend on the parametrization of the curve, provided that the curve is traversed exactly once as t increases from a to b.

The way to remember Formula 3 is to express everything in terms of the parameter t. Use the parametric equations to express x and y in terms of t and recall the arc length formula from Section 9.3:

$$ds = \sqrt{\left(\frac{dx}{dt}\right)^2 + \left(\frac{dy}{dt}\right)^2} \, dt$$

In the special case where C is the line segment that joins $(a, 0)$ to $(b, 0)$, using x as the parameter, we can write the parametric equations of C as follows: $x = x$, $y = 0$, $a \leq x \leq b$. Formula 3 then becomes

$$\int_C f(x, y) \, ds = \int_a^b f(x, 0) \, dx$$

and so the line integral reduces to an ordinary single integral in this case.

Just as for an ordinary single integral, we can interpret the line integral of a *positive* function as an area. In fact, if $f(x, y) \geq 0$, $\int_C f(x, y) \, ds$ represents the area of one side of the "fence" or "curtain" in Figure 2 whose base is C and whose height above the point (x, y) is $f(x, y)$.

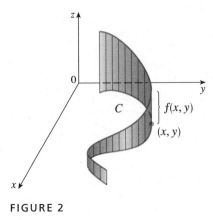

FIGURE 2

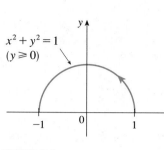

$x^2 + y^2 = 1$
$(y \geqslant 0)$

FIGURE 3

EXAMPLE 1   Evaluate $\int_C (2 + x^2 y)\, ds$, where C is the upper half of the unit circle $x^2 + y^2 = 1$.

SOLUTION   Recall that the unit circle can be parametrized by means of the equations

$$x = \cos t \qquad y = \sin t$$

and the upper half of the circle is described by the parameter interval $0 \leqslant t \leqslant \pi$ (see Figure 3). Therefore, Formula 3 gives

$$\int_C (2 + x^2 y)\, ds = \int_0^\pi (2 + \cos^2 t \sin t) \sqrt{\left(\frac{dx}{dt}\right)^2 + \left(\frac{dy}{dt}\right)^2}\, dt$$

$$= \int_0^\pi (2 + \cos^2 t \sin t) \sqrt{\sin^2 t + \cos^2 t}\, dt$$

$$= \int_0^\pi (2 + \cos^2 t \sin t)\, dt = \left[2t - \frac{\cos^3 t}{3}\right]_0^\pi$$

$$= 2\pi + \tfrac{2}{3} \qquad\blacksquare$$

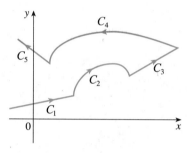

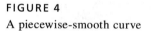

FIGURE 4
A piecewise-smooth curve

Suppose now that C is a **piecewise-smooth curve**; that is, C is a union of a finite number of smooth curves $C_1, C_2, \ldots, C_n$, where, as illustrated in Figure 4, the initial point of C_{i+1} is the terminal point of C_i. Then we define the integral of f along C as the sum of the integrals of f along each of the smooth pieces of C:

$$\int_C f(x, y)\, ds = \int_{C_1} f(x, y)\, ds + \int_{C_2} f(x, y)\, ds + \cdots + \int_{C_n} f(x, y)\, ds$$

EXAMPLE 2   Evaluate $\int_C 2x\, ds$, where C consists of the arc C_1 of the parabola $y = x^2$ from $(0, 0)$ to $(1, 1)$ followed by the vertical line segment C_2 from $(1, 1)$ to $(1, 2)$.

SOLUTION   The curve C is shown in Figure 5. C_1 is the graph of a function of x, so we can choose x as the parameter and the equations for C_1 become

$$x = x \qquad y = x^2 \qquad 0 \leqslant x \leqslant 1$$

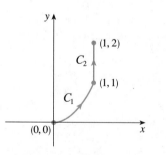

FIGURE 5
$C = C_1 \cup C_2$

Therefore

$$\int_{C_1} 2x\, ds = \int_0^1 2x \sqrt{\left(\frac{dx}{dx}\right)^2 + \left(\frac{dy}{dx}\right)^2}\, dx$$

$$= \int_0^1 2x \sqrt{1 + 4x^2}\, dx = \tfrac{1}{4} \cdot \tfrac{2}{3}(1 + 4x^2)^{3/2}\Big]_0^1 = \frac{5\sqrt{5} - 1}{6}$$

On C_2 we choose y as the parameter, so the equations of C_2 are

$$x = 1 \qquad y = y \qquad 1 \leqslant y \leqslant 2$$

and

$$\int_{C_2} 2x\, ds = \int_1^2 2(1) \sqrt{\left(\frac{dx}{dy}\right)^2 + \left(\frac{dy}{dy}\right)^2}\, dy = \int_1^2 2\, dy = 2$$

Thus

$$\int_C 2x\, ds = \int_{C_1} 2x\, ds + \int_{C_2} 2x\, ds = \frac{5\sqrt{5} - 1}{6} + 2 \qquad\blacksquare$$

Any physical interpretation of a line integral $\int_C f(x, y)\, ds$ depends on the physical interpretation of the function f. Suppose that $\rho(x, y)$ represents the linear density at a point (x, y) of a thin wire shaped like a curve C. Then the mass of the part of the wire from P_{i-1} to P_i in Figure 1 is approximately $\rho(x_i^*, y_i^*)\,\Delta s_i$ and so the total mass of the wire is approximately $\Sigma\, \rho(x_i^*, y_i^*)\,\Delta s_i$. By taking finer partitions of the curve, we obtain the **mass** m of the wire as the limiting value of these approximations:

$$m = \lim_{\|P\| \to 0} \sum_{i=1}^{n} \rho(x_i^*, y_i^*)\,\Delta s_i = \int_C \rho(x, y)\, ds$$

[For example, if $f(x, y) = 2 + x^2 y$ represents the density of a semicircular wire, then the integral in Example 1 would represent the mass of the wire.] The **center of mass** of the wire with density function ρ is at the point $(\overline{x}, \overline{y})$, where

$$(4) \qquad \overline{x} = \frac{1}{m} \int_C x\rho(x, y)\, ds \qquad \overline{y} = \frac{1}{m} \int_C y\rho(x, y)\, ds$$

Other physical interpretations of line integrals will be discussed later in this chapter.

EXAMPLE 3   A wire takes the shape of the semicircle $x^2 + y^2 = 1$, $y \geqslant 0$, and is thicker near its base than near the top. Find the center of mass of the wire if the linear density at any point is proportional to its distance from the line $y = 1$.

SOLUTION   As in Example 1 we use the parametrization $x = \cos t$, $y = \sin t$, $0 \leqslant t \leqslant \pi$, and find that $ds = dt$. The linear density is

$$\rho(x, y) = k(1 - y)$$

where k is a constant, and so the mass of the wire is

$$m = \int_C k(1 - y)\, ds = \int_0^\pi k(1 - \sin t)\, dt$$

$$= k\big[t + \cos t\big]_0^\pi = k(\pi - 2)$$

From Equations 4 we have

$$\overline{y} = \frac{1}{m} \int_C y\rho(x, y)\, ds = \frac{1}{k(\pi - 2)} \int_C yk(1 - y)\, ds$$

$$= \frac{1}{\pi - 2} \int_0^\pi (\sin t - \sin^2 t)\, dt = \frac{1}{\pi - 2}\Big[-\cos t - \tfrac{1}{2}t + \tfrac{1}{4}\sin 2t\Big]_0^\pi$$

$$= \frac{4 - \pi}{2(\pi - 2)}$$

By symmetry we see that $\overline{x} = 0$, so the center of mass is at

$$\left(0, \frac{4 - \pi}{2(\pi - 2)}\right) \approx (0, 0.38)$$

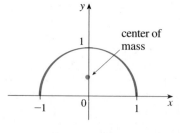

FIGURE 6

See Figure 6.

Two other line integrals are obtained by replacing Δs_i by $\Delta x_i = x_i - x_{i-1}$ or $\Delta y_i = y_i - y_{i-1}$ in Definition 2. They are called the **line integrals of f along C with respect to x and y:**

(5)
$$\int_C f(x, y)\, dx = \lim_{\|P\| \to 0} \sum_{i=1}^{n} f(x_i^*, y_i^*)\, \Delta x_i$$

(6)
$$\int_C f(x, y)\, dy = \lim_{\|P\| \to 0} \sum_{i=1}^{n} f(x_i^*, y_i^*)\, \Delta y_i$$

When we want to distinguish the original line integral $\int_C f(x, y)\, ds$ from those in Equations 5 and 6, we call it the **line integral with respect to arc length.**

The following formulas say that line integrals with respect to x and y can also be evaluated by expressing everything in terms of t: $x = x(t)$, $y = y(t)$, $dx = x'(t)\, dt$, $dy = y'(t)\, dt$.

(7)
$$\int_C f(x, y)\, dx = \int_a^b f(x(t), y(t))\, x'(t)\, dt$$

$$\int_C f(x, y)\, dy = \int_a^b f(x(t), y(t))\, y'(t)\, dt$$

It frequently happens that line integrals with respect to x and y occur together. When this happens, it is customary to abbreviate by writing

$$\int_C P(x, y)\, dx + \int_C Q(x, y)\, dy = \int_C P(x, y)\, dx + Q(x, y)\, dy$$

When you are setting up a line integral, sometimes the most difficult thing is to think of a parametric representation for a curve whose geometric description is given. It is useful to remember that a vector representation of the line segment that starts at $\mathbf{r}_0$ and ends at $\mathbf{r}_1$ is given by

(8)
$$\mathbf{r}(t) = (1 - t)\, \mathbf{r}_0 + t\, \mathbf{r}_1 \qquad 0 \le t \le 1$$

(See Equation 11.5.1 with $\mathbf{v} = \mathbf{r}_1 - \mathbf{r}_0$.)

EXAMPLE 4   Evaluate $\int_C y^2\, dx + x\, dy$, where (a) $C = C_1$ is the line segment from $(-5, -3)$ to $(0, 2)$ and (b) $C = C_2$ is the arc of the parabola $x = 4 - y^2$ from $(-5, -3)$ to $(0, 2)$ (see Figure 7).

SOLUTION
(a) A parametric representation for the line segment is

$$x = 5t - 5 \qquad y = 5t - 3 \qquad 0 \le t \le 1$$

(Use Equation 8 with $\mathbf{r}_0 = \langle -5, -3 \rangle$ and $\mathbf{r}_1 = \langle 0, 2 \rangle$.) Then $dx = 5\, dt$, $dy = 5\, dt$, and Formula 7 gives

$$\int_{C_1} y^2\, dx + x\, dy = \int_0^1 (5t - 3)^2 (5\, dt) + (5t - 5)(5\, dt)$$

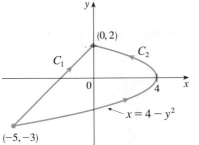

FIGURE 7

$$= 5 \int_0^1 (25t^2 - 25t + 4)\, dt$$

$$= 5 \left[\frac{25t^3}{3} - \frac{25t^2}{2} + 4t \right]_0^1 = -\frac{5}{6}$$

(b) Since the parabola is given as a function of y, let us take y as the parameter and write C_2 as

$$x = 4 - y^2 \qquad y = y \qquad -3 \le y \le 2$$

Then $dx = -2y\, dy$ and by Formula 7 we have

$$\int_{C_2} y^2\, dx + x\, dy = \int_{-3}^2 y^2(-2y)\, dy + (4 - y^2)\, dy$$

$$= \int_{-3}^2 (-2y^3 - y^2 + 4)\, dy$$

$$= \left[-\frac{y^4}{2} - \frac{y^3}{3} + 4y \right]_{-3}^2 = 40\tfrac{5}{6} \qquad ∎$$

Notice that we got different answers in parts (a) and (b) of Example 4 even though the two curves had the same endpoints. Thus, in general, the value of a line integral depends not just on the endpoints of the curve but also on the path. (But see Section 14.3 for conditions under which the integral is independent of the path.)

Notice also that the answers in Example 4 depend on the direction, or orientation, of the curve. If $-C_1$ denotes the line segment from $(0, 2)$ to $(-5, -3)$, you can verify, using the parametrization

$$x = -5t \qquad y = 2 - 5t \qquad 0 \le t \le 1$$

that

$$\int_{-C_1} y^2\, dx + x\, dy = \tfrac{5}{6}$$

In general, a given parametrization $x = x(t)$, $y = y(t)$, $a \le t \le b$, determines an **orientation** of a curve C, with the positive direction corresponding to increasing values of the parameter t. (See Figure 8, where the initial point A corresponds to the parameter value a and the terminal point B corresponds to $t = b$.)

If $-C$ denotes the curve consisting of the same points as C but with the opposite orientation (from initial point B to terminal point A in Figure 8), then we have

$$\int_{-C} f(x, y)\, dx = -\int_C f(x, y)\, dx \qquad \int_{-C} f(x, y)\, dy = -\int_C f(x, y)\, dy$$

But if we integrate with respect to arc length, the value of the line integral does *not* change when we reverse the orientation of the curve:

$$\int_{-C} f(x, y)\, ds = \int_C f(x, y)\, ds$$

(This is because Δs_i is always positive, whereas Δx_i and Δy_i change sign when we reverse the orientation of C.)

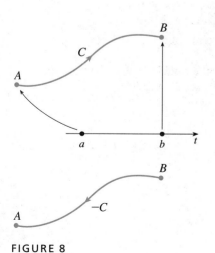

FIGURE 8

LINE INTEGRALS IN SPACE

We now suppose that C is a smooth space curve given by the parametric equations

$$x = x(t) \qquad y = y(t) \qquad z = z(t) \qquad a \le t \le b$$

or by a vector equation $\mathbf{r}(t) = x(t)\,\mathbf{i} + y(t)\,\mathbf{j} + z(t)\,\mathbf{k}$. If f is a function of three variables that is continuous on some region containing C, then we define the **line integral of f along C** (with respect to arc length) in a manner similar to that for plane curves:

$$\int_C f(x, y, z)\, ds = \lim_{\|P\| \to 0} \sum_{i=1}^{n} f(x_i^*, y_i^*, z_i^*)\, \Delta s_i$$

We evaluate it using a formula similar to Formula 3:

(9) $\displaystyle \int_C f(x, y, z)\, ds = \int_a^b f(x(t), y(t), z(t)) \sqrt{\left(\frac{dx}{dt}\right)^2 + \left(\frac{dy}{dt}\right)^2 + \left(\frac{dz}{dt}\right)^2}\, dt$

Observe that the integrals in both Formulas 3 and 9 can be written in the more compact vector notation

$$\int_a^b f(\mathbf{r}(t))\,|\mathbf{r}'(t)|\, dt$$

For the special case $f(x, y, z) \equiv 1$, we get

$$\int_C ds = \int_a^b |\mathbf{r}'(t)|\, dt = L$$

where L is the length of the curve C (see Formula 11.8.3).

Line integrals along C with respect to x, y, and z can also be defined. For example,

$$\int_C f(x, y, z)\, dz = \lim_{\|P\| \to 0} \sum_{i=1}^{n} f(x_i^*, y_i^*, z_i^*)\, \Delta z_i$$

$$= \int_a^b f(x(t), y(t), z(t))\, z'(t)\, dt$$

Therefore, as with line integrals in the plane, we evaluate integrals of the form

(10) $\displaystyle \int_C P(x, y, z)\, dx + Q(x, y, z)\, dy + R(x, y, z)\, dz$

by expressing everything (x, y, z, dx, dy, dz) in terms of the parameter t.

EXAMPLE 5   Evaluate $\int_C y \sin z\, ds$, where C is the circular helix given by the equations $x = \cos t$, $y = \sin t$, $z = t$, $0 \le t \le 2\pi$ (see Figure 9).

SOLUTION   Formula 9 gives

$$\int_C y \sin z\, ds = \int_0^{2\pi} (\sin t) \sin t \sqrt{\left(\frac{dx}{dt}\right)^2 + \left(\frac{dy}{dt}\right)^2 + \left(\frac{dz}{dt}\right)^2}\, dt$$

$$= \int_0^{2\pi} \sin^2 t \sqrt{\sin^2 t + \cos^2 t + 1}\, dt$$

$$= \sqrt{2} \int_0^{2\pi} \tfrac{1}{2}(1 - \cos 2t)\, dt = \frac{\sqrt{2}}{2}\Big[t - \tfrac{1}{2}\sin 2t\Big]_0^{2\pi} = \sqrt{2}\,\pi \qquad \blacksquare$$

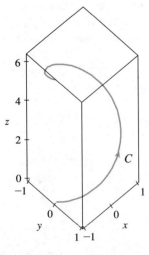

FIGURE 9

EXAMPLE 6   Evaluate $\int_C y\,dx + z\,dy + x\,dz$, where C consists of the line segment C_1 from $(2, 0, 0)$ to $(3, 4, 5)$ followed by the vertical line segment C_2 from $(3, 4, 5)$ to $(3, 4, 0)$.

SOLUTION   The curve C is shown in Figure 10. Using Equation 8, we write C_1 as

$$\mathbf{r}(t) = (1 - t)\langle 2, 0, 0 \rangle + t\langle 3, 4, 5 \rangle = \langle 2 + t, 4t, 5t \rangle$$

or, in parametric form, as

$$x = 2 + t \qquad y = 4t \qquad z = 5t \qquad 0 \le t \le 1$$

Thus

$$\int_{C_1} y\,dx + z\,dy + x\,dz = \int_0^1 (4t)\,dt + (5t)4\,dt + (2 + t)5\,dt$$

$$= \int_0^1 (10 + 29t)\,dt = 10t + 29\frac{t^2}{2}\Big]_0^1 = 24.5$$

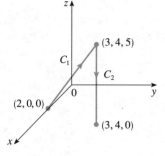

FIGURE 10

Likewise, C_2 can be written in the form

$$\mathbf{r}(t) = (1 - t)\langle 3, 4, 5 \rangle + t\langle 3, 4, 0 \rangle = \langle 3, 4, 5 - 5t \rangle$$

or $\qquad\qquad x = 3 \qquad y = 4 \qquad z = 5 - 5t \qquad 0 \le t \le 1$

Then $dx = 0 = dy$, so

$$\int_{C_2} y\,dx + z\,dy + x\,dz = \int_0^1 3(-5)\,dt = -15$$

Adding the values of these integrals together, we obtain

$$\int_C y\,dx + z\,dy + x\,dz = 24.5 - 15 = 9.5$$

LINE INTEGRALS OF VECTOR FIELDS

Recall from Section 5.4 that the work done by a variable force $f(x)$ in moving a particle from a to b along the x-axis is $W = \int_a^b f(x)\,dx$. Then in Section 11.3 we found that the work done by a constant force $\mathbf{F}$ in moving an object from a point P to another point Q in space is $\mathbf{W} = \mathbf{F} \cdot \mathbf{D}$, where $\mathbf{D} = \vec{PQ}$ is the displacement vector.

Now suppose that $\mathbf{F} = P\mathbf{i} + Q\mathbf{j} + R\mathbf{k}$ is a continuous force field on $\mathbb{R}^3$, such as the gravitational field of Example 4 in Section 14.1 or the electric force field of Example 5 in Section 14.1. (A force field on $\mathbb{R}^2$ could be regarded as a special case where $R = 0$ and P and Q depend only on x and y.) We wish to compute the work done by this force in moving a particle along a smooth curve C.

We partition C into subarcs $P_{i-1}P_i$ with lengths Δs_i by means of a partition of the parameter interval $[a, b]$. (See Figure 1 for the two-dimensional case or Figure 11 for the three-dimensional case.) Choose a point $P_i^*(x_i^*, y_i^*, z_i^*)$ on the ith subarc corresponding to the parameter value t_i^*. If Δs_i is small, then as the particle moves from P_{i-1} to P_i along the curve, it proceeds approximately in the direction of $\mathbf{T}(t_i^*)$, the unit tangent vector at P_i^*. Therefore, the work done by the force $\mathbf{F}$ in moving the particle from P_{i-1} to P_i is approximately

$$\mathbf{F}(x_i^*, y_i^*, z_i^*) \cdot [\Delta s_i\,\mathbf{T}(t_i^*)] = [\mathbf{F}(x_i^*, y_i^*, z_i^*) \cdot \mathbf{T}(t_i^*)]\Delta s_i$$

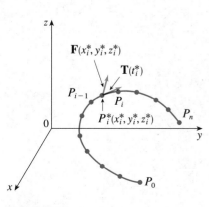

FIGURE 11

and the total work done in moving the particle along C is approximately

$$(11) \qquad \sum_{i=1}^{n} [\mathbf{F}(x_i^*, y_i^*, z_i^*) \cdot \mathbf{T}(x_i^*, y_i^*, z_i^*)] \Delta s_i$$

where $\mathbf{T}(x, y, z)$ is the unit tangent vector at the point (x, y, z) on C. Intuitively, we see that these approximations ought to become better as $\|P\|$ becomes smaller. Therefore, we define the **work** W done by the force field $\mathbf{F}$ as the limit of the Riemann sums in (11), namely,

$$(12) \qquad W = \int_C \mathbf{F}(x, y, z) \cdot \mathbf{T}(x, y, z) \, ds = \int_C \mathbf{F} \cdot \mathbf{T} \, ds$$

Equation 12 says that *work is the integral with respect to arc length of the tangential component of the force.*

If the curve C is given by the vector equation $\mathbf{r}(t) = x(t)\,\mathbf{i} + y(t)\,\mathbf{j} + z(t)\,\mathbf{k}$, then $\mathbf{T}(t) = \mathbf{r}'(t)/|\mathbf{r}'(t)|$, so using Equation 9 we can rewrite Equation 12 in the form

$$W = \int_a^b \left[\mathbf{F}(\mathbf{r}(t)) \cdot \frac{\mathbf{r}'(t)}{|\mathbf{r}'(t)|} \right] |\mathbf{r}'(t)| \, dt$$

$$= \int_a^b \mathbf{F}(\mathbf{r}(t)) \cdot \mathbf{r}'(t) \, dt$$

This latter integral is often abbreviated as $\int_C \mathbf{F} \cdot d\mathbf{r}$ and occurs in other areas of physics as well. Therefore, we make the following definition for the line integral of *any* continuous vector field.

---

(13) DEFINITION   Let $\mathbf{F}$ be a continuous vector field defined on a smooth curve C given by a vector function $\mathbf{r}(t)$, $a \le t \le b$. Then the **line integral of $\mathbf{F}$ along C** is

$$\int_C \mathbf{F} \cdot d\mathbf{r} = \int_a^b \mathbf{F}(\mathbf{r}(t)) \cdot \mathbf{r}'(t) \, dt = \int_C \mathbf{F} \cdot \mathbf{T} \, ds$$

When you use Definition 13, remember that $\mathbf{F}(\mathbf{r}(t))$ is just short for $\mathbf{F}(x(t), y(t), z(t))$, so we simply evaluate $\mathbf{F}(\mathbf{r}(t))$ by putting $x = x(t)$, $y = y(t)$, and $z = z(t)$ in the expression for $\mathbf{F}(x, y, z)$. Notice also that we can formally write $d\mathbf{r} = \mathbf{r}'(t) \, dt$.

EXAMPLE 7   Find the work done by the force field $\mathbf{F}(x, y) = x^2\,\mathbf{i} - xy\,\mathbf{j}$ in moving a particle along the quarter-circle $\mathbf{r}(t) = \cos t\,\mathbf{i} + \sin t\,\mathbf{j}$, $0 \le t \le \pi/2$.

SOLUTION   Since $x = \cos t$ and $y = \sin t$, we have

$$\mathbf{F}(\mathbf{r}(t)) = \cos^2 t\,\mathbf{i} - \cos t \sin t\,\mathbf{j}$$

and

$$\mathbf{r}'(t) = -\sin t\,\mathbf{i} + \cos t\,\mathbf{j}$$

Therefore, the work done is

$$\int_C \mathbf{F} \cdot d\mathbf{r} = \int_0^{\pi/2} \mathbf{F}(\mathbf{r}(t)) \cdot \mathbf{r}'(t) \, dt = \int_0^{\pi/2} (-2 \cos^2 t \sin t) \, dt$$

$$= 2 \left. \frac{\cos^3 t}{3} \right]_0^{\pi/2} = -\frac{2}{3}$$

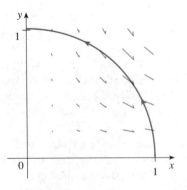

FIGURE 12

Figure 12 shows the force field and the curve in Example 7. The work done is negative because the field impedes movement along the curve.

NOTE: Even though $\int_C \mathbf{F} \cdot d\mathbf{r} = \int_C \mathbf{F} \cdot \mathbf{T}\, ds$ and integrals with respect to arc length are unchanged when orientation is reversed, it is still true that

$$\int_{-C} \mathbf{F} \cdot d\mathbf{r} = -\int_C \mathbf{F} \cdot d\mathbf{r}$$

because the unit tangent vector $\mathbf{T}$ is replaced by its negative when C is replaced by $-C$.

Figure 13 shows the twisted cubic C in Example 8 and some typical vectors acting at three points on C.

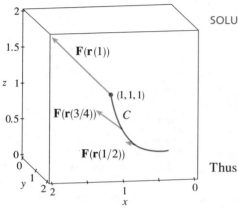

FIGURE 13

EXAMPLE 8   Evaluate $\int_C \mathbf{F} \cdot d\mathbf{r}$, where $\mathbf{F}(x, y, z) = xy\,\mathbf{i} + yz\,\mathbf{j} + zx\,\mathbf{k}$ and C is the twisted cubic given by

$$x = t \qquad y = t^2 \qquad z = t^3 \qquad 0 \le t \le 1$$

SOLUTION   We have

$$\mathbf{r}(t) = t\,\mathbf{i} + t^2\,\mathbf{j} + t^3\,\mathbf{k}$$

$$\mathbf{r}'(t) = \mathbf{i} + 2t\,\mathbf{j} + 3t^2\,\mathbf{k}$$

$$\mathbf{F}(\mathbf{r}(t)) = t^3\,\mathbf{i} + t^5\,\mathbf{j} + t^4\,\mathbf{k}$$

Thus

$$\int_C \mathbf{F} \cdot d\mathbf{r} = \int_0^1 \mathbf{F}(\mathbf{r}(t)) \cdot \mathbf{r}'(t)\, dt$$

$$= \int_0^1 (t^3 + 5t^6)\, dt = \frac{t^4}{4} + \frac{5t^7}{7} \Big]_0^1 = \frac{27}{28} \quad\blacksquare$$

Finally, we note the connection between line integrals of vector fields and line integrals of scalar fields. Suppose the vector field $\mathbf{F}$ on $\mathbb{R}^3$ is given in component form by the equation $\mathbf{F} = P\,\mathbf{i} + Q\,\mathbf{j} + R\,\mathbf{k}$. We use Definition 13 to compute its line integral along C:

$$\int_C \mathbf{F} \cdot d\mathbf{r} = \int_a^b \mathbf{F}(\mathbf{r}(t)) \cdot \mathbf{r}'(t)\, dt$$

$$= \int_a^b (P\,\mathbf{i} + Q\,\mathbf{j} + R\,\mathbf{k}) \cdot (x'(t)\,\mathbf{i} + y'(t)\,\mathbf{j} + z'(t)\,\mathbf{k})\, dt$$

$$= \int_a^b [P(x(t), y(t), z(t))\, x'(t) + Q(x(t), y(t), z(t))\, y'(t) + R(x(t), y(t), z(t))\, z'(t)]\, dt$$

But this last integral is precisely the line integral in (10). Therefore, we have

$$\int_C \mathbf{F} \cdot d\mathbf{r} = \int_C P\, dx + Q\, dy + R\, dz \qquad \text{where } \mathbf{F} = P\,\mathbf{i} + Q\,\mathbf{j} + R\,\mathbf{k}$$

For example, the integral $\int_C y\, dx + z\, dy + x\, dz$ in Example 6 could be expressed as $\int_C \mathbf{F} \cdot d\mathbf{r}$ where

$$\mathbf{F}(x, y, z) = y\,\mathbf{i} + z\,\mathbf{j} + x\,\mathbf{k}$$

EXERCISES 14.2

1–16 ■ Evaluate the line integral, where C is the given curve.

1. $\int_C x \, ds, \quad C: x = t^3, \, y = t, \, 0 \leqslant t \leqslant 1$

2. $\int_C y \, ds, \quad C: x = t^3, \, y = t^2, \, 0 \leqslant t \leqslant 1$

3. $\int_C xy^4 \, ds, \quad C$ is the right half of the circle $x^2 + y^2 = 16$

4. $\int_C xy \, ds, \quad C$ is the line segment joining $(-1, 1)$ to $(2, 3)$

5. $\int_C (x - 2y^2) \, dy,$
C is the arc of the parabola $y = x^2$ from $(-2, 4)$ to $(1, 1)$

6. $\int_C \sin x \, dx,$
C is the arc of the curve $x = y^4$ from $(1, -1)$ to $(1, 1)$

7. $\int_C xy \, dx + (x - y) \, dy,$
C consists of line segments from $(0, 0)$ to $(2, 0)$ and from $(2, 0)$ to $(3, 2)$

8. $\int_C x \sqrt{y} \, dx + 2y \sqrt{x} \, dy,$
C consists of the arc of the circle $x^2 + y^2 = 1$ from $(1, 0)$ to $(0, 1)$ and the line segment from $(0, 1)$ to $(4, 3)$

9. $\int_C xyz \, ds, \quad C: x = 2t, \, y = 3 \sin t, \, z = 3 \cos t, \, 0 \leqslant t \leqslant \pi/2$

10. $\int_C x^2 z \, ds, \quad C: x = \sin 2t, \, y = 3t, \, z = \cos 2t, \, 0 \leqslant t \leqslant \pi/4$

11. $\int_C xy^2 z \, ds, \quad C$ is the line segment from $(1, 0, 1)$ to $(0, 3, 6)$

12. $\int_C xz \, ds, \quad C: x = 6t, \, y = 3 \sqrt{2} t^2, \, z = 2t^3, \, 0 \leqslant t \leqslant 1$

13. $\int_C x^3 y^2 z \, dz, \quad C: x = 2t, \, y = t^2, \, z = t^2, \, 0 \leqslant t \leqslant 1$

14. $\int_C yz \, dy + xy \, dz, \quad C: x = \sqrt{t}, \, y = t, \, z = t^2, \, 0 \leqslant t \leqslant 1$

15. $\int_C z^2 \, dx - z \, dy + 2y \, dz,$
C consists of line segments from $(0, 0, 0)$ to $(0, 1, 1)$, from $(0, 1, 1)$ to $(1, 2, 3)$, and from $(1, 2, 3)$ to $(1, 2, 4)$

16. $\int_C yz \, dx + xz \, dy + xy \, dz,$
C consists of line segments from $(0, 0, 0)$ to $(2, 0, 0)$, from $(2, 0, 0)$ to $(1, 3, -1)$, and from $(1, 3, -1)$ to $(1, 3, 0)$

17–20 ■ Evaluate the line integral $\int_C \mathbf{F} \cdot d\mathbf{r}$, where C is given by the vector function $\mathbf{r}(t)$.

17. $\mathbf{F}(x, y) = x^2 y \, \mathbf{i} - xy \, \mathbf{j},$
$\mathbf{r}(t) = t^3 \mathbf{i} + t^4 \mathbf{j}, \quad 0 \leqslant t \leqslant 1$

18. $\mathbf{F}(x, y, z) = (y + z) \mathbf{i} - x^2 \mathbf{j} - 4y^2 \mathbf{k},$
$\mathbf{r}(t) = t \mathbf{i} + t^2 \mathbf{j} + t^4 \mathbf{k}, \quad 0 \leqslant t \leqslant 1$

19. $\mathbf{F}(x, y, z) = \sin x \, \mathbf{i} + \cos y \, \mathbf{j} + xz \, \mathbf{k},$
$\mathbf{r}(t) = t^3 \mathbf{i} - t^2 \mathbf{j} + t \mathbf{k}, \quad 0 \leqslant t \leqslant 1$

20. $\mathbf{F}(x, y, z) = x^2 \mathbf{i} + xy \mathbf{j} + z^2 \mathbf{k},$
$\mathbf{r}(t) = \sin t \, \mathbf{i} + \cos t \, \mathbf{j} + t^2 \mathbf{k}, \quad 0 \leqslant t \leqslant \pi/2$

21. (a) Evaluate the line integral $\int_C \mathbf{F} \cdot d\mathbf{r}$, where $\mathbf{F}(x, y) = e^{x-1} \mathbf{i} + xy \mathbf{j}$ and C is given by $\mathbf{r}(t) = t^2 \mathbf{i} + t^3 \mathbf{j}, \, 0 \leqslant t \leqslant 1$.

(b) Illustrate part (a) by using a graphing calculator or computer to graph C and the vectors from the vector field corresponding to $t = 0, 1/\sqrt{2}$, and 1 (as in Figure 13).

22. (a) Evaluate the line integral $\int_C \mathbf{F} \cdot d\mathbf{r}$, where $\mathbf{F}(x, y, z) = x \mathbf{i} - z \mathbf{j} + y \mathbf{k}$ and C is given by $\mathbf{r}(t) = 2t \mathbf{i} + 3t \mathbf{j} - t^2 \mathbf{k}, \, -1 \leqslant t \leqslant 1$.

(b) Illustrate part (a) by using a computer to graph C and the vectors from the vector field corresponding to $t = \pm 1$ and $\pm \frac{1}{2}$ (as in Figure 13).

23–24 ■ If your calculator or CAS evaluates integrals numerically, use it to find the integral to three decimal places. Otherwise, use the Midpoint Rule with $n = 4$ to approximate the integral.

23. $\int x \sin y \, ds, \quad C: x = \ln t, \, y = e^{-t}, \, 1 \leqslant t \leqslant 2$

24. $\int z^2 \ln(1 + x^2 + y^2) \, ds,$
$C: x = t, \, y = t^2, \, z = t^3, \, 0 \leqslant t \leqslant 1$

25. Find the exact value of $\int_C x^3 y^5 \, ds$, where C is the part of the astroid $x = \cos^3 t, \, y = \sin^3 t$ in the first quadrant.

26. Find the exact value of $\int_C \mathbf{F} \cdot d\mathbf{r}$, where $\mathbf{F}(x, y, z) = x^4 e^y \mathbf{i} + \ln z \, \mathbf{j} + \sqrt{y^2 + z^2} \, \mathbf{k}$ and C is the line segment from $(1, 2, 1)$ to $(6, 4, 5)$.

27. Let $\mathbf{F}$ be the vector field shown in the figure.
(a) If C_1 is the vertical line segment from $(-3, -3)$ to $(-3, 3)$, determine whether $\int_{C_1} \mathbf{F} \cdot d\mathbf{r}$ is positive, negative, or zero.
(b) If C_2 is the counterclockwise-oriented circle with radius 3 and center the origin, determine whether $\int_{C_2} \mathbf{F} \cdot d\mathbf{r}$ is positive, negative, or zero.

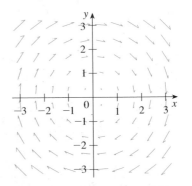

28. (a) Find the work done by the force field $\mathbf{F}(x, y) = x^2 \mathbf{i} + xy \mathbf{j}$ on a particle that moves once around the circle $x^2 + y^2 = 4$ oriented in the counterclockwise direction.
(b) Use a computer algebra system to graph the force field and circle on the same screen. Use the graph to explain your answer to part (a).

29. A thin wire is bent into the shape of a semicircle $x^2 + y^2 = 4, \, x \geqslant 0$. If the linear density is a constant k, find the mass and center of mass of the wire.

30. Find the mass and center of mass of a thin wire in the shape of a quarter-circle $x^2 + y^2 = r^2, \, x \geqslant 0, \, y \geqslant 0$, if the density function is $\rho(x, y) = x + y$.

31. (a) Write the formulas analogous to Equations 4 for the center of mass $(\bar{x}, \bar{y}, \bar{z})$ of a thin wire with density function $\rho(x, y, z)$ in the shape of a space curve C.
    (b) Find the center of mass of a wire in the shape of the helix $x = 2 \sin t$, $y = 2 \cos t$, $z = 3t$, $0 \le t \le 2\pi$, if the density is a constant k.

32. Find the mass and center of mass of a wire in the shape of the helix $x = t$, $y = \cos t$, $z = \sin t$, $0 \le t \le 2\pi$, if the density at any point is equal to the square of the distance from the origin.

33. If a wire with linear density $\rho(x, y)$ lies along a plane curve C, its **moments of inertia** about the x- and y-axes are defined as

$$I_x = \int_C y^2 \rho(x, y)\, ds \qquad I_y = \int_C x^2 \rho(x, y)\, ds$$

Find the moments of inertia for the wire in Exercise 31.

34. If a wire with linear density $\rho(x, y, z)$ lies along a space curve C, its **moments of inertia** about the x-, y-, and z-axes are defined as

$$I_x = \int_C (y^2 + z^2) \rho(x, y, z)\, ds$$

$$I_y = \int_C (x^2 + z^2) \rho(x, y, z)\, ds$$

$$I_z = \int_C (x^2 + y^2) \rho(x, y, z)\, ds$$

Find the moments of inertia for the wire in Example 3.

35. Find the work done by the force field $\mathbf{F}(x, y) = x\,\mathbf{i} + (y + 2)\,\mathbf{j}$ in moving an object along an arch of the cycloid $\mathbf{r}(t) = (t - \sin t)\,\mathbf{i} + (1 - \cos t)\,\mathbf{j}$, $0 \le t \le 2\pi$.

36. Find the work done by the force field $\mathbf{F}(x, y) = x \sin y\,\mathbf{i} + y\,\mathbf{j}$ on a particle that moves along the parabola $y = x^2$ from $(-1, 1)$ to $(2, 4)$.

37. Find the work done by the force field $\mathbf{F}(x, y, z) = xz\,\mathbf{i} + yx\,\mathbf{j} + zy\,\mathbf{k}$ on a particle that moves along the curve $\mathbf{r}(t) = t^2\,\mathbf{i} - t^3\,\mathbf{j} + t^4\,\mathbf{k}$, $0 \le t \le 1$.

38. The force exerted by an electric charge at the origin on a charged particle at a point (x, y, z) with position vector $\mathbf{r} = \langle x, y, z \rangle$ is $\mathbf{F}(\mathbf{r}) = K\mathbf{r}/|\mathbf{r}|^3$ where K is a constant. (See Example 5 in Section 14.1.) Find the work done as the particle moves along a straight line from $(2, 0, 0)$ to $(2, 1, 5)$.

39. A 160-lb man carries a 25-lb can of paint up a helical staircase that encircles a silo with a radius of 20 ft. If the silo is 90 ft high and the man makes exactly three complete revolutions, how much work is done by the man against gravity in climbing to the top?

40. Suppose there is a hole in the can of paint in Exercise 39 and 9 lb of paint leak steadily out of the can during the man's ascent. How much work is done?

41. Experiments show that a steady current I in a long wire produces a magnetic field $\mathbf{B}$ that is tangent to any circle that lies in the plane perpendicular to the wire and whose center is the axis of the wire (as in the figure). *Ampère's Law* relates the electric current to its magnetic effects and states that

$$\int_C \mathbf{B} \cdot d\mathbf{r} = \mu_0 I$$

where I is the net current that passes through any surface bounded by a closed curve C and μ_0 is a constant called the permeability of free space. By taking C to be a circle with radius r, show that the magnitude $B = |\mathbf{B}|$ of the magnetic field at a distance r from the center of the wire is

$$B = \frac{\mu_0 I}{2\pi r}$$

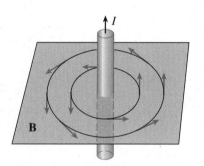

14.3   THE FUNDAMENTAL THEOREM FOR LINE INTEGRALS

Recall from Section 4.4 that Part 2 of the Fundamental Theorem of Calculus can be written as

(1)
$$\int_a^b F'(x)\, dx = F(b) - F(a)$$

where F' is continuous on $[a, b]$. If we think of the gradient vector ∇f of a function f of two or three variables as a sort of derivative of f, then the following theorem can be regarded as a version of the fundamental theorem for line integrals.

> **(2) THEOREM**   Let C be a smooth curve given by the vector function $\mathbf{r}(t)$, $a \leqslant t \leqslant b$. Let f be a differentiable function of two or three variables whose gradient vector ∇f is continuous on C. Then
>
> $$\int_C \nabla f \cdot d\mathbf{r} = f(\mathbf{r}(b)) - f(\mathbf{r}(a))$$

NOTE: Theorem 2 says that we can evaluate the line integral of a conservative vector field (the gradient vector field of the potential function f) simply by knowing the value of f at the endpoints of C. If f is a function of two variables and C is a plane curve with initial point $A(x_1, y_1)$ and terminal point $B(x_2, y_2)$, as in Figure 1, then Theorem 2 becomes

$$\int_C \nabla f \cdot d\mathbf{r} = f(x_2, y_2) - f(x_1, y_1)$$

If f is a function of three variables and C is a space curve joining $A(x_1, y_1, z_1)$ to $B(x_2, y_2, z_2)$, then we have

$$\int_C \nabla f \cdot d\mathbf{r} = f(x_2, y_2, z_2) - f(x_1, y_1, z_1)$$

Let us prove Theorem 2 for the latter case.

PROOF OF THEOREM 2   Using Definition 14.2.13, we have

$$\int_C \nabla f \cdot d\mathbf{r} = \int_a^b \nabla f(\mathbf{r}(t)) \cdot \mathbf{r}'(t)\, dt$$

$$= \int_a^b \left(\frac{\partial f}{\partial x} \frac{dx}{dt} + \frac{\partial f}{\partial y} \frac{dy}{dt} + \frac{\partial f}{\partial z} \frac{dz}{dt} \right) dt$$

$$= \int_a^b \frac{d}{dt} f(\mathbf{r}(t))\, dt \qquad \text{(by the Chain Rule)}$$

$$= f(\mathbf{r}(b)) - f(\mathbf{r}(a))$$

The last step follows from the Fundamental Theorem of Calculus (Equation 1).   □

Although we have proved Theorem 2 for smooth curves, it is also true for piecewise-smooth curves. This can be seen by subdividing C into a finite number of smooth curves and adding the resulting integrals.

EXAMPLE 1   Find the work done by the gravitational field

$$\mathbf{F}(\mathbf{x}) = -\frac{mMG}{|\mathbf{x}|^3} \mathbf{x}$$

in moving a particle with mass m from the point $(3, 4, 12)$ to the point $(2, 2, 0)$ along a piecewise-smooth curve C. (See Example 4 in Section 14.1.)

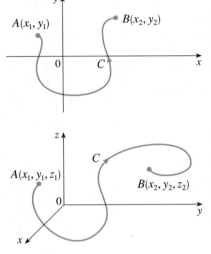

FIGURE 1

SOLUTION   From Section 14.1 we know that **F** is a conservative vector field and, in fact, $\mathbf{F} = \nabla f$, where

$$f(x, y, z) = \frac{mMG}{\sqrt{x^2 + y^2 + z^2}}$$

Therefore, by Theorem 2, the work done is

$$W = \int_C \mathbf{F} \cdot d\mathbf{r} = \int_C \nabla f \cdot d\mathbf{r}$$

$$= f(2, 2, 0) - f(3, 4, 12)$$

$$= \frac{mMG}{\sqrt{2^2 + 2^2}} - \frac{mMG}{\sqrt{3^2 + 4^2 + 12^2}} = mMG\left(\frac{1}{2\sqrt{2}} - \frac{1}{13}\right) \qquad \blacksquare$$

INDEPENDENCE OF PATH

Suppose C_1 and C_2 are two piecewise-smooth curves (which are called **paths**) that have the same initial point A and terminal point B. We know from Example 4 in Section 14.2 that, in general, $\int_{C_1} \mathbf{F} \cdot d\mathbf{r} \neq \int_{C_2} \mathbf{F} \cdot d\mathbf{r}$. But one implication of Theorem 2 is that $\int_{C_1} \nabla f \cdot d\mathbf{r} = \int_{C_2} \nabla f \cdot d\mathbf{r}$ whenever ∇f is continuous. In other words, the line integral of a conservative vector field depends only on the initial point and terminal point of a curve.

In general, if **F** is a continuous vector field with domain D, we say that the line integral $\int_C \mathbf{F} \cdot d\mathbf{r}$ is **independent of path** if $\int_{C_1} \mathbf{F} \cdot d\mathbf{r} = \int_{C_2} \mathbf{F} \cdot d\mathbf{r}$ for any two paths C_1 and C_2 in D that have the same initial and terminal points. With this terminology we can say that **line integrals of conservative vector fields are independent of path.**

A curve is called **closed** if its terminal point coincides with its initial point, that is, $\mathbf{r}(b) = \mathbf{r}(a)$ (see Figure 2). If $\int_C \mathbf{F} \cdot d\mathbf{r}$ is independent of path in D and C is any closed path in D, we can choose any two points A and B on C and regard C as being composed of the path C_1 from A to B followed by the path C_2 from B to A (see Figure 3). Then

$$\int_C \mathbf{F} \cdot d\mathbf{r} = \int_{C_1} \mathbf{F} \cdot d\mathbf{r} + \int_{C_2} \mathbf{F} \cdot d\mathbf{r} = \int_{C_1} \mathbf{F} \cdot d\mathbf{r} - \int_{-C_2} \mathbf{F} \cdot d\mathbf{r} = 0$$

since C_1 and $-C_2$ have the same initial and terminal points.

Conversely, if it is true that $\int_C \mathbf{F} \cdot d\mathbf{r} = 0$ whenever C is a closed path in D, then we demonstrate independence of path as follows. Take any two paths C_1 and C_2 from A to B in D and define C to be the curve consisting of C_1 followed by $-C_2$. Then

$$0 = \int_C \mathbf{F} \cdot d\mathbf{r} = \int_{C_1} \mathbf{F} \cdot d\mathbf{r} + \int_{-C_2} \mathbf{F} \cdot d\mathbf{r} = \int_{C_1} \mathbf{F} \cdot d\mathbf{r} - \int_{C_2} \mathbf{F} \cdot d\mathbf{r}$$

and so $\int_{C_1} \mathbf{F} \cdot d\mathbf{r} = \int_{C_2} \mathbf{F} \cdot d\mathbf{r}$. Thus we have proved the following theorem:

> **(3) THEOREM** $\int_C \mathbf{F} \cdot d\mathbf{r}$ is independent of path in D if and only if $\int_C \mathbf{F} \cdot d\mathbf{r} = 0$ for every closed path C in D.

Since we know that the line integral of any conservative vector field **F** is independent of path, it follows that $\int_C \mathbf{F} \cdot d\mathbf{r} = 0$ for closed paths. The physical interpretation

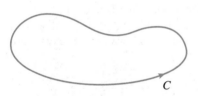

FIGURE 2
A closed curve

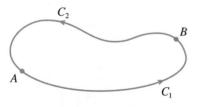

FIGURE 3

is that the work done by a conservative force field (such as the gravitational or electric field in Section 14.1) as it moves an object around a closed path is 0.

The following theorem says that the *only* vector fields that are independent of path are conservative. It is stated and proved for plane curves, but there is a similar version for space curves. We assume that D is **open,** which means that for every point P in D there is a disk with center P that lies entirely in D. (Thus every point in D is an interior point in the sense of Section 12.2.) In addition, we assume that D is **connected.** This means that any two points in D can be joined by a path that lies in D.

---

(4) THEOREM　Suppose **F** is a vector field that is continuous on an open connected region D. If $\int_C \mathbf{F} \cdot d\mathbf{r}$ is independent of path in D, then **F** is a conservative vector field on D; that is, there exists a function f such that $\nabla f = \mathbf{F}$.

PROOF　Let $A(a, b)$ be a fixed point in D. We construct the desired potential function f by defining

$$f(x, y) = \int_{(a, b)}^{(x, y)} \mathbf{F} \cdot d\mathbf{r}$$

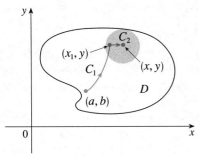

FIGURE 4

for any point (x, y) in D. Since $\int_C \mathbf{F} \cdot d\mathbf{r}$ is independent of path, it does not matter which path C from (a, b) to (x, y) is used to evaluate $f(x, y)$. Since D is open, there exists a disk contained in D with center (x, y). Choose any point (x_1, y) in the disk with $x_1 < x$ and let C consist of any path C_1 from (a, b) to (x_1, y) followed by the horizontal line segment C_2 from (x_1, y) to (x, y) (see Figure 4). Then

$$f(x, y) = \int_{C_1} \mathbf{F} \cdot d\mathbf{r} + \int_{C_2} \mathbf{F} \cdot d\mathbf{r} = \int_{(a, b)}^{(x_1, y)} \mathbf{F} \cdot d\mathbf{r} + \int_{C_2} \mathbf{F} \cdot d\mathbf{r}$$

Notice that the first of these integrals does not depend on x, so

$$\frac{\partial}{\partial x} f(x, y) = 0 + \frac{\partial}{\partial x} \int_{C_2} \mathbf{F} \cdot d\mathbf{r}$$

If we write $\mathbf{F} = P\,\mathbf{i} + Q\,\mathbf{j}$, then

$$\int_{C_2} \mathbf{F} \cdot d\mathbf{r} = \int_{C_2} P\, dx + Q\, dy$$

On C_2, y is constant, so $dy = 0$. Using t as the parameter, where $x_1 \leq t \leq x$, we have

$$\frac{\partial}{\partial x} f(x, y) = \frac{\partial}{\partial x} \int_{C_2} P\, dx + Q\, dy$$

$$= \frac{\partial}{\partial x} \int_{x_1}^{x} P(t, y)\, dt = P(x, y)$$

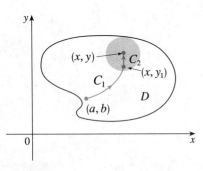

FIGURE 5

by Part 1 of the Fundamental Theorem of Calculus (4.4.2). A similar argument, using a vertical line segment (see Figure 5), shows that

$$\frac{\partial}{\partial y} f(x, y) = \frac{\partial}{\partial y} \int_{C_2} P\, dx + Q\, dy = \frac{\partial}{\partial y} \int_{y_1}^{y} Q(x, t)\, dt = Q(x, y)$$

Thus
$$\mathbf{F} = P\,\mathbf{i} + Q\,\mathbf{j} = \frac{\partial f}{\partial x}\,\mathbf{i} + \frac{\partial f}{\partial y}\,\mathbf{j} = \nabla f$$

which says that **F** is conservative. ◻

The question remains: How is it possible to determine whether or not a vector field **F** is conservative? Suppose it is known that $\mathbf{F} = P\,\mathbf{i} + Q\,\mathbf{j}$ is conservative, where P and Q have continuous first-order partial derivatives. Then there is a function f such that $\mathbf{F} = \nabla f$, that is,

$$P = \frac{\partial f}{\partial x} \quad \text{and} \quad Q = \frac{\partial f}{\partial y}$$

Therefore, by Clairaut's Theorem,

$$\frac{\partial P}{\partial y} = \frac{\partial^2 f}{\partial y\,\partial x} = \frac{\partial^2 f}{\partial x\,\partial y} = \frac{\partial Q}{\partial x}$$

> **(5) THEOREM** If $\mathbf{F}(x, y) = P(x, y)\,\mathbf{i} + Q(x, y)\,\mathbf{j}$ is a conservative vector field, where P and Q have continuous first-order partial derivatives on a domain D, then throughout D we have
> $$\frac{\partial P}{\partial y} = \frac{\partial Q}{\partial x}$$

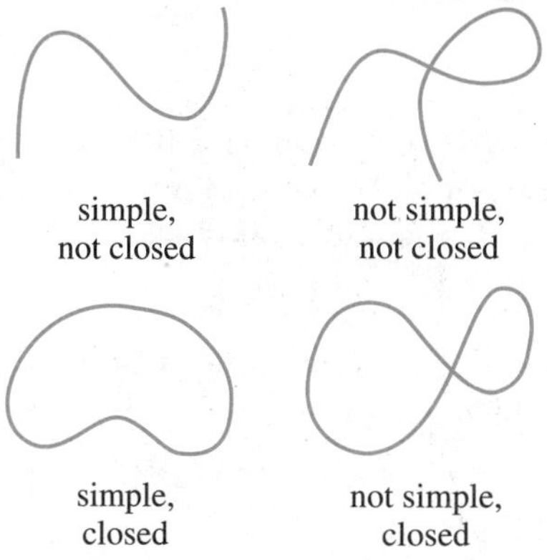

simple,   not simple,
not closed   not closed

simple,   not simple,
closed   closed

FIGURE 6

Types of curves

The converse of Theorem 5 is true only for a special type of region. To explain this, we first need the concept of a **simple curve,** which is a curve that does not intersect itself anywhere between its endpoints. [See Figure 6; $\mathbf{r}(a) = \mathbf{r}(b)$ for a simple closed curve, but $\mathbf{r}(t_1) \neq \mathbf{r}(t_2)$ when $a < t_1 < t_2 < b$.]

In Theorem 4 we needed an open connected region. For the next theorem we need a stronger condition. A **simply-connected region** in the plane is a connected region D such that every simple closed curve in D encloses only points that are in D. Notice from Figure 7 that, intuitively speaking, a simply-connected region contains no hole and cannot consist of two separate pieces.

In terms of simply-connected regions we can now state a partial converse to Theorem 5 that gives a convenient method for verifying that a vector field on $\mathbb{R}^2$ is conservative. The proof will be sketched in the next section as a consequence of Green's Theorem.

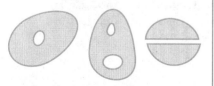

simply-connected region

regions that are not simply-connected

FIGURE 7

> **(6) THEOREM** Let $\mathbf{F} = P\,\mathbf{i} + Q\,\mathbf{j}$ be a vector field on an open simply-connected region D. Suppose that P and Q have continuous first-order derivatives and
> $$\frac{\partial P}{\partial y} = \frac{\partial Q}{\partial x} \qquad \text{throughout } D$$
> Then **F** is conservative.

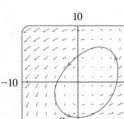

FIGURE 8

Figures 8 and 9 show the vector fields in Examples 2 and 3, respectively. The vectors in Figure 8 that start on the closed curve C all appear to point in roughly the same direction as C. So it looks as if $\int_C \mathbf{F} \cdot d\mathbf{r} > 0$ and therefore $\mathbf{F}$ is not conservative. The calculation in Example 2 confirms this impression. Some of the vectors near the curves C_1 and C_2 in Figure 9 point in approximately the same direction as the curves, whereas others point in the opposite direction. So it appears plausible that line integrals around all closed paths are 0. Example 3 shows that $\mathbf{F}$ is indeed conservative.

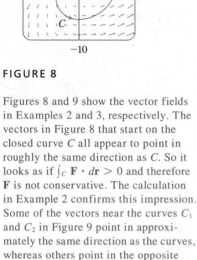

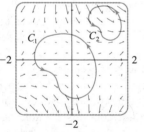

FIGURE 9

EXAMPLE 2   Determine whether or not the vector field

$$\mathbf{F}(x, y) = (x - y)\,\mathbf{i} + (x - 2)\,\mathbf{j}$$

is conservative.

SOLUTION   Let $P(x, y) = x - y$ and $Q(x, y) = x - 2$. Then

$$\frac{\partial P}{\partial y} = -1 \qquad \frac{\partial Q}{\partial x} = 1$$

Since $\partial P/\partial y \neq \partial Q/\partial x$, $\mathbf{F}$ is not conservative by Theorem 5.   ∎

EXAMPLE 3   Determine whether or not the vector field

$$\mathbf{F}(x, y) = (3 + 2xy)\,\mathbf{i} + (x^2 - 3y^2)\,\mathbf{j}$$

is conservative.

SOLUTION   Let $P(x, y) = 3 + 2xy$ and $Q(x, y) = x^2 - 3y^2$. Then

$$\frac{\partial P}{\partial y} = 2x = \frac{\partial Q}{\partial x}$$

Also the domain of $\mathbf{F}$ is the entire plane ($D = \mathbb{R}^2$), which is open and simply-connected. Therefore, we can apply Theorem 6 and conclude that $\mathbf{F}$ is conservative.   ∎

In Example 3, Theorem 6 told us that $\mathbf{F}$ is conservative, but it did not tell us how to find the (potential) function f such that $\mathbf{F} = \nabla f$. The proof of Theorem 4 gives us a clue as to how to find f. We use "partial integration" as in the following example.

EXAMPLE 4
(a) If $\mathbf{F}(x, y) = (3 + 2xy)\,\mathbf{i} + (x^2 - 3y^2)\,\mathbf{j}$, find a function f such that $\mathbf{F} = \nabla f$.
(b) Evaluate the line integral $\int_C \mathbf{F} \cdot d\mathbf{r}$, where C is the curve given by
$\mathbf{r}(t) = e^t \sin t\,\mathbf{i} + e^t \cos t\,\mathbf{j}$, $0 \leq t \leq \pi$.

SOLUTION
(a) From Example 3 we know that $\mathbf{F}$ is conservative and so there exists a function f with $\nabla f = \mathbf{F}$, that is,

(7) $$f_x(x, y) = 3 + 2xy$$

(8) $$f_y(x, y) = x^2 - 3y^2$$

Integrating (7) with respect to x, we obtain

(9) $$f(x, y) = 3x + x^2 y + g(y)$$

Notice that the constant of integration is a constant with respect to x, that is, a function of y, which we have called $g(y)$. Next we differentiate both sides of (9) with respect to y:

(10) $$f_y(x, y) = x^2 + g'(y)$$

Comparing (8) and (10), we see that

$$g'(y) = -3y^2$$

Integrating with respect to y, we have

$$g(y) = -y^3 + K$$

where K is a constant. Putting this in (9), we have

$$f(x, y) = 3x + x^2 y - y^3 + K$$

as the desired potential function.

(b) To use Theorem 2 all we have to know are the initial and terminal points of C, namely, $\mathbf{r}(0) = (0, 1)$ and $\mathbf{r}(\pi) = (0, -e^{\pi})$. In the expression for $f(x, y)$ in part (a), any value of the constant K will do, so let us choose $K = 0$. Then we have

$$\int_C \mathbf{F} \cdot d\mathbf{r} = \int_C \nabla f \cdot d\mathbf{r} = f(0, -e^{\pi}) - f(0, 1)$$

$$= e^{3\pi} - (-1) = e^{3\pi} + 1$$

This method is much shorter than the straightforward method for evaluating line integrals that we learned in Section 14.2. ∎

A criterion for determining whether or not a vector field $\mathbf{F}$ on $\mathbb{R}^3$ is conservative is given in Section 14.5. Meanwhile, the next example shows that the technique for finding the potential function is much the same as for vector fields on $\mathbb{R}^2$.

EXAMPLE 5   If $\mathbf{F}(x, y, z) = y^2 \mathbf{i} + (2xy + e^{3z}) \mathbf{j} + 3ye^{3z} \mathbf{k}$, find a function f such that $\nabla f = \mathbf{F}$.

SOLUTION   If there is such a function f, then

(11) $$f_x(x, y, z) = y^2$$

(12) $$f_y(x, y, z) = 2xy + e^{3z}$$

(13) $$f_z(x, y, z) = 3ye^{3z}$$

Integrating (11) with respect to x, we get

(14) $$f(x, y, z) = xy^2 + g(y, z)$$

where $g(y, z)$ is a constant with respect to x. Then differentiating (14) with respect to y, we have

$$f_y(x, y, z) = 2xy + g_y(y, z)$$

and comparison with (12) gives

$$g_y(y, z) = e^{3z}$$

Thus $g(y, z) = ye^{3z} + h(z)$ and we rewrite (14) as

$$f(x, y, z) = xy^2 + ye^{3z} + h(z)$$

Finally, differentiating with respect to z and comparing with (13), we obtain $h'(z) = 0$ and, therefore, $h(z) = K$, a constant. The desired function is

$$f(x, y, z) = xy^2 + ye^{3z} + K$$

It is easily verified that $\nabla f = \mathbf{F}$. ∎

CONSERVATION OF ENERGY

Let us apply the ideas of this chapter to a continuous force field $\mathbf{F}$ that moves an object along a path C given by $\mathbf{r}(t)$, $a \leq t \leq b$, where $\mathbf{r}(a) = A$ is the initial point and $\mathbf{r}(b) = B$ is the terminal point of C. According to Newton's Second Law of Motion (see Section 11.9), the force $\mathbf{F}(\mathbf{r}(t))$ at a point on C is related to the acceleration $\mathbf{a}(t) = \mathbf{r}''(t)$ by the equation

$$\mathbf{F}(\mathbf{r}(t)) = m\mathbf{r}''(t)$$

So the work done by the force on the object is

$$W = \int_C \mathbf{F} \cdot d\mathbf{r} = \int_a^b \mathbf{F}(\mathbf{r}(t)) \cdot \mathbf{r}'(t)\, dt$$

$$= \int_a^b m\mathbf{r}''(t) \cdot \mathbf{r}'(t)\, dt$$

$$= \frac{m}{2} \int_a^b \frac{d}{dt}[\mathbf{r}'(t) \cdot \mathbf{r}'(t)]\, dt \qquad \text{(Theorem 11.7.5, Formula 4)}$$

$$= \frac{m}{2} \int_a^b \frac{d}{dt}|\mathbf{r}'(t)|^2\, dt$$

$$= \frac{m}{2}\Big[|\mathbf{r}'(t)|^2\Big]_a^b \qquad \text{(Fundamental Theorem of Calculus)}$$

$$= \frac{m}{2}(|\mathbf{r}'(b)|^2 - |\mathbf{r}'(a)|^2)$$

Therefore

(15) $$W = \tfrac{1}{2}m|\mathbf{v}(b)|^2 - \tfrac{1}{2}m|\mathbf{v}(a)|^2$$

where $\mathbf{v} = \mathbf{r}'$ is the velocity.

The quantity $\frac{1}{2}m|\mathbf{v}(t)|^2$, that is, half the mass times the square of the speed, is called the **kinetic energy** of the object. Therefore, we can rewrite Equation 15 as

(16) $$W = K(B) - K(A)$$

which says that the work done by the force field along C is equal to the change in kinetic energy at the endpoints of C.

Now let us further assume that $\mathbf{F}$ is a conservative force field; that is, we can write $\mathbf{F} = \nabla f$. In physics, the **potential energy** of an object at the point (x, y, z) is defined as $P(x, y, z) = -f(x, y, z)$, so we have $\mathbf{F} = -\nabla P$. Then by Theorem 2 we have

$$W = \int_C \mathbf{F} \cdot d\mathbf{r} = -\int_C \nabla P \cdot d\mathbf{r}$$

$$= -[P(\mathbf{r}(b)) - P(\mathbf{r}(a))]$$

$$= P(A) - P(B)$$

Comparing this equation with Equation 16, we see that

$$P(A) + K(A) = P(B) + K(B)$$

which says that if an object moves from one point A to another point B under the influence of a conservative force field, then the sum of its potential energy and its kinetic energy remains constant. This is called the **Law of Conservation of Energy** and it is the reason the vector field is called **conservative.**

EXERCISES 14.3

1–10 ■ Determine whether or not **F** is a conservative vector field. If it is, find a function f such that $\mathbf{F} = \nabla f$.

1. $\mathbf{F}(x, y) = (2x - 3y)\,\mathbf{i} + (2y - 3x)\,\mathbf{j}$

2. $\mathbf{F}(x, y) = (3x^2 - 4y)\,\mathbf{i} + (4y^2 - 2x)\,\mathbf{j}$

3. $\mathbf{F}(x, y) = (x^2 + y)\,\mathbf{i} + x^2\,\mathbf{j}$

4. $\mathbf{F}(x, y) = (x^2 + y)\,\mathbf{i} + (y^2 + x)\,\mathbf{j}$

5. $\mathbf{F}(x, y) = (1 + 4x^3y^3)\,\mathbf{i} + 3x^4y^2\,\mathbf{j}$

6. $\mathbf{F}(x, y) = (y\cos x - \cos y)\,\mathbf{i} + (\sin x + x\sin y)\,\mathbf{j}$

7. $\mathbf{F}(x, y) = (e^{2x} + x\sin y)\,\mathbf{i} + x^2\cos y\,\mathbf{j}$

8. $\mathbf{F}(x, y) = (ye^{xy} + 4x^3y)\,\mathbf{i} + (xe^{xy} + x^4)\,\mathbf{j}$

9. $\mathbf{F}(x, y) = (ye^x + \sin y)\,\mathbf{i} + (e^x + x\cos y)\,\mathbf{j}$

10. $\mathbf{F}(x, y) = (x + y^2)\,\mathbf{i} + (2xy + y^2)\,\mathbf{j}$

11–18 ■ (a) Find a function f such that $\mathbf{F} = \nabla f$ and (b) use part (a) to evaluate $\int_C \mathbf{F} \cdot d\mathbf{r}$ along the given curve C.

11. $\mathbf{F}(x, y) = x\,\mathbf{i} + y\,\mathbf{j}$,
C is the arc of the parabola $y = x^2$ from $(-1, 1)$ to $(3, 9)$

12. $\mathbf{F}(x, y) = y\,\mathbf{i} + x\,\mathbf{j}$,
C is the arc of the curve $y = x^4 - x^3$ from $(1, 0)$ to $(2, 8)$

13. $\mathbf{F}(x, y) = 2xy^3\,\mathbf{i} + 3x^2y^2\,\mathbf{j}$,
C: $\mathbf{r}(t) = \sin t\,\mathbf{i} + (t^2 + 1)\,\mathbf{j}$, $0 \le t \le \pi/2$

14. $\mathbf{F}(x, y) = e^{2y}\,\mathbf{i} + (1 + 2xe^{2y})\,\mathbf{j}$,
C: $\mathbf{r}(t) = te^t\,\mathbf{i} + (1 + t)\,\mathbf{j}$, $0 \le t \le 1$

15. $\mathbf{F}(x, y, z) = y\,\mathbf{i} + (x + z)\,\mathbf{j} + y\,\mathbf{k}$,
C is the line segment from $(2, 1, 4)$ to $(8, 3, -1)$

16. $\mathbf{F}(x, y, z) = 2xy^3z^4\,\mathbf{i} + 3x^2y^2z^4\,\mathbf{j} + 4x^2y^3z^3\,\mathbf{k}$,
C: $x = t$, $y = t^2$, $z = t^3$, $0 \le t \le 2$

17. $\mathbf{F}(x, y, z) = (2xz + \sin y)\,\mathbf{i} + x\cos y\,\mathbf{j} + x^2\,\mathbf{k}$,
C: $\mathbf{r}(t) = \cos t\,\mathbf{i} + \sin t\,\mathbf{j} + t\,\mathbf{k}$, $0 \le t \le 2\pi$

18. $\mathbf{F}(x, y, z) = 4xe^z\,\mathbf{i} + \cos y\,\mathbf{j} + 2x^2e^z\,\mathbf{k}$,
C: $\mathbf{r}(t) = t\,\mathbf{i} + t^2\,\mathbf{j} + t^4\,\mathbf{k}$, $0 \le t \le 1$

19–20 ■ Show that the line integral is independent of path and evaluate the integral.

19. $\int_C 2x\sin y\,dx + (x^2\cos y - 3y^2)\,dy$,
C is any path from $(-1, 0)$ to $(5, 1)$

20. $\int_C (2y^2 - 12x^3y^3)\,dx + (4xy - 9x^4y^2)\,dy$,
C is any path from $(1, 1)$ to $(3, 2)$

21–22 ■ Find the work done by the force field **F** in moving an object from P to Q.

21. $\mathbf{F}(x, y) = x^2y^3\,\mathbf{i} + x^3y^2\,\mathbf{j}$; $P(0, 0)$, $Q(2, 1)$

22. $\mathbf{F}(x, y) = (y^2/x^2)\,\mathbf{i} - (2y/x)\,\mathbf{j}$; $P(1, 1)$, $Q(4, -2)$

23. Is the vector field shown in the figure conservative? Explain.

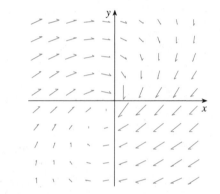

CAS **24–25** ■ From a plot of **F** guess whether it is conservative. Then determine whether your guess is correct.

24. $\mathbf{F}(x, y) = (2xy + \sin y)\,\mathbf{i} + (x^2 + x\cos y)\,\mathbf{j}$

25. $\mathbf{F}(x, y) = \dfrac{(x - 2y)\,\mathbf{i} + (x - 2)\,\mathbf{j}}{\sqrt{1 + x^2 + y^2}}$

26. Let $\mathbf{F} = \nabla f$, where $f(x, y) = \sin(x - 2y)$. Find curves C_1 and C_2 that are not closed and satisfy the equation.

(a) $\displaystyle\int_{C_1} \mathbf{F} \cdot d\mathbf{r} = 0$ (b) $\displaystyle\int_{C_2} \mathbf{F} \cdot d\mathbf{r} = 1$

27. Show that if the vector field $\mathbf{F} = P\,\mathbf{i} + Q\,\mathbf{j} + R\,\mathbf{k}$ is conservative and P, Q, R have continuous first-order partial derivatives, then

$$\frac{\partial P}{\partial y} = \frac{\partial Q}{\partial x} \qquad \frac{\partial P}{\partial z} = \frac{\partial R}{\partial x} \qquad \frac{\partial Q}{\partial z} = \frac{\partial R}{\partial y}$$

28. Use Exercise 27 to show that the line integral $\int_C y\,dx + x\,dy + xyz\,dz$ is not independent of path.

29–32 ■ Determine whether or not the given set is (a) open, (b) connected, and (c) simply-connected.

29. $\{(x, y) \mid x > 0, y > 0\}$

30. $\{(x, y) \mid x \ne 0\}$

31. $\{(x, y) \mid 1 < x^2 + y^2 < 4\}$

32. $\{(x, y) \mid x^2 + y^2 \le 1 \text{ or } 4 \le x^2 + y^2 \le 9\}$

33. Let $\mathbf{F}(x, y) = \dfrac{-y\,\mathbf{i} + x\,\mathbf{j}}{x^2 + y^2}$.

  (a) Show that $\partial P/\partial y = \partial Q/\partial x$.

  (b) Show that $\int_C \mathbf{F} \cdot d\mathbf{r}$ is not independent of path. [*Hint:* Compute $\int_{C_1} \mathbf{F} \cdot d\mathbf{r}$ and $\int_{C_2} \mathbf{F} \cdot d\mathbf{r}$, where C_1 and C_2 are the upper and lower halves of the circle $x^2 + y^2 = 1$ from $(1, 0)$ to $(-1, 0)$.] Does this contradict Theorem 6?

34. (a) Suppose that $\mathbf{F}$ is an inverse square force field, that is,

$$\mathbf{F}(\mathbf{r}) = \frac{c\mathbf{r}}{|\mathbf{r}|^3}$$

  for some constant c, where $\mathbf{r} = x\,\mathbf{i} + y\,\mathbf{j} + z\,\mathbf{k}$. Find the work done by $\mathbf{F}$ in moving an object from a point P_1 along a path to a point P_2 in terms of the distances d_1 and d_2 from these points to the origin.

  (b) An example of an inverse square field is the gravitational field $\mathbf{F} = -(mMG)\mathbf{r}/|\mathbf{r}|^3$ discussed in Example 4 in Section 14.1. Use part (a) to find the work done by the gravitational field when the earth moves from aphelion (at a maximum distance of 1.52×10^8 km from the sun) to perihelion (at a minimum distance of 1.47×10^8 km). (Use the values $m = 5.97 \times 10^{24}$ kg, $M = 1.99 \times 10^{30}$ kg, and $G = 6.67 \times 10^{-11}$ N·m²/kg².)

  (c) Another example of an inverse square field is the electric field $\mathbf{E} = \varepsilon q Q\mathbf{r}/|\mathbf{r}|^3$ discussed in Example 5 in Section 14.1. Suppose that an electron with a charge of -1.6×10^{-19} C is at the origin. A positive unit charge is at a distance 10^{-12} m from the electron and moves to a position half that distance from the electron. Use part (a) to find the work done by the electric field. (Use the value $\varepsilon = 8.985 \times 10^{10}$.)

14.4   GREEN'S THEOREM

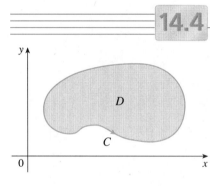

FIGURE 1

Green's Theorem gives the relationship between a line integral around a simple closed curve C and a double integral over the plane region D bounded by C. (See Figure 1. We assume that D consists of all points inside C as well as all points on C.) In stating Green's Theorem we use the convention that the **positive orientation** of a simple closed curve C refers to a single *counterclockwise* traversal of C. Thus if C is given by the vector function $\mathbf{r}(t)$, $a \le t \le b$, then the region D is always on the left as the point $\mathbf{r}(t)$ traverses C (see Figure 2).

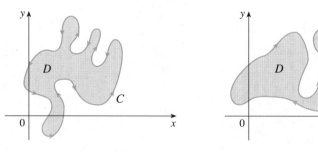

FIGURE 2   (a) Positive orientation   (b) Negative orientation

(1) GREEN'S THEOREM   Let C be a positively oriented, piecewise-smooth, simple closed curve in the plane and let D be the region bounded by C. If P and Q have continuous partial derivatives on an open region that contains D, then

$$\int_C P\,dx + Q\,dy = \iint_D \left(\frac{\partial Q}{\partial x} - \frac{\partial P}{\partial y} \right) dA$$

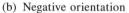

NOTE: The notation

$$\oint_C P\,dx + Q\,dy \qquad \text{or} \qquad \oint_C P\,dx + Q\,dy$$

is sometimes used to indicate that the line integral is calculated using the positive orientation of the closed curve C. Another notation for the positively oriented boundary curve of D is ∂D, so the equation in Green's Theorem can be written as

(2) $$\iint_D \left(\frac{\partial Q}{\partial x} - \frac{\partial P}{\partial y} \right) dA = \int_{\partial D} P\,dx + Q\,dy$$

Green's Theorem should be regarded as the analogue of the Fundamental Theorem of Calculus for double integrals. Compare Equation 2 with the statement of the Fundamental Theorem of Calculus, Part 2, in the following equation:

$$\int_a^b F'(x)\,dx = F(b) - F(a)$$

In both cases there is an integral involving derivatives (F', $\partial Q/\partial x$, and $\partial P/\partial y$) on the left side of the equation. And in both cases the right side involves the values of the original functions (F, Q, and P) only on the *boundary* of the domain. (In the one-dimensional case, the domain is an interval $[a, b]$ whose boundary consists of just two points, a and b.)

Green's Theorem is not easy to prove in the generality stated in Theorem 1, but we can give a proof for the special case where the region is both of type I and of type II (see Section 13.3). Let us call such regions **simple regions.**

PROOF OF GREEN'S THEOREM FOR THE CASE
IN WHICH D IS A SIMPLE REGION

Notice that Green's Theorem will be proved if we can show that

(3) $$\int_C P\,dx = -\iint_D \frac{\partial P}{\partial y}\,dA$$

and

(4) $$\int_C Q\,dy = \iint_D \frac{\partial Q}{\partial x}\,dA$$

We prove Equation 3 by expressing D as a type I region:

$$D = \{(x, y)\,|\,a \leq x \leq b, g_1(x) \leq y \leq g_2(x)\}$$

where g_1 and g_2 are continuous functions. This enables us to compute the double integral on the right side of Equation 3 as follows:

(5) $$\iint_D \frac{\partial P}{\partial y}\,dA = \int_a^b \int_{g_1(x)}^{g_2(x)} \frac{\partial P}{\partial y}(x, y)\,dy\,dx$$

$$= \int_a^b [P(x, g_2(x)) - P(x, g_1(x))]\,dx$$

where the last step follows from the Fundamental Theorem of Calculus.

Green's Theorem is named after the self-taught English scientist George Green (1793–1841). He worked in his father's bakery from an early age and taught himself mathematics from library books. In 1828 he published privately *An Essay on the Application of Mathematical Analysis to the Theories of Electricity and Magnetism,* but only 100 copies were printed and most of those went to friends. What we know as Green's Theorem was in this book, but it did not become widely known at that time. Finally, at age 40, Green entered Cambridge University as an undergraduate but died four years after graduation. In 1846 William Thomson (Lord Kelvin) located a copy of Green's essay, realized its significance, and had it reprinted. Green was the first person to try to formulate a mathematical theory of electricity and magnetism. His work was the basis for the subsequent electromagnetic theories of Thomson, Stokes, Rayleigh, and Maxwell.

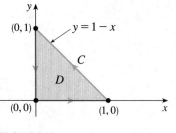

FIGURE 3

Now we compute the left side of Equation 3 by breaking up C as the union of the four curves C_1, C_2, C_3, and C_4 shown in Figure 3. On C_1 we take x as the parameter and write the parametric equations as $x = x$, $y = g_1(x)$, $a \leq x \leq b$. Thus

$$\int_{C_1} P(x, y)\, dx = \int_a^b P(x, g_1(x))\, dx$$

Observe that C_3 goes from right to left but $-C_3$ goes from left to right, so we can write the parametric equations of $-C_3$ as $x = x$, $y = g_2(x)$, $a \leq x \leq b$. Therefore

$$\int_{C_3} P(x, y)\, dx = -\int_{-C_3} P(x, y)\, dx = -\int_a^b P(x, g_2(x))\, dx$$

On C_2 or C_4 (either of which might reduce to just a single point), x is constant, so $dx = 0$ and

$$\int_{C_2} P(x, y)\, dx = 0 = \int_{C_4} P(x, y)\, dx$$

Hence

$$\int_C P(x, y)\, dx = \int_{C_1} P(x, y)\, dx + \int_{C_2} P(x, y)\, dx + \int_{C_3} P(x, y)\, dx + \int_{C_4} P(x, y)\, dx$$

$$= \int_a^b P(x, g_1(x))\, dx - \int_a^b P(x, g_2(x))\, dx$$

Comparing this expression with the one in Equation 5, we see that

$$\int_C P(x, y)\, dx = -\iint_D \frac{\partial P}{\partial y}\, dA$$

Equation 4 can be proved in much the same way by expressing D as a type II region (see Exercise 30). Then, by adding Equations 3 and 4, we obtain Green's Theorem. $\square$

EXAMPLE 1   Evaluate $\int_C x^4\, dx + xy\, dy$, where C is the triangular curve consisting of the line segments from $(0, 0)$ to $(1, 0)$, from $(1, 0)$ to $(0, 1)$, and from $(0, 1)$ to $(0, 0)$.

SOLUTION   Although the given line integral could be evaluated as usual by the methods of Section 14.2, that would involve setting up three separate integrals along the three sides of the triangle, so let us instead use Green's Theorem. Notice that the region D enclosed by C is simple and C has positive orientation (see Figure 4). If we let $P(x, y) = x^4$ and $Q(x, y) = xy$, then we have

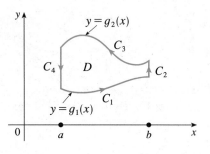

FIGURE 4

$$\int_C x^4\, dx + xy\, dy = \iint_D \left(\frac{\partial Q}{\partial x} - \frac{\partial P}{\partial y} \right) dA = \int_0^1 \int_0^{1-x} (y - 0)\, dy\, dx$$

$$= \int_0^1 \left[\frac{y^2}{2} \right]_{y=0}^{y=1-x} dx = \tfrac{1}{2} \int_0^1 (1 - x)^2\, dx$$

$$= -\tfrac{1}{6}(1 - x)^3 \Big]_0^1 = \tfrac{1}{6}$$

EXAMPLE 2   Evaluate $\oint_C (3y - e^{\sin x})\, dx + (7x + \sqrt{y^4 + 1}\,)\, dy$, where C is the circle $x^2 + y^2 = 9$.

SOLUTION The region D bounded by C is the disk $x^2 + y^2 \le 9$, so let us change to polar coordinates after applying Green's Theorem:

$$\oint_C (3y - e^{\sin x}) \, dx + (7x + \sqrt{y^4 + 1}) \, dy$$

$$= \iint_D \left[\frac{\partial}{\partial x}(7x + \sqrt{y^4 + 1}) - \frac{\partial}{\partial y}(3y - e^{\sin x}) \right] dA$$

$$= \int_0^{2\pi} \int_0^3 (7 - 3) r \, dr \, d\theta$$

$$= 4 \int_0^{2\pi} d\theta \int_0^3 r \, dr = 36\pi \qquad \blacksquare$$

Instead of using polar coordinates, we could simply use the fact that D is a disk of radius 3 and write

$$\iint_D 4 \, dA = 4 \cdot \pi(3)^2 = 36\pi$$

In Examples 1 and 2 we found that the double integral was easier to evaluate than the line integral. (Try setting up the line integral in Example 2 and you will be convinced!) But sometimes it is easier to evaluate the line integral, and Green's Theorem is used in the reverse direction. For instance, if it is known that $P(x, y) = Q(x, y) = 0$ on the curve C, then Green's Theorem gives

$$\iint_D \left(\frac{\partial Q}{\partial x} - \frac{\partial P}{\partial y} \right) dA = \int_C P \, dx + Q \, dy = 0$$

no matter what values P and Q assume in the region D.

Another application of the reverse direction of Green's Theorem is in computing areas. Since the area of D is $\iint_D 1 \, dA$, we wish to choose P and Q so that

$$\frac{\partial Q}{\partial x} - \frac{\partial P}{\partial y} = 1$$

There are several possibilities: $P(x, y) = 0$ and $Q(x, y) = x$; $P(x, y) = -y$ and $Q(x, y) = 0$; or $P(x, y) = -y/2$ and $Q(x, y) = x/2$. Then Green's Theorem gives the following formulas for the area of D:

(6)
$$A = \oint_C x \, dy = -\oint_C y \, dx = \tfrac{1}{2} \oint_C x \, dy - y \, dx$$

EXAMPLE 3 Find the area enclosed by the ellipse $\dfrac{x^2}{a^2} + \dfrac{y^2}{b^2} = 1$.

SOLUTION The ellipse has parametric equations $x = a \cos t$ and $y = b \sin t$, where $0 \le t \le 2\pi$. Using the third formula in Equation 6, we have

$$A = \tfrac{1}{2} \int_C x \, dy - y \, dx$$

$$= \tfrac{1}{2} \int_0^{2\pi} (a \cos t)(b \cos t) \, dt - (b \sin t)(-a \sin t) \, dt$$

$$= \frac{ab}{2} \int_0^{2\pi} dt = \pi ab \qquad \blacksquare$$

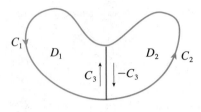

C_1 D_1 D_2 C_2

C_3 $-C_3$

FIGURE 5

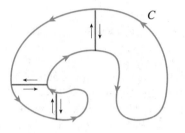

C

FIGURE 6

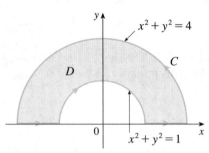

FIGURE 7

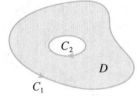

C_2

D

C_1

FIGURE 8

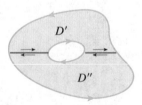

D'

D''

FIGURE 9

Although we have proved Green's Theorem only for the case where D is simple, we can now extend it to the case where D is a finite union of simple regions. For example, if D is the region shown in Figure 5, then we can write $D = D_1 \cup D_2$, where D_1 and D_2 are both simple. The boundary of D_1 is $C_1 \cup C_3$ and the boundary of D_2 is $C_2 \cup (-C_3)$ so, applying Green's Theorem to D_1 and D_2 separately, we get

$$\int_{C_1 \cup C_3} P\,dx + Q\,dy = \iint_{D_1} \left(\frac{\partial Q}{\partial x} - \frac{\partial P}{\partial y} \right) dA$$

$$\int_{C_2 \cup (-C_3)} P\,dx + Q\,dy = \iint_{D_2} \left(\frac{\partial Q}{\partial x} - \frac{\partial P}{\partial y} \right) dA$$

If we add these two equations, the line integrals along C_3 and $-C_3$ cancel, so we get

$$\int_{C_1 \cup C_2} P\,dx + Q\,dy = \iint_{D} \left(\frac{\partial Q}{\partial x} - \frac{\partial P}{\partial y} \right) dA$$

which is Green's Theorem for $D = D_1 \cup D_2$, since its boundary is $C = C_1 \cup C_2$.

The same sort of argument allows us to establish Green's Theorem for any finite union of simple regions (see Figure 6).

EXAMPLE 4   Evaluate $\oint_C y^2\,dx + 3xy\,dy$, where C is the boundary of the semi-annular region D in the upper half-plane between the circles $x^2 + y^2 = 1$ and $x^2 + y^2 = 4$.

SOLUTION   Notice that although D is not simple, the y-axis divides it into two simple regions (see Figure 7). In polar coordinates we can write

$$D = \{(r, \theta) \,|\, 1 \le r \le 2, 0 \le \theta \le \pi\}$$

Therefore, Green's Theorem gives

$$\int_C y^2\,dx + 3xy\,dy = \iint_D \left[\frac{\partial}{\partial x}(3xy) - \frac{\partial}{\partial y}(y^2) \right] dA$$

$$= \iint_D y\,dA = \int_0^\pi \int_1^2 (r\sin\theta)r\,dr\,d\theta$$

$$= \int_0^\pi \sin\theta\,d\theta \int_1^2 r^2\,dr = \left[-\cos\theta \right]_0^\pi \left[\frac{r^3}{3} \right]_1^2$$

$$= \frac{14}{3}$$

Green's Theorem can be extended to apply to regions with holes, that is, regions that are not simply-connected. Observe that the boundary C of the region D in Figure 8 consists of two simple closed curves C_1 and C_2. We assume that these boundary curves are oriented so that the region D is always on the left as the curve C is traversed. Thus the positive direction is counterclockwise for the outer curve C_1 but clockwise for the inner curve C_2. If we divide D into two regions D' and D'' by means of the lines shown in Figure 9 and then apply Green's Theorem to each of D' and D'', we get

$$\iint_D \left(\frac{\partial Q}{\partial x} - \frac{\partial P}{\partial y} \right) dA = \iint_{D'} \left(\frac{\partial Q}{\partial x} - \frac{\partial P}{\partial y} \right) dA + \iint_{D''} \left(\frac{\partial Q}{\partial x} - \frac{\partial P}{\partial y} \right) dA$$

$$= \int_{\partial D'} P\,dx + Q\,dy + \int_{\partial D''} P\,dx + Q\,dy$$

Since the line integrals along the common boundary lines are in opposite directions, they cancel and we get

$$\iint_D \left(\frac{\partial Q}{\partial x} - \frac{\partial P}{\partial y} \right) dA = \int_{C_1} P\,dx + Q\,dy + \int_{C_2} P\,dx + Q\,dy = \int_C P\,dx + Q\,dy$$

which is Green's Theorem for the region D.

EXAMPLE 5   If $\mathbf{F}(x, y) = (-y\,\mathbf{i} + x\,\mathbf{j})/(x^2 + y^2)$, show that $\int_C \mathbf{F} \cdot d\mathbf{r} = 2\pi$ for every simple closed path that encloses the origin.

SOLUTION   Since C is an *arbitrary* closed path that encloses the origin, it is difficult to compute the given integral directly. So let us consider a counterclockwise-oriented circle C' with center the origin and radius a, where a is chosen to be small enough that C' lies inside C (see Figure 10). Let D be the region bounded by C and C'. Then its positively oriented boundary is $C \cup (-C')$ and so the general version of Green's Theorem gives

$$\int_C P\,dx + Q\,dy + \int_{-C'} P\,dx + Q\,dy = \iint_D \left(\frac{\partial Q}{\partial x} - \frac{\partial P}{\partial y} \right) dA$$

$$= \iint_D \left[\frac{y^2 - x^2}{(x^2 + y^2)^2} - \frac{y^2 - x^2}{(x^2 + y^2)^2} \right] dA$$

$$= 0$$

Therefore

$$\int_C P\,dx + Q\,dy = \int_{C'} P\,dx + Q\,dy$$

that is,

$$\int_C \mathbf{F} \cdot d\mathbf{r} = \int_{C'} \mathbf{F} \cdot d\mathbf{r}$$

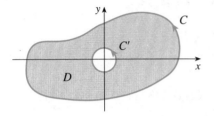

FIGURE 10

We now easily compute the latter integral using the parametrization given by $\mathbf{r}(t) = a \cos t\,\mathbf{i} + a \sin t\,\mathbf{j}$, $0 \le t \le 2\pi$. Thus

$$\int_C \mathbf{F} \cdot d\mathbf{r} = \int_{C'} \mathbf{F} \cdot d\mathbf{r} = \int_0^{2\pi} \mathbf{F}(\mathbf{r}(t)) \cdot \mathbf{r}'(t)\,dt$$

$$= \int_0^{2\pi} \frac{(-a \sin t)(-a \sin t) + (a \cos t)(a \cos t)}{a^2 \cos^2 t + a^2 \sin^2 t}\,dt$$

$$= \int_0^{2\pi} dt = 2\pi \qquad \blacksquare$$

We end this section by using Green's Theorem to discuss a result that was stated in the preceding section.

SKETCH OF PROOF OF THEOREM 14.3.6   We are assuming that $\mathbf{F} = P\,\mathbf{i} + Q\,\mathbf{j}$ is a vector field on an open simply-connected region D, that P and Q have continuous first-order partial derivatives, and that

$$\frac{\partial P}{\partial y} = \frac{\partial Q}{\partial x} \qquad \text{throughout } D$$

If C is any simple closed path in D and R is the region that C encloses, then Green's Theorem gives

$$\oint_C \mathbf{F} \cdot d\mathbf{r} = \oint_C P\,dx + Q\,dy = \iint_R \left(\frac{\partial Q}{\partial x} - \frac{\partial P}{\partial y} \right) dA = \iint_R 0\,dA = 0$$

A curve that is not simple crosses itself at one or more points and can be broken up into a number of simple curves. We have shown that the line integrals of $\mathbf{F}$ around these simple curves are all 0 and, adding these integrals, we see that $\int_C \mathbf{F} \cdot d\mathbf{r} = 0$ for any closed curve C. Therefore $\int_C \mathbf{F} \cdot d\mathbf{r}$ is independent of path in D by Theorem 14.3.3. It follows that $\mathbf{F}$ is a conservative vector field by Theorem 14.3.4. $\square$

EXERCISES 14.4

1–4 ■ Evaluate the line integral by two methods: (a) directly and (b) using Green's Theorem.

1. $\oint_C x^2 y\,dx + xy^3\,dy$,
C is the square with vertices $(0,0)$, $(1,0)$, $(1,1)$, and $(0,1)$

2. $\oint_C x\,dx - x^2 y^2\,dy$,
C is the triangle with vertices $(0,0)$, $(1,1)$, and $(0,1)$

3. $\oint_C (x + 2y)\,dx + (x - 2y)\,dy$,
C consists of the arc of the parabola $y = x^2$ from $(0,0)$ to $(1,1)$ followed by the line segment from $(1,1)$ to $(0,0)$

4. $\oint_C (x^2 + y^2)\,dx + 2xy\,dy$, C consists of the arc of the parabola $y = x^2$ from $(0,0)$ to $(2,4)$ and the line segments from $(2,4)$ to $(0,4)$ and from $(0,4)$ to $(0,0)$

CAS **5–6** ■ Verify Green's Theorem by using a computer algebra system to evaluate both the line integral and the double integral.

5. $P(x, y) = x^4 y^5$, $Q(x, y) = -x^7 y^6$; $C: x^2 + y^2 = 1$

6. $P(x, y) = y^2 \sin x$, $Q(x, y) = x^2 \sin y$;
C is the curve in Exercise 3

7–18 ■ Use Green's Theorem to evaluate the line integral along the given positively oriented curve.

7. $\int_C xy\,dx + y^5\,dy$,
C is the triangle with vertices $(0,0)$, $(2,0)$, and $(2,1)$

8. $\int_C x^2 y\,dx + xy^5\,dy$,
C is the square with vertices $(\pm 1, \pm 1)$

9. $\int_C \left(y + e^{\sqrt{x}} \right) dx + (2x + \cos y^2)\,dy$,
C is the boundary of the region enclosed by the parabolas $y = x^2$ and $x = y^2$

10. $\int_C (y^2 - \tan^{-1} x)\,dx + (3x + \sin y)\,dy$,
C is the boundary of the region enclosed by the parabola $y = x^2$ and the line $y = 4$

11. $\int_C x^2\,dx + y^2\,dy$, C is the curve $x^6 + y^6 = 1$

12. $\int_C x^2 y\,dx - 3y^2\,dy$, C is the circle $x^2 + y^2 = 1$

13. $\int_C xy\,dx + 2x^2\,dy$,
C consists of the line segment from $(-2,0)$ to $(2,0)$ and the top half of the circle $x^2 + y^2 = 4$

14. $\int_C 2xy\,dx + x^2\,dy$,
C is the cardioid $r = 1 + \cos\theta$

15. $\int_C (xy + e^{x^2})\,dx + (x^2 - \ln(1 + y))\,dy$,
C consists of the line segment from $(0,0)$ to $(\pi, 0)$ and the curve $y = \sin x$, $0 \leqslant x \leqslant \pi$

16. $\int_C (x^3 - y^3)\,dx + (x^3 + y^3)\,dy$,
C is the boundary of the region between the circles $x^2 + y^2 = 1$ and $x^2 + y^2 = 9$

17. $\int_C \mathbf{F} \cdot d\mathbf{r}$, where $\mathbf{F}(x, y) = (y^2 - x^2 y)\mathbf{i} + xy^2 \mathbf{j}$ and C consists of the circle $x^2 + y^2 = 4$ from $(2,0)$ to $(\sqrt{2}, \sqrt{2})$ and the line segments from $(\sqrt{2}, \sqrt{2})$ to $(0,0)$ and from $(0,0)$ to $(2,0)$

18. $\int_C \mathbf{F} \cdot d\mathbf{r}$, where $\mathbf{F}(x, y) = x^3 y\mathbf{i} + x^4 \mathbf{j}$ and C is the curve $x^4 + y^4 = 1$

19. Use Green's Theorem to find the work done by the force $\mathbf{F}(x, y) = x(x + y)\mathbf{i} + xy^2 \mathbf{j}$ in moving a particle from the origin along the x-axis to $(1,0)$, then along the line segment to $(0,1)$, and then back to the origin along the y-axis.

20. A particle starts at the point $(-2,0)$, moves along the x-axis to $(2,0)$, and then along the semicircle $y = \sqrt{4 - x^2}$ to the starting point. Use Green's Theorem to find the work done on this particle by the force field $\mathbf{F}(x, y) = \langle x, x^3 + 3xy^2 \rangle$.

21–22 ■ Find the area of the given region using one of the formulas in Equations 6.

21. The region bounded by the hypocycloid with vector equation $\mathbf{r}(t) = \cos^3 t\,\mathbf{i} + \sin^3 t\,\mathbf{j}$, $0 \leqslant t \leqslant 2\pi$

22. The region bounded by the curve with vector equation $\mathbf{r}(t) = \cos t\,\mathbf{i} + \sin^3 t\,\mathbf{j}$, $0 \leqslant t \leqslant 2\pi$

23. (a) If C is the line segment connecting the point (x_1, y_1) to the point (x_2, y_2), show that

$$\int_C x\,dy - y\,dx = x_1 y_2 - x_2 y_1$$

(b) If the vertices of a polygon, in counterclockwise order, are (x_1, y_1), (x_2, y_2), $\ldots$, (x_n, y_n), show that the area of the polygon is

$$A = \tfrac{1}{2}[(x_1 y_2 - x_2 y_1) + (x_2 y_3 - x_3 y_2) + \cdots$$
$$+ (x_{n-1} y_n - x_n y_{n-1}) + (x_n y_1 - x_1 y_n)]$$

(c) Find the area of the pentagon with vertices $(0, 0)$, $(2, 1)$, $(1, 3)$, $(0, 2)$, and $(-1, 1)$.

24. Let D be a region bounded by a simple closed path C in the xy-plane. Use Green's Theorem to prove that the coordinates of the centroid $(\bar{x}, \bar{y})$ of D are

$$\bar{x} = \frac{1}{2A} \oint_C x^2\,dy \qquad \bar{y} = -\frac{1}{2A} \oint_C y^2\,dx$$

where A is the area of D.

25. Use Exercise 24 to find the centroid of the triangle with vertices $(0, 0)$, $(1, 0)$, and $(0, 1)$.

26. Use Exercise 24 to find the centroid of a semicircular region of radius a.

27. A plane lamina with constant density $\rho(x, y) = \rho$ occupies a region in the xy-plane bounded by a simple closed path C.

Show that its moments of inertia about the axes are

$$I_x = -\frac{\rho}{3} \oint_C y^3\,dx$$

$$I_y = \frac{\rho}{3} \oint_C x^3\,dy$$

28. Use Exercise 27 to find the moment of inertia of a circular disk of radius a with constant density ρ about a diameter. (Compare with Example 4 in Section 13.5.)

29. If $\mathbf{F}$ is the vector field of Example 5, show that $\int_C \mathbf{F} \cdot d\mathbf{r} = 0$ for every simple closed path that does not pass through or enclose the origin.

30. Complete the proof of the special case of Green's Theorem by proving Equation 4.

31. Use Green's Theorem to prove the change of variables formula for a double integral (13.9.9) for the case where $f(x, y) = 1$:

$$\iint_R dx\,dy = \iint_S \left| \frac{\partial(x, y)}{\partial(u, v)} \right| du\,dv$$

Here R is the region in the xy-plane that corresponds to the region S in the uv-plane under the transformation given by $x = g(u, v)$, $y = h(u, v)$. [*Hint:* Note that the left side is $A(R)$ and apply the first part of Equation 6. Convert the line integral over ∂R to a line integral over ∂S and apply Green's Theorem in the uv-plane.]

14.5 **CURL AND DIVERGENCE**

In this section we define two operations that can be performed on vector fields and that play a basic role in the applications of vector calculus. Each operation resembles differentiation, but one produces a vector field whereas the other produces a scalar field.

CURL

If $\mathbf{F} = P\mathbf{i} + Q\mathbf{j} + R\mathbf{k}$ is a vector field on $\mathbb{R}^3$ and the partial derivatives of P, Q, and R all exist, then the **curl** of $\mathbf{F}$ is the vector field on $\mathbb{R}^3$ defined by

$$(1) \qquad \text{curl } \mathbf{F} = \left(\frac{\partial R}{\partial y} - \frac{\partial Q}{\partial z} \right) \mathbf{i} + \left(\frac{\partial P}{\partial z} - \frac{\partial R}{\partial x} \right) \mathbf{j} + \left(\frac{\partial Q}{\partial x} - \frac{\partial P}{\partial y} \right) \mathbf{k}$$

As an aid to the memory, let us rewrite Equation 1 using operator notation. We introduce the vector differential operator ∇ ("del") as

$$\nabla = \mathbf{i}\,\frac{\partial}{\partial x} + \mathbf{j}\,\frac{\partial}{\partial y} + \mathbf{k}\,\frac{\partial}{\partial z}$$

It has meaning when it operates on a scalar function to produce the gradient of f:

$$\nabla f = \mathbf{i}\,\frac{\partial f}{\partial x} + \mathbf{j}\,\frac{\partial f}{\partial y} + \mathbf{k}\,\frac{\partial f}{\partial z} = \frac{\partial f}{\partial x}\mathbf{i} + \frac{\partial f}{\partial y}\mathbf{j} + \frac{\partial f}{\partial z}\mathbf{k}$$

If we think of ∇ as a vector with components $\partial/\partial x$, $\partial/\partial y$, and $\partial/\partial z$, we can also consider the formal cross product of ∇ with the vector field **F** as follows:

$$\nabla \times \mathbf{F} = \begin{vmatrix} \mathbf{i} & \mathbf{j} & \mathbf{k} \\ \dfrac{\partial}{\partial x} & \dfrac{\partial}{\partial y} & \dfrac{\partial}{\partial z} \\ P & Q & R \end{vmatrix}$$

$$= \left(\frac{\partial R}{\partial y} - \frac{\partial Q}{\partial z} \right) \mathbf{i} + \left(\frac{\partial P}{\partial z} - \frac{\partial R}{\partial x} \right) \mathbf{j} + \left(\frac{\partial Q}{\partial x} - \frac{\partial P}{\partial y} \right) \mathbf{k}$$

$$= \text{curl } \mathbf{F}$$

Thus the easiest way to remember Definition 1 is by means of the symbolic expression

(2)
$$\boxed{\text{curl } \mathbf{F} = \nabla \times \mathbf{F}}$$

EXAMPLE 1   If $\mathbf{F}(x, y, z) = xz\,\mathbf{i} + xyz\,\mathbf{j} - y^2\,\mathbf{k}$, find curl **F**.

SOLUTION   Using Equation 2, we have

$$\text{curl } \mathbf{F} = \nabla \times \mathbf{F} = \begin{vmatrix} \mathbf{i} & \mathbf{j} & \mathbf{k} \\ \dfrac{\partial}{\partial x} & \dfrac{\partial}{\partial y} & \dfrac{\partial}{\partial z} \\ xz & xyz & -y^2 \end{vmatrix}$$

$$= \left[\frac{\partial}{\partial y}(-y^2) - \frac{\partial}{\partial z}(xyz) \right] \mathbf{i} - \left[\frac{\partial}{\partial x}(-y^2) - \frac{\partial}{\partial z}(xz) \right] \mathbf{j}$$

$$+ \left[\frac{\partial}{\partial x}(xyz) - \frac{\partial}{\partial y}(xz) \right] \mathbf{k}$$

$$= (-2y - xy)\,\mathbf{i} - (0 - x)\,\mathbf{j} + (yz - 0)\,\mathbf{k}$$

$$= -y(2 + x)\,\mathbf{i} + x\,\mathbf{j} + yz\,\mathbf{k} \qquad \blacksquare$$

Most computer algebra systems have commands that compute the curl and divergence of vector fields. If you have access to a CAS, use these commands to check the answers to the examples and exercises in this section.

Recall that the gradient of a function f of three variables is a vector field on $\mathbb{R}^3$ and so we can compute its curl. The following theorem says that the curl of a gradient vector field is **0**. Notice the similarity to Example 2 in Section 11.4: $\mathbf{a} \times \mathbf{a} = \mathbf{0}$ for every three-dimensional vector **a**.

---

(3) THEOREM   If f is a function of three variables that has continuous second-order partial derivatives, then

$$\text{curl}(\nabla f) = \mathbf{0}$$

PROOF   We have

$$\text{curl}(\nabla f) = \nabla \times (\nabla f) = \begin{vmatrix} \mathbf{i} & \mathbf{j} & \mathbf{k} \\ \dfrac{\partial}{\partial x} & \dfrac{\partial}{\partial y} & \dfrac{\partial}{\partial z} \\ \dfrac{\partial f}{\partial x} & \dfrac{\partial f}{\partial y} & \dfrac{\partial f}{\partial z} \end{vmatrix}$$

$$= \left(\frac{\partial^2 f}{\partial y \, \partial z} - \frac{\partial^2 f}{\partial z \, \partial y} \right) \mathbf{i} + \left(\frac{\partial^2 f}{\partial z \, \partial x} - \frac{\partial^2 f}{\partial x \, \partial z} \right) \mathbf{j} + \left(\frac{\partial^2 f}{\partial x \, \partial y} - \frac{\partial^2 f}{\partial y \, \partial x} \right) \mathbf{k}$$

$$= 0\mathbf{i} + 0\mathbf{j} + 0\mathbf{k} = \mathbf{0}$$

by Clairaut's Theorem. □

Since a conservative vector field is one for which $\mathbf{F} = \nabla f$, Theorem 3 can be rephrased as saying that if $\mathbf{F}$ is conservative, then curl $\mathbf{F} = \mathbf{0}$. (Compare this with Exercise 27 in Section 14.3.) This gives us a way of verifying that a vector field is not conservative.

EXAMPLE 2 Show that the vector field $\mathbf{F}(x, y, z) = xz \, \mathbf{i} + xyz \, \mathbf{j} - y^2 \mathbf{k}$ is not conservative.

SOLUTION In Example 1 we showed that

$$\text{curl } \mathbf{F} = -y(2 + x) \mathbf{i} + x \mathbf{j} + yz \, \mathbf{k}$$

This shows that curl $\mathbf{F} \neq \mathbf{0}$ and so, by Theorem 3, $\mathbf{F}$ is not conservative. ∎

The converse of Theorem 3 is not true in general, but the following theorem says the converse is true if $\mathbf{F}$ is defined everywhere. (More generally it is true if the domain is simply-connected, that is, "has no hole.") Theorem 4 is the three-dimensional analogue of Theorem 14.3.6. Its proof requires Stokes' Theorem and is sketched at the end of Section 14.8.

> **(4) THEOREM** If $\mathbf{F}$ is a vector field defined on all of $\mathbb{R}^3$ whose component functions have continuous partial derivatives and curl $\mathbf{F} = \mathbf{0}$, then $\mathbf{F}$ is a conservative vector field.

EXAMPLE 3
(a) Show that $\mathbf{F}(x, y, z) = y^2 z^3 \mathbf{i} + 2xyz^3 \mathbf{j} + 3xy^2 z^2 \mathbf{k}$ is a conservative vector field.
(b) Find a function f such that $\mathbf{F} = \nabla f$.

SOLUTION
(a) We compute the curl of $\mathbf{F}$:

$$\text{curl } \mathbf{F} = \nabla \times \mathbf{F} = \begin{vmatrix} \mathbf{i} & \mathbf{j} & \mathbf{k} \\ \dfrac{\partial}{\partial x} & \dfrac{\partial}{\partial y} & \dfrac{\partial}{\partial z} \\ y^2 z^3 & 2xyz^3 & 3xy^2 z^2 \end{vmatrix}$$

$$= (6xyz^2 - 6xyz^2) \mathbf{i} - (3y^2 z^2 - 3y^2 z^2) \mathbf{j} + (2yz^3 - 2yz^3) \mathbf{k}$$

$$= \mathbf{0}$$

Since curl $\mathbf{F} = \mathbf{0}$ and the domain of $\mathbf{F}$ is $\mathbb{R}^3$, $\mathbf{F}$ is a conservative vector field by Theorem 4.

(b) The technique for finding f was given in Section 14.3. We have

(5) $$f_x(x, y, z) = y^2 z^3$$

(6) $$f_y(x, y, z) = 2xyz^3$$

(7) $$f_z(x, y, z) = 3xy^2 z^2$$

Integrating (5) with respect to x, we obtain

(8)
$$f(x, y, z) = xy^2z^3 + g(y, z)$$

Differentiating (8) with respect to y, we get $f_y(x, y, z) = 2xyz^3 + g_y(y, z)$, so comparison with (6) gives $g_y(y, z) = 0$. Thus $g(y, z) = h(z)$ and

$$f_z(x, y, z) = 3xy^2z^2 + h'(z)$$

Then (7) gives $h'(z) = 0$. Therefore

$$f(x, y, z) = xy^2z^3 + K \qquad\blacksquare$$

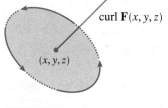

curl $\mathbf{F}(x, y, z)$

(x, y, z)

FIGURE 1

The reason for the name *curl* is that the curl vector is associated with rotations. One connection is explained in Exercise 43. Another occurs when $\mathbf{F}$ represents the velocity field in fluid flow (see Example 3 in Section 14.1). Particles near (x, y, z) in the fluid tend to rotate about the axis that points in the direction of curl $\mathbf{F}(x, y, z)$ and the length of this curl vector is a measure of how quickly the particles move around the axis (see Figure 1). If curl $\mathbf{F} = \mathbf{0}$ at a point P, then the fluid is free from rotations at P and $\mathbf{F}$ is called **irrotational** at P. In other words, there are no whirlpools or eddies. If curl $\mathbf{F} = \mathbf{0}$, then a tiny paddle wheel moves with the fluid but does not rotate about its axis. If curl $\mathbf{F} \neq \mathbf{0}$, the paddle wheel rotates about its axis. We give a more detailed explanation in Section 14.8 as a consequence of Stokes' Theorem.

DIVERGENCE

If $\mathbf{F} = P\,\mathbf{i} + Q\,\mathbf{j} + R\,\mathbf{k}$ is a vector field on $\mathbb{R}^3$ and $\partial P/\partial x$, $\partial Q/\partial y$, and $\partial R/\partial z$ exist, then the **divergence of F** is the function of three variables defined by

(9)
$$\operatorname{div}\mathbf{F} = \frac{\partial P}{\partial x} + \frac{\partial Q}{\partial y} + \frac{\partial R}{\partial z}$$

Observe that curl $\mathbf{F}$ is a vector field but div $\mathbf{F}$ is a scalar field. In terms of the gradient operator $\nabla = (\partial/\partial x)\,\mathbf{i} + (\partial/\partial y)\,\mathbf{j} + (\partial/\partial z)\,\mathbf{k}$, the divergence of $\mathbf{F}$ can be written symbolically as the dot product of ∇ and $\mathbf{F}$:

(10)
$$\operatorname{div}\mathbf{F} = \nabla \cdot \mathbf{F}$$

EXAMPLE 4   If $\mathbf{F}(x, y, z) = xz\,\mathbf{i} + xyz\,\mathbf{j} - y^2\,\mathbf{k}$, find div $\mathbf{F}$.

SOLUTION   By the definition of divergence (Equation 9 or 10) we have

$$\operatorname{div}\mathbf{F} = \nabla \cdot \mathbf{F} = \frac{\partial}{\partial x}(xz) + \frac{\partial}{\partial y}(xyz) + \frac{\partial}{\partial z}(-y^2)$$

$$= z + xz \qquad\blacksquare$$

If $\mathbf{F}$ is a vector field on $\mathbb{R}^3$, then curl $\mathbf{F}$ is also a vector field on $\mathbb{R}^3$. As such, we can compute its divergence. The next theorem shows that the result is 0. [Note the analogy with the scalar triple product: $\mathbf{a} \cdot (\mathbf{a} \times \mathbf{b}) = 0$.]

(11) THEOREM If $\mathbf{F} = P\mathbf{i} + Q\mathbf{j} + R\mathbf{k}$ is a vector field on $\mathbb{R}^3$ and P, Q, and R have continuous second-order partial derivatives, then

$$\text{div curl } \mathbf{F} = 0$$

PROOF Using the definitions of divergence and curl, we have

$$\text{div curl } \mathbf{F} = \nabla \cdot (\nabla \times \mathbf{F})$$

$$= \frac{\partial}{\partial x}\left(\frac{\partial R}{\partial y} - \frac{\partial Q}{\partial z}\right) + \frac{\partial}{\partial y}\left(\frac{\partial P}{\partial z} - \frac{\partial R}{\partial x}\right) + \frac{\partial}{\partial z}\left(\frac{\partial Q}{\partial x} - \frac{\partial P}{\partial y}\right)$$

$$= \frac{\partial^2 R}{\partial x\,\partial y} - \frac{\partial^2 Q}{\partial x\,\partial z} + \frac{\partial^2 P}{\partial y\,\partial z} - \frac{\partial^2 R}{\partial y\,\partial x} + \frac{\partial^2 Q}{\partial z\,\partial x} - \frac{\partial^2 P}{\partial z\,\partial y}$$

$$= 0$$

because the terms cancel in pairs by Clairaut's Theorem. $\square$

EXAMPLE 5 Show that the vector field $\mathbf{F}(x, y, z) = xz\,\mathbf{i} + xyz\,\mathbf{j} - y^2\,\mathbf{k}$ cannot be written as the curl of another vector field, that is, $\mathbf{F} \neq \text{curl } \mathbf{G}$.

SOLUTION In Example 4 we showed that

$$\text{div } \mathbf{F} = z + xz$$

and therefore div $\mathbf{F} \neq 0$. If it were true that $\mathbf{F} = \text{curl } \mathbf{G}$, then Theorem 11 would give

$$\text{div } \mathbf{F} = \text{div curl } \mathbf{G} = 0$$

which contradicts div $\mathbf{F} \neq 0$. Therefore, $\mathbf{F}$ is not the curl of another vector field. ∎

The reason for this interpretation of div $\mathbf{F}$ will be explained at the end of Section 14.9 as a consequence of the Divergence Theorem.

Again, the reason for the name *divergence* can be understood in the context of fluid flow. If $\mathbf{F}(x, y, z)$ is the velocity of a fluid (or gas), then div $\mathbf{F}(x, y, z)$ represents the net rate of change (with respect to time) of the mass of fluid (or gas) flowing from the point (x, y, z) per unit volume. In other words, div $\mathbf{F}(x, y, z)$ measures the tendency of the fluid to diverge from the point (x, y, z). If div $\mathbf{F} = 0$, then $\mathbf{F}$ is said to be **incompressible.**

Another differential operator occurs when we compute the divergence of a gradient vector field ∇f. If f is a function of three variables, we have

$$\text{div}(\nabla f) = \nabla \cdot (\nabla f) = \frac{\partial^2 f}{\partial x^2} + \frac{\partial^2 f}{\partial y^2} + \frac{\partial^2 f}{\partial z^2}$$

and this expression occurs so often that we abbreviate it as $\nabla^2 f$. The operator

$$\nabla^2 = \nabla \cdot \nabla$$

is called the **Laplace operator** because of its relation to **Laplace's equation**

$$\nabla^2 f = \frac{\partial^2 f}{\partial x^2} + \frac{\partial^2 f}{\partial y^2} + \frac{\partial^2 f}{\partial z^2} = 0$$

We can also apply the Laplace operator ∇^2 to a vector field

$$\mathbf{F} = P\,\mathbf{i} + Q\,\mathbf{j} + R\,\mathbf{k}$$

in terms of its components:

$$\nabla^2\mathbf{F} = \nabla^2 P\,\mathbf{i} + \nabla^2 Q\,\mathbf{j} + \nabla^2 R\,\mathbf{k}$$

VECTOR FORMS OF GREEN'S THEOREM

The curl and divergence operators allow us to rewrite Green's Theorem in versions that will be useful in our later work. We suppose that the plane region D, its boundary curve C, and the functions P and Q satisfy the hypotheses of Green's Theorem (14.4.1). Then we consider the vector field $\mathbf{F} = P\,\mathbf{i} + Q\,\mathbf{j}$. Its line integral is

$$\oint_C \mathbf{F} \cdot d\mathbf{r} = \oint_C P\,dx + Q\,dy$$

and its curl is

$$\operatorname{curl} \mathbf{F} = \begin{vmatrix} \mathbf{i} & \mathbf{j} & \mathbf{k} \\ \dfrac{\partial}{\partial x} & \dfrac{\partial}{\partial y} & \dfrac{\partial}{\partial z} \\ P(x, y) & Q(x, y) & 0 \end{vmatrix} = \left(\frac{\partial Q}{\partial x} - \frac{\partial P}{\partial y} \right) \mathbf{k}$$

Therefore

$$(\operatorname{curl} \mathbf{F}) \cdot \mathbf{k} = \left(\frac{\partial Q}{\partial x} - \frac{\partial P}{\partial y} \right) \mathbf{k} \cdot \mathbf{k} = \frac{\partial Q}{\partial x} - \frac{\partial P}{\partial y}$$

and we can now rewrite the equation in Green's Theorem in the vector form

(12)
$$\oint_C \mathbf{F} \cdot d\mathbf{r} = \iint_D (\operatorname{curl} \mathbf{F}) \cdot \mathbf{k}\,dA$$

Equation 12 expresses the line integral of the tangential component of $\mathbf{F}$ along C as the double integral of $(\operatorname{curl} \mathbf{F}) \cdot \mathbf{k}$ over the region D enclosed by C. We now derive a similar formula involving the *normal* component of $\mathbf{F}$.

If C is given by the vector equation

$$\mathbf{r}(t) = x(t)\,\mathbf{i} + y(t)\,\mathbf{j} \qquad a \le t \le b$$

then the unit tangent vector (see Section 11.7) is

$$\mathbf{T}(t) = \frac{x'(t)}{|\mathbf{r}'(t)|}\,\mathbf{i} + \frac{y'(t)}{|\mathbf{r}'(t)|}\,\mathbf{j}$$

You can verify that the outward unit normal vector to C is given by

$$\mathbf{n}(t) = \frac{y'(t)}{|\mathbf{r}'(t)|}\,\mathbf{i} - \frac{x'(t)}{|\mathbf{r}'(t)|}\,\mathbf{j}$$

(See Figure 2.) Then, from Equation 14.2.3, we have

$$\oint_C \mathbf{F} \cdot \mathbf{n}\,ds = \int_a^b (\mathbf{F} \cdot \mathbf{n})(t)\,|\mathbf{r}'(t)|\,dt$$

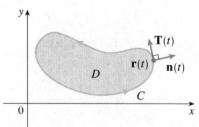

FIGURE 2

$$= \int_a^b \left[\frac{P(x(t), y(t)) \, y'(t)}{|\mathbf{r}'(t)|} - \frac{Q(x(t), y(t)) \, x'(t)}{|\mathbf{r}'(t)|} \right] |\mathbf{r}'(t)| \, dt$$

$$= \int_a^b P(x(t), y(t)) \, y'(t) \, dt - Q(x(t), y(t)) \, x'(t) \, dt$$

$$= \int_C P \, dy - Q \, dx$$

$$= \iint_D \left(\frac{\partial P}{\partial x} + \frac{\partial Q}{\partial y} \right) dA$$

by Green's Theorem. But the integrand in this double integral is just the divergence of **F**. So we have a second vector form of Green's Theorem.

(13)
$$\oint_C \mathbf{F} \cdot \mathbf{n} \, ds = \iint_D \operatorname{div} \mathbf{F}(x, y) \, dA$$

This version says that the line integral of the normal component of **F** along C is equal to the double integral of the divergence of **F** over the region D enclosed by C.

EXERCISES 14.5

1–10 ■ Find (a) the curl and (b) the divergence of the vector field.

1. $\mathbf{F}(x, y, z) = x\mathbf{i} + y\mathbf{j} + z\mathbf{k}$

2. $\mathbf{F}(x, y, z) = x^2 y\mathbf{i} + yz^2\mathbf{j} + zx^2\mathbf{k}$

3. $\mathbf{F}(x, y, z) = yz\mathbf{i} + xz\mathbf{j} + xy\mathbf{k}$

4. $\mathbf{F}(x, y, z) = y^2 z\mathbf{i} - x^2 yz\mathbf{k}$

5. $\mathbf{F}(x, y, z) = xy\mathbf{j} + xyz\mathbf{k}$

6. $\mathbf{F}(x, y, z) = \sin x\mathbf{i} + \cos x\mathbf{j} + z^2\mathbf{k}$

7. $\mathbf{F}(x, y, z) = e^{xz}\mathbf{i} - 2e^{yz}\mathbf{j} + 3xe^y\mathbf{k}$

8. $\mathbf{F}(x, y, z) = (x + 3y - 5z)\mathbf{i} + (z - 3y)\mathbf{j} + (5x + 6y - z)\mathbf{k}$

9. $\mathbf{F}(x, y, z) = xe^y\mathbf{i} - ze^{-y}\mathbf{j} + y\ln z\mathbf{k}$

10. $\mathbf{F}(x, y, z) = e^{xyz}\mathbf{i} + \sin(x - y)\mathbf{j} - \dfrac{xy}{z}\mathbf{k}$

11–18 ■ Determine whether or not the given vector field is conservative. If it is conservative, find a function f such that $\mathbf{F} = \nabla f$.

11. $\mathbf{F}(x, y, z) = y\mathbf{i} + x\mathbf{j} + \mathbf{k}$

12. $\mathbf{F}(x, y, z) = x\mathbf{i} + y\mathbf{j} + x\mathbf{k}$

13. $\mathbf{F}(x, y, z) = yz\mathbf{i} - z^2\mathbf{j} + x^2\mathbf{k}$

14. $\mathbf{F}(x, y, z) = z\mathbf{i} + 2yz\mathbf{j} + (x + y^2)\mathbf{k}$

15. $\mathbf{F}(x, y, z) = \cos y\mathbf{i} + \sin x\mathbf{j} + \tan z\mathbf{k}$

16. $\mathbf{F}(x, y, z) = x\mathbf{i} + e^y \sin z\mathbf{j} + e^y \cos z\mathbf{k}$

17. $\mathbf{F}(x, y, z) = yz\mathbf{i} + (y^2 + xz)\mathbf{j} + xy\mathbf{k}$

18. $\mathbf{F}(x, y, z) = zx\mathbf{i} + xy\mathbf{j} + yz\mathbf{k}$

19. Is there a vector field **G** on $\mathbb{R}^3$ such that curl $\mathbf{G} = xy^2\mathbf{i} + yz^2\mathbf{j} + zx^2\mathbf{k}$? Explain.

20. Is there a vector field **G** on $\mathbb{R}^3$ such that curl $\mathbf{G} = yz\mathbf{i} + xyz\mathbf{j} + xy\mathbf{k}$? Explain.

21. Show that any vector field of the form
$$\mathbf{F}(x, y, z) = f(x)\mathbf{i} + g(y)\mathbf{j} + h(z)\mathbf{k},$$ where f, g, h are differentiable functions, is irrotational.

22. Show that any vector field of the form
$$\mathbf{F}(x, y, z) = f(y, z)\mathbf{i} + g(x, z)\mathbf{j} + h(x, y)\mathbf{k}$$ is incompressible.

23–30 ■ Prove the identity, assuming that the appropriate partial derivatives exist and are continuous. If f is a scalar field and **F**, **G** are vector fields, then $f\mathbf{F}$, $\mathbf{F} \cdot \mathbf{G}$, and $\mathbf{F} \times \mathbf{G}$ are vector fields defined by

$$(f\mathbf{F})(x, y, z) = f(x, y, z)\mathbf{F}(x, y, z)$$

$$(\mathbf{F} \cdot \mathbf{G})(x, y, z) = \mathbf{F}(x, y, z) \cdot \mathbf{G}(x, y, z)$$

$$(\mathbf{F} \times \mathbf{G})(x, y, z) = \mathbf{F}(x, y, z) \times \mathbf{G}(x, y, z)$$

23. $\operatorname{div}(\mathbf{F} + \mathbf{G}) = \operatorname{div} \mathbf{F} + \operatorname{div} \mathbf{G}$

24. $\operatorname{curl}(\mathbf{F} + \mathbf{G}) = \operatorname{curl} \mathbf{F} + \operatorname{curl} \mathbf{G}$

25. $\operatorname{div}(f\mathbf{F}) = f \operatorname{div} \mathbf{F} + \mathbf{F} \cdot \nabla f$

26. $\operatorname{curl}(f\mathbf{F}) = f\operatorname{curl}\mathbf{F} + (\nabla f) \times \mathbf{F}$

27. $\operatorname{div}(\mathbf{F} \times \mathbf{G}) = \mathbf{G} \cdot \operatorname{curl}\mathbf{F} - \mathbf{F} \cdot \operatorname{curl}\mathbf{G}$

28. $\operatorname{div}(\nabla f \times \nabla g) = 0$

29. $\operatorname{curl}\operatorname{curl}\mathbf{F} = \operatorname{grad}\operatorname{div}\mathbf{F} - \nabla^2\mathbf{F}$

30. $\nabla(\mathbf{F} \cdot \mathbf{G}) = (\mathbf{F} \cdot \nabla)\mathbf{G} + (\mathbf{G} \cdot \nabla)\mathbf{F}$
$\qquad\qquad + \mathbf{F} \times \operatorname{curl}\mathbf{G} + \mathbf{G} \times \operatorname{curl}\mathbf{F}$

31. Let f be a scalar field and $\mathbf{F}$ a vector field. State whether each expression is a scalar field, a vector field, or meaningless.
   (a) $\operatorname{curl} f$ (b) $\operatorname{grad} f$
   (c) $\operatorname{div}\mathbf{F}$ (d) $\operatorname{curl}(\operatorname{grad} f)$
   (e) $\operatorname{grad}\mathbf{F}$ (f) $\operatorname{grad}(\operatorname{div}\mathbf{F})$
   (g) $\operatorname{div}(\operatorname{grad} f)$ (h) $\operatorname{grad}(\operatorname{div} f)$
   (i) $\operatorname{curl}(\operatorname{curl}\mathbf{F})$ (j) $\operatorname{div}(\operatorname{div}\mathbf{F})$
   (k) $(\operatorname{grad} f) \times (\operatorname{div}\mathbf{F})$ (l) $\operatorname{div}(\operatorname{curl}(\operatorname{grad} f))$

32–38 ■ Verify the identity, where $\mathbf{r} = x\mathbf{i} + y\mathbf{j} + z\mathbf{k}$ and $r = |\mathbf{r}|$.

32. $\nabla r = \mathbf{r}/r$ **33.** $\nabla \cdot \mathbf{r} = 3$

34. $\nabla \times \mathbf{r} = \mathbf{0}$ **35.** $\nabla(1/r) = -\mathbf{r}/r^3$

36. $\nabla \cdot (r\mathbf{r}) = 4r$ **37.** $\nabla \ln r = \mathbf{r}/r^2$

38. $\nabla^2 r^3 = 12r$

39–40 ■ Is the divergence of the vector field positive, negative, or zero? Explain.

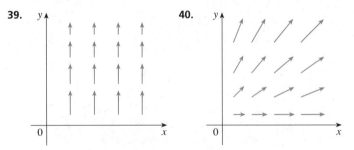

39. **40.**

41. Use Green's Theorem in the form of Equation 13 to prove **Green's first identity:**

$$\iint_D f\nabla^2 g\, dA = \oint_C f(\nabla g) \cdot \mathbf{n}\, ds - \iint_D \nabla f \cdot \nabla g\, dA$$

where D and C satisfy the hypotheses of Green's Theorem and the appropriate partial derivatives of f and g exist and are continuous. (The quantity $\nabla g \cdot \mathbf{n} = D_\mathbf{n}g$ occurs in the line integral. This is the directional derivative in the direction of the normal vector $\mathbf{n}$ and is called the **normal derivative** of g.)

42. Use Green's first identity (Exercise 41) to prove **Green's second identity:**

$$\iint_D (f\nabla^2 g - g\nabla^2 f)\, dA = \oint_C (f\nabla g - g\nabla f) \cdot \mathbf{n}\, ds$$

where D and C satisfy the hypotheses of Green's Theorem and the appropriate partial derivatives of f and g exist and are continuous.

43. This exercise demonstrates a connection between the curl vector and rotations. Let B be a rigid body rotating about the z-axis. The rotation can be described by the vector $\mathbf{w} = \omega\mathbf{k}$, where ω is the angular speed of B, that is, the tangential speed of any point P in B divided by the distance d from the axis of rotation. Let $\mathbf{r} = \langle x, y, z \rangle$ be the position vector of P.
   (a) By considering the angle θ in the figure, show that the velocity field of B is given by $\mathbf{v} = \mathbf{w} \times \mathbf{r}$.
   (b) Show that $\mathbf{v} = -\omega y\mathbf{i} + \omega x\mathbf{j}$.
   (c) Show that $\operatorname{curl}\mathbf{v} = 2\mathbf{w}$.

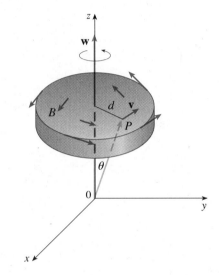

44. Maxwell's equations relating the electric field $\mathbf{E}$ and magnetic field $\mathbf{H}$ as they vary with time in a region containing no charges and no currents can be stated as follows:

$$\operatorname{div}\mathbf{E} = 0 \qquad\qquad \operatorname{div}\mathbf{H} = 0$$

$$\operatorname{curl}\mathbf{E} = -\frac{1}{c}\frac{\partial\mathbf{H}}{\partial t} \qquad \operatorname{curl}\mathbf{H} = \frac{1}{c}\frac{\partial\mathbf{E}}{\partial t}$$

where c is the speed of light. Use these equations to prove the following:

   (a) $\nabla \times (\nabla \times \mathbf{E}) = -\dfrac{1}{c^2}\dfrac{\partial^2\mathbf{E}}{\partial t^2}$

   (b) $\nabla \times (\nabla \times \mathbf{H}) = -\dfrac{1}{c^2}\dfrac{\partial^2\mathbf{H}}{\partial t^2}$

   (c) $\nabla^2\mathbf{E} = \dfrac{1}{c^2}\dfrac{\partial^2\mathbf{E}}{\partial t^2}$ [*Hint:* Use Exercise 29.]

   (d) $\nabla^2\mathbf{H} = \dfrac{1}{c^2}\dfrac{\partial^2\mathbf{H}}{\partial t^2}$

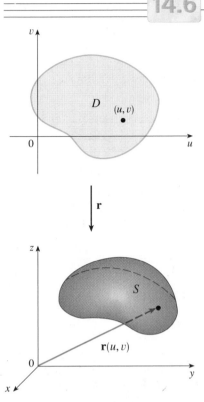

FIGURE 1

A parametric surface

One of the uses of parametric surfaces is in computer graphics. Figure 2 shows the result of trying to graph the sphere $x^2 + y^2 + z^2 = 1$ by solving the equation for z and graphing the top and bottom hemispheres separately. Part of the sphere appears to be missing because of the rectangular grid system used by the computer. The much better picture in Figure 3 was produced by a computer using the parametric equations found in Example 1.

14.6 PARAMETRIC SURFACES AND THEIR AREAS

In Section 8.3 we found the area of a surface of revolution and in Section 13.6 we found the area of a surface with the equation $z = f(x, y)$. Here we discuss more general surfaces, called parametric surfaces, and compute their areas.

In much the same way that we describe a space curve by a vector function $\mathbf{r}(t)$ of a single parameter t, we can describe a surface by a vector function $\mathbf{r}(u, v)$ of two parameters u and v. We suppose that

(1) $\mathbf{r}(u, v) = x(u, v)\,\mathbf{i} + y(u, v)\,\mathbf{j} + z(u, v)\,\mathbf{k}$

is a vector-valued function defined on a region D in the uv-plane and the partial derivatives of x, y, and z with respect to u and v are all continuous. The set of all points (x, y, z) in $\mathbb{R}^3$, such that

(2) $x = x(u, v)$ $y = y(u, v)$ $z = z(u, v)$

and (u, v) varies throughout D, is called a **parametric surface** S and Equations 2 are called **parametric equations** of S. In other words, the surface S is traced out by the position vector $\mathbf{r}(u, v)$ as (u, v) moves throughout the region D (see Figure 1).

EXAMPLE 1   Find a parametric representation of the sphere

$$x^2 + y^2 + z^2 = a^2$$

SOLUTION   The sphere has a simple representation $\rho = a$ in spherical coordinates, so let us choose the angles ϕ and θ in spherical coordinates as the parameters (see Section 11.10). Then, putting $\rho = a$ in the equations for conversion from spherical to rectangular coordinates (Equations 11.10.3), we obtain

$$x = a\sin\phi\cos\theta \qquad y = a\sin\phi\sin\theta \qquad z = a\cos\phi$$

as the parametric equations of the sphere. The corresponding vector equation is

$$\mathbf{r}(\phi, \theta) = a\sin\phi\cos\theta\,\mathbf{i} + a\sin\phi\sin\theta\,\mathbf{j} + a\cos\phi\,\mathbf{k}$$

We have $0 \leq \phi \leq \pi$ and $0 \leq \theta \leq 2\pi$, so the parameter domain is the rectangle $D = [0, \pi] \times [0, 2\pi]$.   ∎

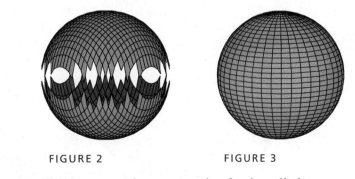

FIGURE 2   FIGURE 3

EXAMPLE 2   Find a parametric representation for the cylinder

$$x^2 + y^2 = 4 \qquad 0 \leq z \leq 1$$

SOLUTION   The cylinder has a simple representation $r = 2$ in cylindrical coordinates, so we choose as parameters θ and z in cylindrical coordinates. Then the parametric equations of the cylinder are

$$x = 2\cos\theta \qquad y = 2\sin\theta \qquad z = z$$

where $0 \le \theta \le 2\pi$ and $0 \le z \le 1$. ∎

EXAMPLE 3 Find a parametric representation for the elliptic paraboloid $z = x^2 + 2y^2$.

SOLUTION   If we regard x and y as parameters, then the parametric equations are simply

$$x = x \qquad y = y \qquad z = x^2 + 2y^2$$

and the vector equation is

$$\mathbf{r}(x, y) = x\,\mathbf{i} + y\,\mathbf{j} + (x^2 + 2y^2)\,\mathbf{k}$$ ∎

In general, a surface given as the graph of a function of x and y, that is, with an equation of the form $z = f(x, y)$, can always be regarded as a parametric surface by taking x and y as parameters and writing the parametric equations as

$$x = x \qquad y = y \qquad z = f(x, y)$$

Surfaces of revolution can be represented parametrically and thus graphed using a computer. For instance, let us consider the surface S obtained by rotating the curve $y = f(x)$, $a \le x \le b$, about the x-axis, where $f(x) \ge 0$ and f' is continuous. Let θ be the angle of rotation as shown in Figure 4. If (x, y, z) is a point on S, then

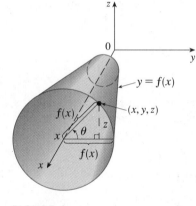

FIGURE 4

(3) $$x = x \qquad y = f(x)\cos\theta \qquad z = f(x)\sin\theta$$

Therefore, we take x and θ as parameters and regard Equations 3 as parametric equations of S. The parameter domain is given by $a \le x \le b$, $0 \le \theta \le 2\pi$.

EXAMPLE 4 Find parametric equations for the surface generated by rotating the curve $y = \sin x$, $0 \le x \le 2\pi$, about the x-axis. Use these equations to graph the surface of revolution.

SOLUTION   From Equations 3, the parametric equations are

$$x = x \qquad y = \sin x \cos\theta \qquad z = \sin x \sin\theta$$

FIGURE 5

and the parameter domain is $0 \le x \le 2\pi$, $0 \le \theta \le 2\pi$. Using a computer to plot these equations and rotate the image, we obtain the graph in Figure 5. ∎

TANGENT PLANES

Let us now find the tangent plane to a parametric surface S given by a vector function $\mathbf{r}(u, v)$ at a point P_0 with position vector $\mathbf{r}(u_0, v_0)$. If we keep u constant by putting $u = u_0$, then $\mathbf{r}(u_0, v)$ becomes a vector function of the single parameter v and defines a curve C_1 lying on S (see Figure 6). The tangent vector C_1 at P_0 is obtained by taking the partial derivative of $\mathbf{r}$ with respect to v:

(4) $$\mathbf{r}_v = \frac{\partial x}{\partial v}(u_0, v_0)\,\mathbf{i} + \frac{\partial y}{\partial v}(u_0, v_0)\,\mathbf{j} + \frac{\partial z}{\partial v}(u_0, v_0)\,\mathbf{k}$$

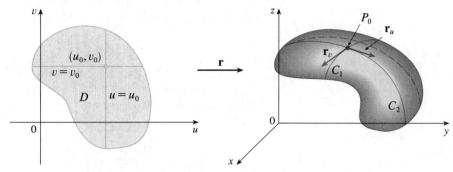

FIGURE 6

Similarly, if we keep v constant by putting $v = v_0$, we get a curve C_2 given by $\mathbf{r}(u, v_0)$ that lies on S, and its tangent vector at P_0 is

(5)
$$\mathbf{r}_u = \frac{\partial x}{\partial u}(u_0, v_0)\,\mathbf{i} + \frac{\partial y}{\partial u}(u_0, v_0)\,\mathbf{j} + \frac{\partial z}{\partial u}(u_0, v_0)\,\mathbf{k}$$

If the **normal vector** $\mathbf{r}_u \times \mathbf{r}_v$ is not $\mathbf{0}$, then the surface S is called **smooth.** (It has no "corners.") In this case the tangent plane to S at P_0 exists and can be found as usual using the normal vector.

EXAMPLE 5 Find the tangent plane to the surface with parametric equations $x = u^2$, $y = v^2$, $z = u + 2v$ at the point $(1, 1, 3)$.

SOLUTION We first compute the tangent vectors:

$$\mathbf{r}_u = \frac{\partial x}{\partial u}\,\mathbf{i} + \frac{\partial y}{\partial u}\,\mathbf{j} + \frac{\partial z}{\partial u}\,\mathbf{k} = 2u\,\mathbf{i} + \mathbf{k}$$

$$\mathbf{r}_v = \frac{\partial x}{\partial v}\,\mathbf{i} + \frac{\partial y}{\partial v}\,\mathbf{j} + \frac{\partial z}{\partial v}\,\mathbf{k} = 2v\,\mathbf{j} + 2\,\mathbf{k}$$

Figure 7 shows the self-intersecting surface in Example 5 and its tangent plane at $(1, 1, 3)$.

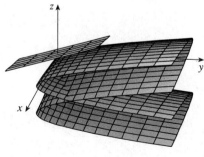

FIGURE 7

Thus the normal vector is

$$\mathbf{r}_u \times \mathbf{r}_v = \begin{vmatrix} \mathbf{i} & \mathbf{j} & \mathbf{k} \\ 2u & 0 & 1 \\ 0 & 2v & 2 \end{vmatrix} = -2v\,\mathbf{i} - 4u\,\mathbf{j} + 4uv\,\mathbf{k}$$

Notice that the point $(1, 1, 3)$ corresponds to the parameter values $u = 1$ and $v = 1$, so the normal vector there is

$$-2\,\mathbf{i} - 4\,\mathbf{j} + 4\,\mathbf{k}$$

Therefore, the tangent plane at $(1, 1, 3)$ is

$$-2(x - 1) - 4(y - 1) + 4(z - 3) = 0$$

or
$$x + 2y - 2z + 3 = 0 \qquad\blacksquare$$

SURFACE AREA

Now we define the surface area of a general parametric surface given by Equation 1. For simplicity we start by considering a surface whose parameter domain D is a rectangle, and we partition it into subrectangles R_{ij}. Let us choose (u_i^*, v_j^*) to be the lower

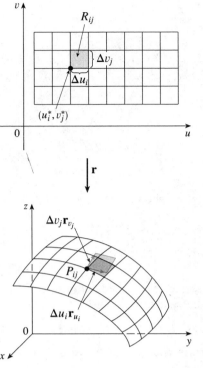

FIGURE 8

left corner of R_{ij} (see Figure 8). The part S_{ij} of the surface S that corresponds to R_{ij} has the point P_{ij} with position vector $\mathbf{r}(u_i^*, v_j^*)$ as one of its corners. Let

$$\mathbf{r}_{u_i} = \mathbf{r}_u(u_i^*, v_j^*) \qquad \text{and} \qquad \mathbf{r}_{v_j} = \mathbf{r}_v(u_i^*, v_j^*)$$

be the tangent vectors at P_{ij} as given by Equations 5 and 4. We approximate S_{ij} by the parallelogram determined by the vectors $\Delta u_i \mathbf{r}_{u_i}$ and $\Delta v_j \mathbf{r}_{v_j}$. (This parallelogram is shown in Figure 8 and lies in the tangent plane to S at P_{ij}.) The area of this parallelogram is

$$|(\Delta u_i \mathbf{r}_{u_i}) \times (\Delta v_j \mathbf{r}_{v_j})| = |\mathbf{r}_{u_i} \times \mathbf{r}_{v_j}| \Delta u_i \Delta v_j$$

and so an approximation to the area of S is

$$\sum_{i=1}^{m} \sum_{j=1}^{n} |\mathbf{r}_{u_i} \times \mathbf{r}_{v_j}| \Delta u_i \Delta v_j$$

Our intuition tells us that this approximation gets better as $\|P\| \to 0$, and we recognize the double sum as a Riemann sum for the double integral $\iint_D |\mathbf{r}_u \times \mathbf{r}_v| \, du \, dv$. This motivates the following definition:

---

(6) DEFINITION   If a smooth parametric surface S is given by the equation

$$\mathbf{r}(u, v) = x(u, v) \, \mathbf{i} + y(u, v) \, \mathbf{j} + z(u, v) \, \mathbf{k} \qquad (u, v) \in D$$

and S is covered just once as (u, v) ranges throughout the parameter domain D, then the **surface area** of S is

$$A(S) = \iint_D |\mathbf{r}_u \times \mathbf{r}_v| \, dA$$

where $\mathbf{r}_u = \dfrac{\partial x}{\partial u} \mathbf{i} + \dfrac{\partial y}{\partial u} \mathbf{j} + \dfrac{\partial z}{\partial u} \mathbf{k}$ $\mathbf{r}_v = \dfrac{\partial x}{\partial v} \mathbf{i} + \dfrac{\partial y}{\partial v} \mathbf{j} + \dfrac{\partial z}{\partial v} \mathbf{k}$

Computing the cross product of the tangent vectors, we get

$$\mathbf{r}_u \times \mathbf{r}_v = \begin{vmatrix} \mathbf{i} & \mathbf{j} & \mathbf{k} \\ \dfrac{\partial x}{\partial u} & \dfrac{\partial y}{\partial u} & \dfrac{\partial z}{\partial u} \\ \dfrac{\partial x}{\partial v} & \dfrac{\partial y}{\partial v} & \dfrac{\partial z}{\partial v} \end{vmatrix}$$

$$= \left(\frac{\partial y}{\partial u} \frac{\partial z}{\partial v} - \frac{\partial z}{\partial u} \frac{\partial y}{\partial v} \right) \mathbf{i} + \left(\frac{\partial z}{\partial u} \frac{\partial x}{\partial v} - \frac{\partial x}{\partial u} \frac{\partial z}{\partial v} \right) \mathbf{j} + \left(\frac{\partial x}{\partial u} \frac{\partial y}{\partial v} - \frac{\partial y}{\partial u} \frac{\partial x}{\partial v} \right) \mathbf{k}$$

$$= \frac{\partial(y, z)}{\partial(u, v)} \mathbf{i} + \frac{\partial(z, x)}{\partial(u, v)} \mathbf{j} + \frac{\partial(x, y)}{\partial(u, v)} \mathbf{k}$$

Notice the similarity between the surface area formula in Equation 7 and the arc length formula in Equation 11.8.2:

$$L = \int_a^b \sqrt{\left(\frac{dx}{dt}\right)^2 + \left(\frac{dy}{dt}\right)^2 + \left(\frac{dz}{dt}\right)^2} \, dt$$

where we have used the notation for Jacobians given in Definition 13.9.7. Thus the formula for surface area in Definition 6 becomes

(7) $$A(S) = \iint_D \sqrt{\left[\frac{\partial(x, y)}{\partial(u, v)}\right]^2 + \left[\frac{\partial(y, z)}{\partial(u, v)}\right]^2 + \left[\frac{\partial(x, z)}{\partial(u, v)}\right]^2} \, dA$$

EXAMPLE 6 Find the surface area of a sphere of radius a.

SOLUTION In Example 1 we found the parametric representation

$$x = a \sin \phi \cos \theta \qquad y = a \sin \phi \sin \theta \qquad z = a \cos \phi$$

where the parameter domain is

$$D = \{(\phi, \theta) \mid 0 \le \phi \le \pi, 0 \le \theta \le 2\pi\}$$

We first compute the cross product of the tangent vectors:

$$\mathbf{r}_\phi \times \mathbf{r}_\theta = \begin{vmatrix} \mathbf{i} & \mathbf{j} & \mathbf{k} \\ \dfrac{\partial x}{\partial \phi} & \dfrac{\partial y}{\partial \phi} & \dfrac{\partial z}{\partial \phi} \\ \dfrac{\partial x}{\partial \theta} & \dfrac{\partial y}{\partial \theta} & \dfrac{\partial z}{\partial \theta} \end{vmatrix} = \begin{vmatrix} \mathbf{i} & \mathbf{j} & \mathbf{k} \\ a \cos \phi \cos \theta & a \cos \phi \sin \theta & -a \sin \phi \\ -a \sin \phi \sin \theta & a \sin \phi \cos \theta & 0 \end{vmatrix}$$

$$= a^2 \sin^2 \phi \cos \theta \, \mathbf{i} + a^2 \sin^2 \phi \sin \theta \, \mathbf{j} + a^2 \sin \phi \cos \phi \, \mathbf{k}$$

Thus

$$|\mathbf{r}_\phi \times \mathbf{r}_\theta| = \sqrt{a^4 \sin^4 \phi \cos^2 \theta + a^4 \sin^4 \phi \sin^2 \theta + a^4 \sin^2 \phi \cos^2 \phi}$$

$$= \sqrt{a^4 \sin^4 \phi + a^4 \sin^2 \phi \cos^2 \phi} = a^2 \sqrt{\sin^2 \phi} = a^2 \sin \phi$$

since $\sin \phi \ge 0$ for $0 \le \phi \le \pi$. Therefore, the area of the sphere is

$$A = \iint_D |\mathbf{r}_\phi \times \mathbf{r}_\theta| \, dA = \int_0^{2\pi} \int_0^\pi a^2 \sin \phi \, d\phi \, d\theta$$

$$= a^2 \int_0^{2\pi} d\theta \int_0^\pi \sin \phi \, d\phi = a^2 (2\pi) 2 = 4\pi a^2 \qquad \blacksquare$$

For the special case of a surface S with equation $z = f(x, y)$, where (x, y) lies in D and f has continuous partial derivatives, we take x and y as parameters. The parametric equations are

$$x = x \qquad y = y \qquad z = f(x, y)$$

so

$$\mathbf{r}_x = \mathbf{i} + \left(\frac{\partial f}{\partial x} \right) \mathbf{k}$$

$$\mathbf{r}_y = \mathbf{j} + \left(\frac{\partial f}{\partial y} \right) \mathbf{k}$$

and

$$(8) \qquad \mathbf{r}_x \times \mathbf{r}_y = \begin{vmatrix} \mathbf{i} & \mathbf{j} & \mathbf{k} \\ 1 & 0 & \dfrac{\partial f}{\partial x} \\ 0 & 1 & \dfrac{\partial f}{\partial y} \end{vmatrix} = -\frac{\partial f}{\partial x} \mathbf{i} - \frac{\partial f}{\partial y} \mathbf{j} + \mathbf{k}$$

Thus the surface area formula in Definition 6 becomes

(9)
$$A(S) = \iint_D \sqrt{1 + \left(\frac{\partial z}{\partial x}\right)^2 + \left(\frac{\partial z}{\partial y}\right)^2}\, dA$$

which agrees with Formula 13.6.2. Equation 9 should be compared with the arc length formula

$$L = \int_a^b \sqrt{1 + \left(\frac{dy}{dx}\right)^2}\, dx$$

EXAMPLE 7   Find the area of the part of the paraboloid $z = x^2 + y^2$ that lies under the plane $z = 9$.

SOLUTION   The plane intersects the paraboloid in the circle $x^2 + y^2 = 9$, $z = 9$. Therefore, the given surface lies above the disk D with center the origin and radius 3 (see Figure 9). Using Formula 9, we have

$$A = \iint_D \sqrt{1 + \left(\frac{\partial z}{\partial x}\right)^2 + \left(\frac{\partial z}{\partial y}\right)^2}\, dA = \iint_D \sqrt{1 + (2x)^2 + (2y)^2}\, dA$$

$$= \iint_D \sqrt{1 + 4(x^2 + y^2)}\, dA$$

Converting to polar coordinates, we obtain

$$A = \int_0^{2\pi} \int_0^3 \sqrt{1 + 4r^2}\, r\, dr\, d\theta = \int_0^{2\pi} d\theta \int_0^3 r\sqrt{1 + 4r^2}\, dr$$

$$= 2\pi \left(\frac{1}{8}\right) \frac{2}{3}(1 + 4r^2)^{3/2}\Big]_0^3 = \frac{\pi}{6}(37\sqrt{37} - 1)$$ ∎

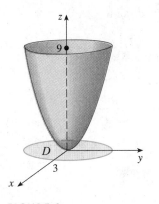

FIGURE 9

The question remains whether our definition of surface area (6) is consistent with the surface area formula from single-variable calculus (8.3.4).

We consider the surface S obtained by rotating the curve $y = f(x)$, $a \le x \le b$, about the x-axis, where $f(x) \ge 0$ and f' is continuous. From Equations 3 we know that parametric equations of S are

$$x = x \qquad y = f(x)\cos\theta \qquad z = f(x)\sin\theta \qquad a \le x \le b \qquad 0 \le \theta \le 2\pi$$

To compute the surface area of S we need the tangent vectors

$$\mathbf{r}_x = \mathbf{i} + f'(x)\cos\theta\,\mathbf{j} + f'(x)\sin\theta\,\mathbf{k}$$

$$\mathbf{r}_\theta = -f(x)\sin\theta\,\mathbf{j} + f(x)\cos\theta\,\mathbf{k}$$

Thus

$$\mathbf{r}_x \times \mathbf{r}_\theta = \begin{vmatrix} \mathbf{i} & \mathbf{j} & \mathbf{k} \\ 1 & f'(x)\cos\theta & f'(x)\sin\theta \\ 0 & -f(x)\sin\theta & f(x)\cos\theta \end{vmatrix}$$

$$= f(x)f'(x)\,\mathbf{i} - f(x)\cos\theta\,\mathbf{j} - f(x)\sin\theta\,\mathbf{k}$$

and so
$$|\mathbf{r}_x \times \mathbf{r}_\theta| = \sqrt{[f(x)]^2[f'(x)]^2 + [f(x)]^2\cos^2\theta + [f(x)]^2\sin^2\theta}$$
$$= \sqrt{[f(x)]^2[1 + (f'(x))^2]} = f(x)\sqrt{1 + [f'(x)]^2}$$

because $f(x) \geqslant 0$. Therefore, the area of S is

$$A = \iint\limits_{D} |\mathbf{r}_x \times \mathbf{r}_\theta|\, dA = \int_0^{2\pi} \int_a^b f(x)\sqrt{1 + [f'(x)]^2}\; dx\, d\theta$$
$$= 2\pi \int_a^b f(x)\sqrt{1 + [f'(x)]^2}\; dx$$

This is precisely the formula that was used to define the area of a surface of revolution in single-variable calculus (8.3.4).

EXERCISES 14.6

1–8 ■ Find a parametric representation for the surface.

1. The upper half of the ellipsoid $3x^2 + 2y^2 + z^2 = 1$

2. The part of the hyperboloid $-x^2 - y^2 + z^2 = 1$ that lies below the rectangle $[-1, 1] \times [-3, 3]$

3. The part of the elliptic paraboloid $y = 6 - 3x^2 - 2z^2$ that lies to the right of the xz-plane

4. The part of the elliptic paraboloid $x + y^2 + 2z^2 = 4$ that lies in front of the plane $x = 0$

5. The part of the sphere $x^2 + y^2 + z^2 = 4$ that lies above the cone $z = \sqrt{x^2 + y^2}$

6. The part of the cylinder $x^2 + z^2 = 1$ that lies between the planes $y = -1$ and $y = 3$

7. The part of the plane $z = 5$ that lies inside the cylinder $x^2 + y^2 = 16$

8. The part of the plane $z = x + 3$ that lies inside the cylinder $x^2 + y^2 = 1$

9–12 ■ Match the equations with the graphs (labeled I–IV). Give reasons for your answer.

9. $\mathbf{r}(u, v) = u\cos v\,\mathbf{i} + u\sin v\,\mathbf{j} + v\,\mathbf{k}$

10. $\mathbf{r}(u, v) = u\cos v\,\mathbf{i} + u\sin v\,\mathbf{j} + u\,\mathbf{k}$

11. $x = (u - \sin u)\cos v, \quad y = (1 - \cos u)\sin v, \quad z = u$

12. $x = (1 - u)(3 + \cos v)\cos 4\pi u,$
    $y = (1 - u)(3 + \cos v)\sin 4\pi u, \quad z = 3u + (1 - u)\sin v$

I   II

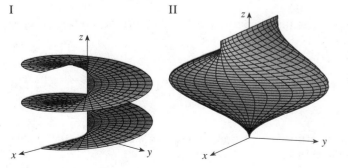

III   IV

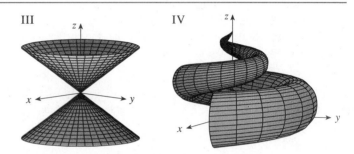

13. Find parametric equations for the surface obtained by rotating the curve $y = e^{-x}$, $0 \leqslant x \leqslant 3$, about the x-axis and use them to graph the surface.

14. Find parametric equations for the surface obtained by rotating the curve $x = 4y^2 - y^4$, $-2 \leqslant y \leqslant 2$, about the y-axis and use them to graph the surface.

15–18 ■ Find an equation of the tangent plane to the given parametric surface at the specified point. If you have software that graphs parametric surfaces, use a computer to graph the surface and the tangent plane.

15. $x = u + v, \quad y = 3u^2, \quad z = u - v; \quad (2, 3, 0)$

16. $x = u^2, \quad y = u - v^2, \quad z = v^2; \quad (1, 0, 1)$

17. $\mathbf{r}(u, v) = uv\,\mathbf{i} + ue^v\,\mathbf{j} + ve^u\,\mathbf{k}; \quad (0, 0, 0)$

18. $\mathbf{r}(u, v) = (u + v)\,\mathbf{i} + u\cos v\,\mathbf{j} + v\sin u\,\mathbf{k}; \quad (1, 1, 0)$

19–27 ■ Find the area of the given surface.

19. The part of the plane $x + 2y + z = 4$ that lies inside the cylinder $x^2 + y^2 = 4$

20. The part of the surface $z = x + y^2$ that lies above the triangle with vertices $(0, 0)$, $(1, 1)$, and $(0, 1)$

21. The part of the hyperbolic paraboloid $z = y^2 - x^2$ that lies between the cylinders $x^2 + y^2 = 1$ and $x^2 + y^2 = 4$

22. The part of the paraboloid $x = y^2 + z^2$ that lies inside the cylinder $y^2 + z^2 = 9$

23. The part of the surface $y = 4x + z^2$ that lies between the planes $x = 0$, $x = 1$, $z = 0$, and $z = 1$

24. The part of the cylinder $x^2 + z^2 = a^2$ that lies inside the cylinder $x^2 + y^2 = a^2$

25. The part of the sphere $x^2 + y^2 + z^2 = a^2$ that lies inside the cylinder $x^2 + y^2 = ax$

26. The helicoid (or spiral ramp) with vector equation $\mathbf{r}(u, v) = u \cos v\, \mathbf{i} + u \sin v\, \mathbf{j} + v\, \mathbf{k}$, $0 \leq u \leq 1$, $0 \leq v \leq \pi$

27. The surface with parametric equations $x = uv$, $y = u + v$, $z = u - v$, $u^2 + v^2 \leq 1$

28. (a) Set up, but do not evaluate, a double integral for the area of the surface with parametric equations $x = au \cos v$, $y = bu \sin v$, $z = u^2$, $0 \leq u \leq 2$, $0 \leq v \leq 2\pi$.
(b) Eliminate the parameters to show that the surface is an elliptic paraboloid and set up another double integral for the surface area.
(c) Use the parametric equations in part (a) with $a = 2$ and $b = 3$ to graph the surface.
(d) For the case $a = 2$, $b = 3$, use a computer algebra system to find the surface area correct to four decimal places.

29. (a) Show that the parametric equations $x = a \sin u \cos v$, $y = b \sin u \sin v$, $z = c \cos u$, $0 \leq u \leq \pi$, $0 \leq v \leq 2\pi$, represent an ellipsoid.
(b) Use the parametric equations in part (a) to graph the ellipsoid for the case $a = 1$, $b = 2$, $c = 3$.

(c) Set up, but do not evaluate, a double integral for the surface area of the ellipsoid in part (b).

30. (a) Show that the parametric equations $x = a \cosh u \cos v$, $y = b \cosh u \sin v$, $z = c \sinh u$, represent a hyperboloid of one sheet.
(b) Use the parametric equations in part (a) to graph the hyperboloid for the case $a = 1$, $b = 2$, $c = 3$.
(c) Set up, but do not evaluate, a double integral for the surface area of the part of the hyperboloid in part (b) that lies between the planes $z = -3$ and $z = 3$.

31. Find a parametric representation for the torus obtained by rotating about the z-axis the circle in the xz-plane with center $(b, 0, 0)$ and radius $a < b$. [*Hint:* Take as parameters the angles θ and α shown in the figure.]

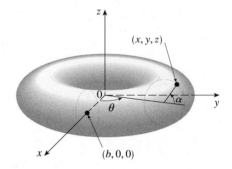

32. Use the parametric representation from Exercise 31 to find the surface area of the torus.

  SURFACE INTEGRALS

The relationship between surface integrals and surface area is much the same as the relationship between line integrals and arc length. Suppose f is a function of three variables whose domain includes a surface S. We divide S into patches S_{ij} with area ΔS_{ij}. We evaluate f at a point P_{ij}^* in each patch, multiply by the area ΔS_{ij}, and form the sum

$$\sum_{i=1}^{m} \sum_{j=1}^{n} f(P_{ij}^*)\, \Delta S_{ij}$$

Then we take the limit as the patch size approaches 0 and define the **surface integral of f over the surface S** as

(1)

$$\iint_S f(x, y, z)\, dS = \lim_{\|P\| \to 0} \sum_{i=1}^{m} \sum_{j=1}^{n} f(P_{ij}^*)\, \Delta S_{ij}$$

Notice the analogy with the definition of a line integral (14.2.2) and also the analogy with the definition of a double integral (13.1.4).

To evaluate the surface integral in Equation 1 we approximate the patch area ΔS_{ij} by the area ΔT_{ij} of an approximating parallelogram in the tangent plane, and the limit becomes a double integral. We now explain the details for two types of surfaces, graphs and parametric surfaces.

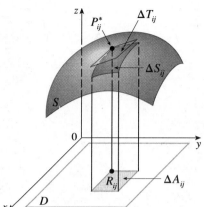

FIGURE 1

GRAPHS   If the surface S is a graph of a function of two variables, then it has an equation of the form $z = g(x, y)$, $(x, y) \in D$, and is illustrated in Figure 1. The patch S_{ij} lies directly above a rectangle R_{ij} in a partition P of D and the point P_{ij}^* in S_{ij} is of the form $(x_i^*, y_j^*, g(x_i^*, y_j^*))$. As with surface area in Section 13.6, we use the approximation

$$\Delta S_{ij} \approx \Delta T_{ij} = \sqrt{[g_x(x_i, y_j)]^2 + [g_y(x_i, y_j)]^2 + 1} \; \Delta A_{ij}$$

It can be shown that, if f is continuous on S and g has continuous derivatives, then Definition 1 becomes

$$\iint_S f(x, y, z) \, dS$$

$$= \lim_{\|P\| \to 0} \sum_{i=1}^{m} \sum_{j=1}^{n} f(x_i^*, y_j^*, g(x_i^*, y_j^*)) \sqrt{[g_x(x_i, y_j)]^2 + [g_y(x_i, y_j)]^2 + 1} \; \Delta A_{ij}$$

$$= \iint_D f(x, y, g(x, y)) \sqrt{[g_x(x, y)]^2 + [g_y(x, y)]^2 + 1} \; dA$$

This formula is usually written in the following form:

(2) $$\iint_S f(x, y, z) \, dS = \iint_D f(x, y, g(x, y)) \sqrt{\left(\frac{\partial z}{\partial x}\right)^2 + \left(\frac{\partial z}{\partial y}\right)^2 + 1} \; dA$$

Similar formulas apply when it is more convenient to project S onto the yz-plane or xz-plane. For instance, if S is a surface with equation $y = h(x, z)$ and D is its projection on the xz-plane, then

$$\iint_S f(x, y, z) \, dS = \iint_D f(x, h(x, z), z) \sqrt{\left(\frac{\partial y}{\partial x}\right)^2 + \left(\frac{\partial y}{\partial z}\right)^2 + 1} \; dA$$

Surface integrals have applications similar to those for the integrals we have previously considered. For example, if a thin sheet (say, of aluminum foil) has the shape of a surface S and the density (mass per unit area) at the point (x, y, z) is $\rho(x, y, z)$, then the total **mass** of the sheet is

$$m = \iint_S \rho(x, y, z) \, dS$$

and the **center of mass** is $(\bar{x}, \bar{y}, \bar{z})$, where

$$\bar{x} = \frac{1}{m} \iint_S x\rho(x, y, z) \, dS \qquad \bar{y} = \frac{1}{m} \iint_S y\rho(x, y, z) \, dS \qquad \bar{z} = \frac{1}{m} \iint_S z\rho(x, y, z) \, dS$$

Moments of inertia can also be defined as before (see Exercise 33).

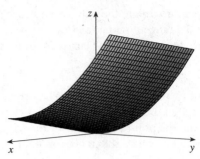

FIGURE 2

EXAMPLE 1   Evaluate $\iint_S y\,dS$, where S is the surface $z = x + y^2$, $0 \leqslant x \leqslant 1$, $0 \leqslant y \leqslant 2$ (see Figure 2).

SOLUTION   Since

$$\frac{\partial z}{\partial x} = 1 \qquad \text{and} \qquad \frac{\partial z}{\partial y} = 2y$$

Formula 2 gives

$$\iint_S y\,dS = \iint_D y\,\sqrt{1 + \left(\frac{\partial z}{\partial x}\right)^2 + \left(\frac{\partial z}{\partial y}\right)^2}\,dA$$

$$= \int_0^1 \int_0^2 y\,\sqrt{1 + 1 + 4y^2}\,dy\,dx$$

$$= \int_0^1 dx\,\sqrt{2}\int_0^2 y\,\sqrt{1 + 2y^2}\,dy$$

$$= \sqrt{2}\,(\tfrac{1}{4})\tfrac{2}{3}(1 + 2y^2)^{3/2}\big]_0^2 = \frac{13\sqrt{2}}{3} \qquad ■$$

PARAMETRIC SURFACES   Suppose that a surface S has a vector equation

$$\mathbf{r}(u, v) = x(u, v)\,\mathbf{i} + y(u, v)\,\mathbf{j} + z(u, v)\,\mathbf{k} \qquad (u, v) \in D$$

We first assume that the parameter domain D is a rectangle. A partition of D into sub-rectangles R_{ij} with dimensions Δu_i and Δv_j determines a partition of S into curvilinear regions S_{ij} that lie on S (see Figure 3). A point (u_i^*, v_j^*) in R_{ij} determines a point on S_{ij} with position vector $\mathbf{r}(u_i^*, v_j^*)$. As in Section 14.6 we approximate S_{ij} by a parallelogram with area

$$\Delta T_{ij} = |\mathbf{r}_{u_i} \times \mathbf{r}_{v_j}|\,\Delta u_i\,\Delta v_j$$

where $\qquad \mathbf{r}_u = \dfrac{\partial x}{\partial u}\,\mathbf{i} + \dfrac{\partial y}{\partial u}\,\mathbf{j} + \dfrac{\partial z}{\partial u}\,\mathbf{k} \qquad \mathbf{r}_v = \dfrac{\partial x}{\partial v}\,\mathbf{i} + \dfrac{\partial y}{\partial v}\,\mathbf{j} + \dfrac{\partial z}{\partial v}\,\mathbf{k}$

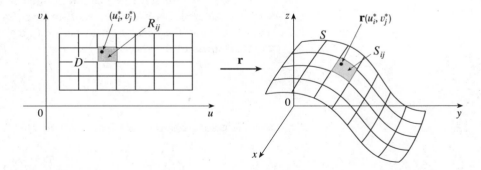

FIGURE 3

If the components are continuous and $\mathbf{r}_u$ and $\mathbf{r}_v$ are nonzero and nonparallel in the interior of D, it can be shown from Definition 1, even when D is not a rectangle, that

We assume that the surface is covered only once as (u, v) ranges throughout D. The value of the surface integral does not depend on the parametrization that is used.

(3)
$$\iint_S f(x, y, z)\, dS = \iint_D f(\mathbf{r}(u, v))\, |\mathbf{r}_u \times \mathbf{r}_v|\, dA$$

This should be compared with the formula for a line integral:

$$\int_C f(x, y, z)\, ds = \int_a^b f(\mathbf{r}(t))\, |\mathbf{r}'(t)|\, dt$$

Formula 3 allows us to compute a surface integral by converting it into a double integral over the parameter domain D. When using this formula, remember that $f(\mathbf{r}(u, v))$ is evaluated by writing $x = x(u, v)$, $y = y(u, v)$, and $z = z(u, v)$ in the formula for $f(x, y, z)$. In fact, by using the same calculation that led to Equation 14.6.7, we can write the right side of Equation 3 as

$$\iint_D f(x(u, v), y(u, v), z(u, v))\, \sqrt{\left[\frac{\partial(x, y)}{\partial(u, v)}\right]^2 + \left[\frac{\partial(y, z)}{\partial(u, v)}\right]^2 + \left[\frac{\partial(x, z)}{\partial(u, v)}\right]^2}\, dA$$

which is the analogue of Equation 14.2.9. Observe that $\iint_S 1\, dS = A(S)$.

Any surface S with equation $z = g(x, y)$ can be regarded as a parametric surface with parametric equations

$$x = x \qquad y = y \qquad z = g(x, y)$$

and from Equation 14.6.8 we have

$$|\mathbf{r}_x \times \mathbf{r}_y| = \sqrt{1 + \left(\frac{\partial z}{\partial x}\right)^2 + \left(\frac{\partial z}{\partial y}\right)^2}$$

Therefore, in this case, Formula 3 becomes Formula 2.

EXAMPLE 2   Compute the surface integral $\iint_S x^2\, dS$, where S is the unit sphere $x^2 + y^2 + z^2 = 1$.

SOLUTION   As in Example 1 in Section 14.6, we use the parametric representation

$$x = \sin\phi \cos\theta \qquad y = \sin\phi \sin\theta \qquad z = \cos\phi \qquad 0 \le \phi \le \pi \qquad 0 \le \theta \le 2\pi$$

that is,
$$\mathbf{r}(\phi, \theta) = \sin\phi \cos\theta\, \mathbf{i} + \sin\phi \sin\theta\, \mathbf{j} + \cos\phi\, \mathbf{k}$$

As in Example 6 in Section 14.6, we can compute that

$$|\mathbf{r}_\phi \times \mathbf{r}_\theta| = \sin\phi$$

Therefore, by Formula 3,

$$\iint_S x^2\, dS = \iint_D (\sin\phi \cos\theta)^2\, |\mathbf{r}_\phi \times \mathbf{r}_\theta|\, dA$$

$$= \int_0^{2\pi} \int_0^\pi \sin^2\phi \cos^2\theta \sin\phi\, d\phi\, d\theta = \int_0^{2\pi} \cos^2\theta\, d\theta \int_0^\pi \sin^3\phi\, d\phi$$

$$= \int_0^{2\pi} \tfrac{1}{2}(1 + \cos 2\theta)\, d\theta \int_0^\pi (\sin\phi - \sin\phi \cos^2\phi)\, d\phi$$

$$= \tfrac{1}{2}\left[\theta + \tfrac{1}{2}\sin 2\theta\right]_0^{2\pi}\left[-\cos\phi + \tfrac{1}{3}\cos^3\phi\right]_0^{\pi} = \frac{4\pi}{3}$$ ∎

If S is a piecewise-smooth surface, that is, a finite union of smooth surfaces S_1, S_2, ..., S_n that intersect only along their boundaries, then the surface integral of f over S is defined by

$$\iint_S f(x, y, z)\, dS = \iint_{S_1} f(x, y, z)\, dS + \cdots + \iint_{S_n} f(x, y, z)\, dS$$

EXAMPLE 3   Evaluate $\iint_S z\, dS$, where S is the surface whose sides S_1 are given by the cylinder $x^2 + y^2 = 1$, whose bottom S_2 is the disk $x^2 + y^2 \leq 1$ in the plane $z = 0$, and whose top S_3 is the part of the plane $z = x + 1$ that lies above S_2.

SOLUTION   The surface S is shown in Figure 4. (We have changed the usual position of the axes to get a better look at S.) For S_1 we use θ and z as parameters (see Example 2 in Section 14.6) and write its parametric equations as

$$x = \cos\theta \qquad y = \sin\theta \qquad z = z$$

where $\qquad 0 \leq \theta \leq 2\pi \quad$ and $\quad 0 \leq z \leq 1 + x = 1 + \cos\theta$

Therefore

$$\mathbf{r}_\theta \times \mathbf{r}_z = \begin{vmatrix} \mathbf{i} & \mathbf{j} & \mathbf{k} \\ -\sin\theta & \cos\theta & 0 \\ 0 & 0 & 1 \end{vmatrix} = \cos\theta\, \mathbf{i} + \sin\theta\, \mathbf{j}$$

and $\qquad |\mathbf{r}_\theta \times \mathbf{r}_z| = \sqrt{\cos^2\theta + \sin^2\theta} = 1$

Thus the surface integral over S_1 is

$$\iint_{S_1} z\, dS = \iint_D z\, |\mathbf{r}_\theta \times \mathbf{r}_z|\, dA$$

$$= \int_0^{2\pi} \int_0^{1+\cos\theta} z\, dz\, d\theta = \int_0^{2\pi} \tfrac{1}{2}(1 + \cos\theta)^2\, d\theta$$

$$= \tfrac{1}{2}\int_0^{2\pi} \left[1 + 2\cos\theta + \tfrac{1}{2}(1 + \cos 2\theta)\right] d\theta$$

$$= \tfrac{1}{2}\left[\tfrac{3}{2}\theta + 2\sin\theta + \tfrac{1}{4}\sin 2\theta\right]_0^{2\pi} = \frac{3\pi}{2}$$

Since S_2 lies in the plane $z = 0$, we have

$$\iint_{S_2} z\, dS = \iint_{S_2} 0\, dS = 0$$

The top surface S_3 lies above the unit disk D and is part of the plane $z = 1 + x$. So, taking $g(x, y) = 1 + x$ in Formula 2 and converting to polar coordinates, we have

$$\iint_{S_3} z\, dS = \iint_D (1 + x) \sqrt{1 + \left(\frac{\partial z}{\partial x}\right)^2 + \left(\frac{\partial z}{\partial y}\right)^2}\, dA$$

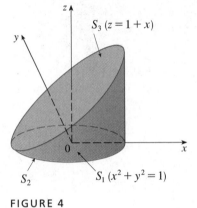

$S_3\ (z = 1 + x)$

S_2

$S_1\ (x^2 + y^2 = 1)$

FIGURE 4

$$= \int_0^{2\pi} \int_0^1 (1 + r\cos\theta)\sqrt{1 + 1 + 0}\; r\, dr\, d\theta$$

$$= \sqrt{2} \int_0^{2\pi} \int_0^1 (r + r^2\cos\theta)\, dr\, d\theta$$

$$= \sqrt{2} \int_0^{2\pi} \left(\tfrac{1}{2} + \tfrac{1}{3}\cos\theta\right) d\theta = \sqrt{2} \left[\frac{\theta}{2} + \frac{\sin\theta}{3}\right]_0^{2\pi} = \sqrt{2}\,\pi$$

Therefore
$$\iint_S z\, dS = \iint_{S_1} z\, dS + \iint_{S_2} z\, dS + \iint_{S_3} z\, dS$$

$$= \frac{3\pi}{2} + 0 + \sqrt{2}\,\pi = \left(\tfrac{3}{2} + \sqrt{2}\right)\pi \qquad\blacksquare$$

ORIENTED SURFACES

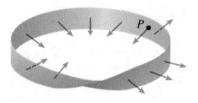

FIGURE 5
A Möbius strip

In order to define surface integrals of vector fields, we need to rule out nonorientable surfaces such as the Möbius strip shown in Figure 5. [It is named after the German geometer August Möbius (1790–1868).] You can construct one for yourself by taking a long rectangular strip of paper, giving it a half-twist, and taping the short edges together as in Figure 6. If an ant were to crawl along the Möbius strip starting at a point P, it would end up on the "other side" of the strip (that is, with its upper side pointing in the opposite direction). Then, if the ant continued to crawl in the same direction, it would end up back at the same point P without ever having crossed an edge. (If you have constructed a Möbius strip, try drawing a pencil line down the middle.) Therefore a Möbius strip really has only one side.

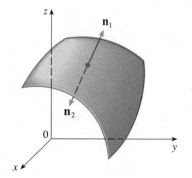

FIGURE 6
Constructing a Möbius strip

From now on we consider only orientable (two-sided) surfaces. We start with a surface S that has a tangent plane at every point (x, y, z) on S (except at any boundary points). There are two unit normal vectors $\mathbf{n}_1$ and $\mathbf{n}_2 = -\mathbf{n}_1$ at (x, y, z) (see Figure 7). If it is possible to choose a unit normal vector $\mathbf{n}$ at every such point (x, y, z) so that $\mathbf{n}$ varies continuously over S, then S is called an **oriented surface** and the given choice of

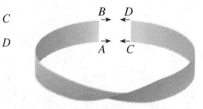

FIGURE 7

n provides S with an **orientation.** There are two possible orientations for any orientable surface (see Figure 8).

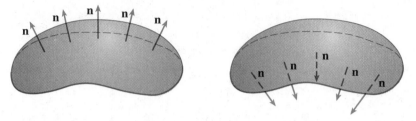

FIGURE 8

The two orientations of an orientable surface

For a surface $z = g(x, y)$ given as the graph of g, we use Equation 14.6.8 and see that the induced orientation is given by the unit normal vector

(4)
$$\mathbf{n} = \frac{-\dfrac{\partial g}{\partial x}\,\mathbf{i} - \dfrac{\partial g}{\partial y}\,\mathbf{j} + \mathbf{k}}{\sqrt{1 + \left(\dfrac{\partial g}{\partial x}\right)^2 + \left(\dfrac{\partial g}{\partial y}\right)^2}}$$

Since the **k**-component is positive, this gives the *upward* orientation of the surface.

If S is a smooth orientable surface given in parametric form by a vector function $\mathbf{r}(u, v)$, then it is automatically supplied with the orientation of the unit normal vector

(5)
$$\mathbf{n} = \frac{\mathbf{r}_u \times \mathbf{r}_v}{|\mathbf{r}_u \times \mathbf{r}_v|}$$

and the opposite orientation is given by $-\mathbf{n}$. For instance, in Example 1 in Section 14.6 we found the parametric representation

$$\mathbf{r}(\phi, \theta) = a \sin\phi \cos\theta\,\mathbf{i} + a \sin\phi \sin\theta\,\mathbf{j} + a \cos\phi\,\mathbf{k}$$

for the sphere $x^2 + y^2 + z^2 = a^2$. Then in Example 6 in Section 14.6 we found that

$$\mathbf{r}_\phi \times \mathbf{r}_\theta = a^2 \sin^2\phi \cos\theta\,\mathbf{i} + a^2 \sin^2\phi \sin\theta\,\mathbf{j} + a^2 \sin\phi \cos\phi\,\mathbf{k}$$

and $$|\mathbf{r}_\phi \times \mathbf{r}_\theta| = a^2 \sin\phi$$

So the orientation induced by $\mathbf{r}(\phi, \theta)$ is defined by the unit normal vector

$$\mathbf{n} = \frac{\mathbf{r}_\phi \times \mathbf{r}_\theta}{|\mathbf{r}_\phi \times \mathbf{r}_\theta|} = \sin\phi \cos\theta\,\mathbf{i} + \sin\phi \sin\theta\,\mathbf{j} + \cos\phi\,\mathbf{k} = \frac{1}{a}\,\mathbf{r}(\phi, \theta)$$

Observe that **n** points in the same direction as the position vector, that is, outward from the sphere (see Figure 9). The opposite (inward) orientation would have been obtained (see Figure 10) if we had reversed the order of the parameters because $\mathbf{r}_\theta \times \mathbf{r}_\phi = -\mathbf{r}_\phi \times \mathbf{r}_\theta$.

For a **closed surface,** that is, a surface that is the boundary of a solid region E, the convention is that the **positive orientation** is the one for which the normal vectors point *outward* from E, and inward-pointing normals give the negative orientation (see Figures 9 and 10).

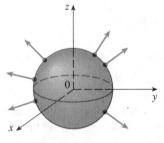

FIGURE 9

Positive orientation

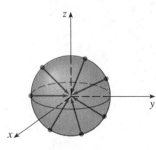

FIGURE 10

Negative orientation

SURFACE INTEGRALS OF VECTOR FIELDS

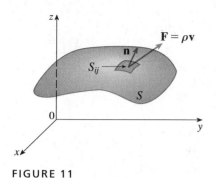

FIGURE 11

Suppose that S is an oriented surface with unit normal vector $\mathbf{n}$, and imagine a fluid with density $\rho(x, y, z)$ and velocity field $\mathbf{v}(x, y, z)$ flowing through S. (Think of S as an imaginary surface so that it does not impede the fluid flow.) Then the rate of flow (mass per unit time) per unit area is $\rho\mathbf{v}$. If we partition S into small parts S_{ij}, as in Figure 11 (compare with Figures 1 and 2), then S_{ij} is nearly planar and so we can approximate the mass of fluid crossing S_{ij} in the direction of the normal $\mathbf{n}$ per unit time by the quantity

$$\rho\mathbf{v} \cdot \mathbf{n} A(S_{ij})$$

where ρ, $\mathbf{v}$, and $\mathbf{n}$ are evaluated at some point on S_{ij}. (Recall that the component of the vector $\rho\mathbf{v}$ in the direction of the unit vector $\mathbf{n}$ is $\rho\mathbf{v} \cdot \mathbf{n}$.) By summing these quantities and taking the limit we get, according to Definition 1, the surface integral of the function $\rho\mathbf{v} \cdot \mathbf{n}$ over S:

$$(6) \qquad \iint_S \rho\mathbf{v} \cdot \mathbf{n} \, dS = \iint_S \rho(x, y, z)\mathbf{v}(x, y, z) \cdot \mathbf{n}(x, y, z) \, dS$$

and this is interpreted physically as the rate of flow through S.

If we write $\mathbf{F} = \rho\mathbf{v}$, then $\mathbf{F}$ is also a vector field on $\mathbb{R}^3$ and the integral in Equation 6 becomes

$$\iint_S \mathbf{F} \cdot \mathbf{n} \, dS$$

A surface integral of this form occurs frequently in physics, even when $\mathbf{F}$ is not $\rho\mathbf{v}$, and is called the *surface integral* (or *flux integral*) of $\mathbf{F}$ over S.

---

(7) DEFINITION If $\mathbf{F}$ is a continuous vector field defined on an oriented surface S with unit normal vector $\mathbf{n}$, then the **surface integral of F over S** is

$$\iint_S \mathbf{F} \cdot d\mathbf{S} = \iint_S \mathbf{F} \cdot \mathbf{n} \, dS$$

This integral is also called the **flux** of $\mathbf{F}$ across S.

In words, Definition 7 says that the surface integral of a vector field over S is equal to the surface integral of its normal component over S (as previously defined).

GRAPHS   In the case of a surface S given by a graph $z = g(x, y)$, we can find $\mathbf{n}$ by noting that S is also the level surface $f(x, y, z) = z - g(x, y) = 0$. We know that the gradient $\nabla f(x, y, z)$ is normal to this surface at (x, y, z) and so a unit normal vector is

$$\mathbf{n} = \frac{\nabla f(x, y, z)}{|\nabla f(x, y, z)|} = \frac{-g_x(x, y)\,\mathbf{i} - g_y(x, y)\,\mathbf{j} + \mathbf{k}}{\sqrt{[g_x(x, y)]^2 + [g_y(x, y)]^2 + 1}}$$

Since the $\mathbf{k}$-component is positive, this is the upward unit normal. If we now use Formula 2 to evaluate the surface integral (7) with

$$\mathbf{F}(x, y, z) = P(x, y, z)\,\mathbf{i} + Q(x, y, z)\,\mathbf{j} + R(x, y, z)\,\mathbf{k}$$

we get

$$\iint_S \mathbf{F} \cdot d\mathbf{S} = \iint_S \mathbf{F} \cdot \mathbf{n} \, dS$$

$$= \iint_D (P\mathbf{i} + Q\mathbf{j} + R\mathbf{k}) \cdot \frac{-\dfrac{\partial g}{\partial x}\mathbf{i} - \dfrac{\partial g}{\partial y}\mathbf{j} + \mathbf{k}}{\sqrt{\left(\dfrac{\partial g}{\partial x}\right)^2 + \left(\dfrac{\partial g}{\partial y}\right)^2 + 1}} \sqrt{\left(\dfrac{\partial g}{\partial x}\right)^2 + \left(\dfrac{\partial g}{\partial y}\right)^2 + 1} \, dA$$

or

(8)
$$\iint_S \mathbf{F} \cdot d\mathbf{S} = \iint_D \left(-P\frac{\partial g}{\partial x} - Q\frac{\partial g}{\partial y} + R \right) dA$$

For a downward unit normal we multiply by -1. Similar formulas can be worked out if S is given by $y = h(x, z)$ or $x = k(y, z)$. (See Exercises 29 and 30.)

EXAMPLE 4   Evaluate $\iint_S \mathbf{F} \cdot d\mathbf{S}$, where $\mathbf{F}(x, y, z) = y\mathbf{i} + x\mathbf{j} + z\mathbf{k}$ and S is the boundary of the solid region E enclosed by the paraboloid $z = 1 - x^2 - y^2$ and the plane $z = 0$.

SOLUTION S consists of a parabolic top surface S_1 and a circular bottom surface S_2 (see Figure 12). Since S is a closed surface, we use the convention of positive (outward) orientation. This means that S_1 is oriented upward and we can use Equation 8 with D being the projection of S_1 on the xy-plane, namely, the disk $x^2 + y^2 \leqslant 1$. Since

$$P(x, y, z) = y \qquad Q(x, y, z) = x \qquad R(x, y, z) = z = 1 - x^2 - y^2$$

on S_1 and

$$\frac{\partial g}{\partial x} = -2x \qquad \frac{\partial g}{\partial y} = -2y$$

we have

$$\iint_{S_1} \mathbf{F} \cdot d\mathbf{S} = \iint_D \left(-P\frac{\partial g}{\partial x} - Q\frac{\partial g}{\partial y} + R \right) dA$$

$$= \iint_D [-y(-2x) - x(-2y) + 1 - x^2 - y^2] \, dA$$

$$= \iint_D (1 + 4xy - x^2 - y^2) \, dA$$

$$= \int_0^{2\pi} \int_0^1 (1 + 4r^2 \cos\theta \sin\theta - r^2)r \, dr \, d\theta$$

$$= \int_0^{2\pi} \int_0^1 (r - r^3 + 4r^3 \cos\theta \sin\theta) \, dr \, d\theta$$

$$= \int_0^{2\pi} \left(\tfrac{1}{4} + \cos\theta \sin\theta \right) d\theta = \tfrac{1}{4}(2\pi) + 0 = \frac{\pi}{2}$$

The disk S_2 is oriented downward, so its unit normal vector is $\mathbf{n} = -\mathbf{k}$ and we have

$$\iint_{S_2} \mathbf{F} \cdot d\mathbf{S} = \iint_{S_2} \mathbf{F} \cdot (-\mathbf{k}) \, dS = \iint_D (-z) \, dA = \iint_D 0 \, dA = 0$$

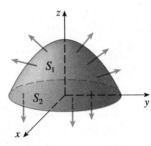

FIGURE 12

since $z = 0$ on S_2. Finally, we compute, by definition, $\iint_S \mathbf{F} \cdot d\mathbf{S}$ as the sum of the surface integrals of $\mathbf{F}$ over the pieces S_1 and S_2:

$$\iint_S \mathbf{F} \cdot d\mathbf{S} = \iint_{S_1} \mathbf{F} \cdot d\mathbf{S} + \iint_{S_2} \mathbf{F} \cdot d\mathbf{S} = \frac{\pi}{2} + 0 = \frac{\pi}{2}$$ ∎

PARAMETRIC SURFACES   If S is given by a vector function $\mathbf{r}(u, v)$, then $\mathbf{n}$ is given by Equation 5, and from Definition 7 and Equation 3 we have

$$\iint_S \mathbf{F} \cdot d\mathbf{S} = \iint_S \mathbf{F} \cdot \frac{\mathbf{r}_u \times \mathbf{r}_v}{|\mathbf{r}_u \times \mathbf{r}_v|} \, dS$$

$$= \iint_D \left[\mathbf{F}(\mathbf{r}(u, v)) \cdot \frac{\mathbf{r}_u \times \mathbf{r}_v}{|\mathbf{r}_u \times \mathbf{r}_v|} \right] |\mathbf{r}_u \times \mathbf{r}_v| \, dA$$

where D is the parameter domain. Thus we have

Compare Equation 9 to the similar expression for evaluating line integrals of vector fields in Definition 14.2.13:

$$\int_C \mathbf{F} \cdot d\mathbf{r} = \int_a^b \mathbf{F}(\mathbf{r}(t)) \cdot \mathbf{r}'(t) \, dt$$

(9)
$$\boxed{\iint_S \mathbf{F} \cdot d\mathbf{S} = \iint_D \mathbf{F} \cdot (\mathbf{r}_u \times \mathbf{r}_v) \, dA}$$

Notice that, in view of Equation 14.6.8, we have

$$\mathbf{r}_x \times \mathbf{r}_y = -\frac{\partial g}{\partial x} \mathbf{i} - \frac{\partial g}{\partial y} \mathbf{j} + \mathbf{k}$$

and so Formula 8 is just a special case of Formula 9.

Figure 13 shows the vector field $\mathbf{F}$ in Example 5 at points on the unit sphere.

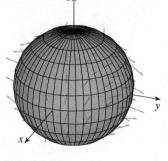

FIGURE 13

EXAMPLE 5   Find the flux of the vector field $\mathbf{F}(x, y, z) = z\,\mathbf{i} + y\,\mathbf{j} + x\,\mathbf{k}$ across the unit sphere $x^2 + y^2 + z^2 = 1$.

SOLUTION   Using the parametric representation

$$\mathbf{r}(\phi, \theta) = \sin\phi \cos\theta\,\mathbf{i} + \sin\phi \sin\theta\,\mathbf{j} + \cos\phi\,\mathbf{k} \qquad 0 \le \phi \le \pi \qquad 0 \le \theta \le 2\pi$$

we have

$$\mathbf{F}(\mathbf{r}(\phi, \theta)) = \cos\phi\,\mathbf{i} + \sin\phi \sin\theta\,\mathbf{j} + \sin\phi \cos\theta\,\mathbf{k}$$

and, from Example 6 in Section 14.6,

$$\mathbf{r}_\phi \times \mathbf{r}_\theta = \sin^2\phi \cos\theta\,\mathbf{i} + \sin^2\phi \sin\theta\,\mathbf{j} + \sin\phi \cos\phi\,\mathbf{k}$$

Therefore

$$\mathbf{F}(\mathbf{r}(\phi, \theta)) \cdot (\mathbf{r}_\phi \times \mathbf{r}_\theta) = \cos\phi \sin^2\phi \cos\theta + \sin^3\phi \sin^2\theta + \sin^2\phi \cos\phi \cos\theta$$

and, by Formula 9, the flux is

$$\iint_S \mathbf{F} \cdot d\mathbf{S} = \iint_D \mathbf{F} \cdot (\mathbf{r}_\phi \times \mathbf{r}_\theta) \, dA$$

$$= \int_0^{2\pi} \int_0^{\pi} (2\sin^2\phi \cos\phi \cos\theta + \sin^3\phi \sin^2\theta) \, d\phi \, d\theta$$

$$= 2 \int_0^{\pi} \sin^2\phi \cos\phi \, d\phi \int_0^{2\pi} \cos\theta \, d\theta + \int_0^{\pi} \sin^3\phi \, d\phi \int_0^{2\pi} \sin^2\theta \, d\theta$$

$$= 0 + \int_0^{\pi} \sin^3\phi \, d\phi \int_0^{2\pi} \sin^2\theta \, d\theta \quad \left(\text{since } \int_0^{2\pi} \cos\theta \, d\theta = 0 \right)$$

$$= \frac{4\pi}{3}$$

by the same calculation as in Example 2. ∎

If, for instance, the vector field in Example 5 is a velocity field describing the flow of a fluid with density 1, then the answer, $4\pi/3$, represents the rate of flow through the unit sphere in units of mass per unit time.

Although we motivated the surface integral of a vector field using the example of fluid flow, this concept also arises in other physical situations. For instance, if **E** is an electric field (see Example 5 in Section 14.1), then the surface integral

$$\iint_S \mathbf{E} \cdot d\mathbf{S}$$

is called the **electric flux** of **E** through the surface S. One of the important laws of electrostatics is **Gauss's Law,** which says that the net charge enclosed by a closed surface S is

$$(10) \qquad Q = \varepsilon_0 \iint_S \mathbf{E} \cdot d\mathbf{S}$$

where ε_0 is a constant (called the permittivity of free space) that depends on the units used. (In the SI system, $\varepsilon_0 \approx 8.8542 \times 10^{-12}$ C^2/N·m^2.) Thus if the vector field **F** in Example 5 represents an electric field, we can conclude that the charge enclosed by S is $Q = 4\pi\varepsilon_0/3$.

Another application of surface integrals occurs in the study of heat flow. Suppose the temperature at a point (x, y, z) in a body is $u(x, y, z)$. Then the **heat flow** is defined as the vector field

$$\mathbf{F} = -K \, \nabla u$$

where K is an experimentally determined constant called the **conductivity** of the substance. The rate of heat flow across the surface S in the body is then given by the surface integral

$$\iint_S \mathbf{F} \cdot d\mathbf{S} = -K \iint_S \nabla u \cdot d\mathbf{S}$$

EXAMPLE 6   The temperature u in a metal ball is proportional to the square of the distance from the center of the ball. Find the rate of heat flow across a sphere S of radius a with center at the center of the ball.

SOLUTION   Taking the center of the ball to be at the origin, we have

$$u(x, y, z) = C(x^2 + y^2 + z^2)$$

where C is the proportionality constant. Then the heat flow is

$$\mathbf{F}(x, y, z) = -K \, \nabla u = -KC(2x\,\mathbf{i} + 2y\,\mathbf{j} + 2z\,\mathbf{k})$$

where K is the conductivity of the metal. Instead of using the usual parametrization of the sphere as in Example 5, we observe that the outward unit normal to the sphere $x^2 + y^2 + z^2 = a^2$ at the point (x, y, z) is

$$\mathbf{n} = \frac{1}{a}(x\,\mathbf{i} + y\,\mathbf{j} + z\,\mathbf{k})$$

and so

$$\mathbf{F} \cdot \mathbf{n} = -\frac{2KC}{a}(x^2 + y^2 + z^2)$$

But on S we have $x^2 + y^2 + z^2 = a^2$, so $\mathbf{F} \cdot \mathbf{n} = -2aKC$. Therefore, the rate of heat flow across S is

$$\iint_S \mathbf{F} \cdot d\mathbf{S} = \iint_S \mathbf{F} \cdot \mathbf{n}\, dS$$

$$= -2aKC \iint_S dS = -2aKCA(S)$$

$$= -2aKC(4\pi a^2) = -8KC\pi a^3 \qquad \blacksquare$$

EXERCISES 14.7

1. Let S be the cube with vertices $(\pm 1, \pm 1, \pm 1)$. Approximate $\iint_S \sqrt{x^2 + 2y^2 + 3z^2}\, dS$ by using a Riemann sum as in Definition 1, taking the patches S_{ij} to be the squares that are the faces of the cube and the points P_{ij}^* to be the centers of the squares.

2. Suppose that $f(x, y, z) = g(\sqrt{x^2 + y^2 + z^2})$, where g is a function of one variable such that $g(2) = -5$. Evaluate $\iint_S f(x, y, z)\, dS$, where S is the sphere $x^2 + y^2 + z^2 = 4$.

3–14 ■ Evaluate the given surface integral.

3. $\iint_S y\, dS$, S is the part of the plane $3x + 2y + z = 6$ that lies in the first octant

4. $\iint_S xz\, dS$,
S is the triangle with vertices $(1, 0, 0)$, $(0, 1, 0)$, and $(0, 0, 1)$

5. $\iint_S x\, dS$,
S is the surface $y = x^2 + 4z$, $0 \leqslant x \leqslant 2$, $0 \leqslant z \leqslant 2$

6. $\iint_S (y^2 + z^2)\, dS$,
S is the part of the paraboloid $x = 4 - y^2 - z^2$ that lies in front of the plane $x = 0$

7. $\iint_S yz\, dS$, S is the part of the plane $z = y + 3$ that lies inside the cylinder $x^2 + y^2 = 1$

8. $\iint_S xy\, dS$,
S is the boundary of the region enclosed by the cylinder $x^2 + z^2 = 1$ and the planes $y = 0$ and $x + y = 2$

9. $\iint_S (x^2 z + y^2 z)\, dS$,
S is the hemisphere $x^2 + y^2 + z^2 = 4$, $z \geqslant 0$

10. $\iint_S xyz\, dS$, S is the part of the sphere $x^2 + y^2 + z^2 = 1$ that lies above the cone $z = \sqrt{x^2 + y^2}$

11. $\iint_S (x^2 y + z^2)\, dS$, S is the part of the cylinder $x^2 + y^2 = 9$ between the planes $z = 0$ and $z = 2$

12. $\iint_S (x^2 + y^2 + z^2)\, dS$, S consists of the cylinder in Exercise 9 together with its top and bottom disks

13. $\iint_S yz\, dS$, S is the surface with parametric equations $x = uv$, $y = u + v$, $z = u - v$, $u^2 + v^2 \leqslant 1$

14. $\iint_S \sqrt{1 + x^2 + y^2}\, dS$, S is the helicoid with vector equation $\mathbf{r}(u, v) = u \cos v\, \mathbf{i} + u \sin v\, \mathbf{j} + v\, \mathbf{k}$, $0 \leqslant u \leqslant 1$, $0 \leqslant v \leqslant \pi$

15–24 ■ Evaluate the surface integral $\iint_S \mathbf{F} \cdot d\mathbf{S}$ for the given vector field $\mathbf{F}$ and the given oriented surface S. In other words, find the flux of $\mathbf{F}$ across S. For closed surfaces, use the positive (outward) orientation.

15. $\mathbf{F}(x, y, z) = e^y\mathbf{i} + ye^x\mathbf{j} + x^2 y\,\mathbf{k}$, S is the part of the paraboloid $z = x^2 + y^2$ that lies above the square $0 \leqslant x \leqslant 1$, $0 \leqslant y \leqslant 1$ and has upward orientation

16. $\mathbf{F}(x, y, z) = x^2 y\,\mathbf{i} - 3xy^2\mathbf{j} + 4y^3\mathbf{k}$,
S is the part of the elliptic paraboloid $z = x^2 + y^2 - 9$ that lies below the square $0 \leqslant x \leqslant 2$, $0 \leqslant y \leqslant 1$ and has downward orientation

17. $\mathbf{F}(x, y, z) = x\,\mathbf{i} + xy\,\mathbf{j} + xz\,\mathbf{k}$,
S is the surface of Exercise 3 with upward orientation

18. $\mathbf{F}(x, y, z) = -x\mathbf{i} - y\mathbf{j} + z^2\mathbf{k}$,
S is the part of the cone $z = \sqrt{x^2 + y^2}$ between the planes $z = 1$ and $z = 2$ with upward orientation

19. $\mathbf{F}(x, y, z) = x\,\mathbf{i} + y\,\mathbf{j} + z\,\mathbf{k}$,
S is the sphere $x^2 + y^2 + z^2 = 9$

20. $\mathbf{F}(x, y, z) = -y\mathbf{i} + x\mathbf{j} + 3z\mathbf{k}$, S is the hemisphere $z = \sqrt{16 - x^2 - y^2}$ with upward orientation

21. $\mathbf{F}(x, y, z) = y\mathbf{j} - z\mathbf{k}$, S consists of the paraboloid $y = x^2 + z^2, 0 \le y \le 1$, and the disk $x^2 + z^2 \le 1, y = 1$

22. $\mathbf{F}(x, y, z) = x\mathbf{i} + y\mathbf{j} + 5\mathbf{k}$, S is the surface of Exercise 8

23. $\mathbf{F}(x, y, z) = x\mathbf{i} + 2y\mathbf{j} + 3z\mathbf{k}$, S is the cube with vertices $(\pm 1, \pm 1, \pm 1)$

24. $\mathbf{F}(x, y, z) = y\mathbf{i} + x\mathbf{j} + z^2\mathbf{k}$, S is the helicoid of Exercise 14

CAS **25.** Evaluate $\iint_S xyz \, dS$ correct to four decimal places, where S is the surface $z = xy, 0 \le x \le 1, 0 \le y \le 1$.

CAS **26.** Find the exact value of $\iint_S x^2yz \, dS$, where S is the surface in Exercise 25.

CAS **27.** Find the value of $\iint_S x^2y^2z^2 \, dS$ correct to four decimal places, where S is the part of the paraboloid $z = 3 - 2x^2 - y^2$ that lies above the xy-plane.

CAS **28.** Find the flux of $\mathbf{F}(x, y, z) = \sin(xyz)\mathbf{i} + x^2y\mathbf{j} + z^2e^{x/5}\mathbf{k}$ across the part of the cylinder $4y^2 + z^2 = 4$ that lies above the xy-plane and between the planes $x = -2$ and $x = 2$ with upward orientation. Illustrate by using a computer algebra system to draw the cylinder and the vector field on the same screen.

29. Find a formula for $\iint_S \mathbf{F} \cdot d\mathbf{S}$ similar to Formula 8 for the case where S is given by $y = h(x, z)$ and $\mathbf{n}$ is the unit normal that points toward the left.

30. Find a formula for $\iint_S \mathbf{F} \cdot d\mathbf{S}$ similar to Formula 8 for the case where S is given by $x = k(y, z)$ and $\mathbf{n}$ is the unit normal that points forward (that is, toward the viewer when the axes are drawn in the usual way).

31. Find the center of mass of the hemisphere $x^2 + y^2 + z^2 = a^2, z \ge 0$, if it has constant density.

32. Find the mass of a thin funnel in the shape of a cone $z = \sqrt{x^2 + y^2}, 1 \le z \le 4$, if its density function is $\rho(x, y, z) = 10 - z$.

33. (a) Give an integral expression for the moment of inertia I_z about the z-axis of a thin sheet in the shape of a surface S if the density function is ρ.
(b) Find the moment of inertia about the z-axis of the funnel in Exercise 32.

34. The conical surface $z^2 = x^2 + y^2, 0 \le z \le a$, has constant density k. Find (a) the center of mass and (b) the moment of inertia about the z-axis.

35. A fluid with density 1200 flows with velocity $\mathbf{v} = y\mathbf{i} + \mathbf{j} + z\mathbf{k}$. Find the rate of flow upward through the paraboloid $z = 9 - (x^2 + y^2)/4, x^2 + y^2 \le 36$.

36. A fluid has density 1500 and velocity field $\mathbf{v} = -y\mathbf{i} + x\mathbf{j} + 2z\mathbf{k}$. Find the rate of flow outward through the sphere $x^2 + y^2 + z^2 = 25$.

37. Use Gauss's Law to find the charge contained in the solid hemisphere $x^2 + y^2 + z^2 \le a^2, z \ge 0$, if the electric field is $\mathbf{E}(x, y, z) = x\mathbf{i} + y\mathbf{j} + 2z\mathbf{k}$.

38. Use Gauss's Law to find the charge enclosed by the cube with vertices $(\pm 1, \pm 1, \pm 1)$ if the electric field is $\mathbf{E}(x, y, z) = x\mathbf{i} + y\mathbf{j} + z\mathbf{k}$.

39. The temperature at the point (x, y, z) in a substance with conductivity $K = 6.5$ is $u(x, y, z) = 2y^2 + 2z^2$. Find the rate of heat flow inward across the cylindrical surface $y^2 + z^2 = 6, 0 \le x \le 4$.

40. The temperature at a point in a ball with conductivity K is inversely proportional to the distance from the center of the ball. Find the rate of heat flow across a sphere S of radius a with center at the center of the ball.

14.8   STOKES' THEOREM

Stokes' Theorem can be regarded as a higher-dimensional version of Green's Theorem. Whereas Green's Theorem relates a double integral over a plane region D to a line integral around its plane boundary curve, Stokes' Theorem relates a surface integral over a surface S to a line integral around the boundary curve of S (which is a space curve). Figure 1 shows an oriented surface with unit normal vector $\mathbf{n}$. The orientation of S induces the **positive orientation of the boundary curve C** shown in the figure. This means that if you walk in the positive direction around C with your head pointing in the direction of $\mathbf{n}$, then the surface will always be on your left.

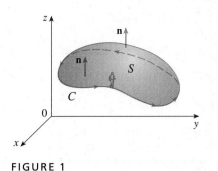

FIGURE 1

(1) STOKES' THEOREM   Let S be an oriented piecewise-smooth surface that is bounded by a simple, closed, piecewise-smooth boundary curve C with positive orientation. Let $\mathbf{F}$ be a vector field whose components have continuous partial derivatives on an open region in $\mathbb{R}^3$ that contains S. Then

$$\int_C \mathbf{F} \cdot d\mathbf{r} = \iint_S \text{curl } \mathbf{F} \cdot d\mathbf{S}$$

Since

$$\int_C \mathbf{F} \cdot d\mathbf{r} = \int_C \mathbf{F} \cdot \mathbf{T}\, ds \qquad \text{and} \qquad \iint_S \text{curl } \mathbf{F} \cdot d\mathbf{S} = \iint_S \text{curl } \mathbf{F} \cdot \mathbf{n}\, dS$$

Stokes' Theorem says that the line integral around the boundary curve of S of the tangential component of $\mathbf{F}$ is equal to the surface integral of the normal component of the curl of $\mathbf{F}$.

The positively oriented boundary curve of the oriented surface S is often written as ∂S, so Stokes' Theorem can be expressed as

(2)
$$\iint_S \text{curl } \mathbf{F} \cdot d\mathbf{S} = \int_{\partial S} \mathbf{F} \cdot d\mathbf{r}$$

There is an analogy among Stokes' Theorem, Green's Theorem, and the Fundamental Theorem of Calculus. As before, there is an integral involving derivatives on the left side of Equation 2 (recall that curl $\mathbf{F}$ is a sort of derivative of $\mathbf{F}$) and the right side involves the values of $\mathbf{F}$ only on the *boundary* of S.

In fact, in the special case where the surface S is flat and lies in the xy-plane with upward orientation, the unit normal is $\mathbf{k}$, the surface integral becomes a double integral, and Stokes' Theorem becomes

$$\int_C \mathbf{F} \cdot d\mathbf{r} = \iint_S \text{curl } \mathbf{F} \cdot d\mathbf{S} = \iint_S (\text{curl } \mathbf{F}) \cdot \mathbf{k}\, dA$$

This is precisely the vector form of Green's Theorem given in Equation 14.5.12. Thus we see that Green's Theorem is really a special case of Stokes' Theorem.

Although Stokes' Theorem is too difficult for us to prove in its full generality, we can give a proof when S is a graph and $\mathbf{F}$, S, and C are well behaved.

PROOF OF A SPECIAL CASE OF STOKES' THEOREM   We assume that the equation of S is $z = g(x, y)$, $(x, y) \in D$, where g has continuous second-order partial derivatives and D is a simple plane region whose boundary curve C_1 corresponds to C. If the orientation of S is upward, then the positive orientation of C corresponds to the positive orientation of C_1 (see Figure 2). We are also assuming that $\mathbf{F} = P\,\mathbf{i} + Q\,\mathbf{j} + R\,\mathbf{k}$, where the partial derivatives of P, Q, and R are continuous.

Since S is a graph, we can apply Formula 14.7.8 with $\mathbf{F}$ replaced by curl $\mathbf{F}$. The result is

(3)
$$\iint_S \text{curl } \mathbf{F} \cdot d\mathbf{S}$$
$$= \iint_D \left[-\left(\frac{\partial R}{\partial y} - \frac{\partial Q}{\partial z} \right) \frac{\partial z}{\partial x} - \left(\frac{\partial P}{\partial z} - \frac{\partial R}{\partial x} \right) \frac{\partial z}{\partial y} + \left(\frac{\partial Q}{\partial x} - \frac{\partial P}{\partial y} \right) \right] dA$$

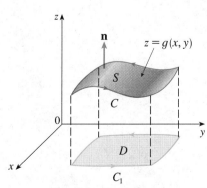

FIGURE 2

where the partial derivatives of P, Q, and R are evaluated at $(x, y, g(x, y))$. If

$$x = x(t) \qquad y = y(t) \qquad a \leqslant t \leqslant b$$

is a parametric representation of C_1, then a parametric representation of C is

$$x = x(t) \qquad y = y(t) \qquad z = g(x(t), y(t)) \qquad a \leqslant t \leqslant b$$

This allows us, with the aid of the Chain Rule, to evaluate the line integral as follows:

$$
\begin{aligned}
\int_C \mathbf{F} \cdot d\mathbf{r} &= \int_a^b \left(P \frac{dx}{dt} + Q \frac{dy}{dt} + R \frac{dz}{dt} \right) dt \\
&= \int_a^b \left[P \frac{dx}{dt} + Q \frac{dy}{dt} + R \left(\frac{\partial z}{\partial x} \frac{dx}{dt} + \frac{\partial z}{\partial y} \frac{dy}{dt} \right) \right] dt \\
&= \int_a^b \left[\left(P + R \frac{\partial z}{\partial x} \right) \frac{dx}{dt} + \left(Q + R \frac{\partial z}{\partial y} \right) \frac{dy}{dt} \right] dt \\
&= \int_{C_1} \left(P + R \frac{\partial z}{\partial x} \right) dx + \left(Q + R \frac{\partial z}{\partial y} \right) dy \\
&= \iint_D \left[\frac{\partial}{\partial x} \left(Q + R \frac{\partial z}{\partial y} \right) - \frac{\partial}{\partial y} \left(P + R \frac{\partial z}{\partial x} \right) \right] dA
\end{aligned}
$$

where we have used Green's Theorem in the last step. Then, using the Chain Rule again and remembering that P, Q, and R are functions of x, y, and z and that z is itself a function of x and y, we get

$$
\begin{aligned}
\int_C \mathbf{F} \cdot d\mathbf{r} = \iint_D &\left[\left(\frac{\partial Q}{\partial x} + \frac{\partial Q}{\partial z} \frac{\partial z}{\partial x} + \frac{\partial R}{\partial x} \frac{\partial z}{\partial y} + \frac{\partial R}{\partial z} \frac{\partial z}{\partial x} \frac{\partial z}{\partial y} + R \frac{\partial^2 z}{\partial x \, \partial y} \right) \right. \\
&\left. - \left(\frac{\partial P}{\partial y} + \frac{\partial P}{\partial z} \frac{\partial z}{\partial y} + \frac{\partial R}{\partial y} \frac{\partial z}{\partial x} + \frac{\partial R}{\partial z} \frac{\partial z}{\partial y} \frac{\partial z}{\partial x} + R \frac{\partial^2 z}{\partial y \, \partial x} \right) \right] dA
\end{aligned}
$$

Four of the terms in this double integral cancel and the remaining six terms can be arranged to coincide with the right side of Equation 3. Therefore

$$\int_C \mathbf{F} \cdot d\mathbf{r} = \iint_S \operatorname{curl} \mathbf{F} \cdot d\mathbf{S} \qquad\qquad \square$$

EXAMPLE 1   Evaluate $\int_C \mathbf{F} \cdot d\mathbf{r}$, where $\mathbf{F}(x, y, z) = -y^2 \mathbf{i} + x \mathbf{j} + z^2 \mathbf{k}$ and C is the curve of intersection of the plane $y + z = 2$ and the cylinder $x^2 + y^2 = 1$. (Orient C to be counterclockwise when viewed from above.)

SOLUTION   The curve C (an ellipse) is shown in Figure 3. Although $\int_C \mathbf{F} \cdot d\mathbf{r}$ could be evaluated directly, it is not easy to parametrize C, so let us use Stokes' Theorem instead. We first compute

$$
\operatorname{curl} \mathbf{F} =
\begin{vmatrix}
\mathbf{i} & \mathbf{j} & \mathbf{k} \\
\dfrac{\partial}{\partial x} & \dfrac{\partial}{\partial y} & \dfrac{\partial}{\partial z} \\
-y^2 & x & z^2
\end{vmatrix}
= (1 + 2y) \mathbf{k}
$$

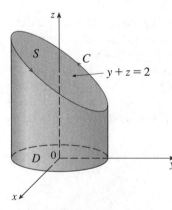

FIGURE 3

Let us choose S to be the elliptical region in the plane $y + z = 2$ that is bounded by C. If we orient S upward, then C has the induced positive orientation. The projection D of S on the xy-plane is the disk $x^2 + y^2 \le 1$ and so using Equation 14.7.8 with $z = g(x, y) = 2 - y$, we have

$$\int_C \mathbf{F} \cdot d\mathbf{r} = \iint_S \text{curl } \mathbf{F} \cdot d\mathbf{S} = \iint_D (1 + 2y) \, dA$$

$$= \int_0^{2\pi} \int_0^1 (1 + 2r \sin \theta) r \, dr \, d\theta$$

$$= \int_0^{2\pi} \left[\frac{r^2}{2} + 2 \frac{r^3}{3} \sin \theta \right]_0^1 d\theta = \int_0^{2\pi} \left(\tfrac{1}{2} + \tfrac{2}{3} \sin \theta \right) d\theta$$

$$= \tfrac{1}{2}(2\pi) + 0 = \pi$$

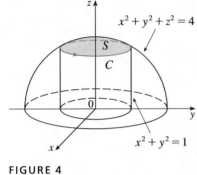

FIGURE 4

EXAMPLE 2   Use Stokes' Theorem to compute the integral $\iint_S \text{curl } \mathbf{F} \cdot d\mathbf{S}$, where $\mathbf{F}(x, y, z) = yz\,\mathbf{i} + xz\,\mathbf{j} + xy\,\mathbf{k}$ and S is the part of the sphere $x^2 + y^2 + z^2 = 4$ that lies inside the cylinder $x^2 + y^2 = 1$ and above the xy-plane (see Figure 4).

SOLUTION   To find the boundary curve C we solve the equations $x^2 + y^2 + z^2 = 4$ and $x^2 + y^2 = 1$. Subtracting, we get $z^2 = 3$ and so $z = \sqrt{3}$ (since $z > 0$). Thus C is the circle given by the equations $x^2 + y^2 = 1$, $z = \sqrt{3}$. The vector equation of C is

$$\mathbf{r}(t) = \cos t\,\mathbf{i} + \sin t\,\mathbf{j} + \sqrt{3}\,\mathbf{k} \qquad 0 \le t \le 2\pi$$

so

$$\mathbf{r}'(t) = -\sin t\,\mathbf{i} + \cos t\,\mathbf{j}$$

Also, we have

$$\mathbf{F}(\mathbf{r}(t)) = \sqrt{3}\,\sin t\,\mathbf{i} + \sqrt{3}\,\cos t\,\mathbf{j} + \cos t \sin t\,\mathbf{k}$$

Therefore, by Stokes' Theorem,

$$\iint_S \text{curl } \mathbf{F} \cdot d\mathbf{S} = \int_C \mathbf{F} \cdot d\mathbf{r} = \int_0^{2\pi} \mathbf{F}(\mathbf{r}(t)) \cdot \mathbf{r}'(t) \, dt$$

$$= \int_0^{2\pi} \left(-\sqrt{3} \sin^2 t + \sqrt{3} \cos^2 t \right) dt$$

$$= \sqrt{3} \int_0^{2\pi} \cos 2t \, dt = 0$$

Note that in Example 2 we computed a surface integral simply by knowing the values of $\mathbf{F}$ on the boundary curve C. This means that if we have a different oriented surface with the same boundary curve C, then we get exactly the same value for the surface integral!

In general, if S_1 and S_2 are oriented surfaces with the same oriented boundary curve C and both satisfy the hypotheses of Stokes' Theorem, then

(4)
$$\iint_{S_1} \text{curl } \mathbf{F} \cdot d\mathbf{S} = \int_C \mathbf{F} \cdot d\mathbf{r} = \iint_{S_2} \text{curl } \mathbf{F} \cdot d\mathbf{S}$$

This fact is useful when it is difficult to integrate over one surface but easy to integrate over the other.

Let us now use Stokes' Theorem to throw some light on the meaning of the curl vector. Suppose that C is an oriented closed curve and **v** represents the velocity field in fluid flow. Consider the line integral

$$\int_C \mathbf{v} \cdot d\mathbf{r} = \int_C \mathbf{v} \cdot \mathbf{T}\, ds$$

and recall that $\mathbf{v} \cdot \mathbf{T}$ is the component of **v** in the direction of the unit tangent vector **T**. This means that the closer the direction of **v** is to the direction of **T**, the larger the value of $\mathbf{v} \cdot \mathbf{T}$. Thus $\int_C \mathbf{v} \cdot d\mathbf{r}$ is a measure of the tendency of the fluid to move around C and is called the **circulation** of **v** around C (see Figure 5).

Now let $P_0(x_0, y_0, z_0)$ be a point in the fluid and let S_a be a small disk with radius a and center P_0. Then (curl $\mathbf{F})(P) \approx$ (curl $\mathbf{F})(P_0)$ for all points P on S_a because curl **F** is continuous. Thus, by Stokes' Theorem, we get the following approximation to the circulation around the boundary circle C_a:

$$\int_{C_a} \mathbf{v} \cdot d\mathbf{r} = \iint_{S_a} \text{curl } \mathbf{v} \cdot d\mathbf{S} = \iint_{S_a} \text{curl } \mathbf{v} \cdot \mathbf{n}\, dS$$

$$\approx \iint_{S_a} \text{curl } \mathbf{v}(P_0) \cdot \mathbf{n}(P_0)\, dS$$

$$= \text{curl } \mathbf{v}(P_0) \cdot \mathbf{n}(P_0)\, \pi a^2$$

This approximation becomes better as $a \to 0$ and we have

(5)
$$\text{curl } \mathbf{v}(P_0) \cdot \mathbf{n}(P_0) = \lim_{a \to 0} \frac{1}{\pi a^2} \int_{C_a} \mathbf{v} \cdot d\mathbf{r}$$

Equation 5 gives the relationship between the curl and the circulation. It shows that curl $\mathbf{v} \cdot \mathbf{n}$ is a measure of the rotating effect of the fluid about the axis **n**. The curling effect is greatest about the axis parallel to curl **v**.

Finally, we mention that Stokes' Theorem can be used to prove Theorem 14.5.4 (which states that if curl $\mathbf{F} = \mathbf{0}$ on all of $\mathbb{R}^3$, then **F** is conservative). From our previous work (Theorems 14.3.3 and 14.3.4), we know that **F** is conservative if $\int_C \mathbf{F} \cdot d\mathbf{r} = 0$ for every closed path C. Given C, suppose we can find an orientable surface S whose boundary is C. (This can be done, but the proof requires advanced techniques.) Then Stokes' Theorem gives

$$\int_C \mathbf{F} \cdot d\mathbf{r} = \iint_S \text{curl } \mathbf{F} \cdot d\mathbf{S} = \iint_S \mathbf{0} \cdot d\mathbf{S} = 0$$

A curve that is not simple can be broken into a number of simple curves, and the integrals around these simple curves are all 0. Adding these integrals, we obtain $\int_C \mathbf{F} \cdot d\mathbf{r} = 0$ for any closed curve C.

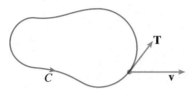

(a) $\int_C \mathbf{v} \cdot d\mathbf{r} > 0$, positive circulation

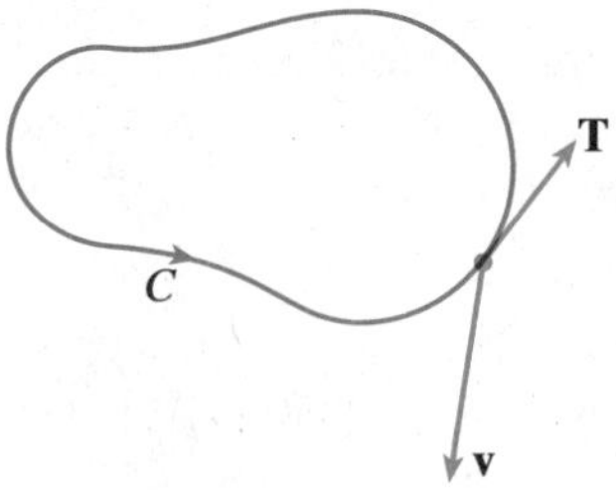

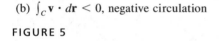

(b) $\int_C \mathbf{v} \cdot d\mathbf{r} < 0$, negative circulation

FIGURE 5

Imagine a tiny paddle wheel placed in the fluid at a point P, as in Figure 6; the paddle wheel rotates fastest when its axis is parallel to curl **v**.

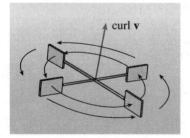

FIGURE 6

EXERCISES 14.8

1–6 ■ Use Stokes' Theorem to evaluate $\iint_S \text{curl } \mathbf{F} \cdot d\mathbf{S}$.

1. $\mathbf{F}(x, y, z) = xyz\,\mathbf{i} + x\,\mathbf{j} + e^{xy}\cos z\,\mathbf{k}$, S is the hemisphere
$x^2 + y^2 + z^2 = 1$, $z \geq 0$, oriented upward

2. $\mathbf{F}(x, y, z) = y^2z\,\mathbf{i} + xz\,\mathbf{j} + x^2y^2\,\mathbf{k}$,
S is the part of the paraboloid $z = x^2 + y^2$ that lies inside
the cylinder $x^2 + y^2 = 1$, oriented upward

3. $\mathbf{F}(x, y, z) = yz^3\,\mathbf{i} + \sin(xyz)\,\mathbf{j} + x^3\,\mathbf{k}$,
S is the part of the paraboloid $y = 1 - x^2 - z^2$ that lies to
the right of the xz-plane, oriented toward the xz-plane

4. $\mathbf{F}(x, y, z) = (x + \tan^{-1}yz)\,\mathbf{i} + y^2z\,\mathbf{j} + z\,\mathbf{k}$,
S is the part of the hemisphere $x = \sqrt{9 - y^2 - z^2}$ that lies
inside the cylinder $y^2 + z^2 = 4$, oriented in the direction
of the positive x-axis

5. $\mathbf{F}(x, y, z) = xyz\,\mathbf{i} + xy\,\mathbf{j} + x^2yz\,\mathbf{k}$,
S consists of the top and the four sides (but not the bottom)
of the cube with vertices $(\pm 1, \pm 1, \pm 1)$, oriented outward
[*Hint:* Use Equation 4.]

6. $\mathbf{F}(x, y, z) = xy\,\mathbf{i} + e^z\,\mathbf{j} + xy^2\,\mathbf{k}$,
S consists of the four sides of the pyramid with vertices
$(0, 0, 0)$, $(1, 0, 0)$, $(0, 0, 1)$, $(1, 0, 1)$, and $(0, 1, 0)$ that lie to the
right of the xz-plane, oriented in the direction of the
positive y-axis   [*Hint:* Use Equation 4.]

7–12 ■ Use Stokes' Theorem to evaluate $\int_C \mathbf{F} \cdot d\mathbf{r}$. In each case
C is oriented counterclockwise as viewed from above.

7. $\mathbf{F}(x, y, z) = xz\,\mathbf{i} + 2xy\,\mathbf{j} + 3xy\,\mathbf{k}$, C is the boundary of
the part of the plane $3x + y + z = 3$ in the first octant

8. $\mathbf{F}(x, y, z) = z^2\,\mathbf{i} + y^2\,\mathbf{j} + xy\,\mathbf{k}$, C is the triangle with
vertices $(1, 0, 0)$, $(0, 1, 0)$, and $(0, 0, 2)$

9. $\mathbf{F}(x, y, z) = 2z\,\mathbf{i} + 4x\,\mathbf{j} + 5y\,\mathbf{k}$,
C is the curve of intersection of the plane $z = x + 4$ and
the cylinder $x^2 + y^2 = 4$

10. $\mathbf{F}(x, y, z) = x\,\mathbf{i} + y\,\mathbf{j} + (x^2 + y^2)\,\mathbf{k}$,
C is the boundary of the part of the paraboloid
$z = 1 - x^2 - y^2$ in the first octant

11. (a) Use Stokes' Theorem to evaluate $\int_C \mathbf{F} \cdot d\mathbf{r}$, where

$$\mathbf{F}(x, y, z) = x^2z\,\mathbf{i} + xy^2\,\mathbf{j} + z^2\,\mathbf{k}$$

and C is the curve of intersection of the plane
$x + y + z = 1$ and the cylinder $x^2 + y^2 = 9$ oriented
counterclockwise as viewed from above.

 (b) Graph both the plane and the cylinder with domains
chosen so that you can see the curve C and the surface
that you used in part (a).

(c) Find parametric equations for C and use them to
graph C.

12. (a) Use Stokes' Theorem to evaluate $\int_C \mathbf{F} \cdot d\mathbf{r}$, where
$\mathbf{F}(x, y, z) = x^2y\,\mathbf{i} + \frac{1}{3}x^3\,\mathbf{j} + xy\,\mathbf{k}$ and C is the curve of
intersection of the hyperbolic paraboloid $z = y^2 - x^2$
and the cylinder $x^2 + y^2 = 1$ oriented counterclockwise as viewed from above.

(b) Graph both the hyperbolic paraboloid and the cylinder
with domains chosen so that you can see the curve C
and the surface that you used in part (a).

(c) Find parametric equations for C and use them to
graph C.

13–16 ■ Verify that Stokes' Theorem is true for the given
vector field $\mathbf{F}$ and surface S.

13. $\mathbf{F}(x, y, z) = 3y\,\mathbf{i} + 4z\,\mathbf{j} - 6x\,\mathbf{k}$,
S is the part of the paraboloid $z = 9 - x^2 - y^2$ that lies
above the xy-plane, oriented upward

14. $\mathbf{F}(x, y, z) = xy\,\mathbf{i} + yz\,\mathbf{j} + xz\,\mathbf{k}$,
S is the hemisphere $z = \sqrt{a^2 - x^2 - y^2}$, oriented upward

15. $\mathbf{F}(x, y, z) = y\,\mathbf{i} + z\,\mathbf{j} + x\,\mathbf{k}$, S is the part of the plane
$x + y + z = 1$ that lies in the first octant, oriented upward

16. $\mathbf{F}(x, y, z) = y\,\mathbf{i} + z\,\mathbf{j} + x\,\mathbf{k}$,
S is the helicoid of Exercise 26 in Section 14.6

17. Calculate the work done by the force field

$$\mathbf{F}(x, y, z) = (x^x + z^2)\,\mathbf{i} + (y^y + x^2)\,\mathbf{j} + (z^z + y^2)\,\mathbf{k}$$

when a particle moves under its influence around the edge
of the part of the sphere $x^2 + y^2 + z^2 = 4$ that lies in the
first octant, in a counterclockwise direction as viewed
from above.

18. Evaluate $\int_C (y + \sin x)\,dx + (z^2 + \cos y)\,dy + x^3\,dz$,
where C is the curve $\mathbf{r}(t) = \langle \sin t, \cos t, \sin 2t \rangle$, $0 \leq t \leq 2\pi$.
[*Hint:* Observe that C lies on the surface $z = 2xy$.]

19. If S is a sphere and $\mathbf{F}$ satisfies the hypotheses of Stokes'
Theorem, show that $\iint_S \text{curl } \mathbf{F} \cdot d\mathbf{S} = 0$.

20. If S and C satisfy the hypotheses of Stokes' Theorem and
f, g have continuous second-order partial derivatives, show
that
(a) $\int_C (f\,\nabla g) \cdot d\mathbf{r} = \iint_S (\nabla f \times \nabla g) \cdot d\mathbf{S}$
(b) $\int_C (f\,\nabla f) \cdot d\mathbf{r} = 0$
(c) $\int_C (f\,\nabla g + g\,\nabla f) \cdot d\mathbf{r} = 0$

14.9 THE DIVERGENCE THEOREM

In Section 14.5 we rewrote Green's Theorem in a vector version as

$$\int_C \mathbf{F} \cdot \mathbf{n} \, ds = \iint_D \text{div } \mathbf{F}(x, y) \, dA$$

where C is the positively oriented boundary curve of the plane region D. If we were seeking to extend this theorem to vector fields on $\mathbb{R}^3$, we might make the guess that

(1)
$$\iint_S \mathbf{F} \cdot \mathbf{n} \, dS = \iiint_E \text{div } \mathbf{F}(x, y, z) \, dV$$

The Divergence Theorem is sometimes called Gauss's Theorem after the great German mathematician Karl Friedrich Gauss (1777–1855) who discovered this theorem during his investigation of electrostatics. In Eastern Europe the Divergence Theorem is known as Ostrogradsky's Theorem after the Russian mathematician Mikhail Ostrogradsky (1801–1862), who published this result in 1826.

where S is the boundary surface of the solid region E. It turns out that Equation 1 is true, under appropriate hypotheses, and is called the Divergence Theorem. Notice its similarity to Green's Theorem and Stokes' Theorem in that it relates the integral of a derivative of a function (div $\mathbf{F}$ in this case) over a region to the integral of the original function $\mathbf{F}$ over the boundary of the region.

At this stage you may wish to review the various types of regions over which we were able to evaluate triple integrals in Section 13.7. We state and prove the Divergence Theorem for regions E that are simultaneously of types 1, 2, and 3 and we call such regions **simple solid regions.** (For instance, regions bounded by ellipsoids or rectangular boxes are simple solid regions.) The boundary of E is a closed surface, and we use the convention, introduced in Section 14.7, that the positive orientation is outward; that is, the unit normal vector $\mathbf{n}$ is directed outward from E.

> **(2) THE DIVERGENCE THEOREM**   Let E be a simple solid region whose boundary surface S has positive (outward) orientation. Let $\mathbf{F}$ be a vector field whose component functions have continuous partial derivatives on an open region that contains E. Then
>
> $$\iint_S \mathbf{F} \cdot d\mathbf{S} = \iiint_E \text{div } \mathbf{F} \, dV$$

PROOF   Let $\mathbf{F} = P\mathbf{i} + Q\mathbf{j} + R\mathbf{k}$. Then

$$\text{div } \mathbf{F} = \frac{\partial P}{\partial x} + \frac{\partial Q}{\partial y} + \frac{\partial R}{\partial z}$$

so

$$\iiint_E \text{div } \mathbf{F} \, dV = \iiint_E \frac{\partial P}{\partial x} \, dV + \iiint_E \frac{\partial Q}{\partial y} \, dV + \iiint_E \frac{\partial R}{\partial z} \, dV$$

If $\mathbf{n}$ is the unit outward normal of S, then the surface integral on the left side of the Divergence Theorem is

$$\iint_S \mathbf{F} \cdot d\mathbf{S} = \iint_S \mathbf{F} \cdot \mathbf{n} \, dS = \iint_S (P\mathbf{i} + Q\mathbf{j} + R\mathbf{k}) \cdot \mathbf{n} \, dS$$

$$= \iint_S P\mathbf{i} \cdot \mathbf{n} \, dS + \iint_S Q\mathbf{j} \cdot \mathbf{n} \, dS + \iint_S R\mathbf{k} \cdot \mathbf{n} \, dS$$

Therefore, to prove the Divergence Theorem, it suffices to prove the three equations:

$$(3) \qquad \iint\limits_{S} P\,\mathbf{i} \cdot \mathbf{n}\,dS = \iiint\limits_{E} \frac{\partial P}{\partial x}\,dV$$

$$(4) \qquad \iint\limits_{S} Q\,\mathbf{j} \cdot \mathbf{n}\,dS = \iiint\limits_{E} \frac{\partial Q}{\partial y}\,dV$$

$$(5) \qquad \iint\limits_{S} R\,\mathbf{k} \cdot \mathbf{n}\,dS = \iiint\limits_{E} \frac{\partial R}{\partial z}\,dV$$

To prove Equation 4 we use the fact that E is a type 1 region:

$$E = \{(x, y, z) \mid (x, y) \in D, \ \phi_1(x, y) \leq z \leq \phi_2(x, y)\}$$

where D is the projection of E onto the xy-plane. By Equation 13.7.6, we have

$$\iiint\limits_{E} \frac{\partial R}{\partial z}\,dV = \iint\limits_{D} \left[\int_{\phi_1(x, y)}^{\phi_2(x, y)} \frac{\partial R}{\partial z}(x, y, z)\,dz \right] dA$$

and, therefore, by the Fundamental Theorem of Calculus,

$$(6) \qquad \iiint\limits_{E} \frac{\partial R}{\partial z}\,dV = \iint\limits_{D} [R(x, y, \phi_2(x, y)) - R(x, y, \phi_1(x, y))]\,dA$$

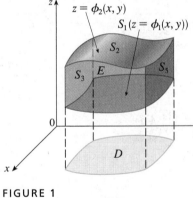

$z = \phi_2(x, y)$

$S_1(z = \phi_1(x, y))$

S_2

S_3 E S_5

D

FIGURE 1

The boundary surface S consists of six pieces: the bottom surface S_1, the top surface S_2, and the four vertical sides S_3, S_4, S_5, and S_6 (see Figure 1). Notice that on each of the vertical sides we have $\mathbf{k} \cdot \mathbf{n} = 0$ and so

$$\iint\limits_{S_i} R\,\mathbf{k} \cdot \mathbf{n}\,dS = \iint\limits_{S_i} 0\,dS = 0 \qquad i = 3, 4, 5, 6$$

This gives

$$(7) \qquad \iint\limits_{S} R\,\mathbf{k} \cdot \mathbf{n}\,dS = \iint\limits_{S_1} R\,\mathbf{k} \cdot \mathbf{n}\,dS = \iint\limits_{S_2} R\,\mathbf{k} \cdot \mathbf{n}\,dS$$

(It may happen that some of the vertical sides do not appear, as in the case of an ellipsoid, but Equation 7 is still true.)

The equation of S_2 is $z = \phi_2(x, y)$, $(x, y) \in D$, and the outward normal $\mathbf{n}$ points upward, so from Equation 14.7.8 (with $\mathbf{F}$ replaced by $R\,\mathbf{k}$) we have

$$\iint\limits_{S_2} R\,\mathbf{k} \cdot \mathbf{n}\,dS = \iint\limits_{D} R(x, y, \phi_2(x, y))\,dA$$

On S_1 we have $z = \phi_1(x, y)$, but here the outward normal $\mathbf{n}$ points downward, so

$$\mathbf{n} = \frac{1}{\sqrt{\left(\dfrac{\partial \phi_1}{\partial x}\right)^2 + \left(\dfrac{\partial \phi_1}{\partial y}\right)^2 + 1}} \left(\frac{\partial \phi_1}{\partial x}\,\mathbf{i} + \frac{\partial \phi_1}{\partial y}\,\mathbf{j} - \mathbf{k}\right)$$

and

$$\iint\limits_{S_1} R\,\mathbf{k} \cdot \mathbf{n}\,dS$$

$$= \iint\limits_{D} R(x, y, \phi_1(x, y)) \frac{(-1)}{\sqrt{\left(\dfrac{\partial\phi_1}{\partial x}\right)^2 + \left(\dfrac{\partial\phi_1}{\partial y}\right)^2 + 1}} \sqrt{1 + \left(\dfrac{\partial\phi_1}{\partial x}\right)^2 + \left(\dfrac{\partial\phi_1}{\partial y}\right)^2}\,dA$$

$$= -\iint\limits_{D} R(x, y, \phi_1(x, y))\,dA$$

Therefore, Equation 7 gives

$$\iint\limits_{S} R\,\mathbf{k} \cdot \mathbf{n}\,dS = \iint\limits_{D} [R(x, y, \phi_2(x, y)) - R(x, y, \phi_1(x, y))]\,dA$$

Comparison with Equation 6 shows that

$$\iint\limits_{S} R\,\mathbf{k} \cdot \mathbf{n}\,dS = \iiint\limits_{E} \frac{\partial R}{\partial z}\,dV$$

Notice that the method of proof of the Divergence Theorem is very similar to that of Green's Theorem.

Equations 3 and 4 are proved in a similar manner using the expressions for E as a type 2 or type 3 region. □

EXAMPLE 1 Find the flux of the vector field $\mathbf{F}(x, y, z) = z\,\mathbf{i} + y\,\mathbf{j} + x\,\mathbf{k}$ over the unit sphere $x^2 + y^2 + z^2 = 1$.

SOLUTION First we compute the divergence of $\mathbf{F}$:

$$\text{div }\mathbf{F} = \frac{\partial}{\partial x}\,(z) + \frac{\partial}{\partial y}\,(y) + \frac{\partial}{\partial z}\,(x) = 1$$

The unit sphere S is the boundary of the unit ball B given by $x^2 + y^2 + z^2 \leq 1$. Thus the Divergence Theorem gives the flux as

The solution in Example 1 should be compared with the solution in Example 5 in Section 14.7.

$$\iint\limits_{S} \mathbf{F} \cdot d\mathbf{S} = \iiint\limits_{B} \text{div }\mathbf{F}\,dV = \iiint\limits_{B} 1\,dV$$

$$= V(B) = \tfrac{4}{3}\pi(1)^3 = \frac{4\pi}{3}$$ ■

EXAMPLE 2 Evaluate $\iint\limits_{S} \mathbf{F} \cdot d\mathbf{S}$, where

$$\mathbf{F}(x, y, z) = xy\,\mathbf{i} + (y^2 + e^{xz^2})\,\mathbf{j} + \sin(xy)\,\mathbf{k}$$

and S is the surface of the region E bounded by the parabolic cylinder $z = 1 - x^2$ and the planes $z = 0$, $y = 0$, and $y + z = 2$ (see Figure 2).

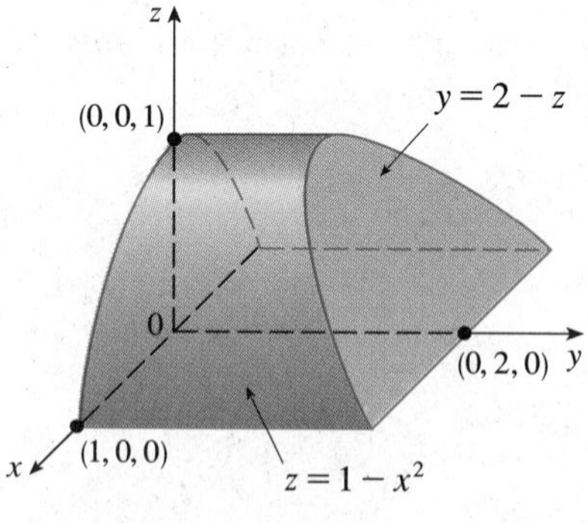

SOLUTION It would be extremely difficult to evaluate the given surface integral directly. (We would have to evaluate four surface integrals corresponding to the four pieces of S.) Furthermore, the divergence of $\mathbf{F}$ is much less complicated than $\mathbf{F}$ itself:

$$\text{div }\mathbf{F} = \frac{\partial}{\partial x}\,(xy) + \frac{\partial}{\partial y}\,(y^2 + e^{xz^2}) + \frac{\partial}{\partial z}\,(\sin xy)$$

$$= y + 2y = 3y$$

FIGURE 2

Therefore, we use the Divergence Theorem to transform the given surface integral into a triple integral. The easiest way to evaluate the triple integral is to express E as a type 3 region:

$$E = \{(x, y, z) \mid -1 \leq x \leq 1, 0 \leq z \leq 1 - x^2, 0 \leq y \leq 2 - z\}$$

Then we have

$$\iint_S \mathbf{F} \cdot d\mathbf{S} = \iiint_E \operatorname{div} \mathbf{F} \, dV = \iiint_E 3y \, dV$$

$$= 3 \int_{-1}^{1} \int_0^{1-x^2} \int_0^{2-z} y \, dy \, dz \, dx$$

$$= 3 \int_{-1}^{1} \int_0^{1-x^2} \frac{(2 - z)^2}{2} \, dz \, dx$$

$$= \frac{3}{2} \int_{-1}^{1} \left[-\frac{(2 - z)^3}{3} \right]_0^{1-x^2} dx$$

$$= -\tfrac{1}{2} \int_{-1}^{1} [(x^2 + 1)^3 - 8] \, dx$$

$$= -\int_0^1 (x^6 + 3x^4 + 3x^2 - 7) \, dx = \frac{184}{35}$$

∎

Although we have proved the Divergence Theorem only for simple solid regions, it can be proved for regions that are finite unions of simple solid regions. (The procedure is analogous to the one we used in Section 14.4 to extend Green's Theorem.)

For example, let us consider the region E that lies between the closed surfaces S_1 and S_2, where S_1 lies inside S_2. Let $\mathbf{n}_1$ and $\mathbf{n}_2$ be outward normals of S_1 and S_2. Then the boundary surface of E is $S = S_1 \cup S_2$ and its normal $\mathbf{n}$ is given by $\mathbf{n} = -\mathbf{n}_1$ on S_1 and $\mathbf{n} = \mathbf{n}_2$ on S_2 (see Figure 3). Applying the Divergence Theorem to S, we get

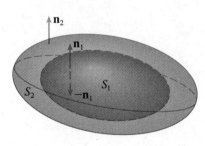

FIGURE 3

(8)
$$\iiint_E \operatorname{div} \mathbf{F} \, dV = \iint_S \mathbf{F} \cdot d\mathbf{S} = \iint_S \mathbf{F} \cdot \mathbf{n} \, dS$$

$$= \iint_{S_1} \mathbf{F} \cdot (-\mathbf{n}_1) \, dS + \iint_{S_2} \mathbf{F} \cdot \mathbf{n}_2 \, dS$$

$$= -\iint_{S_1} \mathbf{F} \cdot d\mathbf{S} + \iint_{S_2} \mathbf{F} \cdot d\mathbf{S}$$

Let us apply this to the electric field (see Example 5 in Section 14.1):

$$\mathbf{E}(\mathbf{x}) = \frac{\varepsilon Q}{|\mathbf{x}|^3} \mathbf{x}$$

where S_1 is a small sphere with radius a and center the origin. You can verify that $\operatorname{div} \mathbf{E} = 0$ (see Exercise 19). Therefore, Equation 8 gives

$$\iint_{S_2} \mathbf{E} \cdot d\mathbf{S} = \iint_{S_1} \mathbf{E} \cdot d\mathbf{S} + \iiint_E \operatorname{div} \mathbf{E} \, dV$$

$$= \iint_{S_1} \mathbf{E} \cdot d\mathbf{S} = \iint_{S_1} \mathbf{E} \cdot \mathbf{n} \, dS$$

The point of this calculation is that we can compute the surface integral over S_1 because S_1 is a sphere. The normal vector at $\mathbf{x}$ is $\mathbf{x}/|\mathbf{x}|$. Therefore

$$\mathbf{E} \cdot \mathbf{n} = \frac{\varepsilon Q}{|\mathbf{x}|^3} \mathbf{x} \cdot \left(\frac{\mathbf{x}}{|\mathbf{x}|} \right) = \frac{\varepsilon Q}{|\mathbf{x}|^4} \mathbf{x} \cdot \mathbf{x}$$

$$= \frac{\varepsilon Q}{|\mathbf{x}|^2} = \frac{\varepsilon Q}{a^2}$$

since the equation of S_1 is $|\mathbf{x}| = a$. Thus we have

$$\iint_{S_2} \mathbf{E} \cdot d\mathbf{S} = \iint_{S_1} \mathbf{E} \cdot \mathbf{n}\, dS$$

$$= \frac{\varepsilon Q}{a^2} \iint_{S_1} dS = \frac{\varepsilon Q}{a^2} A(S_1)$$

$$= \frac{\varepsilon Q}{a^2} 4\pi a^2 = 4\pi \varepsilon Q$$

This shows that the electric flux of $\mathbf{E}$ is $4\pi\varepsilon Q$ through *any* closed surface S_2 that contains the origin. [This is a special case of Gauss's Law (Equation 14.7.10) for a single charge. The relationship between ε and ε_0 is $\varepsilon = 1/(4\pi\varepsilon_0)$.]

Another application of the Divergence Theorem occurs in fluid flow. Let $\mathbf{v}(x, y, z)$ be the velocity field of a fluid with constant density ρ. Then $\mathbf{F} = \rho\mathbf{v}$ is the rate of flow per unit area. If $P_0(x_0, y_0, z_0)$ is a point in the fluid and B_a is a ball with center P_0 and very small radius a, then div $\mathbf{F}(P) \approx$ div $\mathbf{F}(P_0)$ for all points in B_a since div $\mathbf{F}$ is continuous. We approximate the flux over the boundary sphere S_a as follows:

$$\iint_{S_a} \mathbf{F} \cdot d\mathbf{S} = \iiint_{B_a} \text{div } \mathbf{F}\, dV$$

$$\approx \iiint_{B_a} \text{div } \mathbf{F}(P_0)\, dV$$

$$= \text{div } \mathbf{F}(P_0) V(B_a)$$

This approximation becomes better as $a \to 0$ and suggests that

(9)
$$\text{div } \mathbf{F}(P_0) = \lim_{a \to 0} \frac{1}{V(B_a)} \iint_{S_a} \mathbf{F} \cdot d\mathbf{S}$$

Equation 9 says that div $\mathbf{F}(P_0)$ is the net rate of outward flux per unit volume at P_0. (This is the reason for the name *divergence*.) If div $\mathbf{F}(P) > 0$, the net flow is outward near P and P is called a **source**. If div $\mathbf{F}(P) < 0$, the net flow is inward near P and P is called a **sink**.

EXERCISES 14.9

1–2 ■ Verify that the Divergence Theorem is true for the vector field $\mathbf{F}$ on the region E.

1. $\mathbf{F}(x, y, z) = 3x\,\mathbf{i} + xy\,\mathbf{j} + 2xz\,\mathbf{k}$; E is the cube bounded by the planes $x = 0$, $x = 1$, $y = 0$, $y = 1$, $z = 0$, and $z = 1$

2. $\mathbf{F}(x, y, z) = xz\,\mathbf{i} + yz\,\mathbf{j} + 3z^2\,\mathbf{k}$; E is the solid bounded by the paraboloid $z = x^2 + y^2$ and the plane $z = 1$

3–15 ■ Use the Divergence Theorem to calculate the surface integral $\iint_S \mathbf{F} \cdot d\mathbf{S}$; that is, calculate the flux of $\mathbf{F}$ across S.

3. $\mathbf{F}(x, y, z) = 3y^2z^3\mathbf{i} + 9x^2yz^2\mathbf{j} - 4xy^2\mathbf{k}$,
S is the surface of the cube with vertices $(\pm 1, \pm 1, \pm 1)$

4. $\mathbf{F}(x, y, z) = x^2y\mathbf{i} - x^2z\mathbf{j} + z^2y\mathbf{k}$,
S is the surface of the rectangular box bounded by the planes $x = 0$, $x = 3$, $y = 0$, $y = 2$, $z = 0$, and $z = 1$

5. $\mathbf{F}(x, y, z) = -xz\mathbf{i} - yz\mathbf{j} + z^2\mathbf{k}$,
S is the ellipsoid $x^2/a^2 + y^2/b^2 + z^2/c^2 = 1$

6. $\mathbf{F}(x, y, z) = 3xy\mathbf{i} + y^2\mathbf{j} - x^2y^4\mathbf{k}$,
S is the surface of the tetrahedron with vertices $(0, 0, 0)$, $(1, 0, 0)$, $(0, 1, 0)$, and $(0, 0, 1)$

7. $\mathbf{F}(x, y, z) = z\cos y\,\mathbf{i} + x\sin z\,\mathbf{j} + xz\,\mathbf{k}$,
S is the surface of the tetrahedron bounded by the planes $x = 0$, $y = 0$, $z = 0$, and $2x + y + z = 2$

8. $\mathbf{F}(x, y, z) = (x + e^{y\tan z})\mathbf{i} + 3xe^{xz}\mathbf{j} + (\cos y - z)\mathbf{k}$,
S is the surface with equation $x^4 + y^4 + z^4 = 1$

9. $\mathbf{F}(x, y, z) = x^3\mathbf{i} + y^3\mathbf{j} + z^3\mathbf{k}$,
S is the sphere $x^2 + y^2 + z^2 = 1$

10. $\mathbf{F}(x, y, z) = x^3\mathbf{i} + 2xz^2\mathbf{j} + 3y^2z\mathbf{k}$,
S is the surface of the solid bounded by the paraboloid $z = 4 - x^2 - y^2$ and the xy-plane

11. $\mathbf{F}(x, y, z) = ye^z\mathbf{i} + y^2\mathbf{j} + e^{xy}\mathbf{k}$,
S is the surface of the solid bounded by the cylinder $x^2 + y^2 = 9$ and the planes $z = 0$ and $z = y - 3$

12. $\mathbf{F}(x, y, z) = (x^3 + y\sin z)\mathbf{i} + (y^3 + z\sin x)\mathbf{j} + 3z\mathbf{k}$,
S is the surface of the solid bounded by the hemispheres $z = \sqrt{4 - x^2 - y^2}$, $z = \sqrt{1 - x^2 - y^2}$ and the plane $z = 0$

13. $\mathbf{F}(x, y, z) = xy^2\mathbf{i} + yz\mathbf{j} + zx^2\mathbf{k}$, S is the surface of the solid that lies between the cylinders $x^2 + y^2 = 1$ and $x^2 + y^2 = 4$ and between the planes $z = 1$ and $z = 3$

14. $\mathbf{F}(x, y, z) = (x^3 + yz)\mathbf{i} + x^2y\mathbf{j} + xy^2\mathbf{k}$,
S is the surface of the solid bounded by the spheres $x^2 + y^2 + z^2 = 4$ and $x^2 + y^2 + z^2 = 9$

CAS 15. $\mathbf{F}(x, y, z) = e^y\tan z\,\mathbf{i} + y\sqrt{3 - x^2}\,\mathbf{j} + x\sin y\,\mathbf{k}$, S is the surface of the solid that lies above the xy-plane and below the surface $z = 2 - x^4 - y^4$, $-1 \leq x \leq 1$, $-1 \leq y \leq 1$

CAS 16. Use a computer algebra system to plot the vector field $\mathbf{F}(x, y, z) = \sin x\cos^2 y\,\mathbf{i} + \sin^3 y\cos^4 z\,\mathbf{j} + \sin^5 z\cos^6 x\,\mathbf{k}$ in the cube cut from the first octant by the planes $x = \pi/2$, $y = \pi/2$, and $z = \pi/2$. Then compute the flux across the surface of the cube.

17. Use the Divergence Theorem to evaluate $\iint_S \mathbf{F} \cdot d\mathbf{S}$, where $\mathbf{F}(x, y, z) = z^2x\mathbf{i} + (\frac{1}{3}y^3 + \tan z)\mathbf{j} + (x^2z + y^2)\mathbf{k}$ and S is the top half of the sphere $x^2 + y^2 + z^2 = 1$. [*Hint:* Note that S is not a closed surface. First compute integrals over S_1 and S_2, where S_1 is the disk $x^2 + y^2 \leq 1$ oriented downward and $S_2 = S \cup S_1$.]

18. Let $\mathbf{F}(x, y, z) = z\tan^{-1}(y^2)\mathbf{i} + z^3\ln(x^2 + 1)\mathbf{j} + z\mathbf{k}$. Find the flux of $\mathbf{F}$ across the part of the paraboloid $x^2 + y^2 + z = 2$ that lies above the plane $z = 1$ and is oriented upward.

19. Verify that div $\mathbf{E} = 0$ for the electric field $\mathbf{E}(\mathbf{x}) = \varepsilon Q\mathbf{x}/|\mathbf{x}|^3$.

20. Use the Divergence Theorem to evaluate

$$\iint_S (2x + 2y + z^2)\,dS$$

where S is the sphere $x^2 + y^2 + z^2 = 1$.

21–26 ■ Prove each identity, assuming that S and E satisfy the conditions of the Divergence Theorem and the scalar functions and components of the vector fields have continuous second-order partial derivatives.

21. $\iint_S \mathbf{a} \cdot \mathbf{n}\,dS = 0$, where $\mathbf{a}$ is a constant vector

22. $V(E) = \frac{1}{3}\iint_S \mathbf{F} \cdot d\mathbf{S}$, where $\mathbf{F}(x, y, z) = x\mathbf{i} + y\mathbf{j} + z\mathbf{k}$

23. $\iint_S \text{curl } \mathbf{F} \cdot d\mathbf{S} = 0$

24. $\iint_S D_\mathbf{n} f\,dS = \iiint_E \nabla^2 f\,dV$

25. $\iint_S (f\nabla g) \cdot \mathbf{n}\,dS = \iiint_E (f\nabla^2 g + \nabla f \cdot \nabla g)\,dV$

26. $\iint_S (f\nabla g - g\nabla f) \cdot \mathbf{n}\,dS = \iiint_E (f\nabla^2 g - g\nabla^2 f)\,dV$

14.10 SUMMARY

The main results of this chapter are all higher-dimensional analogues of the Fundamental Theorem of Calculus. To help you remember them, we collect them together here (without hypotheses) so that you can see more easily their essential similarity. Notice that in each case we have an integral of a "derivative" over a region on the left side, and the right side involves the values of the original function only on the *boundary* of the region.

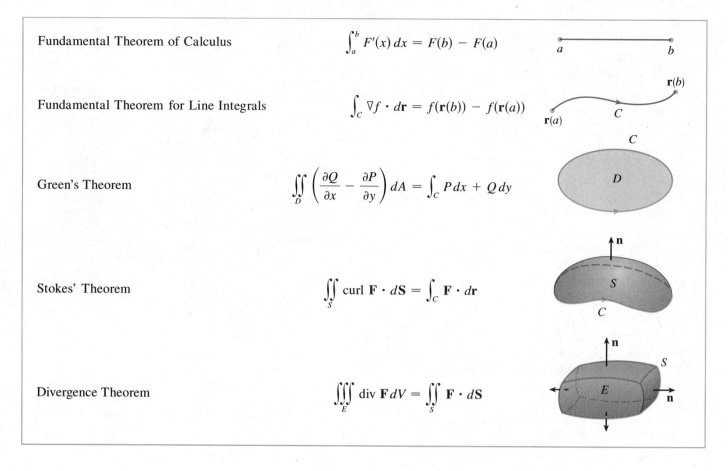

Fundamental Theorem of Calculus	$\int_a^b F'(x)\,dx = F(b) - F(a)$	
Fundamental Theorem for Line Integrals	$\int_C \nabla f \cdot d\mathbf{r} = f(\mathbf{r}(b)) - f(\mathbf{r}(a))$	
Green's Theorem	$\iint_D \left(\dfrac{\partial Q}{\partial x} - \dfrac{\partial P}{\partial y} \right) dA = \int_C P\,dx + Q\,dy$	
Stokes' Theorem	$\iint_S \operatorname{curl} \mathbf{F} \cdot d\mathbf{S} = \int_C \mathbf{F} \cdot d\mathbf{r}$	
Divergence Theorem	$\iiint_E \operatorname{div} \mathbf{F}\,dV = \iint_S \mathbf{F} \cdot d\mathbf{S}$	

14   REVIEW

KEY TOPICS ■ Define, state, or discuss the following.

1. Vector field
2. Conservative vector field
3. Potential function
4. Line integral of a scalar function with respect to arc length
5. Line integral of a scalar function with respect to x, y, and z
6. Line integral of a vector field
7. Work done by a force field
8. Fundamental theorem for line integrals
9. Independence of path
10. Green's Theorem
11. Curl
12. Divergence

13. Vector forms of Green's Theorem
14. Parametric surface
15. Normal vector to a parametric surface
16. Smooth surface
17. Surface area of a parametric surface
18. Formula for the surface area of a graph $z = f(x, y)$
19. Surface integral of a scalar function
20. Oriented surface
21. Surface integral of a vector field
22. Stokes' Theorem
23. The Divergence Theorem

EXERCISES

1–8 ■ Determine whether the statement is true or false.

1. If $\mathbf{F}$ is a vector field, then div $\mathbf{F}$ is a vector field.

2. If $\mathbf{F}$ is a vector field, then curl $\mathbf{F}$ is a vector field.

3. If f has continuous partial derivatives of all orders on $\mathbb{R}^3$, then $\text{div}(\text{curl}\,\nabla f) = 0$.

4. If f has continuous partial derivatives on $\mathbb{R}^3$ and C is any circle, then $\int_C \nabla f \cdot d\mathbf{r} = 0$.

5. If $\mathbf{F} = P\mathbf{i} + Q\mathbf{j}$ and $P_y = Q_x$ in an open region D, then $\mathbf{F}$ is conservative.

6. $\int_{-C} f(x, y)\,ds = -\int_C f(x, y)\,ds$

7. If S is a sphere and $\mathbf{F}$ is a constant vector field, then $\iint_S \mathbf{F} \cdot d\mathbf{S} = 0$.

8. There is a vector field $\mathbf{F}$ such that curl $\mathbf{F} = x\mathbf{i} + y\mathbf{j} + z\mathbf{k}$.

9–17 ■ Evaluate the given line integral.

9. $\int_C y\,ds$,
C is the arc of the parabola $y^2 = 2x$ from $(0, 0)$ to $(2, 2)$

10. $\int_C yz^2\,ds$, C is the line segment from $(-1, 1, 3)$ to $(0, 3, 5)$

11. $\int_C x^3 z\,ds$,
$C: x = 2\sin t, y = t, z = 2\cos t, 0 \le t \le \pi/2$

12. $\int_C xy\,dx + y\,dy$,
C is the sine curve $y = \sin x, 0 \le x \le \pi/2$

13. $\int_C x^3 y\,dx - x\,dy$, C is the circle $x^2 + y^2 = 1$ with counterclockwise orientation

14. $\int_C x\sin y\,dx + xyz\,dz$, C is given by
$\mathbf{r}(t) = t\mathbf{i} + t^2\mathbf{j} + t^3\mathbf{k}, 0 \le t \le 1$

15. $\int_C y\,dx + z\,dy + x\,dz$,
C consists of the line segments from $(0, 0, 0)$ to $(1, 1, 2)$ and from $(1, 1, 2)$ to $(3, 1, 4)$

16. $\int_C \mathbf{F} \cdot d\mathbf{r}$, where $\mathbf{F}(x, y) = x^2 y\mathbf{i} + e^y\mathbf{j}$ and C is given by $\mathbf{r}(t) = t^2\mathbf{i} - t^3\mathbf{j}, 0 \le t \le 1$

17. $\int_C \mathbf{F} \cdot d\mathbf{r}$, where $\mathbf{F}(x, y, z) = (x + y)\mathbf{i} + z\mathbf{j} + x^2 y\mathbf{k}$ and C is given by $\mathbf{r}(t) = 2t\mathbf{i} + t^2\mathbf{j} + t^4\mathbf{k}, 0 \le t \le 1$

18. Find the work done by the force field
$\mathbf{F}(x, y, z) = z\mathbf{i} + x\mathbf{j} + y\mathbf{k}$ in moving a particle from the point $(3, 0, 0)$ to the point $(0, \pi/2, 3)$ (a) along a straight line and (b) along the helix $x = 3\cos t, y = t, z = 3\sin t$.

19–20 ■ Show that $\mathbf{F}$ is a conservative vector field. Then find a function f such that $\mathbf{F} = \nabla f$.

19. $\mathbf{F}(x, y) = \sin y\mathbf{i} + (x\cos y + \sin y)\mathbf{j}$

20. $\mathbf{F}(x, y, z) = (2xy^3 + z^2)\mathbf{i} + (3x^2 y^2 + 2yz)\mathbf{j} + (y^2 + 2xz)\mathbf{k}$

21–22 ■ Show that $\mathbf{F}$ is conservative and use this fact to evaluate $\int_C \mathbf{F} \cdot d\mathbf{r}$ along the given curve.

21. $\mathbf{F}(x, y) = (2x + y^2 + 3x^2 y)\mathbf{i} + (2xy + x^3 + 3y^2)\mathbf{j}$,
C is the arc of the curve $y = x\sin x$ from $(0, 0)$ to $(\pi, 0)$

22. $\mathbf{F}(x, y, z) = yz(2x + y)\mathbf{i} + xz(x + 2y)\mathbf{j} + xy(x + y)\mathbf{k}$,
C is given by $\mathbf{r}(t) = (1 + t)\mathbf{i} + (1 + 2t^2)\mathbf{j} + (1 + 3t^3)\mathbf{k}$,
$0 \le t \le 1$

23. Verify that Green's Theorem is true for the line integral $\int_C xy\,dx + x^2\,dy$, where C is the triangle with vertices $(0, 0)$, $(1, 0)$, and $(1, 2)$.

24. Use Green's Theorem to evaluate

$$\int_C (1 + \tan x)\,dx + (x^2 + e^y)\,dy$$

where C is the positively oriented boundary of the region enclosed by the curves $y = \sqrt{x}$, $x = 1$, and $y = 0$.

25. Use Green's Theorem to evaluate

$$\int_C x^2 y\,dx - xy^2\,dy$$

where C is the circle $x^2 + y^2 = 4$ with counterclockwise orientation.

26. Find curl $\mathbf{F}$ and div $\mathbf{F}$ if

$$\mathbf{F}(x, y, z) = x^2 z\mathbf{i} + 2x\sin y\mathbf{j} + 2z\cos y\mathbf{k}$$

27. Show that there is no vector field $\mathbf{G}$ such that

$$\text{curl}\,\mathbf{G} = 2x\mathbf{i} + 3yz\mathbf{j} - xz^2\mathbf{k}$$

28. Show that, under conditions to be stated on the vector fields $\mathbf{F}$ and $\mathbf{G}$,

$$\text{curl}(\mathbf{F} \times \mathbf{G}) = \mathbf{F}\,\text{div}\,\mathbf{G} - \mathbf{G}\,\text{div}\,\mathbf{F} + (\mathbf{G} \cdot \nabla)\mathbf{F} - (\mathbf{F} \cdot \nabla)\mathbf{G}$$

29. If C is any piecewise-smooth simple closed plane curve and f and g are differentiable functions, show that

$$\int_C f(x)\,dx + g(y)\,dy = 0$$

30. If f and g are twice differentiable functions, show that

$$\nabla^2(fg) = f\nabla^2 g + g\nabla^2 f + 2\nabla f \cdot \nabla g$$

31. If f is a harmonic function, that is, $\nabla^2 f = 0$, show that the line integral $\int f_y\,dx - f_x\,dy$ is independent of path in any simple region D.

32. (a) Sketch the curve C with parametric equations

$$x = \cos t \qquad y = \sin t \qquad z = \sin t \qquad 0 \le t \le 2\pi$$

(b) Find $\int_C 2xe^{2y}\,dx + (2x^2 e^{2y} + 2y\cot z)\,dy - y^2\csc^2 z\,dz$.

33. Find the area of the part of the surface $z = x^2 + 2y$ that lies above the triangle with vertices $(0, 0)$, $(1, 0)$, and $(1, 2)$.

34. (a) Find an equation of the tangent plane at the point
$(4, -2, 1)$ to the parametric surface S given by
$\mathbf{r}(u, v) = v^2\mathbf{i} - uv\mathbf{j} + u^2\mathbf{k}$, $0 \leqslant u \leqslant 3$, $-3 \leqslant v \leqslant 3$.

(b) Use a computer to graph the surface S and the tangent
plane found in part (a).

(c) Set up, but do not evaluate, an integral for the surface
area of S.

(d) If

$$\mathbf{F}(x, y, z) = \frac{z^2}{1 + x^2}\mathbf{i} + \frac{x^2}{1 + y^2}\mathbf{j} + \frac{y^2}{1 + z^2}\mathbf{k}$$

find $\iint_S \mathbf{F} \cdot d\mathbf{S}$ correct to four decimal places.

35–38 ■ Evaluate the surface integral.

35. $\iint_S z\, dS$, where S is the part of the paraboloid $z = x^2 + y^2$
that lies under the plane $z = 4$

36. $\iint_S (x^2z + y^2z)\, dS$, where S is the part of the plane
$z = 4 + x + y$ that lies inside the cylinder $x^2 + y^2 = 4$

37. $\iint_S \mathbf{F} \cdot d\mathbf{S}$, where $\mathbf{F}(x, y, z) = xz\mathbf{i} - 2y\mathbf{j} + 3x\mathbf{k}$ and S is
the sphere $x^2 + y^2 + z^2 = 4$ with outward orientation

38. $\iint_S \mathbf{F} \cdot d\mathbf{S}$, where $\mathbf{F}(x, y, z) = x^2\mathbf{i} + xy\mathbf{j} + z\mathbf{k}$ and S is the
part of the paraboloid $z = x^2 + y^2$ below the plane $z = 1$
with upward orientation

39. Verify that Stokes' Theorem is true for the vector field
$\mathbf{F}(x, y, z) = x^2\mathbf{i} + y^2\mathbf{j} + z^2\mathbf{k}$, where S is the part of the
paraboloid $z = 1 - x^2 - y^2$ that lies above the xy-plane,
and S has upward orientation.

40. Use Stokes' Theorem to evaluate $\iint_S \text{curl } \mathbf{F} \cdot d\mathbf{S}$, where
$\mathbf{F}(x, y, z) = x^2yz\mathbf{i} + yz^2\mathbf{j} + z^3e^{xy}\mathbf{k}$; S is the part of the
sphere $x^2 + y^2 + z^2 = 5$ that lies above the plane $z = 1$,
and S is oriented upward.

41. Use Stokes' Theorem to evaluate $\int_C \mathbf{F} \cdot d\mathbf{r}$, where
$\mathbf{F}(x, y, z) = xy\mathbf{i} + yz\mathbf{j} + zx\mathbf{k}$ and C is the triangle with
vertices $(1, 0, 0)$, $(0, 1, 0)$, and $(0, 0, 1)$, oriented
counterclockwise as viewed from above.

42. Use the Divergence Theorem to calculate the surface
integral $\iint_S \mathbf{F} \cdot d\mathbf{S}$, where $\mathbf{F}(x, y, z) = x^3\mathbf{i} + y^3\mathbf{j} + z^3\mathbf{k}$ and
S is the surface of the solid bounded by the cylinder
$x^2 + y^2 = 1$ and the planes $z = 0$ and $z = 2$.

43. Verify that the Divergence Theorem is true for the vector
field $\mathbf{F}(x, y, z) = x\mathbf{i} + y\mathbf{j} + z\mathbf{k}$, where E is the unit ball
$x^2 + y^2 + z^2 \leqslant 1$.

44. Compute the outward flux of

$$\mathbf{F}(x, y, z) = \frac{x\mathbf{i} + y\mathbf{j} + z\mathbf{k}}{(x^2 + y^2 + z^2)^{3/2}}$$

through the ellipsoid $4x^2 + 9y^2 + 6z^2 = 36$.

45. Let

$$\mathbf{F}(x, y, z) = (3x^2yz - 3y)\mathbf{i} + (x^3z - 3x)\mathbf{j} + (x^3y + 2z)\mathbf{k}$$

Evaluate $\int_C \mathbf{F} \cdot d\mathbf{r}$, where C is the curve with initial point
$(0, 0, 2)$ and terminal point $(0, 3, 0)$ shown in the figure.

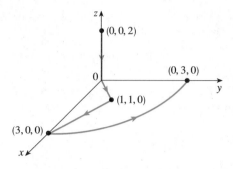

46. Let

$$\mathbf{F}(x, y) = \frac{(2x^3 + 2xy^2 - 2y)\mathbf{i} + (2y^3 + 2x^2y + 2x)\mathbf{j}}{x^2 + y^2}$$

Evaluate $\oint_C \mathbf{F} \cdot d\mathbf{r}$, where C is shown in the figure.

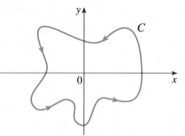

47. Find $\iint_S \mathbf{F} \cdot \mathbf{n}\, dS$, where $\mathbf{F}(x, y, z) = x\mathbf{i} + y\mathbf{j} + z\mathbf{k}$ and S
is the outwardly oriented surface shown in the figure (the
boundary surface of a cube with a unit corner cube
removed).

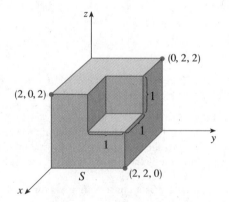

48. If the components of $\mathbf{F}$ have continuous second partial
derivatives and S is the boundary surface of a simple solid
region, show that $\iint_S \text{curl } \mathbf{F} \cdot d\mathbf{S} = 0$.

PROBLEMS PLUS

1. The double integral $\int_0^1 \int_0^1 \dfrac{1}{1 - xy}\, dx\, dy$ is an improper integral and could be defined as the limit of double integrals over the rectangle $[0, t] \times [0, t]$ as $t \to 1^-$. But if we expand the integrand as a geometric series, we can express the integral as the sum of an infinite series. Show that

$$\int_0^1 \int_0^1 \frac{1}{1 - xy}\, dx\, dy = \sum_{n=1}^{\infty} \frac{1}{n^2}$$

2. Leonhard Euler was able to find the exact sum of the series in Problem 1. In 1736 he proved that

$$\sum_{n=1}^{\infty} \frac{1}{n^2} = \frac{\pi^2}{6}$$

   In this problem we ask you to prove this fact by evaluating the double integral in Problem 1. Start by making the change of variables

$$x = \frac{u - v}{\sqrt{2}} \qquad y = \frac{u + v}{\sqrt{2}}$$

   This gives a rotation about the origin through the angle $\pi/4$. You will need to sketch the corresponding region in the uv-plane.

   [*Hint*: If, in evaluating the integral, you encounter either of the expressions $(1 - \sin\theta)/\cos\theta$ or $(\cos\theta)/(1 + \sin\theta)$, you might like to use the identity $\cos\theta = \sin((\pi/2) - \theta)$ and the corresponding identity for $\sin\theta$.]

3. (a) Show that

$$\int_0^1 \int_0^1 \int_0^1 \frac{1}{1 - xyz}\, dx\, dy\, dz = \sum_{n=1}^{\infty} \frac{1}{n^3}$$

   (Nobody has ever been able to find the exact value of the sum of this series.)

   (b) Show that

$$\int_0^1 \int_0^1 \int_0^1 \frac{1}{1 + xyz}\, dx\, dy\, dz = \sum_{n=1}^{\infty} \frac{(-1)^{n-1}}{n^3}$$

   and use this equation to evaluate the triple integral correct to two decimal places.

4. Find the simple closed curve C for which the value of the line integral

$$\int_C (y^3 - y)\, dx - 2x^3\, dy$$

   is a maximum.

5. If $[\![x]\!]$ denotes the greatest integer in x, evaluate the integral

$$\iint_R [\![x + y]\!]\, dA$$

   where $R = \{(x, y) \mid 1 \leqslant x \leqslant 3,\ 2 \leqslant y \leqslant 5\}$.

6. Evaluate the integral

$$\int_0^1 \int_0^1 e^{\max\{x^2,\,y^2\}}\, dy\, dx$$

7. Find the average value of the function $f(x) = \int_x^1 \cos(t^2)\, dt$ on the interval $[0, 1]$.

8. Among all planes that are tangent to the surface $xy^2z^2 = 1$, find the ones that are farthest from the origin.

9. If f is continuous, show that

$$\int_0^x \int_0^y \int_0^z f(t)\, dt\, dz\, dy = \tfrac{1}{2} \int_0^x (x - t)^2 f(t)\, dt$$

10. Show that

$$\int_0^\infty \frac{\arctan \pi x - \arctan x}{x}\, dx = \frac{\pi}{2} \ln \pi$$

by first expressing the integral as an iterated integral.

11. If $\mathbf{a}$, $\mathbf{b}$, and $\mathbf{c}$ are constant vectors, $\mathbf{r}$ is the position vector $x\,\mathbf{i} + y\,\mathbf{j} + z\,\mathbf{k}$, and E is given by the inequalities $0 \le \mathbf{a} \cdot \mathbf{r} \le \alpha$, $0 \le \mathbf{b} \cdot \mathbf{r} \le \beta$, $0 \le \mathbf{c} \cdot \mathbf{r} \le \gamma$, show that

$$\iiint_E (\mathbf{a} \cdot \mathbf{r})(\mathbf{b} \cdot \mathbf{r})(\mathbf{c} \cdot \mathbf{r})\, dV = \frac{(\alpha\beta\gamma)^2}{8\,|\mathbf{a} \cdot (\mathbf{b} \times \mathbf{c})|}$$

12. In this problem we find formulas for the volume enclosed by a hypersphere in n-dimensional space.
  (a) Use a double integral and trigonometric substitution to find the area of a circle with radius r.
  (b) Use a triple integral and trigonometric substitution to find the volume of a sphere with radius r.
  (c) Use a quadruple integral to find the hypervolume enclosed by the hypersphere $x^2 + y^2 + z^2 + w^2 = r^2$ in $\mathbb{R}^4$. (Use only trigonometric substitution and the reduction formulas for $\int \sin^n x\, dx$ or $\int \cos^n x\, dx$.)
  (d) Use an n-tuple integral to find the volume enclosed by a hypersphere of radius r in n-dimensional space $\mathbb{R}^n$.   [*Hint:* The formulas for n even and for n odd are different.]

13. Let S be a smooth parametric surface and let P be a point such that each line that starts at P intersects S at most once. The **solid angle** $\Omega(S)$ subtended by S at P is the set of lines starting at P and passing through S. Let $S(a)$ be the intersection of $\Omega(S)$ with the surface of the sphere with center P and radius a. Then the measure of the solid angle (in steradians) is defined to be

$$|\Omega(S)| = \frac{\text{area of } S(a)}{a^2}$$

Apply the Divergence Theorem to the part of $\Omega(S)$ between $S(a)$ and S to show that

$$|\Omega(S)| = \iint_S \frac{\mathbf{r} \cdot \mathbf{n}}{r^3}\, dS$$

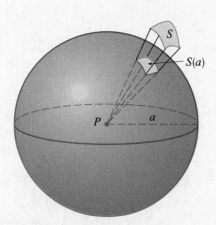

FIGURE FOR PROBLEM 13

where **r** is the radius vector from P to any point on S, $r = |\mathbf{r}|$, and the unit normal vector **n** is directed away from P.

This shows that the definition of the measure of a solid angle is independent of the radius a of the sphere. Thus the measure of the solid angle is equal to the area subtended on a *unit* sphere. (Note the analogy with the definition of radian measure.) The total solid angle subtended by a sphere at its center is thus 4π steradians.

14. (a) Investigate the solid enclosed by the three cylinders $x^2 + y^2 = 1$, $x^2 + z^2 = 1$, and $y^2 + z^2 = 1$. Describe the solid and sketch it carefully. Then compute its volume.

(b) Use a computer algebra system to draw the edges of the solid.

(c) What happens to the solid in part (a) if the radius of the first cylinder is different from 1?

DIFFERENTIAL EQUATIONS

■ Among all the mathematical disciplines the theory of differential equations is the most important. It furnishes the explanation of all those elementary manifestations of nature which involve time.

SOPHUS LIE

In this book we have come into contact now and then with differential equations. For instance, in Section 6.5 we saw that the only solutions of the equation $y' = ky$ are of the form $y = Ce^{kx}$, and in Section 8.1 we learned how to solve separable differential equations. We encountered Laplace's equation and other partial differential equations in Section 12.3. In this chapter we give an introduction to the general study of differential equations by showing how to solve several basic types of equations and how to apply them.

15.1 BASIC CONCEPTS; SEPARABLE AND HOMOGENEOUS EQUATIONS

■ Calculus is the mathematics of change and differential equations are the engines of calculus.

J. M. A. DANBY

There are two kinds of differential equations. An **ordinary differential equation** involves an unknown function of a single variable and some of its derivatives (ordinary derivatives). A **partial differential equation** involves an unknown function of two or more variables and some of its partial derivatives. Since a derivative is a rate of change, it is only natural that differential equations arise in the description of change and motion. In fact, we have already seen how differential equations occur in problems of exponential and logistic growth, radioactive decay, Newton's Law of Cooling, chemical reactions, and the motion of a particle or a planet. Later we will see how differential equations can be applied to the motion of springs and to electric circuits.

The **order** of a differential equation is the order of the highest derivative that appears in the equation. Thus

$$x \frac{dy}{dx} = e^{xy}$$

is an ordinary differential equation of order 1 and

$$y''' - 2xy' + y = \sin x$$

is a third-order ordinary differential equation, whereas

$$\frac{\partial^2 u}{\partial x^2} + 2xy \frac{\partial^2 u}{\partial x \, \partial y} - \frac{\partial^2 u}{\partial y^2} = 0$$

is a second-order partial differential equation. We study only ordinary differential equations in this chapter, so when we talk about a differential equation we mean an ordinary differential equation.

SEPARABLE EQUATIONS

A first-order differential equation has the form

(1)
$$\frac{dy}{dx} = F(x, y)$$

where F is some function of the two variables x and y. The special case in which F can be factored as a function of x times a function of y, that is, $F(x, y) = g(x)f(y)$, is called a **separable equation;** we learned how to solve this type of equation in Section 8.1. We can write Equation 1 in the form

$$\frac{dy}{dx} = \frac{g(x)}{h(y)}$$

and to solve it we rewrite it in differential form:

$$h(y)\,dy = g(x)\,dx$$

Then we integrate both sides:

$$\int h(y)\,dy = \int g(x)\,dx$$

This equation defines y implicitly as a function of x. If possible, we solve for y in terms of x.

An **initial-value problem** for a first-order differential equation consists of finding a solution of Equation 1 that also satisfies an **initial condition** of the form

(2)
$$y(x_0) = y_0$$

Such a **solution** is a function ϕ that satisfies both

$$\phi'(x) = F(x, \phi(x)) \qquad \text{for all } x \text{ in some interval}$$

and
$$\phi(x_0) = y_0$$

It is proved in advanced books on differential equations that if F and $\partial F/\partial y$ are continuous on an open region D of the xy-plane and $(x_0, y_0) \in D$, then there exists a unique solution (defined on some interval containing x_0) to the initial-value problem given by Equations 1 and 2.

EXAMPLE 1

(a) Solve the differential equation $\dfrac{dy}{dx} = e^{-y}\cos x$.

(b) Solve the initial-value problem $y' = e^{-y}\cos x$, $y(0) = 1$.

SOLUTION

(a) The differential equation is separable and we write it as

$$e^y\,dy = \cos x\,dx$$

Now we integrate:

$$\int e^y\,dy = \int \cos x\,dx$$

$$e^y = \sin x + C$$

$$y = \ln(\sin x + C)$$

Figure 1 shows members of the family of solutions in Example 1. The solution of the initial-value problem in part (b) is shown in red.

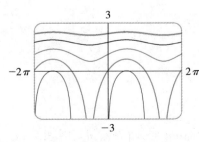

FIGURE 1

This is the general solution. Notice that it involves an arbitrary constant C.

(b) To determine the value of C so that the initial condition $y(0) = 1$ is satisfied, we set $x = 0$ and $y = 1$ in the general solution:

$$y(0) = \ln(0 + C) = \ln C = 1$$

Then $C = e$ and the solution of the initial-value problem is

$$y = \ln(\sin x + e)$$ ■

DIRECTION FIELDS

Even if it is not possible to solve explicitly a differential equation of the form of Equation 1, we can still get a rough picture of how the graphs of the solution look. (These are called **solution curves.**) At any point (x, y) in the domain of F, we know that the number $F(x, y) = dy/dx$ represents the slope of the solution curve. Thus, at each point (x, y), the differential equation tells us the direction in which the curve is proceeding. If we draw a short line segment with slope $F(x, y)$ at a large number of points (x, y), the result is called a **direction field** and allows us to visualize the general shape of the solution curves. The more line segments we draw, the clearer the picture becomes (see Figure 2).

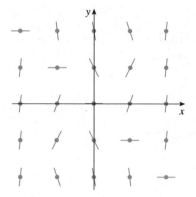
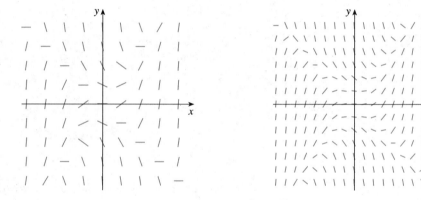

FIGURE 2   Direction fields for $y' = F(x, y)$

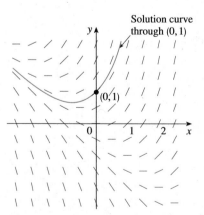

FIGURE 3
Direction field for $y' = x + y$

EXAMPLE 2
(a) Sketch the direction field for the differential equation $y' = x + y$.
(b) Use the direction field to sketch the solution curve that passes through the point $(0, 1)$.

SOLUTION
(a) We first compute the slope at several points in the following chart.

x	0	0	0	0	0	0.5	0.5	0.5	0.5	0.5	1	1	1	$\cdots$
y	0	0.5	1	-0.5	-1	0	0.5	1	-0.5	-1	0	1	-1	$\cdots$
$y' = x + y$	0	0.5	1	-0.5	-1	0.5	1	1.5	0	-0.5	1	2	0	$\cdots$

Then we draw line segments with these slopes at these points in Figure 3.

(b) We sketch the solution curve through $(0, 1)$ in Figure 3 by following the direction field. ■

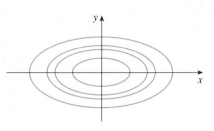

(a) Direction field

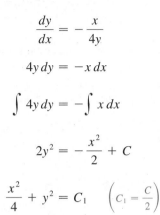

(b) Solution curves

FIGURE 4

$$y' = -\frac{x}{4y}$$

EXAMPLE 3

(a) Sketch the direction field for the differential equation $y' = -x/(4y)$.

(b) Solve the differential equation and sketch the solution curves.

SOLUTION

(a) By computing slopes as in Example 2, we sketch the direction field in Figure 4(a). Notice that the slope is constant on lines through the origin.

(b) The differential equation is separable and we solve it as follows:

$$\frac{dy}{dx} = -\frac{x}{4y}$$

$$4y\,dy = -x\,dx$$

$$\int 4y\,dy = -\int x\,dx$$

$$2y^2 = -\frac{x^2}{2} + C$$

$$\frac{x^2}{4} + y^2 = C_1 \qquad \left(C_1 = \frac{C}{2}\right)$$

We recognize these solutions as being a family of concentric ellipses, which are sketched in Figure 4(b). Notice the similarity to the direction field in Figure 4(a). ∎

EULER'S METHOD

The basic idea behind direction fields can be used to find numerical approximations to solutions of differential equations. We illustrate the method on the initial-value problem in Example 2: $y' = x + y$, $y(0) = 1$. The differential equation tells us that $y'(0) = 0 + 1 = 1$, so the solution curve has slope 1 at the point $(0, 1)$. As a first approximation to the solution we could use the linear approximation $L(x) = x + 1$. In other words, we could use the tangent line at $(0, 1)$ as a rough approximation to the solution curve (see Figure 5).

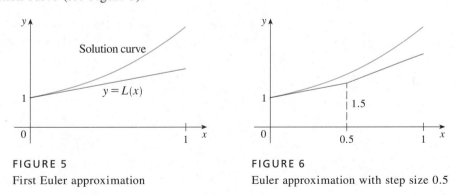

FIGURE 5

First Euler approximation

FIGURE 6

Euler approximation with step size 0.5

Euler's idea was to improve on this approximation by proceeding only a short distance along this tangent line and then making a midcourse correction by changing direction as indicated by the direction field. Figure 6 shows what happens if we start out along the tangent line but stop when $x = 0.5$. (This horizontal distance traveled is called the *step size*.) Since $L(0.5) = 1.5$, we have $y(0.5) \approx 1.5$ and we take $(0.5, 1.5)$ as the starting point for a new line segment. The differential equation tells us that

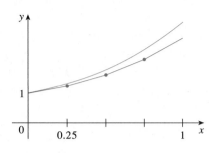

FIGURE 7
Euler approximation with
step size 0.25

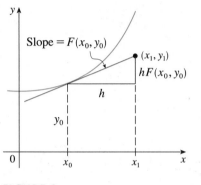

FIGURE 8

$y'(0.5) = 0.5 + 1.5 = 2$, so we use the linear function

$$y = 1.5 + 2(x - 0.5) = 2x + 0.5$$

as an approximation to the solution for $x > 0.5$. If we decrease the step size from 0.5 to 0.25, we get the better Euler approximation shown in Figure 7.

In general, Euler's method says to start at the point given by the initial value and proceed in the direction indicated by the direction field. Stop after a short time, look at the slope at the new location, and proceed in that direction. Keep stopping and changing directions according to the direction field. Euler's method does not produce the exact solution to an initial-value problem—it gives approximations. But by decreasing the step size (and therefore increasing the number of midcourse corrections) we obtain successively better approximations to the exact solution. (Compare Figures 5, 6, and 7.)

For the general first-order initial-value problem $y' = F(x, y)$, $y(x_0) = y_0$, our aim is to find approximate values for the solution at equally spaced numbers x_0, $x_1 = x_0 + h$, $x_2 = x_1 + h$, ..., where h is the step size. The differential equation tells us that the slope at (x_0, y_0) is $y' = F(x_0, y_0)$, so Figure 8 shows that the approximate value of the solution when $x = x_1$ is

$$y_1 = y_0 + hF(x_0, y_0)$$

Similarly,

$$y_2 = y_1 + hF(x_1, y_1)$$

In general,

$$y_n = y_{n-1} + hF(x_{n-1}, y_{n-1})$$

EXAMPLE 4   Use Euler's method with step size 0.1 to construct a table of approximate values for the solution of the initial-value problem $y' = x + y$, $y(0) = 1$.

SOLUTION   We are given that $h = 0.1$, $x_0 = 0$, $y_0 = 1$, and $F(x, y) = x + y$. So we have

$$y_1 = y_0 + hF(x_0, y_0) = 1 + 0.1(0 + 1) = 1.1$$

$$y_2 = y_1 + hF(x_1, y_1) = 1.1 + 0.1(0.1 + 1.1) = 1.22$$

$$y_3 = y_2 + hF(x_2, y_2) = 1.22 + 0.1(0.2 + 1.22) = 1.362$$

This means that if $y(x)$ is the exact solution, then $y(0.3) \approx 1.362$. Proceeding with similar calculations, we get the values in the following table:

The least accurate of the values in Example 4 is the last one: $y(1) = y_{10} \approx 3.1875$. If we reduce the step size to $h = 0.02$, we get $y(1) = y_{50} \approx 3.3832$. Compare these values with the exact value (obtained by solving the differential equation using the methods of Section 15.2): $y(1) = 2(e - 1) \approx 3.4366$.

n	x_n	y_n	n	x_n	y_n
1	0.1	1.100000	6	0.6	1.943122
2	0.2	1.220000	7	0.7	2.197434
3	0.3	1.362000	8	0.8	2.487178
4	0.4	1.528200	9	0.9	2.815895
5	0.5	1.721020	10	1.0	3.187485

HOMOGENEOUS EQUATIONS

A first-order differential equation $y' = f(x, y)$ is called **homogeneous** if $f(x, y)$ can be written as $g(y/x)$, where g is a function of a single variable. For instance, the differential equation

$$\frac{dy}{dx} = \frac{x^2 - xy + y^2}{x^2 - y^2} + \ln x - \ln y + \frac{x + y}{x + 2y}$$

is homogeneous because it can be written as

$$\frac{dy}{dx} = \frac{1 - \left(\dfrac{y}{x}\right) + \left(\dfrac{y}{x}\right)^2}{1 - \left(\dfrac{y}{x}\right)^2} - \ln\left(\dfrac{y}{x}\right) + \frac{1 + \dfrac{y}{x}}{1 + 2\left(\dfrac{y}{x}\right)}$$

However, the differential equation

$$\frac{dy}{dx} = \frac{x^2 - xy^2 + y^2}{x - y^2}$$

is *not* homogeneous because the right side cannot be written as a function of y/x.

A homogeneous differential equation $y' = g(y/x)$ can always be transformed into a separable equation by making the change of variable

$$v = \frac{y}{x}$$

Then

$$y = xv$$

and so

$$y' = v + xv'$$

Thus the differential equation $y' = g(y/x)$ becomes

$$v + xv' = g(v)$$

or

$$v' = \frac{g(v) - v}{x}$$

After we solve this separable differential equation for v as a function of x, we have the solution $y = xv(x)$ of the original differential equation.

EXAMPLE 5   Solve the differential equation

$$y' = \frac{xy + y^2}{x^2}$$

SOLUTION   First we notice that this equation is homogeneous, since we can rewrite it as

$$y' = \left(\frac{y}{x}\right) + \left(\frac{y}{x}\right)^2$$

Therefore, we make the substitution $v = y/x$, which gives

$$y = xv \qquad y' = v + xv'$$

So the differential equation becomes

$$v + xv' = v + v^2$$

or

$$v' = \frac{v^2}{x}$$

We solve this separable differential equation in the usual way:

$$\frac{dv}{dx} = \frac{v^2}{x}$$

$$\int \frac{dv}{v^2} = \int \frac{dx}{x} \qquad v \neq 0$$

$$-\frac{1}{v} = \ln|x| + C$$

$$v = -\frac{1}{\ln|x| + C}$$

Then the solution to the original differential equation is

$$y(x) = xv(x) = -\frac{x}{\ln|x| + C}$$

Notice that in solving the differential equation we had to rule out $v = 0$. But if $v = 0$, then $y = 0$, and we can verify directly that $y = 0$ gives another solution of the differential equation. ∎

Three members of the family of solutions of the differential equation in Example 5 are shown in Figure 9.

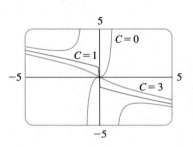

FIGURE 9

ORTHOGONAL TRAJECTORIES

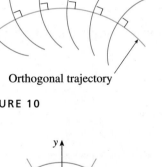

Orthogonal trajectory

FIGURE 10

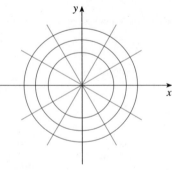

FIGURE 11

An **orthogonal trajectory** of a family of curves is a curve that intersects each curve of the family orthogonally, that is, at right angles (see Figure 10). For instance, each member of the family $y = mx$ of straight lines through the origin is an orthogonal trajectory of the family $x^2 + y^2 = r^2$ of concentric circles with center the origin (see Figure 11). We say that the two families are orthogonal trajectories of each other.

EXAMPLE 6   Find the orthogonal trajectories of the family of curves $x = ky^2$, where k is an arbitrary constant.

SOLUTION   The curves $x = ky^2$ form a family of parabolas whose axis of symmetry is the x-axis. The first step is to find a single differential equation that is satisfied by all members of the family. If we differentiate $x = ky^2$, we get

$$1 = 2ky\frac{dy}{dx} \qquad \text{or} \qquad \frac{dy}{dx} = \frac{1}{2ky}$$

This is a differential equation but it depends on k. To eliminate k we note that, from the equation of the given general parabola $x = ky^2$, we have $k = x/y^2$ and so the differential equation can be written as

$$\frac{dy}{dx} = \frac{1}{2ky} = \frac{1}{2\dfrac{x}{y^2}\,y}$$

or

$$\frac{dy}{dx} = \frac{y}{2x}$$

This means that the slope of the tangent line at any point (x, y) on one of the parabolas is $y' = y/(2x)$. On an orthogonal trajectory the slope of the tangent line must be the negative reciprocal of this slope. Therefore, the orthogonal trajectories must satisfy the differential equation

$$\frac{dy}{dx} = -\frac{2x}{y}$$

This differential equation is separable and we solve it as follows:

$$\int y\,dy = -\int 2x\,dx$$

$$\frac{y^2}{2} = -x^2 + C$$

(3)
$$x^2 + \frac{y^2}{2} = C$$

where C is an arbitrary positive constant. Thus the orthogonal trajectories are the family of ellipses given by Equation 3 and sketched in Figure 12. ∎

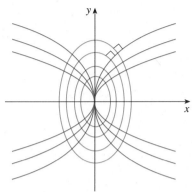

FIGURE 12

Orthogonal trajectories occur in various branches of physics. For example, in an electrostatic field the lines of force are orthogonal to the lines of constant potential. Also, the streamlines in aerodynamics are orthogonal trajectories of the velocity-equipotential curves.

EXERCISES 15.1

1–4 ■ Solve the differential equation.

1. $x^2 y' + y = 0$

2. $(x^2 + 1)\dfrac{dy}{dx} = xy$

3. $\dfrac{dy}{dx} = \dfrac{x\sqrt{x^2 + 1}}{ye^y}$

4. $y' = \dfrac{\ln x}{xy + xy^3}$

5–8 ■ Solve the initial-value problem.

5. $\dfrac{dy}{dx} = y^2 + 1, \quad y(1) = 0$

6. $\dfrac{dy}{dx} = e^{x-y}, \quad y(0) = 1$

7. $\dfrac{du}{dt} = \dfrac{2t + 1}{2(u - 1)}, \quad u(0) = -1$

8. $\dfrac{dy}{dt} = \dfrac{ty + 3t}{t^2 + 1}, \quad y(2) = 2$

9. Solve the initial-value problem $y' = y\sin x$, $y(0) = 1$, and graph the solution.

10. Solve the equation $y' = (3x^2 + e^x)/(4y^3)$ and graph several members of the family of solutions. How does the solution curve change as the constant C varies?

11–14 ■ Match the differential equations with their direction fields (labeled I–IV). Give reasons for your answer.

11. $y' = y - 1$

12. $y' = y - x$

13. $y' = y^2 - x^2$

14. $y' = y^3 - x^3$

I

II

III

IV

15. Use the direction field labeled I (for Exercises 11–14) to sketch the graphs of the solutions that satisfy the given initial conditions.
   (a) $y(0) = 1$ (b) $y(0) = 0$ (c) $y(0) = -1$

16. Repeat Exercise 15 for the direction field labeled III.

17–20 ■ Sketch the direction field of the differential equation. Then use it to sketch a solution curve that passes through the given point.

17. $y' = y^2$, $(0, 1)$ **18.** $y' = x^2 + y$, $(1, 1)$

19. $y' = x^2 + y^2$, $(0, 0)$ **20.** $y' = y(4 - y)$, $(0, 1)$

21–22 ■
(a) Sketch the direction field.
(b) Sketch some solution curves without solving the differential equation.
(c) Solve the differential equation.
(d) Sketch the solutions obtained in part (c) and compare with the graphs from part (b).

21. $y' = 1/y$ **22.** $y' = x^2/y$

CAS **23–24** ■ Use a computer algebra system to draw a direction field for the differential equation. Get a printout and sketch on it the solution curve that passes through $(0, 1)$. Then use the CAS to draw the solution curve and compare with your sketch.

23. $y' = y \sin 2x$ **24.** $y' = \sin(x + y)$

25. Use Euler's method with step size 0.5 to compute the approximate y-values y_1, y_2, y_3, and y_4 of the solution of the initial-value problem $y' = 1 + 3x - 2y$, $y(1) = 2$.

26. Use Euler's method with step size 0.2 to estimate $y(1)$, where $y(x)$ is the solution of the initial-value problem $y' = x + y^2$, $y(0) = 0$.

27. Use Euler's method with step size 0.1 to estimate $y(0.5)$, where $y(x)$ is the solution of the initial-value problem $y' = x^2 + y^2$, $y(0) = 1$.

28. (a) Use Euler's method with step size 0.2 to estimate $y(0.4)$, where $y(x)$ is the solution of the initial-value problem $y' = 2xy^2$, $y(0) = 1$.
(b) Repeat part (a) with step size 0.1.
(c) Find the exact solution of the differential equation and compare the value at 0.4 with the approximations in parts (a) and (b).

29–32 ■ Determine whether the equation is homogeneous.

29. $x^2 + 1 + 2xyy' = 0$

30. $\sqrt{x^2 + y^2}\, dx + y\, dy = 0$, $x > 0$

31. $y' = \ln y - \ln x$

32. $(x^2 + y^2)y' = 2xy + x^2y$

33–38 ■ Solve the homogeneous differential equation.

33. $y' = \dfrac{x - y}{x}$ **34.** $y' = \dfrac{x + y}{x - y}$

35. $\dfrac{dy}{dx} = \dfrac{x^2 + y^2}{xy}$ **36.** $\dfrac{dy}{dx} = \dfrac{y^2 - x^2}{2xy}$

37. $xy' = y + xe^{y/x}$ **38.** $xy' \sin \dfrac{y}{x} = y \sin \dfrac{y}{x} - x$

39–44 ■ Find the orthogonal trajectories of the family of curves. Sketch several members of each family.

39. $y = kx^2$ **40.** $x^2 - y^2 = k$

41. $y = (x + k)^{-1}$ **42.** $y = kx^3$

43. $x^2 - 2y^2 = k$ **44.** $y = ke^{-x}$

45. Let $y(t)$ and $V(t)$ be the height and volume of water in a tank at time t. If water leaks through a hole with area a at the bottom of the tank, then Torricelli's Law says that

$$\frac{dV}{dt} = -a\sqrt{2gy}$$

where g is the acceleration due to gravity.
(a) Suppose the tank is cylindrical with height 6 ft and radius 2 ft and the hole is circular with radius 1 in. If we take $g = 32$ ft/s^2, show that y satisfies the differential equation

$$\frac{dy}{dt} = -\frac{1}{72}\sqrt{y}$$

(b) Solve this equation to find the height of the water at time t, assuming the tank is full at time $t = 0$.
(c) How long will it take for the water to drain completely?

46. Suppose the tank in Exercise 45 is not cylindrical but has cross-sectional area $A(y)$ at height y. Then the volume of water up to height y is $V = \int_0^y A(u)\, du$ and so the Fundamental Theorem of Calculus gives $dV/dy = A(y)$. It follows that

$$\frac{dV}{dt} = \frac{dV}{dy}\frac{dy}{dt} = A(y)\frac{dy}{dt}$$

and so Torricelli's Law becomes

$$A(y)\frac{dy}{dt} = -a\sqrt{2gy}$$

(a) Suppose the tank has the shape of a sphere with radius 2 m and is initially half full of water. If the radius of the circular hole is 1 cm and we take $g = 10$ m/s^2, show that y satisfies the differential equation

$$(4y - y^2)\frac{dy}{dt} = -0.0001\sqrt{20y}$$

(b) How long will it take for the water to drain completely?

15.2 FIRST-ORDER LINEAR EQUATIONS

A first-order **linear** differential equation is one that can be put into the form

(1) $$\frac{dy}{dx} + P(x)y = Q(x)$$

where P and Q are continuous functions on a given interval. This type of equation occurs frequently in various sciences, as we will see.

The standard method for solving Equation 1 is to multiply both sides of the equation by a suitable function $I(x)$ called an *integrating factor*. We try to find I so that the left side of Equation 1, when multiplied by $I(x)$, becomes the derivative of the product $I(x)y$:

(2) $$I(x)\,(\,y' + P(x)y) = (I(x)y)'$$

If we can find such a function I, then Equation 1 becomes

$$(I(x)y)' = I(x)Q(x)$$

Integrating both sides, we would have

$$I(x)y = \int I(x)Q(x)\,dx + C$$

so the solution would be

(3) $$y(x) = \frac{1}{I(x)}\left[\int I(x)Q(x)\,dx + C\right]$$

To find such an I we expand Equation 2 and cancel terms:

$$I(x)y' + I(x)P(x)y = (I(x)y)' = I'(x)y + I(x)y'$$

$$I(x)P(x) = I'(x)$$

This is a separable differential equation for I, which we solve as follows:

$$\int \frac{dI}{I} = \int P(x)\,dx$$

$$\ln|I| = \int P(x)\,dx$$

$$I = Ae^{\int P(x)\,dx}$$

where $A = \pm e^C$. We are looking for a particular integrating factor, not the most general one, so we take $A = 1$ and use

(4) $$I(x) = e^{\int P(x)\,dx}$$

Thus a formula for the general solution to Equation 1 is provided by Equation 3, where I is given by Equation 4. Instead of memorizing this formula, however, we just remember the form of the integrating factor.

> To solve the linear differential equation $y' + P(x)y = Q(x)$, multiply both sides by the **integrating factor** $I(x) = e^{\int P(x)\,dx}$ and integrate both sides.

EXAMPLE 1   Solve the differential equation $\dfrac{dy}{dx} + 3x^2 y = 6x^2$.

SOLUTION   The given equation is linear since it has the form of Equation 1 with $P(x) = 3x^2$ and $Q(x) = 6x^2$. An integrating factor is

$$I(x) = e^{\int 3x^2\,dx} = e^{x^3}$$

Multiplying both sides of the differential equation by e^{x^3}, we get

$$e^{x^3}\,\frac{dy}{dx} + 3x^2 e^{x^3} y = 6x^2 e^{x^3}$$

or

$$\frac{d}{dx}\left(e^{x^3} y\right) = 6x^2 e^{x^3}$$

Figure 1 shows the graphs of several members of the family of solutions in Example 1.

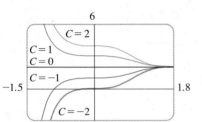

FIGURE 1

Integrating both sides, we have

$$e^{x^3} y = \int 6x^2 e^{x^3}\,dx = 2e^{x^3} + C$$

$$y = 2 + Ce^{-x^3}$$ ∎

EXAMPLE 2   Find the solution of the initial-value problem

$$x^2 y' + xy = 1 \qquad x > 0 \qquad y(1) = 2$$

SOLUTION   We must first divide both sides by the coefficient of y' to put the differential equation into standard form:

(5)
$$y' + \frac{1}{x}y = \frac{1}{x^2} \qquad x > 0$$

The integrating factor is

$$I(x) = e^{\int (1/x)\,dx} = e^{\ln x} = x$$

Multiplication of Equation 5 by x gives

$$xy' + y = \frac{1}{x} \qquad \text{or} \qquad (xy)' = \frac{1}{x}$$

The solution of the initial-value problem in Example 2 is shown in Figure 2.

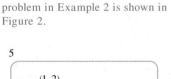
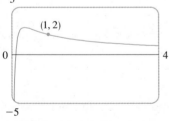

FIGURE 2

Then

$$xy = \int \frac{1}{x}\,dx = \ln x + C$$

and so

$$y = \frac{\ln x + C}{x}$$

Since $y(1) = 2$, we have

$$2 = \frac{\ln 1 + C}{1} = C$$

Therefore, the solution to the initial-value problem is

$$y = \frac{\ln x + 2}{x}$$

■

EXAMPLE 3 Solve $y' + 2xy = 1$.

SOLUTION The given equation is in the standard form for a linear equation. Multiplying by the integrating factor

$$e^{\int 2x\, dx} = e^{x^2}$$

we get

$$e^{x^2}y' + 2xe^{x^2}y = e^{x^2}$$

or

$$(e^{x^2}y)' = e^{x^2}$$

Therefore

$$e^{x^2}y = \int e^{x^2}\, dx + C$$

Recall from Section 7.6 that $\int e^{x^2}\, dx$ cannot be expressed in terms of elementary functions. Nonetheless, it is a perfectly good function and we can leave the answer as

$$y = e^{-x^2}\int e^{x^2}\, dx + Ce^{-x^2}$$

Another way of writing the solution is

$$y = e^{-x^2}\int_0^x e^{t^2}\, dt + Ce^{-x^2}$$

(Any number can be chosen for the lower limit of integration.) ■

Even though the solutions of the differential equation in Example 3 are expressed in terms of an integral, they can still be graphed by a computer algebra system (Figure 3).

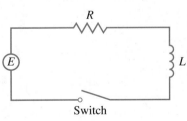

FIGURE 3

APPLICATION TO ELECTRIC CIRCUITS

The simple electric circuit shown in Figure 4 contains an electromotive force (usually a battery or generator) that produces a voltage of $E(t)$ volts (V) and a current of $I(t)$ amperes (A) at time t. The circuit also contains a resistor with a resistance of R ohms (Ω) and an inductor with an inductance of L henries (H).

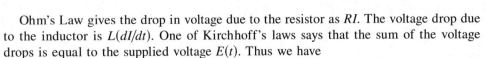

FIGURE 4

Ohm's Law gives the drop in voltage due to the resistor as RI. The voltage drop due to the inductor is $L(dI/dt)$. One of Kirchhoff's laws says that the sum of the voltage drops is equal to the supplied voltage $E(t)$. Thus we have

(6)
$$L\frac{dI}{dt} + RI = E(t)$$

which is a first-order linear differential equation. The solution gives the current I at time t.

EXAMPLE 4   Suppose that in the simple circuit of Figure 4 the resistance is 12 Ω and the inductance is 4 H. If a battery gives a constant voltage of 60 V and the switch is closed when $t = 0$ so the current starts with $I(0) = 0$, find (a) $I(t)$, (b) the current after 1 s, and (c) the limiting value of the current.

SOLUTION
(a) If we put $L = 4$, $R = 12$, and $E(t) = 60$ in Equation 6, we obtain the initial-value problem

$$4 \frac{dI}{dt} + 12I = 60 \qquad I(0) = 0$$

or

$$\frac{dI}{dt} + 3I = 15 \qquad I(0) = 0$$

Multiplying by the integrating factor $e^{\int 3\,dt} = e^{3t}$, we get

$$e^{3t} \frac{dI}{dt} + 3e^{3t}I = 15e^{3t}$$

$$\frac{d}{dt}(e^{3t}I) = 15e^{3t}$$

Figure 5 shows how the current in Example 4 approaches its limiting value.

$$e^{3t}I = \int 15e^{3t}\,dt = 5e^{3t} + C$$

$$I(t) = 5 + Ce^{-3t}$$

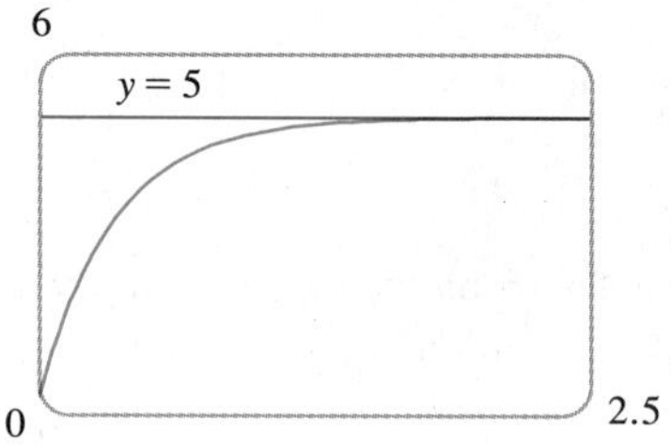

Since $I(0) = 0$, we have $5 + C = 0$, so $C = -5$ and

$$I(t) = 5(1 - e^{-3t})$$

(b) After 1 s the current is

$$I(1) = 5(1 - e^{-3}) \approx 4.75 \text{ A}$$

(c)

$$\lim_{t \to \infty} I(t) = \lim_{t \to \infty} 5(1 - e^{-3t})$$

$$= 5 - 5 \lim_{t \to \infty} e^{-3t}$$

$$= 5 - 0 = 5$$

FIGURE 5

Figure 6 shows the graph of the current when the battery is replaced by a generator.

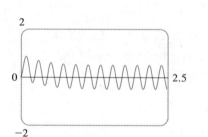

FIGURE 6

EXAMPLE 5   Suppose that the resistance and inductance remain as in Example 4 but, instead of the battery, we use a generator that produces a variable voltage of $E(t) = 60 \sin 30t$ volts. Find $I(t)$.

SOLUTION   This time the differential equation becomes

$$4 \frac{dI}{dt} + 12I = 60 \sin 30t \qquad \text{or} \qquad \frac{dI}{dt} + 3I = 15 \sin 30t$$

The same integrating factor e^{3t} gives

$$\frac{d}{dt}(e^{3t}I) = e^{3t} \frac{dI}{dt} + 3e^{3t}I = 15e^{3t} \sin 30t$$

Using Formula 98 in the Table of Integrals, we have

$$e^{3t}I = \int 15e^{3t}\sin 30t\,dt = 15\,\frac{e^{3t}}{909}(3\sin 30t - 30\cos 30t) + C$$

$$I = \tfrac{5}{101}(\sin 30t - 10\cos 30t) + Ce^{-3t}$$

Since $I(0) = 0$, we get

$$-\tfrac{50}{101} + C = 0$$

so

$$I(t) = \tfrac{5}{101}(\sin 30t - 10\cos 30t) + \tfrac{50}{101}e^{-3t}$$
∎

EXERCISES 15.2

1–4 ■ Determine whether the differential equation is linear.

1. $y' + x^2 y = y^2$

2. $x^2 y' - y + x = 0$

3. $xy' = x - y$

4. $yy' = \sin x$

5–14 ■ Solve the differential equation.

5. $y' - 3y = e^x$

6. $y' + 4y = x$

7. $y' - 2xy = x$

8. $xy' + 2y = e^{x^2}$

9. $y'\cos x = y\sin x + \sin 2x$, $\quad -\pi/2 < x < \pi/2$

10. $1 + xy = xy'$

11. $\dfrac{dy}{dx} + 2xy = x^2$

12. $\dfrac{dy}{dx} = x\sin 2x + y\tan x$, $\quad -\pi/2 < x < \pi/2$

13. $\dfrac{dy}{d\theta} - y\tan\theta = 1$, $\quad -\pi/2 < \theta < \pi/2$

14. $xy' + xy + y = e^{-x}$, $\quad x > 0$

15–20 ■ Solve the initial-value problem.

15. $y' + y = x + e^x$, $\quad y(0) = 0$

16. $xy' - 3y = x^2$, $\quad x > 0$, $\quad y(1) = 0$

17. $y' - 2xy = 2xe^{x^2}$, $\quad y(0) = 3$

18. $(1 + x^2)y' + 2xy = 3\sqrt{x}$, $\quad y(0) = 2$

19. $x^2\dfrac{dy}{dx} + 2xy = \cos x$, $\quad y(\pi) = 0$

20. $x\dfrac{dy}{dx} - \dfrac{y}{x+1} = x$, $\quad y(1) = 0$, $\quad x > 0$

21–22 ■ Solve the differential equation and use a graphing calculator or computer to graph several members of the family of solutions. How does the solution curve change as C varies?

21. $xy' + y = x\cos x$, $\quad x > 0$

22. $y' + (\cos x)y = \cos x$

23. A **Bernoulli differential equation** [named after James Bernoulli (1654–1705)] is of the form

$$\frac{dy}{dx} + P(x)y = Q(x)y^n$$

Observe that, if $n = 0$ or 1, the Bernoulli equation is linear. For other values of n, show that the substitution $u = y^{1-n}$ transforms the Bernoulli equation into the linear equation

$$\frac{du}{dx} + (1 - n)P(x)u = (1 - n)Q(x)$$

24–26 ■ Use the method of Exercise 23 to solve the given differential equation.

24. $xy' + y = -xy^2$

25. $y' + \dfrac{2}{x}y = \dfrac{y^3}{x^2}$

26. $y' + y = xy^3$

27. In the circuit shown in Figure 4 a battery supplies a constant voltage of 40 V, the inductance is 2 H, the resistance is 10 Ω, and $I(0) = 0$.
   (a) Find $I(t)$.
   (b) Find the current after 0.1 s.

28. In the circuit shown in Figure 4 a generator supplies a voltage of $E(t) = 40\sin 60t$ volts, the inductance is 1 H, the resistance is 20 Ω, and $I(0) = 1$ A.
   (a) Find $I(t)$.
   (b) Find the current after 0.1 s.
   (c) Use a graphing device to draw the graph of the current function.

29. The figure shows a circuit containing an electromotive force, a capacitor with a capacitance of C farads (F), and a

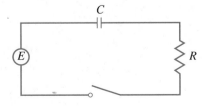

resistor with a resistance of R ohms (Ω). The voltage drop across the capacitor is Q/C, where Q is the charge (in coulombs), so in this case Kirchhoff's Law gives

$$RI + \frac{Q}{C} = E(t)$$

But $I = dQ/dt$ (see Example 3 in Section 2.3), so we have

$$R\frac{dQ}{dt} + \frac{1}{C}Q = E(t)$$

Suppose the resistance is 5 Ω, the capacitance is 0.05 F, a battery gives a constant voltage of 60 V, and the initial charge is $Q(0) = 0$ C. Find the charge and the current at time t.

30. In the circuit of Exercise 29, $R = 2\ \Omega$, $C = 0.01$ F, $Q(0) = 0$, and $E(t) = 10 \sin 60t$. Find the charge and the current at time t.

31. Psychologists interested in learning theory study **learning curves.** A learning curve is the graph of a function $P(t)$, the performance of someone learning a skill as a function of the training time t. The derivative dP/dt represents the rate at which performance improves. If M is the maximum level of performance of which the learner is capable, it is reasonable to assume that dP/dt is proportional to $M - P(t)$. (At first, learning is rapid. Then, as performance increases

and approaches its maximal value, the rate of learning decreases.) Thus

$$\frac{dP}{dt} = k(M - P(t))$$

where k is a positive constant. Solve this linear differential equation and sketch the learning curve.

32. Two new workers were hired for an assembly line. Jim processed 25 units during the first hour and 45 units during the second hour. Mark processed 35 units during the first hour and 50 units the second hour. Using the model of Exercise 31 and assuming that $P(0) = 0$, estimate the maximum number of units per hour that each worker is capable of processing.

33. An object with mass m is dropped from rest and we assume that the air resistance is proportional to the speed of the object. If $s(t)$ is the distance dropped after t seconds, then the speed is $v = s'(t)$ and the acceleration is $a = v'(t)$. If g is the acceleration due to gravity, then the downward force on the object is $mg - cv$, where c is a positive constant, and Newton's Second Law gives

$$m\frac{dv}{dt} = mg - cv$$

(a) Solve this differential equation to find the velocity at time t.
(b) What is the limiting velocity?
(c) Find the distance the object has fallen after t seconds.

15.3   EXACT EQUATIONS

Suppose the equation

(1)
$$f(x, y) = C$$

defines y implicitly as a differentiable function of x. Then $y = y(x)$ satisfies a first-order differential equation obtained by using the Chain Rule (12.5.2) to differentiate both sides of Equation 1 with respect to x:

(2)
$$f_x(x, y) + f_y(x, y)y' = 0$$

A differential equation of the form of Equation 2 is called *exact*.

> **(3) DEFINITION**   A first-order differential equation of the form
>
> $$P(x, y) + Q(x, y)\frac{dy}{dx} = 0$$
>
> is called **exact** if there is a function $f(x, y)$ such that
>
> $$f_x(x, y) = P(x, y) \qquad \text{and} \qquad f_y(x, y) = Q(x, y)$$

If the function f in Definition 3 is known, then the exact differential equation $P(x, y) + Q(x, y)y' = 0$ is easy to solve because

$$\frac{d}{dx} f(x, y(x)) = f_x(x, y) + f_y(x, y)y'$$

$$= P(x, y) + Q(x, y)y' = 0$$

Thus the solution is given implicitly by

(4) $$f(x, y) = C$$

We may be able to solve Equation 4 for y as an explicit function of x.

How can we tell whether a differential equation of the form $P(x, y) + Q(x, y)y' = 0$ is exact? If we consider the vector field $\mathbf{F}(x, y) = \langle P(x, y), Q(x, y) \rangle$, then the condition for exactness in Definition 3 can be written as $\mathbf{F}(x, y) = \nabla f(x, y)$; that is, $\mathbf{F}$ is a conservative vector field. But we know from Theorems 14.3.5 and 14.3.6 that a vector field $\mathbf{F} = P\mathbf{i} + Q\mathbf{j}$ is conservative if and only if $\partial P/\partial y = \partial Q/\partial x$ (assuming these partial derivatives are continuous and the domain is simply-connected). Therefore, we have the following convenient method for testing the exactness of a differential equation.

---

(5) **THEOREM** Suppose P and Q have continuous partial derivatives on a simply-connected domain. Then the differential equation

$$P(x, y) + Q(x, y) \frac{dy}{dx} = 0$$

is exact if and only if

$$\frac{\partial P}{\partial y} = \frac{\partial Q}{\partial x}$$

Furthermore, the procedures that we used in Section 14.3 for determining the (potential) function f such that $\nabla f = \mathbf{F}$ can also be used here to solve an exact differential equation.

EXAMPLE 1 Solve the differential equation

$$4x + 3y + 3(x + y^2)y' = 0$$

SOLUTION Here

$$P(x, y) = 4x + 3y \quad \text{and} \quad Q(x, y) = 3x + 3y^2$$

have continuous partial derivatives on $\mathbb{R}^2$. Also

$$\frac{\partial P}{\partial y} = 3 = \frac{\partial Q}{\partial x}$$

so the differential equation is exact by Theorem 5. Thus there exists a function f such that

(6) $$f_x(x, y) = 4x + 3y$$

(7) $$f_y(x, y) = 3x + 3y^2$$

To determine f we first integrate (6) with respect to x:

$$f(x, y) = 2x^2 + 3xy + g(y) \tag{8}$$

(The constant of integration is a function of y.) Now we differentiate (8) with respect to y:

$$f_y(x, y) = 3x + g'(y) \tag{9}$$

Comparing (7) and (9), we see that

$$g'(y) = 3y^2 \qquad \text{and so} \qquad g(y) = y^3$$

In Figure 1 we used a computer algebra system that is capable of graphing implicit functions to draw members of the family of solutions of the differential equation in Example 1.

(We do not need the arbitrary constant here.) Thus

$$f(x, y) = 2x^2 + 3xy + y^3$$

and by Equation 4 the solution is given implicitly by

$$2x^2 + 3xy + y^3 = C$$

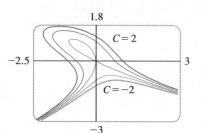

As a check on our work we can verify this as follows:

$$\frac{d}{dx}(2x^2 + 3xy + y^3) = 4x + 3y + 3xy' + 3y^2 y'$$

$$= (4x + 3y) + 3(x + y^2)y' = 0$$

from the differential equation. Thus

$$2x^2 + 3xy + y^3 = C \qquad \blacksquare$$

FIGURE 1

INTEGRATING FACTORS

If the differential equation

$$P(x, y) + Q(x, y)\frac{dy}{dx} = 0$$

is not exact, it may be possible to find an integrating factor $I(x, y)$ such that, after multiplication by $I(x, y)$, the resulting equation

$$I(x, y)P(x, y) + I(x, y)Q(x, y)\frac{dy}{dx} = 0 \tag{10}$$

is exact.

To find such an integrating factor we use Theorem 5. Equation 10 is exact if

$$\frac{\partial}{\partial y}(IP) = \frac{\partial}{\partial x}(IQ)$$

that is,

$$I_y P + IP_y = I_x Q + IQ_x$$

or

$$PI_y - QI_x = I(Q_x - P_y) \tag{11}$$

In general, it is harder to solve this partial differential equation than the original differential equation. But it is sometimes possible to find I that is a function of x or y alone.

For instance, suppose I is a function of x alone. Then $I_y = 0$, so Equation 11 becomes

(12)
$$\frac{dI}{dx} = \frac{P_y - Q_x}{Q} I$$

If $(P_y - Q_x)/Q$ is a function of x alone, then Equation 12 is a first-order linear (and separable) ordinary differential equation that can be solved for $I(x)$. Then Equation 10 is exact and can be solved as in Example 1.

EXAMPLE 2 Solve the equation $2x + y^2 + xyy' = 0$.

SOLUTION Here

$$P(x, y) = 2x + y^2 \qquad Q(x, y) = xy$$

$$\frac{\partial P}{\partial y} = 2y \qquad\qquad \frac{\partial Q}{\partial x} = y$$

Since $\partial P/\partial y \neq \partial Q/\partial x$, the given equation is not exact. But

$$\frac{P_y - Q_x}{Q} = \frac{2y - y}{xy} = \frac{1}{x}$$

is a function of x alone, so by Equation 12 there is an integrating factor I that satisfies

$$\frac{dI}{dx} = \frac{I}{x}$$

Solving this equation as a separable differential equation, we get $I(x) = x$. (We do not need the most general integrating factor, just a particular one.) Multiplying the original equation by x, we get

(13)
$$2x^2 + xy^2 + x^2yy' = 0$$

If we let
$$p(x, y) = 2x^2 + xy^2 \qquad q(x, y) = x^2y$$

then
$$\frac{\partial p}{\partial y} = 2xy = \frac{\partial q}{\partial x}$$

so Equation 13 is now exact. Thus there is a function f such that

$$f_x(x, y) = 2x^2 + xy^2 \qquad f_y(x, y) = x^2y$$

Integrating the first of these equations, we get

$$f(x, y) = \tfrac{2}{3}x^3 + \tfrac{1}{2}x^2y^2 + g(y)$$

so
$$f_y(x, y) = x^2y + g'(y)$$

Comparison then gives $g'(y) = 0$, so g is a constant (which we can take to be 0). Therefore

$$f(x, y) = \tfrac{2}{3}x^3 + \tfrac{1}{2}x^2y^2$$

and the solution is given by

$$\tfrac{2}{3}x^3 + \tfrac{1}{2}x^2y^2 = C$$

Figure 2 shows some of the solutions of the differential equation in Example 2. Notice that if C is positive, then the solution curve consists of three parts.

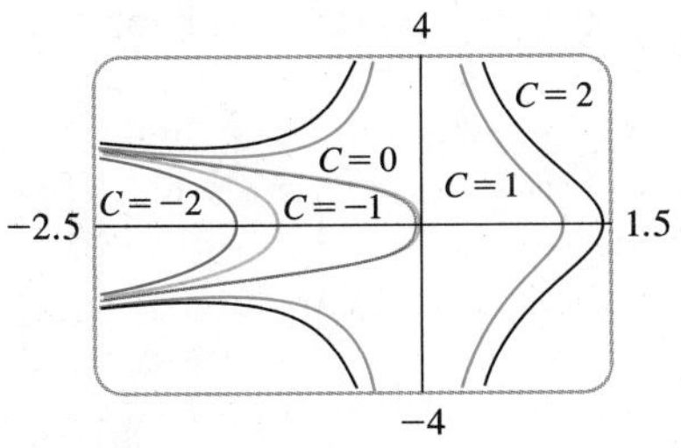

FIGURE 2

EXERCISES 15.3

1. Determine whether each differential equation is exact.
   (a) $x \sin y + (y \cos x)y' = 0$
   (b) $2xy^3 \, dx + 3x^2 y^2 \, dy = 0$
   (c) $(x - y) + (x + y)\dfrac{dy}{dx} = 0$

2. For what values of the number k is the given differential equation exact?
   (a) $2xyy' + y^k = 0$
   (b) $8x^3 y^3 y' + kx^2 y^4 = 0$
   (c) $(xy^2 + kx^2 y) + (x + y)x^2 y' = 0$

3–13 ■ Determine whether the differential equation is exact. If so, solve it.

3. $2x + y + (x + 2y)y' = 0$

4. $2x - y + (x - 2y)y' = 0$

5. $3xy - 2 + (3y^2 - x^2)y' = 0$

6. $3x^2 - 2x + 3y + (3x - 2y)y' = 0$

7. $\sin y + (1 + x \cos y)y' = 0$

8. $y - e^x \cos y + (x + e^x \sin y)y' = 0$

9. $(x + y)e^{x/y} + \left(x - \dfrac{x^2}{y}\right)e^{x/y}y' = 0$

10. $e^x + y \cos x + (e^y - y \sin x)y' = 0$

11. $x \ln y \, dx - (x + y \ln x) \, dy = 0$

12. $\left(2x^3 y^2 - \tfrac{1}{2}e^{2y}\right) dx + (x^4 y - xe^{2y}) \, dy = 0$

13. $\dfrac{1}{y} + \dfrac{2y}{x^3} = \left(\dfrac{x}{y^2} + \dfrac{1}{x^2}\right)\dfrac{dy}{dx}$

CAS 14. If your computer algebra system is capable of graphing implicit functions, use it to graph several members of the family of solutions of the differential equation

$$\frac{dy}{dx} = \frac{\cos y + y \cos x}{x \sin y - \sin x}$$

15–18 ■ Solve the initial-value problem.

15. $3x^2 + 2xy + 3y^2 + (x^2 + 6xy)y' = 0, \quad y(1) = 2$

16. $3x^2 y^2 + 8xy^5 + (2x^3 y + 20x^2 y^4)y' = 0, \quad y(2) = 1$

17. $1 + y \cos xy + (x \cos xy)y' = 0, \quad y(1) = 0$

18. $\ln y + 3y^2 + \left(\dfrac{x}{y} + 6xy\right)y' = 0, y > 0, \quad y(1) = 1$

19–22 ■ Show that the given equation is not exact but becomes exact when multiplied by the specified integrating factor. Then solve the equation.

19. $y^2 + (1 + xy)y' = 0, \quad I(x, y) = e^{xy}$

20. $y(x + y) - x^2 y' = 0, \quad I(x, y) = \dfrac{1}{xy^2}$

21. $y + y^3 + (x + x^3)y' = 0, \quad I(x, y) = \dfrac{1}{(1 + x^2 + y^2)^{3/2}}$

22. $2y \cos x - xy \sin x + (2x \cos x)y' = 0, \quad I(x, y) = xy$

23–25 ■ Find an integrating factor and thus solve the equation.

23. $3xy + 2y^2 + (x^2 + 2xy)y' = 0$

24. $1 - xy + x(y - x)y' = 0, \quad x > 0$

25. $2xy + 3x^2 y + 3y^2 + (x^2 + 2y)y' = 0$

26. Show that if $(Q_x - P_y)/P$ is a function of y alone, then Equation 11 enables us to find an integrating factor for $P + Qy' = 0$. Use this method to solve the equation
$$y - 6x^2 y^3 + (2x - 8x^3 y^2)y' = 0$$

27. Show that every separable differential equation is exact.

15.4   STRATEGY FOR SOLVING FIRST-ORDER EQUATIONS

In solving first-order differential equations we used the technique for separable equations in Sections 8.1 and 15.1 and the method for linear equations in Section 15.2. We also developed methods for solving homogeneous equations in Section 15.1 and exact equations in Section 15.3. In this section we present a miscellaneous collection of first-order differential equations, and part of the problem is to recognize which technique should be used on each equation.

As with the strategy of integration (Section 7.6) and the strategy of testing series (Section 10.7), the main idea is to classify the equation according to its *form*. Here, however, the important thing is not so much the form of the functions involved as it is the form of the equation itself.

Recall that a **separable** equation can be written in the form

(1)
$$\frac{dy}{dx} = g(x)f(y)$$

that is, the expression for dy/dx can be factored as a product of a function of x and a function of y.

A **linear** equation can be put into the form

(2)
$$\frac{dy}{dx} + P(x)y = Q(x)$$

A **homogeneous** equation can be expressed in the form

(3)
$$\frac{dy}{dx} = g\left(\frac{y}{x}\right)$$

An **exact** equation has the form

(4)
$$P(x, y) + Q(x, y)\frac{dy}{dx} = 0$$

where $\partial P/\partial y = \partial Q/\partial x$.

If an equation has none of these forms, we can, as a last resort, attempt to find an **integrating factor** and thus make the equation exact.

In each of these cases, some preliminary algebra may be required in order to put a given equation into one of the preceding forms. (This step is analogous to Step 1 of the strategy for integration: algebraic simplification.)

It may happen that a given equation is of more than one type. For instance, the equation

$$xy' = y$$

is separable because dy/dx can be written as

$$\frac{dy}{dx} = \left(\frac{1}{x}\right)y \qquad \text{(Compare with Equation 1.)}$$

It is also linear, since we can write the equation as

$$\frac{dy}{dx} + \left(-\frac{1}{x}\right)y = 0 \qquad \text{(Compare with Equation 2.)}$$

Furthermore, it is also homogeneous because we can write it as

$$\frac{dy}{dx} = \frac{y}{x} \qquad \text{(Compare with Equation 3.)}$$

In such a case we could solve the equation using any one of the corresponding methods, although one of the methods might be easier than the others.

In the following examples we identify the type of each equation without working out the details of the solutions.

EXAMPLE 1 $y' = 1 - x - y + xy$

Initially, this equation may not appear to be in any of the forms of Equations 1, 2, 3, or 4, but observe that we can factor the right side and therefore write the equation as

$$y' = (1 - x)(1 - y)$$

We now recognize the equation as being separable and we can solve it using the methods of Section 15.1. ∎

EXAMPLE 2 $x^2 - y^2 + 2xyy' = 0$

The equation is clearly not separable, nor is it linear. Since $P_y = -2y$ and $Q_x = 2y$, it is not exact. But if we solve for y', we get

$$y' = \frac{y^2 - x^2}{2xy} = \frac{1}{2}\left[\frac{y}{x} - \frac{1}{y/x}\right]$$

which shows that y' is a function of y/x and the equation is homogeneous (see Equation 3). (We could have anticipated this because the expressions x^2, y^2, and $2xy$ are all of degree 2.) The change of variable $v = y/x$ converts the equation into a separable equation. ∎

EXAMPLE 3 $y' = -\dfrac{3x^2 + 2xy^3}{3x^2y^2}$

This equation is not separable, linear, or homogeneous. We suspect it might be exact, so we write it in the form

$$(3x^2 + 2xy^3) + (3x^2y^2)y' = 0$$

If $P(x, y) = 3x^2 + 2xy^3$ and $Q(x, y) = 3x^2y^2$, then $P_y = 6xy^2 = Q_x$. Therefore, the equation is indeed exact and can be solved by the methods of Section 15.3. ∎

EXAMPLE 4 $(x + 1)y - xy' = x^3 - x^2$

If we put the equation in the form

$$y' - \left(\frac{x + 1}{x}\right)y = x - x^2$$

we recognize it as having the form of Equation 2. It is therefore linear and can be solved using the integrating factor

$$e^{-\int(1 + 1/x)\,dx}$$

∎

EXERCISES 15.4

1–4 ■ Classify the differential equation for each characteristic: (a) separable or not, (b) homogeneous or not, (c) linear or not, (d) exact or not.

1. $yy' + x^2 = 0$

2. $y' + x^2y = 0$

3. $xy' + y = 0$

4. $\sqrt{x}\,y' + \sqrt{y} = 0$

5–24 ■ Classify and solve the differential equation.

5. $y' = y + \sin x$

6. $y' = y \sin x$

7. $x - e^x + 4y^3y' = 0$

8. $3x^2y^2 + 2x^3yy' = 0$

9. $(x^2 - 2xy - y^2)y' = x^2 + 2xy - y^2$

10. $\dfrac{dy}{dx} = \dfrac{e^x - y}{x}$

11. $(2xy - 3\tan x)y' = 3y\sec^2 x - y^2$

12. $xy' + y^2 = 1$

13. $x - (x^2y + y)y' = 0$

14. $y' = \dfrac{\cos y}{x \sin y - 1}$

15. $xy' = \sqrt{1 + x^2} + 2y$

16. $x(x + y)^2 y' = y(x^2 + xy + y^2)$

17. $y^2 \sqrt{1 + x^3} + y' = 0$

18. $xy' = 2 - x^2 + 2y^2 - x^2y^2$

19. $2(x + yy') + e^y(1 + xy') = 0$

20. $xy' = x^2 + x^3 + y$

21. $[\sin(xy) + xy\cos(xy)]y' = 1 - y^2\cos(xy)$

22. $2x^2y + y^2 + (x^3 + 2xy\ln x)y' = 0$

23. $xy' = 2\sqrt{xy} - y$

24. $y' = \dfrac{x^2 + xy}{x^2 - 1}$

25. Let $y' = 3x(y + x^n)$, where n is an integer.
(a) For what value(s) of n is this a separable equation?
(b) For what value(s) of n is this a linear equation?

15.5   SECOND-ORDER LINEAR EQUATIONS

A **second-order linear differential equation** has the form

(1)
$$P(x)\frac{d^2y}{dx^2} + Q(x)\frac{dy}{dx} + R(x)y = G(x)$$

where P, Q, R, and G are continuous functions. We will see in Section 15.7 that equations of this type arise in the study of the motion of a spring as well as in electric circuits.

In this section we study the case where $G(x) = 0$, for all x, in Equation 1. Such equations are called **homogeneous** linear equations. (This use of the word *homogeneous* has nothing to do with the meaning given in Section 15.1.) Thus the form of a second-order linear homogeneous differential equation is

(2)
$$P(x)\frac{d^2y}{dx^2} + Q(x)\frac{dy}{dx} + R(x)y = 0$$

If $G(x) \neq 0$ for some x, Equation 1 is **nonhomogeneous** and is discussed in Section 15.6.

Two basic facts enable us to solve homogeneous linear equations. The first of these says that if we know two solutions y_1 and y_2 of such an equation, then the **linear combination** $y = c_1y_1 + c_2y_2$ is also a solution.

---

(3) THEOREM   If $y_1(x)$ and $y_2(x)$ are both solutions of the linear homogeneous equation (2) and c_1 and c_2 are any constants, then the function

$$y(x) = c_1y_1(x) + c_2y_2(x)$$

is also a solution of Equation 2.

PROOF   Since y_1 and y_2 are solutions of Equation 2, we have

$$P(x)y_1'' + Q(x)y_1' + R(x)y_1 = 0$$

and
$$P(x)y_2'' + Q(x)y_2' + R(x)y_2 = 0$$

Therefore, using the basic rules for differentiation, we have

$$P(x)y'' + Q(x)y' + R(x)y$$

$$= P(x)(c_1 y_1 + c_2 y_2)'' + Q(x)(c_1 y_1 + c_2 y_2)' + R(x)(c_1 y_1 + c_2 y_2)$$

$$= P(x)(c_1 y_1'' + c_2 y_2'') + Q(x)(c_1 y_1' + c_2 y_2') + R(x)(c_1 y_1 + c_2 y_2)$$

$$= c_1[P(x)y_1'' + Q(x)y_1' + R(x)y_1] + c_2[P(x)y_2'' + Q(x)y_2' + R(x)y_2]$$

$$= c_1(0) + c_2(0) = 0$$

Thus $y = c_1 y_1 + c_2 y_2$ is a solution of Equation 2.　□

The other fact we need is given by the following theorem, which is proved in more advanced courses. It says that the general solution is a linear combination of two **linearly independent** solutions y_1 and y_2. This means that neither y_1 nor y_2 is a constant multiple of the other. For instance, the functions $f(x) = x^2$ and $g(x) = 5x^2$ are linearly dependent, but $f(x) = e^x$ and $g(x) = xe^x$ are linearly independent.

---

(4) THEOREM   If y_1 and y_2 are linearly independent solutions of Equation 2, then the general solution is given by

$$y(x) = c_1 y_1(x) + c_2 y_2(x)$$

where c_1 and c_2 are arbitrary constants.

Theorem 4 is very useful because it says that if we know *two* particular linearly independent solutions, then we know *every* solution.

In general, it is not easy to discover particular solutions to a second-order linear equation. But it is always possible to do so if the coefficient functions P, Q, and R are constant functions, that is, if the differential equation has the form

(5)
$$ay'' + by' + cy = 0$$

where a, b, and c are constants.

It is not hard to think of some likely candidates for particular solutions of Equation 5 if we state the equation verbally. We are looking for a function y such that a constant times its second derivative y'' plus another constant times y' plus a third constant times y is equal to 0. We know that the exponential function $y = e^{rx}$ (where r is a constant) has the property that its derivative is a constant multiple of itself: $y' = re^{rx}$. Furthermore, $y'' = r^2 e^{rx}$. If we substitute these expressions into Equation 5, we see that $y = e^{rx}$ is a solution if

$$ar^2 e^{rx} + bre^{rx} + ce^{rx} = 0$$

or
$$(ar^2 + br + c)e^{rx} = 0$$

But e^{rx} is never 0. Therefore, $y = e^{rx}$ is a solution of Equation 5 if r is a root of the equation

(6)
$$ar^2 + br + c = 0$$

Equation 6 is called the **auxiliary equation** (or **characteristic equation**) of the differential equation $ay'' + by' + cy = 0$. Notice that it is an algebraic equation that is obtained from the differential equation by replacing y'' by r^2, y' by r, and y by 1.

Sometimes the roots r_1 and r_2 of the auxiliary equation can be found by factoring. In other cases they are found by using the quadratic formula:

$$(7) \qquad r_1 = \frac{-b + \sqrt{b^2 - 4ac}}{2a} \qquad r_2 = \frac{-b - \sqrt{b^2 - 4ac}}{2a}$$

We distinguish three cases according to the sign of the discriminant $b^2 - 4ac$.

CASE I $b^2 - 4ac > 0$

In this case the roots r_1 and r_2 of the auxiliary equation are real and distinct, so $y_1 = e^{r_1 x}$ and $y_2 = e^{r_2 x}$ are two linearly independent solutions of Equation 5. (Note that $e^{r_2 x}$ is not a constant multiple of $e^{r_1 x}$.) Thus, by Theorem 4, we have the following:

---

(8)   If the roots r_1 and r_2 of the auxiliary equation $ar^2 + br + c = 0$ are real and unequal, then the general solution of $ay'' + by' + cy = 0$ is

$$y = c_1 e^{r_1 x} + c_2 e^{r_2 x}$$

EXAMPLE 1   Solve the equation $y'' + y' - 6y = 0$.

SOLUTION   The auxiliary equation is

$$r^2 + r - 6 = (r - 2)(r + 3) = 0$$

whose roots are $r = 2, -3$. Therefore, by (8) the general solution of the given differential equation is

$$y = c_1 e^{2x} + c_2 e^{-3x}$$ ∎

In Figure 1 the graphs of the basic solutions $f(x) = e^{2x}$ and $g(x) = e^{-3x}$ of the differential equation in Example 1 are shown in red and blue, respectively. Some of the other solutions, linear combinations of f and g, are shown in gold.

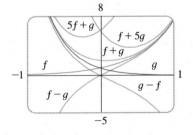

FIGURE 1

EXAMPLE 2   Solve $3\dfrac{d^2 y}{dx^2} + \dfrac{dy}{dx} - y = 0$.

SOLUTION   To solve the auxiliary equation $3r^2 + r - 1 = 0$ we use the quadratic formula:

$$r = \frac{-1 \pm \sqrt{13}}{6}$$

Since the roots are real and distinct, the general solution is

$$y = c_1 e^{(-1+\sqrt{13})x/6} + c_2 e^{(-1-\sqrt{13})x/6}$$ ∎

CASE II $b^2 - 4ac = 0$

In this case $r_1 = r_2$; that is, the roots of the auxiliary equation are real and equal. Let us denote by r the common value of r_1 and r_2. Then, from Equations 7, we have

$$(9) \qquad r = -\frac{b}{2a} \qquad \text{so} \qquad 2ar + b = 0$$

We know that $y_1 = e^{rx}$ is one solution of Equation 5. We now verify that $y_2 = xe^{rx}$ is also a solution:

$$ay_2'' + by_2' + cy_2 = a(2re^{rx} + r^2xe^{rx}) + b(e^{rx} + rxe^{rx}) + cxe^{rx}$$

$$= (2ar + b)e^{rx} + (ar^2 + br + c)xe^{rx}$$

$$= 0(e^{rx}) + 0(xe^{rx}) = 0$$

The first term is 0 by Equations 9; the second term is 0 because r is a root of the auxiliary equation. Since $y_1 = e^{rx}$ and $y_2 = xe^{rx}$ are linearly independent solutions, Theorem 4 provides us with the general solution.

---

(10)   If the auxiliary equation $ar^2 + br + c = 0$ has only one real root r, then the general solution of $ay'' + by' + cy = 0$ is

$$y = c_1e^{rx} + c_2xe^{rx}$$

Figure 2 shows the basic solutions $f(x) = e^{-3x/2}$ and $g(x) = xe^{-3x/2}$ in Example 3 and some other members of the family of solutions. Notice that all of them approach 0 as $x \to \infty$.

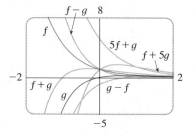

FIGURE 2

EXAMPLE 3   Solve the equation $4y'' + 12y' + 9y = 0$.

SOLUTION   The auxiliary equation $4r^2 + 12r + 9 = 0$ can be factored as

$$(2r + 3)^2 = 0$$

so the only root is $r = -\frac{3}{2}$. By (10) the general solution is

$$y = c_1e^{-3x/2} + c_2xe^{-3x/2}$$ ∎

CASE III $b^2 - 4ac < 0$

In this case the roots r_1 and r_2 of the auxiliary equation are complex numbers. (See Appendix H for information about complex numbers.) We can write

$$r_1 = \alpha + i\beta \qquad r_2 = \alpha - i\beta$$

where α and β are real numbers. [In fact, $\alpha = -b/(2a)$, $\beta = \sqrt{4ac - b^2}/(2a)$.] Then, using Euler's equation

$$e^{i\theta} = \cos\theta + i\sin\theta$$

from Appendix H, we write the solution of the differential equation as

$$y = C_1e^{r_1x} + C_2e^{r_2x}$$

$$= C_1e^{(\alpha+i\beta)x} + C_2e^{(\alpha-i\beta)x}$$

$$= C_1e^{\alpha x}(\cos\beta x + i\sin\beta x) + C_2e^{\alpha x}(\cos\beta x - i\sin\beta x)$$

$$= e^{\alpha x}[(C_1 + C_2)\cos\beta x + i(C_1 - C_2)\sin\beta x]$$

$$= e^{\alpha x}(c_1\cos\beta x + c_2\sin\beta x)$$

where $c_1 = C_1 + C_2$, $c_2 = i(C_1 - C_2)$. This gives all solutions (real or complex) of the differential equation. The solutions are real when the constants c_1 and c_2 are real. We summarize the discussion as follows:

Figure 3 shows the graphs of the solutions in Example 4, $f(x) = e^{3x}\cos 2x$ and $g(x) = e^{3x}\sin 2x$, together with some linear combinations. All solutions approach 0 as $x \to -\infty$.

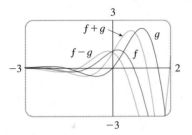

FIGURE 3

> **(11)** If the roots of the auxiliary equation $ar^2 + br + c = 0$ are the complex numbers $r_1 = \alpha + i\beta$, $r_2 = \alpha - i\beta$, then the general solution of $ay'' + by' + cy = 0$ is
>
> $$y = e^{\alpha x}(c_1 \cos \beta x + c_2 \sin \beta x)$$

EXAMPLE 4 Solve the equation $y'' - 6y' + 13y = 0$.

SOLUTION The auxiliary equation is $r^2 - 6r + 13 = 0$. By the quadratic formula, the roots are

$$r = \frac{6 \pm \sqrt{36 - 52}}{2} = \frac{6 \pm \sqrt{-16}}{2} = 3 \pm 2i$$

By (11) the general solution of the differential equation is

$$y = e^{3x}(c_1 \cos 2x + c_2 \sin 2x)$$ ∎

INITIAL-VALUE AND BOUNDARY-VALUE PROBLEMS

An **initial-value problem** for the second-order Equation 1 or 2 consists of finding a solution y of the differential equation that also satisfies initial conditions of the form

$$y(x_0) = y_0 \qquad y'(x_0) = y_1$$

where y_0 and y_1 are given constants. If P, Q, R, and G are continuous on an interval and $P(x) \neq 0$ there, then a theorem found in more advanced books guarantees the existence and uniqueness of a solution to this initial-value problem. Examples 5 and 6 illustrate the technique for solving such a problem.

A **boundary-value problem** for Equation 1 consists of finding a solution y of the differential equation that also satisfies boundary conditions of the form

$$y(x_0) = y_0 \qquad y(x_1) = y_1$$

In contrast with the situation for initial-value problems, a boundary-value problem does not always have a solution. The method is illustrated in Example 7.

EXAMPLE 5 Solve the initial-value problem

$$y'' + y' - 6y = 0 \qquad y(0) = 1 \qquad y'(0) = 0$$

Figure 4 shows the graph of the solution of the initial-value problem in Example 5. Compare with Figure 1.

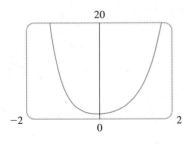

FIGURE 4

SOLUTION From Example 1 we know that the general solution of the differential equation is

$$y(x) = c_1 e^{2x} + c_2 e^{-3x}$$

Differentiating this solution, we get

$$y'(x) = 2c_1 e^{2x} - 3c_2 e^{-3x}$$

To satisfy the initial conditions we require that

(12) $$y(0) = c_1 + c_2 = 1$$

(13) $$y'(0) = 2c_1 - 3c_2 = 0$$

From (13) we have $c_2 = \frac{2}{3}c_1$ and so (12) gives

$$c_1 + \tfrac{2}{3}c_1 = 1 \qquad c_1 = \tfrac{3}{5} \qquad c_2 = \tfrac{2}{5}$$

Thus the required solution of the initial-value problem is

$$y = \tfrac{3}{5}e^{2x} + \tfrac{2}{5}e^{-3x}$$

EXAMPLE 6   Solve the initial-value problem

$$y'' + y = 0 \qquad y(0) = 2 \qquad y'(0) = 3$$

SOLUTION   The auxiliary equation is $r^2 + 1 = 0$, or $r^2 = -1$, whose roots are $\pm i$. Thus $\alpha = 0$, $\beta = 1$, and since $e^{0x} = 1$, the general solution is

$$y(x) = c_1 \cos x + c_2 \sin x$$

Since

$$y'(x) = -c_1 \sin x + c_2 \cos x$$

the initial conditions become

$$y(0) = c_1 = 2 \qquad y'(0) = c_2 = 3$$

Therefore, the solution of the initial-value problem is

$$y(x) = 2 \cos x + 3 \sin x$$

EXAMPLE 7   Solve the boundary-value problem

$$y'' + 2y' + y = 0 \qquad y(0) = 1 \qquad y(1) = 3$$

SOLUTION   The auxiliary equation is

$$r^2 + 2r + 1 = 0 \qquad \text{or} \qquad (r + 1)^2 = 0$$

whose only root is $r = -1$. Therefore, the general solution is

$$y(x) = c_1 e^{-x} + c_2 x e^{-x}$$

The boundary conditions are satisfied if

$$y(0) = c_1 = 1$$

$$y(1) = c_1 e^{-1} + c_2 e^{-1} = 3$$

The first condition gives $c_1 = 1$, so the second condition becomes

$$e^{-1} + c_2 e^{-1} = 3$$

Solving this equation for c_2 by first multiplying through by e, we get

$$1 + c_2 = 3e \qquad c_2 = 3e - 1$$

Thus the solution of the boundary-value problem is

$$y = e^{-x} + (3e - 1)xe^{-x}$$

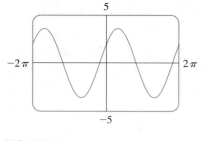

FIGURE 5

The solution to Example 6 is graphed in Figure 5. It appears to be a shifted sine curve and, indeed, you can verify that another way of writing the solution is
$y = \sqrt{13} \sin(x + \phi)$ where $\tan \phi = \tfrac{2}{3}$

Figure 6 shows the graph of the solution of the boundary-value problem in Example 7.

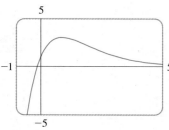

FIGURE 6

SUMMARY: SOLUTIONS OF $ay'' + by' + c = 0$

Roots of $ar^2 + br + c = 0$	General solution
r_1, r_2 real and distinct	$y = c_1 e^{r_1 x} + c_2 e^{r_2 x}$
$r_1 = r_2 = r$	$y = c_1 e^{rx} + c_2 x e^{rx}$
r_1, r_2 complex: $\alpha \pm i\beta$	$y = e^{\alpha x}(c_1 \cos \beta x + c_2 \sin \beta x)$

EXERCISES 15.5

1–17 ■ Solve the differential equation.

1. $y'' - 3y' + 2y = 0$

2. $y'' - y' = 0$

3. $3y'' - 8y' - 3y = 0$

4. $y'' + 9y' + 20y = 0$

5. $y'' + 2y' + 10y = 0$

6. $y'' + 10y' + 41y = 0$

7. $y'' = y$

8. $9y'' - 30y' + 25y = 0$

9. $y'' + 25y = 0$

10. $y'' - 4y' + 13y = 0$

11. $2y'' + y' = 0$

12. $y'' - 2y' - 4y = 0$

13. $y'' - y' + 2y = 0$

14. $y'' = -5y$

15. $\dfrac{d^2 y}{dx^2} + 2\dfrac{dy}{dx} - y = 0$

16. $2\dfrac{d^2 y}{dx^2} + 5\dfrac{dy}{dx} + y = 0$

17. $2\dfrac{d^2 y}{dx^2} + \dfrac{dy}{dx} + 3y = 0$

18–20 ■ Graph the two basic solutions of the differential equation and several other solutions. What features do the solutions have in common?

18. $6\dfrac{d^2 y}{dx^2} - \dfrac{dy}{dx} - 2y = 0$

19. $\dfrac{d^2 y}{dx^2} - 8\dfrac{dy}{dx} + 16y = 0$

20. $\dfrac{d^2 y}{dx^2} - 2\dfrac{dy}{dx} + 5y = 0$

21–28 ■ Solve the initial-value problem.

21. $y'' + 3y' - 4y = 0$, $y(0) = 2$, $y'(0) = -3$

22. $y'' - 4y = 0$, $y(0) = 1$, $y'(0) = 0$

23. $y'' - 2y' + 2y = 0$, $y(0) = 1$, $y'(0) = 2$

24. $y'' + 4y' + 6y = 0$, $y(0) = 2$, $y'(0) = 4$

25. $y'' - 2y' - 3y = 0$, $y(1) = 3$, $y'(1) = 1$

26. $y'' - 2y' + y = 0$, $y(2) = 0$, $y'(2) = 1$

27. $y'' + 9y = 0$, $y(\pi/3) = 0$, $y'(\pi/3) = 1$

28. $y'' + 4y = 0$, $y(\pi/6) = 1$, $y'(\pi/6) = 0$

29–36 ■ Solve each boundary-value problem if possible.

29. $y'' + 4y' + 4y = 0$, $y(0) = 0$, $y(1) = 3$

30. $y'' + 5y' - 6y = 0$, $y(0) = 0$, $y(2) = 1$

31. $y'' + y = 0$, $y(0) = 1$, $y(\pi) = 0$

32. $y'' + 9y = 0$, $y(0) = 1$, $y(\pi/2) = 0$

33. $y'' - y' - 2y = 0$, $y(-1) = 1$, $y(1) = 0$

34. $y'' + 4y' + 3y = 0$, $y(1) = 0$, $y(3) = 2$

35. $y'' + 4y' + 13y = 0$, $y(0) = 2$, $y(\pi/2) = 1$

36. $y'' + 2y' + 5y = 0$, $y(0) = 1$, $y(\pi) = 2$

37. (a) Show that the boundary-value problem $y'' + \lambda y = 0$, $y(0) = 0$, $y(L) = 0$ has only the trivial solution $y = 0$ for the cases $\lambda = 0$ and $\lambda < 0$.
   (b) For the case $\lambda > 0$, find the values of λ for which this problem has a nontrivial solution and give the corresponding solution.

38. If a, b, and c are all positive constants and $y(x)$ is a solution of the differential equation $ay'' + by' + cy = 0$, show that $\lim_{x \to \infty} y(x) = 0$.

15.6 NONHOMOGENEOUS LINEAR EQUATIONS

In this section we learn how to solve second-order nonhomogeneous linear differential equations with constant coefficients, that is, equations of the form

(1) $$ay'' + by' + cy = G(x)$$

where a, b, c are constants and G is a continuous function. The related homogeneous equation

(2)
$$ay'' + by' + cy = 0$$

is called the **complementary equation** and plays an important role in the solution of the original nonhomogeneous equation (1).

---

(3) THEOREM   The general solution of the nonhomogeneous differential equation (1) can be written as

$$y(x) = y_p(x) + y_c(x)$$

where y_p is a particular solution of Equation 1 and y_c is the general solution of the complementary Equation 2.

PROOF   All we have to do is verify that if y is any solution of Equation 1, then $y - y_p$ is a solution of the complementary Equation 2. Indeed

$$a(y - y_p)'' + b(y - y_p)' + c(y - y_p) = ay'' - ay_p'' + by' - by_p' + cy - cy_p$$

$$= (ay'' + by' + cy) - (ay_p'' + by_p' + cy_p)$$

$$= g(x) - g(x) = 0 \qquad \square$$

We know from Section 15.5 how to solve the complementary equation. (Recall that the solution is $y_c = c_1 y_1 + c_2 y_2$, where y_1 and y_2 are linearly independent solutions of Equation 2.) Therefore, Theorem 3 says that we know the general solution of the non-homogeneous equation as soon as we know a particular solution y_p. There are two methods for finding a particular solution. The method of undetermined coefficients is straightforward but works only for a restricted class of functions G. The method of variation of parameters works for every function G but is usually more difficult to apply in practice.

THE METHOD OF UNDETERMINED COEFFICIENTS

We first illustrate the method of undetermined coefficients for the equation

$$ay'' + by' + cy = G(x)$$

where $G(x)$ is a polynomial. It is reasonable to guess that there is a particular solution y_p that is a polynomial of the same degree as G because if y is a polynomial, then $ay'' + by' + cy$ is also a polynomial. We therefore substitute $y_p(x) = $ a polynomial (of the same degree as G) into the differential equation and determine the coefficients.

EXAMPLE 1   Solve the equation $y'' + y' - 2y = x^2$.

SOLUTION   The auxiliary equation of $y'' + y' - 2y = 0$ is

$$r^2 + r - 2 = (r - 1)(r + 2) = 0$$

with roots $r = 1, -2$. So the solution of the complementary equation is

$$y_c = c_1 e^x + c_2 e^{-2x}$$

Since $G(x) = x^2$ is a polynomial of degree 2, we seek a particular solution of the form

$$y_p(x) = Ax^2 + Bx + C$$

Then $y_p' = 2Ax + B$ and $y_p'' = 2A$ so, substituting into the given differential equation, we have

$$(2A) + (2Ax + B) - 2(Ax^2 + Bx + C) = x^2$$

or

$$-2Ax^2 + (2A - 2B)x + (2A + B - 2C) = x^2$$

Polynomials are equal when their coefficients are equal. Thus

$$-2A = 1 \qquad 2A - 2B = 0 \qquad 2A + B - 2C = 0$$

The solution of this system of equations is

$$A = -\tfrac{1}{2} \qquad B = -\tfrac{1}{2} \qquad C = -\tfrac{3}{4}$$

A particular solution is therefore

$$y_p(x) = -\tfrac{1}{2}x^2 - \tfrac{1}{2}x - \tfrac{3}{4}$$

and, by Theorem 3, the general solution is

$$y = y_c + y_p = c_1 e^x + c_2 e^{-2x} - \tfrac{1}{2}x^2 - \tfrac{1}{2}x - \tfrac{3}{4} \qquad \blacksquare$$

Figure 1 shows four solutions of the differential equation in Example 1 in terms of the particular solution y_p and the functions $f(x) = e^x$ and $g(x) = e^{-2x}$.

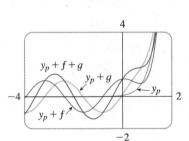

FIGURE 1

If $G(x)$ (the right side of Equation 1) is of the form Ce^{kx}, where C and k are constants, then we take as a trial solution a function of the same form, $y_p(x) = Ae^{kx}$, because the derivatives of e^{kx} are constant multiples of e^{kx}.

EXAMPLE 2 Solve $y'' + 4y = e^{3x}$.

SOLUTION The auxiliary equation is $r^2 + 4 = 0$ with roots $\pm 2i$, so the solution of the complementary equation is

$$y_c(x) = c_1 \cos 2x + c_2 \sin 2x$$

For a particular solution we try $y_p(x) = Ae^{3x}$. Then $y_p' = 3Ae^{3x}$ and $y_p'' = 9Ae^{3x}$. Substituting into the differential equation, we have

$$9Ae^{3x} + 4(Ae^{3x}) = e^{3x}$$

so $13Ae^{3x} = e^{3x}$ and $A = \tfrac{1}{13}$. Thus a particular solution is

$$y_p(x) = \tfrac{1}{13}e^{3x}$$

and the general solution is

$$y(x) = c_1 \cos 2x + c_2 \sin 2x + \tfrac{1}{13}e^{3x} \qquad \blacksquare$$

Figure 2 shows solutions of the differential equation in Example 2 in terms of y_p and the functions $f(x) = \cos 2x$ and $g(x) = \sin 2x$. Notice that all solutions approach ∞ as $x \to \infty$ and all solutions resemble sine functions when x is negative.

FIGURE 2

If $G(x)$ is either $C \cos kx$ or $C \sin kx$, then, because of the rules for differentiating the sine and cosine functions, we take as a trial particular solution a function of the form

$$y_p(x) = A \cos kx + B \sin kx$$

EXAMPLE 3   Solve $y'' + y' - 2y = \sin x$.

SOLUTION   We try a particular solution

$$y_p(x) = A \cos x + B \sin x$$

Then $y_p' = -A \sin x + B \cos x$ $y_p'' = -A \cos x - B \sin x$

so substitution in the differential equation gives

$$(-A \cos x - B \sin x) + (-A \sin x + B \cos x) - 2(A \cos x + B \sin x) = \sin x$$

or $$(-3A + B) \cos x + (-A - 3B) \sin x = \sin x$$

This is true if

$$-3A + B = 0 \qquad \text{and} \qquad -A - 3B = 1$$

The solution of this system is

$$A = -\tfrac{1}{10} \qquad B = -\tfrac{3}{10}$$

so a particular solution is

$$y_p(x) = -\tfrac{1}{10} \cos x - \tfrac{3}{10} \sin x$$

In Example 1 we determined that the solution of the complementary equation is $y_c = c_1 e^x + c_2 e^{-2x}$. Thus the general solution of the given equation is

$$y(x) = c_1 e^x + c_2 e^{-2x} - \tfrac{1}{10}(\cos x + 3 \sin x)$$ ∎

If $G(x)$ is a product of functions of the preceding types, then we take the trial solution to be a product of functions of the same type. For instance, in solving the differential equation

$$y'' + 2y' + 4y = x \cos 3x$$

we would try

$$y_p(x) = (Ax + B) \cos 3x + (Cx + D) \sin 3x$$

If $G(x)$ is a sum of functions of these types, we use the easily verified *principle of superposition*, which says that if y_{p_1} and y_{p_2} are solutions of

$$ay'' + by' + cy = G_1(x) \qquad ay'' + by' + cy = G_2(x)$$

respectively, then $y_{p_1} + y_{p_2}$ is a solution of

$$ay'' + by' + cy = G_1(x) + G_2(x)$$

EXAMPLE 4   Solve $y'' - 4y = xe^x + \cos 2x$.

SOLUTION   The auxiliary equation is $r^2 - 4 = 0$ with roots ± 2, so the solution of the complementary equation is $y_c(x) = c_1 e^{2x} + c_2 e^{-2x}$. For the equation $y'' - 4y = xe^x$ we try

$$y_{p_1}(x) = (Ax + B)e^x$$

Then $y'_{p_1} = (Ax + A + B)e^x$, $y''_{p_1} = (Ax + 2A + B)e^x$, so substitution in the equation gives

$$(Ax + 2A + B)e^x - 4(Ax + B)e^x = xe^x$$

or

$$(-3Ax + 2A - 3B)e^x = xe^x$$

Thus $-3A = 1$ and $2A - 3B = 0$, so $A = -\frac{1}{3}$, $B = -\frac{2}{9}$, and

$$y_{p_1}(x) = \left(-\tfrac{1}{3}x - \tfrac{2}{9}\right)e^x$$

In Figure 3 we show the particular solution y_p of the differential equation in Example 4. The other solutions are given in terms of $f(x) = e^{2x}$ and $g(x) = e^{-2x}$.

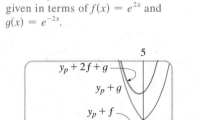

FIGURE 3

For the equation $y'' - 4y = \cos 2x$, we try

$$y_{p_2}(x) = C \cos 2x + D \sin 2x$$

Substitution gives

$$-4C \cos 2x - 4D \sin 2x - 4(C \cos 2x + D \sin 2x) = \cos 2x$$

or

$$-8C \cos 2x - 8D \sin 2x = \cos 2x$$

Therefore, $-8C = 1$, $-8D = 0$, and

$$y_{p_2}(x) = -\tfrac{1}{8} \cos 2x$$

By the superposition principle, the general solution is

$$y = y_c + y_{p_1} + y_{p_2} = c_1 e^{2x} + c_2 e^{-2x} - \left(\tfrac{1}{3}x + \tfrac{2}{9}\right)e^x - \tfrac{1}{8} \cos 2x \quad \blacksquare$$

Finally we note that the recommended trial solution y_p sometimes turns out to be a solution of the complementary equation and therefore cannot be a solution of the nonhomogeneous equation. In such cases we multiply the recommended trial solution by x (or by x^2 if necessary) so that no term in $y_p(x)$ is a solution of the complementary equation.

EXAMPLE 5   Solve $y'' + y = \sin x$.

SOLUTION   The auxiliary equation is $r^2 + 1 = 0$ with roots $\pm i$, so the solution of the complementary equation is

$$y_c(x) = c_1 \cos x + c_2 \sin x$$

Ordinarily, we would use the trial solution

$$y_p(x) = A \cos x + B \sin x$$

but we observe that it is a solution of the complementary equation, so instead we try

$$y_p(x) = Ax \cos x + Bx \sin x$$

Then

$$y'_p(x) = A \cos x - Ax \sin x + B \sin x + Bx \cos x$$

$$y''_p(x) = -2A \sin x - Ax \cos x + 2B \cos x - Bx \sin x$$

Substitution in the differential equation gives

$$y''_p + y_p = -2A \sin x + 2B \cos x = \sin x$$

The graphs of four solutions of the differential equation in Example 5 are shown in Figure 4.

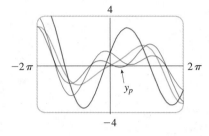

FIGURE 4

so $A = -\frac{1}{2}$, $B = 0$, and

$$y_p(x) = -\frac{1}{2}x \cos x$$

The general solution is

$$y(x) = c_1 \cos x + c_2 \sin x - \frac{1}{2}x \cos x$$

∎

SUMMARY OF UNDETERMINED COEFFICIENTS

$G(x) =$	First try $y_p =$
$C_1 x^n + \cdots + C_1 x + C_0$ Ce^{kx} $C \cos kx + D \sin kx$	$A_1 x^n + \cdots + A_1 x + A_0$ Ae^{kx} $A \cos kx + B \sin kx$
Modification: If any term of y_p is a solution of the complementary equation, multiply y_p by x (or by x^2 if necessary).	

THE METHOD OF VARIATION OF PARAMETERS

Suppose we have already solved the homogeneous equation $ay'' + by' + cy = 0$ and written the solution as

$$(4) \qquad y(x) = c_1 y_1(x) + c_2 y_2(x)$$

where y_1 and y_2 are linearly independent solutions. Let us replace the constants (or parameters) c_1 and c_2 in Equation 4 by arbitrary functions $u_1(x)$ and $u_2(x)$. We look for a particular solution of the nonhomogeneous equation $ay'' + by' + cy = G(x)$ of the form

$$(5) \qquad y_p(x) = u_1(x)y_1(x) + u_2(x)y_2(x)$$

(This method is called **variation of parameters** because we have varied the parameters c_1 and c_2 to make them functions.) Differentiating Equation 5, we get

$$(6) \qquad y_p' = (u_1' y_1 + u_2' y_2) + (u_1 y_1' + u_2 y_2')$$

Since u_1 and u_2 are arbitrary functions, we can impose two conditions on them. One condition is that y_p is a solution of the differential equation; we can choose the other condition so as to simplify our calculations. In view of the expression in Equation 6, let us impose the condition that

$$(7) \qquad u_1' y_1 + u_2' y_2 = 0$$

Then

$$y_p'' = u_1' y_1' + u_2' y_2' + u_1 y_1'' + u_2 y_2''$$

Substituting in the differential equation, we get

$$a(u_1' y_1' + u_2' y_2' + u_1 y_1'' + u_2 y_2'') + b(u_1 y_1' + u_2 y_2') + c(u_1 y_1 + u_2 y_2) = G$$

or

$$(8) \quad u_1(ay_1'' + by_1' + cy_1) + u_2(ay_2'' + by_2' + cy_2) + a(u_1' y_1' + u_2' y_2') = G$$

But y_1 and y_2 are solutions of the complementary equation, so

$$ay_1'' + by_1' + cy_1 = 0 \quad \text{and} \quad ay_2'' + by_2' + cy_2 = 0$$

and Equation 8 simplifies to

(9)
$$a(u_1' y_1' + u_2' y_2') = G$$

Equations 7 and 9 form a system of two equations in the unknown functions u_1' and u_2'. After solving this system we may be able to integrate to find u_1 and u_2 and then the particular solution is given by Equation 5.

EXAMPLE 6   Solve the equation $y'' + y = \tan x$, $0 < x < \pi/2$.

SOLUTION   The auxiliary equation is $r^2 + 1 = 0$ with roots $\pm i$, so the solution of $y'' + y = 0$ is $c_1 \sin x + c_2 \cos x$. Using variation of parameters, we seek a solution of the form

$$y_p(x) = u_1(x) \sin x + u_2(x) \cos x$$

Then
$$y_p' = (u_1' \sin x + u_2' \cos x) + (u_1 \cos x - u_2 \sin x)$$

Set

(10)
$$u_1' \sin x + u_2' \cos x = 0$$

Then
$$y_p'' = u_1' \cos x - u_2' \sin x - u_1 \sin x - u_2 \cos x$$

For y_p to be a solution we must have

(11)
$$y_p'' + y_p = u_1' \cos x - u_2' \sin x = \tan x$$

Solving Equations 10 and 11, we get

$$u_1'(\sin^2 x + \cos^2 x) = \cos x \tan x$$

$$u_1' = \sin x \qquad u_1(x) = -\cos x$$

(We seek a particular solution, so we do not need a constant of integration here.) Then, from Equation 10, we obtain

$$u_2' = -\frac{\sin x}{\cos x} u_1' = -\frac{\sin^2 x}{\cos x} = \frac{\cos^2 x - 1}{\cos x} = \cos x - \sec x$$

So
$$u_2(x) = \sin x - \ln(\sec x + \tan x)$$

(Note that $\sec x + \tan x > 0$ for $0 < x < \pi/2$.) Therefore

$$y_p(x) = -\cos x \sin x + (\sin x - \ln(\sec x + \tan x)) \cos x$$

$$= -\cos x \ln(\sec x + \tan x)$$

and the general solution is

$$y(x) = c_1 \sin x + c_2 \cos x - \cos x \ln(\sec x + \tan x)$$

Figure 5 shows four solutions of the differential equation in Example 6.

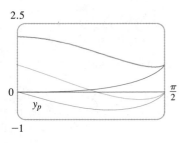

FIGURE 5

EXERCISES 15.6

1–10 ■ Solve the differential equation or initial-value problem using the method of undetermined coefficients.

1. $y'' - y' - 6y = \cos 3x$ **2.** $y'' + 2y' + 2y = x^3 - 1$

3. $y'' - 4y' + 4y = e^{-x}$

4. $y'' - 7y' + 12y = \sin x - \cos x$

5. $y'' + 36y = 2x^2 - x$

6. $y'' + 2y' + y = xe^{-x}$

7. $y'' - 2y' + 5y = x + \sin 3x$, $y(0) = 1$, $y'(0) = 2$

8. $y'' + 2y = e^x \sin x$, $y(0) = 1$, $y'(0) = 1$

9. $y'' - y = xe^{3x}$, $y(0) = 0$, $y'(0) = 1$

10. $y'' - 3y' + 2y = 2x - e^{-2x}$, $y(0) = 1$, $y'(0) = 0$

11–12 ■ Graph the particular solution and several other solutions. What characteristics do these solutions have in common?

11. $4y'' + 5y' + y = e^x$ **12.** $2y'' + 3y' + y = 1 + \cos 2x$

13–16 ■ Write a trial solution for the method of undetermined coefficients. Do not determine the coefficients.

13. $y'' + 2y' + 6y = x^4 e^{2x}$

14. $y'' + 6y' + 2y = x^3 + e^x \sin 2x$

15. $y'' - 2y' + 2y = e^x \cos x$

16. $y'' + 3y' = 1 + xe^{-3x}$

17–20 ■ Solve the differential equation using (a) undetermined coefficients and (b) variation of parameters.

17. $y'' + 4y = x$ **18.** $y'' - 3y' + 2y = \sin x$

19. $y'' - 2y' + y = e^{2x}$ **20.** $y'' - y' = e^x$

21–26 ■ Solve the differential equation using the method of variation of parameters.

21. $y'' + y = \sec x, \ 0 < x < \pi/2$

22. $y'' + y = \cot x, \ 0 < x < \pi/2$

23. $y'' - 3y' + 2y = \dfrac{1}{1 + e^{-x}}$ **24.** $y'' + 3y' + 2y = \sin(e^x)$

25. $y'' - y = 1/x$ **26.** $y'' + 4y' + 4y = \dfrac{e^{-2x}}{x^3}$

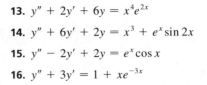

APPLICATIONS OF SECOND-ORDER DIFFERENTIAL EQUATIONS

Second-order linear differential equations have a variety of applications in science and engineering. In this section we explore two of them: the vibration of springs and electric circuits.

VIBRATING SPRINGS

We consider the motion of an object with mass m at the end of a spring that is either vertical (as in Figure 1) or horizontal on a level surface (as in Figure 2).

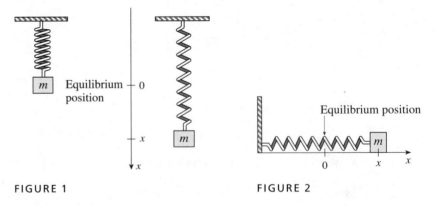

FIGURE 1   FIGURE 2

In Section 5.4 we discussed Hooke's Law, which says that if the spring is stretched (or compressed) x units from its natural length, then it exerts a force that is proportional to x:

$$\text{restoring force} = -kx$$

where k is a positive constant (called the *spring constant*). If we ignore any external resisting forces (due to air resistance or friction) then, by Newton's Second Law (force equals mass times acceleration), we have

$$(1) \qquad m\frac{d^2x}{dt^2} = -kx \qquad \text{or} \qquad m\frac{d^2x}{dt^2} + kx = 0$$

This is a second-order linear differential equation. Its auxiliary equation is $mr^2 + k = 0$ with roots $r = \pm\omega i$, where $\omega = \sqrt{k/m}$. Thus the general solution is

$$x(t) = c_1 \cos \omega t + c_2 \sin \omega t$$

which can also be written as

$$x(t) = A\cos(\omega t + \delta)$$

where

$$\omega = \sqrt{\frac{k}{m}} \qquad \text{(frequency)}$$

$$A = \sqrt{c_1^2 + c_2^2} \quad \text{(amplitude)}$$

$$\cos \delta = \frac{c_1}{A} \qquad \sin \delta = -\frac{c_2}{A} \quad \text{(δ is the phase angle)}$$

(see Exercise 13). This type of motion is called **simple harmonic motion.**

EXAMPLE 1   A spring with a mass of 2 kg has natural length 0.5 m. A force of 25.6 N is required to maintain it stretched to a length of 0.7 m. If the spring is stretched to a length of 0.7 m and then released with initial velocity 0, find the position of the mass at any time t.

SOLUTION   From Hooke's Law, the force required to stretch the spring is

$$k(0.2) = 25.6$$

so $k = 25.6/0.2 = 128$. Using this value of the spring constant k, together with $m = 2$ in Equation 1, we have

$$2\frac{d^2x}{dt^2} + 128x = 0$$

As in the earlier general discussion, the solution of this equation is

$$(2) \qquad x(t) = c_1 \cos 8t + c_2 \sin 8t$$

We are given the initial condition that $x(0) = 0.2$. But, from Equation 2, $x(0) = c_1$. Therefore, $c_1 = 0.2$. Differentiating Equation 2, we get

$$x'(t) = -8c_1 \sin 8t + 8c_2 \cos 8t$$

Since the initial velocity is given as $x'(0) = 0$, we have $c_2 = 0$ and so the solution is

$$x(t) = \tfrac{1}{5} \cos 8t \qquad \blacksquare$$

DAMPED VIBRATIONS

FIGURE 3

We next consider the motion of a spring that is subject to a frictional force (in the case of the horizontal spring of Figure 2) or a damping force (in the case where a vertical spring moves through a fluid as in Figure 3). An example is the damping force supplied by a shock absorber in a car.

We assume that the damping force is proportional to the velocity of the mass and acts in the direction opposite to the motion. (This has been confirmed, at least approximately, by some physical experiments.) Thus

$$\text{damping force} = -c\,\frac{dx}{dt}$$

where c is a positive constant, called the **damping constant.** Thus, in this case, Newton's Second Law gives

$$m\,\frac{d^2x}{dt^2} = \text{restoring force} + \text{damping force}$$

$$= -kx - c\,\frac{dx}{dt}$$

or

(3)
$$m\,\frac{d^2x}{dt^2} + c\,\frac{dx}{dt} + kx = 0$$

This is a second-order linear differential equation and its auxiliary equation is $mr^2 + cr + k = 0$. The roots are

(4) $\qquad r_1 = \dfrac{-c + \sqrt{c^2 - 4mk}}{2m} \qquad\qquad r_2 = \dfrac{-c - \sqrt{c^2 - 4mk}}{2m}$

According to Section 15.5 we need to discuss three cases.

CASE I $c^2 - 4mk > 0$ (overdamping)
In this case r_1 and r_2 are distinct real roots and

$$x = c_1 e^{r_1 t} + c_2 e^{r_2 t}$$

Since c, m, and k are all positive, we have $\sqrt{c^2 - 4mk} < c$, so the roots r_1 and r_2 given by Equations 4 must both be negative. This shows that $x \to 0$ as $t \to \infty$. Typical graphs of x as a function of t are shown in Figure 4. Notice that oscillations do not occur. This is because $c^2 > 4mk$ means that there is a strong damping force (high-viscosity oil or grease) compared with a weak spring or small mass.

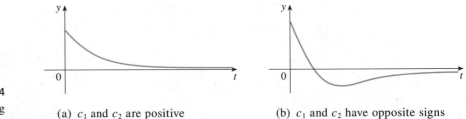

FIGURE 4
Overdamping

(a) c_1 and c_2 are positive

(b) c_1 and c_2 have opposite signs

CASE II $c^2 - 4mk = 0$ **(critical damping)**

This case corresponds to equal roots

$$r_1 = r_2 = -\frac{c}{2m}$$

The solution is given by

$$x = (c_1 + c_2 t)e^{-(c/2m)t}$$

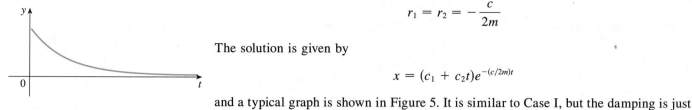

and a typical graph is shown in Figure 5. It is similar to Case I, but the damping is just sufficient to suppress vibrations. Any decrease in the viscosity of the fluid leads to the vibrations of the following case.

CASE III $c^2 - 4mk < 0$ **(underdamping)**

Here the roots are complex:

$$\left.\begin{array}{c} r_1 \\ r_2 \end{array}\right\} = -\frac{c}{2m} \pm \omega i$$

where

$$\omega = \frac{\sqrt{4mk - c^2}}{2m}$$

The solution is given by

$$x = e^{-(c/2m)t}(c_1 \cos \omega t + c_2 \sin \omega t)$$

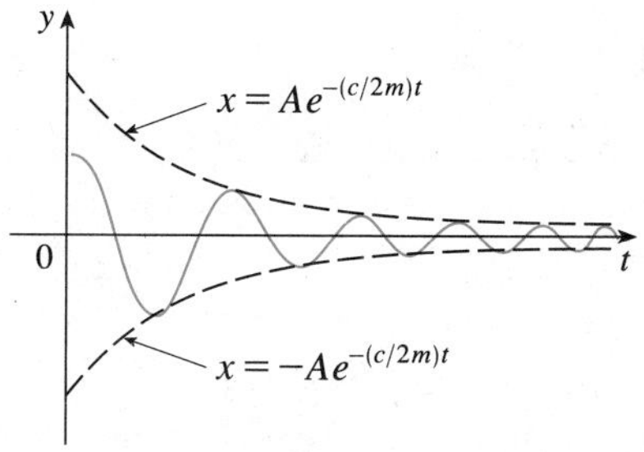

We see that there are oscillations that are damped by the factor $e^{-(c/2m)t}$. Since $c > 0$ and $m > 0$, we have $-(c/2m) < 0$ so $e^{-(c/2m)t} \to 0$ as $t \to \infty$. This implies that $x \to 0$ as $t \to \infty$; that is, the motion decays to 0 as time increases. A typical graph is shown in Figure 6.

EXAMPLE 2   Suppose that the spring of Example 1 is immersed in a fluid with damping constant $c = 40$. Find the position of the mass at any time t if it starts from the equilibrium position and is given a push to start it with an initial velocity of 0.6 m/s.

SOLUTION   From Example 1 the mass is $m = 2$ and the spring constant is $k = 128$, so the differential equation (3) becomes

$$2\frac{d^2x}{dt^2} + 40\frac{dx}{dt} + 128x = 0$$

or

$$\frac{d^2x}{dt^2} + 20\frac{dx}{dt} + 64x = 0$$

The auxiliary equation is $r^2 + 20r + 64 = (r + 4)(r + 16) = 0$ with roots -4 and -16, so the motion is overdamped and the solution is

$$x(t) = c_1 e^{-4t} + c_2 e^{-16t}$$

We are given that $x(0) = 0$, so $c_1 + c_2 = 0$. Differentiating, we get

$$x'(t) = -4c_1 e^{-4t} - 16c_2 e^{-16t}$$

so

$$x'(0) = -4c_1 - 16c_2 = 0.6$$

y

0   t

FIGURE 5

Critical damping

$x = Ae^{-(c/2m)t}$

0

$x = -Ae^{-(c/2m)t}$

FIGURE 6

Underdamping

Figure 7 shows the graph of the position function for the overdamped motion in Example 2.

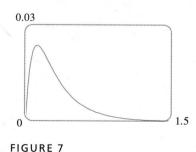

0.03

0   1.5

FIGURE 7

Since $c_2 = -c_1$, this gives $12c_1 = 0.6$ or $c_1 = 0.05$. Therefore

$$x = 0.05(e^{-4t} - e^{-16t})$$

∎

FORCED VIBRATIONS

Suppose that, in addition to the restoring force and the damping force, the motion of the spring is affected by an external force $F(t)$. Then Newton's Second Law gives

$$m\,\frac{d^2x}{dt^2} = \text{restoring force} + \text{damping force} + \text{external force}$$

$$= -kx - c\,\frac{dx}{dt} + F(t)$$

Thus, instead of the homogeneous equation (3), the motion of the spring is now governed by the following nonhomogeneous differential equation:

(5)
$$m\,\frac{d^2x}{dt^2} + c\,\frac{dx}{dt} + kx = F(t)$$

The motion of the spring can be determined by the methods of Section 15.6.

A commonly occurring type of external force is a periodic force function

$$F(t) = F_0 \cos \omega_0 t \qquad \text{where } \omega_0 \neq \omega = \sqrt{\frac{k}{m}}$$

In this case, and in the absence of a damping force ($c = 0$), you are asked in Exercise 7 to use the method of undetermined coefficients to show that

(6)
$$x(t) = c_1 \cos \omega t + c_2 \sin \omega t + \frac{F_0}{m(\omega^2 - \omega_0^2)} \cos \omega_0 t$$

If $\omega_0 = \omega$, then the applied frequency reinforces the natural frequency and the result is vibrations of large amplitude. This is the phenomenon of **resonance** (see Exercise 8).

ELECTRIC CIRCUITS

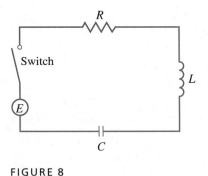

FIGURE 8

In Section 15.2 we were able to use first-order linear equations to analyze electric circuits that contain a resistor and inductor (see Figure 4 in Section 15.2) or a resistor and capacitor (see Exercise 29 in Section 15.2). Now that we know how to solve second-order linear equations, we are in a position to analyze the circuit shown in Figure 8. It contains an electromotive force E (supplied by a battery or generator), a resistor R, an inductor L, and a capacitor C, in series. If the charge on the capacitor at time t is $Q = Q(t)$, then the current is the rate of change of Q with respect to t: $I = dQ/dt$. As in Section 15.2, it is known from physics that the voltage drops across the resistor, inductor, and capacitor are

$$RI \qquad L\,\frac{dI}{dt} \qquad \frac{Q}{C}$$

respectively. Kirchhoff's voltage law says that the sum of these voltage drops is equal to the supplied voltage:

$$L\,\frac{dI}{dt} + RI + \frac{Q}{C} = E(t)$$

Since $I = dQ/dt$, this equation becomes

(7)
$$L\frac{d^2Q}{dt^2} + R\frac{dQ}{dt} + \frac{1}{C}Q = E(t)$$

which is a second-order linear differential equation with constant coefficients. If the charge Q_0 and the current I_0 are known at time 0, then we have the initial conditions

$$Q(0) = Q_0 \qquad Q'(0) = I(0) = I_0$$

and the initial-value problem can be solved by the methods of Section 15.6.

A differential equation for the current can be obtained by differentiating Equation 7 with respect to t and remembering that $I = dQ/dt$:

$$L\frac{d^2I}{dt^2} + R\frac{dI}{dt} + \frac{1}{C}I = E'(t)$$

EXAMPLE 3   Find the charge and current at time t in the circuit of Figure 8 if $R = 40\ \Omega$, $L = 1$ H, $C = 16 \times 10^{-4}$ F, $E(t) = 100\cos 10t$, and the initial charge and current are both 0.

SOLUTION   With the given values of L, R, C, and $E(t)$, Equation 7 becomes

(8)
$$\frac{d^2Q}{dt^2} + 40\frac{dQ}{dt} + 625Q = 100\cos 10t$$

The auxiliary equation is $r^2 + 40r + 625 = 0$ with roots

$$r = \frac{-40 \pm \sqrt{-900}}{2} = -20 \pm 15i$$

so the solution of the complementary equation is

$$Q_c(t) = e^{-20t}(c_1\cos 15t + c_2\sin 15t)$$

For the method of undetermined coefficients we try the particular solution

$$Q_p(t) = A\cos 10t + B\sin 10t$$

Then
$$Q_p'(t) = -10A\sin 10t + 10B\cos 10t$$

$$Q_p''(t) = -100A\cos 10t - 100B\sin 10t$$

Substituting into Equation 8, we have

$$(-100A\cos 10t - 100B\sin 10t) + 40(-10A\sin 10t + 10B\cos 10t)$$
$$+ 625(A\cos 10t + B\sin 10t) = 100\cos 10t$$

or $(525A + 400B)\cos 10t + (-400A + 525B)\sin 10t = 100\cos 10t$

Equating coefficients, we have

$$525A + 400B = 100 \qquad\qquad 21A + 16B = 4$$
$$\text{or}$$
$$-400A + 525B = 0 \qquad\qquad -16A + 21B = 0$$

The solution of this system is $A = \frac{84}{697}$ and $B = \frac{64}{697}$, so a particular solution is

$$Q_p(t) = \tfrac{1}{697}(84 \cos 10t + 64 \sin 10t)$$

and the general solution is

$$
\begin{aligned}
Q(t) &= Q_c(t) + Q_p(t) \\
&= e^{-20t}(c_1 \cos 15t + c_2 \sin 15t) + \tfrac{4}{697}(21 \cos 10t + 16 \sin 10t)
\end{aligned}
$$

Imposing the initial condition $Q(0) = 0$, we get

$$Q(0) = c_1 + \tfrac{84}{697} = 0 \qquad c_1 = -\tfrac{84}{697}$$

To impose the other initial condition we first differentiate to find the current:

$$
\begin{aligned}
I = \frac{dQ}{dt} &= e^{-20t}[(-20c_1 + 15c_2) \cos 15t + (-15c_1 - 20c_2) \sin 15t] \\
&\quad + \tfrac{40}{697}(-21 \sin 10t + 16 \cos 10t)
\end{aligned}
$$

$$I(0) = -20c_1 + 15c_2 + \tfrac{640}{697} = 0 \qquad c_2 = -\tfrac{464}{2091}$$

Thus the formula for the charge is

$$Q(t) = \frac{4}{697}\left[\frac{e^{-20t}}{3}(-63 \cos 15t - 116 \sin 15t) + (21 \cos 10t + 16 \sin 10t)\right]$$

and the expression for the current is

$$I(t) = \tfrac{1}{2091}[e^{-20t}(-1920 \cos 15t + 13{,}060 \sin 15t) + 120(-21 \sin 10t + 16 \cos 10t)]$$

∎

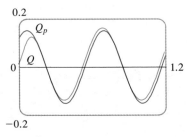

$$0.2$$
$$Q_p$$
$$0$$
$$Q$$
$$1.2$$
$$-0.2$$

FIGURE 9

$$m\frac{d^2x}{dt^2} + c\frac{dx}{dt} + kx = F(t)$$

$$L\frac{d^2Q}{dt^2} + R\frac{dQ}{dt} + \frac{1}{C}Q = E(t)$$

NOTE 1: In Example 3 the solution for $Q(t)$ consists of two parts. Since $e^{-20t} \to 0$ as $t \to \infty$ and both $\cos 15t$ and $\sin 15t$ are bounded functions,

$$Q_c(t) = \tfrac{4}{2091}e^{-20t}(-63 \cos 15t - 116 \sin 15t) \to 0 \qquad \text{as } t \to \infty$$

So, for large values of t,

$$Q(t) \approx Q_p(t) = \tfrac{4}{697}(21 \cos 10t + 16 \sin 10t)$$

and, for this reason, $Q_p(t)$ is called the **steady state solution.** Figure 9 shows how the graph of the steady state solution compares with the graph of Q in this case.

NOTE 2: Comparing Equations 5 and 7, we see that mathematically they are identical. This suggests the analogies given in the following chart between physical situations that, at first glance, are very different.

Spring system		Electric circuit	
x	displacement	Q	charge
dx/dt	velocity	$I = dQ/dt$	current
m	mass	L	inductance
c	damping constant	R	resistance
k	spring constant	$1/C$	elastance
$F(t)$	external force	$E(t)$	electromotive force

We can also transfer other ideas from one situation to the other. For instance, the steady state solution discussed in Note 1 makes sense in the spring system. And the phenomenon of resonance in the spring system can be usefully carried over to electric circuits as electrical resonance.

EXERCISES 15.7

1. A spring with a 3-kg mass is held stretched 0.6 m beyond its natural length by a force of 20 N. If the spring begins at its equilibrium position but a push gives it an initial velocity of 1.2 m/s, find the position of the mass after t seconds.

2. A spring with a 4-kg mass has natural length 1 m and is maintained stretched to a length of 1.3 m by a force of 24.3 N. If the spring is compressed to a length of 0.8 m and then released with zero velocity, find the position of the mass at any time t.

3. A spring with a mass of 2 kg has damping constant 14, and a force of 6 N is required to keep the spring stretched 0.5 m beyond its natural length. The spring is stretched 1 m beyond its natural length and then released with zero velocity. Find the position of the mass at any time t.

4. A spring with a mass of 3 kg has damping constant 30 and spring constant 123.
   (a) Find the position of the mass at time t if it starts at the equilibrium position with a velocity of 2 m/s.
   (b) Graph the position function of the mass.

5. For the spring in Exercise 3, find the mass that would produce critical damping.

6. For the spring in Exercise 4, find the damping constant that would produce critical damping.

7. Suppose a spring has mass m and spring constant k and let $\omega = \sqrt{k/m}$. Suppose that the damping constant is so small that the damping force is negligible. If an external force $F(t) = F_0 \cos \omega_0 t$ is applied, where $\omega_0 \neq \omega$, use the method of undetermined coefficients to show that the motion of the mass is described by Equation 6.

8. As in Exercise 7, consider a spring with mass m, spring constant k, and damping constant $c = 0$, and let $\omega = \sqrt{k/m}$. If an external force $F(t) = F_0 \cos \omega t$ is applied (the applied frequency equals the natural frequency), use the method of undetermined coefficients to show that the motion of the mass is given by

$$x(t) = c_1 \cos \omega t + c_2 \sin \omega t + \frac{F_0}{2m\omega}\, t \sin \omega t$$

9. A series circuit consists of a resistor with $R = 20\ \Omega$, an inductor with $L = 1$ H, a capacitor with $C = 0.002$ F, and a 12-V battery. If the initial charge and current are both 0, find the charge and current at time t.

10. A series circuit contains a resistor with $R = 24\ \Omega$, an inductor with $L = 2$ H, a capacitor with $C = 0.005$ F, and a 12-V battery. The initial charge is $Q = 0.001$ C and the initial current is 0.
   (a) Find the charge and current at time t.
   (b) Graph the charge and current functions.

11. The battery in Exercise 9 is replaced by a generator producing a voltage of $E(t) = 12 \sin 10t$. Find the charge at time t.

12. The battery in Exercise 10 is replaced by a generator producing a voltage of $E(t) = 12 \sin 10t$.
   (a) Find the charge at time t.
   (b) Graph the charge function.

13. Verify that the solution to Equation 1 can be written in the form $x(t) = A \cos(\omega t + \delta)$.

15.8 SERIES SOLUTIONS

So far, the only second-order differential equations that we have been able to solve are linear equations with constant coefficients. Even a simple-looking equation like

(1) $$y'' - 2xy' + y = 0$$

is not easy to solve. But it is important to be able to solve equations such as Equation 1 since they arise from physical problems and, in particular, in connection with the

Schrödinger equation in quantum mechanics. In such a case we use the method of power series; that is, we look for a solution of the form

$$y = f(x) = \sum_{n=0}^{\infty} c_n x^n = c_0 + c_1 x + c_2 x^2 + c_3 x^3 + \cdots$$

The method is to substitute this expression into the differential equation and determine the values of the coefficients $c_0, c_1, c_2, \ldots$. This technique resembles the method of undetermined coefficients discussed in Section 15.6.

Before using power series to solve Equation 1, we illustrate the method on the simpler equation $y'' + y = 0$. It is true that we already know how to solve this equation by the techniques of Section 15.5, but it is easier to understand the power series method when it is applied to this simpler equation.

EXAMPLE 1   Use power series to solve the equation $y'' + y = 0$.

SOLUTION   We assume there is a solution of the form

(2)
$$y = c_0 + c_1 x + c_2 x^2 + c_3 x^3 + \cdots = \sum_{n=0}^{\infty} c_n x^n$$

By Theorem 10.9.2 we can differentiate power series term by term, so

$$y' = c_1 + 2c_2 x + 3c_3 x^2 + \cdots = \sum_{n=1}^{\infty} n c_n x^{n-1}$$

(3)
$$y'' = 2c_2 + 2 \cdot 3c_3 x + \cdots = \sum_{n=2}^{\infty} n(n-1) c_n x^{n-2}$$

In order to compare the expressions for y and y'' more easily, we rewrite y'' as follows:

(4)
$$y'' = \sum_{n=0}^{\infty} (n+2)(n+1) c_{n+2} x^n$$

By writing out the first few terms of Equation 4 you can see that it is the same as Equation 3. To obtain Equation 4 we replaced n by $n+2$ and began the summation at 0 instead of 2. Substituting these expressions into the differential equation, we obtain

$$\sum_{n=0}^{\infty} (n+2)(n+1) c_{n+2} x^n + \sum_{n=0}^{\infty} c_n x^n = 0$$

or

(5)
$$\sum_{n=0}^{\infty} [(n+2)(n+1) c_{n+2} + c_n] x^n = 0$$

If two power series are equal, then the corresponding coefficients must be equal (from Section 10.10). Therefore, the coefficients of x^n in Equation 5 must be 0:

$$(n+2)(n+1) c_{n+2} + c_n = 0$$

(6)
$$c_{n+2} = -\frac{c_n}{(n+1)(n+2)} \qquad n = 0, 1, 2, 3, \ldots$$

Equation 6 is called a recursion relation. If c_0 and c_1 are known, this equation allows us to determine the remaining coefficients recursively by putting $n = 0, 1, 2, 3, \ldots$ in succession.

$$\text{Put } n = 0: \qquad c_2 = -\frac{c_0}{1 \cdot 2}$$

$$\text{Put } n = 1: \qquad c_3 = -\frac{c_1}{2 \cdot 3}$$

$$\text{Put } n = 2: \qquad c_4 = -\frac{c_2}{3 \cdot 4} = \frac{c_0}{1 \cdot 2 \cdot 3 \cdot 4} = \frac{c_0}{4!}$$

$$\text{Put } n = 3: \qquad c_5 = -\frac{c_3}{4 \cdot 5} = \frac{c_1}{2 \cdot 3 \cdot 4 \cdot 5} = \frac{c_1}{5!}$$

$$\text{Put } n = 4: \qquad c_6 = -\frac{c_4}{5 \cdot 6} = -\frac{c_0}{4! \, 5 \cdot 6} = -\frac{c_0}{6!}$$

$$\text{Put } n = 5: \qquad c_7 = -\frac{c_5}{6 \cdot 7} = -\frac{c_1}{5! \, 6 \cdot 7} = -\frac{c_1}{7!}$$

By now we see the pattern:

$$\text{For the even coefficients, } c_{2n} = (-1)^n \frac{c_0}{(2n)!}$$

$$\text{For the odd coefficients, } c_{2n+1} = (-1)^n \frac{c_1}{(2n + 1)!}$$

Putting these values back into Equation 2, we write the solution as

$$y = c_0 + c_1 x + c_2 x^2 + c_3 x^3 + c_4 x^4 + c_5 x^5 + \cdots$$

$$= c_0 \left(1 - \frac{x^2}{2!} + \frac{x^4}{4!} - \frac{x^6}{6!} + \cdots + (-1)^n \frac{x^{2n}}{(2n)!} + \cdots \right)$$

$$+ c_1 \left(x - \frac{x^3}{3!} + \frac{x^5}{5!} - \frac{x^7}{7!} + \cdots + (-1)^n \frac{x^{2n+1}}{(2n + 1)!} + \cdots \right)$$

$$= c_0 \sum_{n=0}^{\infty} (-1)^n \frac{x^{2n}}{(2n)!} + a_1 \sum_{n=0}^{\infty} (-1)^n \frac{x^{2n+1}}{(2n + 1)!}$$

Notice that there are two arbitrary constants c_0 and c_1, which is to be expected. ∎

NOTE 1: We recognize the series obtained in Example 1 as being the Maclaurin series for $\cos x$ and $\sin x$. (See Equations 10.10.16 and 10.10.15.) Therefore, we could write the solution as

$$y(x) = c_0 \cos x + c_1 \sin x$$

But we are not usually able to express power series solutions of differential equations in terms of known functions.

EXAMPLE 2   Solve $y'' - 2xy' + y = 0$.

SOLUTION   We assume there is a solution of the form

$$y = \sum_{n=0}^{\infty} c_n x^n$$

Then

$$y' = \sum_{n=1}^{\infty} n c_n x^{n-1}$$

and

$$y'' = \sum_{n=2}^{\infty} n(n-1)c_n x^{n-2} = \sum_{n=0}^{\infty} (n+2)(n+1)c_{n+2}x^n$$

as in Example 1. Substituting in the differential equation, we get

$$\sum_{n=0}^{\infty} (n+2)(n+1)c_{n+2}x^n - 2x \sum_{n=1}^{\infty} n c_n x^{n-1} + \sum_{n=0}^{\infty} c_n x^n = 0$$

$$\sum_{n=0}^{\infty} (n+2)(n+1)c_{n+2}x^n - \sum_{n=1}^{\infty} 2n c_n x^n + \sum_{n=0}^{\infty} c_n x^n = 0$$

$$\sum_{n=1}^{\infty} 2n c_n x^n = \sum_{n=0}^{\infty} 2n c_n x^n$$

$$\sum_{n=0}^{\infty} [(n+2)(n+1)c_{n+2} - (2n-1)c_n]x^n = 0$$

This is true if the coefficient of x_n is 0:

$$(n+2)(n+1)c_{n+2} - (2n-1)c_n = 0$$

(7)
$$c_{n+2} = \frac{2n-1}{(n+1)(n+2)}c_n \qquad n = 0, 1, 2, 3, \ldots$$

We solve this recursion relation by putting $n = 0, 1, 2, 3, \ldots$ successively in Equation 7:

Put $n = 0$: $c_2 = \dfrac{-1}{1 \cdot 2}c_0$

Put $n = 1$: $c_3 = \dfrac{1}{2 \cdot 3}c_1$

Put $n = 2$: $c_4 = \dfrac{3}{3 \cdot 4}c_2 = -\dfrac{3}{1 \cdot 2 \cdot 3 \cdot 4}c_0 = -\dfrac{3}{4!}c_0$

Put $n = 3$: $c_5 = \dfrac{5}{4 \cdot 5}c_3 = \dfrac{1 \cdot 5}{2 \cdot 3 \cdot 4 \cdot 5}c_1 = \dfrac{1 \cdot 5}{5!}c_1$

Put $n = 4$: $c_6 = \dfrac{7}{5 \cdot 6}c_4 = -\dfrac{3 \cdot 7}{4! \, 5 \cdot 6}c_0 = -\dfrac{3 \cdot 7}{6!}c_0$

Put $n = 5$: $c_7 = \dfrac{9}{6 \cdot 7}c_5 = \dfrac{1 \cdot 5 \cdot 9}{5! \, 6 \cdot 7}c_1 = \dfrac{1 \cdot 5 \cdot 9}{7!}c_1$

Put $n = 6$: $c_8 = \dfrac{11}{7 \cdot 8}c_6 = -\dfrac{3 \cdot 7 \cdot 11}{8!}c_0$

Put $n = 7$: $c_9 = \dfrac{13}{8 \cdot 9}c_7 = \dfrac{1 \cdot 5 \cdot 9 \cdot 13}{9!}c_1$

In general, the even coefficients are given by

$$c_{2n} = -\frac{3 \cdot 7 \cdot 11 \cdot \cdots \cdot (4n - 5)}{(2n)!} c_0$$

and the odd coefficients are given by

$$c_{2n+1} = \frac{1 \cdot 5 \cdot 9 \cdot \cdots \cdot (4n - 3)}{(2n + 1)!} c_1$$

The solution is

$$y = c_0 + c_1 x + c_2 x^2 + c_3 x^3 + c_4 x^4 + \cdots$$

$$= c_0\left(1 - \frac{1}{2!}x^2 - \frac{3}{4!}x^4 - \frac{3 \cdot 7}{6!}x^6 - \frac{3 \cdot 7 \cdot 11}{8!}x^8 - \cdots\right)$$

$$+ c_1\left(x + \frac{1}{3!}x^3 + \frac{1 \cdot 5}{5!}x^5 + \frac{1 \cdot 5 \cdot 9}{7!}x^7 + \frac{1 \cdot 5 \cdot 9 \cdot 13}{9!}x^9 + \cdots\right)$$

or

(8)
$$y = c_0\left(1 - \frac{1}{2!}x^2 - \sum_{n=2}^{\infty} \frac{3 \cdot 7 \cdot \cdots \cdot (4n - 5)}{(2n)!}x^{2n}\right)$$

$$+ c_1\left(x + \sum_{n=1}^{\infty} \frac{1 \cdot 5 \cdot 9 \cdot \cdots \cdot (4n - 3)}{(2n + 1)!}x^{2n+1}\right) \qquad \blacksquare$$

NOTE 2: In Example 2 we had to assume that the differential equation had a series solution. But now we could verify directly that the function given by Equation 8 is indeed a solution.

NOTE 3: Unlike the situation of Example 1, the power series that arise in the solution of Example 2 do not define elementary functions. The functions

$$y_1(x) = 1 - \frac{1}{2!}x^2 - \sum_{n=2}^{\infty} \frac{3 \cdot 7 \cdot \cdots \cdot (4n - 5)}{(2n)!}x^{2n}$$

and

$$y_2(x) = x + \sum_{n=1}^{\infty} \frac{1 \cdot 5 \cdot 9 \cdot \cdots \cdot (4n - 3)}{(2n + 1)!}x^{2n+1}$$

are perfectly good functions but they cannot be expressed in terms of familiar functions. We can use these power series expressions for y_1 and y_2 to compute approximate values of the functions and even to graph them. Figure 1 shows the first few partial sums T_0, T_2, T_4, ... (Taylor polynomials) for $y_1(x)$ and we see how they converge to y_1. In this way we can graph both y_1 and y_2 in Figure 2.

NOTE 4: If we were asked to solve the initial-value problem

$$y'' - 2xy' + y = 0 \qquad y(0) = 0 \qquad y'(0) = 1$$

we would observe from Theorem 10.10.5 that

$$c_0 = y(0) = 0 \qquad c_1 = y'(0) = 1$$

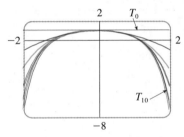

FIGURE 1

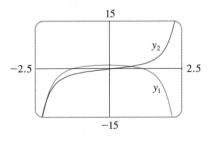

FIGURE 2

This would simplify the calculations in Example 2, since all of the even coefficients would be 0. The solution to the initial-value problem is

$$y(x) = x + \sum_{n=1}^{\infty} \frac{1 \cdot 5 \cdot 9 \cdot \cdots \cdot (4n-3)}{(2n+1)!} x^{2n+1}$$

EXERCISES 15.8

1–11 ■ Use power series to solve the differential equation.

1. $y' = 6y$

2. $y' = xy$

3. $y' = x^2 y$

4. $y'' = y$

5. $y'' + 3xy' + 3y = 0$

6. $y'' = xy$

7. $(x^2 + 1)y'' + xy' - y = 0$

8. $y'' - xy' + 2y = 0$

9. $y'' - xy' - y = 0,\quad y(0) = 1,\quad y'(0) = 0$

10. $y'' + x^2 y = 0,\quad y(0) = 1,\quad y'(0) = 0$

11. $y'' + x^2 y' + xy = 0,\quad y(0) = 0,\quad y'(0) = 1$

12. The solution of the initial-value problem

$$x^2 y'' + xy' + x^2 y = 0$$
$$y(0) = 1 \qquad y'(0) = 0$$

is called a Bessel function of order 0.
(a) Solve the initial-value problem to find a power series expansion for the Bessel function.
(b) Graph several Taylor polynomials until you reach one that looks like a good approximation to the Bessel function on the interval $[-5, 5]$.

15 REVIEW

KEY TOPICS ■ Write the general form of each type of differential equation. Then discuss the method of solution.

1. First-order differential equations:
   (a) separable　　　　(b) homogeneous
   (c) linear　　　　　　(d) exact
   (e) integrating factor

2. Second-order linear differential equations with constant coefficients:
   (a) homogeneous　　　(b) nonhomogeneous

EXERCISES

1–8 ■ Determine whether the statement is true or false.

1. The equation $(xy - y^2)y' = x^2 y$ is homogeneous.

2. The equation $y' + xy^2 = e^x$ is linear.

3. The equation $3(x^2 + 2xy) + (3x^2 + 2y)y' = 0$ is exact.

4. The equation $y' = 3y - 2x + 6xy - 1$ is separable.

5. An integrating factor for the linear equation $xy' + 2y = \sin x$ is x^2.

6. If y_1 and y_2 are solutions of $y'' + 6y' + 5y = x$, then $c_1 y_1 + c_2 y_2$ is also a solution of the equation.

7. The general solution of $y'' - y = 0$ can be written as $y = c_1 \cosh x + c_2 \sinh x$.

8. The equation $y'' - y = e^x$ has a particular solution of the form $y_p = Ae^x$.

9–28 ■ Solve the differential equation.

9. $1 + 2xy^2 + 2x^2 yy' = 0$

10. $y' + 2xy = 2x^3$

11. $xy' - 2y = x^3$

12. $xy' = y + x\cos^2(y/x)$

13. $(2y - 3y^2)y' = x\cos x$

14. $y' = 2 + 2x^2 + y + x^2 y$

15. $(x^2 + xy)y' = x^2 + y^2$

16. $(2x\cos y - 3y^2)y' = 2(x - \sin y)$

17. $1 + y^2 - \sqrt{1 - x^2}\, y' = 0$

18. $yy' = x\sqrt{1 + x^2}\sqrt{1 + y^2}$

19. $y'' - 6y' + 34y = 0$

20. $4y'' - 20y' + 25y = 0$

21. $2y'' + y' = y$

22. $3y'' + 2y' + y = 0$

23. $\dfrac{d^2 y}{dx^2} + 2\dfrac{dy}{dx} + y = \sin 3x$

24. $\dfrac{d^2y}{dx^2} - 2\dfrac{dy}{dx} - 3y = \cos 4x$

25. $4\dfrac{d^2y}{dx^2} + 9y = 2x^2 - 3$

26. $\dfrac{d^2y}{dx^2} - 4\dfrac{dy}{dx} + 20y = xe^x$

27. $\dfrac{d^2y}{dx^2} - 3\dfrac{dy}{dx} + 2y = e^{2x}$

28. $\dfrac{d^2y}{dx^2} + y = \csc x, \quad 0 < x < \pi/2$

29. Solve the equation $y' = y + e^{2x}$ and graph several members of the family of solutions. How does the solution curve change as the constant C varies?

30. (a) Graph the two basic solutions of the differential equation $y'' = y$ and several other solutions. What do the solutions have in common?
   (b) Graph the particular solution of $y'' = y + e^{2x}$ and several other solutions. What do these solutions have in common?

31–36 ■ Solve the initial-value problem.

31. $y' + y = \sqrt{x}\,e^{-x}, \quad y(0) = 3$

32. $1 + x = 2xyy', \quad x > 0, \quad y(1) = -2$

33. $y'' + 6y' = 0, \quad y(1) = 3, \quad y'(1) = 12$

34. $y'' - 6y' + 25y = 0, \quad y(0) = 2, \quad y'(0) = 1$

35. $y'' - 5y' + 4y = 0, \quad y(0) = 0, \quad y'(0) = 1$

36. $9y'' + y = 3x + e^{-x}, \quad y(0) = 1, \quad y'(0) = 2$

37. Use Euler's method with step size 0.1 to estimate $y(0.3)$, where $y(x)$ is the solution of the initial-value problem $y' = x^2 - y^2, y(0) = 1$.

38. Sketch the direction field of the differential equation $y' = y^2 - x$. Then use it to sketch a solution curve that passes through the point $(0, 1)$.

39–40 ■ Find the orthogonal trajectories of the given family of curves.

39. $kx^2 + y^2 = 1$

40. $y = \dfrac{k}{1 + x^2}$

41. Use power series to solve the initial-value problem
$$y'' + xy' + y = 0 \qquad y(0) = 0 \qquad y'(0) = 1$$

42. Use power series to solve the equation
$$y'' - xy' - 2y = 0$$

43. A series circuit contains a resistor with $R = 40\ \Omega$, an inductor with $L = 2$ H, a capacitor with $C = 0.0025$ F, and a 12-V battery. The initial charge is $Q = 0.01$ C and the initial current is 0. Find the charge at time t.

44. A spring with a mass of 2 kg has damping constant 16, and a force of 12.8 N keeps the spring stretched 0.2 m beyond its natural length. Find the position of the mass at time t if it starts at the equilibrium position with a velocity of 2.4 m/s.

45. Consider a population $P = P(t)$ with constant birth and death rates α and β, respectively, and a constant emigration rate m, where α, β, and m are positive constants. Assume that $\alpha > \beta$. Then the rate of change of the population at time t is given by
$$\frac{dP}{dt} = kP - m \qquad k = \alpha - \beta$$
   (a) Find the solution of this differential equation that satisfies the initial condition $P(0) = P_0$.
   (b) What conditions on m will lead to an exponential expansion of the population?
   (c) What conditions on m will result in a constant population? A population decline?

46. There is considerable evidence to support the theory that for some species there is a minimum population level m such that the species will become extinct if the size of the population falls below m. This condition can be incorporated into the logistic model for population "growth" by introducing the factor $(1 - m/P)$. Thus,
$$\frac{dP}{dt} = kP(C - P)\left(1 - \frac{m}{P}\right)$$
In this equation k and C are positive constants with k the "growth" rate and C the carrying capacity, and $C > m$.
   (a) Find the solution of this differential equation that satisfies the initial condition $P(0) = P_0$.
   (b) Show that if $P_0 < m$, then the species will become extinct. What will happen if $P(t) < m$ for some $t > 0$?

APPLICATIONS PLUS

1. Assume that the earth is a solid sphere of uniform density with mass M and radius $R = 3960$ mi. For a particle of mass m within the earth at a distance r from the earth's center, the gravitational force attracting the particle to the center is

$$F_r = \frac{-GM_r m}{r^2}$$

where G is the gravitational constant and M_r is the mass of the earth within the sphere of radius r.

(a) Show that $F_r = \dfrac{-GMm}{R^3} r$.

(b) Suppose a hole is drilled through the earth along a diameter. Show that if a particle of mass m is dropped from rest, at the surface, into the hole, then the distance $y = y(t)$ of the particle from the center of the earth at time t is given by

$$y''(t) = -k^2 y(t)$$

where $k^2 = GM/R^3 = g/R$.

(c) Conclude from part (b) that the particle undergoes simple harmonic motion. Find the period T.

(d) With what speed does the particle pass through the center of the earth?

2. A particle of mass m moves along the x-axis. It is acted on by a force $\mathbf{F}$ that is proportional to its position $x = x(t)$ and acts in a direction opposite to its motion; that is, $\mathbf{F} = -k\mathbf{x}$, where k is a positive constant. (See the discussion of Hooke's Law in Section 15.7.)

(a) Show that the force field is conservative and determine a potential energy function.

(b) Suppose the particle is at the position x_1 at time $t = t_1$ and that it is at the position x_2 at time $t = t_2$. Find the work done in moving the particle from x_1 to x_2.

(c) Suppose that the force field is modified to include a damping force that is proportional to the speed of the particle and opposes the motion. Then the new force field is given by $\mathbf{F} = -k\mathbf{x} - a\mathbf{v}$, where $\mathbf{v} = d\mathbf{x}/dt$ and a is a positive constant. Find the work done in moving the particle from x_1 to x_2 in this force field. (Leave your answer in terms of an integral.) Is this force field conservative?

3. A missile is fired from a point 50 mi due east of its intended target. The speed of the missile is 600 mi/h and it has a homing device that keeps it heading toward its target at all times. A wind is blowing due north at 30 mi/h.

(a) Find the components dx/dt and dy/dt of the missile's velocity in terms of the angle θ in the figure.

(b) Divide dy/dt by dx/dt to obtain an equation for dy/dx. Eliminate θ from this equation.

(c) Determine the general solution of the differential equation found in part (b) and use the initial condition $y(50) = 0$ to find the path of the missile.

(d) Use a graphing device to plot the path of the missile. [*Hint:* First use polar coordinates to obtain parametric equations for the path.]

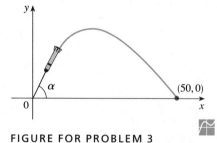

FIGURE FOR PROBLEM 3

4. A rocket is fired straight up, burning fuel at the constant rate of b kilograms per second. Let $v = v(t)$ be the velocity of the rocket at time t and suppose that the velocity u of the exhaust gas is constant. Let $M = M(t)$ be the mass of the rocket at time t and note that M decreases as the fuel burns. If we neglect air resistance, it follows from Newton's Second Law that

$$F = M \frac{dv}{dt} - ub$$

where the force $F = -Mg$. Thus

(1)
$$M \frac{dv}{dt} - ub = -Mg$$

Let M_1 be the mass of the rocket, M_2 the mass of the fuel, and $M_0 = M_1 + M_2$. Then, until the fuel runs out at time $t = M_2b$, the mass is $M = M_0 - bt$.

(a) Substitute $M = M_0 - bt$ into Equation 1 and solve the resulting differential equation for v. Use the initial condition $v(0) = 0$ to evaluate the arbitrary constant.

(b) Determine the velocity of the rocket at time $t = M_2/b$. This is called the *burnout velocity*.

(c) Determine the height of the rocket $y = y(t)$ at the burnout time.

(d) Find the height of the rocket at any time t.

5. In the human bloodstream, potassium ions (K^+) constantly move into and out of the red blood cells; that is, the surfaces of the red blood cells are *permeable* to K ions. The K^+ ions move from the plasma into the red cells at a certain rate, while other ions within the cells move out into the plasma at a certain rate. These rates are called the *permeabilities* of the cells' surfaces to K^+ ions. One technique for determining the permeabilities is the radioactive tracer technique. It is based on the fact that there is a radioactive isotope of K^+, namely K^{42+}, and that the K^{42+} ions are indistinguishable from K^+ ions as far as blood cell permeability is concerned. A fixed quantity S of radioactive K^{42+} ions is introduced into the bloodstream. Initially, all the ions are in the plasma. Let $P = P(t)$ be the amount of K^{42+} ions in the plasma and $C = C(t)$ the amount of K^{42+} ions in the cells at time t $[P(0) = S, C(0) = 0]$. Assume that the system is *closed* in the sense that there is no loss of K^{42+} ions from the system. Thus $P(t) + C(t) = S$ for all $t \geq 0$. Assume, also, that the rate of transfer from the cells to the plasma is proportional to the amount in the cells and that the rate of transfer from the plasma into the cells is proportional to the amount in the plasma. If the respective constants of proportionality are the positive constants k and m, then these assumptions lead to the mathematical model

$$C(t) + P(t) = S \qquad P'(t) = -mP(t) + kC(t)$$

In these equations, S is predetermined by the experimenter and $P(t)$ can be observed empirically. Thus $C(t)$ can easily be computed.

(a) Determine $P(t)$ by solving the initial-value problem $P' = -mP + kC$, $P(0) = S$. Calculate $Q = \lim_{t \to \infty} P(t)$.

(b) Find $C(t)$ and calculate $\lim_{t \to \infty} C(t)$.

(c) The function

$$g(t) = \ln\left(\frac{P(t)}{Q} - 1\right)$$

can be determined from experimental data. Show that $g(t) = -(m + k)t + \ln(m/k)$. If, from the experimental data, g is the straight line $g(t) = at + b$, then

$$a = -(m + k) \qquad b = \ln(m/k)$$

Solve these equations for m and k.

6. When a radioactive substance A decays into a substance B, A and B are referred to as *parent* and *daughter*. If the parent is decaying at the rate k, $k > 0$, then $dA/dt = -kA$ and the amount remaining at time $t > 0$ is given by

$$A(t) = A_0 e^{-kt}$$

where $A_0 = A(0)$. The amount B of the daughter is given by

$$B(t) = A_0 - A(t)$$

(a) Show that

$$t = \frac{1}{k} \ln\left[\frac{B(t)}{A(t)} + 1\right]$$

This formula can be used to determine the age of an object.

(b) Rubidium (Rb^{87}) decays into strontium (Sr^{87}). If the time t is measured in years, then the value of k in this reaction is 1.39×10^{-11}. Suppose a sample of ore contains 250 parts per million of rubidium and 4.35 parts per million of strontium. Determine the age of the sample.

(c) If the daughter is itself radioactive, then it is the parent of the daughter C, and so on. If B were not being formed, then the amount of B remaining at time t could be determined from the differential equation $dB/dt = -rB$, r a positive constant. But B is being replenished; each atom of A that decays becomes an atom of B, and this is happening at the rate $kA(t)$. Thus

$$\frac{dB}{dt} = kA(t) - rB(t) = kA_0 e^{-kt} - rB$$

Show that the solution of this differential equation that satisfies $B(0) = 0$ is

$$B(t) = \frac{kA_0}{k - r}(e^{-rt} - e^{-kt}) \qquad r \neq k$$

Sketch the graph of B.

7. If a mirror has the shape of the solid of revolution formed by rotating $y^2 = 4px$, $0 \leq x \leq c$, about the x-axis, then it has the property that light rays that are parallel to the axis of symmetry are reflected through the focus and, conversely, light rays emanating from the focus are reflected parallel to the axis.

Now assume that a light source is positioned at the origin. Suppose $y = f(x)$ is a function with the following properties:

   (i) f is defined and nonnegative on an interval $[a, b]$, $a < 0$, $b > 0$.

   (ii) If a mirror has the shape of the solid of revolution formed by rotating the graph of f about the x-axis, then the light rays from the source at the origin will be reflected parallel to the x-axis.

(a) Find a first-order differential equation that the function $y = f(x)$ must satisfy.

(b) Determine the general solution of the differential equation found in part (a), and show that the members of the one-parameter family are parabolas, each with focus at the origin.

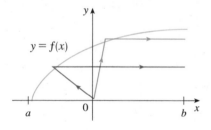

$y = f(x)$

FIGURE FOR PROBLEM 7

8. A certain small country has $10 billion in paper currency in circulation, and each day $50 million comes into the country's banks. The government decides to introduce new currency by having the banks replace old bills with new ones whenever old currency comes into the banks. Let $x = x(t)$ denote the amount of new currency in circulation at time t, with $x(0) = 0$.

(a) Determine a mathematical model in the form of an initial-value problem that represents the "flow" of the new currency into circulation.

(b) Solve the initial-value problem found in part (a).

(c) Find how long it will take for the new bills to account for 90% of the currency in circulation.

In this appendix we present proofs of several theorems that are stated in the main body of the text. The sections in which they occur are indicated in the margin.

In order to prove Theorem 10.8.3 we first need the following results:

THEOREM

1. If a power series $\Sigma\, a_n x^n$ converges when $x = b$ (where $b \neq 0$), then it converges whenever $|x| < |b|$.

2. If a power series $\Sigma\, a_n x^n$ diverges when $x = d$ (where $d \neq 0$), then it diverges whenever $|x| > |d|$.

PROOF OF 1 Suppose that $\Sigma\, a_n b^n$ converges. Then, by Theorem 10.2.6, we have $\lim_{n \to \infty} a_n b^n = 0$. According to Definition 10.1.1 with $\varepsilon = 1$, there is a positive integer N such that $|a_n b^n| < 1$ whenever $n \geq N$. Thus, for $n \geq N$, we have

$$|a_n x^n| = \left| \frac{a_n b^n x^n}{b^n} \right| = |a_n b^n| \left| \frac{x}{b} \right|^n < \left| \frac{x}{b} \right|^n$$

If $|x| < |b|$, then $|x/b| < 1$, so $\Sigma\, |x/b|^n$ is a convergent geometric series. Therefore, by the Comparison Test, the series $\Sigma_{n=N}^{\infty} |a_n x^n|$ is convergent. Thus the series $\Sigma\, a_n x^n$ is absolutely convergent and therefore convergent. □

PROOF OF 2 Suppose that $\Sigma\, a_n d^n$ diverges. If x is any number such that $|x| > |d|$, then $\Sigma\, a_n x^n$ cannot converge because, by part 1, the convergence of $\Sigma\, a_n x^n$ would imply the convergence of $\Sigma\, a_n d^n$. Therefore, $\Sigma\, a_n x^n$ diverges whenever $|x| > |d|$. □

THEOREM For a power series $\Sigma\, a_n x^n$ there are only three possibilities:

1. The series converges only when $x = 0$.

2. The series converges for all x.

3. There is a positive number R such that the series converges if $|x| < R$ and diverges if $|x| > R$.

PROOF Suppose that neither case 1 nor case 3 is true. Then there are nonzero numbers b and d such that $\Sigma\, a_n x^n$ converges for $x = b$ and diverges for $x = d$. Therefore, the set $S = \{x \,|\, \Sigma\, a_n x^n \text{ converges}\}$ is not empty. By the preceding theorem, the series diverges if $|x| > |d|$, so $|x| \leq |d|$ for all $x \in S$. This says that $|d|$ is an upper bound for the set S. Thus, by the Completeness Axiom (see Section 10.1), S has a least upper bound R. If $|x| > R$, then $x \notin S$, so $\Sigma\, a_n x^n$ diverges. If $|x| < R$, then $|x|$ is not an upper bound for S and so there exists $b \in S$ such that $b > |x|$. Since $b \in S$, $\Sigma\, a_n b^n$ converges, so by the preceding theorem $\Sigma\, a_n x^n$ converges. □

---

(3) THEOREM For a power series $\Sigma\, a_n(x - c)^n$ there are only three possibilities:

1. The series converges only when $x = c$.

2. The series converges for all x.

3. There is a positive number R such that the series converges if $|x - c| < R$ and diverges if $|x - c| > R$.

PROOF If we make the change of variable $u = x - c$, then the power series becomes $\Sigma\, a_n u^n$ and we can apply the preceding theorem to this series. In case 3 we have convergence for $|u| < R$ and divergence for $|u| > R$. Thus we have convergence for $|x - c| < R$ and divergence for $|x - c| > R$. □

> **(5) CLAIRAUT'S THEOREM**   Suppose f is defined on a disk D containing the point (a, b). If the functions f_{xy} and f_{yx} are both continuous on D, then $f_{xy}(a, b) = f_{yx}(a, b)$.

PROOF   For small values of h, $h \neq 0$, consider the difference

$$\Delta(h) = [\,f(a + h, b + h) - f(a + h, b)\,] - [\,f(a, b + h) - f(a, b)\,]$$

Notice that if we let $g(x) = f(x, b + h) - f(x, b)$, then

$$\Delta(h) = g(a + h) - g(a)$$

By the Mean Value Theorem, there is a number c between a and $a + h$ such that

$$g(a + h) - g(a) = g'(c)h = h[\,f_x(c, b + h) - f_x(c, b)\,]$$

Applying the Mean Value Theorem again, this time to f_x, we get a number d between b and $b + h$ such that

$$f_x(c, b + h) - f_x(c, b) = f_{xy}(c, d)h$$

Combining these equations, we obtain

$$\Delta(h) = h^2 f_{xy}(c, d)$$

If $h \to 0$, then $(c, d) \to (a, b)$, so the continuity of f_{xy} at (a, b) gives

$$\lim_{h \to 0} \frac{\Delta(h)}{h^2} = \lim_{(c, d) \to (a, b)} f_{xy}(c, d) = f_{xy}(a, b)$$

Similarly, by writing

$$\Delta(h) = [\,f(a + h, b + h) - f(a, b + h)\,] - [\,f(a + h, b) - f(a, b)\,]$$

and using the Mean Value Theorem twice and the continuity of f_{yx} at (a, b), we obtain

$$\lim_{h \to 0} \frac{\Delta(h)}{h^2} = f_{yx}(a, b)$$

It follows that $f_{xy}(a, b) = f_{yx}(a, b)$. $\square$

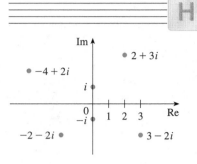

FIGURE 1

Complex numbers as points in
the Argand plane

H   COMPLEX NUMBERS

A **complex number** can be represented by an expression of the form $a + bi$, where a and b are real numbers and i is a symbol with the property that $i^2 = -1$. The complex number $a + bi$ can also be represented by the ordered pair (a, b) and plotted as a point in a plane (called the Argand plane) as in Figure 1. Thus the complex number $i = 0 + 1 \cdot i$ is identified with the point $(0, 1)$.

The **real part** of the complex number $a + bi$ is the real number a and the **imaginary part** is the real number b. Thus the real part of $4 - 3i$ is 4 and the imaginary part is -3. Two complex numbers $a + bi$ and $c + di$ are **equal** if $a = c$ and $b = d$, that is, their real parts are equal and their imaginary parts are equal. In the Argand plane the x-axis is called the real axis and the y-axis is called the imaginary axis.

The sum and difference of two complex numbers are defined by adding or subtracting their real parts and their imaginary parts, respectively:

$$(a + bi) + (c + di) = (a + c) + (b + d)i$$

$$(a + bi) - (c + di) = (a - c) + (b - d)i$$

For instance,

$$(1 - i) + (4 + 7i) = (1 + 4) + (-1 + 7)i = 5 + 6i$$

The product of complex numbers is defined so that the usual commutative and distributive laws hold:

$$(a + bi)(c + di) = a(c + di) + (bi)(c + di)$$

$$= ac + adi + bci + bdi^2$$

Since $i^2 = -1$, this becomes

$$(a + bi)(c + di) = (ac - bd) + (ad + bc)i$$

EXAMPLE 1

$$(-1 + 3i)(2 - 5i) = -(2 - 5i) + 3i(2 - 5i)$$

$$= -2 + 5i + 6i - 15(-1) = 13 + 11i \qquad \blacksquare$$

Division of complex numbers is much like rationalizing the denominator of a rational expression. For the complex number $z = a + bi$, we define its **complex conjugate** to be $\bar{z} = a - bi$. To find the quotient of two complex numbers we multiply numerator and denominator by the complex conjugate of the denominator.

EXAMPLE 2   Express the number $\dfrac{-1 + 3i}{2 + 5i}$ in the form $a + bi$.

SOLUTION   We multiply numerator and denominator by the complex conjugate of $2 + 5i$, namely $2 - 5i$, and we take advantage of the result of Example 1:

$$\frac{-1 + 3i}{2 + 5i} = \frac{-1 + 3i}{2 + 5i} \cdot \frac{2 - 5i}{2 - 5i} = \frac{13 + 11i}{2^2 + 5^2} = \frac{13}{29} + \frac{11}{29}i \qquad \blacksquare$$

The geometric interpretation of the complex conjugate is shown in Figure 2: $\bar{z}$ is the reflection of z in the real axis. We list some of the properties of the complex conjugate in the following box. The proofs follow from the definition and are requested in Exercise 18.

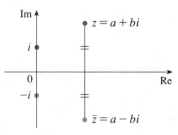

FIGURE 2

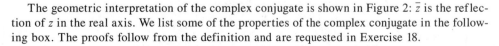

PROPERTIES OF CONJUGATES

$$\overline{z + w} = \bar{z} + \bar{w} \qquad\qquad \overline{zw} = \bar{z}\,\bar{w} \qquad\qquad \overline{z^n} = \bar{z}^n$$

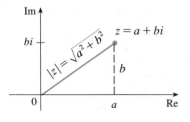

FIGURE 3

The **modulus,** or **absolute value,** $|z|$ of a complex number $z = a + bi$ is its distance from the origin. From Figure 3 we see that if $z = a + bi$, then

$$|z| = \sqrt{a^2 + b^2}$$

Notice that

$$z\bar{z} = (a + bi)(a - bi) = a^2 + abi - abi - b^2i^2 = a^2 + b^2$$

and so

$$z\bar{z} = |z|^2$$

This explains why the division procedure in Example 2 works in general:

$$\frac{z}{w} = \frac{z\bar{w}}{w\bar{w}} = \frac{z\bar{w}}{|w|^2}$$

Since $i^2 = -1$, we can think of i as a square root of -1. But we also have $(-i)^2 = i^2 = -1$ and so $-i$ is also a square root of -1. We say that i is the **principal square root** of -1 and write $\sqrt{-1} = i$. In general, if c is any positive number, we write

$$\sqrt{-c} = \sqrt{c}\, i$$

With this convention the usual derivation and formula for the roots of the quadratic equation $ax^2 + bx + c = 0$ are valid even when $b^2 - 4ac < 0$:

$$x = \frac{-b \pm \sqrt{b^2 - 4ac}}{2a}$$

EXAMPLE 3   Find the roots of the equation $x^2 + x + 1 = 0$.

SOLUTION   Using the quadratic formula, we have

$$x = \frac{-1 \pm \sqrt{1^2 - 4 \cdot 1}}{2} = \frac{-1 \pm \sqrt{-3}}{2} = \frac{-1 \pm \sqrt{3}\, i}{2}$$ ∎

We observe that the solutions of the equation in Example 3 are complex conjugates of each other. In general, the solutions of any quadratic equation $ax^2 + bx + c = 0$ with real coefficients a, b, and c are always complex conjugates. (If z is real, $\bar{z} = z$, so z is its own conjugate.)

We have seen that if we allow complex numbers as solutions, then every quadratic equation has a solution. More generally, it is true that every polynomial equation

$$a_n x^n + a_{n-1} x^{n-1} + \cdots + a_1 x + a_0 = 0$$

of degree at least one has a solution among the complex numbers. This fact is known as the Fundamental Theorem of Algebra and was proved by Gauss.

POLAR FORM

We know that any complex number $z = a + bi$ can be considered as a point (a, b) and that any such point can be represented by polar coordinates (r, θ) with $r \geqslant 0$. In fact,

$$a = r\cos\theta \qquad b = r\sin\theta$$

as in Figure 4. Therefore, we have

$$z = a + bi = (r\cos\theta) + (r\sin\theta)i$$

Thus we can write any complex number z in the form

$$z = r(\cos\theta + i\sin\theta)$$

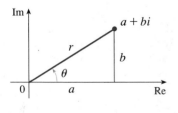

FIGURE 4

where $\qquad$ $r = |z| = \sqrt{a^2 + b^2}$ $\quad$ and $\quad$ $\tan\theta = \dfrac{b}{a}$

The angle θ is called the **argument** of z and we write $\theta = \arg(z)$. Note that $\arg(z)$ is not unique; any two arguments of z differ by an integer multiple of 2π.

EXAMPLE 4 Write the following numbers in polar form:
(a) $z = 1 + i$ $\qquad$ (b) $w = \sqrt{3} - i$

SOLUTION
(a) We have $r = |z| = \sqrt{1 + 1} = \sqrt{2}$ and $\tan\theta = 1$, so we can take $\theta = \pi/4$. Therefore, the polar form is

$$z = \sqrt{2}\left(\cos\frac{\pi}{4} + i\sin\frac{\pi}{4}\right)$$

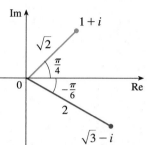

FIGURE 5

(b) Here we have $r = |w| = \sqrt{3 + 1} = 2$ and $\tan\theta = -1/\sqrt{3}$. Since w lies in the fourth quadrant, we take $\theta = -\pi/6$ and

$$w = 2\left[\cos\left(-\frac{\pi}{6}\right) + i\sin\left(-\frac{\pi}{6}\right)\right]$$

The numbers z and w are shown in Figure 5. $\blacksquare$

The polar form of complex numbers gives insight into multiplication and division. Let

$$z_1 = r_1(\cos\theta_1 + i\sin\theta_1) \qquad z_2 = r_2(\cos\theta_2 + i\sin\theta_2)$$

be two complex numbers written in polar form. Then

$$z_1 z_2 = r_1 r_2(\cos\theta_1 + i\sin\theta_1)(\cos\theta_2 + i\sin\theta_2)$$
$$= r_1 r_2[(\cos\theta_1\cos\theta_2 - \sin\theta_1\sin\theta_2) + i(\sin\theta_1\cos\theta_2 + \cos\theta_1\sin\theta_2)]$$

Therefore, using the addition formulas for cosine and sine, we have

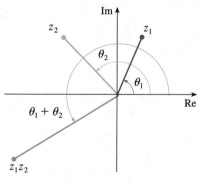

FIGURE 6

(1) $\qquad$ $$z_1 z_2 = r_1 r_2[\cos(\theta_1 + \theta_2) + i\sin(\theta_1 + \theta_2)]$$

This formula says that *to multiply two complex numbers we multiply the moduli and add the arguments.* (See Figure 6.)

A similar argument using the subtraction formulas for sine and cosine shows that *to divide two complex numbers we divide the moduli and subtract the arguments.*

$$\frac{z_1}{z_2} = \frac{r_1}{r_2}[\cos(\theta_1 - \theta_2) + i\sin(\theta_1 - \theta_2)] \qquad z_2 \neq 0$$

In particular, taking $z_1 = 1$ and $z_2 = z$, we have the following, which is illustrated in Figure 7.

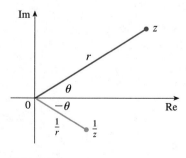

FIGURE 7

$$\text{If} \quad z = r(\cos\theta + i\sin\theta), \quad \text{then} \quad \frac{1}{z} = \frac{1}{r}(\cos\theta - i\sin\theta).$$

EXAMPLE 5   Find the product of the complex numbers $1 + i$ and $\sqrt{3} - i$ in polar form.

SOLUTION   From Example 4 we have

$$1 + i = \sqrt{2}\left(\cos\frac{\pi}{4} + i\sin\frac{\pi}{4}\right)$$

and

$$\sqrt{3} - i = 2\left[\cos\left(-\frac{\pi}{6}\right) + i\sin\left(-\frac{\pi}{6}\right)\right]$$

So, by Equation 1,

$$(1 + i)(\sqrt{3} - i) = 2\sqrt{2}\left[\cos\left(\frac{\pi}{4} - \frac{\pi}{6}\right) + i\sin\left(\frac{\pi}{4} - \frac{\pi}{6}\right)\right]$$

$$= 2\sqrt{2}\left(\cos\frac{\pi}{12} + i\sin\frac{\pi}{12}\right)$$

This is illustrated in Figure 8.

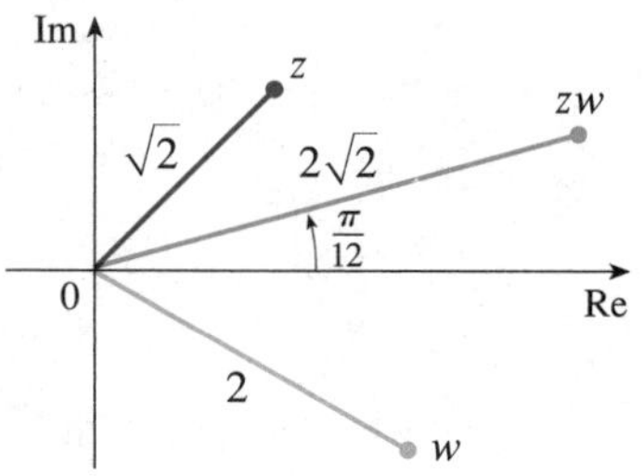

FIGURE 8

Repeated use of Formula 1 shows how to compute powers of a complex number. If

$$z = r(\cos\theta + i\sin\theta)$$

then

$$z^2 = r^2(\cos 2\theta + i\sin 2\theta)$$

and

$$z^3 = zz^2 = r^3(\cos 3\theta + i\sin 3\theta)$$

In general, we obtain the following result, which is named after the French mathematician Abraham De Moivre (1667–1754).

---

(2) DE MOIVRE'S THEOREM   If $z = r(\cos\theta + i\sin\theta)$ and n is a positive integer, then

$$z^n = [r(\cos\theta + i\sin\theta)]^n = r^n(\cos n\theta + i\sin n\theta)$$

This says that *to take the* n*th power of a complex number we take the* n*th power of the modulus and multiply the argument by* n.

EXAMPLE 6   Find $\left(\frac{1}{2} + \frac{1}{2}i\right)^{10}$.

SOLUTION   Since $\frac{1}{2} + \frac{1}{2}i = \frac{1}{2}(1 + i)$, it follows from Example 4(a) that $\frac{1}{2} + \frac{1}{2}i$ has the polar form

$$\frac{1}{2} + \frac{1}{2}i = \frac{\sqrt{2}}{2}\left(\cos\frac{\pi}{4} + i\sin\frac{\pi}{4}\right)$$

So by De Moivre's Theorem,

$$\left(\tfrac{1}{2} + \tfrac{1}{2}i\right)^{10} = \left(\frac{\sqrt{2}}{2}\right)^{10}\left(\cos\frac{10\pi}{4} + i\sin\frac{10\pi}{4}\right)$$

$$= \frac{2^5}{2^{10}}\left(\cos\frac{5\pi}{2} + i\sin\frac{5\pi}{2}\right) = \frac{1}{32}i$$

De Moivre's Theorem can also be used to find the *n*th roots of complex numbers. An *n*th root of the complex number *z* is a complex number *w* such that

$$w^n = z$$

Writing these two numbers in trigonometric form as

$$w = s(\cos\phi + i\sin\phi) \qquad \text{and} \qquad z = r(\cos\theta + i\sin\theta)$$

and using De Moivre's Theorem, we get

$$s^n(\cos n\phi + i\sin n\phi) = r(\cos\theta + i\sin\theta)$$

The equality of these two complex numbers shows that

$$s^n = r \qquad \text{or} \qquad s = r^{1/n}$$

and $$\cos n\phi = \cos\theta \qquad \text{and} \qquad \sin n\phi = \sin\theta$$

From the fact that sine and cosine have period 2π it follows that

$$n\phi = \theta + 2k\pi \qquad \text{or} \qquad \phi = \frac{\theta + 2k\pi}{n}$$

Thus $$w = r^{1/n}\left[\cos\left(\frac{\theta + 2k\pi}{n}\right) + i\sin\left(\frac{\theta + 2k\pi}{n}\right)\right]$$

Since this expression gives a different value of *w* for $k = 0, 1, 2, \ldots, n - 1$, we have the following:

---

(3) ROOTS OF A COMPLEX NUMBER   Let $z = r(\cos\theta + i\sin\theta)$ and let *n* be a positive integer. Then *z* has the *n* distinct *n*th roots

$$w_k = r^{1/n}\left[\cos\left(\frac{\theta + 2k\pi}{n}\right) + i\sin\left(\frac{\theta + 2k\pi}{n}\right)\right]$$

where $k = 0, 1, 2, \ldots, n - 1$.

Notice that each of the *n*th roots of *z* has modulus $|w_k| = r^{1/n}$. Thus all the *n*th roots of *z* lie on the circle of radius $r^{1/n}$ in the complex plane. Also, since the argument of each successive *n*th root exceeds the argument of the previous root by $2\pi/n$, we see that the *n*th roots of *z* are equally spaced on this circle.

EXAMPLE 7   Find the six sixth roots of $z = -8$ and graph these roots in the complex plane.

SOLUTION   In trigonometric form, $z = 8(\cos\pi + i\sin\pi)$. Applying Equation 3 with $n = 6$, we get

$$w_k = 8^{1/6}\left(\cos\frac{\pi + 2k\pi}{6} + i\sin\frac{\pi + 2k\pi}{6}\right)$$

We get the six sixth roots of -8 by taking $k = 0, 1, 2, 3, 4, 5$ in this formula:

$$w_0 = 8^{1/6}\left(\cos\frac{\pi}{6} + i\sin\frac{\pi}{6}\right) = \sqrt{2}\left(\frac{\sqrt{3}}{2} + \frac{1}{2}i\right)$$

$$w_1 = 8^{1/6}\left(\cos\frac{\pi}{2} + i\sin\frac{\pi}{2}\right) = \sqrt{2}\, i$$

$$w_2 = 8^{1/6}\left(\cos\frac{5\pi}{6} + i\sin\frac{5\pi}{6}\right) = \sqrt{2}\left(-\frac{\sqrt{3}}{2} + \frac{1}{2}i\right)$$

$$w_3 = 8^{1/6}\left(\cos\frac{7\pi}{6} + i\sin\frac{7\pi}{6}\right) = \sqrt{2}\left(-\frac{\sqrt{3}}{2} - \frac{1}{2}i\right)$$

$$w_4 = 8^{1/6}\left(\cos\frac{3\pi}{2} + i\sin\frac{3\pi}{2}\right) = -\sqrt{2}\, i$$

$$w_5 = 8^{1/6}\left(\cos\frac{11\pi}{6} + i\sin\frac{11\pi}{6}\right) = \sqrt{2}\left(\frac{\sqrt{3}}{2} - \frac{1}{2}i\right)$$

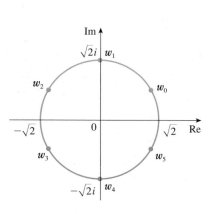

FIGURE 9

The six sixth roots of $z = -8$

All these points lie on the circle of radius $\sqrt{2}$ as shown in Figure 9. ∎

COMPLEX EXPONENTIALS

We also need to give a meaning to the expression e^z when $z = x + iy$ is a complex number. The theory of infinite series as developed in Chapter 10 can be extended to the case where the terms are complex numbers. Using the Taylor series for e^x (10.10.12) as our guide, we define

$$(4) \qquad e^z = \sum_{n=0}^{\infty} \frac{z^n}{n!} = 1 + z + \frac{z^2}{2!} + \frac{z^3}{3!} + \cdots$$

and it turns out that this complex exponential function has the same properties as the real exponential function. In particular, it is true that

$$(5) \qquad e^{z_1 + z_2} = e^{z_1} e^{z_2}$$

If we put $z = iy$, where y is a real number, in Equation 4, and use the facts that

$$i^2 = -1, \quad i^3 = i^2 i = -i, \quad i^4 = 1, \quad i^5 = i, \quad \ldots$$

we get

$$e^{iy} = 1 + iy + \frac{(iy)^2}{2!} + \frac{(iy)^3}{3!} + \frac{(iy)^4}{4!} + \frac{(iy)^5}{5!} + \cdots$$

$$= 1 + iy - \frac{y^2}{2!} - i\frac{y^3}{3!} + \frac{y^4}{4!} + i\frac{y^5}{5!} + \cdots$$

$$= \left(1 - \frac{y^2}{2!} + \frac{y^4}{4!} - \frac{y^6}{6!} + \cdots\right) + i\left(y - \frac{y^3}{3!} + \frac{y^5}{5!} - \cdots\right)$$

$$= \cos y + i\sin y$$

Here we have used the Taylor series for $\cos y$ and $\sin y$ (Equations 10.10.16 and 10.10.15). The result is a famous formula called **Euler's formula:**

$$(6) \qquad \boxed{\; e^{iy} = \cos y + i\sin y \;}$$

Combining Euler's formula with Equation 5, we get

(7)
$$e^{x+iy} = e^x e^{iy} = e^x(\cos y + i \sin y)$$

EXAMPLE 8   Evaluate: (a) $e^{i\pi}$ (b) $e^{-1+i\pi/2}$

SOLUTION
(a) From Euler's equation (6) we have

$$e^{i\pi} = \cos \pi + i \sin \pi = -1 + i(0) = -1$$

(b) Using Equation 7 we get

$$e^{-1+i\pi/2} = e^{-1}\left(\cos \frac{\pi}{2} + i \sin \frac{\pi}{2}\right) = \frac{1}{e}[0 + i(1)] = \frac{i}{e}$$ ∎

Finally, we note that Euler's equation provides us with an easier method of proving De Moivre's Theorem:

$$[r(\cos \theta + i \sin \theta)]^n = (re^{i\theta})^n = r^n e^{in\theta} = r^n(\cos n\theta + i \sin n\theta)$$

EXERCISES H

1–14 ■ Evaluate the expression and write your answer in the form $a + bi$.

1. $(3 + 2i) + (7 - 3i)$ **2.** $(1 + i) - (2 - 3i)$

3. $(3 - i)(4 + i)$ **4.** $(4 - 7i)(1 + 3i)$

5. $\overline{12 + 7i}$ **6.** $\overline{2i(\frac{1}{2} - i)}$

7. $\dfrac{2 + 3i}{1 - 5i}$ **8.** $\dfrac{5 - i}{3 + 4i}$

9. $\dfrac{1}{1 + i}$ **10.** $\dfrac{3}{4 - 3i}$

11. i^3 **12.** i^{100}

13. $\sqrt{-25}$ **14.** $\sqrt{-3}\,\sqrt{-12}$

15–17 ■ Find the complex conjugate and the modulus of each number.

15. $3 + 4i$ **16.** $\sqrt{3} - i$ **17.** $-4i$

18. Prove the following properties of complex numbers.
  (a) $\overline{z + w} = \overline{z} + \overline{w}$
  (b) $\overline{zw} = \overline{z}\,\overline{w}$
  (c) $\overline{z^n} = \overline{z}^n$, where n is a positive integer
     [*Hint:* Write $z = a + bi$, $w = c + di$.]

19–24 ■ Find all solutions of the equation.

19. $4x^2 + 9 = 0$ **20.** $x^4 = 1$

21. $x^2 - 8x + 17 = 0$ **22.** $x^2 - 4x + 5 = 0$

23. $z^2 + z + 2 = 0$ **24.** $z^2 + \frac{1}{2}z + \frac{1}{4} = 0$

25–28 ■ Write the number in polar form with argument between 0 and 2π.

25. $-3 + 3i$ **26.** $1 - \sqrt{3}\,i$

27. $3 + 4i$ **28.** $8i$

29–32 ■ Find polar forms for zw, z/w, and $1/z$ by first putting z and w into polar form.

29. $z = \sqrt{3} + i$, $w = 1 + \sqrt{3}\,i$

30. $z = 4\sqrt{3} - 4i$, $w = 8i$

31. $z = 2\sqrt{3} - 2i$, $w = -1 + i$

32. $z = 4(\sqrt{3} + i)$, $w = -3 - 3i$

33–36 ■ Find the indicated power using De Moivre's Theorem.

33. $(1 + i)^{20}$ **34.** $(1 - \sqrt{3}\,i)^5$

35. $(2\sqrt{3} + 2i)^5$ **36.** $(1 - i)^8$

37–40 ■ Find the indicated roots. Sketch the roots in the complex plane.

37. The eighth roots of 1   **38.** The fifth roots of 32

39. The cube roots of i **40.** The cube roots of $1 + i$

41–46 ■ Write the number in the form $a + bi$.

41. $e^{i\pi/2}$ **42.** $e^{2\pi i}$ **43.** $e^{i3\pi/4}$

44. $e^{-i\pi}$ **45.** $e^{2+i\pi}$ **46.** e^{1+2i}

47. Use De Moivre's Theorem with $n = 3$ to express $\cos 3\theta$ and $\sin 3\theta$ in terms of $\cos \theta$ and $\sin \theta$.

48. Use Euler's formula to prove the following formulas for $\cos x$ and $\sin x$:

$$\cos x = \frac{e^{ix} + e^{-ix}}{2} \qquad \sin x = \frac{e^{ix} - e^{-ix}}{2i}$$

49. If $u(x) = f(x) + ig(x)$ is a complex-valued function of a real variable x and the real and imaginary parts $f(x)$ and $g(x)$ are differentiable functions of x, then the derivative of u is defined to be $u'(x) = f'(x) + ig'(x)$. Use this together with Equation 7 to prove that if $F(x) = e^{rx}$, then $F'(x) = re^{rx}$ when $r = a + bi$ is a complex number.

| | ANSWERS TO ODD-NUMBERED EXERCISES

CHAPTER 10

Exercises 10.1 ■ page 606

1. $\left\{\frac{1}{3}, \frac{2}{5}, \frac{3}{7}, \frac{4}{9}, \frac{5}{11}, \ldots\right\}$ **3.** $\left\{1, \frac{3}{2}, \frac{5}{2}, \frac{35}{8}, \frac{63}{8}, \ldots\right\}$

5. $\{1, 0, -1, 0, 1, \ldots\}$ **7.** $a_n = 1/2^n$ **9.** $a_n = 3n - 2$

11. $a_n = (-1)^{n+1}(3/2)^n$ **13.** 0 **15.** 1

17. Diverges (to ∞) **19.** 0 **21.** Diverges

23. Diverges (to ∞) **25.** 0 **27.** 0 **29.** 0 **31.** 0

33. 1 **35.** 0 **37.** $\frac{1}{2}$ **39.** Diverges (to ∞)

41. Diverges **43.** $\pi/4$ **45.** 0 **47.** 0

49. $-1 < r < 1$ **51.** Decreasing **53.** Increasing

55. Not monotonic **57.** Decreasing **59.** 2

61. $(3 + \sqrt{5})/2$ **63.** (b) $(1 + \sqrt{5})/2$

65. (a) 0 (b) 9, 11

71. (b) $b_n = (1 + 2\cos\theta)/[1 + 2\cos(\theta/2^n)]$

Exercises 10.2 ■ page 616

1. 3.33333, 4.44444, 4.81481, 4.93827, 4.97942, 4.99314, 4.99771, 4.99924, 4.99975, 4.99992
Convergent, sum = 5

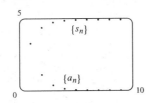

3. 0.50000, 1.16667, 1.91667, 2.71667, 3.55000, 4.40714, 5.28214, 6.17103, 7.07103, 7.98012
Divergent (terms do not approach 0)

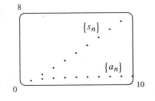

5. 0.64645, 0.80755, 0.87500, 0.91056, 0.93196, 0.94601, 0.95581, 0.96296, 0.96838, 0.97259
Convergent, sum = 1

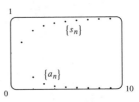

7. $\frac{20}{3}$ **9.** $\frac{1}{2}$ **11.** 8 **13.** $5e/(3 - e)$ **15.** $\frac{8}{3}$

17. Divergent **19.** Divergent **21.** $\frac{1}{3}$ **23.** $\frac{17}{36}$

25. Divergent **27.** $\frac{3}{4}$ **29.** $\frac{3}{2}$ **31.** $\sin 1$ **33.** Divergent

35. Divergent **37.** $\frac{5}{9}$ **39.** $\frac{307}{999}$ **41.** 41,111/333,000

43. $2 < x < 4$, $1/(4 - x)$ **45.** $-5 < x < 5$, $x^2/[5(5 - x)]$

47. $|x - n\pi| < \pi/6$, n any integer, $1/(1 - 2\sin x)$ **49.** $\frac{1}{4}$

51. $a_1 = 0$, $a_n = 2/[n(n + 1)]$ for $n > 1$, sum = 1

53. (a) $S_n = D(1 - c^n)/(1 - c)$ (b) 5 **55.** $(\sqrt{3} - 1)/2$

57. $1/[n(n + 1)]$ **59.** The series is divergent

65. $\{s_n\}$ is bounded and increasing

67. (a) $0, \frac{1}{9}, \frac{2}{9}, \frac{1}{3}, \frac{2}{3}, \frac{7}{9}, \frac{8}{9}, 1$

69. (a) $\frac{1}{2}, \frac{5}{6}, \frac{23}{24}, \frac{119}{120}$; $[(n + 1)! - 1]/(n + 1)!$ (c) 1

Exercises 10.3 ■ page 623

Abbreviations: C, convergent; D, divergent

1. D **3.** C **5.** C **7.** D **9.** C **11.** D **13.** D

15. C **17.** C **19.** $p > 1$ **21.** $p < -1$ **23.** $(1, \infty)$

25. (a) 1.54977, error $\leqslant$ 0.1 (b) 1.64522, error $\leqslant$ 0.005
(c) $n > 1000$

27. 2.6124 **31.** $b < 1/e$

Exercises 10.4 ■ page 628

1. C **3.** C **5.** D **7.** C **9.** D **11.** C **13.** C

15. C **17.** C **19.** D **21.** C **23.** D **25.** C

27. C **29.** C **31.** D **33.** 0.567975, error $\leqslant$ 0.000$\overline{3}$

35. 0.76352, error $<$ 0.001 **45.** Yes

Exercises 10.5 ■ page 633

1. C   **3.** D   **5.** C   **7.** D   **9.** C   **11.** C   **13.** D
15. C   **17.** C   **19.** D   **21.** $\{b_n\}$ is not decreasing
23. p is not a negative integer   **25.** 0.82
27. 0.13 (or 0.137)   **29.** 0.8415   **31.** 0.6065
33. An underestimate

Exercises 10.6 ■ page 639

Abbreviations: AC, absolutely convergent;
CC, conditionally convergent

1. AC   **3.** D   **5.** CC   **7.** AC   **9.** CC   **11.** D
13. AC   **15.** AC   **17.** AC   **19.** D   **21.** AC   **23.** D
25. AC   **27.** AC   **29.** D   **31.** AC   **33.** D
35. (a) and (d)
39. (a) $\frac{661}{960} \approx 0.68854$, error < 0.00521 (b) $n \geq 11$, 0.693109

Exercises 10.7 ■ page 642

1. C   **3.** C   **5.** C   **7.** C   **9.** C   **11.** D   **13.** C
15. C   **17.** C   **19.** C   **21.** D   **23.** D   **25.** C
27. D   **29.** C   **31.** C   **33.** C   **35.** C   **37.** C
39. C

Exercises 10.8 ■ page 647

1. (a) Yes   (b) No   **3.** 1, $[-1, 1)$ **5.** 1, $(-1, 1)$
7. ∞, $(-\infty, \infty)$ **9.** 2, $(-2, 2]$ **11.** $\frac{1}{3}$, $[-\frac{1}{3}, \frac{1}{3}]$
13. 1, $[-1, 1)$ **15.** 2, $(-\frac{3}{2}, \frac{5}{2})$ **17.** 1, $(0, 2]$
19. ∞, $(-\infty, \infty)$ **21.** 0.5, $[2.5, 3.5)$ **23.** 0, $\{-6\}$
25. $\frac{1}{2}$, $[0, 1]$ **27.** ∞, $(-\infty, \infty)$ **29.** k^k
31. (a) $(-\infty, \infty)$
(b), (c)

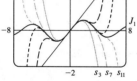

33. $(-1, 1)$, $f(x) = (1 + 2x)/(1 - x^2)$ **37.** 2

Exercises 10.9 ■ page 652

1. $\sum\limits_{n=0}^{\infty} (-1)^n x^n$, $(-1, 1)$ **3.** $\sum\limits_{n=0}^{\infty} (-1)^n 4^n x^{2n}$, $(-\frac{1}{2}, \frac{1}{2})$

5. $\sum\limits_{n=0}^{\infty} \dfrac{(-1)^n}{4^{n+1}} x^{2n}$, $(-2, 2)$ **7.** $-\sum\limits_{n=1}^{\infty} \left(\dfrac{x}{3}\right)^n$, $(-3, 3)$

9. $-\sum\limits_{n=0}^{\infty} (2^n + 1) x^n$, $(-\frac{1}{2}, \frac{1}{2})$ **11.** $\sum\limits_{n=0}^{\infty} (-1)^n (n + 1) x^n$, $R = 1$

13. $\frac{1}{2} \sum\limits_{n=0}^{\infty} (-1)^n (n + 2)(n + 1) x^n$, $R = 1$

15. $\ln 5 - \sum\limits_{n=1}^{\infty} \dfrac{x^n}{n5^n}$, $R = 5$ **17.** $\sum\limits_{n=0}^{\infty} \dfrac{2x^{2n+1}}{2n + 1}$, $R = 1$

19. $\ln 3 + \sum\limits_{n=1}^{\infty} \dfrac{(-1)^{n-1}}{n3^n} x^n$, $R = 3$
The partial sums approximate f
better (on the interval of
convergence).

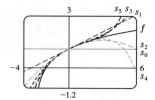

21. $C + \sum\limits_{n=0}^{\infty} \dfrac{(-1)^n x^{4n+1}}{4n + 1}$ **23.** $C + \sum\limits_{n=0}^{\infty} (-1)^n \dfrac{x^{2n+1}}{(2n + 1)^2}$
25. 0.199936   **27.** 0.000065   **29.** 0.09531
31. (b) 0.920   **35.** $[-1, 1]$, $[-1, 1)$, $(-1, 1)$

Exercises 10.10 ■ page 663

1. $\sum\limits_{n=0}^{\infty} (-1)^n \dfrac{x^{2n}}{(2n)!}$, $R = \infty$ **3.** $\sum\limits_{n=0}^{\infty} (-1)^n (n + 1) x^n$, $R = 1$

5. $\sum\limits_{n=0}^{\infty} \dfrac{x^{2n+1}}{(2n + 1)!}$, $R = \infty$

7. $\sum\limits_{n=0}^{\infty} \dfrac{(-1)^{n(n-1)/2}(x - \pi/4)^n}{\sqrt{2}\, n!}$, $R = \infty$

9. $\sum\limits_{n=0}^{\infty} (-1)^n (x - 1)^n$, $R = 1$ **11.** $\sum\limits_{n=0}^{\infty} \dfrac{e^3}{n!} (x - 3)^n$, $R = \infty$

17. $\sum\limits_{n=0}^{\infty} \dfrac{3^n x^n}{n!}$, $R = \infty$ **19.** $\sum\limits_{n=0}^{\infty} \dfrac{(-1)^n x^{2n+2}}{(2n)!}$, $R = \infty$

21. $\sum\limits_{n=0}^{\infty} \dfrac{(-1)^n x^{2n+2}}{2^{2n+1}(2n + 1)!}$, $R = \infty$

23. $\sum\limits_{n=1}^{\infty} \dfrac{(-1)^{n+1} 2^{2n-1} x^{2n}}{(2n)!}$, $R = \infty$

25. $\sum\limits_{n=0}^{\infty} \dfrac{(-1)^n x^{2n}}{(2n + 1)!}$, $R = \infty$

27. $1 + \dfrac{x}{2} + \sum\limits_{n=2}^{\infty} (-1)^{n-1} \dfrac{1 \cdot 3 \cdot 5 \cdot \cdots \cdot (2n - 3)}{2^n n!} x^n$, $R = 1$

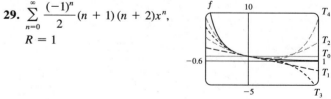

29. $\sum\limits_{n=0}^{\infty} \dfrac{(-1)^n}{2} (n + 1)(n + 2) x^n$,
$R = 1$

31. $\sum\limits_{n=1}^{\infty} (-1)^{n-1} \dfrac{x^n}{n}$, 0.09531   **33.** $C + \sum\limits_{n=0}^{\infty} \dfrac{(-1)^n x^{4n+3}}{(4n + 3)(2n + 1)!}$

35. $C + x + \dfrac{x^4}{8} + \sum\limits_{n=2}^{\infty} (-1)^{n-1} \dfrac{1 \cdot 3 \cdot 5 \cdot \cdots \cdot (2n - 3)}{2^n n! \, (3n + 1)} x^{3n+1}$

37. 0.310   **39.** 0.09998750   **41.** $1 - \frac{3}{2}x^2 + \frac{25}{4}x^4$

43. $-x + \frac{1}{2}x^2 - \frac{1}{3}x^3$ **45.** e^{-x^4} **47.** $1/\sqrt{2}$ **49.** $e^x - 1$

53. $1/120$

Exercises 10.11 ■ page 667

1. $1 + \dfrac{x}{2} + \displaystyle\sum_{n=2}^{\infty} (-1)^{n-1} \dfrac{1 \cdot 3 \cdot 5 \cdot \cdots \cdot (2n-3)}{2^n n!} x^n$, $R = 1$

3. $\displaystyle\sum_{n=0}^{\infty} (-1)^n \dfrac{(n+1)(n+2)(n+3)2^n}{6} x^n$, $R = \dfrac{1}{2}$

5. $x + \displaystyle\sum_{n=1}^{\infty} \dfrac{1 \cdot 3 \cdot 5 \cdot \cdots \cdot (2n-1)}{2^n n!} x^{n+1}$, $R = 1$

7. $1 - \dfrac{x^4}{4} - \displaystyle\sum_{n=2}^{\infty} \dfrac{3 \cdot 7 \cdot 11 \cdot \cdots \cdot (4n-5)}{4^n n!} x^{4n}$, $R = 1$

9. $\displaystyle\sum_{n=0}^{\infty} \dfrac{(n+4)!}{4! \cdot n!} x^{n+5}$, $R = 1$

11. $\dfrac{1}{2} + \dfrac{1}{2}\displaystyle\sum_{n=1}^{\infty} \dfrac{(-1)^n 1 \cdot 4 \cdot 7 \cdot \cdots \cdot (3n-2)}{24^n n!} x^n$, $R = 8$

13. (a) $1 + \displaystyle\sum_{n=1}^{\infty} \dfrac{1 \cdot 3 \cdot 5 \cdot \cdots \cdot (2n-1)}{2^n n!} x^{2n}$

(b) $x + \displaystyle\sum_{n=1}^{\infty} \dfrac{1 \cdot 3 \cdot 5 \cdot \cdots \cdot (2n-1)}{2^n n!} \dfrac{x^{2n+1}}{2n+1}$

15. (a) $1 + \displaystyle\sum_{n=1}^{\infty} (-1)^n \dfrac{1 \cdot 3 \cdot 5 \cdot \cdots \cdot (2n-1)}{2^n n!} x^n$ (b) 0.953

17. (a) $\displaystyle\sum_{n=1}^{\infty} n x^n$ (b) 2

19. (a) $1 + \dfrac{x^2}{2} + \displaystyle\sum_{n=2}^{\infty} (-1)^{n-1} \dfrac{1 \cdot 3 \cdot 5 \cdot \cdots \cdot (2n-3)}{2^n n!} x^{2n}$

(b) 99,225

Exercises 10.12 ■ page 673

1. $\dfrac{1}{2} + \dfrac{\sqrt{3}}{2}\left(x - \dfrac{\pi}{6}\right) - \dfrac{1}{4}\left(x - \dfrac{\pi}{6}\right)^2 - \dfrac{\sqrt{3}}{12}\left(x - \dfrac{\pi}{6}\right)^3$

3. $x + \dfrac{1}{3}x^3$ **5.** $x + x^2 + \dfrac{1}{3}x^3$

7. $\dfrac{1}{2} - \dfrac{1}{48}(x-8) + \dfrac{1}{576}(x-8)^2 - \dfrac{7}{41,472}(x-8)^3$

9. $T_1(x) = 1$, $T_2(x) = 1 - \frac{1}{2}x^2$,
$T_3(x) = 1 - \frac{1}{2}x^2$,
$T_4(x) = 1 - \frac{1}{2}x^2 + \frac{1}{24}x^4$

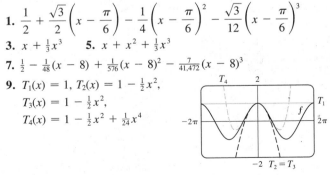

11. $T_8(x) = 1 + \frac{1}{2}x^2 + \frac{5}{24}x^4 + \frac{61}{720}x^6 + \frac{277}{8064}x^8$

13. (a) $1 + \frac{1}{2}x$ (b) 0.00125

15. (a) $\dfrac{1}{\sqrt{2}} + \dfrac{1}{\sqrt{2}}\left(x - \dfrac{\pi}{4}\right) - \dfrac{1}{2\sqrt{2}}\left(x - \dfrac{\pi}{4}\right)^2 -$

$\dfrac{1}{6\sqrt{2}}\left(x - \dfrac{\pi}{4}\right)^3 + \dfrac{1}{24\sqrt{2}}\left(x - \dfrac{\pi}{4}\right)^4 + \dfrac{1}{120\sqrt{2}}\left(x - \dfrac{\pi}{4}\right)^5$

(b) 0.00033   **17.** (a) $x + \frac{1}{3}x^3$ (b) 0.06

19. (a) $1 + x^2$ (b) 0.00006

21. (a) $8 + \frac{3}{8}(x - 16) - \frac{3}{1024}(x-16)^2 + \frac{5}{65,536}(x-16)^3$

(b) 0.0000034   **23.** 0.57358   **25.** 3

27. $-1.037 < x < 1.037$ **29.** 21 m, no

Review Exercises for Chapter 10 ■ page 675

1. False   **3.** False   **5.** False   **7.** False   **9.** False

11. True   **13.** True   **15.** False   **17.** True

19. C, $\frac{1}{2}$ **21.** D   **23.** D   **25.** C, e^{12} **27.** 2

29. D   **31.** C   **33.** C   **35.** C   **37.** C   **39.** D

41. CC   **43.** AC   **45.** 8   **47.** $\pi/4$ **49.** $\frac{4111}{3330}$

51. 0.9721   **53.** 0.18976224, error $< 6.4 \times 10^{-7}$

57. 3, $[-3, 3]$ **59.** 0.5, $[2.5, 3.5)$

61. $\dfrac{1}{2} + \dfrac{\sqrt{3}}{2}\left(x - \dfrac{\pi}{6}\right) - \dfrac{1}{2}\dfrac{1}{2!}\left(x - \dfrac{\pi}{6}\right)^2$

$- \dfrac{\sqrt{3}}{2}\dfrac{1}{3!}\left(x - \dfrac{\pi}{6}\right)^3 + \cdots$

$= \dfrac{1}{2}\displaystyle\sum_{n=0}^{\infty} (-1)^n\left[\dfrac{1}{(2n)!}\left(x - \dfrac{\pi}{6}\right)^{2n} + \dfrac{\sqrt{3}}{(2n+1)!}\left(x - \dfrac{\pi}{6}\right)^{2n+1}\right]$

63. $\displaystyle\sum_{n=0}^{\infty} (-1)^n x^{n+2}$, 1   **65.** $-\displaystyle\sum_{n=1}^{\infty} \dfrac{x^n}{n}$, 1

67. $\displaystyle\sum_{n=0}^{\infty} (-1)^n \dfrac{x^{8n+4}}{(2n+1)!}$, ∞

69. $\dfrac{1}{2} + \displaystyle\sum_{n=1}^{\infty} \dfrac{1 \cdot 5 \cdot 9 \cdot \cdots \cdot (4n-3)}{n! \, 2^{6n+1}} x^n$, 16

71. $\ln|x| + C + \displaystyle\sum_{n=1}^{\infty} \dfrac{x^n}{n \cdot n!}$

73. (a) $1 + \frac{1}{2}(x-1) - \frac{1}{8}(x-1)^2 + \frac{1}{16}(x-1)^3$

(b)   (c) 0.000006   **75.** 1

PROBLEMS PLUS ■ page 678

1. $15!/5! = 10{,}897{,}286{,}400$

3. (b) 0 if $x = 0$, $(1/x) - \cot x$ if $x \neq n\pi$, n an integer

5. (a) $s_n = 3 \cdot 4^n$, $l_n = 1/3^n$, $p_n = 4^n/3^{n-1}$ (c) $2\sqrt{3}/5$

7. $2\pi/3 - \sqrt{3}/2$ **11.** $(-1,1)$, $(x^3 + 4x^2 + x)/(1 - x)^4$

25. (b)

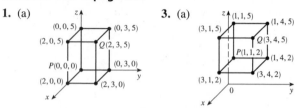

29. 100 **31.** $\frac{7}{4}$ **33.** (b) 1 (c) $\sqrt{3}$

CHAPTER 11

Exercises 11.1 ■ page 688

1. (a)

3. (a)

(b) $|PQ| = \sqrt{38}$ (b) $|PQ| = \sqrt{22}$

5. $|AB| = \sqrt{21}$, $|BC| = \sqrt{6}$, $|CA| = 3\sqrt{3}$; right triangle

7. $|AB| = \sqrt{69}$, $|BC| = \sqrt{158}$, $|CA| = 13$; neither

9. Not collinear   **11.** $x^2 + (y - 1)^2 + (z + 1)^2 = 16$

13. $(x + 6)^2 + (y + 1)^2 + (z - 2)^2 = 12$ **15.** $(-1, -4, 2)$, 7

17. $(-\frac{1}{2}, 1, -3)$, $\frac{7}{2}$ **19.** $(\frac{1}{2}, 0, 0)$, $\frac{1}{2}$ **23.** $\sqrt{85}/2$, $\frac{5}{2}$, $\sqrt{94}/2$

25. (a) $(x - 2)^2 + (y + 3)^2 + (z - 6)^2 = 36$

(b) $(x - 2)^2 + (y + 3)^2 + (z - 6)^2 = 4$

(c) $(x - 2)^2 + (y + 3)^2 + (z - 6)^2 = 9$

27. $14x - 6y - 10z = 9$, a plane perpendicular to AB

29. A plane parallel to the yz-plane and 9 units in front of it

31. A half-space consisting of all points to the right of the plane $y = 2$

33. A plane perpendicular to the xz-plane and intersecting it in the line $x = z$, $y = 0$

35. Circular cylinder, radius 1, axis the z-axis

37. All points outside the sphere with radius 1 and center O

39. All points inside the sphere with radius 2 and center $(0, 0, 1)$ **41.** Hyperbolic cylinder

43. All points on and between the horizontal planes $z = 2$ and $z = -2$

45. $y < 0$ **47.** $r^2 < x^2 + y^2 + z^2 < R^2$

49. (a) $(2, 1, 4)$ (b)

Exercises 11.2 ■ page 695

1. $\mathbf{a} = \langle 3, 1 \rangle$ **3.** $\mathbf{a} = \langle 0, -2 \rangle$

5. $\mathbf{a} = \langle 2, 0, -2 \rangle$ **7.** $\langle 5, -1 \rangle$

9. $\langle 1, 0, 2 \rangle$ **11.** 13, $\langle 3, -4 \rangle$, $\langle 7, -20 \rangle$, $\langle 10, -24 \rangle$, $\langle 7, -4 \rangle$

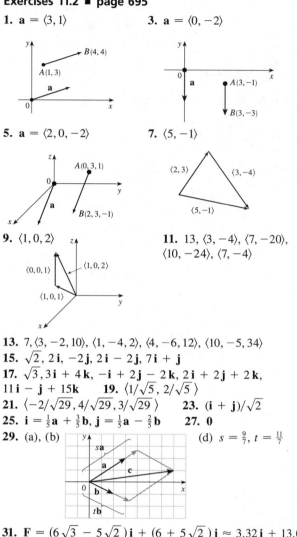

13. 7, $\langle 3, -2, 10 \rangle$, $\langle 1, -4, 2 \rangle$, $\langle 4, -6, 12 \rangle$, $\langle 10, -5, 34 \rangle$

15. $\sqrt{2}$, $2\mathbf{i}$, $-2\mathbf{j}$, $2\mathbf{i} - 2\mathbf{j}$, $7\mathbf{i} + \mathbf{j}$

17. $\sqrt{3}$, $3\mathbf{i} + 4\mathbf{k}$, $-\mathbf{i} + 2\mathbf{j} - 2\mathbf{k}$, $2\mathbf{i} + 2\mathbf{j} + 2\mathbf{k}$, $11\mathbf{i} - \mathbf{j} + 15\mathbf{k}$ **19.** $\langle 1/\sqrt{5}, 2/\sqrt{5} \rangle$

21. $\langle -2/\sqrt{29}, 4/\sqrt{29}, 3/\sqrt{29} \rangle$ **23.** $(\mathbf{i} + \mathbf{j})/\sqrt{2}$

25. $\mathbf{i} = \frac{1}{5}\mathbf{a} + \frac{3}{5}\mathbf{b}$, $\mathbf{j} = \frac{1}{5}\mathbf{a} - \frac{2}{5}\mathbf{b}$ **27.** $\mathbf{0}$

29. (a), (b)   (d) $s = \frac{9}{7}$, $t = \frac{11}{7}$

31. $\mathbf{F} = (6\sqrt{3} - 5\sqrt{2})\mathbf{i} + (6 + 5\sqrt{2})\mathbf{j} \approx 3.32\mathbf{i} + 13.07\mathbf{j}$, $|\mathbf{F}| \approx 13.5$ lb, $\theta \approx 76°$ **33.** $\sqrt{493} \approx 22.2$ mi/h N8°W

35. A sphere with radius 1, centered at (x_0, y_0, z_0)

Exercises 11.3 ■ page 700

1. -1 **3.** -5 **5.** -11 **7.** 3 **11.** $\cos^{-1}\frac{11}{15} \approx 43°$

13. $\cos^{-1}(2/(13\sqrt{5})) \approx 86°$ **15.** $\cos^{-1}(1/(7\sqrt{3})) \approx 85°$

17. $114°$, $33°$, $33°$ **19.** Parallel   **21.** Neither

23. Orthogonal   **25.** ± 4 **27.** $-\frac{4}{5}$

29. $(\mathbf{i} - \mathbf{j} - \mathbf{k})/\sqrt{3}$ [or $(-\mathbf{i} + \mathbf{j} + \mathbf{k})/\sqrt{3}$]

31. $\frac{1}{3}$, $\frac{2}{3}$, $\frac{2}{3}$; $71°$, $48°$, $48°$

33. $-8/\sqrt{77}$, $3/\sqrt{77}$, $2/\sqrt{77}$; $156°$, $70°$, $77°$

35. $5/\sqrt{38}$, $3/\sqrt{38}$, $2/\sqrt{38}$; $36°$, $61°$, $71°$

37. $11/\sqrt{13}$, $\langle 22/13, 33/13 \rangle$ **39.** $3/\sqrt{5}$, $\langle 6/5, 3/5, 0 \rangle$

41. $1/\sqrt{2}$, $(\mathbf{i} + \mathbf{k})/2$

45. $\langle 0, 0, -2\sqrt{10} \rangle$ or any vector of the form $\langle s, t, 3s - 2\sqrt{10} \rangle$, $s, t \in \mathbb{R}$ **47.** 38 J   **49.** $250 \cos 20° \approx 235$ ft-lb

51. (a), (e), (f)   **53.** $\frac{13}{5}$ **55.** $\cos^{-1}(1/\sqrt{3}) \approx 55°$

Exercises 11.4 ■ page 707

1. $\langle -1, 0, 1 \rangle$ **3.** $\langle 3, 14, -9 \rangle$ **5.** $-2\mathbf{i} + 2\mathbf{j}$
7. $2\mathbf{i} - \mathbf{j} + 4\mathbf{k}$ **9.** $\langle -2, 6, -3 \rangle$, $\langle 2, -6, 3 \rangle$
11. $\langle -2/\sqrt{6}, -1/\sqrt{6}, 1/\sqrt{6} \rangle$, $\langle 2/\sqrt{6}, 1/\sqrt{6}, -1/\sqrt{6} \rangle$ **19.** 4
21. (a) $\langle 6, 3, 2 \rangle$ (b) $\frac{7}{2}$ **23.** (a) $\langle -10, -3, 7 \rangle$ (b) $\sqrt{158}/2$
25. 226   **27.** 21   **31.** $10.8 \sin 100° \approx 10.6$ J
33. (b) $\sqrt{97/3}$ **39.** (a) No   (b) No   (c) Yes

Exercises 11.5 ■ page 716

1. $\mathbf{r} = \langle 3, -1, 8 \rangle + t\langle 2, 3, 5 \rangle$,
$x = 3 + 2t$, $y = -1 + 3t$, $z = 8 + 5t$
3. $\mathbf{r} = (\mathbf{j} + 2\mathbf{k}) + t(6\mathbf{i} + 3\mathbf{j} + 2\mathbf{k})$,
$x = 6t$, $y = 1 + 3t$, $z = 2 + 2t$
5. $x = 2 + 4t$, $y = 1 - t$, $z = 8 - 5t$;
$(x - 2)/4 = (y - 1)/(-1) = (z - 8)/(-5)$
7. $x = 3$, $y = 1 + t$, $z = -1 - 5t$;
$x = 3$, $y - 1 = (z + 1)/(-5)$
9. $x = (-1 + t)/3$, $y = 1 + 4t$, $z = 1 - 9t$;
$(x + \frac{1}{3})/(\frac{1}{3}) = (y - 1)/4 = (z - 1)/(-9)$
13. (a) $x/2 = (y - 2)/3 = (z + 1)/(-7)$
(b) $(-\frac{2}{7}, \frac{11}{7}, 0)$, $(-\frac{4}{3}, 0, \frac{11}{3})$, $(0, 2, -1)$
15. Skew   **17.** Parallel
19. $7x + y + 4z = 31$ **21.** $5x + 3y - 4z + 1 = 0$
23. $x + y - z = 13$ **25.** $3x - 4y - 6z = 33$
27. $x - 2y + z = 0$ **29.** $17x - 6y - 5z = 32$
31. $25x + 14y + 8z = 77$ **33.** $x - 2y + z = 0$
35. $(1, 0, 0)$ **37.** $(-3, -1, -2)$ **39.** $1, 0, -1$
41. Neither, $60°$ **43.** Perpendicular   **45.** Parallel
47. (a) $x - 2 = y/(-8) = z/(-7)$
(b) $\cos^{-1}(-\sqrt{6}/5) \approx 119°$ (or $61°$)
49. $x = 6t$, $y = -\frac{1}{6} + t$, $z = -\frac{1}{6} + 7t$ **51.** $x = z$
53. $x - 2y + 4z + 1 = 0$ **55.** $(x/a) + (y/b) + (z/c) = 1$
57. $x = 3t$, $y = 1 - t$, $z = 2 - 2t$
59. P_1 and P_3 are parallel, P_2 and P_4 are identical
61. $\sqrt{22/5}$ **63.** $\frac{25}{3}$ **65.** $7\sqrt{6}/18$ **69.** $1/\sqrt{6}$

Exercises 11.6 ■ page 722

1. $x = k$, $z^2 - y^2 = 1 - k^2$, hyperbola
$y = k$, $x^2 + z^2 = 1 + k^2$, circle
$z = k$, $x^2 - y^2 = 1 - k^2$,
hyperbola
Hyperboloid of one sheet
with axis the y-axis

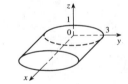

3. $x = k$, $y^2 + 4z^2 = 4 - 4k^2/9$, ellipse $(|k| < 3)$
$y = k$, $x^2 + 9z^2 = 9 - 9k^2/4$, ellipse $(|k| < 2)$
$z = k$, $4x^2 + 9y^2 = 36(1 - k^2)$, ellipse $(|k| < 1)$
Ellipsoid

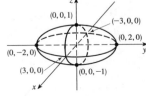

5. $x = k$, $4z^2 - y^2 = 1 + k^2$, hyperbola
$y = k$, $4z^2 - x^2 = 1 + k^2$, hyperbola
$z = k$, $x^2 + y^2 = 4k^2 - 1$, circle $(|k| > \frac{1}{2})$
Hyperboloid of two sheets

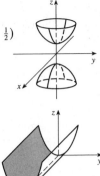

7. $x = k$, $z = y^2$, parabola
$y = k$, $z = k^2$, line
$z = k$, $y = \pm\sqrt{k}$, lines $(k > 0)$
Parabolic cylinder

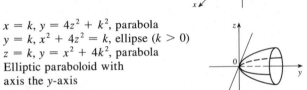

9. $x = k$, $y^2 - z^2 = k^2$, hyperbola $(k \neq 0)$
$y = k$, $x^2 + z^2 = k^2$, circle $(k \neq 0)$
$z = k$, $y^2 - x^2 = k^2$,
hyperbola $(k \neq 0)$
$x = 0$, $y = \pm z$, lines
Cone

11. $x = k$, $y = 4z^2 + k^2$, parabola
$y = k$, $x^2 + 4z^2 = k$, ellipse $(k > 0)$
$z = k$, $y = x^2 + 4k^2$, parabola
Elliptic paraboloid with
axis the y-axis

13. $x = k$, $y^2 + 9z^2 = 9$, ellipse
$y = k$, $z = \pm\sqrt{1 - (k^2/9)}$, pairs of lines $(|k| < 3)$
$z = k$, $y = \pm 3\sqrt{1 - k^2}$, pairs of lines $(|k| < 1)$
Elliptic cylinder with axis the x-axis

15. $x = k$, $y = z^2 - k^2$, parabola
$y = k$, $z^2 - x^2 = k$, hyperbola $(k \neq 0)$
$z = k$, $y = k^2 - x^2$, parabola
Hyperbolic paraboloid

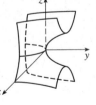

17. VII   **19.** II   **21.** VI   **23.** VIII
25. $(x^2/4) + (y^2/3) - (z^2/12) = 1$
Hyperboloid of one sheet with
axis the z-axis

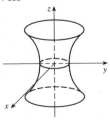

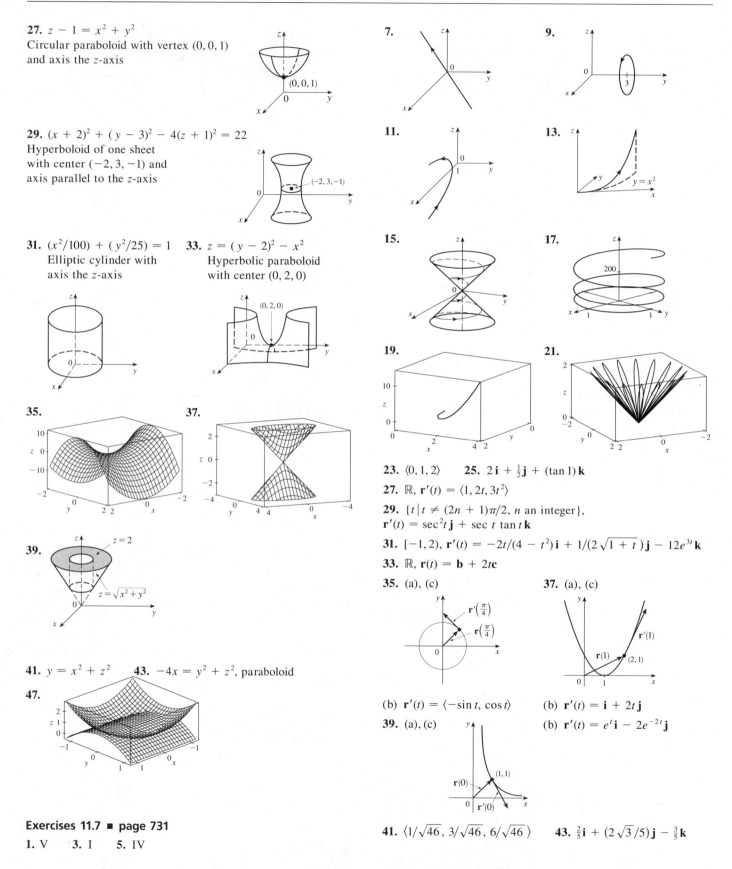

27. $z - 1 = x^2 + y^2$
Circular paraboloid with vertex $(0, 0, 1)$
and axis the z-axis

29. $(x + 2)^2 + (y - 3)^2 - 4(z + 1)^2 = 22$
Hyperboloid of one sheet
with center $(-2, 3, -1)$ and
axis parallel to the z-axis

31. $(x^2/100) + (y^2/25) = 1$
Elliptic cylinder with
axis the z-axis

33. $z = (y - 2)^2 - x^2$
Hyperbolic paraboloid
with center $(0, 2, 0)$

35.

37.

39.

41. $y = x^2 + z^2$ **43.** $-4x = y^2 + z^2$, paraboloid

47.

Exercises 11.7 ■ page 731

1. V   **3.** I   **5.** IV

7.

9.

11.

13.

15.

17.

19.

21.

23. $\langle 0, 1, 2 \rangle$ **25.** $2\mathbf{i} + \frac{1}{2}\mathbf{j} + (\tan 1)\mathbf{k}$

27. $\mathbb{R}$, $\mathbf{r}'(t) = \langle 1, 2t, 3t^2 \rangle$

29. $\{t \mid t \neq (2n + 1)\pi/2, n \text{ an integer}\}$,
$\mathbf{r}'(t) = \sec^2 t\,\mathbf{j} + \sec t \tan t\,\mathbf{k}$

31. $[-1, 2)$, $\mathbf{r}'(t) = -2t/(4 - t^2)\mathbf{i} + 1/(2\sqrt{1 + t})\mathbf{j} - 12e^{3t}\mathbf{k}$

33. $\mathbb{R}$, $\mathbf{r}(t) = \mathbf{b} + 2t\mathbf{c}$

35. (a), (c)

37. (a), (c)

(b) $\mathbf{r}'(t) = \langle -\sin t, \cos t \rangle$
39. (a), (c)

(b) $\mathbf{r}'(t) = \mathbf{i} + 2t\,\mathbf{j}$
(b) $\mathbf{r}'(t) = e^t\mathbf{i} - 2e^{-2t}\mathbf{j}$

41. $\langle 1/\sqrt{46}, 3/\sqrt{46}, 6/\sqrt{46} \rangle$ **43.** $\frac{2}{5}\mathbf{i} + (2\sqrt{3}/5)\mathbf{j} - \frac{3}{5}\mathbf{k}$

45. $x = 1 + t$, $y = 1 + 2t$, $z = 1 + 3t$

47. $x = -(\pi/2)t$, $y = \frac{1}{4} + t$, $z = 1 + 4t$

49. $x = (\pi/4) + t$, $y = 1 - t$, $z = 1 + t$ **51.** $66°$

55. $x = 2 \cos t$, $y = 2 \sin t$, $z = 4 \cos^2 t$ **57.** $\frac{1}{2}\mathbf{i} + \frac{1}{3}\mathbf{j} + \frac{1}{4}\mathbf{k}$

59. $\frac{1}{2}\mathbf{i} + \frac{1}{2}\mathbf{j} + (4 - \pi)/(4\sqrt{2})\,\mathbf{k}$

61. $(t^3/3)\mathbf{i} + (t^4 + 1)\mathbf{j} - (t^3/3)\mathbf{k}$

69. $1 - 4t\cos t + 11t^2 \sin t + 3t^3 \cos t$

Exercises 11.8 ■ page 739

1. $\sqrt{13}\,(b - a)$ **3.** 8 **5.** 9.5706

7. $\mathbf{r}(t(s)) = (1 + s/\sqrt{2})\sin[\ln(1 + s/\sqrt{2})]\,\mathbf{i}$
$+ (1 + s/\sqrt{2})\cos[\ln(1 + s/\sqrt{2})]\,\mathbf{j}$

9. $\mathbf{r}(t(s)) = 3\sin(s/5)\mathbf{i} + (4s/5)\mathbf{j} + 3\cos(s/5)\mathbf{k}$

11. (a) $\langle \frac{4}{5}\cos 4t, \frac{3}{5}, -\frac{4}{5}\sin 4t \rangle$, $\langle -\sin 4t, 0, -\cos 4t \rangle$ (b) $\frac{16}{25}$

13. (a) $\langle -\sqrt{2}\sin t, \cos t, \cos t \rangle/\sqrt{2}$,
$\langle -\sqrt{2}\cos t, -\sin t, -\sin t \rangle/\sqrt{2}$ (b) $1/\sqrt{2}$

15. $2/(4t^2 + 1)^{3/2}$

17. $\sqrt{t^4 + 4t^2 + 1}/[6(t^4 + t^2 + 1)^{3/2}]$ **19.** $\sqrt{2}/(1 + \cos^2 t)^{3/2}$

21. $6|x|/(1 + 9x^4)^{3/2}$ **23.** $|\sin x|/(1 + \cos^2 x)^{3/2}$

25. $(-\frac{1}{2}\ln 2, 1/\sqrt{2})$

27. $y = \kappa(x)$ 2.5   **29.** $6/[\,|t|(9t^2 + 4)^{3/2}]$

$y = x^4$

-1.2 1.2

-0.5

31. $\langle \frac{2}{3}, \frac{2}{3}, \frac{1}{3} \rangle$, $\langle -\frac{1}{3}, \frac{2}{3}, -\frac{2}{3} \rangle$, $\langle -\frac{2}{3}, \frac{1}{3}, \frac{2}{3} \rangle$

33. $y = 6x + \pi$, $x + 6y = 6\pi$

35. $(x + \frac{5}{2})^2 + y^2 = \frac{81}{4}$,
$x^2 + (y - \frac{5}{3})^2 = \frac{16}{9}$

    5

-7.5 2.5

-5

37. $(-1, -3, 1)$ **47.** $2/(t^4 + 4t^2 + 1)$ **49.** 2 m

Exercises 11.9 ■ page 747

1. $\mathbf{v}(t) = \langle 2t, 1 \rangle$ **3.** $\mathbf{v}(t) = e^t\mathbf{i} - e^{-t}\mathbf{j}$
$\mathbf{a}(t) = \langle 2, 0 \rangle$ $\mathbf{a}(t) = e^t\mathbf{i} + e^{-t}\mathbf{j}$
$|\mathbf{v}(t)| = \sqrt{4t^2 + 1}$ $|\mathbf{v}(t)| = \sqrt{e^{2t} + e^{-2t}}$

5. $\mathbf{v}(t) = \cos t\,\mathbf{i} + \mathbf{j} - \sin t\,\mathbf{k}$
$\mathbf{a}(t) = -\sin t\,\mathbf{i} - \cos t\,\mathbf{k}$
$|\mathbf{v}(t)| = \sqrt{2}$

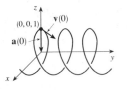

7. $\langle 3t^2, 2t, 3t^2 \rangle$, $\langle 6t, 2, 6t \rangle$, $\sqrt{18t^4 + 4t^2}$

9. $-t^{-2}\mathbf{i} + 2t\mathbf{k}$, $2t^{-3}\mathbf{i} + 2\mathbf{k}$, $\sqrt{t^{-4} + 4t^2}$

11. $e^t[(\cos t - \sin t)\mathbf{i} + (\sin t + \cos t)\mathbf{j} + (t + 1)\mathbf{k}]$,
$e^t[-2\sin t\,\mathbf{i} + 2\cos t\,\mathbf{j} + (t + 2)\mathbf{k}]$, $e^t\sqrt{t^2 + 2t + 3}$

13. $\mathbf{v}(t) = \mathbf{i} - \mathbf{j} + t\mathbf{k}$, $\mathbf{r}(t) = t\mathbf{i} - t\mathbf{j} + (t^2/2)\mathbf{k}$

15. (a) $\mathbf{r}(t) = (1 + t^2/2)\mathbf{i} + t^2\mathbf{j} + (1 + t^3/3)\mathbf{k}$

(b)

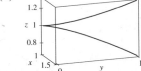

17. $t = 4$ **19.** $\mathbf{r}(t) = t\mathbf{i} - t\mathbf{j} + \frac{5}{2}t^2\mathbf{k}$, $|\mathbf{v}(t)| = \sqrt{25t^2 + 2}$

21. (a) 22 km   (b) 3.2 km   (c) 500 m/s   **23.** 30 m/s

27. $(\sin t)/\sqrt{2(1 - \cos t)}$, $\sqrt{(1 - \cos t)/2}$

29. $(18t^3 + 4t)/\sqrt{9t^4 + 4t^2 + 1}$,
$2\sqrt{9t^4 + 9t^2 + 1}/\sqrt{9t^4 + 4t^2 + 1}$ **31.** $e^t - e^{-t}$, $\sqrt{2}$

33. $t = 1$ **37.** 2.99×10^8 km

Exercises 11.10 ■ page 753

1. $(0, 3, 1)$ **3.** $(-1, -\sqrt{3}, 8)$ **5.** $(3, 0, -6)$ **7.** $(1, \pi, 0)$

9. $(2, \pi/6, 4)$ **11.** $(4\sqrt{2}, \pi/4, 4)$ **13.** $(0, 0, 1)$

15. $(\sqrt{3}/4, \frac{1}{4}, \sqrt{3}/2)$ **17.** $(\sqrt{2}, \sqrt{2}, 2\sqrt{3})$ **19.** $(3, \pi, \pi/2)$

21. $(2, 0, \pi/3)$ **23.** $(2, 7\pi/4, 3\pi/4)$ **25.** $(\sqrt{2}, \pi/4, \pi/2)$

27. $(4\sqrt{2}, \pi/3, \pi/4)$ **29.** $(0, 0, 2)$ **31.** $(8, \pi/6, 0)$

33. Cylinder with radius 3   **35.** Half-cone

37. Circular paraboloid   **39.** Horizontal plane

41. Positive z-axis

43. Circular cylinder, radius 1, axis parallel to the z-axis

45. Sphere, radius 5, center the origin

47. Circular cylinder, radius 2, axis the y-axis

49. Cylinder, radius 1, together with the z-axis

51. (a) $r^2 + z^2 = 16$ (b) $\rho = 4$

53. (a) $r\cos\theta + 2r\sin\theta + 3z = 6$
(b) $\rho(\sin\phi\cos\theta + 2\sin\phi\sin\theta + 3\cos\phi) = 6$

55. (a) $2z^2 = r^2\cos 2\theta - 4$
(b) $\rho^2(\sin^2\phi\cos 2\theta - 2\cos^2\phi) = 4$

57. (a) $r = 2\sin\theta$ (b) $\rho\sin\phi = 2\sin\theta$

59.   **61.**

63. $0 \le \phi \le \pi/4, 0 \le \rho \le \cos\phi$ **65.**

Review Exercises for Chapter 11 ■ page 754

1. True **3.** True **5.** True **7.** True **9.** True
11. False **13.** False
15. $|AB| = 13, |BC| = \sqrt{38}, |AC| = \sqrt{65}$
17. $(-2, -3, 5), 6$ **19.** $6\mathbf{i} + \mathbf{j} + 13\mathbf{k}$ **21.** -1
23. $3\sqrt{35}$ **25.** 0 **27.** $96°$ **29.** $-1/\sqrt{6}$ **31.** 4
33. (a) 2 (b) -2 (c) -2 (d) 0
35. If $\vec{PR}$ is a scalar multiple of $\vec{PQ}$ (or if $\vec{PQ} \times \vec{PR} = \mathbf{0}$),
then P, Q, R are collinear.
37. $\cos^{-1}\frac{1}{3} \approx 71°$ (or $109°$) **39.** (a) $\langle 4, -3, 4 \rangle$ (b) $\sqrt{41}/2$
41. 166 N, 114 N **43.** $x = 1 + 2t, y = 2 - t, z = 4 + 3t$
45. $x = 1 + 4t, y = -3t, z = 1 + 5t$ **47.** $x + 2y + 5z = 8$
49. $(1, 4, 4)$ **51.** Skew **53.** $22/\sqrt{26}$
55. Plane **57.** Circular paraboloid

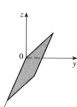 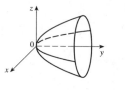

59. Cone **61.** Hyperboloid of two sheets

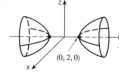 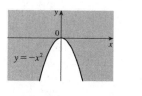

63. $4x^2 + y^2 + z^2 = 16$
65. (a) (b) $\cos t\,\mathbf{j} - \sin t\,\mathbf{k}$, $-\sin t\,\mathbf{j} - \cos t\,\mathbf{k}$

67. $\frac{5}{6}\mathbf{i} + \frac{9}{4}\mathbf{j} + \frac{1}{5}\mathbf{k}$ **69.** 15.9241 **71.** $\pi/2$
73. (a) $\langle t^2, t, 1 \rangle/\sqrt{t^4 + t^2 + 1}$
(b) $\langle 2t, 1 - t^4, -2t^3 - t \rangle/\sqrt{t^8 + 4t^6 + 2t^4 + 5t^2}$
(c) $\sqrt{t^8 + 4t^6 + 2t^4 + 5t^2}/(t^4 + t^2 + 1)^2$ **75.** $12/17^{3/2}$
77. $x - 2y + 2\pi = 0$ **79.** $\langle t + t^3/6, 2t + t^2/2, t + t^4/12 \rangle$
81. $(\sqrt{3}, 1, 2), (2\sqrt{2}, \pi/6, \pi/4)$
83. $(1, \sqrt{3}, 2\sqrt{3}), (2, \pi/3, 2\sqrt{3})$ **85.** A half-plane
87. The horizontal plane $z = 3$ **89.** $r^2 + z^2 = 4, \rho = 2$
91. $z = 4r^2$ **93.** (c) $-2e^{-t}\mathbf{v}_d + e^{-t}\mathbf{R}$

APPLICATIONS PLUS ■ page 757

1. (a) $\mathbf{v} = \omega R(-\sin\omega t\,\mathbf{i} + \cos\omega t\,\mathbf{j})$ (c) $\mathbf{a} = -\omega^2\mathbf{r}$
3. (a) $90°, v_0^2/(2g)$
5. (b) $\mathbf{R}(t) = (m/k)(1 - e^{-kt/m})\mathbf{v}_0 + (gm/k)[(m/k)(1 - e^{-kt/m}) - t]\mathbf{j}$
7. $3x^5 - 8x^4 + 6x^3$, no

CHAPTER 12

Exercises 12.1 ■ page 768

1. (a) 7 (b) -45
(c) $x^2 + 2xh + h^2 - y^2 + 4xy + 4hy - 7x - 7h + 10$
(d) $x^2 - y^2 - 2yk - k^2 + 4xy + 4xk - 7x + 10$
(e) $4x^2 - 7x + 10$
3. (a) 1 (b) $-\frac{2}{3}$ (c) $3t/(t^2 + 2)$
(d) $-3y/(1 + 2y^2)$ (e) $3x/(1 + 2x^2)$
5. $\mathbb{R}^2; \mathbb{R}$ **7.** $\{(x, y) \mid x + y \ne 0\}; \{z \mid z \ne 0\}$
9. $\mathbb{R}^2; \{z \mid z > 0\}$ **11.** $\{(x, y, z) \mid x - y + z > 0\}; \mathbb{R}$
13. $\mathbb{R}^3; \mathbb{R}$
15. $\{(x, y) \mid y \ge 2x\}$ **17.** $\{(x, y) \mid x^2 + y^2 \le 9, x \ne -2y\}$

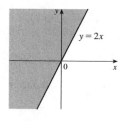 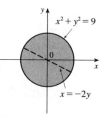

19. $\{(x, y) \mid y \ge -x^2\}$ **21.** $\{(x, y) \mid xy > 1\}$

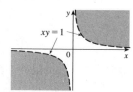 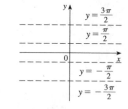

23. $\{(x, y) \mid y \ne (2n + 1)\pi/2, n \text{ an integer}\}$

 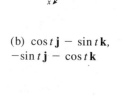

25. $\{(x, y) \mid -1 \le x + y \le 1\}$

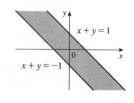

27. $\{(x, y) \mid x > 0, 2n\pi < y < (2n + 1)\pi, n \text{ an integer}\}$

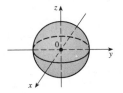

49. $\sqrt{x + y} = k$

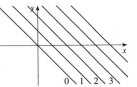

51. $x = y^2 + k$

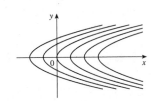

29. $\{(x, y, z) \mid x^2 + y^2 + z^2 \leqslant 1\}$ **31.** $z = 3$, horizontal plane

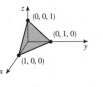   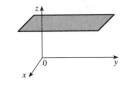

53. Family of parallel planes

55. Family of hyperboloids of one or two sheets with axis the y-axis

57.

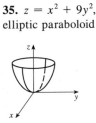

59. (a) B   (b) III
61. (a) F   (b) V
63. (a) D   (b) IV

33. $x + y + z = 1$, plane   **35.** $z = x^2 + 9y^2$, elliptic paraboloid

65.   **67.**

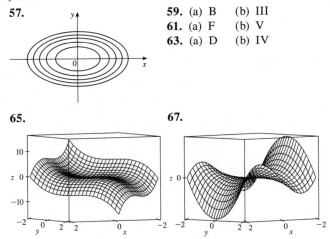

37. $z = \sqrt{x^2 + y^2}$, top half of cone

69. The level curves are elliptic if a and b have the same sign, hyperbolic if they have opposite signs. For both a and b positive, the graph increases rapidly as we leave the origin. If one of a and b is 0, the graph is a cylinder. If one of a and b is positive and the other is negative, the graph is saddle-shaped near the origin. If both a and b are negative, there is a bulge at the origin, which is elongated in the y-direction if a/b is large and in the x-direction if a/b is small.

39. $z = y^2 - x^2$, hyperbolic paraboloid. See Figures 6 and 7(b) in Section 11.6.

41. $z = 1 - x^2$, parabolic cylinder

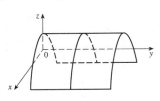

43. $xy = k$

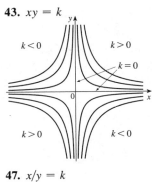

Exercises 12.2 ■ page 777

1. -927 **3.** $-\frac{5}{2}$ **5.** π **7.** Does not exist
9. Does not exist   **11.** Does not exist   **13.** 0
15. Does not exist   **17.** 2   **19.** Does not exist
21. $-\frac{3}{5}$ **23.** Does not exist   **25.** Does not exist
27. The graph shows that the function approaches different numbers along different lines.
29. $h(x, y) = e^{-(x^4+x^2y^2+y^4)} \cos(x^4 + x^2y^2 + y^4); \mathbb{R}^2$
31. $h(x, y) = 4x^2 + 9y^2 + 12xy - 24x - 36y + 36 + \sqrt{2x + 3y - 6}; \{(x, y) \mid 2x + 3y \geqslant 6\}$
33. $\{(x, y) \mid x^2 + y^2 \neq 1\}$
35. $\{(x, y) \mid x^4 - y^4 \neq (2n + 1)\pi/2, n \text{ an integer}\}$ **37.** $\mathbb{R}^2$
39. $\{(x, y) \mid x \geqslant |y|\}$ **41.** $\{(x, y, z) \mid yz > 0\}$
43. $\{(x, y) \mid (x, y) \neq (0, 0)\}$ **45.** $\{(x, y) \mid (x, y) \neq (0, 0)\}$
47. 0   **49.** 0

45. $x^2 + 9y^2 = k$

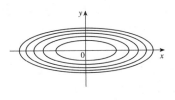

47. $x/y = k$

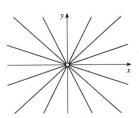

Exercises 12.3 ▪ page 784

1. $f_x(1, 2) = -8 =$ slope of C_1, $f_y(1, 2) = -4 =$ slope of C_2

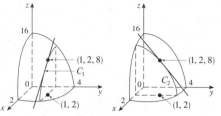

3. -27 **5.** 2

7. $(x^4 + 3x^2y^2 - 2xy^3)/(x^2 + y^2)^2$,
$(y^4 + 3x^2y^2 - 2x^3y)/(x^2 + y^2)^2$

9. $(y - z)/(x - y)$, $(x + z)/(x - y)$

11. $(x - y - z)/(x + z)$, $(y - x)/(x + z)$ **13.** 0

15. $y + z, x + z, x + y$

17. $f_x(x, y) = 3x^2y^5 - 4xy + 1$, $f_y(x, y) = 5x^3y^4 - 2x^2$

19. $f_x(x, y) = 4x^3 + 2xy^2$, $f_y(x, y) = 2x^2y + 4y^3$

21. $f_x(x, y) = 2y/(x + y)^2$, $f_y(x, y) = -2x/(x + y)^2$

23. $f_x = e^x[\tan(x - y) + \sec^2(x - y)]$, $f_y = -e^x \sec^2(x - y)$

25. $f_u = v/(u^2 + v^2)$, $f_v = -u/(u^2 + v^2)$

27. $g_x = 2xy^4 \sec^2(x^2y^3)$, $g_y = \tan(x^2y^3) + 3x^2y^3 \sec^2(x^2y^3)$

29. $\partial z/\partial x = 1/\sqrt{x^2 + y^2}$, $\partial z/\partial y = y/(x^2 + y^2 + x\sqrt{x^2 + y^2})$

31. $f_x = -e^{x^2}$, $f_y = e^{y^2}$

33. $f_x = 2xyz^3 + y$, $f_y = x^2z^3 + x$, $f_z = 3x^2yz^2 - 1$

35. $f_x = yzx^{yz-1}$, $f_y = zx^{yz} \ln x$, $f_z = yx^{yz} \ln x$

37. $u_x = -\dfrac{yz \cos(y/(x + z))}{(x + z)^2}$, $u_y = \dfrac{z \cos(y/(x + z))}{x + z}$,

$u_z = \sin\left(\dfrac{y}{x + z}\right) - \dfrac{yz \cos(y/(x + z))}{(x + z)^2}$

39. $u_x = y^2z^3[\ln(x + 2y + 3z) + x/(x + 2y + 3z)]$,
$u_y = 2xyz^3[\ln(x + 2y + 3z) + y/(x + 2y + 3z)]$,
$u_z = 3xy^2z^2[\ln(x + 2y + 3z) + z/(x + 2y + 3z)]$

41. $f_x = 1/(z - t)$, $f_y = 1/(t - z)$, $f_z = (y - x)/(z - t)^2$,
$f_t = (x - y)/(z - t)^2$

43. $\partial u/\partial x_i = x_i/\sqrt{x_1^2 + x_2^2 + \cdots + x_n^2}$

45. $f_x(x, y) = 2x - y$, $f_y(x, y) = 4y - x$

47. $f_x = 2x + 2xy$, $f_y = 2y + x^2$

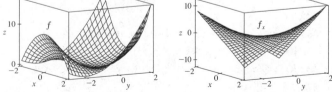

49. $c = f, b = f_x, a = f_y$

51. $f'(x), g'(y)$ **53.** $f'(x + y), f'(x + y)$

55. $f'(x/y)/y, -xf'(x/y)/y^2$

57. $f_{xx} = 2y, f_{xy} = 2x + 1/(2\sqrt{y}) = f_{yx}$,
$f_{yy} = -x/(4y\sqrt{y})$

59. $z_{xx} = \dfrac{3(2x^2 + y^2)}{\sqrt{x^2 + y^2}}$, $z_{xy} = \dfrac{3xy}{\sqrt{x^2 + y^2}} = z_{yx}$,

$z_{yy} = \dfrac{3(x^2 + 2y^2)}{\sqrt{x^2 + y^2}}$

61. $z_{tt} = 0$, $z_{tx} = \dfrac{1}{2\sqrt{x - x^2}} = z_{xt}$, $z_{xx} = \dfrac{t(2x - 1)}{4(x - x^2)^{3/2}}$

67. $-48xy$ **69.** $48x^3y^3z^2$ **71.** $-\sin y$

73. $72yz^2/(x + 2y^2 + 3z^3)^3$ **85.** R^2/R_1^2 **89.** No

91. $x = 1 + t, y = 2, z = 2 - 2t$ **95.** -2

97. (a)

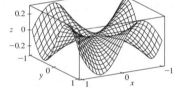

(b) $f_x(x, y) = \dfrac{x^4y + 4x^2y^3 - y^5}{(x^2 + y^2)^2}$, $f_y(x, y) = \dfrac{x^5 - 4x^3y^2 - xy^4}{(x^2 + y^2)^2}$

(c) $0, 0$ **(e)** No, since f_{xy} and f_{yx} are not continuous.

Exercises 12.4 ▪ page 793

1. $4x + 8y - z = 8$ **3.** $2x + 4y - z + 6 = 0$

5. $2x + y - z = 1$

7.   **9.**

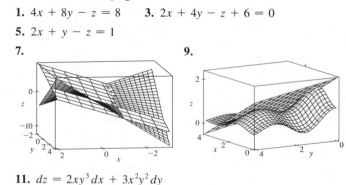

11. $dz = 2xy^3 dx + 3x^2y^2 dy$

13. $du = e^x(\cos xy - y \sin xy) dx - xe^x \sin xy\, dy$

15. $dw = 2xy\, dx + (x^2 + 2yz) dy + y^2 dz$

17. $dw = (x^2 + y^2 + z^2)^{-1}(x\, dx + y\, dy + z\, dz)$

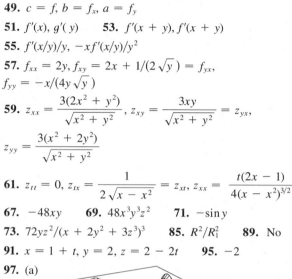
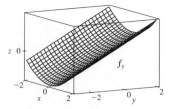

19. $\Delta z = 0.9225$, $dz = 0.9$ **21.** $2.84\overline{6}$ **23.** 65.88
25. 26.76 **27.** 1.015 **29.** 5.4 cm^2 **31.** 16 cm^3
33. 150 **35.** $\frac{1}{17} \approx 0.059 \text{ ohm}$ **37.** $\varepsilon_1 = \Delta x$, $\varepsilon_2 = \Delta y$

Exercises 12.5 ■ page 800

1. $6t^5 + 4t^3 + 4t$ **3.** $(x + y^2)^{-1}(1/(2\sqrt{1+t}) + y/\sqrt{t})$
5. $y^2z^3\cos t - 2xyz^3\sin t + 6xy^2z^2e^{2t}$
7. $\partial z/\partial s = 4sx\sin y + 2tx^2\cos y$,
$\partial z/\partial t = 4tx\sin y + 2sx^2\cos y$
9. $\partial z/\partial s = 2x(1 - 3y^3)e^t - 9x^2y^2e^{-t}$,
$\partial z/\partial t = 2x(1 - 3y^3)se^t + 9x^2y^2se^{-t}$
11. $\partial z/\partial s = 2^{x-3y}(\ln 2)(2st - 3t^2)$,
$\partial z/\partial t = 2^{x-3y}(\ln 2)(s^2 - 6st)$
13. $\dfrac{\partial u}{\partial r} = \dfrac{\partial u}{\partial x}\dfrac{\partial x}{\partial r} + \dfrac{\partial u}{\partial y}\dfrac{\partial y}{\partial r}$, $\dfrac{\partial u}{\partial s} = \dfrac{\partial u}{\partial x}\dfrac{\partial x}{\partial s} + \dfrac{\partial u}{\partial y}\dfrac{\partial y}{\partial s}$,
$\dfrac{\partial u}{\partial t} = \dfrac{\partial u}{\partial x}\dfrac{\partial x}{\partial t} + \dfrac{\partial u}{\partial y}\dfrac{\partial y}{\partial t}$
15. $\dfrac{\partial v}{\partial x} = \dfrac{\partial v}{\partial p}\dfrac{\partial p}{\partial x} + \dfrac{\partial v}{\partial q}\dfrac{\partial q}{\partial x} + \dfrac{\partial v}{\partial r}\dfrac{\partial r}{\partial x}$,
$\dfrac{\partial v}{\partial y} = \dfrac{\partial v}{\partial p}\dfrac{\partial p}{\partial y} + \dfrac{\partial v}{\partial q}\dfrac{\partial q}{\partial y} + \dfrac{\partial v}{\partial r}\dfrac{\partial r}{\partial y}$,
$\dfrac{\partial v}{\partial z} = \dfrac{\partial v}{\partial p}\dfrac{\partial p}{\partial z} + \dfrac{\partial v}{\partial q}\dfrac{\partial q}{\partial z} + \dfrac{\partial v}{\partial r}\dfrac{\partial r}{\partial z}$
17. $2, 0$ **19.** $0, 0, 4$
21. $\partial u/\partial p = 2(z - x)/(y + z)^2 = -t/p^2$, $\partial u/\partial r = 0$,
$\partial u/\partial t = 2/(y + z) = 1/p$
23. $(y - 2x)/(3y^2 - x)$
25. $(18x - x^{-2/3}y^{1/3})/(12y + x^{1/3}y^{-2/3})$
27. $(y - z)/(x - y)$, $(x + z)/(x - y)$
29. $(x - y - z)/(x + z)$, $(y - x)/(x + z)$
31. $-(e^y + ze^x)/(y + e^x)$, $-(xe^y + z)/(y + e^x)$
33. $-9600\pi \text{ cm}^3/\text{s}$
35. (a) $6 \text{ m}^3/\text{s}$ (b) $10 \text{ m}^2/\text{s}$ (c) 0 m/s **37.** -0.27 L/s
39. (a) $\partial z/\partial r = (\partial z/\partial x)\cos\theta + (\partial z/\partial y)\sin\theta$,
$\partial z/\partial\theta = -(\partial z/\partial x)r\sin\theta + (\partial z/\partial y)r\cos\theta$
45. $4rs\,\partial^2z/\partial x^2 + (4r^2 + 4s^2)\,\partial^2z/\partial x\,\partial y + 4rs\,\partial^2z/\partial y^2 + 2\,\partial z/\partial y$

Exercises 12.6 ■ page 810

1. $7\sqrt{3} - 16$ **3.** 1
5. (a) $\nabla f(x, y) = \langle 3x^2 - 8xy, -4x^2 + 2y\rangle$ (b) $\langle 0, -2\rangle$
(c) $-\frac{8}{5}$
7. (a) $\nabla f(x, y, z) = \langle y^2z^3, 2xyz^3, 3xy^2z^2\rangle$ (b) $\langle 4, -4, 12\rangle$
(c) $20/\sqrt{3}$
9. $\frac{7}{52}$ **11.** $29/\sqrt{13}$ **13.** $\frac{1}{6}$ **15.** $-\pi/(4\sqrt{3})$
17. $\sqrt{5}$, $\langle 1, 2\rangle$ **19.** $\sqrt{17}/6$, $\langle 4, 1\rangle$ **21.** $\sqrt{13}/2$, $\langle -3, -2\rangle$
25. (a) $-40/(3\sqrt{3})$
27. (a) $32/\sqrt{3}$ (b) $\langle 38, 6, 12\rangle$ (c) $2\sqrt{406}$ **29.** $\frac{327}{13}$
35. (a) $4x + y + z = 12$ (b) $(x - 2)/4 = y - 2 = z - 2$
37. (a) $3x - y + z = 4$ (b) $(x - 1)/3 = -y = z - 1$
39. (a) $x + y - z = 1$ (b) $x - 1 = y = -z$

41.

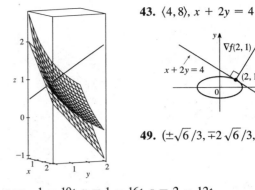

43. $\langle 4, 8\rangle$, $x + 2y = 4$

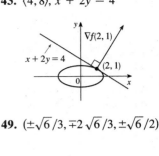

49. $(\pm\sqrt{6}/3, \mp 2\sqrt{6}/3, \pm\sqrt{6}/2)$

55. $x = -1 - 10t$, $y = 1 - 16t$, $z = 2 - 12t$
59. If $\mathbf{u} = \langle a, b\rangle$ and $\mathbf{v} = \langle c, d\rangle$, then $af_x + bf_y$ and $cf_x + df_y$
are known, so we solve linear equations for f_x and f_y.

Exercises 12.7 ■ page 819

1. Minimum $f(-2, 3) = -13$
3. Minimum $f(0, -1) = -1$
5. Minimum $f(0, 0) = 4$, saddle points $(\pm\sqrt{2}, -1)$
7. Minimum $f(1, 1) = -1$, saddle point $(0, 0)$
9. Saddle point $(1, 2)$ **11.** Maximum $f(-\frac{1}{2}, 4) = -6$
13. None   **15.** Saddle points $(0, n\pi)$, n an integer
17. Maximum $f(0, 0) = 2$, minimum $f(0, 2) = -2$,
saddle points $(\pm 1, 1)$
19. Maximum $f(\pi/3, \pi/3) = 3\sqrt{3}/2$,
minimum $f(5\pi/3, 5\pi/3) = -3\sqrt{3}/2$
21. Minima $f(-1.714, 0) \approx -9.200$, $f(1.402, 0) \approx 0.242$,
saddle point $(0.312, 0)$, lowest point $(-1.714, 0, -9.200)$
23. Maxima $f(-1.267, 0) \approx 1.310$, $f(1.629, \pm 1.063) \approx 8.105$,
saddle points $(-0.259, 0)$, $(1.526, 0)$,
highest points $(1.629, \pm 1.063, 8.105)$
25. Minimum $f(4, 0) = -7$, maximum $f(4, 5) = 13$
27. Maximum $f(\pm 1, 1) = 7$, minimum $f(0, 0) = 4$
29. Maximum $f(2, 4) = 3$, minimum $f(-2, 4) = -9$
31. Maximum $f(1, 0) = 2$, minimum $f(-1, 0) = -2$
33.

35. $\left(\frac{2}{7}, \frac{4}{7}, \frac{6}{7}\right)$

37. $7/\sqrt{61}$ **39.** $(0, 0, 1)$, $(0, 0, -1)$ **41.** $\frac{100}{3}, \frac{100}{3}, \frac{100}{3}$
43. $16/\sqrt{3}$ **45.** $\frac{4}{3}$ **47.** Cube, edge length $c/12$
49. Square base of side 40 cm, height 20 cm
53. $y = 6.67x - 308$, 172.2 lb

Exercises 12.8 ▪ page 826

1. Maxima $f(\pm 1, 0) = 1$, minima $f(0, \pm 1) = -1$

3. Maxima $f(\pm\sqrt{2}/3, \pm\sqrt{2}) = \frac{2}{3}$,
minima $f(\pm\sqrt{2}/3, \mp\sqrt{2}) = -\frac{2}{3}$

5. Maximum $f(1/\sqrt{35}, 3/\sqrt{35}, 5/\sqrt{35}) = \sqrt{35}$,
minimum $f(-1/\sqrt{35}, -3/\sqrt{35}, -5/\sqrt{35}) = -\sqrt{35}$

7. Maximum $2/\sqrt{3}$, minimum $-2/\sqrt{3}$

9. Maximum $\sqrt{3}$, minimum 1

11. Maximum $f(\frac{1}{2}, \frac{1}{2}, \frac{1}{2}, \frac{1}{2}) = 2$,
minimum $f(-\frac{1}{2}, -\frac{1}{2}, -\frac{1}{2}, -\frac{1}{2}) = -2$

13. Maximum $f(1, \sqrt{2}, -\sqrt{2}) = 1 + 2\sqrt{2}$,
minimum $f(1, -\sqrt{2}, \sqrt{2}) = 1 - 2\sqrt{2}$

15. Maximum $\frac{3}{2}$, minimum $\frac{1}{2}$

17. Maxima $f(\pm 1/\sqrt{2}, \mp 1/(2\sqrt{2})) = e^{1/4}$,
minima $f(\pm 1/\sqrt{2}, \pm 1/(2\sqrt{2})) = e^{-1/4}$

19. (b) Minimum $f(3 - \frac{3}{2}\sqrt{2}, 3 - \frac{3}{2}\sqrt{2}) = (351 - 243\sqrt{2})/2 \approx 3.7$,
maximum $f(3 + \frac{3}{2}\sqrt{2}, 3 + \frac{3}{2}\sqrt{2}) = (351 + 243\sqrt{2})/2 \approx 347$

25.–39. See Exercises 35–49 in Section 12.7.

41. Nearest $(\frac{1}{2}, \frac{1}{2}, \frac{1}{2})$, farthest $(-1, -1, 2)$

Review Exercises for Chapter 12 ▪ page 828

1. True   **3.** False   **5.** False   **7.** True   **9.** False
11. True

13. $\{(x, y) \mid y > -x - 1, x \neq 1\}$ **15.** $\{(x, y) \mid -1 \leq x \leq 1\}$

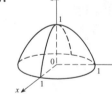

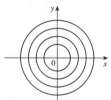

17.

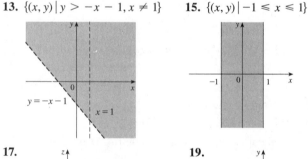

19.

21. 0   **23.** $f_x = 12x^3 - \sqrt{y}, f_y = -\frac{1}{2}x/\sqrt{y}$

25. $f_s = 2e^{2s}\cos \pi t, f_t = -\pi e^{2s}\sin \pi t$

27. $f_x = y^z, f_y = xzy^{z-1}, f_z = xy^z \ln y$

29. $f_{xx} = 2y^3 - 24x^2, f_{xy} = 6xy^2, f_{yy} = 6x^2y + 2$

31. $f_{xx} = 0, f_{yy} = 2xz^3, f_{zz} = 6xy^2z, f_{xy} = 2yz^3$,
$f_{xz} = 3y^2z^2, f_{xz} = 6xyz^2$

35. $6y - z = 1$ **37.** $x + 3y + 9z = 18$

39. $3x + 4y + 3z = 14$ **41.** $(\pm\sqrt{2/7}, \pm 1/\sqrt{14}, \mp 3/\sqrt{14})$

43. 38.656   **45.** $e^t + 2(y/z)(3t^2 + 4) - 2t(y^2/z^2)$

51. $ze^{x\sqrt{y}}\langle z\sqrt{y}, xz/(2\sqrt{y}), 2\rangle$ **53.** $\frac{43}{5}$ **55.** $\sqrt{145}/2, \langle 4, \frac{9}{2}\rangle$

57. $\pi/2$ **59.** Minimum $f(-4, 1) = -9$

61. Maximum $f(1, 1) = 1$, saddle points $(0, 0), (0, 3), (3, 0)$

63. Maximum $f(1, 2) = 4$, minimum $f(2, 4) = -64$

65. Maximum $f(-1, 0) = 2$, minima $f(1, \pm 1) = -3$,
saddle points $(-1, \pm 1), (1, 0)$

67. Maximum $f(\pm\sqrt{2/3}, 1/\sqrt{3}) = 2/(3\sqrt{3})$,
minimum $f(\pm\sqrt{2/3}, -1/\sqrt{3}) = -2/(3\sqrt{3})$

69. Maximum $f(3, 3, 3) = 9$,
minimum $f(1, 1, -1) = f(1, -1, 1) = f(-1, 1, 1) = 1$

71. $(\pm 3^{-1/4}, 3^{-1/4}\sqrt{2}, \pm 3^{1/4}), (\pm 3^{-1/4}, -3^{-1/4}\sqrt{2}, \pm 3^{1/4})$

73. $P(2 - \sqrt{3}), P(3 - \sqrt{3})/6, P(2\sqrt{3} - 3)/3$

PROBLEMS PLUS ▪ page 830

1. $(\sqrt{3} - 1.5)$ m   **5.** $L^2W^2, \frac{1}{4}L^2W^2$ **13.** $\sqrt{6}/2, 3\sqrt{2}/2$

15. (a) $(x + 1)/(-2c) = (y - c)/(c^2 - 1) = (z - c)/(c^2 + 1)$
(b) $x^2 + y^2 = t^2 + 1, z = t$ (c) $4\pi/3$

CHAPTER 13

Exercises 13.1 ▪ page 837

1. (a) -17.75 (b) -15.75 (c) -8.75 (d) -6.75

3. $63, \sqrt{2}$ **5.** $0, \sqrt{5}$ **7.** $\frac{247}{8}, \sqrt{5}$ **9.** $U < V < L$

11. 0.6065, 0.5694, 0.5606, 0.5585, 0.5579, 0.5578   **13.** 3

Exercises 13.2 ▪ page 842

1. $4x^2, y^3/3$ **3.** $(e^2 - 1)xe^x, e^y$ **5.** $\frac{32}{3}$ **7.** 0

9. $\frac{4}{15}(31 - 9\sqrt{3})$ **11.** $3[1 - (1/\sqrt{2})]$ **13.** 6

15. $-\frac{585}{8}$ **17.** $\frac{15}{4}(2 - \sqrt{3})$ **19.** $[(\sqrt{3} - 1)/2] - (\pi/12)$

21. $\ln \frac{27}{16}$ **23.**

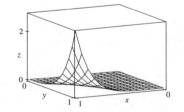

25. 37.5   **27.** $\frac{166}{27}$ **29.** $\frac{4}{15}(2\sqrt{2} - 1)$ **31.** 36

33. $21e - 57$

35. $\frac{5}{6}$

37. Fubini's Theorem does not apply. The integrand has an infinite discontinuity at the origin.

23. 0.43, $\sqrt{3}/2$ **25.**

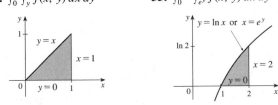

Exercises 13.3 ■ page 850

1. $\frac{1}{6}$ **3.** $16\left[1 - (\sqrt{2}/7)\right]$ **5.** $(1 - \cos 1)/2$ **7.** $\frac{1}{12}$
9. $-\frac{19}{42}$ **11.** $\frac{1}{2}e^4 - 2e$ **13.** $(1 - \cos 1)/2$ **15.** $\frac{500}{3}$
17. $\frac{1}{8}$ **19.** $\frac{6}{35}$ **21.** $\frac{31}{8}$ **23.** $\frac{1}{6}(11\sqrt{5} - 27) + \frac{9}{2}\sin^{-1}\frac{2}{3}$
25. $\frac{1}{6}$ **27.** $\frac{2}{3}$ **29.** 0, 1.213, 0.713
31. 13,984,735,616/14,549,535
33. $\int_0^1 \int_y^1 f(x, y)\, dx\, dy$ **35.** $\int_0^{\ln 2} \int_{e^y}^2 f(x, y)\, dx\, dy$

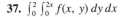

37. $\int_0^2 \int_0^{2x} f(x, y)\, dy\, dx$

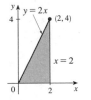

39. $(e^9 - 1)/6$ **41.** $\frac{1}{4}\sin 81$ **43.** $(2\sqrt{2} - 1)/3$ **45.** 1
47. $0 \le \iint_D \sqrt{x^3 + y^3}\, dA \le \sqrt{2}$ **51.** 8π

Exercises 13.4 ■ page 856

1. 0   **3.** $\frac{609}{8}$ **5.** 2   **7.** $24\pi^5$ **9.** $\pi/12$ **11.** 4
13. π **15.** $81\pi/2$ **17.** $5\pi/2$ **19.** $(2\pi/3)[1 - (1/\sqrt{2})]$
21. $(8\pi/3)(64 - 24\sqrt{3})$ **23.** $\frac{4}{3}\pi a^3$
25. $(\pi/4)(e - 1)$ **27.** $4\pi/3$ **29.** 0.587   **31.** $\frac{15}{16}$
33. (a) $\sqrt{\pi}/4$ (b) $\sqrt{\pi}/2$

Exercises 13.5 ■ page 862

1. $\frac{50}{3}$ C   **3.** $\frac{2}{3}$, $(0, \frac{1}{2})$ **5.** 6, $(\frac{3}{4}, \frac{3}{2})$ **7.** $\frac{1}{6}$, $(\frac{4}{7}, \frac{3}{4})$
9. $\frac{27}{2}$, $(\frac{8}{5}, \frac{1}{2})$ **11.** $\pi/4$, $(\pi/2, 16/(9\pi))$ **13.** $(\frac{3}{8}, 3\pi/16)$
15. $(2a/5, 2a/5)$ if vertex is $(0, 0)$ and sides are along positive axes
17. $\frac{1}{10}$, $\frac{1}{16}$, $\frac{13}{80}$ **19.** $\frac{189}{20}$, $\frac{1269}{28}$, $\frac{1917}{35}$
21. $\rho a^4/3$, $\rho a^4/3$; $a/\sqrt{3}$, $a/\sqrt{3}$

Exercises 13.6 ■ page 864

1. $4\sqrt{6}\,\pi$ **3.** $12\sin^{-1}\frac{2}{3}$ **5.** $(\pi/6)(17\sqrt{17} - 5\sqrt{5})$
7. $(2\pi/3)(2\sqrt{2} - 1)$ **9.** $a^2(\pi - 2)$ **11.** 1.83
13. $\frac{3}{2} + \frac{5}{8}\ln 5$ **15.** 3.3213

Exercises 13.7 ■ page 872

3. $\frac{1}{48}$ **5.** $\frac{16}{3}$ **7.** $\frac{7}{5}$ **9.** $\frac{5}{28}$ **11.** $\frac{1}{10}$ **13.** $\frac{1}{12}$
15. $16\pi/3$ **17.** 8   **19.** $\frac{8}{15}$
21. (a) $\int_0^1 \int_0^x \int_0^{\sqrt{1-y^2}} dz\, dy\, dx$ (b) $\frac{1}{4}\pi - \frac{1}{3}$

27. $\int_{-2}^2 \int_0^6 \int_{-\sqrt{4-x^2}}^{\sqrt{4-x^2}} f(x, y, z)\, dz\, dy\, dx$
$= \int_0^6 \int_{-2}^2 \int_{-\sqrt{4-x^2}}^{\sqrt{4-x^2}} f(x, y, z)\, dz\, dx\, dy$
$= \int_{-2}^2 \int_0^6 \int_{-\sqrt{4-z^2}}^{\sqrt{4-z^2}} f(x, y, z)\, dx\, dy\, dz$
$= \int_0^6 \int_{-2}^2 \int_{-\sqrt{4-z^2}}^{\sqrt{4-z^2}} f(x, y, z)\, dx\, dz\, dy$
$= \int_{-2}^2 \int_{-\sqrt{4-x^2}}^{\sqrt{4-x^2}} \int_0^6 f(x, y, z)\, dy\, dz\, dx$
$= \int_{-2}^2 \int_{-\sqrt{4-z^2}}^{\sqrt{4-z^2}} \int_0^6 f(x, y, z)\, dy\, dx\, dz$

29. $\int_{-1}^1 \int_0^{1-x^2} \int_0^y f(x, y, z)\, dz\, dy\, dx$
$= \int_0^1 \int_{-\sqrt{1-y}}^{\sqrt{1-y}} \int_0^y f(x, y, z)\, dz\, dx\, dy$
$= \int_0^1 \int_z^1 \int_{-\sqrt{1-y}}^{\sqrt{1-y}} f(x, y, z)\, dx\, dy\, dz$
$= \int_0^1 \int_0^y \int_{-\sqrt{1-y}}^{\sqrt{1-y}} f(x, y, z)\, dx\, dz\, dy$
$= \int_{-1}^1 \int_0^{1-x^2} \int_z^{1-x^2} f(x, y, z)\, dy\, dz\, dx$
$= \int_0^1 \int_{-\sqrt{1-z}}^{\sqrt{1-z}} \int_z^{1-x^2} f(x, y, z)\, dy\, dx\, dz$

31. $\int_0^1 \int_{\sqrt{x}}^1 \int_0^{1-y} f(x, y, z)\, dz\, dy\, dx = \int_0^1 \int_0^{y^2} \int_0^{1-y} f(x, y, z)\, dz\, dx\, dy$
$= \int_0^1 \int_0^{1-z} \int_0^{y^2} f(x, y, z)\, dx\, dy\, dz = \int_0^1 \int_0^{1-y} \int_0^{y^2} f(x, y, z)\, dx\, dz\, dy$
$= \int_0^1 \int_0^{1-\sqrt{x}} \int_{\sqrt{x}}^{1-z} f(x, y, z)\, dy\, dz\, dx$
$= \int_0^1 \int_0^{(1-z)^2} \int_{\sqrt{x}}^{1-z} f(x, y, z)\, dy\, dx\, dz$

33. $\int_0^1 \int_0^x \int_0^y f(x, y, z)\, dz\, dy\, dx = \int_0^1 \int_z^1 \int_y^1 f(x, y, z)\, dx\, dy\, dz$
$= \int_0^1 \int_0^y \int_y^1 f(x, y, z)\, dx\, dz\, dy = \int_0^1 \int_0^x \int_z^x f(x, y, z)\, dy\, dz\, dx$
$= \int_0^1 \int_z^1 \int_z^x f(x, y, z)\, dy\, dx\, dz$

35. $\frac{9}{10}$, $\left(\frac{22}{27}, \frac{25}{63}, \frac{152}{189}\right)$ **37.** a^5, $(7a/12, 7a/12, 7a/12)$
39. (a) $m = \int_0^1 \int_0^{\sqrt{1-x^2}} \int_0^y (1 + x + y + z)\, dz\, dy\, dx$
(b) $(\bar{x}, \bar{y}, \bar{z})$, where
$\bar{x} = (1/m) \int_0^1 \int_0^{\sqrt{1-x^2}} \int_0^y x(1 + x + y + z)\, dz\, dy\, dx$,
$\bar{y} = (1/m) \int_0^1 \int_0^{\sqrt{1-x^2}} \int_0^y y(1 + x + y + z)\, dz\, dy\, dx$,
$\bar{z} = (1/m) \int_0^1 \int_0^{\sqrt{1-x^2}} \int_0^y z(1 + x + y + z)\, dz\, dy\, dx$
(c) $\int_0^1 \int_0^{\sqrt{1-x^2}} \int_0^y (x^2 + y^2)(1 + x + y + z)\, dz\, dy\, dx$
41. (a) $m = \int_{-1}^1 \int_{-\sqrt{1-y^2}}^{\sqrt{1-y^2}} \int_{4y^2+4z^2}^4 (x^2 + y^2 + z^2)\, dx\, dz\, dy$
(b) $(\bar{x}, \bar{y}, \bar{z})$, where
$\bar{x} = (1/m) \int_{-1}^1 \int_{-\sqrt{1-y^2}}^{\sqrt{1-y^2}} \int_{4y^2+4z^2}^4 x(x^2 + y^2 + z^2)\, dx\, dz\, dy$,
$\bar{y} = (1/m) \int_{-1}^1 \int_{-\sqrt{1-y^2}}^{\sqrt{1-y^2}} \int_{4y^2+4z^2}^4 y(x^2 + y^2 + z^2)\, dx\, dz\, dy$,
$\bar{z} = (1/m) \int_{-1}^1 \int_{-\sqrt{1-y^2}}^{\sqrt{1-y^2}} \int_{4y^2+4z^2}^4 z(x^2 + y^2 + z^2)\, dx\, dz\, dy$
(c) $\int_{-1}^1 \int_{-\sqrt{1-y^2}}^{\sqrt{1-y^2}} \int_{4y^2+4z^2}^4 (x^2 + y^2)(x^2 + y^2 + z^2)\, dx\, dz\, dy$

43. (a) $\frac{3}{32}\pi + \frac{11}{24}$ (b) $(\bar{x}, \bar{y}, \bar{z})$, where $\bar{x} = 28/(9\pi + 44)$,
$\bar{y} = 2(15\pi + 64)/[5(9\pi + 44)]$,
$\bar{z} = (45\pi + 208)/[15(9\pi + 44)]$
(c) $(68 + 15\pi)/240$
45. $I_x = I_y = I_z = \frac{2}{3}kL^5$ **47.** $L^3/8$
49. The region bounded by the ellipsoid $x^2 + 2y^2 + 3z^2 = 1$

Exercises 13.8 ■ page 878

1. 8π **3.** $\pi/6$

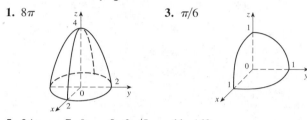

5. 24π **7.** 0   **9.** $2\pi/5$ **11.** 162π
13. $\pi Ka^2/8$, $(0, 0, 2a/3)$ **15.** $4\pi/5$ **17.** $\pi/30$
19. $4\pi(2 - \sqrt{3})$ **21.** 10π
23. $K\pi a^4/2$, where K is the proportionality constant
25. $2\pi Ka^6/9$ **27.** $4K\pi a^5/15$
29. $(2\pi/3)[1 - (1/\sqrt{2})]$, $(0, 0, 3/[8(2 - \sqrt{2})])$ **31.** $5\pi/6$
33. $8\pi/35$ **35.** $243\pi/5$

Exercises 13.9 ■ page 886

1. 3   **3.** $2e^{4u}$ **5.** -4
7. The parallelogram bounded by the lines $y = 2x$,
$y = 2x + 3$, $x = 2y$, $x = 2y - 6$
9. The triangular region with vertices $(0, 0)$, $(4, 0)$, $(3, 4)$
11. $\frac{11}{3}$ **13.** 6π **15.** $2\ln 3$ **17.** $\frac{4}{3}\pi abc$ **19.** $-\frac{66}{125}$
21. $\frac{3}{2}\sin 1$ **23.** $e - e^{-1}$

Review Exercises for Chapter 13 ■ page 887

1. True   **3.** True   **5.** False   **7.** 1152   **9.** $\ln(\frac{27}{4}) - 1$
11. $\frac{1}{110}$
13. The region outside the circle $r = 1$ and inside
the cardioid $r = 1 + \sin\theta$
15. $(e - 1)/2$ **17.** $\ln\frac{3}{2}$ **19.** $\frac{1}{40}$ **21.** $\frac{41}{30}$ **23.** $81\pi/5$
25. $\frac{32}{3}$ **27.** $\pi/96$ **29.** $\frac{64}{15}$ **31.** 176   **33.** $\frac{2}{3}$
35. $2ma^3/9$ **37.** $\frac{1}{4}, (\frac{1}{3}, \frac{8}{15})$ **39.** $(4a/(3\pi), 4a/(3\pi))$
41. $(0, 0, h/4)$ **43.** $\frac{7}{2}$ **45.** $(\pi/8)\ln 5$ **47.** 0.0512
49. $(\frac{4}{21}, \frac{11}{21}, \frac{8}{7})$ **51.** $\int_0^1 \int_0^{1-z} \int_{-\sqrt{y}}^{\sqrt{y}} f(x, y, z) \, dx \, dy \, dz$
53. $-\ln 2$ **55.** 0

APPLICATIONS PLUS ■ page 890

1. (a) $x = w/3$, base $= w/3$ (b) Yes
3. (a) $\iint_D (k/20) \left[20 - \sqrt{(x - x_0)^2 + (y - y_0)^2} \right] dA$, where D is
the disk with radius 10 mi centered at the center of the city
(b) $200\pi k/3 \approx 209k$, $200(\pi/2 - \frac{8}{9})k \approx 136k$, on the edge
5. (a) $\iiint_C h(P)g(P) \, dV$, where C is the cone
(b) $\approx 3.1 \times 10^{19}$ ft-lb
7. (f) $\frac{2}{3}(1 - b^5)/(1 - b^3)$; this approaches $\frac{2}{5}$ as $a \to 0$ and
$\frac{2}{3}$ as $a \to r$

CHAPTER 14

Exercises 14.1 ■ page 896

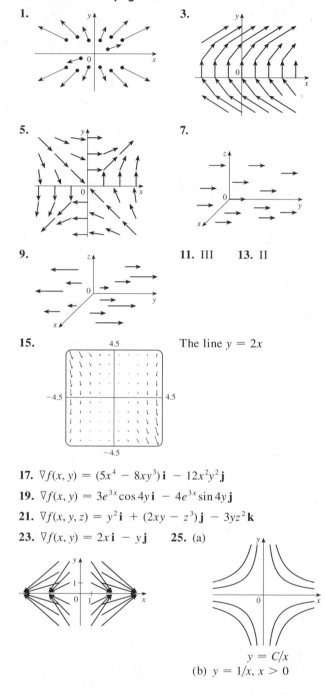

11. III   **13.** II

15.   The line $y = 2x$

17. $\nabla f(x, y) = (5x^4 - 8xy^3)\mathbf{i} - 12x^2y^2\mathbf{j}$
19. $\nabla f(x, y) = 3e^{3x}\cos 4y\,\mathbf{i} - 4e^{3x}\sin 4y\,\mathbf{j}$
21. $\nabla f(x, y, z) = y^2\mathbf{i} + (2xy - z^3)\mathbf{j} - 3yz^2\mathbf{k}$
23. $\nabla f(x, y) = 2x\mathbf{i} - y\mathbf{j}$ **25.** (a)

$y = C/x$
(b) $y = 1/x$, $x > 0$

Exercises 14.2 ■ page 906

1. $(10\sqrt{10} - 1)/54$ **3.** 1638.4   **5.** 48   **7.** $\frac{17}{3}$
9. $9\sqrt{13}\,\pi/4$ **11.** $3\sqrt{35}$ **13.** $\frac{16}{11}$ **15.** $\frac{77}{6}$ **17.** $-\frac{19}{143}$
19. $\frac{6}{5} - \cos 1 - \sin 1$

21. (a) $\frac{11}{8} - 1/e$
(b)

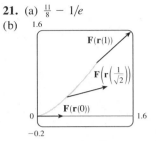

23. 0.052
25. $\frac{945}{16,777,216}\,\pi$
27. (a) Positive
 (b) Negative
29. $2\pi k, (4/\pi, 0)$

31. (a) $\bar{x} = (1/m) \iint_C x\rho(x, y, z)\, ds$,
$\bar{y} = (1/m) \int_C y\rho(x, y, z)\, ds$,
$\bar{z} = (1/m) \int_C z\rho(x, y, z)\, ds$, where $m = \int_C \rho(x, y, z)\, ds$
(b) $2\sqrt{13}\,k\pi, (0, 0, 3\pi)$
33. $I_x = k((\pi/2) - \frac{4}{3})$, $I_y = k((\pi/2) - \frac{2}{3})$ **35.** $2\pi^2$ **37.** $\frac{23}{88}$
39. 1.67×10^4 ft-lb

Exercises 14.3 ■ page 915

1. $f(x, y) = x^2 - 3xy + y^2 + K$ **3.** Not conservative
5. $f(x, y) = x^4y^3 + x + K$ **7.** Not conservative
9. $f(x, y) = ye^x + x\sin y + K$
11. (a) $f(x, y) = (x^2 + y^2)/2$ (b) 44
13. (a) $f(x, y) = x^2y^3$ (b) $[1 + (\pi^2/4)]^3$
15. (a) $f(x, y, z) = xy + yz$ (b) 15
17. (a) $f(x, y, z) = x^2z + x\sin y$ (b) 2π
19. $25\sin 1 - 1$ **21.** $\frac{8}{3}$ **23.** No   **25.** No
29. (a) Yes   (b) Yes   (c) Yes
31. (a) Yes   (b) Yes   (c) No

Exercises 14.4 ■ page 922

1. $-\frac{1}{12}$ **3.** $-\frac{1}{6}$ **7.** $-\frac{4}{3}$ **9.** $\frac{1}{3}$ **11.** 0   **13.** 0
15. π **17.** $\pi + \frac{16}{3}[(1/\sqrt{2}) - 1]$ **19.** $-\frac{1}{12}$ **21.** $3\pi/8$
23. (c) $\frac{9}{2}$ **25.** $(\frac{1}{3}, \frac{1}{3})$

Exercises 14.5 ■ page 929

1. (a) **0**   (b) 3   **3.** (a) **0**   (b) 0
5. (a) $xz\,\mathbf{i} - yz\,\mathbf{j} + y\,\mathbf{k}$ (b) $x + xy$
7. (a) $(3xe^y + 2ye^{yz})\,\mathbf{i} + (xe^{xz} - 3e^y)\,\mathbf{j}$ (b) $ze^{xz} - 2ze^{yz}$
9. (a) $(\ln z + e^{-y})\,\mathbf{i} - xe^y\,\mathbf{k}$ (b) $e^y + ze^{-y} + (y/z)$
11. $f(x, y, z) = xy + z + K$ **13.** Not conservative
15. Not conservative   **17.** $f(x, y, z) = xyz + (y^3/3) + K$
19. No
31. (a) Meaningless   (b) Vector field   (c) Scalar field
(d) Vector field   (e) Meaningless   (f) Vector field
(g) Scalar field   (h) Meaningless   (i) Vector field
(j) Meaningless   (k) Meaningless   (l) Scalar field
39. Negative

Exercises 14.6 ■ page 937

1. $x = x, y = y, z = \sqrt{1 - 3x^2 - 2y^2}$
[or $x = (1/\sqrt{3})\sin\phi\cos\theta, y = (1/\sqrt{2})\sin\phi\sin\theta, z = \cos\phi$,
$0 \le \phi \le \pi/2, 0 \le \theta \le 2\pi$]
3. $x = x, y = 6 - 3x^2 - 2z^2, z = z, 3x^2 + 2z^2 \le 6$

5. $x = 2\sin\phi\cos\theta, y = 2\sin\phi\sin\theta$,
$z = 2\cos\phi, 0 \le \phi \le \pi/4, 0 \le \theta \le 2\pi$
7. $x = r\cos\theta, y = r\sin\theta, z = 5, 0 \le r \le 4, 0 \le \theta \le 2\pi$
[or $x = x, y = y, z = 5, x^2 + y^2 \le 16$]
9. I   **11.** II
13. $x = x, y = e^{-x}\cos\theta$,
$z = e^{-x}\sin\theta, 0 \le x \le 3$,
$0 \le \theta \le 2\pi$

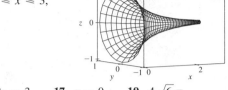

15. $3x - y + 3z = 3$ **17.** $x = 0$ **19.** $4\sqrt{6}\,\pi$
21. $(\pi/6)(17\sqrt{17} - 5\sqrt{5})$
23. $(\sqrt{21}/2) + \frac{17}{4}[\ln(2 + \sqrt{21}) - \ln\sqrt{17}]$
25. $2a^2(\pi - 2)$ **27.** $\pi(2\sqrt{6} - \frac{8}{3})$
29. (b)

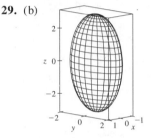

(c) $\int_0^{2\pi} \int_0^\pi \sqrt{36\sin^4 u\cos^2 v + 9\sin^4 u\sin^2 v + 4\cos^2 u\sin^2 u}\ du\,dv$
31. $x = (b + a\cos\alpha)\cos\theta, y = (b + a\cos\alpha)\sin\theta, z = a\sin\alpha$,
$0 \le \theta \le 2\pi, 0 \le \alpha \le 2\pi$

Exercises 14.7 ■ page 949

1. $8(1 + \sqrt{2} + \sqrt{3}) \approx 33.17$ **3.** $3\sqrt{14}$
5. $(33\sqrt{33} - 17\sqrt{17})/6$ **7.** $\pi\sqrt{2}/4$ **9.** 16π **11.** 16π
13. 0   **15.** $(11 - 10e)/6$ **17.** 12   **19.** 108π **21.** 0
23. 48   **25.** 0.1642   **27.** 3.4895
29. $\iint_S \mathbf{F} \cdot d\mathbf{S} = \iint_D [P(\partial h/\partial x) - Q + R(\partial h/\partial z)]\, dA$,
where D = projection on xz-plane
31. $(0, 0, a/2)$
33. (a) $I_z = \iint_S (x^2 + y^2)\rho(x, y, z)\, dS$ (b) $4329\sqrt{2}\,\pi/5$
35. $194,400\pi$ **37.** $8\pi a^3\varepsilon_0/3$ **39.** 1248π

Exercises 14.8 ■ page 955

1. π **3.** $3\pi/4$ **5.** 0   **7.** 3.5   **9.** -4π
11. (a) $81\pi/2$ (b)

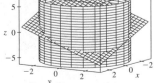

(c) $x = 3\cos t$, $y = 3\sin t$,
$z = 1 - 3(\cos t + \sin t)$,
$0 \le t \le 2\pi$

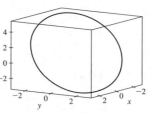

17. 16

Exercises 14.9 ▪ page 960

3. 8   **5.** 0   **7.** $\frac{1}{6}$ **9.** $12\pi/5$ **11.** $-81\pi/2$
13. 27π **15.** $341\sqrt{2}/60 + \frac{81}{20}\arcsin(\sqrt{3}/3)$ **17.** $13\pi/20$

Review Exercises for Chapter 14 ▪ page 963

1. False   **3.** True   **5.** False   **7.** True
9. $(5\sqrt{5} - 1)/3$ **11.** $4\sqrt{5}$ **13.** $-\pi$ **15.** $\frac{17}{2}$
17. 5   **19.** $f(x, y) = x\sin y - \cos y$ **21.** π^2 **25.** -8π
33. $\frac{2}{3}(6\sqrt{6} - 5\sqrt{5})$ **35.** $\pi(391\sqrt{17} + 1)/60$
37. $-64\pi/3$ **41.** $-\frac{1}{2}$ **45.** -4 **47.** 21

PROBLEMS PLUS ▪ page 965

3. (b) 0.90   **5.** 30   **7.** $\frac{1}{2}\sin 1$

CHAPTER 15

Exercises 15.1 ▪ page 975

1. $y = Ce^{1/x}$ **3.** $(y - 1)e^y = \frac{1}{3}(x^2 + 1)^{3/2} + C$
5. $y = \tan(x - 1)$ **7.** $u = 1 - \sqrt{t^2 + t + 4}$
9. $y = e^{1-\cos x}$ **11.** IV   **13.** III

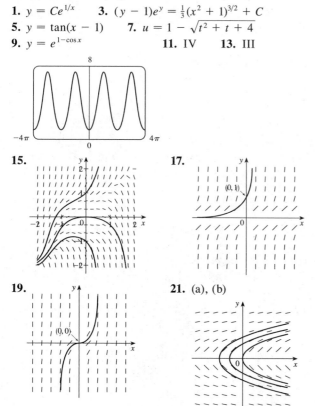

15.   **17.**

19.   **21.** (a), (b)

23.

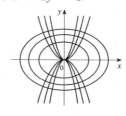

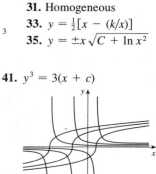

25. 2, 2.75, 3.5, 4.25
27. 1.8371
29. Not homogeneous
31. Homogeneous
33. $y = \frac{1}{2}[x - (k/x)]$
35. $y = \pm x\sqrt{C + \ln x^2}$

37. $y = -x\ln(C - \ln|x|)$
39. $x^2 + 2y^2 = C$ **41.** $y^3 = 3(x + c)$

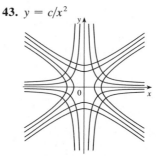

43. $y = c/x^2$ **45.** (b) $y(t) = (\sqrt{6} - \frac{1}{144}t)^2$
                            (c) $144\sqrt{6}$ s $\approx$ 5 min 53 s

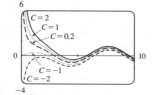

Exercises 15.2 ▪ page 981

1. Not linear   **3.** Linear   **5.** $y = Ce^{3x} - \frac{1}{2}e^x$
7. $y = Ce^{x^2} - \frac{1}{2}$ **9.** $y = \frac{1}{2}\sec x - \cos x + C\sec x$
11. $y = \frac{1}{2}x + Ce^{-x^2} - \frac{1}{2}e^{-x^2}\int e^{x^2}dx$
13. $y = \tan\theta + C\sec\theta$
15. $y = x - 1 + \frac{1}{2}(e^x + e^{-x}) = x - 1 + \cosh x$
17. $y = (x^2 + 3)e^{x^2}$
19. $y = (\sin x)/x^2$
21. $y = \sin x + (\cos x)/x + C/x$

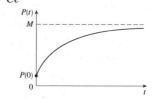

25. $y = \pm[Cx^4 + 2/(5x)]^{-1/2}$
27. (a) $I(t) = 4 - 4e^{-5t}$ (b) $4 - 4e^{-1/2} \approx 1.57$ A
29. $Q(t) = 3(1 - e^{-4t})$, $I(t) = 12e^{-4t}$
31. $P(t) = M + Ce^{-kt}$

33. (a) $(mg/c)(1 - e^{-ct/m})$ (b) mg/c
(c) $(mg/c)[t + (m/c)e^{-ct/m}] - m^2g/c^2$

Exercises 15.3 ■ page 986

1. (a) Not exact   (b) Exact   (c) Not exact
3. $y = \frac{1}{2}(-x \pm \sqrt{4c - 3x^2}\,)$ **5.** Not exact
7. $x \sin y + y = C$ **9.** $xye^{x/y} = C$ **11.** Not exact
13. $xy^{-1} - yx^{-2} = C$ **15.** $y = (-x^2 + \sqrt{180x - 11x^4}\,)/(6x)$
17. $y = [\sin^{-1}(1 - x)]/x$ **19.** $ye^{xy} = C$
21. $xy/\sqrt{1 + x^2 + y^2} = C$ **23.** $x^3y + x^2y^2 = C$
25. $e^{3x}(x^2y + y^2) = C$

Exercises 15.4 ■ page 988

1. (a) Separable   (b) Not homogeneous   (c) Not linear
(d) Exact
3. (a) Separable   (b) Homogeneous   (c) Linear
(d) Exact
5. $y = Ce^x - \frac{1}{2}(\sin x + \cos x)$
7. $y = \pm\sqrt[4]{e^x - \frac{1}{2}x^2 + C}$ **9.** $x + y = k(x^2 + y^2)$
11. $xy^2 - 3y \tan x = C$ **13.** $y^2 = \ln(x^2 + 1) + K$
15. $y = Cx^2 - \frac{1}{2}[\sqrt{1 + x^2} - x^2\ln|(\sqrt{1 + x^2} + 1)/x|]$
17. $y = 1/(k + \int \sqrt{1 + x^3}\,dx)$ **19.** $x^2 + y^2 + xe^y = C$
21. $y\sin(xy) - x = C$ **23.** $y = x[1 - (C/x)]^2$ **25.** (a) 0
(b) All values

Exercises 15.5 ■ page 995

1. $y = c_1e^x + c_2e^{2x}$ **3.** $y = c_1e^{-x/3} + c_2e^{3x}$
5. $y = e^{-x}(c_1\cos 3x + c_2\sin 3x)$ **7.** $y = c_1e^x + c_2e^{-x}$
9. $y = c_1\cos 5x + c_2\sin 5x$ **11.** $y = c_1 + c_2e^{-x/2}$
13. $y = e^{x/2}[c_1\cos(\sqrt{7}x/2) + c_2\sin(\sqrt{7}x/2)]$
15. $y = c_1e^{(-1+\sqrt{2})x} + c_2e^{(-1-\sqrt{2})x}$
17. $y = e^{-x/4}[c_1\cos(\sqrt{23}x/4) + c_2\sin(\sqrt{23}x/4)]$
19.

All solutions approach 0 as $x \to -\infty$ and approach $\pm\infty$ as
$x \to \infty$.
21. $y = e^x + e^{-4x}$ **23.** $y = e^x(\cos x + \sin x)$
25. $y = 2e^{1-x} + e^{3(x-1)}$ **27.** $y = -\frac{1}{3}\sin 3x$
29. $y = 3xe^{-2x+2}$ **31.** No solution
37. (b) $\lambda = n^2\pi^2/L^2$, n a positive integer; $y = C\sin(n\pi x/L)$

Exercises 15.6 ■ page 1002

1. $y = c_1e^{3x} + c_2e^{-2x} - \frac{5}{78}\cos 3x - \frac{1}{78}\sin 3x$
3. $y = e^{2x}(c_1x + c_2) + \frac{1}{9}e^{-x}$
5. $y = c_1\cos 6x + c_2\sin 6x + (x^2/18) - (x/36) - \frac{1}{324}$
7. $y = e^x[\frac{523}{650}\cos 2x + \frac{797}{1300}\sin 2x] + \frac{1}{5}x + \frac{2}{25} + \frac{3}{26}\cos 3x$
$- \frac{1}{13}\sin 3x$

9. $y = \frac{5}{8}e^x - \frac{17}{32}e^{-x} + e^{3x}[\frac{1}{8}x - \frac{3}{32}]$

11.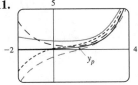
The solutions are all
asymptotic to $y_p = e^x/10$ as
$x \to \infty$. Except for y_p,
all solutions approach
either ∞ or $-\infty$ as $x \to -\infty$.

13. $y_p = (Ax^4 + Bx^3 + Cx^2 + Dx + E)e^{2x}$
15. $y_p = xe^x(A\cos x + B\sin x)$
17. $y = c_1\cos 2x + c_2\sin 2x + \frac{1}{4}x$
19. $y = c_1e^x + c_2xe^x + e^{2x}$
21. $y = (c_1 + x)\sin x + (c_2 + \ln\cos x)\cos x$
23. $y = [c_1 + \ln(1 + e^{-x})]e^x + [c_2 - e^{-x} + \ln(1 + e^{-x})]e^{2x}$
25. $y = [c_1 - \frac{1}{2}\int(e^x/x)\,dx]e^{-x} + [c_2 + \frac{1}{2}\int(e^{-x}/x)\,dx]e^x$

Exercises 15.7 ■ page 1009

1. $x = 0.36\sin(10t/3)$ **3.** $x = -\frac{1}{5}e^{-6t} + \frac{6}{5}e^{-t}$ **5.** $\frac{49}{12}$ kg
9. $Q(t) = (-e^{-10t}/250)(6\cos 20t + 3\sin 20t) + \frac{3}{125}$,
$I(t) = \frac{3}{5}e^{-10t}\sin 20t$
11. $Q(t) = e^{-10t}[\frac{3}{250}\cos 20t - \frac{3}{500}\sin 20t]$
$- \frac{3}{250}\cos 10t + \frac{3}{125}\sin 10t$

Exercises 15.8 ■ page 1014

1. $\sum_{n=0}^{\infty} a_0\frac{6^n}{n!}x^n = a_0e^{6x}$ **3.** $\sum_{n=0}^{\infty} \frac{a_0}{3^nn!}x^{3n} = a_0e^{x^3/3}$
5. $a_0\sum_{n=0}^{\infty}\left(-\frac{3}{2}\right)^n\frac{1}{n!}x^{2n} + a_1\sum_{n=0}^{\infty}\frac{(-6)^nn!}{(2n + 1)!}x^{2n+1}$
7. $a_0 + a_1x + a_0\frac{x^2}{2} + a_0\sum_{n=2}^{\infty}\frac{(-1)^{n-1}(2n - 3)!}{2^{2n-2}n!(n - 2)!}x^{2n}$
9. $\sum_{n=0}^{\infty}\frac{x^{2n}}{2^nn!} = e^{x^2/2}$
11. $x + \sum_{n=0}^{\infty}\frac{(-1)^n2^25^2\cdots(3n - 1)^2}{(3n + 1)!}x^{3n+1}$

Review Exercises for Chapter 15 ■ page 1014

1. False   **3.** True   **5.** True   **7.** True
9. $x + x^2y^2 = C$ **11.** $y = Cx^2 + x^3$
13. $y^2 - y^3 = x\sin x + \cos x + C$
15. $(y/x) + \ln(x - y)^2 = \ln|x| + C$
17. $y = \tan(\sin^{-1}x + C)$
19. $y = e^{3x}(c_1\cos 5x + c_2\sin 5x)$ **21.** $y = c_1e^{-x} + c_2e^{x/2}$
23. $y = e^{-x}(c_1 + c_2x) - \frac{3}{50}\cos 3x - \frac{2}{25}\sin 3x$
25. $y = c_1\cos(3x/2) + c_2\sin(3x/2) + \frac{2}{9}x^2 - \frac{43}{81}$
27. $y = c_1e^x + c_2e^{2x} + xe^{2x}$

29. $y = e^{2x} + Ce^x$

All solutions approach 0 as $x \to -\infty$ and approach ∞ as $x \to \infty$. For $C \ge 0$ all solutions are increasing. For $C < 0$ the solutions have a minimum point, which moves downward and to the right as $C \to -\infty$.

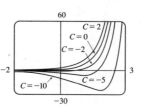

31. $y = e^{-x}\left[\frac{2}{3}x^{3/2} + 3\right]$ **33.** $y = 5 - 2e^{-6(x-1)}$

35. $y = (e^{4x} - e^x)/3$ **37.** 0.7568

39. $y^2 - 2\ln|y| + x^2 = K$ **41.** $\displaystyle\sum_{n=0}^{\infty} \frac{(-2)^n n!}{(2n+1)!}x^{2n+1}$

43. $Q(t) = -0.02e^{-10t}(\cos 10t + \sin 10t) + 0.03$

45. (a) $P(t) = (m/k) + (P_0 - m/k)e^{kt}$ (b) $m < kP_0$

(c) $m = kP_0, \ m > kP_0$

APPLICATIONS PLUS ■ page 1016

1. (c) $2\pi/k \approx 85$ min   (d) $\approx 17{,}600$ mi/h

3. (a) $-600\cos\theta, \ 30 - 600\sin\theta$

(b) $dy/dx = (30\sqrt{x^2 + y^2} - 600y)/(-600x)$

(c) $y + \sqrt{x^2 + y^2} = 50^{1/20}x^{19/20}$

(d)

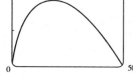

5. (a) $kS/(m + k) + [mS/(m + k)]e^{-(m+k)t}, \ kS/(m + k)$

(b) $[mS/(m + k)](1 - e^{-(m+k)t}), \ mS/(m + k)$

(c) $-ae^b/(e^b + 1), \ -a/(e^b + 1)$

7. (a) $y' = (-x \pm \sqrt{x^2 + y^2})/y$ (b) $y^2 = 2Cx + C^2$

APPENDIXES

Exercises H ■ page A9

1. $10 - i$ **3.** $13 - i$ **5.** $12 - 7i$ **7.** $-\frac{1}{2} + \frac{1}{2}i$

9. $\frac{1}{2} - \frac{1}{2}i$ **11.** $-i$ **13.** $5i$ **15.** $3 - 4i, 5$ **17.** $4i, 4$

19. $\pm\frac{3}{2}i$ **21.** $4 \pm i$ **23.** $-\frac{1}{2} \pm (\sqrt{7}/2)i$

25. $3\sqrt{2}[\cos(3\pi/4) + i\sin(3\pi/4)]$

27. $5\{\cos[\tan^{-1}\frac{4}{3}] + i\sin[\tan^{-1}\frac{4}{3}]\}$

29. $4[\cos(\pi/2) + i\sin(\pi/2)], \ \cos(-\pi/6) + i\sin(-\pi/6),$
$\frac{1}{2}[\cos(-\pi/6) + i\sin(-\pi/6)]$

31. $4\sqrt{2}[\cos(7\pi/12) + i\sin(7\pi/12)],$
$(2\sqrt{2})[\cos(13\pi/12) + i\sin(13\pi/12)], \frac{1}{4}[\cos(\pi/6) + i\sin(\pi/6)]$

33. -1024 **35.** $-512\sqrt{3} + 512i$

37. $\pm 1, \ \pm i, \ (1/\sqrt{2})(\pm 1 \pm i)$ **39.** $\pm(\sqrt{3}/2) + \frac{1}{2}i, \ -i$

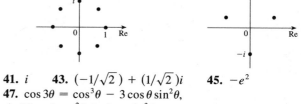

41. i **43.** $(-1/\sqrt{2}) + (1/\sqrt{2})i$ **45.** $-e^2$

47. $\cos 3\theta = \cos^3\theta - 3\cos\theta\sin^2\theta,$
$\sin 3\theta = 3\cos^2\theta\sin\theta - \sin^3\theta$

INDEX

TABLE OF INTEGRALS

BASIC FORMS

1. $\displaystyle\int u\,dv = uv - \int v\,du$

2. $\displaystyle\int u^n\,du = \frac{1}{n+1}u^{n+1} + C,$
$n \neq -1$

3. $\displaystyle\int \frac{du}{u} = \ln|u| + C$

4. $\displaystyle\int e^u\,du = e^u + C$

5. $\displaystyle\int a^u\,du = \frac{1}{\ln a}a^u + C$

6. $\displaystyle\int \sin u\,du = -\cos u + C$

7. $\displaystyle\int \cos u\,du = \sin u + C$

8. $\displaystyle\int \sec^2 u\,du = \tan u + C$

9. $\displaystyle\int \csc^2 u\,du = -\cot u + C$

10. $\displaystyle\int \sec u \tan u\,du = \sec u + C$

11. $\displaystyle\int \csc u \cot u\,du = -\csc u + C$

12. $\displaystyle\int \tan u\,du = \ln|\sec u| + C$

13. $\displaystyle\int \cot u\,du = \ln|\sin u| + C$

14. $\displaystyle\int \sec u\,du = \ln|\sec u + \tan u| + C$

15. $\displaystyle\int \csc u\,du = \ln|\csc u - \cot u| + C$

16. $\displaystyle\int \frac{du}{\sqrt{a^2 - u^2}} = \sin^{-1}\frac{u}{a} + C$

17. $\displaystyle\int \frac{du}{a^2 + u^2} = \frac{1}{a}\tan^{-1}\frac{u}{a} + C$

18. $\displaystyle\int \frac{du}{u\sqrt{u^2 - a^2}} = \frac{1}{a}\sec^{-1}\frac{u}{a} + C$

19. $\displaystyle\int \frac{du}{a^2 - u^2} = \frac{1}{2a}\ln\left|\frac{u+a}{u-a}\right| + C$

20. $\displaystyle\int \frac{du}{u^2 - a^2} = \frac{1}{2a}\ln\left|\frac{u-a}{u+a}\right| + C$

FORMS INVOLVING $\sqrt{a^2 + u^2}$, $a > 0$

21. $\displaystyle\int \sqrt{a^2 + u^2}\,du = \frac{u}{2}\sqrt{a^2 + u^2} + \frac{a^2}{2}\ln(u + \sqrt{a^2 + u^2}) + C$

22. $\displaystyle\int u^2\sqrt{a^2 + u^2}\,du = \frac{u}{8}(a^2 + 2u^2)\sqrt{a^2 + u^2} - \frac{a^4}{8}\ln(u + \sqrt{a^2 + u^2}) + C$

23. $\displaystyle\int \frac{\sqrt{a^2 + u^2}}{u}\,du = \sqrt{a^2 + u^2} - a\ln\left|\frac{a + \sqrt{a^2 + u^2}}{u}\right| + C$

24. $\displaystyle\int \frac{\sqrt{a^2 + u^2}}{u^2}\,du = -\frac{\sqrt{a^2 + u^2}}{u} + \ln(u + \sqrt{a^2 + u^2}) + C$

25. $\displaystyle\int \frac{du}{\sqrt{a^2 + u^2}} = \ln(u + \sqrt{a^2 + u^2}) + C$

26. $\displaystyle\int \frac{u^2\,du}{\sqrt{a^2 + u^2}} = \frac{u}{2}\sqrt{a^2 + u^2} - \frac{a^2}{2}\ln(u + \sqrt{a^2 + u^2}) + C$

27. $\displaystyle\int \frac{du}{u\sqrt{a^2 + u^2}} = -\frac{1}{a}\ln\left|\frac{\sqrt{a^2 + u^2} + a}{u}\right| + C$

28. $\displaystyle\int \frac{du}{u^2\sqrt{a^2 + u^2}} = -\frac{\sqrt{a^2 + u^2}}{a^2 u} + C$

29. $\displaystyle\int \frac{du}{(a^2 + u^2)^{3/2}} = \frac{u}{a^2\sqrt{a^2 + u^2}} + C$

FORMS INVOLVING $\sqrt{a^2 - u^2}$, $a > 0$

30. $\displaystyle\int \sqrt{a^2 - u^2}\,du = \frac{u}{2}\sqrt{a^2 - u^2} + \frac{a^2}{2}\sin^{-1}\frac{u}{a} + C$

31. $\displaystyle\int u^2\sqrt{a^2 - u^2}\,du = \frac{u}{8}(2u^2 - a^2)\sqrt{a^2 - u^2} + \frac{a^4}{8}\sin^{-1}\frac{u}{a} + C$

32. $\displaystyle\int \frac{\sqrt{a^2 - u^2}}{u}\,du = \sqrt{a^2 - u^2} - a\ln\left|\frac{a + \sqrt{a^2 - u^2}}{u}\right| + C$

33. $\displaystyle\int \frac{\sqrt{a^2 - u^2}}{u^2}\,du = -\frac{1}{u}\sqrt{a^2 - u^2} - \sin^{-1}\frac{u}{a} + C$

34. $\displaystyle\int \frac{u^2\,du}{\sqrt{a^2 - u^2}} = -\frac{u}{2}\sqrt{a^2 - u^2} + \frac{a^2}{2}\sin^{-1}\frac{u}{a} + C$

35. $\displaystyle\int \frac{du}{u\sqrt{a^2 - u^2}} = -\frac{1}{a}\ln\left|\frac{a + \sqrt{a^2 - u^2}}{u}\right| + C$

36. $\displaystyle\int \frac{du}{u^2\sqrt{a^2 - u^2}} = -\frac{1}{a^2 u}\sqrt{a^2 - u^2} + C$

37. $\displaystyle\int (a^2 - u^2)^{3/2}\,du = -\frac{u}{8}(2u^2 - 5a^2)\sqrt{a^2 - u^2} + \frac{3a^4}{8}\sin^{-1}\frac{u}{a} + C$

38. $\displaystyle\int \frac{du}{(a^2 - u^2)^{3/2}} = \frac{u}{a^2\sqrt{a^2 - u^2}} + C$

FORMS INVOLVING $\sqrt{u^2 - a^2}$, $a > 0$

39. $\displaystyle\int \sqrt{u^2 - a^2}\,du = \frac{u}{2}\sqrt{u^2 - a^2} - \frac{a^2}{2}\ln|u + \sqrt{u^2 - a^2}| + C$

40. $\displaystyle\int u^2\sqrt{u^2 - a^2}\,du = \frac{u}{8}(2u^2 - a^2)\sqrt{u^2 - a^2} - \frac{a^4}{8}\ln|u + \sqrt{u^2 - a^2}| + C$

41. $\displaystyle\int \frac{\sqrt{u^2 - a^2}}{u}\,du = \sqrt{u^2 - a^2} - a\cos^{-1}\frac{a}{u} + C$

42. $\displaystyle\int \frac{\sqrt{u^2 - a^2}}{u^2}\,du = -\frac{\sqrt{u^2 - a^2}}{u} + \ln|u + \sqrt{u^2 - a^2}| + C$

43. $\displaystyle\int \frac{du}{\sqrt{u^2 - a^2}} = \ln|u + \sqrt{u^2 - a^2}| + C$

44. $\displaystyle\int \frac{u^2\,du}{\sqrt{u^2 - a^2}} = \frac{u}{2}\sqrt{u^2 - a^2} + \frac{a^2}{2}\ln|u + \sqrt{u^2 - a^2}| + C$

45. $\displaystyle\int \frac{du}{u^2\sqrt{u^2 - a^2}} = \frac{\sqrt{u^2 - a^2}}{a^2 u} + C$

46. $\displaystyle\int \frac{du}{(u^2 - a^2)^{3/2}} = -\frac{u}{a^2\sqrt{u^2 - a^2}} + C$

TABLE OF INTEGRALS

FORMS INVOLVING $a + bu$

47. $\displaystyle\int \frac{u\,du}{a + bu} = \frac{1}{b^2}(a + bu - a\ln|a + bu|) + C$

48. $\displaystyle\int \frac{u^2\,du}{a + bu} = \frac{1}{2b^3}[(a + bu)^2 - 4a(a + bu) + 2a^2\ln|a + bu|] + C$

49. $\displaystyle\int \frac{du}{u(a + bu)} = \frac{1}{a}\ln\left|\frac{u}{a + bu}\right| + C$

50. $\displaystyle\int \frac{du}{u^2(a + bu)} = -\frac{1}{au} + \frac{b}{a^2}\ln\left|\frac{a + bu}{u}\right| + C$

51. $\displaystyle\int \frac{u\,du}{(a + bu)^2} = \frac{a}{b^2(a + bu)} + \frac{1}{b^2}\ln|a + bu| + C$

52. $\displaystyle\int \frac{du}{u(a + bu)^2} = \frac{1}{a(a + bu)} - \frac{1}{a^2}\ln\left|\frac{a + bu}{u}\right| + C$

53. $\displaystyle\int \frac{u^2\,du}{(a + bu)^2} = \frac{1}{b^3}\left(a + bu - \frac{a^2}{a + bu} - 2a\ln|a + bu|\right) + C$

54. $\displaystyle\int u\sqrt{a + bu}\,du = \frac{2}{15b^2}(3bu - 2a)(a + bu)^{3/2} + C$

55. $\displaystyle\int \frac{u\,du}{\sqrt{a + bu}} = \frac{2}{3b^2}(bu - 2a)\sqrt{a + bu} + C$

56. $\displaystyle\int \frac{u^2\,du}{\sqrt{a + bu}} = \frac{2}{15b^3}(8a^2 + 3b^2u^2 - 4abu)\sqrt{a + bu} + C$

57. $\displaystyle\int \frac{du}{u\sqrt{a + bu}} = \frac{1}{\sqrt{a}}\ln\left|\frac{\sqrt{a + bu} - \sqrt{a}}{\sqrt{a + bu} + \sqrt{a}}\right| + C, \quad \text{if } a > 0$

$\displaystyle\qquad = \frac{2}{\sqrt{-a}}\tan^{-1}\sqrt{\frac{a + bu}{-a}} + C, \qquad \text{if } a < 0$

58. $\displaystyle\int \frac{\sqrt{a + bu}}{u}\,du = 2\sqrt{a + bu} + a\int \frac{du}{u\sqrt{a + bu}}$

59. $\displaystyle\int \frac{\sqrt{a + bu}}{u^2}\,du = -\frac{\sqrt{a + bu}}{u} + \frac{b}{2}\int \frac{du}{u\sqrt{a + bu}}$

60. $\displaystyle\int u^n\sqrt{a + bu}\,du = \frac{2}{b(2n + 3)}\left[u^n(a + bu)^{3/2} - na\int u^{n-1}\sqrt{a + bu}\,du\right]$

61. $\displaystyle\int \frac{u^n\,du}{\sqrt{a + bu}} = \frac{2u^n\sqrt{a + bu}}{b(2n + 1)} - \frac{2na}{b(2n + 1)}\int \frac{u^{n-1}\,du}{\sqrt{a + bu}}$

62. $\displaystyle\int \frac{du}{u^n\sqrt{a + bu}} = -\frac{\sqrt{a + bu}}{a(n - 1)u^{n-1}} - \frac{b(2n - 3)}{2a(n - 1)}\int \frac{du}{u^{n-1}\sqrt{a + bu}}$

TRIGONOMETRIC FORMS

63. $\displaystyle\int \sin^2 u\,du = \tfrac{1}{2}u - \tfrac{1}{4}\sin 2u + C$

64. $\displaystyle\int \cos^2 u\,du = \tfrac{1}{2}u + \tfrac{1}{4}\sin 2u + C$

65. $\displaystyle\int \tan^2 u\,du = \tan u - u + C$

66. $\displaystyle\int \cot^2 u\,du = -\cot u - u + C$

67. $\displaystyle\int \sin^3 u\,du = -\tfrac{1}{3}(2 + \sin^2 u)\cos u + C$

68. $\displaystyle\int \cos^3 u\,du = \tfrac{1}{3}(2 + \cos^2 u)\sin u + C$

69. $\displaystyle\int \tan^3 u\,du = \tfrac{1}{2}\tan^2 u + \ln|\cos u| + C$

70. $\displaystyle\int \cot^3 u\,du = -\tfrac{1}{2}\cot^2 u - \ln|\sin u| + C$

71. $\displaystyle\int \sec^3 u\,du = \tfrac{1}{2}\sec u\tan u + \tfrac{1}{2}\ln|\sec u + \tan u| + C$

72. $\displaystyle\int \csc^3 u\,du = -\tfrac{1}{2}\csc u\cot u + \tfrac{1}{2}\ln|\csc u - \cot u| + C$

73. $\displaystyle\int \sin^n u\,du = -\frac{1}{n}\sin^{n-1}u\cos u + \frac{n - 1}{n}\int \sin^{n-2}u\,du$

74. $\displaystyle\int \cos^n u\,du = \frac{1}{n}\cos^{n-1}u\sin u + \frac{n - 1}{n}\int \cos^{n-2}u\,du$

75. $\displaystyle\int \tan^n u\,du = \frac{1}{n - 1}\tan^{n-1}u - \int \tan^{n-2}u\,du$

76. $\displaystyle\int \cot^n u\,du = \frac{-1}{n - 1}\cot^{n-1}u - \int \cot^{n-2}u\,du$

77. $\displaystyle\int \sec^n u\,du = \frac{1}{n - 1}\tan u\sec^{n-2}u + \frac{n - 2}{n - 1}\int \sec^{n-2}u\,du$

78. $\displaystyle\int \csc^n u\,du = \frac{-1}{n - 1}\cot u\csc^{n-2}u + \frac{n - 2}{n - 1}\int \csc^{n-2}u\,du$

79. $\displaystyle\int \sin au\sin bu\,du = \frac{\sin(a - b)u}{2(a - b)} - \frac{\sin(a + b)u}{2(a + b)} + C$

80. $\displaystyle\int \cos au\cos bu\,du = \frac{\sin(a - b)u}{2(a - b)} + \frac{\sin(a + b)u}{2(a + b)} + C$

81. $\displaystyle\int \sin au\cos bu\,du = -\frac{\cos(a - b)u}{2(a - b)} - \frac{\cos(a + b)u}{2(a + b)} + C$

82. $\displaystyle\int u\sin u\,du = \sin u - u\cos u + C$

83. $\displaystyle\int u\cos u\,du = \cos u + u\sin u + C$

84. $\displaystyle\int u^n\sin u\,du = -u^n\cos u + n\int u^{n-1}\cos u\,du$

85. $\displaystyle\int u^n\cos u\,du = u^n\sin u - n\int u^{n-1}\sin u\,du$

86. $\displaystyle\int \sin^n u\cos^m u\,du = -\frac{\sin^{n-1}u\cos^{m+1}u}{n + m} + \frac{n - 1}{n + m}\int \sin^{n-2}u\cos^m u\,du$

$\displaystyle\qquad = \frac{\sin^{n+1}u\cos^{m-1}u}{n + m} + \frac{m - 1}{n + m}\int \sin^n u\cos^{m-2}u\,du$